Nutraceutical and Specialty Lipids and their Co-Products

NUTRACEUTICAL SCIENCE AND TECHNOLOGY

Series Editor

FEREIDOON SHAHIDI, PH.D., FACS, FCIC, FCIFST, FIFT, FRSC
University Research Professor
Department of Biochemistry
Memorial University of Newfoundland
St. John's, Newfoundland, Canada

Nutraceutical and Specialty Lipids and their Co-Products

Edited by

Fereidoon Shahidi

CRC Press
Taylor & Francis Group
Boca Raton London New York

CRC Press is an imprint of the
Taylor & Francis Group, an **informa** business

A TAYLOR & FRANCIS BOOK

CRC Press
Taylor & Francis Group
6000 Broken Sound Parkway NW, Suite 300
Boca Raton, FL 33487-2742

First issued in paperback 2019

© 2006 by Taylor & Francis Group, LLC
CRC Press is an imprint of Taylor & Francis Group, an Informa business

No claim to original U.S. Government works

ISBN-13: 978-1-57444-499-5 (hbk)
ISBN-13: 978-0-367-39105-8 (pbk)
Library of Congress Card Number 2005054946

Library of Congress Cataloging-in-Publication Data

Neutraceutical lipids and co-products / edited by Fereidoon Shahidi
      p. cm. -- (Neutraceutical science and technology ; 5)
    Includes bibliographical references and index.
    ISBN 1-57444-499-9
    1. Functional foods. 2. Food--Biotechnology. I. Series.

QP144.F85N84 2006
612.23'97--dc22   2005054946

Visit the Taylor & Francis Web site at
http://www.taylorandfrancis.com

and the CRC Press Web site at
http://www.crcpress.com

Preface

Interest in food lipids has grown dramatically in recent years as a result of findings related to their health effects. Fats and oils have often been condemned because of their high energy value and due to potential health problems associated with certain saturated fatty acids as well as *trans* fats. However, lipids are important in that they provide essential fatty acids and fat-soluble vitamins as well as flavor, texture, and mouthfeel to foods. In addition, the beneficial health effects and essentiality of long-chain omega-3 fatty acids such as eicosapentaenoic acid (EPA), and docosahexaenoic acid (DHA) and that of omega-6 fatty acids such as arachidonic acid (AA) and γ-linolenic acid (GLA) have been recognized. Recently, the role of EPA and/or DHA in heart health, mental health, and brain and retina development has been well documented. In this connection, there has been a surge in the public interest and thus inclusion of these fatty acids into foods such as spreads, bread and cereal products, orange juice, and dairy products, among others. In addition, novel sources of edible oils with specific characteristics such as those of fruit seed oils, nut oils, algal oils, and medium-chain fatty acids as well as diacylglycerols have been explored. The role of minor components in fats and oils and their effects on oil stability have been acknowledged. Minor components such as phospholipids, tocopherols and tocotrienols, carotenoids, and sterols as well as phenolic compounds may be procured from the oil or the leftover meal and used as nutraceuticals and functional food ingredients.

It is the purpose of this book to present a comprehensive assessment of the current state of the chemistry, nutrition, and health aspects of specialty fats and oils and their co-products and to address stability issues and their potential application and delivery in functional foods and geriatric and other formulations. This book provides valuable information for senior undergraduate and graduate students as well as scientists in academia, government laboratories, and industry. I am indebted to the participating authors for their hard work and dedication in providing a state-of-the-art contribution and for their authoritative views resulting from their latest investigations on different aspects of nutraceutical lipids and co-products.

Fereidoon Shahidi

Editor

Fereidoon Shahidi, Ph.D., FACS, FCIC, FCIFST, FIFT, FRSC, is a University Research Professor, the highest academic level, in the Department of Biochemistry, Memorial University of Newfoundland (MUN), Canada. He is also cross-appointed to the Department of Biology, Ocean Sciences Centre, and the aquaculture program at MUN. Dr. Shahidi is the author of over 550 scientific papers and book chapters and has authored or edited over 40 books. He has given over 350 presentations at different scientific meetings and conferences. His research has led to a number of industrial developments around the globe.

Dr. Shahidi's current research interests include different areas of nutraceuticals and functional foods and particularly work on specialty and structured lipids, lipid oxidation, food phenolics, and natural antioxidants, among others. Dr. Shahidi is the editor-in-chief of the *Journal of Food Lipids*, an editor of *Food Chemistry*, and a member on the editorial boards of the *Journal of Food Science*, *Journal of Agricultural and Food Chemistry*, *International Journal of Food Properties*, *Journal of Food Science and Nutrition*, and *Current Food Science and Nutrition*. He is the editor of the sixth edition of *Bailey's Industrial Oils and Fats* in six volumes. Dr. Shahidi has been the recipient of numerous awards, the latest of which was the Stephen S. Chang Award from the Institute of Food Technologists (IFT) in 2005, for his outstanding contributions to food lipids and flavor chemistry, and was also recognized by IFT as a Fellow in 2005.

Dr. Shahidi is a founding member and a past chair of the Nutraceutical and Functional Food Division of IFT and a councilor of IFT. He has also served in the past as chairs for the Agricultural and Food Chemistry Division of the American Chemical Society (ACS) and the Lipid Oxidation and Quality of the American Oil Chemists' Society (AOCS). Dr. Shahidi served as a member of the Expert Advisory Panel of Health Canada on Standards of Evidence for Health Claims for Foods, the Standards Council of Canada on Fats and Oils, the Advisory Group of Agriculture and Agri-Food Canada on Plant Products, and the Nutraceutical Network of Canada. He also served as a member of the Washington-based Council of Agricultural Science and Technology on Nutraceuticals.

Contributors

R.O. Adlof
Food and Industrial Oil Research, National
   Center for Agricultural
Utilization Research
Peoria, Illinois, USA

Scott Bloomer
Archer Daniels Midland Company
James R. Randall Research Center
Decatur, Illinois, USA

Yaakob B. Che Man
Department of Food Technology, Faculty
   of Food Science and Technology
Universiti Putra Malaysia
Serdang, Selangor, Malaysia

Grace Chen
United States Department of Agriculture
Agricultural Research Service
Albany, California, USA

Hang Chen
Department of Food Science
Center for Advanced Food Technology
Rutgers University
New Brunswick, New Jersey, USA

Armand B. Christophe
Department of Internal Medicine
Ghent University Hospital
Ghent, Belgium

Yasushi Endo
Graduate School of Agricultural Science
Tohoku University
Sendai, Japan

Fang Fang
Department of Food Science
Center for Advanced Food Technology
Rutgers University
New Brunswick, New Jersey, USA

Paul Fedec
POS Pilot Plant Corporation
Saskatoon, Saskatchewan, Canada

Jaouad Fichtali
Martek Biosciences Corporation
Winchester, Kentucky, USA

Brent D. Flickinger
Archer Daniels Midland Company
Decatur, Illinois, USA

Kenshiro Fujimoto
Graduate School of Agricultural Science
Tohoku University
Sendai, Japan

Frank D. Gunstone
Scottish Crop Research Institute
Invergowrie, Dundee, Scotland, U.K.

Xiaohua He
United States Department of Agriculture
Agricultural Research Institute
Albany, California, USA

Chi-Tang Ho
Department of Food Science
Center for Advanced Food Technology
Rutgers University
New Brunswick, New Jersey, USA

Masashi Hosokawa
Laboratory of Biofunctional Material
   Chemistry
Hokkaido University
Hakodate, Japan

Chung-yi Huang
Department of Food Science and Technology
University of Georgia
Athens, Georgia, USA

Yao-wen Huang
Department of Food Science and Technology
University of Georgia
Athens, Georgia, USA

Charlotte Jacobsen
Department of Seafood Research
Danish Institute for Fisheries Research
Lyngby, Denmark

J.W. King
Food and Industrial Oil Research
National Center for Agricultural
    Utilization Research
Peoria, Illinois, USA

Yong Li
Center for Enchancing Food to Protect Health
Lipid Chemistry and Molecular Biology
    Laboratory
Purdue University
West Lafayette, Indiana, USA

Jiann-Tsyh Lin
United States Department of Agriculture
Agricultural Research Service
Albany, California, USA

G.R. List
Food and Industrial Oil Research
National Center for Agricultural
    Utilization Research
Peoria, Illinois, USA
and
Food Science and Technology Consultants
Germantown, Tennessee, USA

Hu Liu
School of Pharmacy
Memorial University of Newfoundland
St. John's, Newfoundland, Canada

Marina Abdul Manaf
Department of Food Technology
Faculty of Food Science and Technology
Universiti Putra Malaysia
Serdang, Selangor, Malaysia

Thomas A. McKeon
United States Department of Agriculture
Agricultural Research Service
Albany, California, USA

H. Miraliakbari
Department of Biochemistry
Memorial University of Newfoundland
St. John's, Newfoundland, Canada

Kazur Miyashita
Laboratory of Biofunctional Material Chemistry
Hokkaido University
Hakodate, Japan

Karlene S.T. Mootoosingh
School of Nutrition
Ryerson University
Toronto, Ontario, Canada

Kumar D. Mukherjee
Institute for Lipid Research
Federal Research Centre for Nutrition and Food
Münster, Germany

Toshihiro Nagao
Osaka Municipal Technical Research Institute
Osaka, Japan

Bhaskar Narayan
Laboratory of Biofunctional Material Chemistry
Hokkaido University
Hakodate, Japan

Nina Skall Nielsen
Department of Seafood Research
Danish Institute for Fisheries Research
Lyngby, Denmark

Frank T. Orthoefer
Food Science and Technology Consultants
Germantown, Tennessee, USA

Andreas M. Papas
YASOO Health, Inc.
Johnson City, Tennessee, USA

Si-Bum Park
Graduate School of Agricultural Science
Tohoku University
Sendai, Japan

J.W. Parry
Department of Nutrition and Food Science
University of Maryland
College Park, Maryland, USA

Roman Przybylski
Department of Chemistry and Biochemistry
University of Lethbridge
Alberta, Canada

Robert D. Reichert
Industrial Research Assistance Program
National Research Council of Canada
Ottawa, Ontario, Canada

Robert T. Rosen
Department of Food Science
Center for Advanced Food Technology
Rutgers University
New Brunswick, New Jersey, USA

Dérick Rousseau
School of Nutrition
Ryerson University
Toronto, Ontario, Canada

Karen Schaich
Department of Food Science
Rutgers University
New Brunswick, New Jersey, USA

S.P.J.N. Senanayake
Department of Biochemistry
Memorial University of Newfoundland
St. John's, Newfoundland, Canada
and
Martek Biosciences Corporation
Winchester, Kentucky, USA

Fereidoon Shahidi
Department of Biochemistry
Memorial University of Newfoundland
St. John's, Newfoundland, Canada

Yuji Shimada
Osaka Municipal Technical Research Institute
Osaka, Japan

Barry G. Swanson
Food Science and Human Nutrition
Washington State University
Pullman, Washington, USA

Maike Timm-Heinrich
Department of Seafood Research
Danish Institute for Fisheries Research
Lyngby, Denmark

Charlotta Turner
United States Department of Agriculture
Agricultural Research Service
Albany, California, USA

Udaya Wanasundara
POS Pilot Plant Corporation
Saskatoon, Saskatchewan, Canada

Lili Wang
School of Pharmacy
Memorial University of Newfoundland
St. John's, Newfoundland, Canada

Yomi Watanabe
Osaka Municipal Technical Research Institute
Osaka, Japan

Bruce A. Watkins
Center for Enhancing Food to Protect Health
Lipid Chemistry and Molecular Biology
    Laboratory
Purdue University
West Lafayette, Indiana, USA

Nikolaus Weber
Institute for Lipid Research
Federal Research Centre for Nutrition and Food
Münster, Germany

Liangli Yu
Department of Nutrition and Food Science
University of Maryland,
College Park, Maryland, USA

Kequan Zhou
Department of Nutrition and Food Science
University of Maryland
College Park, Maryland, USA

Contents

1 Nutraceutical and Specialty Lipids

Fereidoon Shahidi and S.P.J.N. Senanayake[*]
Department of Biochemistry, Memorial University of Newfoundland,
St. John's, Newfoundland, Canada

CONTENTS

---

[*]Current address: Martek Biosciences Corporation, 555 Rolling Hills Lane, Winchester, Kentucky.

1.1 INTRODUCTION

Lipids are organic substances that are insoluble or sparingly soluble in water. They are important components in determining the sensory attributes of foods. Lipids contribute to mouthfeel and textural properties in the foods. They have several important biological functions, which include: (1) serving as structural components of membranes; (2) acting as storage and transport forms of metabolic fuel; (3) serving as the protective coating on the surface of many organisms; (4) acting as carriers of fat-soluble vitamins A, D, E, and K and helping in their absorption; and (5) being involved as cell-surface components concerned with cell recognition, species specificity, and tissue immunity. Ironically, overconsumption of lipids is associated with a number of diseases, namely artherosclerosis, hypertension, and breast and colon cancer, and in the development of obesity.

There are several classes of lipids, all having similar and specific characteristics due to the presence of a major hydrocarbon portion in their molecules. Over 80 to 85% of lipids are generally in the form of triacylglycerols (TAGs). These are esters of glycerol and fatty acids. The TAGs occur in many different types, according to the identity and position of the three fatty acid components. Those with a single type of fatty acid in all three positions are called simple TAGs and are named after their fatty acid component. However, in some cases the trivial names are more commonly used. An example of this is trioleylglycerol, which is usually referred to as triolein. The TAGs with two or more different fatty acids are named by a more complex system.

Lipids, and particularly TAGs, are integrated components of our diet and are a major source of caloric intake from foods. The caloric value of lipids is much higher than other food components and about 2.25 times greater than that of proteins and carbohydrates. While a certain amount of fat in the diet is required for growth and maintenance of the body functions, excessive intake of lipids has its own implications. While our body can synthesize saturated and monoenoic acids, polyunsaturated fatty acids (PUFAs) must be provided in the diet. Deficiency of linoleic acid and n-3 fatty acids results in dermatitis and a variety of other disease conditions. The role of n-3 fatty acids in lowering of blood cholesterol level and other benefits has been appreciated. The ratio of the intake of linoleic to α-linolenic acid in our diet should be approximately 2 and our daily caloric intake should have a contribution of 3.0 to 6.0% and 2.0 to 2.5% of each of these fatty acids, respectively.

1.2 CHEMISTRY AND COMPOSITION OF LIPIDS

1.2.1 THE FATTY ACIDS

Fatty acids are divided into saturated and unsaturated groups, the latter being further subdivided into monounsaturated and PUFAs. The PUFAs are divided into main categories depending on the position of the first double bond in the fatty acid carbon chain from the methyl end group of the molecules and are called n-3, n-6, and n-9 families.

1.2.2 SATURATED FATTY ACIDS

Saturated fatty acids contain only single carbon–carbon bonds in the aliphatic chain and all other available bonds are taken up by hydrogen atoms. The most abundant saturated fatty acids in animal and plant tissues are straight-chain compounds with 14, 16, and 18 carbon atoms. In general, saturated fats are solid at room temperature. They are predominantly found in butter, margarine, shortening, coconut and palm oils, as well as foods of animal origin[1]. The most common saturated fatty acids in foods are lauric (12:0), myristic (14:0), palmitic (16:0), and stearic (18:0) acids[2]. The common nomenclature for some saturated fatty acids is given in Table 1.1.

Fatty acids containing 4 to 14 carbon atoms occur in milk fat and in some vegetable oils. For example, cow's milk fat contains butyric acid (4:0) at a level of 4%. In addition, fatty acids containing 6 to 12 carbon atoms are also present in small quantities. The short-chain fatty acids are

TABLE 1.1
Nomenclature of Some Common Saturated Fatty Acids

Common name	Systematic name	No. of carbon atoms	Shorthand notation
Acetic	Ethanoic	2	2:0
Butyric	Butanoic	4	4:0
Caproic	Hexanoic	6	6:0
Caprylic	Octanoic	8	8:0
Capric	Decanoic	10	10:0
Lauric	Dodecanoic	12	12:0
Myristic	Tetradecanoic	14	14:0
Palmitic	Hexadecanoic	16	16:0
Stearic	Octadecanoic	18	18:0
Arachidic	Eicosanoic	20	20:0
Behenic	Docosanoic	22	22:0

usually present in butter and in other milk fat-based products. For example, the short-chain fatty acids from butyric to capric are characteristic of ruminant milk fat.

Tropical fruit oils, such as those from coconut and palm kernel, contain very high amounts (approximately 50%) of lauric acid (12:0). They also contain significant amounts of caprylic (8:0), capric (10:0), and myristic (14:0) acids. Canola oil is another example of a lauric acid-rich oil.

Palmitic acid (16:0) is the most widely occurring saturated fatty acid. It is found in almost all vegetable oils, as well as in fish oils and body fat of land animals. The common sources of palmitic acid include palm oil, cottonseed oil, as well as lard and tallow, among others.

Stearic acid (18:0) is less common compared to palmitic acid. However, it is a major component of cocoa butter. This fatty acid may be produced by hydrogenation of oleic, linoleic, and linolenic acids. Palmitic and stearic acids are employed in food and nonfood (personal hygiene products, cosmetics, surfactants, etc.) products.

1.2.3 UNSATURATED FATTY ACIDS

Unsaturated fatty acids contain carbon–carbon double bonds in the aliphatic chain. In general, these fats are soft at room temperature. When the fatty acids contain one carbon–carbon double bond in the molecule, it is called monounsaturated. Monounsaturated fatty acids are synthesized within the human body[3]. Oleic acid (18:1n-9) is the most common dietary monounsaturated fatty acid and found in most animal fats[1,2]. The common nomenclature for some unsaturated fatty acids is given in Table 1.2.

PUFAs contain two or more carbon–carbon double bonds. The PUFAs are liquid at room temperature. In general, they have low melting points and are susceptible to oxidation. They are found in grains, nuts, vegetables, and seafood (Table 1.3). The PUFAs of animal origin can be categorized into different families according to their derivation from specific biosynthetic precursors. In each case, the families contain from two up to a maximum of six double bonds, separated by methylene-interrupted groups and they have the same terminal structure. Linoleic acid (LA; 18:2n-6) is the most common fatty acid of this type. This fatty acid is found in all vegetable fats and is required for normal growth, reproduction, and health. It is the most predominant PUFA in the Western diet[4]. LA serves as a precursor or "parent" compound of n-6 family of fatty acids that is formed by desaturation and chain elongation, in which the terminal (n-6) structure is retained. Thus, LA can be metabolized into γ-linolenic acid (GLA; 18:3n-6), dihomo-γ-linolenic acid (DGLA; 20:3n-6), and arachidonic acid (AA; 20:4n-6). Of these, AA is particularly important as an essential

TABLE 1.2
Nomenclature of Some Common Unsaturated Fatty Acids

Common name	Systematic name	No. of carbon atoms	Shorthand notation
Myristoleic	Tetradec-9-enoic	14	14:1
Palmitoleic	Hexadec-9-enoic	16	16:1
Oleic	Octadec-9-enoic	18	18:1n-9
Linoleic	Octadeca-9,12-dienoic	18	18:2n-6
α-Linolenic	Octadeca-9,12,15-trienoic	18	18:3n-3
γ-Linolenic	Octadeca-6,9, 12-trienoic	18	18:3n-6
Elaeostearic	Octadeca-9,11,13-trienoic	20	20:3
Gadoleic	Eicosa-9-enoic	20	20:1
Arachidonic	Eicosa-5,8,11,14-tetraenoic	20	20:4n-6
EPA	Eicosa-5,8,11,14,17-pentaenoic	20	20:5n-3
Erucic	Docosa-13-enoic	22	22:1
DHA	Docosa-4,7,10,13,16,19-hexaenoic	22	22:6n-3

component of the membrane phospholipids and as a precursor of the eicosanoids. GLA, an important intermediate in the biosynthesis of AA from LA, is a constituent of certain seed oils and has been a subject of intensive study. α-Linolenic acid (ALA; 18:3n-3) is a precursor of n-3 family of fatty acids. It is found in appreciable amounts in green leaves, stems, and roots. It is a major component of flaxseed oil (45 to 60%) (Table 1.3). When ALA is absorbed into an animal body through the diet, it forms long-chain PUFAs with an n-3 terminal structure. ALA can be metabolized into eicosapentaenoic acid (EPA; 20:5n-3) and docosahexaenoic acid (DHA; 22:6n-3). They have special functions in the membrane phospholipids. In addition, EPA is a precursor of a series of eicosanoids. The major sources of EPA and DHA are algal, fish, and other marine oils.

1.2.4 ACYLGLYCEROLS

Edible fats and oils are composed primarily of TAGs. Partial acylglycerols, such as mono- and diacylglycerols, may also be present as minor components. The TAGs consist of a glycerol moiety, each hydroxyl group of which is esterified to a fatty acid. These compounds are synthesized by enzyme systems in nature. A stereospecific numbering (*sn*) system has been recognized to describe various enantiomeric forms (e.g., different fatty acyl groups in each positions in the glycerol backbone) of TAGs. In a Fischer projection of a natural L-glycerol derivative, the secondary hydroxyl group is shown to the left of carbon-2; the carbon atom above this becomes carbon-1 and that below becomes carbon-3 (Figure 1.1). The prefix "*sn*" is placed before the stem name of the compound.

Partial acylglycerols, namely diacylglycerols (DAGs) and monoacylglycerols (MAGs), are important intermediates in the biosynthesis and catabolism of TAGs and other classes of lipids. For example, 1,2-DAGs are important as intermediates in the biosynthesis of TAGs and other lipids. 2-MAGs are formed as intermediates or end products of the enzymatic hydrolysis of TAGs. The DAGs are fatty acid diesters of glycerol while the MAGs are fatty acid monoesters of glycerol. The MAGs and DAGs are produced on a large scale for use as surface-active agents. Acyl migration may occur with partial acylglycerols, especially on heating, in alcoholic solvents or when protonated reagents are present.

1.2.5 PHOSPHOLIPIDS

In phospholipids, one or more of the fatty acids in the TAG is replaced by phosphoric acid or its derivatives. Phospholipids are major constituents of cell membranes and thus regarded as structural

TABLE 1.3
Dietary Sources of Selected Fatty Acids

Source	Total fatty acid (%)
18:3n-3	
Flaxseed	45–60
Green leaves	56
Rapeseed	10–11
20:5n-3	
Herring	3–5
Mackerel	7–8
Sardine/pilchard	3–17
Pacific anchovy	18
Cod	17
Halibut	13
Menhaden	14
22:6n-3	
Herring	2–3
Mackerel	8
Sardine/pilchard	9–13
Pacific anchovy	9–11
Cod	30
Halibut	38
Menhaden	8
18:2n-6	
Borage	38
Evening primrose	70–75
Blackcurrant	44
Corn	34–62
Soybean	44–62
Sunflower seed	20–75
Safflower seed	55–81
Sesame seed	35–50
Cotton seed	33–59
Groundnut	13–45
18:3n-6	
Borage	20
Evening primrose	10
Blackcurrant	17–20

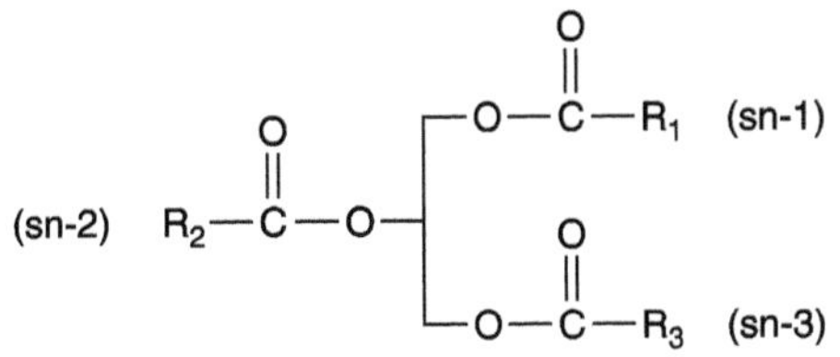

FIGURE 1.1 Stereospecific numbering of triacylglycerols.

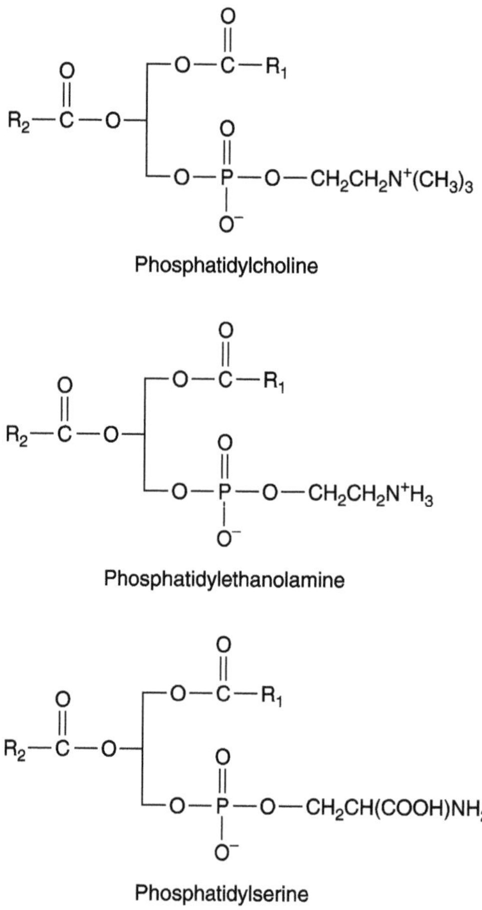

FIGURE 1.2 Chemical structures of the major phospholipids.

lipids in living organisms. The acyl groups in phospholipids occur in the *sn*-1 and *sn*-2 positions of the glycerol moiety while a polar head group involving a phosphate is present in the *sn*-3 position of the molecule. There are several types of phospholipids (Figure 1.2). These are based on the phosphatidic acids (monoesters of the tribasic phosphoric acid), which themselves are diacyl derivatives of 3-glycerophosphoric acid. The major types of phospholipids include phosphatidylcholine (PC), phosphatidylethanolamine (PE), phosphatidylserine (PS), phosphatidylinositol (PI), and phosphatidylglycerol (PG), among others.

Phospholipids of various types are present as minor components (0.5 to 3.0%) in most crude oils. However, these compounds are mainly removed during the refining process. They may be recovered as a distillate byproduct during deodorization and are generally referred to as lecithin, which is a mixture of phospholipids. The major phospholipids in crude lecithin are usually PC, PE, PI, and phosphatidic acids. Lecithin is found in many sources of vegetable oils. Commercial lecithin is generally produced from soybean oil during the degumming process. Lecithin is also available from sunflower, rapeseed, and corn oils. These are important surface-active compounds used extensively in the food, pharmaceutical, and cosmetic applications.

The hydrolysis of phospholipids gives rise to various products. For example, hydrolysis of PC occurs with aqueous acids and the products are glycerol, fatty acids, phosphoric acid, and choline. However, enzyme-assisted hydrolysis is more selective and gives rise to a variety of products.

Phospholipase A_1 causes deacylation at the *sn*-1 position, liberating fatty acids from this position and leaving behind a lysophosphatidylcholine. Phospholipase A_2 behaves in a similar manner at the *sn*-2 position.

There are differences in the natural distribution of fatty acids associated with lipids such as phospholipids and TAGs. For example, it is generally believed that phospholipids, such as lecithin and cephalin, contain more PUFAs than do the TAGs. Lovern[5] reported that phospholipids present in marine species are generally unsaturated and esterified mainly with EPA and DHA. Menzel and Olcott[6] studied PC and PE constituents of menhaden oil and found that their PUFAs were located mainly in the *sn*-2 position of the TAG molecules. The *sn*-2 position of PC in menhaden oil contained 29 and 42% EPA and DHA, respectively, whereas the *sn*-1 position contained only 1.7 and 12.5%, respectively, of these fatty acids. However, the phospholipid content of refined, bleached, and deodorized oils is very low[7], due to the removal of polar compounds during the degumming process.

Phospholipids are usually extracted with the total lipids when using the Bligh and Dyer[8] fat extraction procedure. Silicic acid column chromatography with methanol after eluting neutral lipids can be used to recover phospholipids. Two-dimensional thin-layer chromatography (TLC)[9] and Iatroscan[10] provide a means for separating individual phospholipids.

1.2.6 Fat-Soluble Vitamins and Tocopherols

Vitamins A and D are stored in large amounts in the liver of fish. Therefore, fish liver oils are considered as exceptionally rich sources of vitamins A and D. After vitamin A was synthesized and produced commercially, production of the liver oils for vitamins A and D became a minor industry in North America. Vitamin E or α-tocopherols are also present in marine lipids. Though present in lower amounts, the tocopherols and tocotrienols attract attention because of their vitamin E and antioxidant properties. The tocopherols are a series of benzopyranols with one, two, or three methyl groups attached to the phenolic ring. The molecules also have a 16-carbon side chain moiety on the pyran ring. In tocopherols the side chain is saturated, whereas in tocotrienols the side chain is unsaturated and contains three double bonds. There are four tocopherols and tocotrienols designated α, β, γ, and δ. The tocopherols exhibit different vitamin E activity in the order $\alpha > \beta > \gamma > \delta$. However, the antioxidant activity is generally in the reverse order ($\delta > \gamma > \beta > \alpha$). Thus, various oils do not follow a similar sequence of vitamin E and antioxidant activity.

1.2.7 Sterols

Most vegetable oils contain 0.1 to 0.5% sterols. They may exist as free sterols and esters with long-chain fatty acids. Sitosterol is generally the major phytosterol, contributing 50 to 80% to the total content of sterols. Campesterol and stigmasterol may also be present in significant levels. Cholesterol is generally considered to be an animal sterol. It is not present in plant systems at any significant level. The sterol content of some fats and oils is given in Table 1.4.

1.2.8 Waxes

Waxes include a variety of long-chain compounds occurring in both plants and animals. These are generally water-resistant materials made up of mixtures of fatty alcohols and their esters. They differ from the long-chain fatty acids in the TAG molecules. These include compounds of higher molecular weight (up to 60 carbon atoms and beyond) and are frequently branched with one or more methyl groups. Even though they may be unsaturated they do not generally exhibit a methylene-interrupted unsaturation pattern. Waxes find useful applications in the food, pharmaceutical, and cosmetic industries.

TABLE 1.4
Sterol Content of Fats and Oils

Source	Sterol (%)
Soybean oil	0.7–0.9
Canola oil	0.4–0.5
Corn oil	1.0–2.3
Coconut oil	0.08
Mustard oil	0.06
Milk fat	0.3–0.35
Lard	0.12
Beef tallow	0.08–0.10
Herring	0.4

1.2.9 BIOCHEMISTRY AND METABOLISM OF SHORT-CHAIN FATTY ACIDS (SCFAS)

Short-chain fatty acids (SCFA) are saturated fatty acids with 2 to 4 carbon atoms. This family of fatty acids includes acetic acid (2:0), propionic acid (3:0), and butyric acid (4:0). They are commonly referred to as the volatile fatty acids and are produced in the human gastrointestinal tract via bacterial fermentation of dietary carbohydrates[11]. SCFAs are present in the diet in small amounts, for example acetic acid in vinegar and butyric acid in bovine milk and butter. They may also be present in fermented foods. In humans, SCFAs contribute 3% of the total energy expenditure[12]. SCFAs are more easily absorbed in the stomach and provide fewer calories than MCFAs and LCFAs.

In nutritional applications, there has been a growing interest in the use of SCFAs as an alternative or additional source of energy to the medium- (MCFA) and long-chain fatty acid (LCFA) counterparts. The SCFAs, acetic, propionic, and butyric acids, are easily hydrolyzed from a single TAG structure and are rapidly absorbed by the intestinal mucosa[13]. These fatty acids go directly into the portal vein for transport to the liver where they are broken down to acetate via β-oxidation. The acetate can then be metabolized for energy or use in new fatty acid synthesis.

SCFAs may be incorporated into enteral nutritional formulas. Kripke et al.[14] have shown that a chemically synthesized diet containing 40% (w/w) of nonprotein as short-chain triacylglycerols (1:1, triacetin and tributyrin) maintained body weight, improved nitrogen balance and liver function, and enhanced jejunal and colonic mucosal adaptation in rats after 60% distal small-intestine resection with cecectomy, when compared to short-intestine animals receiving a diet without supplemental lipid calories from medium-chain triacylglycerol (MCT). SCFAs affect gastrointestinal function by stimulating pancreatic enzyme secretion[15] and increasing sodium and water absorption in the intestine[16].

1.2.10 BIOCHEMISTRY AND METABOLISM OF MCFAS

MCFAs are saturated fatty acids with 6 to 12 carbon atoms[17]. The sources of MCFAs include lauric oils such as coconut and palm kernel oils[18]. For example, coconut oil naturally contains some 65% MCFAs[19]. MCFAs, being saturated fatty acids, are resistant to oxidation and stable at high and low temperatures[20]. One of the first medical foods developed, as an alternative to conventional lipids, was MCT. MCT is an excellent source of MCFAs for production of structured and specialty lipids. Pure MCTs have a caloric value of 8.3 calories per gram. However, they do not provide essential fatty acids[21,22]. MCFAs are more hydrophilic than their LCFA counterparts, and hence their solubilization as micelles is not a prerequisite for absorption[23]. MCTs can also be directly incorporated into mucosal cells without hydrolysis and may readily be oxidized in the cell. MCTs pass directly into the portal vein and are readily oxidized in the liver to serve as an energy source. Thus, they are less likely to be deposited in adipose tissues[20] and are more susceptible to oxidation in tissues[24]. MCTs are metabolized as quickly as glucose and have twice the energy value of carbohydrates[25]. Johnson et al.[26] found that MCTs were oxidized much more rapidly than LCTs, with 90% conversion to carbon dioxide in 24 h.

MCTs are liquid or solid products at room temperature. They have a smaller molecular size, lower melting point, and greater solubility than their LCFA counterparts. These characteristics account for their easy absorption, transport, and metabolism compared to LCTs[27]. MCTs are hydrolyzed by pancreatic lipase more rapidly and completely than are LCTs[18]. They may be directly absorbed by the intestinal mucosa with minimum pancreatic or biliary function. They are transported predominantly by the portal vein to the liver for oxidation[28] rather than through the intestinal lymphatics. In addition, MCFAs are more rapidly oxidized to produce acetyl-CoA and ketone bodies and are independent of carnitine for entry into the mitochondria.

MCTs need to be used with LCTs to provide a balanced nutrition in enteral and parenteral products[29,30]. In many medical foods, a mixture of MCTs and LCTs is used to provide both rapidly metabolized and slowly metabolized fuel as well as essential fatty acids. Clinical nutritionists have taken advantage of MCTs' simpler digestion to nourish individuals who cannot utilize LCTs. Any abnormality in the numerous enzymes or processes involved in the digestion of LCTs can cause symptoms of fat malabsorption. Thus, patients with certain diseases have shown improvement when MCTs are included in their diet[31]. MCTs are also increasingly utilized in the feeding of critically ill or septic patients who presumably gain benefits in the setting of associated intestinal dysfunction. Further investigation should clarify potential roles for MCTs in patients with lipid disorders associated with lipoprotein lipase and carnitine deficiencies. MCTs may be used in confectioneries and in other functional foods as carriers for flavors, colors, and vitamins[20]. MCTs have clinical applications in the treatment of fat malabsorption, maldigestion, obesity, and metabolic difficulties related to cystic fibrosis, Crohn's disease, colitis, and enteritis[31,32].

1.2.11 BIOCHEMISTRY AND METABOLISM OF ESSENTIAL FATTY ACIDS (EFAs)

The EFAs are PUFAs which means that they have two or more double bonds in their backbone structure. There are two groups of EFAs, the n-3 fatty acids and the n-6 fatty acids. They are defined by the position of the double bond in the molecule nearest to the methyl end of the chain. In the n-3 group of fatty acids it is between the third and fourth carbon atoms and in the n-6 group of fatty acids it is between the sixth and seventh carbon atoms. The parent compounds of the n-6 and n-3 groups of fatty acids are LA and ALA, respectively. LA and ALA are considered to be essential fatty acids for human health because humans cannot synthesize them and must obtain them from the diet. Within the body, these parent compounds are metabolized by a series of alternating desaturations (in which an extra double bond is inserted by removing two hydrogen atoms) and elongations (in which two carbon atoms are added) as shown in Figure 1.3. This requires a series of special enzymes called desaturases and elongases. It is believed that the enzymes metabolizing both n-6 and n-3 fatty acids are identical[33], resulting in competition between the two PUFA families for these enzymes[3]. Chain elongation and desaturation occurs only at the carboxyl end of the fatty acid molecule[34].

The potential health benefits of n-3 fatty acids include reduced risk of cardiovascular disease, inflammation, hypertension, allergies, and immune and renal disorders[35–37]. Epidemiological studies have linked the dietary intake of n-3 PUFAs in Greenland Eskimos to their lower incidence of coronary heart disease[38,39]. Research has shown that DHA is essential for proper function of central nervous system and visual acuity of infants[40]. The n-3 fatty acids are essential for normal growth and development throughout the life cycle of humans and therefore should be included in the diet. Fish and marine oils are rich sources of n-3 fatty acids, especially EPA and DHA. Cod liver, menhaden, and sardine oils contain approximately 30% EPA and DHA.

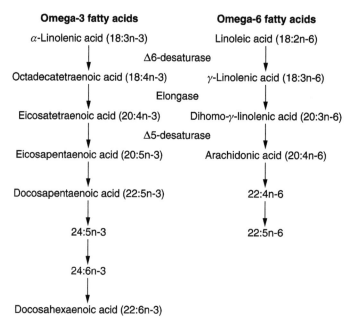

FIGURE 1.3 Metabolic pathways of the omega-3 and omega-6 fatty acids.

The n-6 fatty acids exhibit various physiological functions in the human body. The main functions of these fatty acids are related to their roles in the membrane structure and in the biosynthesis of short-lived derivatives (eicosanoids) which regulate many aspects of cellular activity. The n-6 fatty acids are involved in maintaining the integrity of the water impermeability barrier of the skin. They are also involved in the regulation of cholesterol transport in the body.

GLA, a desaturation product of linoleic acid, has shown therapeutic benefits in a number of diseases, notably atopic eczema, cyclic mastalgia, premenstrual syndrome, cardiovascular disease, inflammation, diabetes, and cancer[33]. Arachidonic acid is found in meats, egg yolk, and human milk. GLA is found in oats, barley, and human milk. GLA is also found in higher amounts in plant seed oils such as those from borage, evening primrose, and blackcurrant. Algae such as *Spirulina* and various species of fungi also seem to be desirable sources of GLA.

1.2.12 EICOSANOIDS

Much attention has been paid to the role of EFAs as precursors of a wide variety of short-lived hormone-like substances called eicosanoids. They are 20-carbon endogenous biomedical mediators derived from EFAs, notably AA and DGLA of the n-6 family and EPA of the n-3 family[41]. DGLA, AA, and EPA are precursors for eicosanoid series 1, 2, and 3, respectively. The members of the eicosanoid cascade include the prostaglandins, prostacyclins, thromboxanes, leukotrienes, and hydroxy fatty acids. They play a major role in regulating the cell-to-cell communication involved in cardiovascular, reproductive, respiratory, renal, endocrine, skin, nervous, and immune system actions. Arachidonic acid is derived from linoleic acid, which gives rise to series-2 prostaglandins, series-2 prostacyclins, series-2 thromboxanes, and series-4 leukotrienes. These end products of n-6 fatty acid metabolism induce inflammation and immunosuppression. Prostanoids (collective name for prostaglandins, prostacyclins, thromboxanes) of series-1 and leukotrienes of series-3 are produced from DGLA. When n-3 fatty acids are processed in the eicosanoid cascade, series-3 prostaglandins, series-3 prostacyclins, series-3 thromboxanes, and series-5 leukotrienes are formed.

The biological activities of the eicosanoids derived from n-3 fatty acids differ from those produced from n-6 fatty acids. For example, series-2 prostaglandins formed from AA may impair the immune functions while series-3 prostaglandins produced from EPA ameliorate immunodysfunction. Thromboxane A_2 produced by AA is a potent vasoconstrictor and platelet aggregator[42]. Thromboxane A_3 synthesized from EPA is a mild vasoconstrictor and has shown antiaggregatory properties[43,44]. Furthermore, n-3 fatty acids competitively inhibit the formation of eicosanoids derived from the n-6 family of fatty acids. In general, the eicosanoids derived from the n-3 PUFAs are less powerful in their effects than those derived from the n-6 PUFAs[42].

1.3 MAJOR SOURCES OF NUTRACEUTICAL AND SPECIALTY LIPIDS

1.3.1 FISH OILS

Fish oils and fish meal are the most convenient sources of n-3 fatty acids. Fish oils contribute about 2% to the total production of fats and oils. They come from various fish species such as menhaden, sardine, herring, anchovy, capelin, and sand eel. Fish oils are generally characterized by a rather large group of saturated and unsaturated fatty acids, which are commonly associated with mixed TAGs. In addition to TAGs, fish oils usually include small amounts of fatty acids as substituents of phospholipids and other lipids. The fatty acids derived from fish oils are of three principal types: saturated, monounsaturated, and polyunsaturated. The saturated fatty acids have carbon chain lengths that generally range from 12 to 24 (mainly 14:0, 16:0, and 18:0). Traces of eight- and ten-carbon fatty acids may also be found in some fish oils. The carbon chain lengths of the unsaturated fatty acids range generally from 14 to 22 (mainly 16:1, 18:1, 20:1, and 22:1). Small amounts of 10- to 12-carbon monounsaturated fatty acids have been found in some fish oils. The major n-3 PUFAs are generally 18:4, 20:5, and 22:6. The lipid of most common fish is 8 to12% EPA and 10 to 20% DHA. The fatty acid composition of menhaden oil, as an example of a fish oil, is given in Table 1.5.

Traditionally, fish oils have been used after a partial hydrogenation process. This is essential if the oil is to be used as a component of fat spreads. The process of partial hydrogenation provides a more desirable product with increased oxidative stability with a required melting behavior. However, the nutritional value of the product is compromised. During hydrogenation, n-3 PUFAs may be converted to fatty acids with lower unsaturation with some double bonds of *trans* configuration.

TABLE 1.5
Fatty Acid Composition of Seal Blubber and Menhaden Oils

Fatty acid (wt%)	Seal blubber oil	Menhaden oil
10:0	—	—
12:0	—	—
14:0	3.4	8.3
14:1	1.0	0.4
16:0	5.0	17.1
16:1n-7	15.1	11.4
18:1n-9 and n-11	26.4	12.1
18:2n-6	1.3	1.4
20:1n-9	15.0	1.4
20:5n-3	5.4	13.2
22:1n-11	3.6	0.1
22:5n-3	4.9	2.4
22:6n-3	7.9	10.1

The n-3 PUFAs, especially EPA and DHA, present in fish and fish oils have an imputed major positive role in human health and disease. It has been shown that n-3 PUFAs in fish oils have an inhibitory effect on platelet aggregation, and this reduces the risk of thrombosis, which is a major cause of stroke and heart attack[42]. Furthermore, the n-3 PUFAs in fish oils are very effective in lowering serum TAGs. DHA is essential for the development of brain and retina in infants[40]. As a consequence awareness of these possible beneficial effects, increased consumption of n-3 fatty acids in the form of either dietary fish or fish oil capsules is a recognized change in the nutritional habits of many individuals.

1.3.2 SEAL BLUBBER OIL (SBO)

Seal blubber is mostly (98.9%) composed of neutral lipids[45]. The blubber oil is rich in long-chain PUFAs, especially those of the n-3 family. The fatty acid composition of SBO is quite similar to that of fish oils as it contains a large proportion of highly unsaturated fatty acids. Table 1.5 summarizes the fatty acid profile of seal blubber and menhaden oils. A comparison of the fatty acid composition of these oils indicated that menhaden oil had a higher amount of EPA and DHA than SBO, but the latter had a higher content of docosapentaenoic acid (DPA, 4.7%) than fish oils, and contained 6.4% EPA and 7.6% DHA.

SBO and fish oils differ from one another in the dominance and distribution of fatty acids in the TAG molecules. The n-3 fatty acids, such as EPA, DPA, and DHA, in SBO are mainly located in the primary positions (sn-1 and sn-3) of TAG molecules. Thus, the proportions in the sn-1 and sn-3 positions were EPA, 8.4 and 11.2%; DPA, 4.0 and 8.2%; and DHA, 10.5 and 17.9%, respectively. However, in fish oils these fatty acids are preferentially esterified at the sn-2 position of the TAGs (17.5% EPA, 3.1% DPA, and 17.2% DHA). In the sn-2 position of SBO, EPA, DPA, and DHA were present at 1.6, 0.8, and 2.3%, respectively[46]. During digestion, the fatty acids are liberated from the primary positions of TAGs via hydrolysis by pancreatic lipase. The rate of hydrolysis at the sn-2 position of TAGs is very slow, and as a result the fatty acids at this position remain intact as 2-MAGs during digestion and absorption.

Processing steps of SBO are similar to those of vegetable oils. The basic processing steps for the manufacturing of SBO involve rendering to release the oil followed by degumming, alkali refining, bleaching, and deodorization. Each processing step has a specific function and may affect the quality of the resultant oil by removing certain major and minor components. During processing, impurities such as free fatty acids, mono- and diacylglycerols, phospholipids, sterols, vitamins, hydrocarbons, pigments, protein and their degradation products, suspended mucilagenous and colloidal materials, and oxidation products are removed from the oil. Heating may be required to denature the residual flesh proteins and to break the cell walls so that the oil and water can be easily removed[47]. Alkali refining removes free fatty acids, phospholipids, metals, as well as some colored compounds[48]. The refined oil is washed with water to remove any remaining traces of soap. Refined oil is then heated and mixed with bleaching clay to remove various colored compounds as well as phospholipids, metals, soap, and oxidation products. This process is generally carried out under vacuum to minimize oxidation. Subsequently, the oil is deodorized in order to remove off-odor volatiles.

Different procedures may be explored for improving the oxidative stability of SBO. Particular emphasis may be placed on the use of natural antioxidants, especially dechlorophyllized green tea extracts (DGTE) and individual tea catechins[49]. The results showed strong antioxidant activity for DGTE as well as individual tea catechins, namely epicatechin (EC), epigallocatechin (EGC), epicatechin gallate (ECG), and epigallocatechin gallate (EGCG), when added to SBO. The potency of catechins in retarding oxidation of SBO was in the decreasing order of ECG > EGCE > EGC > EC. Therefore, DGTE and isolated tea catechins may be used as effective natural antioxidants for stabilization of SBO. tert-Butylhydroquinone (TBHQ), a synthetic antioxidant, was also highly effective in retarding oxidation of SBO.

Microencapsulation provides an alternative method for stabilization of edible oils, possibly together with the use of antioxidants. Among the encapsulating materials tested for encapsulation and hence stabilization of SBO, β-cyclodextrin was most effective and retained 89% of total PUFA content of SBO, after storing the encapsulated oil at room temperature for 49 days[50]. Changes in the n-3 fatty acid content of stored SBO, both in the encapsulated and unencapsulated forms, have been determined (data not shown). The n-3 fatty acid content of unencapsulated SBO decreased by 50%. However, in β-cyclodextrin, the encapsulated SBO remained nearly unchanged even after 49 days of storage. The total PUFA content in β-cyclodextrin-encapsulated SBO decreased marginally after 49 days of storage while that for the control sample changed from 22.6 to 11.5%. The progression of peroxide values in unencapsulated and encapsulated SBO stored at room temperature has been evaluated (data not shown). β-Cyclodextrin served better in controlling the formation of peroxides than the control. The peroxide value of control samples increased from 2.1 to 29.8 meq/kg oil over 49 days of storage. The corresponding values for β-cyclodextrin-encapsulated oil were smaller, changing from 3.0 to 10.2 meq/kg oil. These results suggest that microencapsulation of SBO, using starch-based wall materials, improved the oxidative stability of the oil and preserved the integrity of nutritionally important n-3 fatty acids.

The major use of marine oils has traditionally been for the production of margarine and other edible oil products following their hydrogenation. Because of potential health benefits of unaltered PUFAs, incorporation of these fatty acids into the diet has shown promise for both food manufacturers and nutritionists. A wide variety of foods such as bread, baby foods, margarine, and salad dressings have been produced with n-3 fatty acids obtained from marine oils[51]. Microencapsulated fish oil has been produced and may be used in health food formulations. These products are in powdered forms. They can be incorporated mainly into milk powders, reduced fat products, fruit drinks, salad dressings, soups, cakes, and biscuits[52]. SBO may be used in the manufacturing of the above products. Furthermore, this oil may also be used in infant formulas.

Refined marine oils or their n-3 fatty acid concentrates have been used in pharmaceuticals such as EPA and DHA capsules as well as skin and other personal care products. For nutraceutical applications, SBO may be used in the form of a liquid, as soft gel capsules, or in the microencapsulated form. SBO may provide a very good starting material for preparation of n-3 fatty acid concentrates, as discussed earlier. Marine oils have also been used for topical applications to treat various skin disorders.

SBO may also lend itself to nonedible applications. Oleochemicals (fatty acids, fatty alcohols, esters of alcohols, and nitrogen derivatives) derived from marine oils find a wide range of industrial applications. Marine oils may be used in the production of lubricants, corrosion inhibitors, textile, leather, and paper additives, cleaners, and personal care products. Marine oils have been used as an alternative fuel to petroleum-based products. SBO has traditionally been used for industrial purposes. It was used as a major fuel source for lighthouses in Newfoundland. Other industrial uses of marine oils are in the production of polyurethane resins, cutting oils, printing ink formulations, insecticides, and leather treatment, among others.

1.3.3 BORAGE, EVENING PRIMROSE, AND BLACKCURRANT OILS

Borage (*Borago officinalis* L.) is an annual herbaceous plant and is commercially grown in North America. Borage oil (BO), extracted from seeds of the blue, star shaped borage flower, is attracting the attention of alternative health practitioners and mainstream medicine alike for its profound medicinal properties. Although the oil is getting all of the credit, it is actually the oil's active component, GLA, that has drawn the interest of researchers. GLA is the first intermediate in the bioconversion of LA to AA, and the first step of Δ-6 desaturation (synthesis of GLA from LA) is known to be rate limiting. The seeds of borage contain approximately 38% oil with a GLA content of 20 to 25%[53] (Table 1.6). The level of GLA in the seeds is approximately 7% and this is about three times that in evening primrose seeds. The oil is made up of 95.7% neutral lipids, 2.0%

TABLE 1.6
Fatty Acid Composition of Borage, Evening Primrose, and Blackcurrant Oils

Fatty acid (wt%)	Borage oil	Evening primrose oil	Blackcurrant oil
16:0	9.8	6.2	7.3
18:0	3.1	1.7	2.8
18:1n-9	15.2	8.7	13.3
18:2n-6	38.4	73.6	31.2
18:3n-6	24.4	9.9	19.3
20:1n-9	4.1	—	3.3
22:1n-9	—	—	8.4
24:1	2.5	—	2.1

glycolipids, and 2.3% phospholipids[54]. Neutral lipids of BO are composed of TAGs (99.1%), DAGs (0.06%), MAGs (0.02%), FFAs (0.91%), and sterols (0.02%)[54].

The oil from evening primrose (*Oenothera biennis* L.) is another commercial source of GLA. Evening primrose is a biennial plant and is a common weed that is native to North America. Interest in evening primrose oil (EPO) has intensified in recent years because of its GLA content. Although the evening primrose plant does not produce a high yield of seeds compared to the well-known commercial oilseeds, it is preferred to other sources of GLA because it is easy to produce and does not contain any ALA. At present, EPO is the most important source of GLA, which is in growing demand for its clinical and pharmaceutical applications[55]. EPO is currently available in over 30 countries as a nutritional supplement or as a constituent of specialty foods. In a number of countries, certain nutritional products require governmental registration before they can be marketed. Several large organizations have been able to establish moderately large-scale extraction facilities for oils. The EPO capsules contain 10 to 12% GLA and in Canada were marketed by the Efamol company. The oil content of seeds was 17 to 25%[53,56], of which 7 to 10% was GLA[57,58]. EPO is very high in LA (70 to 75%) (Table 1.6). The total GLA content of the seeds is approximately 2.5%[56]. The oil, as marketed, is made up of 97 to 98% TAG, 1.5 to 2.0% unsaponifiable matter, and 0.5 to 1.0% polar lipids[55]. EPO is generally obtained by mechanical pressing followed by extraction with hexane[59]. There are preliminary indications that EPO may be more effective in some of its physiological effects than other oils in which GLA occurs. One possible explanation is that GLA is present in EPO almost entirely as molecular species of TAG in which one GLA is combined with two LA molecules[58]. Another possibility is that minor components of EPO, not GLA, are responsible for some of the effects. GLA from other oils (borage, blackcurrant, and fungal) may also be biologically less effective than that from EPO, partly because of the other fatty acids present and partly because of the different TAG structure of the oils[33]. The TAG stereospecific structure of EPO is distinct, with GLA being concentrated in the *sn*-3 position[60].

Blackcurrant is a perennial berry crop and is mainly cultivated in Europe and Asia. Blackcurrant is a round, dull black berry[59] and its seeds contain about 30% oil[61] which may be extracted by hexane[59]. The oil differs from BO and EPO in that it contains significant levels of two n-3 fatty acids, namely ALA (18:3n-3) and stearidonic acid (18:4n-3). Blackcurrant oil (BCO), having a GLA content of 15 to 19%[61,62] (Table 1.6), also contains a potent GLA inhibitor, erucic acid (22:1n-9) which reduces its advantage as a medicinal oil[62]. LA is the major PUFA found in BCO[60]. The main uses of BCO, as in the case of borage and evening primrose, are generally based on claims concerning pharmacological properties of GLA[59].

1.3.4 CONCENTRATION OF n-3 FATTY ACIDS FROM MARINE OILS

Several techniques have been explored for the concentration of PUFAs from marine oils. Methods traditionally employed for the concentration of PUFAs in oils make use of differences in physical and chemical properties between saturated and unsaturated fatty acids. For example, the melting points of fatty acids are dependent on their degree of unsaturation. EPA and DHA melt at –54 and –44.5°C compared to 13.4 and 69.6°C for 18:1 and 18:0, respectively[63]. As the temperature of a mixture of a saturated and unsaturated fatty acids decreases, the saturated fatty acids, having a higher melting point, start to crystallize out first and the liquid phase becomes enriched in the unsaturated fatty acids. However, as the number and type of fatty acid components in the mixture increases, the crystallization process becomes more complex and repeated crystallization and separation of fractions must be carried out to obtain purified fractions. In the case of marine oils, not only is there a very wide spectrum of fatty acids but the fatty acids exist, not in the FFA form, but esterified in TAGs. However, the principle of low-temperature crystallization can still be applied to marine oils partially to concentrate TAGs rich in n-3 PUFAs[64]. SBO in the TAG and FFA forms were subjected to low-temperature crystallization using solvents such as hexane and acetone in order to obtain n-3 fatty acid concentrates. When SBO TAGs were dissolved in acetone, the total n-3 fatty acid content of the oil was increased to 48% at –70°C. Meanwhile, when SBO was used in the FFA form, in the presence of hexane, the total n-3 fatty acid content was increased to 66.7% at –70°C.

The ease of complexation of straight-chain saturated fatty acids with urea in comparison with PUFA is well established, and conventional urea complexation techniques using ethanol or methanol as a solvent can be applied to the fatty acids of oils or their methyl or ethyl esters to produce a fraction rich in PUFAs. Initially, the TAGs of the oil are hydrolyzed into their constituent fatty acids via alkaline hydrolysis using alcoholic KOH or NaOH. The resultant free fatty acids are then mixed with ethanolic solution of urea for complex formation. The saturated and monounsaturated fatty acids are readily complexed with urea and crystallize out on cooling and may be removed by filtration. The liquid fraction is enriched with n-3 fatty acids. Urea complex formation of fatty acids has been extensively used for enriching marine oils in n-3 PUFAs[65]. Urea complexation of fatty acids of BO, using methanol, can increase the GLA content from 23.6 to 94%[66]. Haagsma et al.[67] described a urea complexation method for enriching the EPA and DHA levels of cod liver oil from 12 to 28% and 11 to 45%, respectively.

Supercritical fluid extraction is a relatively novel technique which has found use in food and pharmaceutical applications. The process makes use of the fact that at a combined temperature and pressure above a critical point, a gas such as CO_2 has a liquid-like density and possesses a high solvation capacity[64]. This method is mild and, because it uses CO_2, minimizes autoxidation. A number of gases are known to have good solvent properties at pressures above their critical values. For food applications, CO_2 is the solvent of choice because it is inert, inexpensive, nonflammable, environmentally acceptable, safe, readily available, and has a moderate critical temperature (31.1°C) and pressure (1070 psig). It separates fatty acids most effectively on the basis of chain length; hence the method works best for oils with low levels of long-chain fatty acids. Fish oils in the form of free fatty acids and fatty acid esters have been extracted with supercritical gaseous CO_2 to yield concentrates of EPA and DHA. The drawbacks of this method include the use of extremely high pressure and high capital costs.

For the concentration of PUFAs on a large scale, each of the above physical and chemical separation methods has some disadvantages in terms of low yield, a requirement for large volumes of solvent or sophisticated equipment, a risk of structural changes in the fatty acid products, or high operational costs. Lipases work under mild conditions of temperature and pH[68], a factor that favors their potential use for the enrichment of PUFAs in oils. Lipases (EC 3.1.1.3) are enzymes that catalyze the hydrolysis, esterification, interesterification, acidolysis, and alcoholysis reactions. The common feature among lipases is that they are activated by an interface. Lipases have been used for many years to modify the structure and composition of foods. Lipases that act on neutral lipids generally hydrolyze the esters of PUFAs at a slower rate than those of more saturated fatty acids[69]. Use has been made of this relative substrate specificity to increase the concentration of n-3 PUFAs in seal blubber and menhaden oils by subjecting them to hydrolysis by a number of microbial lipases[64]. Concentration of n-3 fatty acids by enzyme-assisted reactions involves mild reaction conditions and provides an alternative to the traditional concentration methods such as distillation and chromatographic separation. Furthermore, concentration via enzymatic means may also produce n-3 fatty acids in the acylglycerol form, which is nutritionally preferred.

1.3.5   APPLICATION OF LIPASES IN SYNTHESIS OF SPECIALTY LIPIDS

Enzymes have been used for production of nutraceutical lipids used for confectionery fat formulations and nutritional applications. Interesterification of high oleic (18:1) sunflower oil and stearic acid using immobilized *Rhizomucor miehei* lipase produces mainly1,3-distearoyl-2-monolein (StOSt). Other reactants may also be used for production of specialty lipids useful as confectionery fats. In particular, there are many reports on enzymatic interesterification of mixtures of palm oil fractions and stearic acid or stearic acid esters to produce fats containing high concentrations of StOSt and 1-palmitoyl-2-oleoyl-3-stearoyl-glycerol (POSt)[70]. These products are the main components of cocoa butter, and enzymatic interesterification processes can produce fats with compositions and physical properties very similar to cocoa butter[71].

The enzymes may also be used to synthesize a human milk fat substitute for use in infant formulas[72]. Acidolysis reaction of a mixture of tripalmitin and unsaturated fatty acids using a *sn*-1,3-specific lipase as biocatalyst afforded TAGs derived entirely from vegetable oils rich in 2-position palmitate with unsaturated fatty acyl groups in the *sn*-1 and *sn*-3 positions. These TAGs closely mimic the fatty acid distribution found in human milk fat, and when they are used in infant formulas instead of conventional fats the presence of palmitate in the *sn*-2 position of the TAGs has been shown to improve digestibility of the fat and absorption of other important nutrients such as calcium.

The possible application of enzyme-assisted reactions for production of lower value nonspecialty lipids such as margarine hardstocks and cooking oils has been investigated. When nonspecific lipases such as those of *Candida cylindraceae* and *C. antarctica* are used as biocatalysts for interesterification of oil blends, the TAG products are very similar to those obtained by chemical interesterification[70]. Therefore, replacement of chemical interesterification by an enzyme process giving similar products is technically feasible, although it has not been adopted on a commercial scale to date, largely because of the comparatively high process and catalyst costs.

Enzymatic reactions can also be used for production of fats and oils containing nutritionally important PUFAs, such as EPA and DHA. For example, various vegetable and fish oils have been enriched in the EPA and DHA by enzyme-catalyzed reactions[73–75]. Use of this technique to produce structured lipids with MCFAs and PUFAs located specifically in either the *sn*-2 or *sn*-1,3 positions of the TAG has been described. Enzymatic processes are particularly suitable for the production and modification of lipids containing PUFAs, because these unstable fatty acids are susceptible to damage under the more severe conditions used for chemical processing.

Interesterification of blends of palm and hydrogenated canola oils and cottonseed and hydrogenated soybean oils using *sn*-1,3-specific lipases as catalysts gave fats with a low *trans* fatty acid content that were effective as margarine hardstocks[76]. Reaction of mixtures of palm stearine and lauric fats using immobilized *Rhizomucor miehei* as catalyst also produced fats that were functional as margarine hardstocks[77]. With these enzymatically interesterified fats, margarine could be formulated without using hydrogenated fats.

1.3.6   STRUCTURED LIPIDS

Structured lipids (SLs) are TAGs containing short- and/or medium- as well as long-chain fatty acids preferably located in the same glycerol molecule. They can be produced by chemical or enzymatic processes and may be prepared as nutraceutical lipids for nutritional, pharmaceutical, and medical applications. These TAGs have been modified by incorporation of desired fatty acids, or by changing the fatty acid profiles from their native state in order to produce novel TAGs. SLs are designed for use as nutraceutical or functional lipids. These specialty lipids may be synthesized via direct esterification, acidolysis, alcoholysis, or interesterification reactions. However, the common methods reported in the literature for the synthesis of SLs are based on reactions between two TAG molecules (interesterification) or between a TAG and an acid (acidolysis). These specialty lipids have been developed to optimize fully the benefit of various fatty acid moieties. SLs have been reported to have beneficial effects on a range of metabolic parameters including immune function, nitrogen balance, and improved lipid clearance from the bloodstream[72]. SLs are also synthesized to improve or change the physical and/or chemical properties of TAGs. Research on SLs remains an interesting area that holds great promise for the future.

Nutraceutical is a term used to describe foods that provide health benefits beyond those ascribed to their nutritional effect benefits[78]. These products may be referred to as functional foods or functional lipids if they are incorporated into products that have the usual appearance of food, but to which they may be added and provide specific health benefits[78]. SLs can be designed for use as medical or functional foods as well as nutraceuticals, depending on the form of use.

Lipids can be modified to incorporate specific fatty acids of interest in order to achieve desired functionality. SLs may be synthesized via the hydrolysis of fatty acyl groups from a mixture of TAGs followed by random reesterification onto the glycerol backbone[27]. Various fatty acids are used in this process, including different classes of saturated, monounsaturated, and n-3 and n-6 PUFAs, depending on the desired metabolic effect. Thus, a mixture of fatty acids is incorporated onto the same glycerol molecule. SLs containing MCFAs and LCFAs have modified absorption rates because MCFAs are rapidly oxidized for energy and LCFAs are oxidized very slowly. SLs are expected to be less toxic than physical mixtures of oils. These specialty lipids are structurally and metabolically different from the simple physical mixtures of MCTs and LCTs. SLs containing MCFAs at the *sn*-1,3 positions and LA at the *sn*-2 position may have beneficial effects both as an energy source and as a source of essential fatty acid[23].

1.3.6 SYNTHESIS OF STRUCTURED LIPIDS FROM VEGETABLE OILS AND n-3 FATTY ACIDS

Borage, evening primrose, blackcurrant, and fungal oils are predominant sources of GLA (18:3n-6). GLA has been used in the treatment of atopic eczema, dermatitis, hypertension, and premenstrual syndrome. Also, n-3 PUFAs have potential for prevention of cardiovascular disease, arthritis, hypertension, immune and renal disorders, diabetes, and cancer[17]. SLs containing both GLA and n-3 PUFAs may be of interest because of their desired health benefits. We have successfully produced SLs containing GLA, EPA, and DHA in the same glycerol backbone using borage and evening primrose oils as the main substrates[73,75]. In this study, a number of commercially available enzymes, namely lipases from *Candida antarctica* (Novozym-435), *Mucor miehei* (Lipozyme-IM), and *Pseudomonas* sp. (Lipase PS-30), were used as biocataysts with free EPA and DHA as acyl donors. Higher incorporation of EPA + DHA (34.1%) in borage oil was obtained with *Pseudomonas* sp. lipase, compared to 20.7 and 22.8% EPA + DHA, respectively, with *Candida antarctica* and *Mucor miehei* lipases. Similarly, in evening primrose oil *Pseudomonas* sp. lipase gave the highest degree of EPA + DHA incorporation (31.4%) followed by lipases from *Mucor miehei* (22.8%) and *Candida antarctica* (17.0%). The modified borage and evening primrose oils thus obtained may have potential health benefits.

Recently, EPA and capric acid (10:0) have been incorporated into borage oil using two immobilized lipases, SP435 from *Candida antarctica* and IM60 from *Rhizomucor miehei*, as biocatalysts[79]. Higher incorporation of EPA (10.2%) and 10:0 (26.3%) was obtained with IM60 lipase, compared to 8.8 and 15.5%, respectively, with SP435 lipase.

Huang and Akoh[80] used immobilized lipases IM60 from *Mucor miehei* and SP435 from *Candida antarctica* to modify the fatty acid composition of soybean oil by incorporation of n-3 fatty acids. The transesterification reaction was carried out with free fatty acid and ethyl esters of EPA and DHA as acyl donors. With free EPA as acyl donor, *Mucor miehei* lipase gave a higher incorporation of EPA than *Candida antarctica* lipase. However, when ethyl esters of EPA and DHA were the acyl donors, *Candida antarctica* lipase gave a higher incorporation of EPA and DHA than *Mucor miehei* lipase.

Akoh and Sista[81] have shown that the fatty acid composition of borage oil can be modified using EPA ethyl ester with an immobilized lipase from *Candida antarctica*. The highest incorporation (31%) was obtained with 20% *Candida antarctica* lipase. At a substrate mole ratio of 1:3, the ratio of n-3 to n-6 fatty acids was 0.64. Under similar conditions, Akoh et al.[82] were able to increase the n-3 fatty acid content (up to 43%) of evening primrose oil with a corresponding increase in the n-3/n-6 ratio from 0.01 to 0.6. Sridhar and Lakshminarayana[83] modified the fatty acid composition of groundnut oil by incorporating EPA and DHA using a *sn*-1,3-specific lipase from *Mucor miehei* as the biocatalyst. The modified groundnut oil had 9.5% EPA and 8.0% DHA.

Ju et al.[84] incorporated n-3 fatty acids into the acylglycerols of borage oil. They have selectively hydrolyzed borage oil using immobilized *Candida rugosa* lipase and then used this product with n-3 fatty acids for the acidolysis reaction. The total content of n-3 and n-6 fatty acids in acylglycerols was 72.8% following acidolysis. The contents of GLA, EPA, and DHA in the SL so prepared were 26.5, 19.8, and 18.1%, respectively. The n-3/n-6 ratio increased from 0 to 1.09, following the acidolysis.

Huang et al.[85] incorporated EPA into crude melon seed oil by two immobilized lipases, IM60 from *Mucor miehei* and SP435 from *Candida antarctica* as biocatalysts. Higher EPA incorporation was obtained using EPA ethyl ester than using EPA free fatty acid for both enzyme-catalyzed reactions.

1.3.7 SYNTHESIS OF STRUCTURED LIPIDS FROM MARINE OILS AND MEDIUM-CHAIN FATTY ACIDS

Lipase-catalyzed acidolysis may be used to produce SLs containing MCFAs in the *sn*-1 and *sn*-3 positions. We used an immobilized *sn*-1,3-specific lipase from *Mucor miehei* to incorporate capric acid (10:0; a MCFA) into SBO containing EPA and DHA. After modification, the fatty acid composition of SBO was different from that of the unaltered oil. Under optimum reaction conditions (500 mg oil, 331 mg capric acid, 45°C, 24 h, 1% water, 83.1 mg lipase, and 3 mL hexane), a SL containing 27.1% capric acid, 2.3% EPA, and 7.6% DHA was obtained. Positional distribution of fatty acids in the SL revealed that *Mucor miehei* lipase incorporated capric acid predominantly at the *sn*-1,3- positions of TAG molecules.

Jennings and Akoh[86] were able to incorporate capric acid (10:0) into fish oil TAGs using immobilized lipase from *Rhizomucor miehei* (IM 60). The fish oil (produced by Pronova Biocare Inc., Sandefjord, Norway) originally contained 40.9% EPA and 33.0% DHA. After a 24 h incubation in hexane, there was an average of 43% incorporation of capric acid into fish oil, while EPA and DHA decreased to 27.8 and 23.5%, respectively. Akoh and Moussata[79] used acidolysis reaction to incorporate capric acid (10:0) and EPA into borage oil using lipase from *Candida antarctica* and *Rhizomucor miehei* as biocatalysts. Higher incorporation of EPA (10.2%) and 10:0 (26.3%) was obtained with *Rhizomucor miehei* lipase, compared to 8.8 and 15.5%, respectively, with *Candida antarctica* lipase.

Iwasaki et al.[87] produced a SL from of a single-cell oil (produced by a marine microorganism, *Schizochytrium* sp.) containing docosapentaenoic acid (DPA; 22:5n-6) and DHA and caprylic acid (8:0) using lipases from *Rhizomucor miehei* and *Pseudomonas* sp. The targeted products were SL containing caprylic acid at the *sn*-1 and *sn*-3 positions and DHA or DPA at the *sn*-2 position of glycerol. When *Pseudomonas* sp. was used, more than 60% of fatty acids in single-cell oil were exchanged with caprylic acid. With *Rhizomucor miehei* lipase, the incorporation of caprylic acid was only 23%. Their results suggested that the difference in the degree of acidolysis by the two enzymes were due to their different selectivity toward DPA and DHA as well as the difference in their positional specificities.

1.3.8   Synthesis of SBO-Based Structured Lipids

We have produced SLs via acidolysis of SBO and GLA (18:3n-6) with lipases PS-30 from *Pseudomonas* sp. and Lipozyme IM from *Mucor miehei* as the biocatalysts. The highest incorporation of GLA (37%) into SBO was achieved with lipase PS-30 (data not shown). The modified SBO contained 37% GLA, 3.8% EPA, and 4.3% DHA[88]. Thus, SLs containing GLA (a n-6 fatty acid), EPA, and DHA (n-3 fatty acids) are produced and may have potential health benefits. The oils containing both n-3 and n-6 fatty acids are considered important for specific clinical as well as nutritional applications.

The fatty acid composition of SBO was also modified by incorporating a MCFA, capric acid (10:0), using a *sn*-1,3-specific lipase from *Mucor miehei*. The content of capric acid incorporated into SBO was 25.4%. Stereospecific analysis of modified oils revealed that capric acid was preferentially esterified at the *sn*-1,3 positions of TAG of SBO. Even though EPA (8.8%) and DHA (10.8%) were mainly located at *sn*-1,3 positions, *sn*-2 position also contained significant amounts of these fatty acids (4.7% EPA and 4.1% DHA). Structured lipids containing n-3 PUFAs at the *sn*-2 position and MCFAs at the *sn*-1,3 positions are expected to supply quick energy to individuals with lipid malabsorption disorders and enhance the absorption of the n-3 PUFAs.

1.3.9   Low-Calorie Structured and Specialty Lipids

The synthesis of low-calorie lipids, which are characterized by a combination of SCFAs and/or MCFAs and LCFAs in the same glycerol backbone, is an interesting area in the field of structured and specialty lipids. Interest in these types of products emerged from the fact that they contain 5 to 7 kcal/g caloric value compared to the 9 kcal/g of conventional fats and oils because of the lower caloric content of SCFAs or MCFAs compared to their LCFA counterparts. Reduced-calorie specialty lipids are designed for use in baking chips, coatings, dips, and bakery and dairy products, or as cocoa butter substitutes (Table 1.7). Currently, such products are synthesized by random chemical interesterification between a short-chain TAGs (SCTs) and LCTs, typically a hydrogenated vegetable oil such as soybean or canola oil[89].

Caprenin, composed of one molecule of a very long-chain saturated fatty acid, behenic acid ($C_{22:0}$), and two molecules of MCFAs, caprylic acid ($C_{8:0}$) and capric acid ($C_{10:0}$), is a commercially

TABLE 1.7
Examples of Reduced-Calorie Lipids and Fat Replacers

Type	Ingredients	Applications
Caprenin	Capric, caprylic, behenic acids, and glycerol	Confections; soft candies
Salatrim	Soybean oil, canola oil	Dairy products, baked goods, confections, margarine, spreads
Olestra	Sucrose core with 6 to 8 fatty acids	Baked goods, fried foods, savory snacks, salad dressing
Sorbestrin	Sorbitol, methyl or ethyl esters of fatty acids	Baked goods, fried foods
Simplesse	Milk and/or egg/white proteins, pectin, sugar, citric acid	Baked goods, ice cream, butter, sour cream, cheese, yogurt

available reduced-calorie SL. It provides 5 kcal/g[90] compared to 9 kcal/g of conventional fats and oils. This product was originally produced by Procter and Gamble Company (Cincinnati, OH). The constituent fatty acids for caprenin synthesis are obtained from natural food sources. For example, caprylic and capric acids are obtained by fractionation of palm kernel and coconut oils while behenic acid is produced from rapeseed oil. Behenic acid, being a very long-chain saturated fatty acid, is poorly absorbed regardless of its position on the glycerol moiety. The MCFAs provide fewer calories than absorbable LCFAs. Caprenin reportedly has functional characteristics similar to cocoa butter and can be used as a cocoa butter substitute in selected confectionery products. It is digested, absorbed, and metabolized by the same pathway as other TAGs[91]. Caprenin is a liquid or semisolid product at room temperature, has a bland taste, and is fairly heat stable. The U.S. Food and Drug Administration (FDA) has received a Generally Recognized As Safe (GRAS) petition for caprenin for use in soft candy bars and in confectionery coatings for nuts, fruits, and cookies.

Salatrim is also a reduced-calorie SL, which is composed of a mixture of very short-chain fatty acids ($C_{2:0}$–$C_{4:0}$) and LCFAa (predominantly $C_{18:0}$)[89]. The SCFAs are chemically transesterified with vegetable oils such as highly hydrogenated canola or soybean oil. The very short-chain fatty acids reduce the caloric value to about 5 kcal/g[90] and LCFAs provide the lipid functionality. Salatrim was developed by Nabisco Foods Group (East Hanover, NJ) and is now marketed under the brand name Benefat™ by Cultor Food Science, Inc. (New York, NY). It has the taste, texture, and functional characteristics of conventional fats. It can be produced to have different melting profiles by adjusting the amounts of SCFAs and LCFAs used in the chemical synthesis. Reduced-fat baking chips are one of the products in the market that contain Salatrim and were introduced in 1995 by Hershey Food Corporation (Hershey, PA). Salatrim received FDA GRAS status in 1994 and can also be used as a cocoa butter substitute. It is intended for use in chocolate-flavored coatings, chips, caramel, fillings for confectionery and baked goods, peanut spreads, savory dressings, dips and sauces, and dairy products[92].

Neobee, another example of a low-calorie SL, is composed of capric and caprylic acids and produced by Stepan Company (Maywood, NJ). This class of specialty lipids includes different products. For example, Neobee® 1053 and Neobee® M-5 contain both capric and caprylic acids while Neobee® 1095 is made up of only capric acid[28]. Neobee® 1095 is a solid product. Therefore, in certain applications, which require solid fats, this product may be suitable. Neobee® 1814 is a MCT derivative made by interesterifying MCT with butter oil[93]. This product contains half of the long-chain saturated fatty acid found in conventional butter oil and is suitable to replace butter oil in a variety of applications. Neobee® 1814 may serve as a flavor carrier and functions as a textural component for low-fat food products[28].

1.4 LOW-CALORIE FAT SUBSTITUTES

Consumers are demanding low-fat and even nonfat products with sensory qualities similar to those of the regular products. As a result, a number of fat substitutes have been developed.

1.4.1 OLESTRA (SUCROSE POLYESTER)

A slightly different carbohydrate-based approach is behind a lipid substitute known chemically as sucrose polyester. The commercial name of this product is Olestra® or Olean®, manufactured by Procter and Gamble Company (Cincinnati, OH). It is made by combining the disaccharide sucrose with six to eight fatty acids via ester linkages, to produce large polymer molecules. These fatty acids are derived from vegetable oils such as soybean or corn oil. It received FDA approval in 1996, based on numerous studies concerning its safety and nutritional effects. Olestra has been approved to replace fully conventional lipids in the preparation of savory or salty snack foods such as chips and crackers. However, it is anticipated that in the future it will be used in salad dressings, shortening, table spreads, and dairy products.

Olestra contains no available calories because the ester linkages hold the sucrose and fatty acid molecules together in a way that cannot be broken down by digestive enzymes and therefore is not absorbed. The taste and mouthfeel of olestra are similar to those of conventional fat[90]. The color, heat stability, and shelf-life stability of oil made with Olestra are comparable to those of conventional fat. It has been extensively tested to determine its safety. Olestra behaves in the mouth much like fat. It travels undigested through the gastrointestinal tract without any of the energy locked up in its structure being made available to the body. Olestra is a fat-free and cholesterol-free substance, is stable under ambient and high-temperature storage conditions, has acceptable flavor, and is suitable for deep frying and baking because it decomposes into the same byproducts as regular fat[94]. One potential drawback of the consumption of Olestra is that it causes a decreased absorption of fat-soluble vitamins. Thus, vitamins A, D, E, and K have been added to Olestra to compensate this. It may also cause flatulence, abdominal cramping, diarrhea, increased bowl movement, reduce absorption of cholesterol, and depletion of carotenoids.

1.4.2 SIMPLESSE

A chemically different approach has been obtained to create protein-based substances with the desired properties of fats. A low-calorie, cholesterol-free fat substitute, called Simplesse, is commercially available at present. It was introduced by the NutraSweet Co. (Chicago, IL) in 1988[95]. The caloric value of Simplesse is about 1 to 2 kcal/g. It can be used in a variety of food applications. It can be added to dairy products such as ice cream, yogurt, cream cheese, and sour cream as well as oil-based foods such as mayonnaise, salad dressings, and margarine[95,96]. It is produced from milk and egg white protein[89] by a process of microparticulation. During this process, proteins in solution are deaerated and heated to a temperature just below the coagulation point of proteins. The solution will then be homogenized and sheared at elevated temperatures. Under heat and shear, the proteins coagulate and shape into small spheroidal particles ranging in size from 0.1 to 2.0 μm. The protein aggregates are so small that the mouth cannot perceive them individually. Once ingested, it is digested and absorbed by the body as protein. The final product provides the rich, creamy mouthfeel properties of fats and oils. Simplesse cannot be used to cook foods because heat causes the protein to gel and lose its creamy quality. The U.S. FDA approved the use of Simplesse in frozen desserts in 1990.

1.4.3 SORBESTRIN (SORBITOL POLYESTER)

This is a low-calorie, thermally stable, liquid lipid substitute composed of fatty acid esters of sorbitol and sorbitol anhydrides. Sorbestrin belongs to the family of carbohydrate-based fatty acid polyesters. Sorbitol and sorbitol anhydrides serve as the backbone of this compound, which is esterified with fatty acids of varying chain length and degree of saturation. It has a caloric value of approximately 1.5 kcal/g and provides a bland oil-like taste. Sorbestrin is suitable for use in all vegetable oil applications including fried foods, salad dressing, mayonnaise, and baked goods. It was discovered in the late 1980s by Pfizer Inc. (New York, NY) and is currently under development by Danisco Cultor America Inc. (Ardsley, NY).

1.4.4 ESTERIFIED PROPOXYLATED GLYCEROLS (EPGS)

EPGs are analogs of TAGs in which a propoxyl group has been introduced between the glycerol backbone and the fatty acids to replace the ester linkage with an ether linkage. Glycerol is first reacted with propylene oxide to form a polyether glycol and then esterified with fatty acids to yield an oil-like product. These fatty acids may be obtained from edible oils such as lard, tallow, corn, canola, soybean, and cottonseed oils. The physical properties of the finished product depend on the type of fatty acids esterified. EPGs are thermally stable. It is manufactured by ARCO Chemical Co. (Newton Square, PA) and suitable for use in formulated products as well as baking and frying applications.

1.4.5 Paselli

This product has been developed by Avebe America Inc. (Princeton, NJ). It is a potato starch-based ingredient[90] that can replace fats and oils in a variety of products such as dips, sources, salad dressings, dessert toppings, dairy products, and bakery goods[95]. Paselli has a caloric value of about 3.8 kcal/g. Under appropriate temperature conditions, this product forms a thermostable gel with a smooth, fat-like texture and neutral taste. It has been used commercially in ice cream, puddings, and meat products.

1.4.6 N-Oil

This product is composed of a tapioca dextrin[90] and can partially or totally replace fats and oils in foods, giving the illusion of high fat content[95]. It has been commercially available since 1984 and is marketed by National Starch and Chemical Corp. (Bridgewater, NJ). It has a caloric value of about 1 kcal/g. This product is resistant to high temperature, shear, and acidic conditions[90]. It can be used in frozen desserts, salad dressings, puddings, spreads, dairy products, and soups[96].

REFERENCES

1. Lee, R.M.K.W., Fish oil, essential fatty acids, and hypertension, *Can. J. Physiol. Pharmacol.*, 72, 945–953, 1994.
2. Nettleton, J.A., n-3 Fatty acids: comparison of plant and seafood sources in human nutrition, *J. Am. Diet. Assoc.*, 91, 331–337, 1991.
3. Simopoulos, A.P., Omega-3 fatty acids in health and disease and in growth and development, *Am. J. Clin. Nutr.*, 54, 438–463, 1991.
4. Newton, I.S., Long chain fatty acids in health and nutrition. *J. Food Lipids*, 3, 233–249, 1996.
5. Lovern, J.A., The lipids of marine organisms, *Ocenogr. Mar. Biol. Ann. Rev.*, 2, 169–191, 1964.
6. Menzel, D.B. and Olcott, H.S., Positional distribution of fatty acids in fish and other animal lecithins, *Biochem. Biophys. Acta*, 84, 133–137, 1964.
7. Young, F.V.K., The production and use of fish oils, in *Nutritional Evaluation of Long-Chain Fatty Acids*, Barlow, S.M. and Stansby, M.E., Academic Press, London, 1982, pp. 1–23.
8. Bligh, E.G. and Dyer, W.J., A rapid method of total lipid extraction and purification, *Can. J. Biochem. Physiol.*, 37, 911–917, 1959.
9. Christie, W.W., Analysis of complex lipids, in *Lipid Analysis*, 2nd ed., Pergamon Press, New York, 1982, pp. 107–134.
10. Parrish, C.C., Separation of aquatic lipid classes by chromarod thin-layer chromatography with measurement by iatroscan flame ionization detection, *Can. J. Fish Aquat. Sci.*, 44, 722–731, 1987.
11. Stein, J., Chemically defined structured lipids: current status and future directions in gastrointestinal diseases, *Int. J. Colorect. Dis.*, 14, 79–85, 1999.
12. Hashim, A. and Babayan, V.K., Studies in man of partially absorbed dietary fats, *Am. J. Clin. Nutr.*, 31, 5273–5276, 1978.
13. Ruppin, H., Bar-Meir, S., Soergel, K.H., Wood, C.M., and Schmitt, M.G., Absorption of short chain fatty acids by the colon, *Gastroenterology*, 78, 1500–1507, 1980.
14. Kripke, S.A., De Paula, J.A., Berman, J.M., Fox, A.D., Rombeau, J.L., and Settle, R.G., Experimental short-bowel syndrome: effect of an elemental diet supplemented with short-chain triglycerides, *Am. J. Clin. Nutr.*, 53, 954–962, 1991.
15. Harada, E. and Kato, S., *Am J. Physiol.*, 244, G284–G290, 1983.
16. Roediger, W.E.W. and Rae, D.A., *Brit. J. Surg.*, 69, 23–25, 1982.

17. Senanayake, S.P.J.N. and Shahidi, F., Structured lipids containing long-chain omega-3 polyunsaturated fatty acids, in *Seafood in Health and Nutrition. Transformation in Fisheries and Aquaculture: Global Perspectives*, Shahidi, F., Ed., ScienceTech, St. John's, NF, Canada, 2000, pp. 29–44.

18. Bell, S.J., Mascioli, E.A., Bistrian, B.R., Babayan, V.K., and Blackburn, G.L., Alternative lipid sources for enteral and parenteral nutrition: long- and medium-chain triglycerides, structured triglycerides, and fish oils. *J. Am. Diet. Assoc.*, 91, 74–78, 1991.

19. Young, F.V.K., *J. Am. Oil Chem. Soc.*, 60, 374–379, 1983.

20. Megremis, C.L., Medium-chain triglycerides: a nonconventional fat, *Food Technol.*, 45, 108–110, 1991.

21. Heird, W.C., Grundy, S.M., and Hubbard, V.S., Structured lipids and their use in clinical nutrition, *Am. J. Clin. Nutr.*, 43, 320–324, 1986.

22. Lee, T.W. and Hastilow, C.I., Quantitative determination of triacylglycerol profile of structured lipid by capillary supercritical fluid chromatography and high-temperature gas chromatogaphy, *J. Am. Oil Chem. Soc.*, 76, 1405–1413, 1999.

23. Ikeda, I., Tomari, Y., Sugano, M., Watanabe, S., and Nagata, J., Lymphatic absorption of structured glycerolipids containing medium-chain fatty acids and linoleic acid, and their effect on cholesterol absorption in rats, *Lipids*, 26, 369–373, 1991.

24. Mascioli, E.A., Bistrian, B.R., Babayan, V.K., and Blackburn, G.L., Medium-chain triglycerides and structured lipids as unique nonglucose energy sources in hyperalimentation, *Lipids*, 22, 421–423, 1987.

25. Bell, S.J., Bradley, D., Forse, R.A., and Bistrian, B.R., The new dietary fats in health and disease, *J. Am. Diet. Assoc.*, 97, 280–286, 1997.

26. Johnson, R.C., Young, S.K., Cotter, R., Lin, L., and Rowe, W.B., Medium-chain-triglyceride lipid emulsion: metabolism and tissue distribution, *Am. J. Clin. Nutr.*, 52, 502–508, 1990.

27. Babayan, V.K., Medium-chain triglycerides and structured lipids, *Lipids*, 22, 417–420, 1987.

28. Heydinger, J.A. and Nakhasi, D.K., Medium chain triacylglycerols, *J. Food Lipids*, 3, 251–257, 1996.

29. Haumann, B.F., Structured lipids allow fat tailoring, *INFORM*, 8, 1004–1011, 1997.

30. Ulrich, H., Pastores, S.M., Katz, D.P., and Kvetan, V., Parenteral use of medium-chain triglycerides: a reappraisal, *Nutrition*, 12, 231–238, 1996.

31. Kennedy, J.P., Structured lipids: fats for the future, *Food Technol.*, 11, 76–83, 1991.

32. Bach, A.C. and Babayan, V.K., Medium-chain triglycerides: an update, *Am. J. Clin. Nutr.*, 36, 950–962, 1982.

33. Horrobin, D.F., Gamma linolenic acid: an intermediate in essential fatty acid metabolism with potential as an ethical pharmaceutical and as a food, *Rev. Contemp. Pharmacother.*, 1, 1–41, 1990.

34. Salem, N., Jr. and Ward, G.R., Are n-3 fatty acids essential nutrients for mammals?, *World Rev. Nutr. Diet.*, 72, 128–147, 1993.

35. Kinsella, J.E., Lokesh, B., and Stone, R.A., Dietary n-3 polyunsaturated fatty acids and amelioration of cardiovascular disease: possible mechanisms, *Am. J. Clin. Nutr.*, 52, 1–28, 1990.

36. Weber, P.C. and Leaf, A., Cardiovascular effects of n-3 fatty acids, *World Rev. Nutr. Diet.*, 66, 218–232, 1991.

37. Endres, S., De Caterina, R., Schmidt, E.B., and Kristensen, S.D., n-3 Polyunsaturated fatty acids: update 1995. *Eur. J. Clin. Invest.*, 25, 629–638, 1995.

38. Bang, H.O. and Dyerberg, J., Plasma lipids and lipoproteins in Greenlandic West-coast Eskimos, *Acta Med. Scand.*, 192, 85–94, 1972.

39. Bang, H.O. and Dyerberg, J., Lipid metabolism and ischemic heart disease in Greenland Eskimos, *Adv. Nutr. Res.*, 3, 1–21, 1986.

40. Carlson, S., The role of PUFA in infant nutrition, *INFORM* 6, 940–946, 1995.

41. Branden, L.M. and Carroll, K.K., Dietary polyunsaturated fats in relation to mammary carcinogenesis in rats, *Lipids*, 21, 285–288, 1986.

42. Groom, H., Oil-rich fish, *Nutr. Food Sci.*, Nov./Dec., 4–8, 1993.

43. Rice, R., *Lipid Technol.*, 23, 112–116, 1991.

44. Bajpai, P. and Bajpai, P.K., Eicosapentaenoic acid (EPA) production from microorganisms: a review, *J. Biotech.*, 30, 161–183, 1993.

45. Shahidi, F., Seal blubber, in *Seal Fishery and Product Development*, ScienceTech, St. John's, Canada, 1998, pp. 99–146.

46. Wanasundara U.N. and Shahidi, F., Positional distribution of fatty acids in triacylglycerols of seal blubber oil, *J. Food Lipids*, 4, 51–64, 1997.

47. Bimbo, A.P.. The emerging marine oil industry, *J. Am. Oil Chem. Soc.*, 64, 706–715, 1987.
48. Carr, R.A., Refining and degumming systems for edible fats and oils, *J. Am. Oil Chem. Soc.*, 55, 765–771, 1978.
49. Wanasundara, U.N. and Shahidi, F., Stabilization of seal blubber and menhaden oils with green tea catechins, *J. Am. Oil Chem. Soc.*, 73, 1183–1190, 1996.
50. Wanasundara, U.N. and Shahidi, F., Storage stability of microencapsulated seal blubber oil, *J. Food Lipids*, 2, 73–86, 1995.
51. Bimbo, A.P., Technology of production and industrial utilization of marine oils, in *Marine Biogenic Lipids, Fats and Oils*, Ackman, R.G., Ed., CRC Press, Boca Raton, FL, 1989, pp. 401–433.
52. Newton, I.S., Food enrichment with long-chain n-3 PUFA, *Food Technol.*, 7, 169–177, 1996.
53. Beaubaire, N.A. and Simon, J.E., Production potential of *Borago officinalis* L., *Acta Horticulturae*, 208, 101–103, 1987.
54. Senanayake, S.P.J.N. and Shahidi, F., Lipid components of borage (*Borago officinalis* L.) seeds and their changes during germination, *J. Am. Oil Chem. Soc.*, 77, 55–61, 2000.
55. Hudson, B.J.F., Evening primrose (*Oenothera* spp.) oil and seed, *J. Am. Oil Chem. Soc.*, 61, 540–543, 1984.
56. Wolf, R.B., Kleiman, R., and England, R.E., New sources of γ-linolenic acid, *J. Am. Oil Chem. Soc.*, 60, 1858–1860, 1983.
57. Gibson, R.A., Lines, D.R., and Neumann, M.A., Gamma linolenic acid (GLA) content of encapsulated evening primrose oil products, *Lipids*, 27, 82–84, 1992.
58. Fieldsend, A., Evening primrose-from garden flower to oilseed crop, *The Horticulturist*, 5, 2–5, 1996.
59. Helme, J.P., Evening primrose, borage, and blackcurrant seeds, in *Oils and Fats Manual. A Comprehensive Treatise. Properties, Productions and Applications*, Karleskind, A., Ed., Intercept, Andover, UK, 1996, pp. 168–179.
60. Lawson, L.D. and Hughes, B.G., Triacylglycerol structure of plant and fungal oils containing γ-linolenic acid, *Lipids*, 23, 313–317, 1988.
61. Traitler, H., Winter, H., Richli, U., and Ingenbleek, Y., Characterization of γ-linolenic acid in *Ribes* seed, *Lipids*, 19, 923–928, 1984.
62. Walker, M., Medicinal oils, *Health Foods Business*, 37, 95–96, 1991.
63. *Merck Index., An Encyclopedia of Chemical, Drugs and Biologicals*, 10th ed., Merck and Co., Rahway, NJ, 1983.
64. Shahidi, F. and Wanasundara, U.N., Omega-3 fatty acid concentrates: nutritional aspects and production technologies, *Trends Food Sci. Technol.*, 9, 230–240, 1998.
65. Hayes, D.G., Alstine, J.M.V., and Setterwall, F., Urea-based fractionation of seed oil samples containing fatty acids and acylglycerols of polyunsaturated and hydroxy fatty acids, *J. Am. Oil Chem. Soc.*, 77, 207–213, 2000.
66. Spurvey, S.A. and Shahidi, F., Concentration of gamma linolenic acid (GLA) from borage oil by urea complexation: optimization of reaction conditions, *J. Food Lipids*, 7, 163–174, 2000
67. Haagsma, N., Gent, C.M., Luten, J.B., Jong, R.W., and Doorn, E., Preparation of an ω3 fatty acid concentrate from cod liver oil, *J. Am. Oil Chem. Soc.*, 59, 117–118, 1982.
68. Gandhi, N.N., Applications of lipase, *J. Am. Oil Chem. Soc.*, 74, 621–634, 1997.
69. Villeneuve, P. and Foglia, T.A., Lipase specificities: potential application in lipid bioconversions, *INFORM*, 8, 640–650, 1997.
70. Macrae, A.R., Lipase-catalyzed interesterification of oils and fats, *J. Am. Oil Chem. Soc.*, 60, 291–294, 1983.
71. Macrae, A.R., Interesterification of fats and oils, in *Biocatalysis in Organic Syntheses*, Tramper, H.C., van der Plas, and Linko, P., Eds., Elsevier, Amsterdam, 1985, pp. 195–208.
72. Quinlan, P. and Moore, S., Modification of triglycerides by lipases: process technology and its application to the production of nutritionally improved fats, *INFORM*, 4, 580–585, 1993.
73. Senanayake, S.P.J.N. and Shahidi, F., Enzyme-assisted acidolysis of borage (*Borago officinalis* L.) and evening primrose (*Oenothera biennis* L.) oils: incorporation of omega-3 polyunsaturated fatty acids, *J. Agric. Food Chem.*, 47, 3105–3112, 1999.
74. Senanayake, S.P.J.N. and Shahidi, F., Enzymatic incorporation of docosahexaenoic acid into borage oil, *J. Am. Oil Chem. Soc.*, 76, 1009–1015, 1999.
75. Senanayake, S.P.J.N. and Shahidi, F., Modified oils containing highly unsaturated fatty acids and their stability, in *Omega-3 Fatty Acids. Chemistry, Nutrition and Health Effects*, Shahidi, F. and Finley, J.W., Eds., American Chemical Society, Washington, DC, 2001, pp. 162–173.

76. Mohamed, H.M.A. and Larsson, K., Modification of fats by lipase interesterification: 2. Effect on crystallisation behaviour and functional properties, *Fat Sci. Technol.*, 96, 56–59, 1994.

77. Posorske, L.H., LeFebvre, G.K., Miller, C.A., Hansen, T.T., and Glenvig, B.L., Process considerations of continuous fat modification with an immobilised lipase, *J. Am. Oil Chem. Soc.*, 65, 922–926, 1988.

78. Scott, F.W. and Lee, N.S., Recommendations for defining and dealing with functional foods, *Report of the Bureau of Nutritional Science Committee on Functional Foods*, Food Directorate, Health Protection Branch, Ottawa, 1996.

79. Akoh, C.C. and Moussata, C.O., Lipase-catalyzed modification of borage oil: incorporation of capric and eicosapentaenoic acids to form structured lipids, *J. Am. Oil Chem. Soc.*, 75, 697–701, 1998.

80. Huang, K. and Akoh, C.C., Lipase-catalyzed incorporation of n-3 polyunsaturated fatty acids into vegetable oils, *J. Am. Oil Chem. Soc.*, 71, 1277–1280, 1994.

81. Akoh, C.C. and Sista, R.V., Enzymatic modification of borage oil: incorporation of eicosapentaenoic acid, *J. Food Lipids*, 2, 231–238, 1995.

82. Akoh, C.C., Jennings, B.H., and Lillard, D.A., Enzymatic modification of evening primrose oil: incorporation of n-3 polyunsaturated fatty acids, *J. Am. Oil Chem. Soc.*, 73, 1059–1062, 1996.

83. Sridhar, R. and Lakshminarayana, G., Incorporation of eicosapentaenoic and docosahexaenoic acids into groundnut oil by lipase-catalyzed ester interchange, *J. Am. Oil Chem. Soc.*, 69, 1041–1042, 1992.

84. Ju, Y., Huang, F., and Fang, C., The incorporation of n-3 polyunsaturated fatty acids into acylglycerols of borage oil via lipase-catalyzed reactions, *J. Am. Oil Chem. Soc.*, 75, 961–965, 1998.

85. Huang, K., Akoh, C.C., and Erickson, M.C., Enzymatic modification of melon seed oil: incorporation of eicosapentaenoic acid, *J. Agric. Food Chem.*, 42, 2646–2648, 1994.

86. Jennings, B.H. and Akoh, C.C., Enzymatic modification of triacylglycerols of high eicosapentaenoic and docosahexaenoic acids content to produce structured lipids, *J. Am. Oil Chem. Soc.*, 76, 1133–1137, 1999.

87. Iwasaki, Y., Han, J.J., Narita, M., Rosu, R., and Yamane, T., Enzymatic synthesis of structured lipids from single cell oil of high docosahexaenoic acid content, *J. Am. Oil Chem. Soc.*, 76, 563–569, 1999.

88. Spurvey, S.A., Senanayake, S.P.J.N., and Shahidi, F., Enzyme-assisted acidolysis of menhaden and seal blubber oils with gamma-linolenic acid, *J. Am. Oil Chem. Soc.*, 78, 1105–1112, 2001.

89. Smith, R.E., Finley, J.W., and Leveille, G.A., Overview of SALATRIM, a family of low-calorie fats, *J. Agric. Food. Chem.*, 42, 432–434, 1994.

90. Schaich, K.M., Rethinking low-fat formulations: matching fat functionality to molecular characteristics, *The Manufacturing Confectioner*, June, 109–122, 1997.

91. Artz, W.E. and Hansen, S.L., Current developments in fat replacers, in *Food Lipids and Health*, McDonald, R.E. and Min, D.B., Eds., Marcel Dekker, New York, 1996, pp. 385–415.

92. Kosmark, R., Salatrim: properties and applications, *Food Technol.*, 50, 98–101, 1996.

93. Babayan, V.K., Blackburn, G.L., and Bistrian, B.R., Structured Lipid Containing Dairy Fat, US Patent 4, 952, 606, 1990.

94. Sandrou, D.K. and Arvanitoyannis, I.S., Low-fat/calorie foods: current state and perspectives, *Crit. Rev. Food Sci. Nutr.*, 40, 427–447, 2000.

95. Dziezak, J.D., Fats, oils and fat substitutes, *Food Technol.*, July, 66–74, 1989.

96. McClements, D.J. and Demetriades, K., An integrated approach to the development of reduced-fat food emulsions, *Crit. Rev. Food Sci. Nutr.*, 38, 511–536, 1998.

2 Medium-Chain Triacylglycerols

Yaakob B. Che Man and Marina Abdul Manaf
Department of Food Technology, Faculty of Food Science and
Technology, Universiti Putra Malaysia, Serdang, Selangor

CONTENTS

2.1 INTRODUCTION

Over the years, much information has been revealed regarding the consumption of fats and oils. The increasing knowledge and understanding accumulated on the effects of chain length, position, and metabolism of fatty acids on health have directed the use of dietary lipids towards prevention as well as treatment of diseases in order to improve health status.

Medium-chain triacylglycerols (MCTs) are recognized for their benefits. Although they fall in the category of saturated fat, MCTs surpass other saturated fats and fatty acids due to their unique properties. Medium-chain fatty acids (MCFAs) make up the MCTs. MCTs were introduced into the clinical arena approximately 50 years ago. MCTs have found uses in various fields from medical to cosmetic applications. They are now being utilized in many food applications as well. In 1994

MCTs were approved as GRAS (Generally Recognized As Safe) for oral or enteral use by the U.S. Food and Drug Administration (USFDA), confirming the good tolerance of MCTs in human nutrition[1].

MCTs are used in the clinical arena for enteral and parenteral nutrition in diverse medical conditions for treatment of patients suffering from fat malabsoprtion. MCTs are included in infant formulas to aid in fat digestion and absorption of the immature digestive system of infants. Supplements are sold as MCT oil with the aim of increasing metabolic rate for weight reduction and providing additional energy for sport activities.

With the diverse applications and roles in disease prevention and optimizing health, MCTs serve as ideal functional fats. Functional food is defined as "any food or food ingredient that may provide a health benefit beyond the traditional nutrients it contains"[2]. MCTs are certainly beneficial and have proved to be valuable in promoting the health of humans.

2.2   SOURCES OF MCTS

2.2.1   COCONUT OIL

The primary source of MCFA is lauric oils, which consist mostly of coconut oil and palm kernel oil[3]. They are called "lauric" due to the higher lauric acid content in these oils (see Table 2.1)[4]. World coconut oil production in 2000/2001 was 3.4 MMT or 3% of the total oils and fats production. The Philippines and Indonesia account for 66% of production and export of coconut oil[5].

Coconut oil is commercially derived from copra, which is the dried kernel or "meat" of coconut. It is a colorless to pale brownish yellow oil. Coconut oil is fluid in warm tropical climates but changes into solid fat in temperate climates. Coconut oil contains a high level of low molecular weight saturated fatty acids, the distinctive characteristic of lauric oils. Coconut oil has a sharp and low melting point, ranging from 23 to 26°C. This property made coconut oil useful for synthetic creams, hard butter, and other similar products. Coconut oil contains a low level of unsaturation and therefore is stable to oxidation[6]. However, the stability of the refined oil is lower due to the loss of natural antioxidants during the refining process. Thus, citric acid is added in the last stage of deodorization to give protection from metal-catalyzed oxidative rancidity[7].

Compared to palm kernel oil, coconut oil is higher in saponification and Reichert–Meissll and Polenske values (indicative of short-chain fatty acids present) due to the higher level of low molecular weight fatty acids in the oil. The Reichert–Meissl value is the amount of sodium hydroxide required to neutralize volatile fatty acids (mainly C4 and C6) in an oil, whereas the Polenske value reflects the amount of sodium hydroxide needed to neutralize insoluble fatty acids, mainly C6, C10, and C12, in an oil. The Reichert values for coconut oil and palm kernel oil are 6 to 8.5 ml and 4 to 7 ml, respectively, and 13 to 18 ml for the Polenske values[8]. The odor and taste of coconut oil is largely contributed to δ- and γ-lactones, which are also responsible for the strong smell of crude palm kernel oil[9].

Coconut oil is usually extracted from copra by pressing in screw presses (expellers) and generally followed by solvent extraction[10]. However, other methods of extraction such as using aqueous enzymes[11] and pure culture of *Lactobacillus plantarum*[12] are also able to afford a higher yield and quality of coconut oil compared to the traditional wet process. Enzymatic treatment offers a high yield due to the mild conditions employed[13].

The chemical composition of coconut oil made its use possible in a wide range of edible and nonedible products (Table 2.1). Coconut oil has unique characteristics such as bland flavor, pleasant odor, high resistance to rancidity, narrow temperature range of melting, easy digestibility and absorbability, high gross for spray oil use, and superior foam retention capacity for whip-topping use[14]. Unlike other oils, coconut oil passes rather abruptly from butter solid to liquid within a narrow temperature range rather than exhibiting a gradual softening with increasing temperature. Some of

TABLE 2.1
Coconut Oil Composition and Physical Characteristics

Characteristic	Typical	Range
Specific gravity, 30/30°C	—	0.915–0.920
Refractive index, 40°C	—	1.448–1.449
Iodine value	10.0	7.5–10.5
Saponification number	—	248–264
Unsaponifiable number	—	0.1–0.8
Titer (°C)	—	20.0–24.0
Melting point (°C)	26.5	25.0–28.0
Solidification point (°C)	—	14.0–22.0
AOM stability (h)	150	30–250
Tocopherol content (ppm):		
γ-Tocopherol	6.0	3–9
Tocotrienol content (ppm):		
α-Tocotrienol	49.0	27–71
Fatty acid composition (%):		
Caproic (C6:0)	0.5	0.4–0.6
Caprylic (C8:0)	7.8	6.9–9.4
Capric (C10:0)	6.7	6.2–7.8
Lauric (C12:0)	47.5	45.9–50.3
Myristic (C14:0)	18.1	16.8–19.2
Palmitic (C16:0)	8.8	7.7–9.7
Stearic (C18:0)	2.6	2.3–3.2
Oleic (C18:1)	6.2	5.4–7.4
Linoleic (C18:2)	1.6	1.3–2.1
Arachidic acid (C20:1)	0.1	<0.2
Gadoleic (C20:1)	Trace	<0.2
Triacylglycerol composition (%):		
Trisaturated	84.0	—
Disaturated	12.0	—
Monounsaturated	4.0	—
Triunsaturated	0.0	—
Crystal habit	β'	—
Solids fat index (%) at:		
10.0°C	54.5	—
21.1°C	26.6	—
26.7°C	0.0	—

Source: O'Brien, R.D., *Fats and Oils: Formulating and Processing for Applications*, CRC Press, New York, 2004, pp. 40–43.

the edible products made from coconut oil include frying oil, shortening, margarine, and confectionary. Coconut oil is an excellent material for nonedible purposes due to desirable properties such as good resistance to rancidity, biodegradability, nonoily character, and mildness to skin. Some of the important nonedible uses of coconut oil are in the production of soaps, plastics, rubbers, and chemical products[15].

2.2.2 PALM KERNEL OIL

Palm kernel oil is the other important lauric oil, besides coconut oil. It is obtained from the kernel of palm fruit, usually enclosed in a hard woody shell. It differs greatly from palm oil in appearance, characteristics, and composition although originating from the same fruit (Table 2.2). Palm kernel

TABLE 2.2
Palm Kernel Oil Composition and Physical Characteristics

Characteristic	Typical	Range
Specific gravity, 40/20°C	—	0.860–0.873
Refractive index, 40°C	1.451	1.448–1.452
Iodine value	17.8	16.2–19.2
Saponification number	245.0	243–249
Unsaponifiable number	0.3	0.3–0.5
Titer (°C)	—	20.0–29.0
Melting point (°C)	28.3	26.8–29.8
Solidification point (°C)	—	20.0–24.0
AOM stability (h)	100+	15–100+
Tocopherol content (ppm)	3.0	3–10
Fatty acid composition (%):		
Caproic (C6:0)	0.2	0.1–0.5
Caprylic (C8:0)	3.3	3.4–5.9
Capric (C10:0)	3.4	3.3–4.4
Lauric (C12:0)	48.2	46.3–51.1
Myristic (C14:0)	16.2	14.3–16.8
Palmitic (C16:0)	8.4	6.5–8.9
Stearic (C18:0)	2.5	1.6–2.6
Oleic (C18:1)	15.3	13.2–16.4
Linoleic (C18:2)	2.3	2.2–3.4
Arachidic acid (C20:1)	0.1	Trace–0.9
Gadoleic (C20:1)	0.1	Trace–0.9
Crystal habit	β'	—
Solids fat index (%) at:		
10.0°C	48.0	—
21.1°C	31.0	—
26.7°C	11.0	—

Source: O'Brien, R.D., *Fats and Oils: Formulating and Processing for Applications*, CRC Press, New York, 2004, pp. 40–43.

oil constitutes about 45% of the palm nut of palm oil fruit. The kernel contains about 45 to 50% of oil on a wet basis[16]. In general, palm kernel oil resembles coconut oil, being a colorless to brownish yellow oil, solidifying in temperate climates to a white or yellowish fat. Coconut and palm kernel oils are derived from the fruit of palm trees but from different species. Coconut palm belongs to *Cocos nucifera* and palm kernel belongs to *Elais guineensis*. World palm kernel oil production in 2000/2001 was 2.9 MMT or 2.5% of total oils and fats production. Malaysia and Indonesia contribute 78% of total production and 90% of the export of palm kernel oil[5].

Palm kernel oil has a similar fatty acid composition to coconut oil. The difference is that coconut oil contains about twice as much caprylic and capric acids. Compared to coconut oil, palm kernel oil has a higher degree of unsaturation. Palm kernel oil contains about 48% of lauric acid, 16% of myristic acid, and 15% of oleic acid[17]. Besides triacylglycerol (TAG) and free fatty acid (FFA), palm kernel oil contains unsaponifiable matter such as sterols, tocols, triterpene alcohols, hydrocarbons, and lactones[18].

Palm kernel oil is more useful as a raw material for edible products than coconut oil. This is due to the slight difference in their chemical composition[19]. Palm kernel oil contains a significantly higher content of oleic acid than coconut oil, which aids in the production of specialty fats because hydrogenation can change this TAG from being liquid to solid at end-use temperature (20 to 24°C), thus permitting more control of the properties of the final fat. It melts at a much higher temperature

(28°C) than coconut oil (24°C) due to the slight difference in their chemical composition. Palm kernel oil requires refining (neutralization, bleaching, filtering, and deodorization) but the bleaching step is not as drastic as that for palm oil[20].

Applications of palm kernel oil in both edible and nonedible fields are quite similar to those of coconut oil because of similarities in their chemical composition and properties. The difference is that palm kernel oil is more unsaturated, which makes its use possible in many types of products for the food industry by means of hydrogenation. It is a valuable component of margarine formulation, giving a rapid melt in the mouth. Its high solid content (15 to 20°C) makes it useful in confectionary products. Palm kernel oil is also used in frying oils, specialty fats (cocoa butter substitute), filling cream, nondairy whipping cream, and ice cream[21]. Palm kernel oil has also found uses in the oleochemical industry. It is used in the production of short fatty acids, fatty alcohols, methyl esters, and amides for use in detergents and cosmetics and commercially fractionated into olein and stearin[22]. Palm kernel oil along with coconut oil is used in the production of MCTs, which are then used for other applications such as health and infant food products.

2.2.3 OTHER SOURCES OF MCTs

Besides coconut oil and palm kernel oil, there are other plants that are rich in MCFAs such as babassu, cuphea, tukum, murumuru, ouricuri, and cohume[5]. They do not enter international trade and are produced only in small quantities. Native babassu palm is found over a very large area in Brazil. This natural resource is only partially exploited to produce oil. However, many other fuels and chemicals can be produced from babassu coconut. Babassu has been integrated in producing lauric oil, charcoal, animal feed, and ethanol[23]. Many species from the genus cuphea have potential as sources of MCTs. Cuphea is a unique genus in the plant kingdom with the diversity of major fatty acids produced in its seeds. Unlike other flowering plants for which their seeds constitute mostly linoleic acid (C18:2), cuphea is rich in MCFAs, which are used for manufacturing soap and pharmaceuticals[24]. However, cuphea generally exhibit some wild plant characteristics such as sticky glandular hairs on stem, leaves, and flowers, indeterminate pattern of growth, and flowering and seed dormancy. These characteristics limit its application in domestication and production[25].

2.3 PHYSICOCHEMICAL PROPERTIES OF MCTS

Since MCTs have short chain length and full saturation, MCT oils have different chemical and physiological properties from long-chain triacylglycerols (LCTs). Compared to LCTs, they are smaller molecules and have a lower melting point, making them liquid at room temperature[26]. The melting points of MCFAs are 16.7 and 31.3°C for C8:0 and C10:0, respectively. MCFAs are soluble in water (0.68 mg/ml for C8:0 versus 0.72 mg/ml for C10:0). They are weak electrolytes and highly ionized at neutral pH, which enhances their solubility in biological fluids. These properties of MCTs affect the way they are absorbed and metabolized. These same properties make them useful ingredients in health, sport, and infant food products.

MCTs consist of a mixture of caproic acid (C6:0, 1 to 2%), caprylic acid (C8:0, 35 to 75%), capric acid (C10:0, 25 to 35%), and lauric acid (C12:0, 1 to 2%)[27]. Since 1950 MCTs have been synthesized by hydrolysis of lauric oils to MCFAs and glycerol. The glycerol is drawn off from the resultant mixture and the MCFAs are fractionally distilled. The desired fatty acids (caprylic and capric acids mixture) are finally reesterified to glycerol with or without a catalyst to form the TAG[5].

2.4 MECHANISM OF FAT DIGESTION AND ABSORPTION

Differences in digestion, absorption, and metabolism of MCTs and LCTs and their corresponding fatty acids, MCFAs and LCFAs, respectively, arise largely from differences in their physiochemical properties. These differences affect both the rate and fate of carbon metabolism of MCTs relative to LCTs[28].

MCTs are more rapidly and completely hydrolyzed compared to LCTs. Fat digestion occurs in the stomach, which is catalyzed partially by lingual or gastric lipases[29]. MCTs are then rapidly hydrolyzed by pancreatic lipase within the intestinal lumen. The product of MCT hydrolysis, monoacylglycerol and MCFAs are then rapidly absorbed through the stomach mucosa into the hepatic portal vein after ingestion[1]. LCTs, however, must be emulsified and hydrolyzed in the gut lumen before absorption. The digestion products of the gastric phase are diacylglycerol and free fatty acids, which facilitate the intestinal phase of digesting as emulsifying agents[30]. Pancreatic lipase is needed to cleave the fatty acid moieties from the *sn*-1 and *sn*-3 positions of TAGs resulting in 2-monoacylglycerols and free fatty acids, which are then mixed with bile salt to form mixed micelles, where they are absorbed through the stomach mucosa[26].

In the mucosa, the LCFAs are resynthesized to give newly formed TAGs. Newly formed TAGs are combined with phospholipids and apolipoproteins to form chylomicron. The LCFAs are transported as chylomicron and enter the lymphatic system[31]. In contrast, the MCFAs are bound to serum albumin and transported in the soluble form of fatty acids and enter the systemic circulation through the portal vein directly to the liver, without being incorporated in the chylomicron. They do not accumulate in adipose tissue or muscle. Since MCFAs leave the intestinal mucosa by the portal vein system, they reach the liver more rapidly than the longer molecules[32].

The metabolism of MCFAs and LCFAs also differs in the transport of the fatty acids into the mitochondria (see Figure 2.1). LCFAs require enzyme transport, carnitine, in order to cross the mitochondrial wall. LCFAs are transformed into acylcarnitines in the presence of carnitine palmityl transferase-I and cross the membrane and regenerate long-chain acyl-coenzyme A in the matrix by the action of carnitine palmityl transferase-II[27]. In contrast, MCFAs are transported across the mitochondrial membrane of the liver rapidly and independently of the acylcarnitine transfer system[33]. In the mitochondrial matrix, MCFAs are acylated by means of an octanoyl-coenzyme A synthetase.

The acyl coenzyme A in mitochondria from MCFAs and LCFAs undergoes β-oxidation, which yields acetyl coenzyme A. However, few LCFAs reach this stage at the same time because fatty acids are prone to be incorporated into the lipids synthesized by the liver, which results in inactivation of palmityl transferase complex. In contrast, MCTs are available and subjected to rapid oxidation, resulting primarily in increased ketone production[28]. Therefore, though MCTs are fats, they behave more like carbohydrate. MCTs deliver fewer calories and are metabolized in an eighth of the time required for LCTs[34]. The products of MCTs are absorbed as fast as glucose. Their rapid transport and oxidation resemble more the carbohydrates than other fats and have less tendency to be stored as fat[27]. Table 2.3 summarizes the oxidative pathway of medium- and long-chain fatty acids.

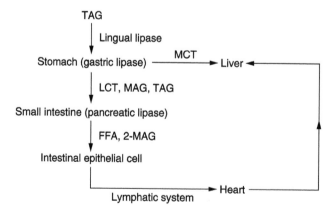

FIGURE 2.1 Transport metabolism of medium- and long-chain triacylglycerols.

TABLE 2.3
Summary of Oxidative Pathway of Medium- and Long-Chain Fatty Acids

	Medium chain (C6–C12)	Long chain (C14–C22)
Digestion	Easier, pancreatic lipase not essential	Pancreatic lipase essential
Absorption	Faster, due to smaller molecular size and greater water solubility	Slower
Transport	Via portal circulation direct to the liver	Via the lymphatic and systemic circulation, required enzyme transport, carnitine in order to cross mitochondria wall
Metabolism	Undergo faster and more complete oxidation	Oxidize slowly, incorporated into chylomicron and transferred to circulation via lymph system
Deposition of adipose tissue	Less adipose tissue deposition	More adipose tissue deposition

2.5 NUTRITIONAL BENEFITS

2.5.1 POTENTIAL USE IN THE TREATMENT OF OBESITY

The increasing incidence of obesity is becoming a medical problem in developed countries[35]. Relative adiposity is also considered as a problem affecting quite a number of individuals in western countries[36]. Weight loss can be achieved but the maintenance after weight loss in the long term is rarely shown[37]. Therefore, identification of substances that can improve or prevent obesity remains a requirement.

Dietary restrictions involving lipids are recognized as being the most important approach for people who intend to prevent any weight gain[38]. MCTs have been the subject of much research lately. MCTs are being promoted as potential agents in the prevention of obesity[39].

2.5.1.1 Effect of MCTs on Energy Expenditure

Research conducted in both animal and human studies on energy expenditure with respect to MCT consumption mostly resulted in increased energy expenditure. In one of the experiments using animal models to determine the energy expenditure of rats fed by intravenous and intragastric nutrition solution, Lasekan et al.[40] found lower weight gain and greater energy expenditure with MCT than with LCT supplemented nutrition. Other animal studies have also shown that MCT consumption increased energy expenditure compared with a meal containing LCTs[41,42]. In a study on the effect of preinfusion with total parenteral nutrition in rats undergoing gastrectomy, Lin et al.[43] found that MCT/LCT preinfusion had beneficial effects in improving liver lipid metabolism and reducing oxidative stress in those rats.

Human studies have also shown an increase in energy expenditure with regard to consuming MCT-containing meals[44,45]. In a single meal or single day experiment, Seaton et al.[46] compared the thermic effect of 400 kcal meals containing MCTs and LCTs in seven healthy men. Mean postprandial oxygen consumption was 12% higher than the basal oxygen consumption after the MCT meal, compared to that after the LCT meal that was only 4% higher than the basal oxygen consumption after 6 hours. This suggested more energy was generated by MCTs. It is possible that long-term substitution of MCTs for LCTs would produce weight loss if energy intake remained constant. Similar results were obtained for a much longer time after the meal (24 hours)[47].

In another single day meal experiment, Scalfi et al.[48] examined the effect of mixed meal containing MCTs on postprandial thermogenesis in lean and obese men. Total energy expenditure was 48 and 65% greater in lean and obese individuals, respectively, after MCT compared to LCT consumption. Another study discovered that dietary LCTs were less oxidized in obese individuals while MCT oxidation was not affected. A negative correlation was found between the amounts of LCTs oxidized with the fat mass content. Therefore, obesity may result in incomplete oxidation of dietary LCTs[49].

Bendixen et al.[50] examined the short-term effect of three modified fats containing mixtures of fatty acids of varying chain length on energy expenditure and substrate oxidation in healthy men. The modified fat mostly contained MCFAs and LCFAs attached to the same glycerol backbone. The results indicated that structured fat produced higher postprandial energy expenditure and fat oxidation than conventional fat.

There have also been studies conducted for a much longer duration (7 to 14 days) which generally compared the effect of MCTs versus LCTs. Hill et al.[44] overfed liquid formula diets containing MCTs (61% octanoate, 32% decanoate) or LCTs (32% oleate, 51% linoleate) for 7 days, in an attempt to examine the energy balance. The results showed that the thermic effect of food (TEF) was 8% of ingested energy after MCT consumption compared with 5.8% after LCT consumption on day 1. TEF was 12 and 6.6% of ingested energy with MCT and LCT consumption, respectively, after 6 days. This suggests that the difference in energy expenditure between MCTs and LCTs continued even after a week of overfeeding.

White et al.[45] determined the changes in energy expenditure or substrate oxidation in 12 nonobese, premenopausal women, after consuming MCT- or LCT-enriched diets. Each meal contained 40% energy as fat (80% of which was the treatment fat). On day 7, postprandial total energy expenditure (TEE) was significantly greater with the MCT than the LCT diet. However, TEE was not significantly different between MCT and LCT diet on day 14, although TEE was still greater for the MCT diet. This result shows that short-time feeding of MCT-enriched diet increased TEE but the effect could be transient with continued feeding. A similar result was observed by Papamandjaris et al.[51], in which increases seen in energy expenditure in healthy lean women, following MCT feeding, were of short duration. It has been suggested that compensatory mechanisms might exist which blunt the effect of MCTs on energy components over the long term. Kasai et al.[52] reported that intake of 5 to 10 g of MCTs caused larger diet-induced thermogenesis than LCTs, regardless of the form of meal containing MCTs.

Besides healthy individuals, there have also been studies done on obese individuals to compare the effect of MCT versus LCT diet consumption. St-Onge et al.[53] studied the effect of a diet rich in MCTs or LCTs on energy expenditure and substrate oxidation. Twenty-four healthy overweight men with body mass index from 25 to 31 kg/m^2 were given a diet rich in MCTs or LCTs for 28 days in a crossover and randomized controlled trial. The results showed that energy expenditure as well as fat oxidation was greater with MCT intake than with LCT consumption. The authors proposed that MCTs could be considered as agents that aid in the prevention of obesity or potentially stimulate weight loss.

Given that consumption of MCTs has been shown to increase energy expenditure, continued work has been done to examine the relationship between body composition and thermogenic responsiveness to MCT treatment. St-Onge and Jones[54] reported that fat oxidation was greater in men with lower body weight than in men with greater body weight. This indicated that overweight men have lower responsiveness to rapidly oxidized fat.

To determine whether the effect of greater increase in oxidation for MCT consumption applied for the longer term, St-Onge et al.[55] performed a study comparing LCT versus LCT consumption on healthy obese women for 27 days. The result was positive, in which MCT consumption enhanced energy expenditure and fat oxidation in the subjects compared to LCT consumption. The authors suggested that substitution of MCTs for LCTs may prevent long-term weight gain via increased energy expenditure. Nevertheless, Yost and Eckel[56] reported that a liquid containing 24% energy as MCTs did not lead to greater weight loss in obese women compared to an isocaloric diet containing LCTs.

Donnel et al.[57] studied the effect of consumption of MCTs on infants after surgery. Stable infants were given total parenteral nutrition of 4 g/kg/day of either pure LCT fat emulsion or 50/50 MCT/LCT fat emulsion. The result showed that net fat oxidation increased when LCTs were partially replaced with MCTs in intravenous fat emulsion, provided that carbohydrate calorie was below resting energy expenditure. In another study on infants, which was designed to investigate the effect of MCTs on linoleic metabolism, Rodriguez et al.[58] reported that oral MCTs were effective in reducing polyunsaturated fatty acid and long-chain polyunsaturated fatty acid oxidation in preterm infants without compromising endogenous n-6 long-chain polyunsaturated fatty acid synthesis.

All the studies summarized above demonstrated that MCTs caused a greater increase in the energy expenditure in animal studies[40–42]. Similar results have been shown in human studies in a single day[48] and several days[44] experiments. The results appear positive with healthy subjects[50], nonobese and premenopausal women[45], and obese individuals[55]. Thus, such convincing results suggest that MCTs can be useful agents in the prevention of obesity[49,53].

2.5.1.2   Effect of MCTs on Fat Deposition

Since MCTs increase energy expenditure in human and animal trials, further work has been done to examine the effect of MCTs on body fat accumulation in order to establish a correlation between energy expenditure and weight loss. Several studies on animals have shown promising results. Baba et al.[59] reported that increased metabolic rate and thermogenesis resulted in decreased body fat after overfeeding rats with an MCT diet. A similar result was obtained by Geliebter et al.[60]. MCT-fed rats gained 20% less weight and had fat depots weighing 23% less than LCT-fed rats. Hill et al.[61], however, did not observe any greater weight reduction in animals fed MCT than those consuming lard, corn oil, or fish oil.

In a more recent study, Han et al.[62] successfully demonstrated that adipose tissue was one of the primary targets on which MCFAs exert their metabolic influence. Rats fed a control of high-fat diet were compared with rats fed an isocaloric diet rich in MCTs. The results showed that MCT-fed animals had smaller fat pads, which contained a considerable amount of MCFAs in both TAGs and phospholipids. The adipose tissue lipoprotein lipase activity also reduced along with improved insulin sensitivity and glucose tolerance in MCT-fed animals.

Few studies have been conducted on humans to parallel the animal trials. In a double-blind, controlled trial, 78 healthy men and women consumed 9218 kJ/day and 60g/day total fat of either MCT or LCT diet[63]. The results showed that subjects with body mass index (BMI) of more than 23 kg/m^2 in the MCT group had significantly greater weight reduction than the LCT group. The MCT group with the same BMI also had significantly greater decrease in subcutaneous fat. Thus, MCT diet might reduce body weight and fat in individuals with a BMI of more than 23 kg/m^2 to a greater degree than LCT diet.

St-Onge et al.[53] also observed the same trend towards reduced subcutaneous adipose tissue with MCT compared to LCT consumption. Another study by Matsuo et al.[64] compared the effects of a liquid diet supplement containing structured fat composed of 10% MCFAs and 90% LCFAs versus LCTs alone on body fat accumulation in 13 healthy male volunteers. The result showed that body fat percentage was significantly lower for the formula diet containing structured medium- and long-chain TAGs compared to the LCT group. Krotkiewski[65] reported that the replacement of LCTs by MCTs through a very low-calorie diet increased the rate of decrease of body fat and body weight and had a sparing effect on fat-free mass.

The consumption of MCTs is proved to affect fat deposition in accordance with the metabolic rate. Fat deposition in the body caused by MCT consumption decreased in all human studies[53,63–65]. Animal trials generally follow the same pattern in some trials[60,62] but there is also a study that does not conform to the majority findings[61].

2.5.1.3   Effect of MCTs on Food Intake

Besides being able to increase energy expenditure and decrease fat deposition in both human and animal studies, MCT consumption also affects food intake. Few researchers have tried to explore the underlying mechanism of MCT metabolism that leads to reduced body weight.

Bray et al.[66] demonstrated that LCTs rendered greater feed intake when included in the diet of rats compared with diets containing MCTs. The study showed that rats fed either corn oil or corn oil with MCTs had higher body weight compared to rats fed MCTs alone. It has been proposed that the difference in feed intake between MCT- and corn oil-fed rats might be contributed by β-hydroxybutyrate.

Maggio and Koopmans[67] investigated the source of signal that controlled the short-term intake of mixed meals containing TAGs with fatty acids of different chain length. The authors found no changes in food intake when the chain length was shifted from medium to long in equicaloric infusions consisting of 21% of energy as fat. The conclusion was reached that satiety may be related to the amount of energy ingested instead of the physical characteristics of the nutrients. However, Furuse et al.[68] reported that satiety is affected by carbon chain length in dietary TAG sources.

In a study on humans, Stubbs and Harbon[69] reported that food and energy intakes were suppressed when two thirds of the fat content of a high-fat diet was derived from MCTs without affecting body weight. Rolls et al.[70] conducted a study to determine the intake of liquid meal containing MCTs among dieters and nondieters and found that nondieters consuming MCTs at all doses had significantly lower caloric intake in a lunch, while dieters were unresponsive to the diet preload.

Wymelbeke et al.[71] investigated the influence of medium- and long-chain TAGs on the control of food intake in men. Four high-carbohydrate breakfasts (1670 kJ) were supplemented with either fat substitute (70 kJ) or monounsaturated long-chain TAGs (1460 kJ), saturated long-chain TAGs (1460 kJ), and MCTs (1460 kJ). The results showed that MCTs did not delay the request of the next meal but decreased the amount of food ingested at the next meal. The authors believed that MCTs decreased food intake by a postabsorptive mechanism. Kovacks et al.[72] combined MCTs with hydroxycitrate (HCA), which was hypothesized to induce hepatic fatty acid oxidation, in an attempt to investigate the effects on satiety and energy intake. Two weeks of supplementation of HCA combined with MCTs failed to increase satiety or decrease energy intake compared to placebo in subjects.

Wylmelbeke et al.[73] examined the role of glucose metabolism in the control of food intake in men by using MCTs to spare carbohydrate oxidation. The studies showed that the carbohydrate oxidation was lower while fat oxidation increased after MCT and LCT lunches. The request time for dinner was significantly delayed after carbohydrate lunch but not after MCT lunch. Nevertheless, food intake at dinner was significantly lower after MCT lunch than after carbohydrate lunch. It was concluded that MCTs played a greater role in satiation of the next meal in the control of food intake but carbohydrate affected more of the duration of the satiety than fat.

2.5.1.4 Effect of MCTs on Lipid Profile

The rapid postingestive oxidation of MCTs appears to be accompanied by a substantial increase in energy expenditure. This greater energy expenditure slows down body weight gain and depot size. However, the effect of MCTs on serum cholesterol also deserves attention. Several researchers have attempted to look into this perspective.

Hill et al.[74] compared the effect of overfeeding of MCTs versus LCTs on blood lipid for six days. The authors observed a reduction in fasting serum total cholesterol concentration with consumption of LCTs but no changes with MCT consumption. There was also a threshold increase in fasting serum TAG concentration with MCT but not with LCT diet. The lipid effects of natural food diet supplemented with MCTs, palm oil, or sunflower oil were compared in hypercholesterolmic men by Carter et al.[75]. MCT oil was equal to palm oil in producing total cholesterol content, and significantly higher than total cholesterol produced by high-oleic sunflower oil. According to this study, MCTs potentially increase total and low-density lipoprotein (LDL) cholesterol concentration by about half that of the palmitic acid.

Asakura et al.[76] fed hypertriacylglycerol subjects with MCTs and corn oil for 12 weeks to examine the changes in plasma lipid. Compared with corn oil, MCTs showed higher mean of total cholesterol concentration (6.39 ± 1.14 versus 5.51 ± 0.98 mmol/l, respectively). Subsequently, Tholstrup et al.[77] demonstrated that MCT fat inconveniently affected the lipid profile in healthy young men by increasing plasma LDL cholesterol and TAGs. A study by Swift et al.[78] also indicated that MCTs produced a significantly higher plasma concentration of TAGs than LCTs.

Several researchers found an improvement in the lipid profile when MCTs were combined with other stimulating factors. In most studies, MCTs were combined with other fatty acids in structured lipids to achieve a specific fatty acid profile with desired benefits. Such structured TAGs were produced by interesterifying a mixture of conventional fats and oils, usually with MCTs which resulted in TAGs containing combinations of short-, medium-, and long-chain fatty acids on a single glycerol backbone[79].

In animal studies, Rao and Lokesh[80] employed a structured lipid consisting of omega-6 polyunsaturated fatty acids (n-6 PUFAs) synthesized from safflower oil. Rats were fed coconut oil, coconut oil–safflower blend, or the structured lipid for 60 days. The structured lipid lowered serum cholesterol levels by 10.3 and 10.5%, respectively, in comparison to coconut oil and blended oil. Compared to coconut oil and blended oil, the structured lipid also showed a decrease in liver cholesterol level by 35.9 and 26.6%, respectively.

In another study on healthy overweight men, St-Onge et al.[81] evaluated the effect of a functional oil (MCTs, phytosterols, and flaxseed oil) on plasma lipid concentration. The diet given to subjects was composed of 40% of energy as fat, 75% of which was added fat, either functional oil or olive oil. The functional oil, which was a cooking oil containing a blend of MCTs and n-3 PUFA structured lipids, resulted in a decrease of total cholesterol concentration by 12.5% compared to 4.7% for olive oil, and lowered the LDL concentration by 13.9% whereas no change was observed for olive oil. Therefore, it was concluded that subjects who consumed the functional oil had a better lipid profile than those consuming olive oil.

In a similar study designed by Bourque et al.[82] on overweight women, mean plasma total concentration was lower by 9.1% for functional oil versus beef tallow diet. Subsequently, mean plasma LDL was also lower for functional oil diet with a 10% difference from beef tallow diet. The study also showed an increase in the ratio of high-density lipoprotein (HDL) to LDL and HDL to total cholesterol concentration by 22.0 and 11.0%, respectively, for functional oil diet compared to beef tallow diet. The authors concluded that consumption of the functional oil improved the overall cardiovascular risk profile in overweight women.

Another study by Beerman et al.[83] combined dietary MCTs with n-3 long-chain fatty acids, which proved to be favorable in healthy volunteers. The subjects were fed diet formula containing 72% MCFAs with 22% n-3 PUFAs versus isoenergetic formula. The result showed that the plasma TAG and cholesterol content decreased in the group fed the formula diet compared to the isoenergetic formula. Manuel-y-Keenoy et al.[84] reported a two-fold increase in serum α-tocopherol in patients supplemented with total parenteral nutrition (TPN) containing MCTs and α-tocopherol. There was also a decrease in the susceptibility of LDL and very low-density lipoprotein (VLDL) to peroxidation *in vitro*.

Although MCTs prove to be beneficial in increasing energy expenditure and potentially aid in weight reduction, the effect they have on plasma lipid seems to be divergent. MCTs produced a significant increase in plasma concentration of TAGs in healthy subjects[78,74], increased total cholesterol in hypertriacylglycerolicdemic individuals[76], and caused a greater frequency of cholesterolemia in mildly hypercholesterolmic men[75]. In contrast, combining MCTs with phytosterols (cholesterol lowering), n-3 PUFAs (TAG suppressing), n-6 PUFAs (essential fatty acid), and α-tocopherol (antioxidant) appears to lower LDL[81], decrease serum cholesterol level[80], improve cardiovascular risk profile[82], and increase stability of LDL towards peroxidation *in vitro*[84].

2.5.2 Clinical Uses

2.5.2.1 Treatment for Fat Malabsorption

Malabsorption is the clinical term for defects occurring during the digestion and absorption of food nutrients by the gastrointestinal tract. MCTs have become an established treatment for many types of malabsorption cases such as steatorrhea, chyluria, and hyperlipoproteinemia[85]. Due to its unique metabolism, MCT oil has proved to be an important source of energy in a variety of clinical conditions.

Glucose was used intravenously in patients before the introduction of lipid emulsions as the only source of calorie[86]. However, application of glucose often led to hepatic lipogenesis and increased respiratory work to expire the excess carbon dioxide produced during lipogenesis[87]. Therefore, researchers have looked for other sources for calorie, and LCTs have been proposed as ideal nonglucose fuel that can provide energy. Intravenous fat emulsions made from soy or safflower oils have been used since they contain linoleic acid, an essential fatty acid. Fat is used as the energy source; therefore the body can use amino acids as protein and not as a caloric source[88]. Moreover, LCTs inhibit lipogenesis from carbohydrate, thereby decreasing fatty livers. Lipid supplementation, when substituted for glucose, can benefit diabetic patients by decreasing insulin requirement. Fat is digested at a slower and lower respiratory quotient than glucose, thereby producing less carbon dioxide for the same amount of oxygen intake. This condition can benefit patients with pulmonary compromise who have problems expiring all the carbon dioxide.

Lipid supplementation using LCTs, however, has some disadvantages. LCT emulsion is slowly cleared from the blood stream[89]. Clearance is not synonymous with oxidation of the fatty acids which is the main reason for using the emulsion. There is also concern that these emulsions are less ideal because of relative carnitine deficiency, which occurs in sepsis, which blocks their entry into mitochondria for β-oxidation[85]. Therefore, attention has been focused on MCT emulsion for clinical intravenous use. MCFAs such as caprylic and capric acids have been used as components for infant feeding and as nutritional supplements for patients suffering from fat malabsorption[90].

MCTs have been used therapeutically since the 1950s in the treatment of malabsorption. MCTs are easily hydrolyzed in the intestine and the fatty acids are transported directly to the liver via the portal venous system, whereas LCTs are incorporated into chylomicrons for transport through the lymphatic system. MCFAs do not require carnitine to cross the double mitochondrial membrane of the hepatocyte; thus they quickly enter mitochondria and undergo rapid β-oxidation[91].

When a diffuse disorder affects the intestine, the absorption of almost all elements is impaired. Digestion of macronutrients occurs mostly via enzymatic hydrolysis into smaller absorbable molecules[92]. Pancreatic enzymes play an important role in macronutrient digestion. Destruction of pancreatic tissue and obstruction of the ducts that lead into the small intestine prevent pancreatic secretions from reaching the small intestine and result in weight loss, abdominal distention, and changes in the appearance and frequency of stools[93]. Effective reduction of nutrient malabsorption in pancreatic insufficiency requires delivery of sufficient enzymatic activity into the duodenal lumen simultaneously with meal nutrients.

Nutrient delivery into the proximal small bowel has been demonstrated to be the most important stimulus of exocrine pancreatic secretion. Small intestine transit is significantly accelerated in patients with pancreatic insufficiency compared with healthy subjects, resulting in a 50% reduction of intestinal transit time[94]. Thus, the available time for digestion and absorption is markedly decreased. As a result of severe lipase and protease deficiency, unabsorbed lipids and protein reach the colon and may induce steatorrhea. Pancreatic enzyme must be reduced to less than 10% of normal secretion before fat absorption is impaired, proving that the pancreas secretes a large surplus of enzymes[95].

The purpose of dietary intervention in patients with maldigestion is to provide sufficient calories and protein to maintain weight while limiting fat intake to an amount that the patient can tolerate[93]. The reaction of pancreatic lipase in the small intestines is enhanced by the relatively low molecular weight of MCTs, which results in rapid and near complete hydrolysis of the fatty acids[91]. MCT oil significantly accelerates small-bowel transit time compared with that seen in control subjects[96]. Shea et al.[97] demonstrated that enteral supplements containing MCTs and hydrolyzed peptides stimulate the exocrine pancreas by blunting cholecystokinin release and thus reducing postprandial pain associated with chronic pancreatitis.

Pancreatic lipase hydrolyzes ester bonds at the *sn*-1 and *sn*-3 positions in TAGs and shows higher activity towards MCFAs. LCTs are hydrolyzed to 2-monoacylglycerols and fatty acids by lipase and the hydrolysis products are absorbed into the intestinal mucosa[98]. Thus, TAGs with MCFAs at the *sn*-1 and *sn*-3 positions and with functional fatty acids at the *sn*-2 position are rapidly hydrolyzed with pancreatic lipase and are absorbed efficiently into mucosal cells. These highly absorptive TAGs are known as structured lipids. According to Straarup et al.[99], a combination of LCFAs and MCFAs is advantageous to provide both energy and essential fatty acids.

Fat malabsorption of maldigestion may occur due to mucosal damage or atrophy. Severe fat malabsorption is evident as steatorrhea. In most patients, steatorrhea is the late event in the course of progressive chronic pancreatic problems[92]. For these patients, supplementation with MCTs is useful because they are hydrolyzed rapidly by pancreatic enzymes, do not require bile acid micelles for absorption, and are primarily directed to the portal rather than lymphatic circulation. MCTs have also been applied in therapy for small bowel resection[100]. Dietary MCTs are also useful in the treatment of infants with short bowel syndrome. The dietary fat aims at restoring the intestinal continuity and at improving the physiological process of gut adaptation[101].

MCTs have been suggested for use as a dietary source in patients with AIDS[102]. The malnutrition of fat, carbohydrate, specific micronutrients, and protein has been reported in patients with HIV[103]. MCTs are readily absorbed from the small bowel under conditions in which the absorption of LCTs is impaired. Wanke et al.[104] reported that HIV patients with chronic diarrhea, fat malabsorption, and weight loss benefit symptomatically from a diet composed of MCT-based supplement. MCT-enriched formula has also been shown to decrease fat and nitrogen losses in patients with AIDS compared to LCT-enriched formula[102].

In summary, the goal of nutritional requirement for patients with fat malabsorption is to provide sufficient nutrients using dietary sources which can compensate for the lack of proper digestion to ensure adequate nutrients are received. MCTs with their unique properties fit into the criteria to suit the needs of fat malabsorption patients where no other fats can.

2.5.2.2   Parenteral Nutrition Formulation for Surgery and Compromised Patients

MCTs have been used as alternative calorie sources for compromised patients. The goal of nutrition support during critical illness is to maintain organ function and prevent dysfunction of the cardiovascular, respiratory, and immune systems. The nutrition supplied should be able to reduce starvation effects and nutritional deficiencies[105].

Feeding through the gastrointestinal tract is the preferred supplementation but when a patient cannot tolerate enteral feeding or when there is a need to supplement enteral intake to meet basic nutritional requirements, parenteral nutrition becomes the main course of treatment[106]. MCTs are formulated for parenteral nutrition either as physical mixtures or as a component of structured lipids. Physical mixtures involve partial replacement of LCT with MCT lipid emulsion while structured lipids are TAGs containing mixtures of medium- and long-chain fatty acids based on the same glycerol molecule in a predetermined proportion[107].

Gastrectomy is a major abdominal surgical procedure, which usually causes stress to patients. Under stress conditions, there is bound to be glucose intolerance, thus making fat the primary substrate for oxidation[108]. Parenteral MCT emulsions can benefit patients with gastrectomy because MCTs are more rapidly cleared from plasma and more completely oxidized than LCTs[53,91]. Lin et al.[43] observed that preinfusion with MCTs along with LCTs has beneficial effects in improving liver lipid metabolism and reducing oxidative stress in rats with gastrectomy.

Major catabolic stresses such as trauma or sepsis cause accelerated breakdown of muscle protein, increase nitrogen excretion, and negative nitrogen balance[109–111]. It has been suggested that MCT emulsions be used as nonglucose fuel because they may have a protein-sparing effect[106]. Denison et al.[112] reported a significant improvement in nitrogen balance among critically ill surgical patients receiving MCT and LCT mixtures compared to those receiving only LCTs. A similar result was obtained by Jiang et al.[113]. Improvement in nitrogen balance in surgical patients was postulated to be associated with the increased ketone and insulin level in accordance with MCT supplementation.

Generally, lipid-based nutritional support is recommended for patients with chronic respiratory illness due to high carbon dioxide production. Oxidation of lipids results in a lower carbon dioxide production compared to carbohydrates. However, clinical studies have shown some deleterious effects associated with intravenous fat emulsions based solely on LCTs in patients with respiratory insufficiently[114]. These adverse effects are present as a result of increased production of eicosanoids (prostaglandin and thromboxanes)[115]. In contrast, MCT emulsions may be advantageous in patients with respiratory problems because they do not influence eicosanoid synthesis[31].

Lipid emulsions containing MCTs and LCTs have also been proposed as an ideal source of fat for patients with chronic hepatic failure. MCTs may cause a reduction in hepatic side effects during parenteral nutrition, such as cholestasis and steatosis[116]. Sepsis is characterized by increased oxidation of all fatty acids regardless of chain length[106]. Critically ill septic patients have decreased carnitine stores[117]. Supplementation with MCT emulsion can provide nutritional support in a beneficial manner because it does not require carnitine for entry into mitochondria, and thus can be rapidly and more efficiently oxidized.

The unique characteristics of MCT emulsion therefore make it possible for use in various clinical applications. MCTs provide beyond basic nutrition to critically ill and compromised patients. MCT emulsions are far superior substrates for parenteral use than LCTs, but their combination through structured TAGs can greatly enhance the efficacy of delivering nutrients to patients.

2.5.3 APPLICATION FOR IMPROVED NUTRITION

2.5.3.1 Incorporation into Infant Formula

Fats are essential components in the diet for neonates and have a profound effect on the growth and development of infants. Dietary fat provides the major energy (~50% of the calories) during infancy[118]. Fats also serve as integral constituents of neural and retinal tissues[119]. Manufacturing infant formula requires both a knowledge of lipids and information on the digestion, absorption, and transport of lipids in infants.

Basically, there is a difference in fat digestion between infants and adults. The difference is due to the immaturity of the digestive system, which includes a low level of pancreatic lipase and bile salts[31]. Fat digestion and absorption in infants depend on the development pattern of lipase[120]. Hydrolysis of lipids begins with lingual lipase, which is important in infant digestion. Lingual lipase accounts for 50 to 70% of the hydrolysis of fat[121]. Once in the stomach, fats come in contact with gastric lipase, which accounts for 10 to 30% of fat digestion[122]. Much of the fat in infants is hydrolyzed in the stomach by lingual and gastric lipases.

It is well established and accepted that human milk is the best food for infants and provides the nutritional requirement for the newborn. It is an ideal formula because of its nutrient balance, ease of digestion, supply of immune-enhancing components, and growth stimulation[105]. Therefore, infant formula should reflect as much as possible the fat composition of human milk. Human milk consists of 98% TAGs, 1% phospholipids, and 0.5% cholesterol and cholesterol esters[123]. New infant formulas are continually being developed as more components of human milk are characterized and the nutritional requirements of infants are identified[124]. Infant formula consists of a mixture of several oils including corn, coconut, soy, canola, sunflower, safflower, and palm oils.

Due to the immature condition of their digestive systems, infants require special nutrition that differs from that of adults. Infants use about 25% of the caloric intake for growth[125]. MCTs can provide a concentrated source of energy needed by infants. Compared to LCTs, MCTs are more efficiently absorbed in the digestive tract[126] and metabolized as quickly as glucose but with twice the energy density of carbohydrates[79]. According to Borum[127], MCFA concentration in infant formula can reach 40 to 50% of the total fatty acids. MCFA oxidation is associated with a ketogenic effect which provides an alternative source and is considered harmless for infants provided that ketone concentration does not exceed values observed in breast-fed infants[58].

In contrast to LCTs, MCFAs do not promote the synthesis of eicosanoid or radical formation, both of which are involved in inflammatory responses[31]. In addition, infants can make use of the metabolism of MCFAs to their advantage. MCFAs do not require pancreatic or biliary secretion for absorption, and thus are best suited for infants, whose level of pancreatic enzyme and bile salt are limited. According to Sann et al.[128], oral lipid supplementation containing a high percentage of MCTs was shown to prevent the occurrence of hypoglycemia in low-birth-weight infants. Isaacs et al.[129] reported that MCFAs or monoacylglycerols added to infant formula provide antimicrobial protection against viral and bacterial pathogens prior to digestion.

The increasing importance of MCFAs in fat digestion in neonates has prompted their inclusion in preterm infant formulas. Preterm infant formulas are considered the best substitute for premature infants who cannot receive sufficient amounts of human milk[58]. According to Klien[130], the nutrient composition of term formula was based on mature breast milk; thus, formulas suitable for term infants are inadequate for premature infants. The preterm formula contains up to 50% MCFAs (C8:0 and C10:0). In an experiment to compare energy expenditure of MCT oil with canola oil, Cohen et al.[131] discovered that MCT oil was better absorbed than canola oil in growing preterm infants.

Despite being beneficial to infants, consumption of MCTs has some drawbacks. Consumption of MCTs may result in deficiency in essential fatty acids such as linoleic acid[132]. High doses of MCFAs may lead to metabolic acidosis, a condition in which the MCT emulsion produces large amounts of ketone bodies[133]. To overcome this problem, MCTs can be given with LCTs as a physical mixture to reduce any potential adverse effects and to ensure adequate supply of essential fatty acids.[134]. Telliez et al.[135] reported that the ratio of MCTs to LCTs is important in neonates' feeding to modify the physiologic functions involved in energy balance regulation.

Differences in the nutritional requirement between infants and adults create different provision in the dietary fat. While adults aim at maintaining health and preventing diseases, infants strive for physical growth. Special requirements for term and preterm infants have promoted the use of MCTs as a rapid source of energy in infant formulas. Modification in dietary MCTs, such as MCT emulsions, is pursued in order to improve and optimize the product to suit the needs of infants.

2.5.3.2   Energy Supplement for Athletes

Many individuals engage in exercise and sport for health purposes. Firm, toned, and developed muscle is considered desirable and is associated with physical fitness. Athletic performance is greatly influenced by the dietary cost of performing a sport and the proficiency of the metabolic system to provide the maximum amount of energy needed.

Carbohydrate consumption has been recognized as an important fuel source during exercise since the beginning of the 20th century. Ingestion of carbohydrates can improve endurance performance, but their oxidation is rather limited[136]. Carbohydrate is stored in the form of glycogen in the liver and muscle tissues. According to Jeukendrup et al.[137], the body glycogen stores are very small (~8 to 16 MJ) which can be depleted within one hour. Depletion of body glycogen has been associated with fatigue during constant load exercise[138]. Furthermore, a two-fold increase in the amount of carbohydrate ingested caused only a slight increase in the rate of oxidation during exercise at 70% maximal oxygen consumption[139,140]. Thus, research has focused on finding ways to improve endurance capacity and to supply additional energy sources.

Hickson et al.[141] found that rats that were fed TAGs and later infused with heparin (intravenous) had elevated circulating FFA consumption. Heparin released lipoprotein lipase from the vascular walls to promote hydrolysis of plasma TAGs, resulting in a marked elevation of the concentration of fatty acids and glycerol in the plasma[26]. Elevated plasma fatty acid has been associated with decreased muscle glycogen utilization and improved exercise performance[142]. A similar study has shown a reduction in muscle glycogen utilization in humans during exercise[143].

Most TAGs are stored in the adipose tissue (~17,500 mmol in lean adult men) and also in skeletal muscle (~300 mmol). There is ~560 MJ of energy stored as TAGs, which is more than 60 times the amount stored as glycogen (~9 MJ)[144]. It has been hypothesized that if fat intake is increased while maintaining sufficient carbohydrate intake, it is possible that endurance exercise time could be improved, provided that enhanced fat oxidation allowed the muscle to spare glycogen[145]. This effect has been related to an increase in FFA availability which reduces muscle glycogen utilization and results in delayed exhaustion[146].

Though high-fat diet enhances endurance, eating a high fat meal that is mainly composed of LCTs before exercise is not a practical approach as a direct source of fat during exercise due to delayed and limited availability of ingested fat for skeletal muscle oxidation[144]. The digestion and absorption of fat are rather slow. LCTs also slow gastric emptying and must be packaged into chylomicrons which are not believed to be a very important source of energy during exercise[147]. LCTs enter the blood only 3 to 4 hours after ingestion[137]. Thus, LCT ingestion is not very effective as an energy source. In contrast, MCTs are more rapidly digested and absorbed as MCFAs that directly enter the blood through the portal system[27]. In addition, MCFAs can diffuse into mitochondria independent of carnitine[148]. These facts suggest that MCTs are better in modulating energy metabolism and enhancing physical performance activity than LCTs[149].

Since MCTs are a readily available energy source for the working muscle, it has been suggested that they serve as an additional substrate during prolonged endurance exercise[136]. MCT ingestion may improve exercise performance by elevating plasma acid level and sparing muscle glycogen[137]. Beckers et al.[150] reported that MCTs added to carbohydrate (CHO) drinks emptied faster from the stomach than an isocaloric CHO drink. In addition, MCTs did not inhibit gastric emptying.

Many studies have been done to examine the effect of ingestion of MCTs on physical performance before and during exercise. Theoretically, MCT ingestion should provide a way to increase plasma fatty acid levels[137]. However, Ivy et al.[151] observed no elevation in plasma fatty acid concentrations of subjects who ingested MCTs one hour before exercise. It was postulated that the large amount of CHO ingested with the MCTs masked the result by favoring CHO metabolism. Consequently, Decombaz et al.[152] conducted a study on energy metabolism by giving preingestion of 25 g of MCTs without added CHO or 50 g of CHO an hour before exercise. There was a slight elevation in plasma fatty acid concentration and an increase in ketone bodies upon MCT feeding. However, there was no decrease in muscle glycogen breakdown during exercise. Other studies with preingestion of MCTs involving low- to moderate-intensity exercise between 60 and 70% of maximal oxygen consumption did not show any improvement in endurance performance[146,153].

Research has also been conducted on MCT ingestion during exercise. It has been suggested that supplying exogenous energy besides CHO during exercise can have ergogenic effects[26]. Jeukendrup at al.[136] demonstrated that carbohydrates coingested with MCTs accelerated the oxidation of MCTs during the first 90 minutes of exercise. This confirms the hypothesis that oral MCTs can serve as additional fuel for the working muscle. However, although rapidly oxidized, MCTs contributed only a slight amount to the total energy expenditure (between 3 and 7%). The amount of MCTs ingested is relatively small (30 g) which limits their contribution to the total energy expenditure. In subsequent studies to determine the effect of MCTs on muscle glycogen breakdown Jeakendrup et al.[154] showed that MCTs did not reduce the use of muscle glycogen, even under the condition where the reliance on blood substrates was maximal such as in glycogen-depleted state[155]. Van Zeyl et al.[156] proposed that the amount of MCTs ingested was too small (~30 g) to render any effect on exercise performance. Therefore, they gave 86 g of MCTs to subjects during 2 hours of exercise. Interestingly, a reduction in the rate of muscle glycogen oxidation and an increase in performance were obtained when MCTs were added to a 10% CHO solution. There was no mention of any gastrointestinal discomfort. A similar experiment was repeated by Jeukendrup et al.[137] in which subjects were given 85 g of MCTs in a 10% CHO solution during exercise for 2 hours at 60% of maximal oxygen consumption. Surprisingly, the researchers did not find any reduction in the use of muscle glycogen oxidation and no improvement in the performance. They reported that subjects developed gastrointestinal problems which led to a decrease in exercise performance. Goedecke et al.[157] adopted the same experiment of Van Zeyl et al.[156] using low and high doses of MCTs (28 and 55 g in 2 hours) but also failed to reproduce the findings reported by Van Zeyl et al.[156].

Recent studies conducted on MCT ingestion during exercise showed no additional improvement in performance[58,159] and no significant rise in plasma MCFA concentration[160]. Oopik et al.[149] gave a daily supplement of MCTs to subjects instead of during exercise and observed an increase in the availability of ketone bodies for oxidation in working muscle during high-intensity endurance exercise but this metabolic adaptation did not improve endurance performance capacity in well-trained runners.

The results of these studies suggest that MCT oil does not reduce the use of muscle glycogen or improve performance. The amount of MCTs that can be tolerated at one time is limited to 25 to 30g; larger amounts may cause adverse effects on gastrointestinal function[151]. However, ingesting 25 to 30g of MCTs produces only ~0.2 to 0.3 MJ/h energy[136]. Thus, it is not surprising that endurance performance is not enhanced by MCT ingestion.

2.6 STRUCTURED LIPIDS CONTAINING MCTs

Considerable advances have been made over the last 10 years that have enhanced our knowledge and understanding of the basic chemistry, physiology, and metabolism of lipids. Recently, structured TAGs made an appearance, which integrate advantages from conventional fats with those used for special purposes. Structured lipids are developed in order to attain the optimal lipid source for improved nutritional or physical properties over conventional fats. Structured lipids are defined as TAGs containing mixtures of fatty acids, either short chain and/or medium chain with long chain, which are incorporated into the same glycerol backbone. Structured lipids have been of much interest in recent years because of their potential nutraceutical application. Nutraceutical food is generally food or parts of food that provide beneficial health effects, which aid in the prevention and/or treatment of diseases[107].

MCTs demonstrate additional characteristics of considerable advantage. MCTs are known to provide rapid energy with little tendency to deposit as stored fat[85]. MCTs are not dependent on carnitine for transport into mitochondria and are preferentially transported via the portal vein to the liver because of their smaller size and greater solubility than LCTs[27]. Emulsions containing MCTs have been used as an alternative source of lipid in addition to LCTs during the past decade. Numerous studies have been reported on the proficiency of MCTs in the treatment of fat malabsorption, metabolism difficulties related to the gastrointestinal tract, and parentral nutrition for severely malnourished patients[3]. Nevertheless, MCTs do not contain linoleic and linolenic acids, which are essential fatty acids (EFAs)[134]. High doses of MCFAs can produce adverse effects such as metabolic acidosis[133] and neurologic effects[161]. Therefore, an appropriate amount of LCTs must be included in MCT emulsions to ensure that the adverse effects related to the sole use of MCTs are avoided. Structured TAG emulsions containing both MCFAs and LCFAs on the same glycerol backbone have been developed to utilize the positive effects of MCFAs while circumventing the side effects as well as satisfying the requirement for EFAs[162].

Structured lipids can be produced either by chemical or enzymatic methods. Chemical methods involve hydrolysis of a mixture of MCTs and LCTs and then reesterification after random mixing of the MCFAs and LCFAs by transesterification[163]. The reaction is catalyzed by alkali metals or alkali metal alkylates. Chemical interesterification is inexpensive and easy to scale up. The reaction requires high temperature and anhydrous conditions. According to Willis et al.[31], chemical interesterification is not an effective method to produce high concentrations of MCFAs due to randomness and lack of positional or fatty acid selectivity inherent, even when lipids with fatty acids of different chain lengths and molecular weights are interesterified.

Fat can also be modified by enzymatic interesterification using lipase, which is derived from yeast, bacterial, and fungal sources. Lipase hydrolyzes TAGs to monoacylglycerols (MAGs), diacylglycerols (DAGs), FFAs, and glycerol. Hydrolysis of TAGs can be carried out through direct esterification or acidolysis. Direct esterification involves preparation of structured lipids by reacting FFAs with glycerol. However, water must be continuously removed from the reaction medium to prevent the products from hydrolyzing back to reactants[107]. Acidolysis reaction can be achieved by exchanging acyl groups between an ester and a free acid. However, in the transesterification one acyl group is exchanged between one ester and another ester[164]. The main application of lipase-catalyzed transesterification is the enrichment of high-EFA oils with MCFAs[31]. Compared to chemical synthesis, the enzymatic reaction offers advantages such as better control over the positional distribution of fatty acids in the final product. This is due to the unique property of lipase on selectivity and regiospecificity of fatty acids. Generally, enzymatic synthesis is carried out under mild reaction conditions and products are easily purified and there is less waste.

The purpose of developing structured TAGs consisting of MCFAs and LCFAs is to complement each other. According to Nagata[165], such structured lipids can retain most of the desired qualities of fatty acids while minimizing the adverse effects. The influence of TAG structure on the absorption of fatty acids has also been of interest[166–169]. Generally, physical blending of MCTs and LCTs does not improve their overall absorption since each of the individual fatty acids maintains its original uptake rate[170]. According to Lepine et al.[171], some of the properties of MCTs, such as digestive, absorptive, and metabolic characteristics, are retained in MCT structured TAGs. The effectiveness of intestinal absorption of MCT structured TAGs is intermediate between that of MCTs and LCTs.

During digestion, TAGs are hydrolyzed to sn-2 MAGs and FFAs in the small intestine by pancreatic lipase[172]. The metabolism of a structured TAG is determined by the nature and position of the constituent fatty acids on the glycerol backbone. The chain length of the MCFAs is important because the distribution of octanoic acid (C8:0) and decanoic acid (C10:0) between the lymphatic system and the portal vein depends on the chain length, with a higher proportion of C8:0 than C10:0 absorbed directly to the portal vein[99]. Typically, C2:0 to C12:0 fatty acids are transported via the portal system and C12:0 to C24:0 via the lymphatic system[107]. However, there is evidence showing that MCFAs may also be absorbed via the lymphatic route[78]. The presence of C10:0 has been reported in the lymph of a canine model fed a structured lipid containing MCTs and fish oil versus their physical mixture[173]. The level of MCFAs detected in lymph lipids increased with an increase in chain length of MCFAs[174]. It has been proposed that not only the types of fatty acid present but also their relative order in the sn-1, sn-2, or sn-3 position on the glycerol moiety can influence the metabolism of TAGs and fatty acids[169].

New developments in the lipid area provide the possibility of supplying EFAs in MCT emulsion. Enhanced absorption of linoleic acid (C18:2, n-6) was demonstrated in cystic fibrosis patients fed structured TAGs containing long- and medium-chain fatty acids[175,176]. Lymphatic absorption of linoleic acid in the sn-2 position and MCFAs in the sn-1 and sn-3 positions (MLM) (M = medium chain and L = long chain) is more rapid than absorption of LML and physical mixtures of LLL and MMM in rats[177,178]. Kishi et al.[179] reported that the structural difference in TAG containing MCFAs and linoleic acid can alter the rate of lipid clearance in the serum of rats.

Continued improvement in the development of structured TAGs has made possible the emergence of fats as functional foods. Combination of MCT and n-3 PUFA emulsions provides unique properties which enhance nutritional and metabolic support physically and chemically[180]. High intake of n-3 PUFAs, especially eicosapentaenoic acid (EPA, C20:5n-3) and docosahexaenoic (DHA, C22:6n-3), has been linked with low incidence of coronary heart disease in Greenland Eskimos[181]. The n-3 fatty acids have been demonstrated to decrease plasma TAGs, platelet aggregation, and possibly VLDL cholesterol. However, n-3 PUFAs are more slowly released in vitro by pancreatic lipase than other fatty acids[182]. Structured TAGs made from fish oil (source of n-3 PUFAs) and MCFAs with specific location of n-3 PUFAs at sn-2 and MCFAs at the sn-1 and sn-3 positions maybe hydrolyzed faster than TAGs in which fatty acids are randomly distributed[99]. Fish oil fatty acids such as EPA and DHA have been shown to be more readily absorbed when esterified in the *sn*-2 position[183]. Therefore, applying structured TAGs can vastly improve the efficiency and functional value of n-3 PUFAs.

With the ability to combine the beneficial properties of their component fatty acids, structured lipids provide a means of enhancing the role of fat in health and medical applications. A structured lipid made from fish oil is reported to improve the nitrogen balance in thermally injured rats[184,185] and inhibit tumor growth in Yoshida sarcoma-bearing rats[186]. Structured TAGs containing MCTs and fish oil fatty acids were absorbed more quickly in rats with intestinal injury and impaired lymph transport compared to a physical mix of the components[187]. Straarup and Hoy[99] demonstrated an increased recovery of EPA and DHA when administered in the form of structured lipids through interesterification of fish oil with MCFAs in both normal and malabsorption rats compared to the fish oil itself.

Structured lipids comprised of both LCFAs and MCFAs have emerged as the favored alternative to physical mixtures for treatment of patients with a number of disease conditions[3]. The TAGs in TPN are typically administered as an emulsion, which is suspected as suppressing the immune function since pneumonia and wound infection often occur in patients treated with TPN. This is probably due to the production of oxygen radicals in physical mixtures compared to LCTs and structured lipid emulsions[188]. A study conducted on postoperative patients revealed no hepatic disturbances in patients given structured lipids, which is typically observed with TPN[189]. Structured lipid diets made from fish oil and MCTs were reported to be significantly better tolerated, reduced the number of infections, and improved hepatic and renal function in patients undergoing surgery for upper gastrointestinal malignancies[190].

Development of structured TAGs is also making its way into the manufacturing of infant formula. The ideal fat component for infant formula should contain fatty acids such as MCFAs, linoleic acid, linolenic acid, and PUFAs in the same positions and amounts as those contained in human milk[3]. In order to increase caloric intake, the saturated fatty acids of the fat in infant formula should be at the *sn*-2 position. Absorption of fat has been associated with the amount of palmitic acid in the *sn*-2 position[166]. Palmitic acid can be better absorbed at the *sn*-2 position[174] and 68% of the palmitic acid molecules in human milk are located at the *sn*-2 position[172]. Lien et al.[191] reported that chemical interesterification of coconut oil and palm olein increased the amounts of palmitic acid in the *sn*-2 position and hence increased the fat absorption in rats compared with a simple mixture of the two fats. Thus, application of structured TAGs can be useful in manufacturing infant formulas which closely resemble human milk in terms of fat composition.

2.7 ANTIMICROBIAL PROPERTIES OF MCFAs

The antimicrobial properties of fatty acids and fatty acid salts or esters have been known for many decades. It has been reported that medium-chain saturated fatty acids and their derivatives (MAGs) are effective against various microorganisms[192]. Those microorganisms that are inactivated are bacteria, yeast, fungi, and enveloped viruses.

The demand for fresher and minimally processed food has led to the development of multiple-barrier food preservation systems[193]. Chemical preservation is widely used to extend food shelf life and inhibit the growth of pathogenic microorganisms. However, the use of chemical preservatives is questionable due to public concerns about food safety[194]. It has been suspected that some food additives such as nitrite and sorbates may give adverse health effects[193]. Therefore, there is a need to develop natural substances as alternatives to commonly used preservatives. Fatty acids and their esters are naturally occurring substances in foods[195] and hence may be used as natural antimicrobial preservatives.

Lauric acid (C12:0) is a MCFA which will turn into monolaurin in the human or animal body. Glycerol monolaurate (monolaurin), a food-grade glycerol monoester of lauric acid, is approved as an emulsifier in foods by the USFDA[194]. Monolaurin inhibited pathogenic bacteria such as *Listeria monocytogenes*, a gram-positive bacterium, that cause foodborne diseases in humans and animals[196,197]. Monolaurin has been reported to be a 5000 times more effective inhibitor against *L. monocytogenes* than ethanol[194] which is potentially used as an inhibitory agent in dairy foods[198] and minimally processed refrigerated foods[199]. Inhibition of monolaurin against *L. monocytogenes* is further enhanced by combination with organic acids[200].

Antimicrobial activity of monolaurin has been demonstrated against *Staphylococcus aureus* strains[201]. Surfactant glycerol monolaurate inhibits the production of the toxin that is responsible for toxic shock syndrome, produced by *S. aureus*[202,203]. Monolaurin is also reported to inactivate pathogenic bacteria, *Helicobacter pylori*, the etiologic agent that plays a role in chronic gastritis and peptic ulcer disease[204]. In addition, monolaurin is capable of inhibiting bacterial spores and vegetative cells from *Bacillus cereus* and *Clostridium botulinum*[193,205–208]; it also exhibited antifungal activity against toxin-producing species *Apergillus niger*[209]. Monolaurin has been identified as a potential cure for HIV/AIDS. The first clinical trial of monolaurin on HIV-infected patients showed positive results by the third month, in which 50% of the patients showed a reduced viral load[210]. Other MCFAs such as capric acid and monocaprin also demonstrated antimicrobial properties. A rapid inactivation of a large number of *Chlamydia trachomatis* by monocaprin suggested that it may be useful as a microbicidal agent against *C. trachomatis*[211].

Some of the viruses inactivated by lauric acid are vesicular stomatitis virus (VSV) and cytomegalovirus (CMV). There is evidence that CMV is involved in the development of athero-sclerosis[212]. Prevalence of CMV is significantly higher in patients with serious atherosclerosis than in patients with minimal atherosclerosis[213,214] and in diabetic mellitus patients with atherosclerosis than in diabetic patients without atherosclerosis[215]. It is suggested that the antimicrobial effect is due to the lauric acid's interference with virus assembly and maturation[216]. Lauric acid also inhibited the late maturation stage of Junin virus (JUNV), the virus that can cause the severe disease in humans known as Argentine hemorrhagic fever[217].

2.8   CONCLUSIONS

The beneficial effects of MCTs are largely contributed by the physicochemical properties of their MCFAs which make them important in dietetic management. The chain length and solubility of MCFAs contribute greatly to the ease of their absorption. This in turn affects fat metabolism and resolves complications pertinent to most fat maldigestion and malabsorption conditions. MCTs behave more like carbohydrates than conventional fats and have been proved to increase energy expenditure and hence may be useful in weight reduction and prevention of obesity. Although some findings report an increase in serum TAGs with MCT consumption, incorporation of structured lipids with other functional substances may improve the lipid profile. Lipid supplementation with MCTs through enteral and parenteral nutrition has provided life support for patients from infants to HIV-infected individuals. The valuable properties of MCTs make them very suitable for incorporation into products such as infant foods. While MCTs may not have been proved to enhance endurance during exercise, they do serve as additional fuel for working muscles. MCFAs, especially lauric acid, provide excellent antimicrobial protection, which enhances the immune system of the body. The performance of MCTs is improved by incorporation in structured TAGs of MCTs and LCTs. Thus structured lipids containing MCFAs and essential fatty acids as well as other fatty acids may serve as important components with nutraceutical benefits.

REFERENCES

1. Traul, K.A., Driedger, A., Ingle, D.L., and Nakhasi, D., Review of the toxicology properties of medium-chain triglycerides, *Food Chem. Toxicol.*, 38, 79–98, 2000.
2. Patch, C.S., Tapsell, L.C., and Williams, P.G., Dietetics and functional foods. *Nutr. Diet.*, 61, 22–29, 2004.

3. Osborn, H.T. and Akoh, C.C., Structured lipids, novel fat, with medical, nutraceutical and food applications, *Comprehensive Rev. Food Sci. Food Safety,* 3, 93–103, 2002.

4. O'Brien, R.D., *Fats and Oils: Formulating and Processing for Applications,* CRC Press, New York, 2004, pp. 40–43.

5. Pantzaris, T.P. and Basiron. Y., The lauric (coconut and palm kernel) oils, In *Vegetable Oils in Food Technology: Composition, Properties and Uses,* Gunstone, F.D., Ed., Blackwell and CRC Press, Boca Raton, 2002, pp. 157–202.

6. Guarte, R.C., Muhlbauer, W., and Kellert, M., Drying characteristics of copra and quality of copra and coconut oil, *Postharvest Biol. Tech.,* 9, 361–372, 1996.

7. Canapi, E.C., Agustin, Y.T.V., Moro, E.A., Pedrosa, E., and Bendano, M.L.J., Coconut oil, in *Bailey's Industrial Oil and Fat Products,* Hui, Y.H, Ed., John Wiley, New York, 1996, pp. 97–124.

8. Gunstone, F.D., Harwood, J.L., and Padley, F.B., *The Lipid Handbook,* 2nd ed., Chapman and Hall, London, 1994.

9. Young, F.V.K., Palm kernel and coconut oils: analytical characteristics, process technology and uses, *J. Am. Oil Chemist Soc.,* 60, 374–379, 1983.

10. Bockisch, M., *Fats and Oils Handbook,* AOCS Press, Champaign, IL, 1998.

11. Che Man, Y.B., Suhardiyono, Asbi, A.B., Azudin, M.N., and Wei, L.S., Aqueous enzymatic extraction of coconut oil, *J. Am. Oil Chem. Soc.,* 73, 683–686, 1996.

12. Che Man, Y.B., Abdul Karim, M.I.B., and Teng, C.T., Extraction of coconut oil with *Lactobacillus plantarum* 1041 IAM, *J. Am. Oil Chem. Soc.,* 74, 1115–1119, 1997.

13. Dominguez, H., Nunez, M.J., and Lema, J.M., Enzymatic pretreatment to enhance oil extraction from fruits and oilseeds: a review, *Food Chem.,* 49, 271–286, 1994.

14. Persley, G.J., *Replanting the Tree of Life,* CAB International, Oxfordshire, UK, 1992.

15. Laureles, L.R., Rodriguez, F.M., Reano, C.E., Santos, G.A., Laurena, A.C., and Mendoza, E.M.T., Variability in fatty acid and triacylglycerol composition of the oil of coconut (Cocos nucifera L.) hybrids and their parentals, *J. Agric. Food Chem.,* 50, 1581–1586, 2002.

16. Nik Norulaini, N.A., Md Zaidul, I.S., Anuar, O., and Mohd Omar, A.K., Supercritical enhancement for separation of lauric acid and oleic acid in palm kernel oil (PKO), *Separat. Purif. Tech.,* 35, 55–60, 2004.

17. Codex. Alinorm 01/17, draft report of the 17th Session of the Codex Committee on Fats and Oils, 2001.

18. Basiron, Y., Coconut oil, in *Bailey's Industrial Oil and Fat Products,* Hui, Y.H, Ed., John Wiley, New York, 1996, pp. 271–375.

19. Teah, Y.K. and Ong, A.S.H., Advantages of palm kernel oil over coconut oils in foods, *Palm Oil Develop.,* no. 9, 1988.

20. Cornelius, J.A., Palm oil and palm kernel oil. *Prog. Chem. Fats Other Lipids,* 15, 5–27, 1977.

21. de Man, J.M. and de Man, L., *Specialty Fats Based on Palm Oil and Palm Kernel Oil,* Malaysian Palm Oil Promotion Council, Kuala Lumpur, 1994.

22. Tang, T.S., Chong, C.L., and Yusoff, M.S.A., Malaysian palm kernel stearin, palm kernel olein and their hydrogenated products, *PORIM Technol.,* no. 16, 1993.

23. Filho, E.A.B., Baruque, M.D.G., and Sant'Anna, Jr., G.L., Babassu coconut starch liquefaction: an industrial scale approach to improve conversion yield, *Bioresource Tech.,* 75, 49–55, 2000.

24. Roath, W.W., Widrlechner, M.P., and Kleinman, R., Variability in Cuphea viscosissima jacq collected in east-central United States, *Ind. Crops Prod.,* 3, 217–223, 1994.

25. Moscheni, E., Angelini, L.G., and Macchia, M., Agronomic potential and seed oil composition of Cuphea lutea and C. laminuligera, *Ind. Crops Prod.,* 3, 3–9, 1994.

26. Jeukendrup, A.E. and Aldred, S., Fat supplementation, health and endurance performance, *Nutrition,* 20, 678–688, 2004.

27. Bach, A.C. and Babayan, V.K., Medium-chain triglycerides: an update, *Am. J. Clin. Nutr.,* 36, 950–962, 1982.

28. Odle, J., New insight into the utilization of medium-chain triglycerides by the neonate: observation from a piglet model, *J. Nutr.,* 127, 1061–1067, 1997.

29. Ramirez, M., Amate, L., and Gil, A., Absorption and distribution of dietary fatty acids from different sources, *Early Human Dev.,* S95–S101, 2001.

30. Lieu, E.L., The role of fatty acid composition and positional distribution in fat absorption in infants, *J. Pediat.,* 125, S62–S68, 1994.

31. Willis, W.M., Lencki, R.W., and Marangoni, A.G., Lipid modification strategies in the production of nutritionally functional fats and oils, *Crit. Rev. Food Sci. Nutr.,* 38, 639–674, 1998.

32. Papamandjaris, A.A., MacDougall, D.E., and Jones, P.J.H., Medium chain fatty acid metabolism and energy expenditure: obesity treatment implications, *Life Sci.*, 62, 1203–1215, 1998.
33. Jong-Yeon, K., Hickner, R.C., Dohm, G.L., and Houmard, J.A., Long- and medium-chain fatty acid oxidation is increased in exercise-trained human skeletal muscle, *Metabolism*, 51, 460–464, 2002.
34. Pszczola, D.E., Fats: in transition, *Food Technol.*, 58, 52–63, 2004.
35. Seidell, J.C., Obesity in Europe, *Obesity Res.*, 3 (Suppl. 2), 249–259, 1995.
36. Atkinson, R.L., Treatment of obesity. *Nutr. Rev.*, 50, 338–345, 1992.
37. Pasman W.J., Saris, W.H.M., and Westerterp-Plantega, M.S., Predictors of weight maintenance, *Obesity Res.*, 7, 43–50, 1999.
38. Bach, A.C., Ingenbleek, Y., and Frey, A., The usefulness of dietary medium-chain triglycerides in body weight control: fact or fancy?, *J. Lipid Res.*, 37, 708–726, 1996.
39. St-Onge, M.P. and Jones, P.J., Physiological effects of medium-chain triglycerides: potential agents in the prevention of obesity, *J. Nutr.*, 132, 329–332, 2002.
40. Lasekan, J.B., Rivera, J., Hirnoven, M.D., Keesey, R.E., and Ney, D.M., Energy expenditure in rats maintained with intravenous or intragastric solutions containing medium or long chain triglyceride emulsions. *J. Nutr.*, 122, 1483–1492, 1992.
41. Mabayo, R.T., Furuse, M., Murai, A., and Okumura, J.I., Interaction between medium-chain and long chain triacylglycerols in the lipid and energy metabolism in growing chicks, *Lipids*, 29, 139–144, 1994.
42. Rothwell, N.J. and Stock, M.J., Stimulation of thermogenesis and brown fat activity in rats fed medium chain triglyceride, *Metabolism*, 36, 128–130, 1987.
43. Lin, M.T., Yeh, S.L., Kuo, M.L., Liaw, K.Y., Lee, P.H., Chang, K.J., and Chen, W.J., Effects on medium-chain triglyceride in parenteral nutrition on rats undergoing gastrectomy, *Clin. Nutr.*, 21, 39–43, 2002.
44. Hill, J.O., Peters, J.C., Yang, D., Sharp, T., Kaler, M., Abumrad, N.N., and Greene, H.L., Thermogenesis in humans during overfeeding with medium-chain triglycerides, *Metabolism*, 38, 641–648, 1989.
45. White, M.D., Papamandjaris, A.A., and Jones, P.J.H., Enhanced postprandial energy expenditure with medium-chain fatty acid feeding is attenuated after 14d in premenopausal women, *Am. J. Clin. Nutr.,* 69, 883–889, 1999.
46. Seaton, T.B., Welle, S.L., Warenko, M.K., and Campbell, R.G., Thermic effect of medium-chain and long chain triglycerides in man, *Am. J. Clini.Nutr.*, 44, 630–634, 1986.
47. Dulloo, A.G. Fathi, M., and Mensi, N., Twenty four hour energy expenditure and urinary catecholamines of humans consuming low to moderate amounts of medium chain triglycerides: a dose dependent study in a human respiratory chamber, *Eur. J. Clini. Nutr.*, 50, 152–158, 1996.
48. Scalfi, L., Coltorti, A., and Contaldo, F., Postprandial thermogenesis in lean and obese subjects after meals supplemented with medium-chain and long-chain triglycerides, *Am. J. Clinic. Nutr.*, 53, 1130–1133, 1991.
49. Binnert, C., Paschiaud, C., Beylot, M., Hans, D., Vandermander, J., Chantre, P., Riou, J.D., and Laville, M., Influence of human obesity on the metabolic fate of dietary long and medium-chain triacylglycerols, *Am. J. Clin. Nutr.*, 67, 595–601, 1998.
50. Bendixen, H., Flint, A., Raben, A., Hoy, C-E., Mu, H., Xu, X., Bartels, E.M., and Astrup, A., Effect of 3 modified fats and a conventional fat on appetite, energy intake, energy expenditure, and substrate oxidation in healthy men, *Am. J. Clin. Nutr.*, 75, 47–56, 2002.
51. Papamandjaris, A.A., White, M.D., and Jones, P.J., Components of total energy expenditure in healthy young women are not affected after 14 days of feeding with medium versus long-chain triglycerides, *Obesity Res.,* 7, 273–280, 1999.
52. Kasai, M.J., Nosaka, N., Maki, H., Sizuki, Y., Takeuchi, H., Aoyama, T., Oh, A., Harada, Y., Okazaki, M., and Kondo, K., Comparison of diet-induced thermogenesis of foods containing medium versus long-chain triacylglycerols, *J. Nutr. Sci. Vit.* (Tokyo), 48, 539–540, 2002.
53. St-Onge, M.P., Ross, R., Parsons, W.D., and Jones, P.J.H., Medium-chain triglycerides increase energy expenditure and decrease adiposity in overweight men, *Obesity Res.*, 11, 395–402, 2003c.
54. St-Onge, M.P. and Jones, P.J., Greater rise in fat oxidation with medium-chain triglyceride consumption relative to long-chain triglyceride is associated with lower initial body weight and greater loss of subcutaneous adipose tissue, *Int. J. Obesity Related Metabolic Disord.*, 27, 1565–1571, 2003.
55. St-Onge, M.P., Bourque, C., Jones, P.J.H., Ross, R., and Parsons, W.E., Medium versus long-chain triglycerides for 27 days increases fat oxidation and energy expenditure without resulting in changes in body composition in overweight women, *Int. J. Obesity*, 27, 95–102, 2003.

56. Yost, T.J. and Eckel, R.H., Hypocaloric feeding in obese women: metabolic effects of medium chain triglyceride substitution, *Am. J. Clin. Nutr.*, 49, 326–330, 1989.

57. Donnel, S.C., Lyoyd, D-A., Eaton, S., and Pierro, A., The metabolic response to intravenous medium-chain triglycerides in infants after surgery, *J. Pediat.*, 141, 689–94, 2002.

58. Rodriguez, M., Funke, S., Fink, M., Demmelmair, H., Turini, M., Crozier, G., and Koletsko, Plasma fatty acids and[13C] linoleic acid metabolism in preterm infants fed a formula with medium chain triglycerides, *J. Lipid Res.*, 44, 41–48, 2003.

59. Baba, N., Bracco, E.F., and Hashim, S.A., Enhanced thermogenesis and diminished deposition of fat in response to overfeeding with diet containing medium-chain triglyceride, *Am. J. Clin. Nutr.*, 35, 678–682, 1982.

60. Geliebter, A., Torbay, N., Bracco, E., Hashim, S.A., and Itallie, V., Overfeeding with medium-chain triglyceride diet results in diminished deposition of fat, *Am. J. Clin. Nutr.*, 37, 1–4, 1983.

61. Hill, J.O., Peters, J.C., Lin, D., Yakubu, F., Greene, H., and Swift, L., Lipid accumulation and body fat distribution is influenced by type of dietary fat fed to rats, *Int. J. Obesity*, 17, 223–236, 1993.

62. Han, J., Hamilton, J.A., Kirkland, J.L., Corkey, B.E., and Guo, W., Medium-chain oil reduce fat mass and down regulates expression of adipogenic genes in rats. *Obesity Res.*, 11, 734–744, 2003.

63. Tsuji, H., Kasai, M., Takeuchi, H., Nakamura, M., Mitsuko, O., and Kondo, K., Dietary medium-chain triacylglycerols suppress accumulation of body fat in a double-blind, controlled trial in healthy men and women, *J. Nutr.*, 131, 2853–2859, 2001.

64. Matsuo, T., Matsuo, M., Kasai, M., and Takeuchi, H., Effects of a liquid diet supplement containing structured medium and long-chain triacylglycerols on body fat accumulation in healthy young subjects, *Asia Pacif. J. Clini. Nutr.*, 10, 46–50, 2001.

65. Krotkiewski, M., Value of VLCD supplementation with medium-chain triglycerides, *Int. J. Obesity Related Metabolic Disord.*, 25, 1393–1400, 2001.

66. Bray, G.A., Lee, M., and Bray, T.L., Weight gain of rats fed medium-chain triglycerides is less than long chain triglycerides, *Int. J. Obesity*, 4, 27–32, 1980.

67. Maggio, C.A., and Koopmans, H.S., Food intake after intragastric meals of short, medium or long-chain triglyceride, *Physiological Behavior*, 28, 921–926, 1982.

68. Furuse, M., Choi, Y.H., Mabayo, R.T., and Okumura, J.L., Feeding behavior in rats fed diets containing medium-chain triglyceride, *Physiological Behavior*, 52, 815–817, 1992.

69. Stubbs, R.J. and Harbron, C.G., Covert manipulation of the ration of medium to long chain triglycerides in isoenergetically dense diets: effect on food intake in ad libitum feeding men, *Int. J. Obesity*, 20, 435–444, 1996.

70. Rolls, B.J., Gnizak, N., Summerfelt, A., and Laster, J., Food intake in dieters and nondieters after a liquid meal containing medium-chain triglycerides, *Am. J. Clini. Nutr.*, 48, 66–71, 1988.

71. Wymelbeke, V.V., Himaya, A., Louis-Sylvester, J., and Fantino, M., Influence of medium-chain and long-chain triacylglycerols on the control of food intake in men, *Am. J. Clin. Nutr.*, 68, 226–34, 1998.

72. Kovacks, E.M.R., Westerterp-Plantenga, M.S., de Vries, M., Brouns, F., and Saris, W.H.M., Effects of 2-week ingestion of (−)-hydroxycitrate combined with medium-chain triglycerides on satiety and food intake, *Physiol. Behavior*, 74, 543–549, 2001.

73. Wymelbeke, V.V., Louis-Sylvestre, J., and Fantino, M., Substrate oxidation and control of food intake in men after a fat substitute meal compared with meals supplemented with an isoenergetic load of carbohydrate, long chain triacylglycerols on medium-chain triacylglycerols, *Am. J. Clinical Nutr.*, 74, 620–30, 2001.

74. Hill, J.O, Peters, J.C., Swift, L.L., Yang, D., Sharp, T., Abumrad, N., and Greene, H.L., Changes in blood lipids during six days of overfeeding with medium or long-chain triglycerides, *J. Lipid Res.*, 31, 407–416, 1990.

75. Carter, N.B., Heller, H.J., and Denke, M.A., Comparison of the effects of medium-chain triacylglycerols, palm oil and high oleic acid sunflower oil on plasma triacylglycerol fatty acids and lipid and lipoproteins concentrations in humans, *Am. J. Clin. Nutr.*, 65, 41–45, 1997.

76. Asakura, L., Lottenberg, A.M.P., Neves, M.Q.T.S., Nunes, V.S., Rocha, J.C., and Passarelli, Dietary medium-chain triacylglycerol prevents the postprandial rise of plasma triacylglycerols but induces hypercholesterolemia in primary hypertriglyceridemia subjects, *Am. J. Clin. Nutr.*, 71, 701–705, 2000.

77. Tholstrup, T., Ehnholm, C., Jauhiainen, M., Peterson, M., Hoy, C.E., Lund, P., and Sanstrom, B., Effects of medium-chain fatty acids and oleic acid on blood lipids, lipoproteins, glucose, insulin and lipid transfer protein activities, *Am. J. Clin. Nutr.*, 79, 564–569, 2004.

78. Swift, L.L., Hill, J.O., Peters, J.C., and Greene, H.L., Plasma lipids and lipoprotein during 6 d of maintenance feeding with long-chain, medium-chain and mixed-chain triglycerides, *Am. J. Clini. Nutr.*, 56, 881–886, 1992.

79. Bell, S.J., Bradley, D., Forse, R.A., and Bistrian, B.R., The new dietary fats in health and disease, *J. Am. Diet Assoc.*, 97, 280–286, 1997.

80. Rao, R. and Lokesh, B.R., Nutritional evaluation of structured lipid containing omega 6 fatty acid synthesized from coconut oil in rats, *Mol. Cellular Biochem.*, 248, 25–33, 2003.

81. St-Onge, M.P., Lamarche, B., Mauger, J.F., and Jones, P.J.H., Consumption of functional oil rich in phytosterol and medium-chain triglyceride oil improves plasma lipid profiles in men, *J. Nutr.*, 133, 1815–1820, 2003.

82. Bourque, C., St-Onge, M.P., Papamandjaris, A.A., Cohn, J.S., and Jones, P.J., Consumption of an oil composed of medium-chain triacylglycerols, phytosterol and N-3 fatty acids improves cardiovascular risk profile in overweight women, *Metabolism*, 52, 771–7, 2003.

83. Beermann, C.J., Jelinek, J., Reinecker, T., Hauenschild, A., Boehm, G., and Klor, H.U., Short term effects of dietary medium-chain fatty acids and. n-3 long chain polyunsaturated fatty acids on the fat metabolism of healthy volunteers, *Lipids Health Dis.*, 2, 10, 2003.

84. Manuel-y-Keenoy, B., Nonneman, L., De Bosscher, H., Vertomen, J., Schrans, S., Klutsch, K., and De Leeuw, I., Effects of intravenous supplementation with alpha-tocopherol in patients receiving total parenteral nutrition containing medium and long-chain triglycerides, *Eur. J. Clini. Nutr.*, 56, 121–128, 2002.

85. Babayan, V.K., Medium chain triglycerides and structured lipids, *Lipids*, 22, 417–420, 1987.

86. Mascioli, E.A., Bistrian, B.R., Babayan, V.K., and Blackburn, G.L., Medium chain triglycrides and structured lipid as unique nonglucose energy sources in hyperalimentation, *Lipids*, 22, 421–423, 1987.

87. Wolfe, R.R., O'Donnell, R.F.J., Stone, M.D., Richmand, D.A., and Burke, J.F., Investigation of factors determining the optimal glucose infusion rate in total parenteral nutrition, *Metabolism*, 20, 892–900, 1980.

88. Jeejebhoy, K.N., Anderson, G.H., Nakhooda, A.F., Greenberg, G.R., Sanderson, I., and Marliss, E.B., Metabolic studies in total parenteral nutrition with lipid in man, comparison with glucose, *J. Clin. Invest.*, 57, 125–136, 1976.

89. Yeh, S.L., Chao, C.Y., Lin, M.T., and Chen, W.J., Effects of parenteral infusion with medium-chain trigycerides and safflower oil emulsions on hepatic lipids, plasma amino acids and inflammatory mediators in septic rats, *Clin. Nutr.*, 19, 115–120, 2000.

90. Kwon, D.Y., Song, H.N., and Yoon, S.H., Synthesis of medium-chain glycerides by lipase in organic solvent, *J. Am. Oil Chem. Soc.*, 73, 1521–1525, 1996.

91. Calabrese. C., Myer, S., Munson, S., Turet, P., and Birdsall, T.C., A cross-over study of the effect of a single oral feeding of medium-chain triglycerides oil versus canola oil on post-ingestion plasma triglyceride levels in healthy men, *Alter. Med. Rev.*, 4, 23–28, 1999.

92. Peter, L. and Jutta, K., Lipase supplementation therapy: standards, alternatives and perspectives, *Pancreas*, 26, 1–7, 2003.

93. Perry, R.S. and Gallagher, J., Management of maldigestion associated with pancreatic insufficiency, *Clin. Pharm.*, 4, 161–169, 1985.

94. Layer, P. Von der One, M.R., Holst, J.J., Jansen, J.B.M.J., Granat, D., Holtmann, G., and Goebell, H., Altered postprandial motility in chronic pancreatitis: role of malabsorption, *Gastroenterology*, 112, 1624–1634, 1997.

95. Doty, J.E., Fink, A.S., and Meyer, J.H., Alterations in digestive function caused by pancreatic disease, *Surg. Clin. N. Am.*, 69, 447–65, 1989.

96. Ledeboer, M., Masclee, A.A., Jansen, J.B., and Lamers, C.B., Effect of equimolar amounts of long-chain triglycerides and medium-chain triglycerides on small bowel transit time in humans, *J. Parenteral Enteral Nutr.*, 19, 5–8, 1995.

97. Shea, J.C., Bishop, M.D., Parker, E.M., Gelrud, A., and Freedman, S.D., An enteral therapy containing medium-chain triglyceride and hydrolyzed peptides reduces postprandial pain associated with chronic pancreatitis, *Pancreatology*, 3, 36–40, 2003.

98. Shimada, Y., Sugihara, A., Nakano, H., Yokota, T., Nagao, T., Komemushi, S., and Tominaga, Y., Production of structured lipids containing essential fatty acids by immobilized Rhizopus delemar lipase, *J. Am. Oil Chemist Soc.*, 73, 1415–1420, 1996.

99. Straarup, E.M. and Hoy, C.-E., Lymphatic transport of fat in rats with normal and malabsorption following intake of fats made from fish oil and decanoic acid, effect of triacylglycerol structure, *Nutr. Res.*, 21, 1001–1013, 2001.

100. Jeppesen, P.B. and Mortensen, P.B., The influence of a preserved colon on the absorption of medium chain fat in patients with small resection, *Gut*, 43, 478–483, 1998.

101. Goulet, O., Lipid requirements in infants with digestive diseases with references to short bowel syndrome, *Eur. J. Med. Res.*, 21, 79–83, 1997.

102. Craig, C.B., Darnell, B.E., Weinsier, R.L., Saag, M.S., Epps, L., Mullins, L., Lapidus, W.I., Ennis, D.M., Akrabawi, S.S., Cornwell, P.E., and Sauberlich, H.E., Decreased fat and nitrogen losses in patients with AIDS receiving medium-chain-triglyceride-enriched formula vs those receiving long-chain-triglyceride containing formula, *J. Am. Diet Assoc.*, 97, 605–611, 1997.

103. Miller, T.L., Orav, E.J., and Martin, S.R., Malnutrition and carbohydrate malabsorption in children with vertically transmitted human immunodeficiency virus 1 infection. *Gastroenterology*, 11, 1296, 1991.

104. Wanke, C.A., Pleskow, D., Degirolami, P.C., Lambl, B.B., Merkel, K., and Akrabawi, S., A medium chain triglyceride-based diet in patients with HIV and chronic diarrhea reduces diarrhea and malabsortion: a perspective, controlled trial, *Nutrition*, 12, 766–771, 1996.

105. Irving, S.Y., Simone, S.D., Hicks, F.W., and Verger, J.T., Nutrition for the critically ill child: enteral and parenteral support, *Adv. Practice Acute Crit. Care*, 11, 541–558, 2000.

106. Ulrich, H., Pastores, S.M., Katz, D.P., and Kveton, V., Parenteral use of medium-chain triglycerides: a reappraisal, *Nutrition*, 12, 231–238, 1996.

107. Akoh, C.C., Structured lipid, in *Food Lipids, Chemistry, Nutrition and Biotechnology*, Akoh, C.C. and Min, D.B., Eds., Marcel Dekker, New York, 2002, pp. 877–908.

108. Clowes, G.H.A., O'Donnel, G.F., and Ryan, N.T., Energy metabolism in sepsis: treatment based on different patterns in shock and high output stage, *Ann. Surg.*, 141, 321–326, 1974.

109. Powanda, M.C., Changes in body balances of nitrogen and other key nutrients: description and underlying mechanisms, *Am. J. Clini. Nutr.*, 30, 1254, 1977.

110. Clowes, G.H.A., Randall, H.T., and Cha, C.J., Amino acid and energy metabolism in septic and traumatized patients, *J. Parenteral Enteral Nutr.*, 4, 195, 1980.

111. Schneeweiss, B., Graninger, W., and Ferenci, P., Short term energy balance in patients with infections: carbohydrate-based versus fat-based diets, *Metabolism*, 41, 125, 1992.

112. Dennison, A.R., Ball, M., and Hands, L.J., Total parenteral nutrition using conventional and medium chain trigycerides: effects on liver function tests, complement and nitrogen balance, *J. Parent. Enteral Nutr.*, 12, 15, 1988.

113. Jiang, Z., Zhang, S., and Wang, X., A comparison of medium-chain and long-chain trigycerides in surgical patients, *Ann. Surg.*, 217, 175, 1993.

114. Venus, B., Prager, R., and Patel, C.B., Cardiopulmonary effects of intralipid infusion in critically ill patients, *Crit. Care Med.*, 16, 587, 1988.

115. Hageman, J.R., McCulloch, K., and Gora, P., Intralipid alterations in pulmonary prostaglandin metabolism and gas exchange, *Crit. Care Med.*, 7, 69, 1983.

116. Muscaritoli, M., Cangiano, C., and Cascino, A., Exogenous lipid clearance in compensated liver cirrhosis, *J. Parenteral Enteral Nutr.*, 10, 599–603, 1986.

117. Border, J.R., Burns, G.P., and Rumph, C., Carnitine levels in severe infections and starvation: a possible key to the prolonged catabolic state, *Surgery*, 68, 175, 1970.

118. Lien, E.L., The role of fatty acids composition and positional distribution in fat absorption in infants, *J. Pediat.*, 125, S62–S68, 1994.

119. Manson, W.G. and Weaver, L.T., Fat digestion in the neonate, *Arch. Dis. Childhood.*, 76, F206–F211, 1997.

120. Birk, R.Z., Regan, K.S., Boyle-Roden, E., and Brannon, P.M., Pancreatic lipase and its related protein 2 are regulated by dietary polyunsaturated fat during the postnatal development of rats, *Pediat. Res.*, 56, 256–262, 2004.

121. Watkins, J.B., Lipid digestion in the developing infant, in *Dietary Fat Requirements in Health and Development*, Beare-Rogers, J., Ed., American Oil Chemists Society, Library of Congress, USA, 1988. pp. 29–42.

122. Carriere, F., Barrowman, J.A., Verger, R., and Laugier, R., Secretion and contribution to lipolysis of gastric and pancreatic lipase during a treatment in humans, *Gastroenterology*, 105, 876–888, 1993.

123. Carnielli, V.P., Verlato, G., Pederzini, F., Luijendijk, I., Boerlage, A., and Pedrotti, D., Intestinal absorption of long-chain polyunsaturated fatty acids in preterm infants fed breast milk formula, *Am. J. Clin. Nutr.*, 67, 97–103, 1998.

124. Carver, J.D., Advances in nutritional modifications of formulas, *Am. J. Clin. Nutr.*, 77 (Suppl.), 1550S–1554S, 2003.

125. Hardy, S.C. and Kleinman, R.E., Fat and cholesterol in the diet of infants and young children: implications for growth, development and long-term health, *J. Ped.*, 125, S69–S77, 1994.

126. Jensen, C., Buist, N.R., and Wilson, T., Absorption of individual fatty acids from long-chain or medium-chain triglycerides in very small infants, *Am. J. Clin. Nutr.*, 43, 745–751, 1986.

127. Borum, P.R., Medium-chain trigycerides in formula for preterm neonates: implications for hepatic and extrahepatic metabolism, *J. Pediat.*, 120, S139–S145, 1992.

128. Sann, L., Mousson, B., Rousson, M., Maire, I., and Bethenod, M., Prevention of neonatal hypoglycemia by oral lipid supplementation in low birth weight infants, *Eur. J. Pediat.*, 147, 158–161, 1998.

129. Isaacs, C.E., Litov, R.E., and Thormar, H., Antimicrobial activity of lipids added to human milk, infant formula and bovine milk, *J. Nutr. Biochem.*, 6, 362–366, 1995.

130. Klein, C., Nutrient requirement for preterm infant formula, *J. Nutr.*, 132, 1395S–1577S, 2002.

131. Cohen, S., Mimouni, F.B., Migdalovich, M., Peled, Y., and Dollberg, S., P1140 energy expenditure in preterm infants supplemented with canola oil versus medium-chain triglyceride oil, *J. Ped. Gastro. Nutr.*, 39, S490, 2004.

132. Pettei, M.J., Daftary, S., and Levine, J.J., Essential fatty acid deficiency with the use of a medium-chain triglyceride infant formula in pediatric hepatobiliary disease, *Am. J. Clini. Nutr.*, 53, 1217–21, 1991.

133. Cotter, R., Taylor, C.A., and Johnson, R., A metabolic comparison of a pure long-chain triglyceride (LCT) and various medium-chain triglyceride (MCT)-LCT combination emulsion in dogs, *Am. J. Clin. Nutr.*, 45, 927, 1987.

134. Rubin, M., Moser, A., Vaserberg, N., Greig, F., Levy, Y., Spivak, H., Ziv, Y., and Lelcuk, S., Structured triacylglycerol emulsion, containing both medium and long-chain fatty acids, in long-term home parenteral nutrition: a double-blind randomized cross-over study, *Nutrition*, 16, 95–100, 2000.

135. Telliez, F., Bach, V., Leke, A., Chardon, K., and Libert, J.P., Feeding behavior in neonates whose diet contained medium-chain triacylglycerols: short-term effects on thermoregulation and sleep, *Am. J. Clini. Nutr.*, 76, 1091–1095, 2002.

136. Jeukendrup, A.E., Saris, W.H.M., Schrauwen, P., Brouns, F and Wagenmakers, A.J.M., Metabolic availability of medium-chain triglycerides coingested with carbohydrates during prolonged exercise: diet related differences between Europe and America, *J. Appl. Physoil.*, 74, 2353–2357, 1995.

137. Jeukendrup, A.E., Saris, W.H.M., and Wagenmakers, A.J.M., Fat metabolism during exercise: a review, *Int. J. Sports Med.*, 19, 371–379, 1998.

138. Coyle, E.F., Coggan, A.R., Hemmert, M.K., and Ivy, J.L., Muscle glycogen utilization during prolonged strenuous exercise when fed carbohydrate, *J. Appl. Physiol.*, 61, 165–172, 1986.

139. Wagenmakers, A.J.M., Brouns, F., Saris, W.H.M., and Halliday, D., Oxidation rates of orally ingested carbohydrates during prolonged exercises in man, *J. Appl. Physiol.*, 75, 2774–2780, 1993.

140. Rehrer, N.J., Wagenmakers, A.J.M., Beckers, E.J., Halliday, D., Leiper, J.B., Brouns, F., Maugham, R.J., Westerterp, K., and Saris, W.H.M., Gastric emptying absorption and carbohydrate oxidation during prolonged exercise, *J. Appl. Physiol.*, 63, 1752–1732, 1992.

141. Hickson, R.C., Rennie, M.J., Conlee, R.K., Winder, W.W., and Holloszy, J.O., Effects of increased plasma free fatty acids on muscle glycogen utilization and endurance, *J. Appl. Physiol.*, 43, 829–833, 1977.

142. Costill, D.L. Coyle, E., Dalkys, G., Evans, W., Fink, W., and Hoopes, D., Effects of elevated plasma FFA and insulin on muscle glycogen usage during exercise, *J. Physiol.*, 43, 695–9, 1977.

143. Vukovich, M.D., Costill, D.L., Hickey, M.S., Trappe, S.W., Cole, K.J., and Fink, W.J., Effect of fat emulsion infusion and fat feeding on muscle glycogen utilization during cycle exercise, *J. Appl. Physiol.*, 75, 1513–1518, 1993.

144. Horowitz, J.F., and Klein, S., Lipid metabolism during endurance exercise, *Am. J. Clin. Nutr.*, 72 (Suppl.), 558S–563S, 2000.

145. Pendergast, D.R., Leddy, J.J., and Venkatrama, J.T., A perspective on fat intake in athletes, *J. Am. Coll. Nutr.*, 19, 345–350, 2000.

146. Satabin, P., Portero, P., Defer, G., Bricout, J., and Guezennec, C.-Y., Metabolic and hormonal responses to lipid and carbohydrate diets during exercise in man, *Med. Sci. Sports Exercise*, 19, 218–223, 1987.

147. Kiens, B. and Lithell, H., Lipoprotein metabolism influenced by training induced changes in human skeletal muscle, *J. Clin. Invest.*, 83, 558–564, 1989.

148. Bremer, J., Carnitine-metabolism and functions, *Physiol. Rev.*, 63, 1420–1479, 1983.

149. Oopik, V., Timpmann, S., Medijainen, L., and Lemberg, H., Effects of daily medium-chain trigyceride ingestion on energy metabolism and endurance performance capacity in well-trained runners, *Nutr. Res.*, 21, 1125–1135, 2001.

150. Beckers, E.J., Jeukendrup, A.E., Brouns, F., Wagenmakers, A.J.M., and Saris, W.H.M., Gastric empty-ing of carbohydrate medium chain triglyceride suspensions at rest, *Int. J. Sports Med.*, 13, 581–584, 1992.

151. Ivy, J.L., Costill, D.L., Fink, W.J., and Maglischo, E., Contribution of medium and long chain triglyc-eride intake to energy metabolism during prolonged exercise, *Int. J. Sports Med.*, 1, 15–20, 1980.

152. Decombaz, J., Arnaud, M.J., Milan, H., Moesch, H., Philippossian, G., Thelin, A.L., and Howald, H., Energy metabolism of medium-chain triglycerides versus carbohydrates during exercise, *Eur. J. Appl. Physiol. Occup. Physiol.*, 52, 9–14, 1983.

153. Massicotte, D., Peronnet, F., Brisson, G.R., and Hillaire-Marcel, C., Oxidation of exogenous medium-chain free fatty acids during prolonged exercise-comparison with glucose, *J. Appl. Physiol.*, 73, 1334, 1992.

154. Jeukendrup, A.E., Wagenmakers, A.J.M., Brouns, F., Halliday, D., and Saris, W.H.M., Effect of carbo-hydrate (CHO) and fat supplementation on cho metabolism during prolonged exercise, *Metabolism*, 45, 915, 1996b.

155. Jeukendrup, A.E., Saris, W.H.M., Van Diesen, R., Brouns, F., and Wagenmakers, A.J.M., Effect of edogenous carbohydrate availability on oral medium-chain triglyceride oxidation during prolonged exercise, *J. Appl. Physiol.*, 80, 949, 1996.

156. Van Zyl, C.G., Lambert, E.V., Hawley, J.A., Noakes, T.D., and Dennis, S.C., Effects of medium-chain triglyceride ingestion on fuel metabolism and cycling performance, *J. Appl. Physiol.*, 88, 113–119, 1996.

157. Goedecke, J.H., Elmer-English, R., Dennis, S.C., Scholass, I., Noakes, T.D., and Lambert, E.V., Effects of medium-chain triacylglycerol ingested with carbohydrate on metabolism and exercise performance, *Int. J. Sport Nutr.*, 9, 35–47, 1999.

158. Angus, D.J., Hargreaves, M., Dancey, J., and Febbraio, M.A., Effect of carbohydrate or carbohydrate plus medium-chain triglyceride ingestion on cycling time trial performance, *J. Appl. Physiol.*, 88, 113–119, 2000.

159. Misell, L.M., Lagomarcino, N.D., Schuster, V., and Kern, M., Chronic medium-chain triacylglycerol consumption and endurance performance in trained runners, *J. Sports Med. Phys. Fit.*, 41, 210–215, 2001.

160. Vistisen, B., Nybo, L., Xu, X., Hoy, C.E., and Kiens, B., Minor amounts of plasma medium-chain fatty acids and no improved time trial performance after consuming lipids, *J. Physiol.*, 95, 2434, 2003.

161. Miles, J., Cattalini, M., and Sharbrough, F., Metabolic and neurologic effects of an intravenous medium-chain triglyceride emulsion, *J. Parenteral Enteral Nutr.*, 15, 37, 1991.

162. Lingren, B.F., Ruokonen, E., and Takala, J., Nitrogen sparing effect of structured triglycerides contain-ing both medium and long chain fatty acids in critically ill patients, a double blind randomized con-trolled trial, *Clin. Nutr.*, 20, 43–48, 2001.

163. Simoens, C., Deckelbaum, R.J., and Carpentier, Y.A., Metabolism of defined structured triglyceride particles compared to mixtures of medium and long chain triglycerides intravenously infused in dogs, *Clin. Nutr.*, 23, 665–672, 2004.

164. Li, Z-Y. and Ward, O.P., Lipase-catalyzed esterification of glycerol and n-3 polyunsaturated fatty acid concentrate in organic solvent, *J. Am. Oil Chem. Soc.*, 70, 745–748, 1993.

165. Nagata. J-I, Kasai, M., Watanabe, S., Ikeda, I., and Saito, M., Effects of highly purified structured lipids containing medium-chain fatty acids and linoleic acid on lipid profiles in rats. *Biosci., Biotech. Biochem.*, 67, 1937–1943, 2003.

166. Kubow, S., The influence of positional distribution of fatty acids in native, interesterified and structure-specific lipids on lipoprotein metabolism and atherogenesis, *J. Nutr. Biochem.*, 7, 530–541, 1996.

167. Small, D.M., The effects of glyceride structure on absorption and metabolism, *Annu. Rev. Nutr.*, 11, 413, 1991.

168. Bracco, U., Effect of triglyceride structure on fat absorption, *Am. J. Clin. Nutr.*, 60, 1002S, 1994.

169. Hayes, K.C., Synthetic and modified glycerides: effect on plasma lipids, *Curr. Opin. Lipidology*, 12, 55–60, 2001.

170. Roy, C.C., Bouthillier, J., Seidman, E., and Levy, E., New lipids in enteral feeding, *Curr Opin. Clin. Nutr. Metabolic Care*, 7, 117–122, 2004.

171. Lepine, A.J., Garleb, K.A., Reinhart, G.A., and Kresty, L.A., Plasma and tissue fatty acid profiles of growing pigs fed structured or non-structured triacylglycerides containing medium-chain and marine oil fatty acids, *J. Nutr. Biochem.*, 4, 362–372, 1993.

172. Decker, E.A., The role of stereospecific saturated fatty acid positions on lipid nutrition, *Nutr. Rev.*, 54, 108–110, 1996.

173. Jensen, G.L., McGarvey, N., Taraszew, R., Wixson, S.K., Seiner, D.L., Pai, T., Yeh, Y.Y., Lee, T.W., and DeMichele, S.J., Lymphatic absorption of enterally fed structured triacylglycerol vs physical in a canine model, *Am. J. Clin. Nutr.*, 60, 518, 1994.

174. Mu, H. and Hoy, C.-E., Intestinal absorption of specific structured triacylglycerols, *J. Lipid Res.*, 42, 792–798, 2001.

175. Hubbard, V.S. and McKenna, Absorption of safflower oil and structured lipid preparations in patients with cystic fibrosis, *Lipids*, 22, 424–428, 1987.

176. McKenna, M.C., Hubbard, V.S., and Pieri, J.G., Linoleic acid absorption from lipid supplements in patients with cystic fibrosis with pancreatic insufficiency and in control subjects, *J. Pediat. Gastro. Nutr.*, 4, 45–51, 1985.

177. Jandacek, R.J., Whiteside, J.A., Holcombe, B.N., Volpenhein, R.A., and Taulbee, J.D., The rapid hydrolysis and efficient absorption of triglyceride with octanoic acid in the land 3 positions and long chain fatty acids in the 2 position, *Am. J. Clin. Nutr.*, 45, 940–945, 1987.

178. Ikeda, I., Tamari, Y., Sugano, M., Watanabe, S., and Nagata, J., Lymphatic absorption of structured glycerolipids containing medium-chain fatty acids and linoleic acid, and their effect on cholesterol absorption in rats, *Lipids*, 26, 369–373, 1991.

179. Kishi, T., Carvajal, O., Tomoyori, H., Ikeda, I., Sugano, M., and Imaizumi, K., Structured trigycerides containing medium-chain fatty acids and linoleic acid differently influence clearance rate in serum of triglycerides in rats, *Nutr. Res.*, 22, 1343–1351, 2002.

180. Chan, S., McCowen, K.C., and Bistrian, B., Medium-chain triglyceride and n-3 polyunsaturated fatty acid containing emulsion in intravenous nutrition, *Lipid Metabolism Therapy*, 1, 163–169, 1998.

181. Dyerberg, J., Bang, H.O., and Hjorne, N., Fatty acid composition of the plasma lipids in Greenland Eskimos, *Am. J. Clin. Nutr.*, 28, 958–66, 1975.

182. Yang, Y-L., Kuksis, A., and Myher, J.J., Lipolysis of menhaden oil triacylglycerols and the corresponding fatty acids alkyl esters by pancreatic lipase in vitro: a reexamination, *J. Lipid Res.*, 31, 137–48, 1990.

183. Christensen, M.S., Hoy, C.E., Becker, C.C., and Redgrave, T.G., Intestinal absorption and lymphatic transport of eicosapentaenoic (EPA) docosahexaenoic (DHA), and decanoic acids: dependence on intramolecular triacylglycerol structure, *Am. J. Clin. Nutr.*, 61, 56–61, 1995.

184. Selleck, K.J., Wan, J.M-F., Gollaher, C.J., Babayan, V.K., and Bistrian, B.R., Effect of low and high amounts of structured lipid containing fish oil on protein metabolism in enterally fed burned rats, *Am. J. Clini. Nutr.*, 60, 216–222, 1994.

185. Swenson, E.S., Selleck, K.M., Babayan, V.K., Blackburn, G.L., and Bistrian, B.R., Persistence of metabolic effects after long term oral feeding of a structured triglyceride derived from medium-chain triglyceride and fish oil in burned and normal rats, *Metabolism*, 40, 484–90, 1991.

186. Ling, R.T., Istfan, N.W., Lopes, S.M., Babayan, V.K., Blackburn, G.L., and Bistrian, B.R., Structured lipid made from fish oil and medium-chain triglycerides alters tumor and host metabolism in Yoshida-sarcoma-bearing rats, *Am. J. Clin. Nutr.*, 53, 1177–1184, 1991.

187. Tso, P., Lee, T., and Stephen, J.D., Lymphatic absorption of structured triglycerides versus physical mix in a rat model of fat malabsorption, *Am. J. Physiol.*, 277, G333–G340, 1999.

188. Kruimel, J.W., Naber, A.H., Curfs, J.H., Wenker, M.A., and Jansen, J.B., With medium-chain triglycerides, higher, and faster oxygen radical production by stimulated polymorphonuclear leukocytes occur, *J. Parenteral Enteral Nutr.*, 24, 107–112, 2000.

189. Chambrier, C., Guiraud, M., Gibault, I.P., Labrosse, H., and Bouletreau, Medium and long-chain triacylgycerols in post operative patients: structured lipids versus a physical mixture, *Nutrition*, 15, 274–277, 1999.

190. Kenler, A.S., Swails, W.S., Driscoll, D.S., DeMichele, S.J., Daley, B., and Babinead, T.J., Early enteral feeding in postsurgical cancer patients: fish oil structured lipid-based polymeric formula versus a standard polymeric formula, *Ann. Surg.*, 223, 316–333, 1996.

191. Lien, E.L., Yuhas, R.J., Boyle, F.G., and Tomarelli, R.M., Corandomization of fats improves absorption in rats, *J. Nutr.*, 123, 1859–1867, 1993.

192. Boddie, R.L. and Nickerson, S.C., Evaluation of postmilking teat germicides containing lauricidian, saturated fatty acids and lactic acid, *J. Dairy Sci.*, 75, 1725–1730, 1992.

193. Chaibi, A., Lahsen, H.A., and Busta, F.F., Inhibition of bacterial spores and vegetative cells by glycerides, *J. Food Prot.*, 59, 716–722, 1996.

194. Oh, D-H. and Marshall, D.L., Antimicrobial activity of ethanol, glycerol monolaurate or lactic acid against *Listeria monocytogenes*, *Int. J. Food Microbiol.*, 20, 239–246, 1993.

195. Oh, D.-H. and Marshall, D.L., Influence of temperature, pH and glycerol monolaurate on growth and survival of *Listeria monocytogenes*, *J. Food Prot.*, 56, 4–749, 1993.

196. Marth, E.H., Disease characteristics of *Listeria monocytogenes*, *Food Technol.*, 42, 165–168, 1988.

197. Bal'a, M.F.A. and Marshall, D.L., Testing matrix, inoculum size and incubation temperature affect monolaurin activity against *Listeria monocytogenes*, *Food Microbiol.*, 13, 467–473, 1996.

198. Wang, L.-L. and Johnson, E.A., Inhibition of *Listeria monocytogenes* by fatty acids and monoglycerides, *Appl. Envir. Microbiol.*, 58, 624–629, 1992.

199. Wang, L.-L. and Johnson, E.A., Control of *Listeria monocytogenes* by monoglycerides in foods, *J. Food Prot.*, 60, 131–138, 1997.

200. Oh, D.-H. and Marshall, D.L., Enhanced inhibition of *Listeria monocytogenes* by glycerol monolaurate with organic acids, *J. Food Sci.*, 59, 1258–1261, 1994.

201. Kabara, J.J., Inhibition of *Staphylococcus Aureus* in a model agar-meat system by monolaurin: a research note, *J. Food Safety*, 6, 197–201, 1983.

202. Projan, S.J., Brown-Skrobot, S., Schlievert, P.M., Vandenesch, F., and Novick, R.P., Glycerol monolaurate inhibits the production of β-lactamase, toxic shock syndrome toxind-1 and other Staphylococcal exoproteins by interfering with signal transduction, *J. Bacteriol.*, 176, 4204–4209, 1994.

203. Witcher, K.J., Novick., and Schlievert, P.M., Modulation of immune cell proliferation by glycerol monolaurate, *Clin. Diagnostic Lab. Immunol.*, 3, 10–13, 1996.

204. Petschow, B.W., Batema, R.P., and Ford, L.L., Susceptibility of *Helicobacter pylori* to bactericidal properties of medium-chain monoglycerides and free fatty acids, *Antimicrob. Agents Chemother.*, 40, 302–306, 1996.

205. Chaibi, A., Lahsen, H.A., and Busta, F.F., Inhibition by monoglycerides of L-alanine-triggered *Bacillus aureus* and *Clostridium botulinum* spore germination and outgrowth, *J. Food Prot.*, 59, 832–837, 1996.

206. Ababouch, L.H., Bouqartacha, F., and Busta, F.F., Inhibition of *Bacillus cereus* respores and vegetative cells by fatty acids and glyceryl monododecanoate, *Food Microbiol.*, 11, 327–336, 1994.

207. Ababouch, L., Chaibi, A., and Busta, F.F., Inhibition of bacterial spore growth by fatty acids and their sodium salts, *J. Food Prot.*, 55, 980–984, 1992.

208. Chaibi, A., Ababouch, L.H., Ghoila, M.R., and Busta, F.F., Effect of monoglycerides on the thermal inactivation kinetics of *Bacillus cereus* F4165/75 spores, *Food Microbiol.*, 15, 527–537, 1998.

209. Rihakova, Z., Plockova, M., and Filip, V., Antifungal activity of lauric acid derivatives against *Aspergillus niger*, *Eur. Food Res. Technol.*, 213, 448–490, 2001.

210. Dayrit, C.S., Coconut oil in heath and disease: its and monolaurin's potential as cure for HIV/AIDS, 37th Cocotech Meeting, India, 2000.

211. Bergsson, G., Arnfinnson, J., Karlsson, S.M., Steingrimsson, O., and Thormar, H., In vitro of *Chlamydia trachomatis* by fatty acids and monoglycerides, *Antimicrob. Agents Chemother.*, 42, 2290–2294, 1998.

212. Zhou, Y.F., Guetta, E., Yu, Z.X., Finkel, T., and Epstien, S.E., Human cytomegalovirus increases modified low density lipoprotein uptake and scavenger receptor mRNA expression in vascular smooth muscle cells, *J. Clin. Invest.*, 98, 2129–2138, 1996.

213. Hendrix, M.G.R., Dormans, P.H.J., Kitslaar, P., Bosman, F., and Bruggeman, C.A., The presence of CMV nucleic acids in arterial walls of atherosclerotic and non-atheroslcerotic patients, *Am. J. Path.*, 134, 1151–1157, 1989.
214. Ellis, R.W., Infection and coronary heart disease, *J. Med. Micro.*, 46, 535–539, 1997.
215. Visseren, F.L.J., Bouler, K.P., Pon, M.J., Hoekstra, B.L., Erkelens, D.W., and Diepersloot, R.J.A., Patients with diabetes mellitus and atherosclerosis; a role for cytomegalovirus?, *Diabetes Res. Clin. Prac.*, 36, 46–55, 1997.
216. Hornung, B., Amtmann, E., and Sauer, G., Lauric acid inhibits the maturation of vesicular stomatitis virus, *J. Gen. Virology*, 75, 353–361, 1994.
217. Bartolotta, S., Garcia, C.C., Candurra, N.A., and Damonte, E.B., Effect of fatty acids on arenavirus replication: inhibition of virus production by lauric acid, *Arch. Virology*, 146, 777–790, 2001.

3 Cereal Grain Oils

Roman Przybylski
Department of Chemistry and Biochemistry, University of Lethbridge,
Alberta, Canada

CONTENTS

3.1   INTRODUCTION

Interest in specialty lipids with specific health-affecting components is growing. Consumers demand these oils as food ingredients, and as health supplements and nutraceuticals to control and protect health and support well being. Specialty oils play a different role from mainstream commodity oils, where nutritional and health value dictates their use and market prices. Consumers are willing to pay a higher price for specific nutritional and health properties of compounds present in these oils. With the nutritional and health value of these oils comes processing, which should protect bioactive components, and prevent "processing contamination." Processing procedures should have a natural connotation when applied to the isolation of oils from oilseeds or other matrices such as cereals. As regards to standard extraction and processing technologies used for mainstream commodity oils, where organic solvents and excessive processing is a common practice, consumers perceive that specialty oils processed in this way will have poor quality and bioactive components will be degraded. Additionally, using organic solvents and chemicals in processing may cause "contamination" with harmful chemicals. Extensive processing of commodity oils may trigger the formation of potentially harmful components, such as polymers formed in deodorization processes. Additionally, extensive processing causes a reduction of the amount of bioactive components such as tocopherols, sterols, and others.

Interest in cereal lipids is driven by the presence of specific compounds, which are usually absent in mainstream oils, particularly natural antioxidants such as tocotrienols, phenolic components, and other bioactive compounds specific to the cereal grain. Additionally, the presence of cereal phytoestrogens such as lignans, some phenolic compounds, for which nutritional and health properties are established, attracts interest.

In this chapter sources, composition, processing, and utilization of wheat, barley, oat, and rye lipids are discussed in terms of possible applications as nutraceuticals and functional food ingredients and components of other products such as cosmetics.

3.2 CEREAL GRAIN LIPIDS

3.2.1 DISTRIBUTION

Wheat, barley, rye, and oat contain small amounts of lipids compared to oilseeds, usually in the range of a few percentage. Oat is an exception, because new varieties bred for their oil content contain up to 15% of oil[1]. That amount is still considerably less than in typical oilseed where the average oil content is usually above 20%. For all cereal grains the amounts of lipid are below 10% with the exception of oat, which contains about twice the amount present in other cereals (Table 3.1). The increased amount of oil in oat is attributed by the higher contribution of free lipids, mainly triacylglycerols.

Lipids in cereal grains are divided in two major groups: (1) nonstarch lipids and (2) starch lipids. The second group includes only a small portion of total lipids, which are bound to starch granules, and can only be isolated by extraction with a boiling mixture of *n*-butanol and water. Typical low-polarity solvents such as hexane and supercritical carbon dioxide, which are usually used for extraction of oils, extract minimal amounts of polar lipids. Bound lipids, starch lipids, consist of phospholipids and galactolipids, the main physiologically active and functional lipids present in cells.

Nonstarch lipids consist of two types, free and bound, the latter requiring polar solvents for their isolation (Table 3.1). Free lipids are extracted with nonpolar solvents as discussed above and consist mainly of acylglycerols and free fatty acids.

Lipids are distributed in the grain kernel unevenly and dispersal is dependent on the particular cereal grain. Table 3.2 gives the distribution of nonstarch lipids in grain kernels.

TABLE 3.1
Nonstarch Lipids of Cereal Grains (%)

Grain	Free[a]	Bound[b]	Total	Ref.
Wheat	1.4–2.6	0.7–1.2	2.1–3.8	2
Barley	1.7–1.9	1.5–1.7	3.2–3.4	3
Oat	5.5–8.0	1.4–1.6	6.9–9.6	2, 7
Rye	1.9–2.0	1.5–1.6	3.4–3.6	4

[a] Lipids extracted with nonpolar solvents such as diethyl or petroleum ethers.
[b] Lipids extracted with polar solvents, generally alcohols with water.

TABLE 3.2
Distribution of Lipids in Kernel Parts of Cereal Grains (%)

Cereal grain	Embryo/germ	Hull/bran	Endosperm	Ref.
Wheat	66	15	19	5
Barley	4	9	87	6
Oat	9	38	53	7
Rye	34	13	53	8, 12

Knowledge of the distribution of lipids in cereal kernel is important from a technological point of view. As in wheat, most of the lipids are located in germ, and it is sufficient to use this part of a seed to extract wheat germ oil; however, this will not guarantee that all bioactive lipid components will be present in this oil, as discussed below. Most of the phenolic derivatives of lipid components are present in the aleurone layer (bran) because they are part of the protection mechanism for the seed[1].

For barely and oat, whole seed need to be extracted to obtain the majority of oil because lipids are dispersed in the endosperm, the starchy and largest part of the seed, with small amounts in bran and germ. Generally, to isolate oil containing all bioactive components, extraction of whole seed needs to be performed because of the unequal distribution of components in the kernel anatomical parts.

3.2.2 Composition of Grain Lipids

The distribution of lipid components within seed is not uniform and different classes of lipid compounds are present in different parts of the seed where they play their physiological and protection roles. As shown in Table 3.3, the composition of lipid classes is usually unique to specific cereal grains.

The neutral lipids group consists of a variety of components, which have different physiological functions. Among them are sterols, tocopherols, and intermediate metabolic components such as discylglycerols (DGs), monoacylglycerols (MGs), and free fatty acids (FFAs). The main components of neutral lipids are triacylglycerols (TGs), which are the final metabolites in lipid anabolism, and in plants they are storage lipids and sources of energy.

The content of galactolipids and phospholipids in wheat and rye oils is the highest (Table 3.3). Barley oil contains the lowest amounts of these polar lipids, while oat oil has a comparable amount of phospholipids with almost half the galactolipids of the other cereal oils. Galactolipids and phospholipids are the main functional lipid components, having specific physiological functions in cells and defining the properties of cell membranes. Their physiological function is dependent on the type of fatty acid present in their structure; typically they contain unsaturated fatty acids that make cell membranes fluid and functional[2]. Phospholipids are also important ingredients of functional foods and nutraceuticals. Phospholipids and mono- and diacylglycerols are used in foods as emulsifiers and depending on their composition they can also affect health. Among this diverse group are phosphatidylcholine, phosphatidylethanolamine, phosphatidylserine, phosphatidylinositol, monogalactosyl mono- and diacylglycerols, and digalactosyl mono- and diacylglycerols.

Typical cereal grain neutral lipids contain mainly TGs (Table 3.4). The presence of DGs, MGs, and FFAs in the oils is considered detrimental to the quality and stability of the oils. To eliminate these components, the oils go through a full refining process including deodorization. Fully processed oil contains TGs as the main component (>95 to 98%) with very small amounts of other compounds present such as tocopherols and sterols. From a nutritional and health point of view these components detrimental to oil quality have a positive health effect and are expected to be present in cereal grain oils. The presence of DGs, MGs, and FFAs in cereal oils is normal because all these components are intermediates in the formation of TGs, storage lipid components. It has been

TABLE 3.3
Composition of Lipid Classes in Cereal Grains (% of Total Lipids)

Grain	Neutral lipids	Galactolipids	Phospholipids	Ref.
Wheat	60–61	17–18	24–26	4
Barley	65–80	7–26	9–18	9
Oat	66–80	6–10	12–26	10
Rye	63–68	10–12	22–25	11

TABLE 3.4
Composition of Neutral Lipids in Cereal Grains (%)

Grain	TG	DG	MG	FFA	Sterols	SE	Ref.
Wheat	70	3	1–3	5	3	7	12
Barley	32–51	2–3	1–9	2–11	1–2	1–2	13
Oat	32–51	2–3	3–9	2–11	1–2	1–2	7
Rye	51–57	8–11	3–7	11–14	3	5–7	4

Note: TG, triacylglycerols; DG, diacylglycerols; MG, monoacylglycerols; FFA, free fatty acids; SE, sterol esters.

TABLE 3.5
Fatty Acid Composition of Nonstarch Grain Lipids and Selected Vegetable Oils (%)

Grain	16:0	18:0	18:1	18:2	18:3	Ref.
Wheat	17–24	1–2	8–21	55–60	3–5	15
Barley	19–26	1–2	13–18	51–60	4–6	16
Oat	15–26	2–4	27–48	31–47	1–4	1, 10
Rye	12–19	1–2	12–16	57–65	3–12	17
Canola oil	3–6	1–3	52–67	16–25	6–14	18
Soybean oil	10–13	3–6	18–29	50–58	6–12	18

established that DGs can act as blood cholesterol lowering agents, also reducing body fat content in humans[14]. Both MGs and DGs go through different absorption and metabolic pathways in the human body from TGs and are direct sources of energy. In contrast, TGs are mainly transferred into storage fats, and less often are used as energy sources in the human body[14].

The fatty acid composition of cereal grain lipids is given in Table 3.5. The content of the main fatty acids is comparable to commercial oils such as soybean, sunflower, corn, and canola oils. However, wheat, barley, and oat oils have less linolenic acid than soybean and canola oils. The exception is rye oil that contains a comparable amount of linolenic acid to soybean and canola oils. Among the cereal oils, oat oil has an almost equal amount of oleic and linoleic acids. Vegetable oils are usually high in linoleic acid with the exception of canola and olive oils that are high in oleic acid. Cereal grain oils contain large amounts of palmitic acid, which is implied as the important factor causing elevated levels of low-density lipoprotein (LDL) and reduced amounts of high-density lipoprotein (HDL) in human blood[19].

The TG composition of oils is governed by the composition of fatty acids. the partial composition of cereal oil TGs was published where not all components were separated; the exception is oat oil where the composition has been established (Table 3.6)[2,20].

3.2.3 CHROMANOLS AND STEROLS IN GRAIN LIPIDS

Tocopherols and tocotrienols, often called chromanols as a group, are one of the most efficient natural antioxidants produced by plants. Additionally, chromanols are the major components of unsaponifiable matter in crude oils. These components are present in cell membranes where they protect cell components from free radicals and the cell from oxidative stress. The efficiency of

TABLE 3.6
Content of Triacylglycerides in Oat Oil (%)[20]

Triacylglyceride	Content
SOO	2.1
POO	10.3
OOO	3.8
POL	20.7
OOL	14.3
PLL	9.4
OLL	2.4
OLLn	5.9
LLL	13.6

Note: S, stearic; O, oleic; P, palmitic; L, linoleic, Ln, linolenic.

TABLE 3.7
Composition of Chromanols in Cereal Grain Oils (ppm)

Grain	αT	βT	γT	δT	αT3	βT3	γT3	δT3	Total	Ref.
Wheat	407	240	—	—	230	1383	—	—	2253	23
Barley	261	27	170	21	1221	264	315	21	2264	23
Oat	186	38	5	—	705	68	—	—	901	23
Rye	386	186	—	—	177	914	—	—	1951	23
Canola	272	1	423	—	—	—	—	—	800	24
Soybean	116	34	737	230	—	—	—	—	1200	24

Note: T, tocopherol; T3, tocotrienol. Canola and soybean oils are fully processed.

tocopherols in cells is very high; one molecule of tocopherol can protect 10^3 to 10^8 molecules of polyunsaturated fatty acids (PUFAs) at low levels of peroxides[21]. This may explain why the small ratio of α-tocopherol to PUFAs in cell membranes is sufficient to disrupt free radical chain reactions[22]. Probably, this last statement is a simplification when considering that in the cells are also other protective components such as enzymes (catalase, glutathione peroxidase, and superoxide dismutase) among a variety of other components that control the amount of free radicals. In addition, tocopherols in the cell system are recuperated from the oxidized form to active antioxidant by ascorbic acid and enzymes mentioned previously. The α-tocopherol can operate as a secondary line of defense to deal with any unusual "flood" of free radicals, destructive components formed as products of normal metabolism[23].

Cereal grain lipids are unique in the plant kingdom because they contain both tocopherol and tocotrienol isomers in significant amounts (Table 3.7). Commodity oils contain only tocopherols; cereal grain oils have both groups of these antioxidants, with higher contribution of tocotrienols. The antioxidant activity of tocotrienols has not been fully established yet; however, some data show their better effectiveness than tocopherols[22,24].

Cereal oils contain about two to three times more chromanols than typical commercial oils (Table 3.7). This difference has to be related to the fact that commercial oils are fully processed and during deodorization about 20 to 50% of tocopherols are removed from the oils[24].

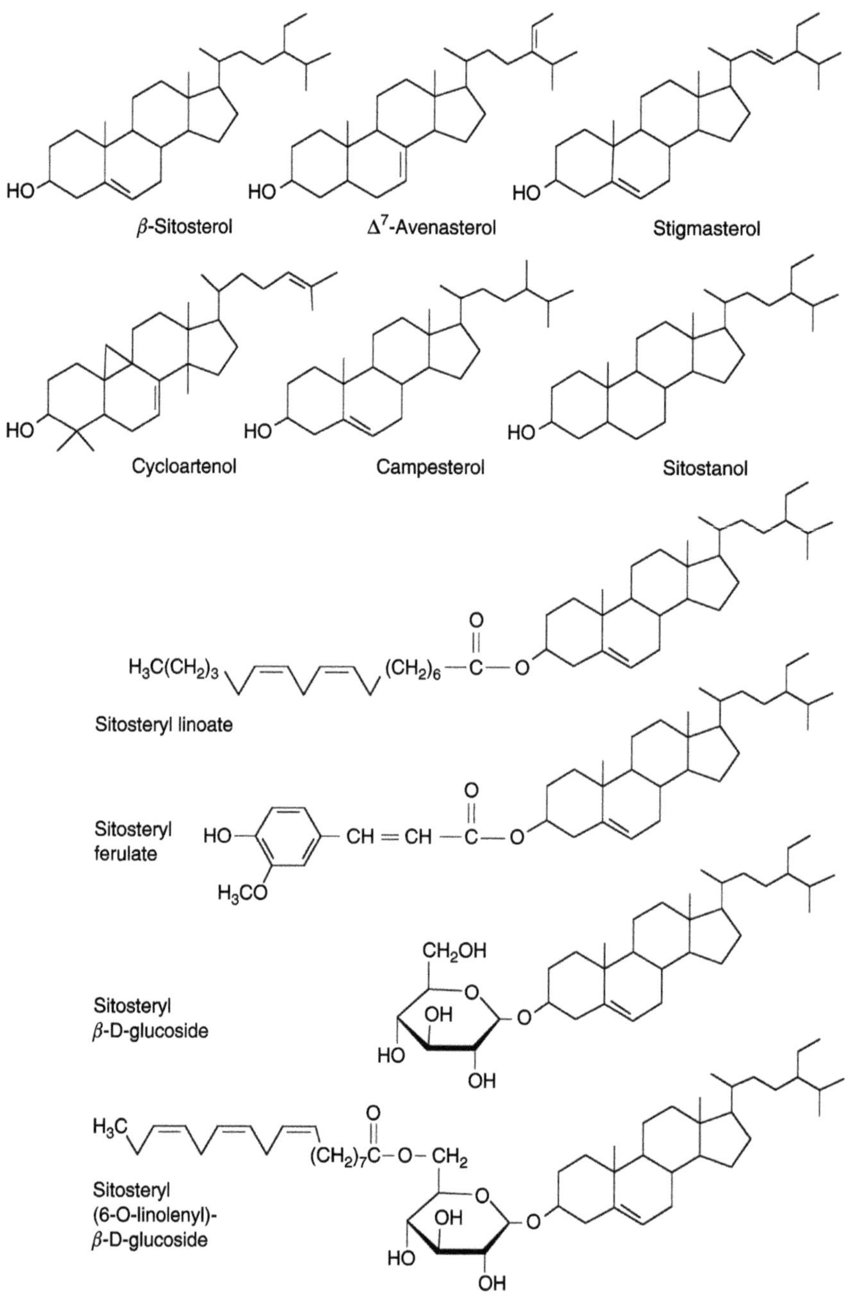

FIGURE 3.1 Structures of plant sterols and their derivatives.

Sterols together with chromanols are the main components of unsaponifiable matter present in the oils. These components contribute very often more than half of the total amount of unsaponifiables. Sterols, also called phytosterols, are similarly affected by processing as discussed for chromanols. Table 3.8 contains data for the composition of sterols in cereal oils and commodity oils. The most common sterols in cereals are β-sitosterol and campesterol. Among minor sterols, stanols, a group of saturated sterols, and avenasterols contribute significantly to the total amount of sterols (Figure 3.1; Table 3.8). Oils of plant origin are rich sources of phytosterols; cereal grain oils contain

TABLE 3.8
Composition of Sterols in Cereals and Commodity Oils (%)[25,26]

Sterol	Wheat	Barley	Oat	Rye	Canola[27]	Soybean[27]
Brassicasterol	3	2	8	1	8	1
Campesterol	17	23	9	18	25	17
Stigmasterol	3	4	4	3	3	20
β-Sitosterol	50	58	60	50	57	55
Δ^5-Avenasterol	1	7	10	2	3	2
Δ^7-Avenasterol	1	1	3	2	1	1
Sitostanol	15	1	2	13	1	2
Campestanol	8	1	0	8	1	1
Others	1	2	6	3	1	1
Amount (g/kg oil)	21.7	26.7	5.6	28.9	7.8	3.5

Note: Sterols contents in canola and soybean oils represent values in crude oils[27].

TABLE 3.9
Composition of Steryl Ferulates in Cereal Grains and Cereal Oils (mg/100 g)

Grain	Campesteryl ferulate		Sitosteryl ferulate		Sitostanyl ferulate		Total steryl ferulates		Total sterols	
	Grain[28]	Oil[29]	Grain[28]	Oil[29]	Grain[28]	Oil[29]	Grain[28]	Oil[29]	Grain[28]	Oil[29]
Wheat	1.2	—	3.3	30	1.9	80	6.3	380	63.4	1080
Barley[30]							0.4	44		
Rye	1.1	—	3.3	30	2.1	210	6.4	830	84.5	3274

Note: Data for oil represent bran oil of the specific grain.

3 to 4 times more sterols than commodity oils (Table 3.8). The exception is oat oil, which contains a similar amount of sterols as commodity oils.

The main places where phytosterols are located in cereal kernel are the germ and bran parts of the seed. Chemical structures of plant sterols are similar to that of cholesterol, the main animal-origin sterol, differing only in the side-chain structure (Figure 3.1). Oils of plant origin consist of phytosterols as free compounds and as a variety of esters (Figure 3.1). The single hydroxyl group at the 3 position in the ring can be esterified by fatty acids to form steryl esters and by phenolic acids, e.g. ferulic acid, to form steryl ferulates, which are common derivatives of phytosterols present in cereals (Table 3.9; Figure 3.1). Interest in different forms of sterol esters started with the assessment of rice bran oil, known as the commercial product γ-oryzanol, which contains significant amounts of different forms of esterified sterols[31].

The health benefits associated with sterol ferulates include lowering effect on plasma cholesterol level. However, these components are less effective in this function than fatty acid sterol esters[32].

Recently published work showed that steryl ferulates might have anticarcinogenic activity for selected forms of cancer. Ferulates act as antitumor promoters by inhibiting ear inflammatory edema and skin carcinogenesis in mice[33,34]. Experiments with Epstein–Barr virus antigens, a known model used for a primary screening for antitumor promoters, showed that cholestanol and stigmastanol ferulates are very potent inhibitors of these antigens[29].

Furthermore, sterol ferulates are efficient antioxidants due to the phenolic moiety, which is a proficient hydrogen donor. The large molecular size and nonpolar structure of sterol ferulates can affect their partitioning in matrices and make them active in hydrophobic and hydrophilic environments as potent antioxidants[35,36]. In cereal grains derivatives of ferulic, p-coumaric, and other phenolic acids were also identified, although their health effect and antioxidant potential has not yet been established[37].

A molecule of glucose can be attached to the same hydroxyl group in the sterol ring, forming steryl glycosides, which are in plants further esterified by attaching a fatty acid residue to the glucose molecule forming acylated steryl glycosides (Figure 3.1)[38]. All the discussed forms of sterols and their derivatives are physiologically active components in plant and animal cells, performing an important role in the structure and function of cell membranes. Plant sterol esters are located intracellularly and they are storage forms of sterols in plants and are precursors of a diverse group of plant growth factors[38].

Phytosterols as free compounds are solid at room temperature, with melting points at 140, 158, and 179°C for β-sitosterol, campesterol, and stigmasterol, respectively. Esterified sterols with fatty acids are hydrophobic and their solubility in oil increases when the length of the fatty acid chain increases. Free sterols and their fatty acid esters are soluble in nonpolar solvents such as hexane. In contrast, steryl glycosides will dissolve better in polar solvents due to the presence of the glucose moiety in the molecule, which is polar and hydrophilic[39]. Those sterol derivatives may play an important role as antioxidants in hydrophobic and hydrophilic environments.

In the 1990 it was established that fat-soluble stanol esters, saturated forms of plant sterols, when consumed with fat products such as margarines, reduced blood serum cholesterol levels by 10 to 15% (Figure 3.1)[40]. A diet rich in plant materials is recommended for the public due to the presence of endogenous and exogenous plant sterols, stanols, and their derivatives, together with a variety of natural and effective antioxidants such as chromanols and phenolic components. This renewed interest in the development of functional foods and nutraceuticals with plant sterols, their derivatives, and antioxidants is expanding fast in terms of controlling blood cholesterol levels and natural protection from oxidative stress[41].

3.2.4   CAROTENOIDS IN CEREAL GRAIN

Carotenoids are polyisoprenoid compounds, where hydrocarbons are known as carotenes while their oxygenated derivatives are xanthophylls. These compounds contain a conjugated polyene chain, which is responsible for color and sensitivity to light[2,4]. Some carotenoids such as carotenes are precursors of vitamin A and they are converted into it in intestinal mucosa and liver[4].

The highest amount of these pigments is found in barley and oat oils with much smaller amounts in wheat and rye oils. The main components among the carotenoids are carotenes and xanthophylls for barley, oat, and rye; the exception is wheat with xanthophylls and their esters being the main pigments (Table 3.10). In cereal grain oils other carotenoids, although in lower amounts, have been identified[2,4].

3.2.5   OTHER FUNCTIONAL COMPONENTS OF GRAIN LIPIDS

Cereal lipids are drastically different from commodity oils, because they contain a variety of compounds that can affect our health. Most of these bioactive compounds will be extracted with oils from grain, due to the complex structures, hydrophobic properties, and better solubility in organic solvents than in water.

3.2.5.1   Avenanthramides

Avenanthramides is a trivial name given to the group of unique components found in oat seed, which are substituted hydroxycinnamic acid conjugates (Figure 3.2). The amount of avenanthramides in

TABLE 3.10
Composition and Content of Carotenoids in Cereal Grain Oils (mg/kg)[2]

Grain	Carotenoids	Carotene	Xanthophylls	Xanthophyll esters
Wheat	43–99	10	55	35
Barley	100–150	55	41	1
Oat	100–310	62	38	2
Rye	15–28	78	6	1

Note: Data for specific carotenoids are percentages of total amount.

oat seed is in the range of 80 to 240 mg/kg; whereas several-fold greater amounts are present in hull than in groats[42]. There is a lack of published data about the presence of these components in other plant sources. More than 25 individual compounds of avenanthramide have been separated and identified in oat seed. The same authors established that these components of oat are very potent antioxidants[43]. These amphiphilic compounds are lipid soluble and can be potent antioxidants in hydrophobic and hydrophilic conditions (Figure 3.2)[42]. The presence of avenanthramides in oat oil was not established; however, based on their structure it is expected that they will be present in oil. Even if present in small amounts, they can be a potent addition to the antioxidant capacity of oat oil. It has already been established that avenanthramides, due to the presence of a nitrogen atom in the structure, are more potent antioxidants than any individual phenolic acids and combinations of them[44].

3.2.5.2 Alkylresorcinols

Alkylresorcinols are amphiphilic phenolic lipids present in considerable amounts in many plant sources, including cereals. Cereal alkylresorcinols are phenolic components in which saturated, unsaturated, and substituted chains of hydrocarbon are attached to position 5 on the phenolic ring (Figure 3.2). The hydrogens of the hydroxyl groups on the ring can also be substituted by methyl groups and sugar residues forming a variety of resorcinols with different functions (Figure 3.2)[45]. Resorcinols have been of interest for a long time because a wide range of activities has been attributed to these complex components. Kozubek and Tyman[45] described antimicrobial, antiparasitic, antitumor, and antioxidant activities of resorcinols in a comprehensive review. Antioxidant activity of resorcinols has been questioned by Kamal-Eldin et al.[46], who found that they have low potency as hydrogen donors and radical scavengers. Wheat and rye bran oils contain 19 and 38 g/kg of 5-alkenylresorcinols, respectively[29]. In whole wheat and rye seeds, alkylresorcinols were found at the level of 476 and 559 mg/kg of dry matter, respectively. The amounts of alkylresorcinols in the bran of these seeds were 2.6 and 2.2 g/kg of dry matter, respectively[47,48]. This fact clearly indicates that these compounds play an important role in protection of the seed. Based on these two grain oils, it is expected that these compounds would be an important component of other cereal oils.

3.2.5.3 Lignans

Lignans were first identified in plants and later in the physiological fluids of mammals. The characteristic chemical structure of lignans is a dibenzylbutane skeleton (Figure 3.3)[49]. The best known phytoestrogenic lignans are secoisolariciresinol and matairesinol, the main components of flax seed (Figure 3.3). Lariciresinol, syringaresinol, isolariciresinol, and pinoresinol were recently identified in cereals (Figure 3.3)[50]. Plant lignans are transformed by gut microflora into mammalian lignans, enterodiol and enterolactone, through complex enzymatic reactions (Figure 3.4). Physiological and

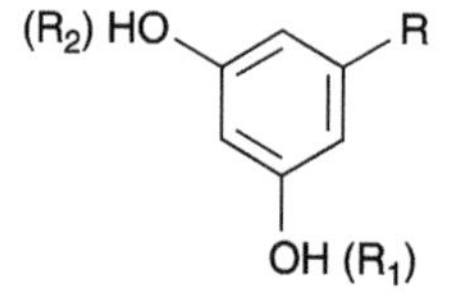

Side chain (R)   Group name
C_5-C_{29} 1, 3-dihydroxy-5-n-alkylresorcinols
C_{15}-C_{29} 1, 3-dihydroxy-5-mono-and polyenesresorcinols
R_1; R_2 H;-CH_3; glucose; galactose

Structure of resorcinol lipids

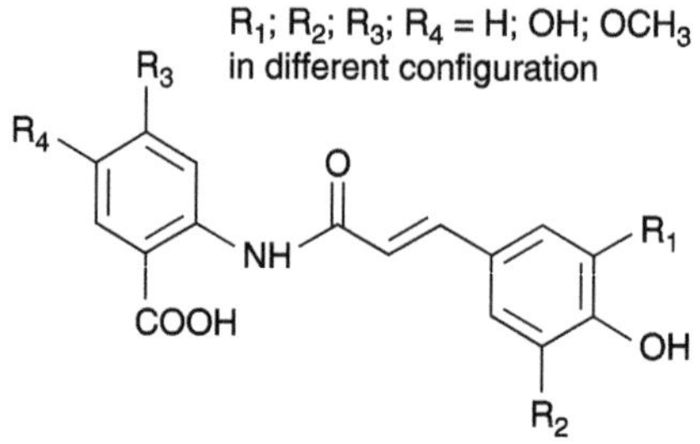

R_1; R_2; R_3; R_4 = H; OH; OCH_3
in different configuration

Structure of avenanthramides

FIGURE 3.2 Structures of avenanthramides and resorcinols.

health functions of lignans are well documented, showing activity such as phytoestrogens, anticancer, lowering blood cholesterol levels, and antioxidants[49]. Lignans as phenolics are amphiphilic components and their presence in oils has been identified (Figure 3.5)[51]. Oils such as corn, soybean and flaxseed contain important amounts of these components, which are transformed into mammalian lignans. Since cereals contain measurable amounts of lignans in seeds, it is expected that their oils will contain them in physiologically important quantities.

3.3 CEREAL GRAIN OIL EXTRACTION AND UTILIZATION

Wheat germ oil is only available on the market as cereal oil, mainly because the large amounts of wheat germ are produced during the milling process. Oil is separated from germs commercially by pressure expulsion or solvent extraction. The former avoids organic solvent extraction as a possible source of oil contamination, but only half of oil available is removed by expelling[52]. Solvent extraction is more efficient, where 99% of oil is removed, with the process being performed below 60°C to protect nutritionally important components present in this oil. Solvents used for extraction of wheat germ oil are hexane, dichloromethane, and ethanol[53]. Recently, supercritical carbon dioxide was applied to extract wheat germ oil. This solvent is considered "green" and not toxic. Similar yields were obtained using this solvent compared to hexane extraction. Composition and physical properties of supercritical- and hexane-extracted oils are similar indicating that this technology can provide good-quality oil with environmentally friendly connotations[54]. Similar extraction solvents have been applied to oat oil, showing comparable quality of oils extracted by hexane and supercritical carbon dioxide[55]. Refining of wheat germ oil has been done to provide oil with quality criteria similar to mainstream oils but many functional components were lost[53].

   Cereal oils, particularly wheat germ oil, have not found many applications, mainly due to the different composition and stability problems. For oils to be used in food applications specific properties have to be fulfilled, which are set for mainstream commodity oils. These specifications are

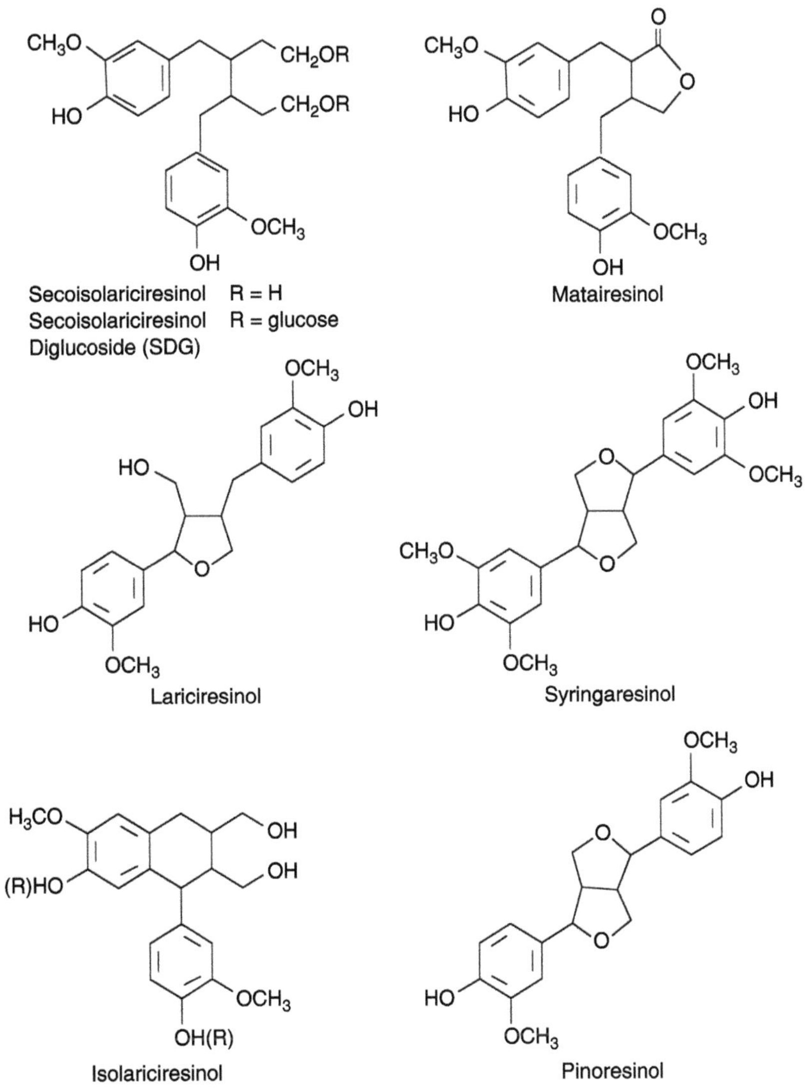

FIGURE 3.3 Sturcture of plant lignans.

very stringent in many aspects and do not allow the presence of free fatty acids, and acylglycerols other than triacylglycerols in the amounts found in cereal oils. That is why these oils have to be refined to fulfill specifications for food application. Recent trends to produce nutraceutical oils and functional food ingredients necessitate development of different quality criteria to be used for these specialty oils, to allow them to be applied in functional foods and nutraceuticals.

Historically, wheat germ oil was used as a food supplement for farm animals, racehorses, pets, and mink, as a fertility agent, as an antioxidant, as an agent to control insects, and as an ingredient in shampoos, conditioners, and lotions[53]. Oat oil has been applied as a dough improver in bread baking to improve dough quality, but only at the experimental stage[56].

There are very limited applications of cereal oils. The main limiting factor in food application is quality specifications that limit the use of nutritionally advantageous oils due to the need to refine them to remove components to the level specified for commodity oils. The second limiting factor is the price, particularly when compared to commodity oils. However, consumers are willing to pay

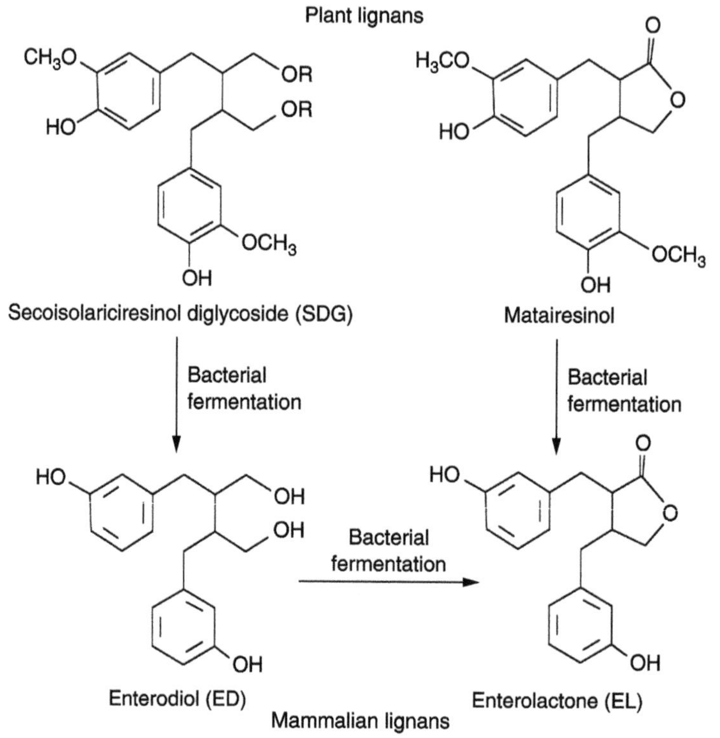

Plant lignans

Secoisolariciresinol diglycoside (SDG)   Matairesinol

Bacterial fermentation   Bacterial fermentation

Bacterial fermentation

Enterodiol (ED)   Enterolactone (EL)

Mammalian lignans

FIGURE 3.4 Formation of mammalian lignans in the digestive tract from their plant precursors.

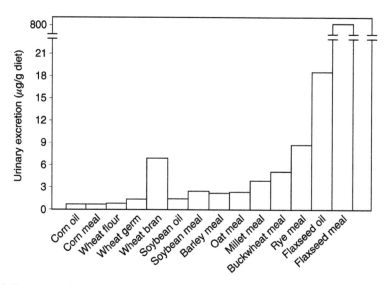

FIGURE 3.5 Excretion of human lignans after supplementing diet with various foods. (Adapted from Thompson, L.U., Robb, P. Serraino, M., and Cheung, F., *Nutr. Cancer*, 16, 43–52, 1991.)

more for products: nutraceuticals with guaranteed nutritional and health quality. Most of these "problematic" components are compounds with specific nutritional and health functions. New approaches to functional foods and nutraceuticals should open opportunities for oils loaded with functional and health compounds. Was our food ever not functional?

REFERENCES

1. Holland, J.B., Frey, K.J., and Hammond, E.G., Correlated response of fatty acid composition, grain quality and agronomic traits to nine cycles of recurrent selection for increased oil content in oat, *Euphytica*, 122, 69–79, 2001.
2. Morrison, W.R., Cereal lipids, in *Advances in Cereal Sciences and Technology*, Vol. II, Pomeranz, Y., Ed., American Association of Cereal Chemists, St. Paul, MN, 1978, pp. 221–348.
3. Pomeranz, Y., Ke, H., and Ward, A.B., Composition and utilization of milled barley products: I. Gross composition of rolled, milled and air-separated fractions, *Cereal Chem.*, 48, 47–58, 1971.
4. Chung, O.K. and Tsen, C.C., Triticale lipids, in *Triticale: First Man-Made Cereal,* Tsen, C.C., Ed., American Association of Cereal Chemists, St. Paul, MN, 1974, pp. 191–200.
5. Inglett, G.H., Kernel structure and composition, in *Wheat: Production and Itilization*, Inglett, G.E., Ed., AVI, Westport, 1974, pp. 108–118.
6. Price, P.B. and Parsons, J.G., Distribution of lipids in embryonic axis, bran-endosperm and hull fractions of hulless barley and hulless oat grain, *J. Agric. Food Chem.*, 27, 813–815, 1979.
7. Youngs, V.L., Paskulcu, H., and Smith, R.R., Oat lipids: I. Composition and distribution of lipid components in two oat cultivars, *Cereal Chem.*, 54, 803–812, 1977.
8. Price, P.B. and Parsons, J.G., Lipids of seven cereal grains, *J. Am. Oil Chem. Soc.*, 52, 490–-493 (1975).
9. Bhatty, R.S. and Rossnagel, B.G., Lipids and fatty acid composition of Risco 1508 and normal barley, *Cereal Chem.*, 57, 382–386, 1980.
10. Sahasrabudhe, M.R., Lipid composition of oats (Avena sativa L.). *J. Am. Oil Chem. Soc.*, 54, 80–84, 1979.
11. Price, P.B. and Parsons, J.G., Lipids of seven cereal grains, *J. Am. Oil Chem. Soc.*, 52, 490–493, 1975.
12. Nechaev, A.P. and Sandler, Z.Y., in *Grain Lipids* (*Lipidy Zerna*), Kolos, Moscow, 1975, p. 159.
13. Price, P.B. and Parsons, J.G., Neutral lipids of barley grain, *J. Agric. Food Chem.*, 28, 875–877, 1980.
14. Nagao, T., Watanabe, H., Naohiro, G., Onizawa, K., Taguchi, H., Matsuo, N., Yasukawa, T., Tsushima, R., Shimasaki, H., and Itakura, H., Dietary diacylglycerol suppresses accumulation of body fat compared to triacylglycerols in men in a double-blind controlled trial, *J. Nutr.*, 130, 792–797, 2000.
15. Morrison, W.R., Cereal lipids, in *Advances in Cereal Science and Technology*, Vol. II, Pomeranz, Y., Ed., American Association of Cereal Chemists, St. Paul, MN, 1978, pp. 221–348.
16. Price, P.B. and Parsons, J.G., Lipids of six cultivated barley (*Hordeum vulgare* L.) varieties, *Lipids*, 9, 560–565, 1974.
17. Weihrauch, J.L. and Matthews, R.H., Lipid content of selected cereal grains and their milled and baked products, *Cereal Chem.*, 54, 444–453, 1977.
18. Eskin, N.A.M., McDonald, B.E., Przybylski, R., Malcomson, L.J., Scarth, R., Mag, T., Ward K., and Adolph, D., Canola oil, in *Bailey's Industrial Oil & Fat Products; Edible Oil and Fat Products: Oils and Oilseeds*, Vol. 2, Hui, Y.H., Ed., John Wiley, New York, 1996, pp. 1–95.
19. Honstra, G., Lipids in functional foods in relation to cardiovascular disease, *Fett/Lipid*, 101, 456–466, 1999.
20. Nechaev, A.P., Novozilova, G.N., and Novitskaya, G.V., Composition of oat triglycerides, *Maslo-Zhir. Prom.*, 38, 16–17, 1972.
21. Patterson, L.K., Studies of radiation-induced peroxidation in fatty acids micelles, in *Oxygen and Oxy-Radicals in Chemistry and Biology,* Rodgers, M.A.J. and Powers, E.L., Eds., Academic Press, New York, 1981, pp. 89–95.
22. Diplock, A.T., and Lucy, J., The biochemical modes of action of vitamin E and selenium: a hypothesis, *FEBS Lett.*, 29, 205–210, 1973.
23. Chance, B., Sies, H., and Boveris, A., Hydroperoxide metabolism in mammalian organs, *Physiol. Rev.*, 59, 527–605, 1979.
24. Panfili, G., Fratianni. A., and Irano, M., Normal phase high-performance liquid chromatography method for the determination of tocopherols and tocotrienols in cereals, *J. Agric. Food Chem.*, 51, 3940–3944, 2003.

25. Piironen, V., Toivo, J., and Lampi, A.M., Plant sterols in cereals and cereal products, *Cereal Chem.*, 79, 148–154, 2002.

26. Piironen, V., Lindsay, D.G., Miettinen, T.A., Toivo, J., and Lampi, A.M., Plant sterols: biosynthesis, biological function and their importance to human nutrition, *J. Sci. Food Agric.*, 80, 939–966, 2000.

27. Phillips, K.M., Ruggio, D.M., Toivo, J., Swank, M.A., and Simpkins, A.H., Free and esterified sterol composition of edible oils and fats, *J. Food Comp. Anal.*, 15, 123–142, 2002.

28. Hakala, P., Lampi, A.M., Ollilainen, V., Werner, U., Murkovic, M., Wahala, K., Karkola, S., and Piironen, V., Steryl phenolic acid esters in cereals and their milling fractions, *J. Agric. Food Chem.*, 50, 5300–5307, 2002.

29. Iwatsuki, K., Akihisa, T., Tokuda, H., Ukiya, M., Higashihara, H., Mukainaka, T., Iizuka, M., Hayashi, Y., Kimura, Y., and Nishino, H., Steryl ferulates, sterols, and 5-alk(en)ylresorcinols from wheat, rye, and corn bran oils and their inhibitory effects on Epstein–Barr virus activation, *J. Agric. Food Chem.*, 51, 6683–6688, 2003.

30. Moreau, R.A., Powell, M.J. Hicks, K.B., and Norton, R.A., A comparison of the levels of ferulate-phytosterol esters in corn and other seeds, in *Advances in Plant Lipid Research*, Sanches, J., Cerda-Olmeda, E., and Matinez-Force, E., Eds., Universidad de Sevilla, Seville, Spain, 1998, pp. 472–474.

31. Rogers, E.J., Rice, S.M., Nicolosi, R.J., Carpenter, D.R., McClelland, C.A., and Romanczyk, L.J., Identification and quantitation of γ-oryzanol components and simultaneous assessment of tocols in rice bran oil, *J. Am. Oil Chem. Soc.*, 70, 301–307, 1993.

32. Weststrate, J.A. and Meijer, G.W., Plant sterol-enriched margarines and reduction of plasma total- and LDL-cholesterol concentrations in normocholesterolaemic and mildly hypercholesterolaemic subject, *Eur. J. Clin. Nutr.*, 52, 334–343, 1998.

33. Akihisa, T., Yakusawa, K., Yamaura, M., Ukiya, M., Kimura, Y., Shimizu, N., and Arai, K., Triterpene alcohols and sterol ferulates from rice bran and their anti-inflammatory effects, *J. Agric. Food Chem.*, 48, 2313–2319, 2000.

34. Yasukawa, K., Akihisa, T., Kimura, Y., Tamura, T., and Takiod, M., Inhibitory effect of cycloartenol ferulate, a component of rice bran, on tumor promotion in two-stage carcinogenesis in mouse skin, *Biol. Pharm. Bull.*, 21, 1072–1078, 1998.

35. Xu, Z. and Godber, J.S., Antioxidant activities of major components of γ-oryzanol from rice bran using a linoleic acid model, *J. Am. Oil. Chem. Soc.*, 78, 645–649, 2001.

36. Yagi, K. and Ohishi, N., Action of ferulic acid and its derivatives as antioxidants, *J. Nutr. Vitaminol.*, 25, 127–130, 1979.

37. Seitz, L.M., Stanol and sterol esters of ferulic and *p*-coumaric acids in wheat, corn, rye and triticale, *J. Agric. Food Chem.*, 37, 662–667, 1989.

38. Akihisa, T., Kokke, W., and Tamura, T., Naturally occurring sterols and related compounds from plants, in *Physiology and Biochemistry of Sterols*, Patterson, G.W. and Nes, W.D., Eds., AOCS Press, Champaign, IL, 1991, pp. 172–228.

39. Hartmann, M.A. and Benveniste, P., Plant membrane sterols: isolation, identification and biosynthesis, *Methods Enzymol.*, 148, 632–650, 1987.

40. Miettinen, T.A. and Gylling, H., Regulation of cholesterol metabolism by dietary plant sterols, *Curr. Opin. Lipidol.*, 10, 9–14, 1998.

41. Plat, J. and Mensik, R.P., Vegetable oil based versus wood based stanol mixtures: effects on serum lipids and homeostatic factors in non-hypercholesterolemic subjects, *Atherosclerosis*, 148, 101–112, 2000.

42. Paterson, D.M., Oat antioxidants, *J. Cereal Sci.*, 33, 115–129, 2001.

43. Dimberg, L.H., Theander, O., and Lingnert, H., Avenanthramides: a group of phenolic antioxidants in oats, *Cereal Chem.*, 42, 637–641, 1993.

44. Martinez-Tome, M., Murcia, M.A., Frega, N., Ruggieri, S., Jimenez, A.M., Rose, F., and Parras, P., Evaluation of antioxidant capacity of cereal brans, *J. Agric. Food Chem.*, 52, 4690–4699, 2004.

45. Kozubek, A. and Tyman, J.H.P., Resorcinolic lipids, the natural non-isoprenoid phenolic amphiphiles and their biological activity, *Chem. Rev.*, 99, 1–25, 1999.

46. Kamal-Eldin, A., Pouru, A., Eliasson, C., and Aman, P., Alkylresorcinols as antioxidants: hydrogen donation and peroxy radical-scavenging effects, *J. Sci. Food Agric.*, 81, 353–356, 2000.

47. Mullin, W.J. and Emery, P.H., Determination of alkylresorcinols in cereal-based foods, *J. Agric. Food Chem.*, 40, 2127–2130, 1992.

48. Ross, A.B., Kamal-Eldin, A., Jung, C., Shepherd, M.J., and Aman, P., Gas chromatographic analysis of alkylresorcinols in rye (*Secale cereale* L) grains, *J. Sci. Food Agric.*, 81, 1405–1411, 2001.

49. Setchell, K.D.R. and Adlercreutz, H., Mammalian lignans and phyto-oestrogens recent studies on their formation, metabolism and biological role in health and disease, in *Role of Gut Flora in Toxicity and Cancer*, Rowland, I.R., Ed., Academic Press, San Diego, 1988, pp. 315–345.

50. Heinonen, S., Nurmi, T., Liukkonen, K., Poutanen, K., Wahala, K., Deyama, T., Nishibe, S., and Adlercreutz, H., In vitro metabolism of plant lignans: new precursors of mammalian lignans enterolactone and enterodiol, *J. Agric. Food Chem.*, 49, 3178–3186, 2001.

51. Thompson, L.U., Robb, P. Serraino, M., and Cheung, F., Mammalian lignan production from various foods, *Nutr. Cancer*, 16, 43–52, 1991.

52. Barnes, P.J., Wheat germ oil, in *Lipids in Cereal Technology*, Barnes, P.J., Ed., Academic Press, London, 1983, pp. 64–72.

53. Kahlon, T.S., Nutritional implications and uses of wheat and oat kernel oils, *Cereal Food World*, 34, 872–875, 1989.

54. Gomez, A.M. and de la Ossa, E.M., Quality of wheat germ oil extracted by liquid and supercritical carbon dioxide, *J. Am. Oil Chem. Soc.*, 77, 969–974, 2000.

55. Fors, S.M. and Eriksson, C.E., Characterization of oils extracted from oats by supercritical carbon dioxide, *Lebensm. Wiss. Technol.*, 23, 390–395, 1990.

56. Erazo-Castrejon, S., Doehlert, D.C., and D'Appolonia, B.L., Application of oat oil in breadbaking, *Cereal Chem.*, 78, 243–248, 2001.

4 Fruit Seed Oils

Liangli Yu, John W. Parry, and Kequan Zhou
Department of Nutrition and Food Science, University of Maryland, College Park, Maryland

CONTENTS

4.1   INTRODUCTION

Commercial edible seed oils are mixtures of lipids including triacylglycerols, diacylgylcerols, monoacylglycerols, free fatty acids, and other minor components. The demand for edible oils has been increasing with the growing world population and consumers' preference for vegetable oils over animal fats[1,2]. Generally, plant oils are liquid at ambient temperature and obtained from oil seeds. Oils are important food ingredients that contribute to sensory properties of food and energy. The fatty acid components of edible oils may also be metabolized and incorporated into cell membranes and are required for cell integrity and human health. The quality, stability, safety, and nutritional value of edible seed oils are determined by their chemical composition, such as fatty acid profile, level of natural antioxidants, and fat-soluble vitamins.

Fatty acids are classified as saturated, monounsaturated (MUFA), and polyunsaturated (PUFA) according to the presence and number of double bonds. PUFAs are further classified as, $\omega 6$ (n-6) and $\omega 3$ (n-3) fatty acids. Linoleic (18:2n-6) and α-linolenic (18:3n-3) acids are essential fatty acids that cannot be synthesized by humans *in vivo* and have to be obtained through diet. Growing evidence suggests the potential application of dietary fatty acid composition on disease risks and general human health[3–15]. Long-chain ω3 fatty acids, including α-linolenic, eicosapentaenoic (EPA), and docosahexaenoic (DHA) acids, have been shown to exert beneficial effects in the prevention of cancer, heart disease, hypertension, and autoimmune disorders[12,13,16–20]. In addition, MUFAs have been recognized for their potential antiatherosclerotic and immunomodulating effects[21–23]. The consumer demands for novel edible oils with desired physicochemical properties and potential health benefits for different applications have driven the characterization and development of new edible oils. Some fruit seeds that are byproducts of processing have been evaluated for their potential utilization in preparing edible oils. Recent research has shown that fruit seed oils may serve as specialty oils for health promotion and disease prevention due to their special fatty acid composition and other beneficial components. This chapter summarizes and discusses the recent research on fatty acid composition, presence of phytochemicals, and potential health benefits of fruit seed oils. A total of 16 fruit seed oils are discussed: those of red raspberry, black raspberry, goldenberry, cranberry, blueberry, marionberry, boysenberry, citrus, gac, watermelon, pumpkin, carob, apple, grape, pomegranate, and currant.

4.2   RED RASPBERRY SEED OIL

The common red raspberry, *Rubus idaeus* L., is cultivated throughout the world, and a very large percentage of it is processed to juice. Raspberry seed is a major byproduct of juice production, constituting about 10% of the total fresh weight[24]. The annual average raspberry production in Canada is approximately 18,000 metric tons of fresh berries and the annual production worldwide is about 312,000 metric tons. Assuming 10% of the fruit's weight is seeds and 23% of the seed's weight is oil, and if all raspberries were processed to juice, this would provide more than 7000 metric tons of raspberry seed oil annually[24,25].

Raspberry seed oil is used as an ingredient in cosmetic products due to its antiinflammatory properties. It has been shown to prevent gingivitis, eczema, rashes, and other skin lesions. It may also be used by the sunscreen industry because of its UV absorbing properties[24].

Two different methods are used to extract the oil from raspberry seeds: hexane[24] and cold pressing[26]. Seed oils prepared by both methods were found to exhibit similar fatty acid compositions (Table 4.1). Red raspberry seed oil contained a significantly higher level of α-linolenic acid, 29 to 32% of the total fatty acids, and the ratio of n-6 to n-3 fatty acids was 1.6:1 to 1.9:1. In addition, PUFAs accounted for 82 to 87% of the total fatty acids in red raspberry seed oil, whereas MUFAs accounted for 12%. These data suggest that red raspberry seed oil may serve as a dietary source for α-linolenic acid, and may reduce the overall ratio of n-6 to n-3 fatty acids. During human evolution, the n-6 to n-3 ratio was estimated to be approximately 1:1, and the current dietary ratio of n-6 to n-3 is 10:1 or higher[27,28]. The n-6 fatty acids suppress the desaturation and elongation of α-linolenic acid to EPA and DHA, and compete with n-3 fatty acids for incorporation in cellular phospholipids. Growing evidence suggests that the ratio of n-6 to n-3 fatty acids might play a role in cancer development and bone health, and a reduction in the ratio of n-6 to n-3 fatty acid may lower the risks of cancer and heart disease[16,29,30]. Also, red raspberry oil displayed a very similar fatty acid composition to that of the cold-pressed black raspberry seed oil[26] as shown in Table 4.1.

Oil-soluble vitamins and other beneficial components were also detected in red raspberry seed oil. Red raspberry seed oil had a carotenoid concentration of 23 mg/100 g oil[24]. The tocopherol concentration was 97 mg/100 g oil in hexane-extracted red raspberry seed oil, and was 61 mg/100 g oil in the corresponding cold-pressed oil, both of which were greater than that in safflower (57 mg/100g oil), and grapeseed (12 mg/100 g oil) oils. The large difference in the tocopherol content of hexane-extracted and cold-pressed raspberry seed oils could possibly be explained by the presence of nonlipid contaminants from the cold-pressing procedure that dilute vitamin E in the oil sample[24], and better extraction of vitamin E by hexane.

The evaluation of red raspberry seed oil demonstrated that its *p*-anisidine value, a measure of aldehydes or secondary oxidation products, was 14.3. This value was significantly higher than those

TABLE 4.1
Fatty Acid Composition (g/100g Fatty Acids) of Cold-Pressed Seed Oils[a]

Fatty acid	Cranberry[43]	Cranberry[12]	Goldenberry[36]	Red raspberry[24,31]	Black raspberry[26]	Blueberry[31]	Marionberry[31]	Boysenberry[31]
12:0	nd	nd	0.4	nd	nd	nd	nd	nd
14:0	nd	nd	1.0	nd	nd	nd	nd	nd
16:0	5.0–6.0	7.8	7.3	1.2–2.7	1.2–1.6	5.7	3.3	4.2
16:1	nd	nd	0.5	nd	nd	nd	nd	nd
18:0	1.0–2.0	1.9	2.5	1.0	trace	2.8	3.1	4.5
18:1	20.0–25.0	22.7	11.7	12.0–12.4	6.2–7.7	22.8	15.1	17.9
18:2n-6	35–40	44.3	76.1	53.0–54.5	55.9–57.9	43.5	62.8	53.8
18:3n-3	30.0–35.0	22.3	0.3	29.1–32.4	35.2–35.3	25.1	15.7	19.5
20:0	0.1	nd	0.2	nd	nd	nd	nd	nd
20:2	nd	1.0	nd	nd	nd	nd	nd	nd
20:5n-3	1.1	nd	nd	nd	nd	nd	nd	nd
Sat	6.1–8.1	9.7	11.3	2.2–3.7	1.2–1.6	8.5	6.4	8.7
MUFA	20.0–25.0	22.7	12.2	12.0–12.4	6.2–7.7	22.8	15.1	17.9
PUFA	66.1–76.1	67.6	76.4	82.1–86.9	91.1–93.1	68.6	78.5	73.3
n-6	35.0–40.0	44.3	76.1	53.0–54.5	55.9–57.8	43.5	62.8	53.8
n-3	31.1–36.1	22.3	0.3	29.1–32.4	35.2–35.3	25.1	15.7	19.5
n-6/n-3 ratio	1.0–1.3	2.0	253.7	1.6–1.9	1.6	1.7	4.0	2.76

[a] Sat, saturated fatty acids; MUFA, monounsaturated fatty acids; PUFA, polyunsaturated fatty acids; n-6, polyunsaturated n-6 fatty acids; n-3, polyunsaturated n-3 fatty acids; nd, not detected.

of safflower seed (5.4) and grapeseed (10.5) oils[24]. The antioxidative capacity of the oil, measured by the oxygen radical absorbance capacity (ORAC) test, using Trolox as the standard, was 48.8 μmol trolox equivalents per gram of oil (TE/g oil). Trolox, 6-hydroxy-2,5,7,8-tetramethylchroman-2-carboxylic acid, is a water-soluble vitamin E analog. This value was significantly higher than that of blueberry seed oil (36.0 TE/g oil)[31], which is known to contain high concentrations of antioxidants[32]. The peroxide value of the oil was determined to be between 4.4[31] and 8.25[24] meq O-OH/kg oil. Meanwhile, the iodine value of the oil was 191 g iodine/100 g oil and its viscosity was 26 mPa s at 25°C[24].

4.3   BLACK RASPBERRY SEED OIL

Black raspberries are phenotypically similar to red raspberries with the exception of their color. In 2000 and 2001 the production of black raspberries in Oregon was 3.83 and 3.81 million pounds, respectively (http://www.nass.usda.gov/or/annsum2002.htm). Recent research indicated that black raspberries might contain a higher level of natural antioxidants than red raspberries[33,34]. According to Wang and Lin[33], black raspberry juice exhibited a higher ORAC, anthocyanin concentration, and total phenolic content than all tested strawberry and blackberry cultivars on a fresh weight basis. Black raspberry (*Rubus occidentalis* L., cv Jewel) juice also had a higher ORAC, and greater levels of both anthocyanin and total phenolics than other red raspberry cultivars[33]. Furthermore, Wang and Jiao[34] reported that black raspberry juice had stronger or comparable scavenging activities against hydroxyl radical, superoxide radical, hydrogen peroxide, and singlet oxygen. In 2002, Huang and others[35] found that black raspberry extracts might impair signal transduction pathways leading to activation of AP-1 and NF-kappaB, and, therefore, inhibit tumor development. These data suggested the potential health benefits of consuming black raspberry and black raspberry-based products, including the juice, against several aging-related diseases such as cancer and heart disease.

More than 99% of the total production of black raspberry is processed, while about 1% of fruits are consumed fresh (http://www.nass.usda.gov/or/annsum2002.htm). Black raspberry seeds are the byproduct from juice manufacture. Parry and Yu[26] evaluated the fatty acid composition, antioxidant properties, oxidative stability, iodine value, total phenolic content, and color perception of cold-pressed black raspberry seed oils. The fatty acid composition of the cold-pressed black raspberry seed oil is presented in Table 4.1. The cold-pressed black raspberry seed oils contained about 91 to 93% PUFAs, and were rich (35%) in α-linolenic acid (18:3n-3), along with 56 to 58% linoleic acid (18:2n-6), suggesting that black raspberry seed oil may serve as a dietary source for essential fatty acids. The ratio of n-6 to n-3 fatty acids was 1.6:1 to 1.7:1, suggesting the potential application of cold-pressed black raspberry seed oil in improving or reducing the current dietary ratio of n-6 to n-3 fatty acids[26]. In addition, the cold-pressed black raspberry seed oil exhibited a significant radical scavenging capacity, and had a total phenolic content of 35 to 93 μg gallic acid equivalent (GE) per gram of oil. The physicochemical properties, including the oxidative stability index, peroxide value, and iodine value, of the cold-pressed black raspberry seed oils were reported[26]. Cold-pressed black raspberry seed oil had a better oxidative stability than cold-pressed hemp seed oil, but was less stable than cold-pressed cranberry seed oil and commercial corn and soybean oils under the same experimental conditions[12,26].

4.4   GOLDENBERRY SEED OIL

Goldenberry, also known as cape gooseberry (*Physalis peruviana* L.), is a shrub native to the Andes. It is related to both tomatoes and potatoes and prefers the same growing conditions as tomatoes. The fruit is round, golden, and about the size of a marble. It is grown throughout the world and has major commercial possibilities, but is not yet a major food source. It is described as having a pleasant flavor that is similar to tomatoes. The fruit is eaten in many ways including in salads, cooked dishes, chocolate-covered desserts, jams, preserves, and natural snacks[36].

Goldenberry has also been used to treat many ailments including asthma, edema, optic nerve disorders, throat afflictions, intestinal parasites, and amoebic dysentery. The fruit is an excellent source of vitamins A and C as well as minerals. Previous work on goldenberry has shown that the fruit contains between 0.16 and 1.30% oil, on a fresh weight basis[37–39]. Goldenberry seed oil was prepared by extracting lyophilized ground seed meal with chloroform–methanol and was characterized for its fatty acid composition using a GC equipped with a FID detector[36]. Table 4.1 shows that the oil had total saturated fatty acids of 11.3%, monounsaturated fatty acids of 12.2%, linoleic acid of 76.1%, linolenic acid of 0.33%, and total polyunsaturated fatty acids of 76.4%. The ratio of saturated to unsaturated fatty acids was 1:7.8 (w/w). These data indicate that goldenberry seed oil may serve as an excellent dietary source for α-linoleic acid, and may be a good choice for a higher overall intake of total unsaturated fatty acids.

The fat-soluble vitamins E and K, carotene, and phytosterols were detected in goldenberry seed oil[36]. Total vitamin E, including α-, β-, γ-, and δ-tocopherols, was 29.7 mg/g oil, comprising of 0.9 mg α-, 11.3 mg β-, 9.1 mg γ-, and 8.4 mg δ-tocopherols. The total vitamin K content was 0.12 mg/g oil, and the β-carotene concentration was 1.30 mg/g oil. In addition, significant levels of phytosterols were detected in goldenberry seed oil. The major phytosterol in goldenberry seed oil was campesterol having a concentration of 6.5 mg/g oil. Other phytosterols included ergosterol, stigmasterol, lanosterol, β-sitosterol, Δ5-avenosterol, and Δ7-avenosterol, at levels of 1.04, 1.32, 2.27, 5.73, 4.70, and 1.11 mg/g oil, respectively.

4.5 CRANBERRY SEED OIL

Cranberry is grown and harvested in the Northeast, Northwest, and Great Lakes regions in the United States. In 2002 total U.S. cranberry harvest reached 5.6 million barrels, 6% above the 2001 production (http://www.nass.usda.gov/nh/cran03.htm). Cranberry contains vitamin C, flavonols, anthocyanins, and procyanins, and has a number of health benefits related to reducing the risk of cardiovascular disease[40–42]. Cranberry seeds are byproducts of cranberry juice production. Cranberry seed oil has been produced and is commercially available[12,43,44]. Heeg and others[43] have determined that cranberry seed oil contains 35 to 40% linoleic acid (18:2n-6) and 30 to 35% α-linolenic acid (18:3n-3), along with 20 to 25% oleic acid, 5 to 6% palmitic acid (16:0), 1 to 2% stearic acid (18:0), a trace amount of arachidonic acid (20:4n-6), and possibly EPA, on an oil weight basis. In 2003, Parker and others[12] reported 44.3% linoleic acid and 22.3% α-linolenic acid, along with 22.7% oleic acid, 7.8% palmitic acid, and 1.9% stearic acid in the total fatty acids of the cold-pressed cranberry seed oil (Table 4.1). Heeg and others[43] also detected significant levels of β-sitosterol (1.3 g/kg oil), and α- and γ-tocopherols at 341 and 110 mg/kg oil, respectively, in cranberry seed oil. Significant antioxidant activities were also detected in the cold-pressed cranberry seed oil extract[31]. The cold-pressed cranberry seed oil contained radical scavengers that could directly react with and quench stable DPPH radicals and ABTS$^{\bullet+}$, and had a total phenolic content of 1.6 mg gallic acid equivalent per gram of oil[31]. The ORAC value of the oil ranged from 3.5 to 3.9 μmol TE/g oil[31]. Parry and Yu[31] also reported the potential effects of cranberry seed oil components in the reduction of human low-density lipoprotein (LDL) oxidation, suggesting possible benefits of cranberry seed oil in the prevention of heart disease. In addition, cold-pressed cranberry seed oil showed similar oxidative stability as commercial soybean and corn oils[12].

In summary, cranberry seed oil is an excellent dietary source of α-linolenic and linoleic acids, and may be used to improve the dietary ratio of n-6/n-3 fatty acids. Cranberry seed oil also provides a significant level of natural antioxidants including phenolic compounds and tocopherols.

4.6 BLUEBERRY SEED OIL

Blueberries are rich in antioxidative phenolic compounds, particularly anthocyanins[45]. Consumption of blueberries may improve serum antioxidant status and reduce the risk of many chronic degenerative diseases[46]. Recently, cold-pressed blueberry seed oil has become commercially available, and has also been examined for its fatty acid composition and antioxidant properties[31]. The fatty acids of blueberry seed oil consisted of 69% PUFAs, 23% MUFAs, and about 8.5% saturated fatty acids (Table 4.1) and may serve as an excellent source of essential fatty acids, linoleic and α-linolenic acids, with a very good ratio of n-6 to n-3 fatty acids of 1.7:1. This suggests the potential application of blueberry seed oil in improving the dietary ratio of n-6 to n-3 fatty acids for optimal human health. Significant antioxidant activity was detected in the cold-pressed blueberry seed oil[31] with an ORAC value of 36.0 μmol TE/g oil, which is significantly higher than that of marionberry, black raspberry, cranberry, and pumpkin seed oils[31].

4.7 MARIONBERRY SEED OIL

The antioxidant activity and phenolic content of marionberry (*Rubus ursinus*) were investigated by Wada and Ou[47]. Marionberry had an ORAC value of 28 μmol TE/g fresh fruit and contained 5.8 mg total phenolics and 1.6 mg anthocyanins per gram on a fresh fruit weight basis[47]. Cyanidin 3-(6N-*p*-coumaryl)glucoside was the major anthocyanin that consisted of 95% of total anthocyanins[47]. Total phenolic and ascorbic acid contents of freeze-dried and air-dried marionberry grown either conventionally or organically have also been evaluated[48]. The corresponding total phenolic content was 350 to 410 and 500 to 620 mg gallic acid equivalents (GE)/100 g fresh fruit for conventionally and organically grown marionberries. No ascorbic acid was detected in marionberries. Recently, cold-pressed marionberry seed oil was evaluated for its fatty acid composition, peroxide value, and antioxidant properties[31]. The major fatty acids in the cold-pressed marionberry seed oil were linoleic, α-linolenic, and oleic acids (Table 4.1). Linoleic acid comprised approximately 63% of the total fatty acids, whereas α-linolenic acid was present at 16% (Table 4.1), suggesting the potential use of marionberry seed oil as a dietary source for essential fatty acids. Cold-pressed marionberry seed oil also contained a significant level of oxygen radical absorbing agents and exhibited an ORAC value of 17.2 μmol TE/g oil.

4.8 BOYSENBERRY SEED OIL

Boysenberry (*Rubus ursinus* × *idaeus*), a commercially available caneberry in the United States, has been investigated for its antioxidant capacity, and phenolic and anthocyanin contents[47]. Boysenberry had an ORAC value of 42 μmol TE/g in fresh fruit, which is higher than that of red raspberry on a dry weight basis[47]. Wada and Ou[47] also reported that boysenberry contained 6.0 and 1.3 mg of phenolics and anthocyanins, respectively, per gram, on a fresh fruit weight basis[31,47]. They evaluated the antioxidant properties and fatty acid profile of cold-pressed boysenberry seed oil and found that it contained about 20% α-linolenic acid and 54% linoleic acid, suggesting that it may serve as a good dietary source for both n-6 and n-3 essential fatty acids (Table 4.1). The ORAC value of the oil was 77.9 μmol TE/g, a value significantly higher than those of five other tested fruit seed oils, including blueberry that is known to contain high concentrations of antioxidants[31].

4.9 CITRUS SEED OILS

In 1994–1995 the world citrus production was approximately 66 million metric tons (MMT) with almost 28 MMTs going into processing (www.fas.usda.gov/htp2/circular/1997/97/jul97cov.html).

The projection for world production of citrus in 2010 is expected to rise by 8.4% to 95.5 MMT compared to the 1996–1998 yield of 88.1 MMT. Oranges are projected to account for 64.0 MMT, tangerines 15.4 MMT, lemons and limes 10.6 MMT, and grapefruit 5.5 MMT. From the total count, 28.3 MMT of the oranges and 2 MMT of the grapefruits are expected to be processed (www.fao.org/docrep/003/x6732e/x6732e02.htm).

In 1986, Habib and others[49] examined the fatty acid composition and other physicochemical properties of four Egyptian citrus fruit seed oils: orange, mandarin, lime, and grapefruit. The seed oils from citrus fruits were also compared to commercial edible cottonseed and soybean oils to determine similarities that could lead to a viable commercial product. The lime seed oil contained 42% α-linolenic acid, which was much greater than the rest (Table 4.2). The n-6 to n-3 fatty acid ration was 1:2.3 suggesting the potential use of lime seed oil in improving n-6 to n-3 fatty acid ratio. The grapefruit seed oil contained 43% palmitic acid (16:0), which was the most in all tested citrus seed oils. It also had a low concentration of total unsaturated fatty acids. Grapefruit seed oil contained about 1.4% of α-linolenic acid and 34.5% linoleic acid (Table 4.2). Interestingly, the mandarin seed oil consisted of 8.5% medium-chain fatty acids (C8 to C12) and 22.7% of unidentified compounds in the total fatty acids (Table 4.2). Most of the unidentified compounds were not detected in other citrus seed oils under the same experimental conditions, except two of them which were present in orange seed oils. No n-3 fatty acid was detected in the mandarin seed oil. The major fatty acids in orange seed oil were linoleic, palmitic, and oleic acids, along with 6.5% α-linolenic acid (Table 4.2). Compared to soybean oil, all citrus seed oils had higher levels of palmitic acid and less oleic acid. Lime and orange seed oils contained a higher proportion of α-linolenic acid than soybean oil[49].

In 1988, Lazos and Servos[50] investigated nutritional and chemical compositions of orange seed oil from seeds obtained in the Argos region of Greece. Less α-linolenic acid was detected in the orange seed oil with a greater level of linoleic acid (Table 4.2). In addition, seed oil was prepared from bitter orange (*Citrus aurantium* L.) by Soxhlet extraction and analyzed for fatty acid composition and other physicochemical properties[51]. The bitter orange seed oil consisted of 35.6% linoleic, 27% oleic, 24% palmitic, and 8.6% α-linolenic acids (Table 4.2). In 1993, Ajewole[52] characterized Nigerian citrus seed oils and reported the fatty acid profiles and other chemical properties for six citrus species including sweet and sour oranges, grapefruit, lime, tangerine, and tangelo. The major fatty acids were palmitic, ranging from 12 to 28%, oleic (26 to 45%), and linoleic (29 to 38%) acids, along with stearic and α-linolenic acids (Table 4.2). In contrast to Habib and others' observation, lime seed oil contained only 3.4% α-linolenic acid, but was rich in oleic acid (34% of total fatty acids). The highest oleic acid content was detected in the tangelo seed oil (45%). Sterols, including β-sitosterol and squalene, and a terpinene were detected in the citrus seed oils[49,50]. The physicochemical properties of citrus seed oils are summarized in Table 4.3.

4.10   GAC SEED OIL

Gac (*Momordica cochinchinensis* Spreng) is a member of the gourd family (Curcubitaceae). Other gourds include fruits such as cantaloupe, pumpkin, squash, cucumber, and watermelon. Gac is round and approximately 20 cm in diameter. It is indigenous to Asia where it is consumed as food and has also been used in Chinese traditional medicine for many years. The seeds comprise approximately 16% of the total fresh weight of the fruit and are surrounded by an aryl that is oily, red, and fleshy, and has a bland to nutty flavor[53]. Possible beneficial effects may be due to the fact that the aryl has very high concentrations of both β-carotene and lycopene[53]. In fact, the lycopene concentration in gac aryl has been shown to be 13 to 20 times higher than that of field-grown tomatoes by weight[54]. In 2002, Vuong and others[54] found that gac significantly increased plasma β-carotene and retinol levels and marginally increased hemoglobin level in children with low hemoglobin levels compared to control subjects.

TABLE 4.2
Fatty Acid Composition (g/100g Fatty Acids) of Citrus Seed Oils[a]

Fatty acid	Ora_Sw[52]	Grap1[52]	Ora_So	Tangerine[52]	Lime1[52]	Tangelo[52]	Ora1[49]	Mandarin[49]	Lime2[49]	Grap2[49]	Ora2[50]	Ora_B[51]
8:0	nd	nd	nd	nd	nd	nd	nd	2.2	nd	nd	nd	nd
10:0	nd	nd	nd	nd	nd	nd	5.4	3.2	0.9	8.4	nd	nd
12:0	nd	nd	nd	nd	nd	nd	nd	3.1	1.0	nd	nd	nd
14:0	nd	nd	nd	nd	nd	nd	nd	3.5	nd	nd	trace	nd
15:0	nd	nd	nd	nd	nd	nd	nd	2.4	nd	nd	nd	nd
15:1	nd	nd	nd	nd	nd	nd	nd	3.2	nd	nd	nd	nd
16:0	25.2	28.0	24.8	27.5	24.6	12.1	28.8	18.1	19.1	42.6	25.4	23.7
16:1	nd	nd	nd	nd	nd	nd	0.7	2.1	0.3	nd	0.3	nd
18:0	4.2	2.9	3.0	1.2	8.6	1.7	2.6	7.8	0.8	0.8	5.3	5.0
18:1	26.1	26.9	27.2	29.0	33.9	45.3	23.7	16.8	17.1	12.2	24.6	27.1
18:2n-6	37.8	34.5	37.6	29.0	30.0	36.4	31.0	13.5	18.6	34.5	39.3	35.6
18:3n-3	6.7	7.7	7.4	13.2	3.4	4.5	6.5	nd	42.3	1.4	4.5	8.6
20:0	nd	nd	nd	nd	nd	nd	nd	nd	nd	nd	0.4	nd
20:1	nd	nd	nd	nd	nd	nd	nd	nd	nd	nd	0.1	nd
Sat	29.4	30.9	27.8	28.7	33.2	13.8	36.8	40.3	21.8	51.8	31.1	28.7
MUFA	26.1	26.9	27.2	29.0	33.9	45.3	24.4	22.1	17.4	12.2	25.0	27.1
PUFA	43.5	42.2	45.0	42.2	33.4	40.9	37.5	13.5	60.9	35.9	43.8	44.2
n-6	37.8	34.5	37.6	29.0	30.0	36.4	31.0	13.5	18.6	34.5	39.3	35.6
n-3	6.7	7.7	7.4	13.2	3.4	4.5	6.5	N/A	42.3	1.4	4.5	8.6
n-6/n-3 ratio	5.6	4.5	5.1	2.2	8.8	8.1	4.8	N/A	0.4	24.6	8.7	4.1

[a] Sat, saturated fatty acids; MUFA, monounsaturated fatty acids; PUFA, polyunsaturated fatty acids; n-6, polyunsaturated n-6 fatty acids; n-3, polyunsaturated n-3 fatty acids; Ora_Sw, sweet orange; Ora_So, sour orange; Ora1, orange1; Ora2, organge2; Grap1, grapefruit1; Grap2, grapefruit2; Ora_B, bitter orange; nd, not detected; N/A, not applicable.

TABLE 4.3
Physical and Chemical Characteristics of Citrus Seed Oils[a]

Source of oil	Density (g/ml)	Refractive index	Iodine value (g iodine/100 g oil)	Acid value[b]	Saponification value[c]
Cottonseed[49]	0.921	1.4641	108.3	0.74	196.0
Mandarin[49]	0.912	1.4650	82.5	0.65	186.2
Lime[49,52]	0.922	1.4671	89.3–100.0	1.2	191.3–196.0
Grapefruit[49,52]	0.913	1.4662	91.4–101.0	0.90	189.6–192.0
Sweet orange[49]	N/A	N/A	102.0	N/A	186.0
Sour orange[52]	N/A	N/A	109.0	N/A	196.0
Tangerine[52]	N/A	N/A	108.0	N/A	188.0
Hybrid[52]	N/A	N/A	114.0	N/A	193.0
Orange[49,50]	0.92–0.933	1.4624–1.4681	99.2–99.5	0.21–1.51	195.8–196.8
Bitter orange[51]	0.92	1.4710	102.0	N/A	190.4

[a] N/A, not applicable.
[b] mg KOH necessary to neutralize fatty acid in 1 g oil.
[c] mg KOH required to saponify 1 g oil.

TABLE 4.4
Fatty Acid Composition (g/100 g Fatty Acids) of Gac Seed Oils[a]

Fatty acid	Test 1	Test 2	Test 3	Average
Palmitic (16:0)	6.2	5.2	5.3	5.6
Palmitoleic (16:1 Δ^9)	0.01	nd	nd	0.1
Stearic (18:0)	71.7	55.2	54.5	60.5
Oleic (18:1 Δ^9)	4.8	11.2	11.0	9.0
cis-Vaccenic (18:1 Δ^{11})	0.4	nd	0.7	0.5
Linoleic (18:2 $\Delta^{9,12}$)	11.2	24.8	25.0	20.3
α-Linolenic (18:3 $\Delta^{9,12,15}$)	0.5	0.6	0.4	0.5
Arachidic (20:0)	1.3	2.0	1.7	1.6
Eicosa-11-enoic (20:1 Δ^{11})	0.8	1.0	1.4	1.1
Eisoa-13-enoic (20:1 Δ^{13})	3.0	nd	nd	3.0

[a] nd, not detected.
Source: Ishida, B., Turner, C., Chapman, M., and McKeon, T., *J. Agric. Food Chem.*, 52, 274–279, 2004.

Gac seed oil contained over 65% saturated fatty acids with stearic acid comprising over 60% of the total fat. Linoleic acid was the primary unsaturated fatty acid present at 20% (Table 4.4). It is interesting to note that the gac aryl oil was very different in its fatty acid composition from gac seed oil containing high concentrations of palmitic, oleic, and linoleic acids, and nearly eight times less stearic acid (Table 4.4)[53].

4.11 WATERMELON SEED OIL

Watermelon, *Citrullus vulgaris*, is a warm-weather crop grown throughout the world where conditions permit. Over 1200 varieties have been cultivated and 200 to 300 varieties are grown in North America (http://www.watermelon.org/index.asp?a=dsp&htype=about&pid=39). In 2002, world production of watermelon was approximately 89 MMT with over 63 MMT grown in China. The

TABLE 4.5
Fatty Acid Composition (g/100 g Fatty Acids) of Pumpkin, Carob, and Watermelon Seed Oils[a]

Fatty acid	Pumpkin[55]	Carob[62]	Watermelon[55]
Caproic	0.2	nd	nd
Palmitic	13.4	17.6	11.3
Palmitoleic	0.4	nd	0.3
Stearic	10.0	nd	10.2
Oleic	20.4	trace	18.1
Linoleic	55.6	22.6	59.6
α-Linolenic	nd	nd	0.4
10-Octadecenoic	nd	40.3	nd
Margaric	nd	6.4	0.1
Oxiraneoctanoic	nd	trace	nd
Ricinoleic	nd	5.6	nd
9-OH-octadecanoic	nd	7.6	nd
Docosanoic	nd	trace	nd
Saturated	23.5	31.6	21.6
MUFA	20.8	45.9	18.4
PUFA	55.6	22.3	60.0
Total unsaturated	76.4	68.2	78.4

[a] MUFA, monounsaturated fatty acids; PUFA, polyunsaturated fatty acids; nd, not detected.

next largest producer was Turkey with a production of 8 MMT. In the U.S., production in both 2002 and 2003 was over 1.9 million tons (http://usda.mannlib.cornell.edu/reports/nassr/fruit/pvg-bban/vgan0104.txt). Watermelon seeds are eaten as a snack food worldwide, and they are also used as a source of oil in some countries.

The fatty acid composition of an unknown variety of *Citrullus vulgaris* is shown in Table 4.5. The primary fatty acid in the watermelon seeds was linoleic acid, which comprised nearly 60% of the total. Oleic acid was the only other compositionally relevant unsaturated fatty acid[55].

Some physicochemical properties of watermelon seed oil were compared to pumpkin seed oil including refractive index, acid value, peroxide value, free fatty acids, saponification value, and iodine value. The refractive index, acid value, free fatty acids, and peroxide value of watermelon seed oil were determined to be 1.4696 (25°C), 2.82 mg of KOH/g of oil, 3.40 meq of O-OH/kg of oil, and 1.41% as oleic acid, respectively, and these were not different from those of pumpkin seed oil. The saponification value of watermelon seed oil was 201 mg of KOH/g oil and was significantly lower than that of pumpkin seed oil at 206 mg of KOH/g oil. The iodine value of watermelon seed oil was 115 g iodine/100 g oil, and it was significantly higher than that of pumpkin seed oil at 109 g iodine/100 g oil[55].

4.12 PUMPKIN SEED OIL

Pumpkin, *Curcubita* spp., is a member of the gourd family, Curcubitaceae, that also includes melons, cucumbers, squash, and the previously mentioned gac. In 2003 the U.S. production of pumpkins was approximately 335,000 MMT (http://usda.mannlib.cornell.edu/reports/nassr/fruit/pvg-bban/vgan0104.txt). In Sudan and Ethiopia, dried pumpkin seeds have been used to treat tapeworm when eaten on an empty stomach[56]. Also, for many years, pumpkin seeds have been used as a remedy for micturition in Europe. Pumpkin seed oil has also shown possible beneficial affects in retarding the progression of hypertension[57], potential antiinflammatory activity in arthritis[58], and may be effective in reducing the risk of bladder-stone disease[59].

The fatty acid compositions of the seed oils prepared from two different pumpkin species (*pepo* and *mixta*) are shown in Table 4.5[55,60]. The seed oils from both species were similar in their composition, and all were fairly high in unsaturated fats that ranged from 76% in the pepo to 80.3% in the mixta. The iodine value (IV) for pumpkin seed oils ranged from 103 to 123 g of iodine absorbed per 100 g oil, and saponification value ranged from 132 to 207 mg of KOH/g oil[50,55,56,60]; unsaponifiable matter for *pepo* was 0.67% and 0.9% for *mixta* on an oil weight basis, and the refractive indices were 1.4616 (40°C)[50], 1.4706 (25°C)[55], and 1.4695 (23°C) for the *pepo*, and 1.4615 (60°C)[60] for the *mixta*[50,55,56,60]. The ORAC value of roasted pumpkin seed oil was 1.1 μmol TE/g oil, which is the lowest in comparison to blueberry, red raspberry, black raspberry, boysenberry, and cranberry seed oils[31].

4.13 CAROB SEED OIL

Carob, *Ceratonia siliquia*, is a tree native to the Mediterranean region but has now been disseminated to warm-climate locations throughout the world. The fruit, similar in appearance to a flattened bean, is approximately 20 cm long by 2.5 cm wide, and seeds comprise from 10 to 20% of the total weight[37]. The world production in the late 1980s was about 330,000 MMT with approximately 45% grown in Spain. Both the seedpod and seed have been used as food and fodder since antiquity[61]. The flesh has a similar flavor to chocolate and is used as a substitute in products such as cakes, cereals, cocoa, coffee, and candy, and it is also used to make other products including gums, sugar, and alcohol. Carob seed oil was obtained from dried seed powder with *n*-hexane using a Soxhlet extractor. The seeds contained 1.13% total oil with major unsaturated fatty acid being 10-octadecenoic (18:1) and linoleic acids at about 40 and 23%, respectively. Palmitic acid was the primary saturated fatty acid constituting about 18% of the total fat (Table 4.5)[62].

4.14 APPLE SEED OIL

In 1997, the world production of apples was 44.7 MMT, and 84% of those were processed (www.geocities.com/perfectapple/prod.html). The world production of apples in 2000–2001 reached a record high of 48 MMT (www.fas.usda.gov/htp/circular/2003/3-7-03%2520Web%2520 Art.%Updates//World%2520Apple%2520Situation%25202002-03.pdf). Apple seed is a byproduct of apple processing. In 1971 Morice and others[63] investigated the seed oils from three different varieties of apples, Granny Smith, Sturmer, and Dougherty, and compared them with the seed oils prepared from other apple varieties. The fatty acid profiles of the apple seed oils showed similarities among the varieties (Table 4.6). Oleic and linoleic acids comprised 85 to 95% of the total fatty acids in all tested apple seed oils[63]. The physicochemical properties of apple seed oils were also examined. The seed oils of Granny Smith apple had an IV of 127, the Sturmer had an IV of 122.4, and the Dougherty's IV was 119 g I/100 g oil. The physicochemical properties of the apple seed oils are summarized in Table 4.7. Apple seed oils may be useful as a dietary source for linoleic and oleic acids.

4.15 GRAPESEED OIL

World grape production in 2001 was 61.2 MMT (www.winetitles.com.au/awol.overview/world.asp). Grapeseeds are byproducts from the manufacturing of grape juice, jam, jelly, and wine. The seed oils from 41 grape varieties showed a similar fatty acid profile[64].

In 1998, Abou Rayan[65] and others investigated the characteristics and composition of Egyptian-grown Cabarina red grapeseed oil. Crude grapeseed oil was extracted with hexane at room temperature. The major fatty acids present were palmitic, stearic, linoleic, and linolenic acids, similar to

TABLE 4.6
Fatty Acid Composition (g/100 g Fatty Acids) of Apple Seed Oils[a]

Fatty acid	Granny Smith	Sturmer	Dougherty	Golden Delicious
16:0	6.8–7.1	4.8–6.4	5.7–6.8	8.5
16:1	0.1–0.2	0.1	0.1–0.2	0.5
18:0	1.0–2.1	1.5–2.5	1.3–2.1	nd
18:1	24.4–27.4	32.8–36.6	34.6–42.1	31
18:2	62.1–64.1	52.1–58.3	48.2–56.1	59
18:3	0.2–0.4	≤0.5	≤0.6	0.5
20:0	0.6–1.1	0.7–1.7	0.6–0.9	0.5
20:1	0.2–0.3	0.2–0.4	0.2–0.3	nd
20:2	0.1–0.7	0.1–0.7	0.0–0.3	nd
22:0	0.1–0.2	0.1–0.3	0.1	trace

[a] nd, not detected; Granny Smith, Sturmer, Dougherty, and Golden Delicious represent varieties of apple.
Source: Morice, I.M., Shorland, F.B., and Williams, E., *J. Sci. Food Agric.*, 22, 186–188, 1971.

TABLE 4.7
Physical and Chemical Characteristics of Apple Seed Oils[a]

Source of oil	Iodine value (g iodine/100 g oil)	Acid value[b]	Saponification value[c]	Unsaponifiable matter (% weight of fat)
Granny Smith	127	1.0–1.8	302–311	1.8–4.4
Sturmer	116–126	0.9–1.7	299– 302	1.5–4.2
Dougherty	118–121	N/A	295–297	1.3–1.4
Golden Delicious 1	122	2.3	299	1.1
Golden Delicious 2	122	2.6	285	0.9

[a] N/A, not applicable. Granny Smith, Sturmer, Dougherty, and Golden Delicious represent varieties of apple.
[b] mg KOH necessary to neutralize fatty acid in 1 g oil.
[c] mg KOH required to sponify 1 g oil.
Source: Morice, I.M., Shorland, F.B., and Williams, E., *J. Sci. Food Agric.*, 22, 186–188, 1971.

other grape varieties (Table 4.8). The measured IV was 130gI/100g oil, and the peroxide value was 2.92 meq O-OH/kg oil. Other characteristics determined were refractive index, specific gravity, saponification value, percentage of unsaponifiable matter, and acid value (Table 4.9).

4.16   POMEGRANATE SEED OIL

Pomegranate (*Punica granatum*), of the Punicaceae family, is a small tree grown in Iran, India and the United States, as well as in most Near and Far Eastern countries[66]. Pomegranate is used as a table fruit and is also processed to juice. Pomegranate preparations, including the juice of the fruit, the dried pericarp, the bark, and the roots, have been used in folk medicine to treat colic, colitis, dysentery, diarrhea, menorrhagia, oxyuriasis, parasis, and headache, and as a vermifugal, carminative, antispasmodic, taenicidal, and emmenagogue[66–69]. Seeds are byproducts from juice manufacture. Cold-pressed pomegranate seed oil was prepared and analyzed for its fatty acid composition, inhibitory effects against both cyclooxygenase and lipoxygenase, antioxidant properties, and total phenolic content[66]. The seed oil contained about 150 ppm total phenolics on an oil weight basis. The

TABLE 4.8
Fatty Acid Composition (g/100 g Fatty Acids) of Grapeseed Oils[a]

Fatty acid	Grape[65]	Palomino grape[75]	Grape[74]
Lauric	0.18	nd	nd
Myristic	0.32	trace	0.0–0.1
Palmitic	10.08	7.61–14.22	5.8–12.8
Palmitoleic	3.34	0.11–3.90	0.0–4.2
Stearic	6.12	4.63–8.56	0.0–6.2
Oleic	17.08	18.34–26.50	13.7–31.9
Linoleic	62.07	50.07–66.96	53.3–77.8
Linolenic	nd	≤4.97	≤0.7
Arachidic	0.21	nd	nd
Behenic	nd	nd	nd
22:1	0.40	nd	nd
20:2	0.21	nd	nd

[a] nd, not detected.

TABLE 4.9
Identity Characteristics of Grapeseed Oil

Characteristic	Value
Refractive index	1.5
Specific gravity (g/cm³)	0.9
Saponification number (mg KOH/g oil)	185.6
Iodine value (g iodine/100 g oil)	130.3
Unsaponifiable matter (%)	1.7
Peroxide value (meq O-OH/kg oil)	2.9
Acid value (mg KOH/g oil)	3.8
Free fatty acids (% oleic acid)	1.9

Source: Abou Rayan, M.A., Abdel-Nabey, A.A., Abou Samaha, O.R., and Mohamed, M.K., *J. Agric. Res.*, 43, 67–79, 1998.

oil extract, at a concentration of 5 μg total phenolics per ml, exhibited 37% inhibition of the sheep cyclooxygenase activity under experimental conditions[66]. The oil extract resulted in 75% inhibition of the soybean lipoxygenase activity, whereas butylated hydroxyanisole (BHA) had a 92% inhibition under the same experimental conditions. The oil extract also showed strong antioxidant activity in the coupled oxidation system of β-carotene and linoleic acid, and its antioxidant capacity is comparable to that of BHA and green tea extract on the same weight basis[66]. The major fatty acid was punicic acid (18:3n-5), which comprised 65% of total fatty acids, along with linoleic, oleic, palmitic, and stearic acids (Table 4.10). These data suggest the potential application of pomegranate seed oil as an antiinflammatory agent and for general health promotion.

4.17   SEED OILS OF BLACKCURRANT AND OTHER RIBES SPECIES

Blackcurrant (*Ribes nigrum*) is cultivated for berry production, and is mainly consumed in the form of juice[70]. Blackcurrant is rich in ascorbic acid and exhibits strong antioxidant activity. Lister and

TABLE 4.10
Fatty Acid Composition (g/100 g Fatty Acids) of Pomegranate and Blackcurrant Seed Oils[a]

Fatty acid	Pomegranate[66]	Blackcurrant[75]	Blackcurrant[74]
16:0	4.8	5.3	6.0–6.3
16:1 n-7	nd	nd	0.1
18:0	2.3	1.5	1.3–1.6
18:1 n-9	6.3	14.7	8.9–9.6
18:1 n-7	nd	0.7	0.7–0.8
18:2 n-6	6.6	47.0	42.7–43.5
18:3 n-3	nd	13.2	10.0–11.5
18:3 n-6	nd	12.2	22.0–24.6
18:3 n-5	65.3	nd	nd
18:4 n-3	nd	2.7	3.2–3.4
20:0	nd	0.1	0.1–0.2
20:1 n-11	nd	nd	0.1
20:1 n-9	nd	1.0	0.8–1.4
20:2 n-6	nd	0.2	0.4

[a] nd, not detected.

TABLE 4.11
Oil Content, Tocopherol Composition, Total Tocopherol, and γ-Linolenic Acid Contents in Seeds of Ribes Species

Ribes species	Oil[a] (%)	Tocopherols[b] (%)				Total-T[c]	γ-18:3[d]
		α-T	β-T	γ-T	δ-T		
Grossularia	16.2–26.4	16.3–37.6	0.0–0.0	59.4–79.4	3.0–15.4	559–1191	5.6–11.3
Nigrum (blackcurrants)	17.2–22.3	29.2–43.7	0.0–0.0	53.6–65.1	2.5–8.6	1228–2458	11.9–15.8
Rubrum	11.2–23.6	9.1–35.8	0.0–4.3	44.5–68.8	12.0–31.2	857–2481	3.3–7.2
Nigrum × *hirtellum*	18.5	43.5	0.0	53.6	3.0	1360	8.3

[a] Oil content, expressed as wt%.
[b] Tocopherols, expressed as % of total tocopherols. α-T, α-tocopherol; β-T, β-tocopherol; γ-T, γ-tocopherol; δ-T, δ-tocopherol.
[c] Total tocopherol content, as mg/kg oil.
[d] γ-Linolenic acid, as % of total fatty acids.
Source: Goffman, F.D. and Galletti, S., *J. Agric. Food Chem.*, 49, 349–354, 2001.

others[71] summarized the health benefits of blackcurrants. Blackcurrant seed oils were analyzed for fatty acid composition, tocopherols, and their potential application in reducing prostaglandin E_2 production[70,72–74]. Blackcurrant seed oil is an excellent dietary source of both γ-linolenic (18:3n-6) and α-linolenic (18:3n-3) acids. γ-Linolenic acid constituted 12 to 25% of the total fatty acids, while α-linolenic acid comprised the other 10 to 13% (Table 4.10). The fatty acid composition depended on genotype and differences in growing conditions. The blackcurrant seed oils also had significant levels of tocopherols[73]. The total tocopherol content was 1.2 to 2.5 mg/g oil, with a mean value of 1.7 mg/g oil for 10 oil samples. The major tocopherol in the blackcurrant seed oil was γ-tocopherol, and no β-tocopherol was detected in the blackcurrant seed oil (Table 4.11). In 1999, Wu and others[72] investigated the effect of dietary supplementation with blackcurrant seed oil on the immune response of healthy elderly subjects. They concluded that the oil may moderately enhance the immune function through reducing the production of prostaglandin E_2. It has been suggested by these researchers that blackcurrant seed oil may have a number of health benefits against cancer, cardiovascular disease, and other health problems.

Other Ribes species, including *R. grossularia* (red-black gooseberries), *R. grossularia* (yellow gooseberries), *R. nigrum* (blackcurrants) *R. rubrum* (red currants), and *R. nigrum* × *R. hirtellum* (jostaberries), were also examined for γ-linolenic acid concentration and tocopherol content in the seed oils. Among the tested samples, blackcurrant seed oil had the greatest level of γ-linolenic acid, and all had a total tocopherol content of over 1.0 mg/g oil (Table 4.11).

4.18 SUMMARY

A number of studies have been conducted to evaluate the chemical composition and potential nutraceutical applications of fruit seed oils. Among the discussed fruit seed oils, some have unique fatty acid compositions, such as lime and cranberry seed oils rich in α-linolenic acid whereas blackcurrant seed oil is rich in γ-linolenic acid. Fruit seed oils may also contain significant levels of tocopherols, carotenoids, phytosterols, and natural antioxidants. The chemical composition of the fruit seed oil determines the potential nutraceutical application of the oil. Individual fruit seed oils may be preferred by special groups of consumers for preventing and treating a selected health problem or for general health promotion. Fruit seeds are one of the byproducts from fruit processing. New applications of the seed-based products may add value to fruit-processing industries. Developing novel utilizations of fruit seed oils may also improve the farm gate value of the fruits and benefit the growers and the agricultural economy in general. Great opportunities are available in the research and development of specialty fruit seed oils and oil-based nutraceutical products. More research is required to screen and characterize the fatty acids and bioactive ingredients in the fruit seeds to develop value-added utilization of fruit seed oils as nutraceuticals.

REFERENCES

1. Fils, J., The production of oils, in *Edible Oil Processing*, Hamm, W. and Hamilton, R.J., Eds., CRC Press, Boca Raton, FL, 2000, pp. 47–78.
2. Gunstone, F.D., Composition and properties of edible oils, in *Edible Oil Processing*, Hamm, W. and Hamilton, R.J., Eds., CRC Press, Boca Raton, FL, 2000, pp. 1–33.
3. Mattson, F.H. and Grundy, S.M., Comparison of effects of dietary saturated, monounsaturated, and polyunsaturated fatty acids on plasma lipids and lipoproteins in man, *J. Lipid Res.*, 26, 194–202, 1985.
4. Mensink, R.P. and Katan, M.S., Effects of a diet enriched with monounsauraged or polyunsauraged fatty acids on levels of low-density and high-density lipoprotein cholesterol in healthy women and men, *New Engl. J. Med.*, 321, 436–441, 1989.
5. Dreon, D.M., Varnizan, K.M., Drauss, R.M., Austin, M.A., and Wood, P.D., The effects of polyunsaturaged fat vs monounsaturated fat on plasma lipoproteins, *J. Am. Med.*, 263, 2462–2466, 1990.
6. Jamal, G.A., The use of gamma linolenic acid in the prevention and treatment of diabetic neuropathy, *Diabetic Med.*, 11, 145–149, 1994.
7. Keen, H., Payan, J., Allawi, J., Walker, J., Jamal, G.A., and Weir, A.I., Treatment of diabetic neuropathy with γ-linolenic acid, *Diabetes Care*, 14, 8–15, 1993.
8. Wright, S. and Burton, J.L., Oral evening primrose oil improves atopic eczema, *Nature*, 2, 1120–1122, 1982.
9. Das, U.N., in *γ-Linolenic Acid: Metabolism and Its Roles in Nutrition and Medicine*, Huang, Y.S. and Mill, D.E., Eds., American Oil Chemists' Society, IL, 1995, pp. 282–292.
10. De Antueno, R.J., Elliot, M., Jenkins, K., Ells, G.W., and Horrobin, D.F., in *γ-Linolenic Acid: Metabolism and Its Roles in Nutrition and Medicine*, Huang, Y.S. and Mill, D.E., Eds., American Oil Chemists' Society, IL, 1995, pp. 293–303.

11. Layne, K.S., Goh, Y.K., Jumpsen, J.A., Ryan, E.A., Chow, P., and Clandinin, M.T., Normal subjects comsuming physiological levels of 18:3(n-3) and 20:5(n-3) from flaxseed or fish oils have characteristic differences in plasma lipid and lipoprotein fatty acid levels, *J. Nutr.*, 126, 2130–2140, 1996.

12. Parker, T.D., Adams, D.A., Zhou, K., Harris, M., and Yu, L., Fatty acid composition and oxidative stability of cold-pressed edible seed oils, *J. Food Sci.*, 68, 1240–1243, 2003.

13. Connor, W.E., Importance of n-3 fatty acid regulation of gene transcription: a molecular mechanism to improve the metabolic syndrome, *J. Nutr.*, 131, 1129–1132, 2000.

14. Wu, F.C., Ting, Y.Y., and Chen, H.Y., Docosahexaenoic acid is superior to eicosapentaenoic acid as the essential fatty acid for growth of grouper, *Epinephelus malabaricus*, *J. Nutr.*, 132, 72–79, 2002.

15. Ringbom, T., Huss, U., Flock, S., Skattebol, L., Perera, P., and Bohlin, L., COX-2 inhibitory effects of naturally occurring and modified fatty acids, *J. Nat. Prod.*, 64, 745–749, 2001.

16. Aronson, W.J., Glaspy, J.A., Reddy, S.T., Reese, D., Heber, D., and Bagga, D., Modulation of omega-3/ omega-6 polyunsaturated ratios with dietary fish oils in men with prostate cancer, *Urology*, 58, 283–288, 2001.

17. Harel, Z., Gascon, G., Riggs, S., Vaz, R., Brown, W., and Exil, G., Supplementation with omega-3 polyunsaturated fatty acids in the management of recurrent migraines in adolescents, *J. Adolescent Health*, 31, 154–161, 2002.

18. Iso, H., Sato, S., Umemura, U., Kudo, M., Koike, K., Kitamura, A., Imano, H., Okamura, T., Naito, Y., and Shimamoto, T., Linoleic, other fatty acids, and the risk of stroke, *Stroke*, 33, 2086–2093, 2002.

19. Tapiero, H., Ba, G.N., Couvreur, P., and Tew, K.D., Polyunsaturated fatty acids (PUFA) and eicosanoids in human health and pathologies, *Biomed. Pharmacother.*, 56, 215–222, 2002.

20. Villa, B., Calbresi, L., Chiesa, G., Rise, P., Galli, C., and Sirtori, C., Omega-3 fatty acid ethyl esters increase heart rate variability in patients with coronary disease, *Pharmacol. Res.*, 45, 475, 2002.

21. Hargrove, R.L., Etherton, T.D., Pearson, T.A., Harrison, E.H., and Kris-Etherton, P.M., Low fat and high monounsaturated fat diets decrease human low density lipoptotein oxidative susceptibility in vitro, *J. Nutr.*, 131, 1758–63, 2001.

22. Yaqoob, P., Monounsaturaged fatty acids and immune function, *Eur. J. Clin. Nutr.*, 56, S9–S13, 2002.

23. Nicolosi, R.J., Wilson, T.A., Handelman, G., Foxall, T., Keaney, J.F., and Vita, J.A., Decreased aortic early atherosclerosis in hepercholesterolemic hamsters fed oleic acid-rich trisun oil compared to linoleic acid-rich sunflower oil, *J. Nutr. Biochem.*, 13, 392–402, 2002.

24. Oohmah, B.D., Ladet, S., Godfrey, D.V., Liang, J., and Benoit, G., Characteristics of raspberry (Rubus idaeus) seed oil, *Food Chem.*, 69, 187–193, 2000.

25. Johansson, A., Laakso, P., and Kallio, H., Characterization of seed oils of wild, edible Finnish berries, *Zeitschrift fur Lebensmitte-luntersuchung und-Forchung A*, 204, 300–307, 1997.

26. Parry, J.W. and Yu, L., Fatty acid content and antioxidant properties of cold-pressed black raspberry seed oil and meal, *J. Food Sci.*, 69, 189–193, 2004.

27. Eaton, S.B., Eaton, S.B., III, Sinclair, A.J., Cordain, L., and Mann, N.J., Dietary intake of long-chain polyunsaturated fatty acids during the Paleolithic, *World Rev. Nutr. Diet.*, 83, 12–23, 1988.

28. Kris-Etherton, P.M., Taylor, D., Poth, S., Huth, P., Moriarty, K., Fishell, V., Hargrove, R., Zhao, G., and Etherton, T., Polyunsaturated fatty acids in the food chain in the United States, *Am. J. Clin. Nutr.*, 71, 179S–188S, 2000.

29. Maillard, V., Bougnoux, P., Ferrari, P., Jourdan, M.L., Pinault, M., Lavillonniere, F., Body, G., Le Flouch, O., and Chajes, V., n-3 and n-6 fatty acids in breast adipose tissue and relative risk of breast cancer in a case-control study in Tours, France, *Int. J. Cancer*, 98, 78–83, 2002.

30. Adams, D.A., Fatty Acid Composition of Commercial Raw Fish Products, MS thesis, Colorado State University, 2003, pp. 47–49.

31. Parry, J.W. and Yu, L., Phytochemical Composition and Free Radical Scavenging Capacities of Selected Cold-Pressed Edible Seed Oils, abstracts of papers, 228th National Meeting of the American Chemical Society, Philadelphia, PA, Aug. 22–26, 2004.

32. Parry, J.W., The Effect of Pycnogenol on Vitamin C Status and Antioxidant Capacity in Humans, thesis, California State University, 2001.

33. Wang, S.Y. and Lin, H.-S., Antioxidant activity in fruits and leaves of blackberry, raspberry, and strawberry varieties with cultivar and developmental stage, *J. Agric. Food Chem.*, 48, 140–146, 2000.

34. Wang, S.Y. and Jiao, H., Scavenging capacity of berry crops on superoxide radicals, hydrogen peroxide, hydroxyl radicals, and singlet oxygen, *J. Agric. Food Chem.*, 48, 5677–5684, 2000.

35. Huang, C., Huang, Y., Li, J., Hu, W., Aziz, R., Tang, M.S., Sun, N., Cassady, J., and Stoner, G.D., Inhibition of benzo(a)pyrene diol-epoxide-induced transactivation of activated protein 1 and nuclear factor kappaB by black raspberry extracts, *Cancer Res.*, 62, 6857–6863, 2002.

36. Ramadan, M.F. and Morsel, J.T., Oil goldenberry (*Physalis peruviana* L.), *J. Agric. Food Chem.*, 51, 969–974, 2003.

37. Morton, J.F., in *Fruits of Warm Climates*, Morton, J.F., Ed., Creative Resource System, Winterville, NC, 1987, pp. 430–434.

38. Popenoe, H., King, S.R., Leon, J., and Kalinowski, L.S., in *Lost Crops of the Incas Little-Known Plants of the Andes with Promise for Worldwide Cultivation*, National Research Council, National Academy Press, Washington, DC, 1990, pp. 241–252.

39. Rehm, S. and Espig, G., in *The Cultivated Plants of The Tropics and Subtropics, Cultivation, Economic Value, Utilization*, Sigmund, R. and Gustav, E., Eds., Verlag Josef Margraf, Weiker-sheim, Germany, 1991, pp. 169–245.

40. Kalt, W., Forney, C.F., Martin, A., and Prior, R., Antioxidant capacity, vitamin C, phenolics, and anthocyanins after fresh storage of small fruits, *J. Agric. Food Chem.*, 47, 4638–4644, 1999.

41. Prior, R.L., Lazarus, S.A., Cao, G., Muccitelli, H., and Hammerstone, J.F., Identification of procyanidins and anthocyanins in blueberries and cranberries (*Vaccinium* spp.) using high-performance liquid chromatography/mass spectrometry, *J. Agric. Food Chem.*, 49, 1270–1276, 2001.

42. Vvedenskaya, I.O., Rosen, R.T., Guido, J.E., Russell, D.J., Mills, K.A., and Vorsa, N., Characterization of flavonols in cranberry (*Vaccinium macrocarpon*) powder, *J. Agric. Food Chem.*, 52, 188–195, 2004.

43. Heeg, T., Lager, H., and Bernard, G., Cranberry Seed Oil, Cranberry Seed Flour and a Method for Making, U.S. Patent 6,391,345, 2002.

44. Yu, L., Zhou, K., and Parry, J.W., Antioxidant properties of cold-pressed black caraway, carrot, cranberry, and hemp seed oils, *Food Chem.*, 91, 723–729, 2005.

45. Prior, R.L., Cao, G., Martin, A., Sofic, E., McEwan, J., O'Brein, C., Lischner, N., Ehlenfeldt, M., Kalt, W., Krewer, G., and Mainland, C.M., Antioxidant capacity as influenced by total phenolic and anthocyanin content, maturity and variety of *Vaccinium* species, *J. Agric. Food Chem.*, 46, 2686–2693, 1998.

46. Kay, C.D. and Holub, B.J., The effect of wild blueberry (Vaccinium angustifolium) consumption on postprandial serum antioxidant status in human subjects, *Br. J. Nutr.*, 88, 389–397, 2002.

47. Wada, L. and Ou, B., Antioxidant activity and phenolic content of Oregon cranberries, *J. Agric. Food Chem.*, 50, 3495–3500, 2002.

48. Asami, D.K., Hong, Y., Barrett, D.M., and Mitchell, A.E., Comparison of the total phenolic and ascorbic acid content of freeze-dried and air-dried marionbery, strawberry, and corn grown using conventional, organic, and sustainable agricultural practices, *J. Agric. Food Chem.*, 51, 1237–1241, 2003.

49. Habib, M.A., Hammam, M.A., Sakr, A.A., and Ashoush, Y.A., Chemical evaluation of Egyptian citrus seeds as potential sources of vegetable oils, *J. Am. Oil Chem. Soc.*, 63, 1192–1197, 1986.

50. Lazos, E.S. and Servos, D.C., Nutritional and chemical characteristics of orange seed oil, *Grasas Aceites.*, 39, 232–234, 1988.

51. Romero, F., Doblado, J., and Cota, J., Characterization of bitter orange (*Citrus aurantium* L) seed oil, *Grasas Aceites.*, 39, 353–358, 1988.

52. Ajewole, K., Characterization of Nigerian citrus seed oils, *Food Chem.*, 47, 77–78, 1993.

53. Ishida, B., Turner, C., Chapman, M., and McKeon, T., Fatty acid and carotenoid compostion of gac (*Momordica cochinensis* Spreng) fruit, *J. Agric. Food Chem.*, 52, 274–279, 2004.

54. Vuong, L.T., Dueker, S.R., and Murphy, S.P., Plasma *β*-carotene and retinal concentrations of children increase after a 30-d supplementation with the fruit *Momordica cochinchinensis* (gac), *Am. J. Clin. Nutr.*, 75, 872–879, 2002.

55. El-Adaway, T.K. and Taha, K.M., Characteristics and composition of watermelon, pumpkin, and paprika seed oils and flours, *J. Agric. Food Chem.*, 49, 1253–1259, 2001.

56. Younis, Y.M.H., Ghirmay, S., and Al-Shihry, S.S., African *Curcubita pepo* L.: properties of seed and variability is fatty acid composition of seed oil, *Phytochemistry*, 54, 71–75, 2000.

57. Zuhair, H.A., Abd El-Fattah, A.A., and El-Sayed, M.I., Pumpkin-seed oil modulates the effect of felodipine and captopril in spontaneously hypertensive rats, *Parmacol. Res.*, 41, 555–563, 2000.

58. Fahim, A.T., Abd-El Fattah, A.A., Agha, A.M., and Gad, M.Z., Effect of pumpkin-seed oil on the level of free radical scavengers induced during adjuvant-arthritis in rats, *Pharmacol. Res.*, 31, 73–79, 1994.

59. Suphakarn, V.S., Yarnnon, C., and Ngunboonsri, B.S., The effect of pumpkin seeds on oxalcrystalluria and urinary compositions of children in hyperendemic area, *Am. J. Clin. Nutr.*, 45, 115–121, 1987.

60. Kamel, B., DeMan, J., and Blackman, B., Nutritional, fatty acid and oil characteristics of different agricultural seeds, *J. Food. Technol.*, 17, 263–269, 1982.

61. Tous, J. and Ferguson, L., in *Progress in New Crops*, Janick, J., Ed., ASHS Press, Arlington, VA, 1996, pp. 416–430.

62. Orhan, I. and Sener, B., Fatty acid content of selected seed oils, *J. Herbal Pharmacol.*, 2, 29–33, 2002.

63. Morice, I.M., Shorland, F.B., and Williams, E., Seed oils of apples (*Malus pumila*), *J. Sci. Food Agric.*, 22, 186–188, 1971.

64. Massanet, M.G., Montiel, J.A., Pando, E., and Rodriguez Luis, F., Study of agricultural by-products: II. Fatty acid composition of palomino grape seed oil, *Grasas Aceites.*, 37, 233–236, 1986.

65. Abou Rayan, M.A., Abdel-Nabey, A.A., Abou Samaha, O.R., and Mohamed, M.K., Characteristics and composition of grape seed oil, *J. Agric. Res.*, 43, 67–79, 1998.

66. Schubert, S.Y., Lansky, E.P., and Neeman, I., Antioxidant and eicosanoid enzyme inhibition properties of pomegranate seed oil and fermented juice flavonoids, *J. Ethnopharmacol.*, 66, 11–17, 1999.

67. Bianchini, F. and Corbetta, F., Health Plants of the World, *Newsweek*, New York, 1979.

68. Aynesu, S.E., *Medicinal Plants of the West Indies*, Reference Publication, Algonac, MI, 1981.

69. Duke, A.J. and Ayensu, S.E., *Medicinal Plants of China*, Reference Publication, Algonac, MI, 1985.

70. Ruiz Del Castillo, R.L., Dobson, G., Brennan, R., and Gordon, S., Fatty acid content and juice characteristics in black currant (Ribes nigrum L.) genotypes, *J. Agric. Food Chem.*, 52, 948–952, 2004.

71. Lister, C.E., Wilson, P.E., Sutton, K.H., and Morrison, S.C., Understanding the health benefits of black-currants, *Acta Hort.*, 585, 443–449, 2002.

72. Wu, D., Meydani, M., Leka, L.S., Nightingale, Z., Handelman, G.J., Blumberg, J. B., and Meydani, S.N., Effect of dietary supplementation with black currant seed oil on the immune response of healthy elderly subjects, *Am. J. Clin. Nutr.*, 70, 536–43, 1999.

73. Goffman, F.D. and Galletti, S., Gamma-linolenic acid and tocopherol contents in the seed oil of 47 accessions from several Ribes species, *J. Agric. Food Chem.*, 49, 349–354, 2001.

74. Ruiz Del Castillo, R.L., Dobson, G., Brennan, R., and Gordon, S., Genotype variation in fatty acid content of blackcurrant seeds, *J. Agric. Food Chem.*, 50, 332–335, 2002.

75. Olsson, U., Kaufmann, P., Herslof, B.G., Multivariate optimization of a gas-liquid chromatographic analysis of fatty acid methyl esters of black currant seed oil, *J. Chromatogr.*, 505, 385–394, 1990.

5 Minor Specialty Oils

Frank D. Gunstone
Scottish Crop Research Institute, Invergowrie, Dundee, Scotland, U.K.

CONTENTS

5.1   INTRODUCTION

There is no accepted definition of "minor oil" so it is necessary to indicate how this term will be interpreted. Some fatty oils are produced in such large amounts that they are recognized as commodity oils. They are produced and used on a large scale with internationally quoted prices and subject to import and export. The four largest in the harvest year 2004/05 were the vegetable oils from soybean (32.6 million metric tons, MMT), palm (32.5 MMT), rapeseed/canola (16.1 MMT), and sunflower (9.1 MMT), and at the bottom end of a list of 13 vegetable oils are those such as sesame and linseed at annual production levels of 0.5 to 1.0 MMT[1]. In addition to these there is a wide range of oils produced, sold, and used in still lower quantities. The list of these is almost endless and the author has made his own selection based on the frequency with which they are reported in the literature and their appearance in lists of specialist oil suppliers. These are presented in alphabetical order after some general points have been made.

These oils are generally of interest because they contain a fatty acid or other component, which gives the oil interesting dietary or technical properties, or they are oils available in modest quantities that can be used in a niche market. Such oils are usually available in only limited quantities and if they are to be marketed it is essential to ensure that the sources located will provide a reliable and adequate supply of good-quality material. As the oils are to be used as dietary supplements, as health foods, as gourmet oils, or in the cosmetics industry[2], it is important that the seeds be handled, transported, and stored under conditions that will maintain quality. It may also be necessary to consider growing the crops in such a way as to minimize the level of pesticides.

Many fruits are now processed at centralized facilities. This means that larger quantities of "waste products" are available at one center and can be more easily treated to recover oil and other valuable byproducts. This is particularly relevant in the fruit industry where pips, stones, and kernels are available in large quantities.

Extraction can be carried out in several ways including cold pressing (at temperatures not exceeding 45°C), pressing at higher temperatures, and/or solvent extraction. Solvent extraction is not favored for high-quality gourmet oils. Supercritical fluid extraction with carbon dioxide is an acceptable possibility but only limited use is made of this. A further possibility is to use enzymes to break down cell walls followed by extraction under the mildest possible conditions.

Some specialty oils such as walnut, virgin olive, hazelnut, pistachio, and sesame can be used as expressed, merely after filtering, but for others some refining is generally necessary. If the oil has a characteristic flavor of its own it may be desirable to retain this and high-temperature deodorization must then be excluded or reduced to a minimum. Once obtained in its final form the oil must be protected from deterioration — particularly by oxidation. This necessitates the use of stainless steel equipment, blanketing with nitrogen, and avoiding unnecessary exposure to heat and light. At the request of the customer natural and/or synthetic antioxidants can be added to provide further protection.

Useful information related to this topic can be found in references[3-20]. Further web sites also furnish information on many of the individual oils.

For all these oils a fatty acid composition has been reported and information about the minor components (tocopherols, sterols, carotenes, etc.) is also sometimes available. Based on this information, claims are frequently made for the superior properties of these oils. These may be valid, but there are few, if any, where tests have been carried out to support the claims.

Most vegetable oils contain only three acids at levels exceeding 10% with these [palmitic (16:0), oleic (18:1), and linoleic (18:2)] frequently having a combined level of 90% or more. This means that other acids such as Δ9-hexadecenoic, stearic, or linolenic acids are present at low levels, if at all. The large number of oils of this type can generally be subdivided into those in which oleic acid dominates, those in which linoleic dominates, and those in which these two acids are present at similar levels. Palmitic acid, though always present, is seldom the dominant component. Beyond these are some oils with less-common acids, sometimes at quite high concentration.

Short- and medium-chain acids. While most oils contain virtually only C_{16} and C_{18} fatty acids a small number are characterized by a dominance of acids of shorter chain lengths. Two commodity oils (coconut and palmkernel) are known collectively as lauric oils because they contain around 50% of lauric acid (12:0) accompanied by 8:0, 10:0, and 14:0 at lower levels[1]. Among the minor oils are some that have a similar fatty acid composition (e.g., babassu) and some in which the shorter-chain acids dominate as in the cuphea oils.

Stearic acid. Stearic acid is more significant in fats from domesticated land animals (especially sheep) than in vegetable oils. Nevertheless, there are some minor oils in which stearic acid accompanies palmitic and oleic acids as a major component. This holds for cocoa butter (palmitic acid ~26%, stearic acid ~34%, and oleic acid ~35%) and for a range of tropical fats which have a similar chemical composition and similar physical properties. An example is Borneo tallow.

Hexadecenoic and erucic acids. Oleic acid (18:1) is the most common monounsaturated acid and also the most common acid produced in nature, a position that it shares with linoleic acid (18:2). Despite this, there are some other monounsaturated acids that become significant in certain vegetable fats. These may be isomers of oleic acid with the unsaturated center different from the common Δ9 (such as petroselinic and *cis*-vaccenic) or they may be acids of different chain length of which the most common are hexadecenoic (16:1), present in macadamia oil and sea buckthorn oil, and erucic acid (22:1) in some forms of rapeseed oil and in crambe oil.

Petroselinic acid. Petroselinic acid ($\Delta 6c$-18:1) is an uncommon isomer of oleic acid, present at high levels in a restricted range of seed oils, especially those from plants of the Umbelliferae family. Oleic acid is usually present also at lower levels. With unsaturation starting on an even carbon atom the $\Delta 6$ acid has a higher melting point than isomers in which unsaturation starts on an odd carbon atom. Petroselinic acid melts at 29°C compared with 11°C for oleic acid. It is formed in seeds by an unusual biosynthetic pathway. The unsaturated center is introduced at the C_{16} stage by a $\Delta 4$-desaturase and this step is followed by chain elongation:

$$16:0 \rightarrow \Delta 4\text{-}16:1 \rightarrow \Delta 6\text{-}18:1$$

γ-Linolenic acid (GLA). The most common polyunsaturated fatty acids occurring in seed oils are linoleic acid ($\Delta 9,12$-18:2) and α-linolenic acid ($\Delta 9,12,15$-18:3) but in a few species the α-linolenic acid is accompanied or replaced by GLA ($\Delta 6,9,12$-18:3) that is now recognized as an interesting material with beneficial health properties. Claims have been made for its use in the treatment of multiple sclerosis, arthritis, eczema, premenstrual syndrome, and other diseases. It is a biological intermediate in the conversion of freely available linoleic acid to the important but less readily available arachidonic acid. This change is a three-step process involving $\Delta 6$-desaturation, elongation, and $\Delta 5$-desaturation of which the first step is considered to be rate determining:

$$9,12\text{-}18:2 \text{ (linoleic)} \rightarrow 6,9,12\text{-}18:3 \rightarrow 8,11,14\text{-}20:3 \rightarrow 5,8,11,14\text{-}20:4 \text{ (arachidonic)}$$

A similar sequence of changes converts α-linolenic acid to eicosapentaenoic acid (20:5) and docosahexaenoic acid (22:6) with the first metabolite being stearidonic acid (6,9,12,15-18:4). Echium oils serve as a source of stearidonic acid.

GLA is present in a number of seed oils of which three are commercially available (black-currant, borage, and evening primrose). A case has been made for incorporating this acid into our dietary intake and businesses have developed to grow the required seeds and to produce the oils from these (see borage).

Acids with conjugated unsaturation. As indicated in the previous paragraph the most common polyunsaturated fatty acids in vegetable oils have methylene-interrupted patterns of unsaturation. However, acids with conjugated unsaturation are present at high levels in a small number of seed oils. These are mainly 18:3 acids with unsaturation at $\Delta 9,11,13$ or $\Delta 8,10,12$, all derived from linoleic acid. There are also some tetraene acids ($\Delta 9,11,13,15$-18:4) derived from α-linolenic acid. Conjugated diene acids occur only very rarely in seed oils. The intensive study of the animal-derived conjugated linoleic acids (18:2) has led to consideration of the potential value of the plant-derived conjugated trienes and tetraenes.

Cocoa butter alternatives. Cocoa butter is an important commodity which carries a premium price. Cheaper alternatives with similar physical properties such as materials derived from lauric oils can be used but products containing these fats cannot be called chocolate and are generally described as confectionery fats. However, in some European countries up to 5% of a product can be fats taken from a prescribed list and the product still be designated chocolate. These include palm mid-fraction and five tropical fats listed in Table 5.2 under Borneo tallow.

5.2 MINOR OILS

Summarizing Tables 8 and 9 are found at the end of this section.

5.2.1 ACEITUNO (*SIMAROUBA GLAUCA*, QUASSIA)

This oil comes from trees grown in Central and South America. The nuts contain about 30% of oil rich in oleic acid (~58%) and with significant levels of stearic (~28%) and palmitic (12%) acids. With this fatty acid composition it is not surprising that its major triacylglycerols are SOO (42%), SOS (29%), and OOO (15%), where S and O represent saturated acids and oleic acid, respectively[11,13].

5.2.2   ALFALFA (*MEDICAGO SATIVA* AND *M. FALCATE*)

Alfalfa seeds (*M. sativa*) contain only 7.8% of oil. The major component acids are linoleic (34%) and α-linolenic (25%) along with lower levels of saturated and monounsaturated acids. The oil is rich in carotenes and in lutein. It has been claimed that the seeds lower low-density lipoprotein (LDL) cholesterol in patients with hyperlipoproteinemia and that the oil reduces erythema caused by sunburn[12,18].

5.2.3   ALMOND (*PRUNUS DULCIS, P. AMYGDALIS, AMYGDALIS COMMUNIS*)

Almond oil is an oleic-rich oil (65 to 70%) accompanied by linoleic, palmitic, and minor acids, though its fatty acid composition can vary widely. The triacylglycerol composition of the oil has also been reported[21] and as expected the major triacylglycerols have three oleic chains (38%) or two oleic chains with linoleic (24%) or palmitic (11%) acids. In common with other low-saturated, high-monounsaturated oils, almond oil shows high oxidative and cold-weather stability (slow to deposit crystals). The oil is commonly used in skincare and massage products because of its non-greasy nature, good skin feel, reasonable price, and consumer appeal. Almond nuts are reported to lower cholesterol levels and the U.S. Food and Drug Administration (USFDA) permits the following claim for a limited range of nuts including almonds: "Scientific evidence suggests but does not prove, that eating 1.5 ounces per day of most nuts as part of a diet low in saturated fat and cholesterol may reduce the risk of heart disease"[11–13,95].

5.2.4   AMARANTHUS (*AMARANTHUS CRUENTUS*)

Amaranthus is a grain containing only low levels (6 to 9%) of oil. A study of 21 accessions gave the following results: oil content 5 to 8% (mean 6.5), palmitic 8 to 22% (mean 19), stearic 1 to 4% (mean 3), oleic 16 to 25% (mean 22), linoleic 41 to 61% (mean 45), tocopherols 2.8 to 7.8 mg/100 g of seed. The average content of tocopherols is 4.94 mg/100 g of seed with the major components being β- and α-tocopherols at 2.17 and 1.66 mg/100 g seed, respectively[22]. The high level of β-tocopherol is unusual and in contrast to the results of an earlier study[23]. A more recent study of five accessions shows palmitic (21 to 24%), oleic (23 to 31%), and linoleic acid (39 to 48%) as the major components and gives details of the triacylglycerol composition[24]. Amaranthus oil is unusual among vegetable oils in that it has a relatively high level (6 to 8%) of squalene and this concentration can be raised 10-fold by short-path high-vacuum distillation. There is no other convenient vegetable source of this C_{30} hydrocarbon other than olive oil which has a squalene level of 0.3 to 0.7% rising to 10 to 30% in deodorizer distillate[25].

5.2.5   ARGEMONE (*ARGEMONE MEXICANA*)

These seeds contain about 39% of oil with palmitic (12 to 15%), oleic (28 to 29%), and linoleic (~55%) acids as major component acids[13].

5.2.6   APRICOT (*PRUNUS ARMENIACA*)

Apricot seed oil is used in cosmetics, particularly as a skin-conditioning agent, and is also available as a specialty oil for food use. It generally contains oleic (58 to 74%) and linoleic acids (20 to 34%) with one study giving values of palmitic 5%, stearic 1%, oleic 66%, and linoleic acid 29%. With its low content of saturated acids it shows excellent cold-weather stability. The fatty acid composition of the phospholipids has been reported and tocopherol levels are given as 570 to 900 mg/kg[13,26–28].

5.2.7 ARGANE (*ARGANIA SPINOSA*)

The argan tree grows mainly in Morocco and also in Israel. Its seeds contain about 50% of an oil rich in oleic (42 to 47%) and linoleic (31 to 37%) acids. Sterols, phenols, tocopherols, and carotenoids are present in the unsaponifiable portion of the oil (~1.0%) and give the oil high oxidative stability. It is used by women in Morocco to protect and soften the skin[13,29].

5.2.8 ARNEBIA (*ARNEBIA GRIFFITHII*)

The seed oil is highly unsaturated. In addition to significant levels of α-linolenic acid (~45%) it also contains γ-linolenic acid (3%) and stearidonic acid (4%) at low levels. Palmitic (7%), oleic (14%), and linoleic (23%) acids are also present[13].

5.2.9 AVOCADO (*PERSEA AMERICANA* AND *P. GRATISSIMA*)

The avocado grows in tropical and subtropical countries between 40°N and 40°S and is available particularly from California, Florida, Israel, New Zealand, and South Africa. Like the palm and the olive, lipid is concentrated in the fruit pulp (4 to 25%) from which it can be pressed. There is very little oil in the seed (2%). The oil is used widely in cosmetic products as it is easily absorbed by the skin and its unsaponifiable material is reported to provide some protection from the sun. It has been claimed that mixtures of avocado and soybean oil may help osteoarthritis. It is also available as a high-oleic specialty oil for food use and is being produced and marketed in New Zealand as a local alternative to olive oil. It is rich in chlorophylls, making it green before processing. It contains 16:0 (10 to 20%), 18:1 (60 to 70%), and 18:2 (10 to 15%) as its major fatty acids. Its unsaponifiable matter (~1%), total sterol which is mainly β-sitosterol, and tocopherol levels (130 to 200 mg/kg, mainly α-tocopherol) have been reported[11–13,30–35].

5.2.10 BABASSU (*ORBIGNYA MARTIANA* AND *O. OLEIFERA*)

This palm, grown in South and Central America, contains a lauric oil in its kernel. Annual production is small and uncertain (100 to 300 kt) but Codex values have been established. In line with other lauric oils it contains 8:0 (6%), 10:0 (4%), 12:0 (45%), 14:0 (17%), 16:0 (9%), 18:0 (3%), 18:1 (13%), and 18:2 (3%) acids. It is used as a skin cosmetic and is being considered in Brazil as a biofuel either as the oil or as its methyl esters, alone or mixed with mineral diesel[11–13].

5.2.11 BAOBAB (*ADANSONIA DIGITATA*)

An African tree whose seeds are eaten raw or roasted by the local population provides an oil of long shelf life which is used in cosmetics and is reported to be edible. The seed oil is reported to contain palmitic (25 to 46%), oleic (21 to 39%), and linoleic acids (12 to 29%) along with minor amounts of stearic and cyclopropene acids. If the oil does contain this last type of acid then it is probably unwise to use it for food and cosmetic purposes. However, one supplier of the oil[19] gives a specification which does not include cyclopropene acids with palmitic (22%), oleic (34%), and linoleic (30%) as the major acids[12,13,36].

5.2.12 BASIL (*OCIMUM* SPP)

Basil seed oil is obtained in a yield of 300 to 400 kg/hectare. The seeds contain 18 to 36% of a highly unsaturated oil with typical levels of palmitic 6 to 11%, oleic 9 to 13%, linoleic 18 to 31%, and linolenic 44 to 65% acids[13,37].

5.2.13 BLACKCURRANT (*RIBES NIGER*)

Blackcurrant seed oil is of interest and of value because it contains γ-linolenic acid (18:3 n-6) and stearidonic acid (18:4 n-3) which are important metabolites of linoleic and linolenic acids, respectively. Blackcurrant seed oil is also a rich source of tocopherols (1700 mg/kg)[38]. More general information about oils containing these acids is included in the entry for borage oil. They are used in cosmetics and also as dietary supplements. Blackcurrant oil is extracted from the seeds, themselves a byproduct of the production of juice from the berries[11–13].

5.2.14 BORAGE (*BORAGO OFFICINALIS*, STARFLOWER)

GLA (Δ 6,9,12-18:3) is an interesting material with beneficial health properties. Claims have been made for its use in the treatment of multiple sclerosis, arthritis, eczema, premenstrual syndrome, and other diseases[39]. It is a biological intermediate in the conversion of freely available linoleic acid to the important but less readily available arachidonic acid. This change is a three-step process involving Δ6-desaturation, elongation, and Δ 5-desaturation of which the first step is considered to be rate determining:

$$9,12\text{-}18{:}2 \rightarrow 6,9,12\text{-}18{:}3 \rightarrow 8,11,14\text{-}20{:}3 \rightarrow 5,8,11,14\text{-}20{:}4$$

GLA is present in a number of seed oils of which three (blackcurrant, borage, evening primrose) are commercially available. The production and use of these oils has been reviewed by Clough[40,41]. Borage oil with just below 25% is the richest source of GLA and there are several reports on ways to isolate the pure acid or to enhance its level in the oil by enzymatic and other methods[42–44]. There are many other plant sources of GLA including hop (*Humulus lupulus*, 3 to 4%), hemp (*Cannabis sativa*, 3 to 6%), redcurrant (*Ribes rubrum*, 4 to 6%), and gooseberry seeds (*Ribes uva crispa*, 10 to 12%)[45]. The value of these GLA-containing oils is such that a genetically modified canola oil rich in GLA (43%) has been developed[46]. The nature of the sterols and alkaloids in borage oil has been described[47,48] and there are general reviews on GLA[49–53].

Table 5.1 lists the component acids of oils containing GLA and stearidonic acid.

5.2.15 BORNEO TALLOW (*SHOREA STENOPTERA*)

This solid fat, also known as illipe butter, contains palmitic (18%), stearic (46%), and oleic (35%) acids. It is one of a group of tropical fats that generally resemble cocoa butter in the proportions of these three acids and therefore have similar triacylglycerol composition and display similar melting

TABLE 5.1

Component Acids of Oils Containing γ-Linolenic Acid (γ-18:3) and Stearidonic Acid (18:4) (Typical Results, wt%)

	16:0	18:0	18:1	18:2	γ-18:3	18:4	Other
Evening primrose	6	2	9	72	10	Tr	1
Borage	10	4	16	38	23	Tr	9[a]
Blackcurrant	7	2	11	47	17	3	13[b]
Echium	6	3	14	13	12	17	35[c]

[a] Including 20:1 (4.5), 22:1 (2.5), and 24:1 (1.5).
[b] Including α-18:3 (13).
[c] Including α-18:3 (33).

behavior. Its major triacylglycerols are POP (7%), POSt (34%), and StOSt (47%). Along with palm oil, kokum butter, sal fat, shea butter, and mango kernel fat, it is one of six permitted tropical fats which can partially replace cocoa butter in chocolate (Table 5.2). An interesting account of the commercial development of illipe, shea, and sal fats has been provided by Campbell[54] and further information is available in articles and books devoted to cocoa butter and to chocolate[55,56].

5.2.16 BRAZIL (*BERTHOLLETIA EXCELSA*)

These nuts come from long-living trees in the Brazilian rain forest. They are rich in oil (66%) and have similar levels of saturated (24%), monounsaturated (35%), and polyunsaturated (36%) acids[11–13].

5.2.17 BUFFALO GOURD (*CUCURBITA FOETIDISSIMA*)

The buffalo gourd is a vine-like plant growing in semiarid regions of the USA, Mexico, Lebanon, and India. The seed contains good-quality oil (32 to 39%) and protein. The oil is very variable in its fatty acid composition thus lending itself to seed breeding. A typical sample contains 16:0 (9%), 18:0 (2%), 18:1 (25%), and 18:2 (62%)[12,13,57].

5.2.18 CALENDULA (*CALENDULA OFFICINALIS*, MARIGOLD)

Interest in this seed oil is based on the fact that it contains significant levels of calendic acid (53 to 62%) along with linoleic acid (28 to 34%). Calendic acid ($\Delta 8t,10t,12c$-18:3) is a conjugated trienoic acid and this makes the oil an effective drying agent. Its alkyl esters can be used as a reactive diluent in alkyd paints replacing volatile organic compounds. Calendula oil is also a rich source of γ-tocopherol (1820 ppm in the crude oil). The crop is being studied particularly in Europe to improve its agronomy. A soybean has been genetically modified to contain 15% of calendic acid[11,13,58,59].

5.2.19 CAMELINA

See gold of pleasure.

TABLE 5.2
Tropical Fats That May Partially Replace Cocoa Butter in Some Countries

		Major triacylglycerols (%)		
Common name	**Botanical name**	**POP**	**POSt**	**StOSt**
Cocoa butter	*Theobroma cacao*	16	38	23
Palm mid fraction	*Elaies guinensis*	57	11	2
Borneo tallow (illipe)	*Shorea stenoptera*	6	37	49
Kokum butter	*Garcinia indica*	1	5	76
Mango kernel stearin	*Mangifer indica*	2	13	55
Sal stearin	*Shorea robusta*	1	10	57
Shea stearin	*Butyrospermim parkii*	1	7	71

Major SOS triacylglycerols are shown as typical values (P = palmitic, O = oleic, St = stearic, S = saturated).

5.2.20   CANDLENUT (ALEURITES MOLUCCANA, LUMBANG, KEMIRI, KUKUI)

This is a tropical tree whose nuts contain a very unsaturated oil: 16:0 (6 to 8%), 18:0 (2 to 3%), 18:1 (17 to 25%), 18:2 (38 to 45%), and 18:3 (25 to 30%). Its iodine value, however, is not as high as that of linseed oil. It is used for cosmetic purposes and has been recommended for the treatment of burns[11–13].

5.2.21   CARAWAY (CARUM CARVII)

This is one of a group of plants whose seed oils contain petroselinic acid (Δ6-18:1). This reaches levels of 35 to 43% in caraway, 66 to 73% in carrot, 31 to 75% in coriander, and ~80% in parsley[11,13]. This isomer of oleic acid has some potential use as a source of lauric and adipic acids, produced by oxidative cleavage. The latter is an important component of many polyamides (nylons) and is usually made from cyclohexane by a reaction that is reported to be environmentally unfriendly. The use of petroselinic acid in food and in skincare products has been described in two patents[60].

5.2.22   CARROT (DAUCUS CAROTA)

See caraway.

5.2.23   CASHEW (ANACARDIUM OCCIDENTALE)

Toschi et al.[61] have given details of the fatty acids, triacylglycerols, sterols, and tocopherols in cashew nut oil. The major fatty acids are palmitic (9 to 14%), stearic (6 to 12%), oleic (57 to 65%), and linoleic (16 to 18%), and the major triacylglycerols are OOO, POO, OOSt, OOL, and POL. The oil contains α- (2 to 6 mg/100 mg of oil), γ- (45 to 83), and δ-tocopherols (3 to 8). The oil is used in cosmetic preparations[11–13,61].

5.2.24   CHAULMOOGRA (HYDNOCARPUS KURZII)

Seed oils of the Flacourtiaceae are unusual in that they contain high levels of cyclopentenyl fatty acids, $C_5H_7(X)COOH$, in which X is a saturated or unsaturated alkyl chain. The most common are hydnocarpic (16:1) and chaulmoogric (18:1) acids. Chaulmoogra oil has been used in folk medicine for the treatment of leprosy but there is no scientific evidence to support this claim[13].

5.2.25   CHERRY (PRUNUS CERASUS)

Obtained by cold pressing and filtering, this oil is sold in the unrefined state for use as a specialty oil for salad dressings, baking, and shallow frying and also in the production of skincare products. Its fatty acid composition is unusual in that in addition to oleic (30 to 40%) and linoleic (40 to 50%) acids it also contains α-eleostearic acid (6 to 12%, Δ9c11t13t-18:3)[11,13,26,27,62]. Some of these potential uses are perhaps surprising for an oil containing a conjugated triene acid. The fatty acid composition of the phospholipids has been reported[28].

5.2.26   CHESTNUT (CASTANEA MOLLISMA)

Chestnut oil contains the usual three major component acids: palmitic (15%), oleic (54%), and linoleic (25%)[11].

5.2.27 CHIA (*SALVIA HISPANICA*)

Chia seeds contain 32 to 38% of a highly unsaturated oil. The fatty acid composition for five samples from Argentina is saturated (9 to 11%), oleic (7 to 8%), linoleic (20 to 21%), and linolenic (52 to 63%) acids[11-13,63].

5.2.28 CHINESE VEGETABLE TALLOW AND STILLINGIA OIL (*SAPIUM SEBIFERUM*, *STILLINGIA SEBIFERA*)

This seed is unusual in that it yields lipids of differing composition from its outer seed coating (Chinese vegetable tallow, 20 to 30%) and from its kernel (stillingia oil, 10 to 17%)[11,13,64]. The former, with ~75% palmitic acid and 20 to 25% oleic acid, is mainly a mixture of PPP (20 to 25%) and POP (~70%) triacylglycerols and is a potential confectionery fat. However, it is difficult to obtain the fat free from stillingia oil (the kernel oil) which is considered to be nutritionally unacceptable. Stillingia oil is quite different in composition, with oleic (13%), linoleic (23%), and linolenic (47%) acids and novel C_8 hydroxy allenic and C_{10} conjugated dienoic acids combined as a C_{18} estolide attached to glycerol at the *sn*-3 position thus:

$$Glyc-OCOCH=C=CH(CH_2)_4OCOCH=CHCH=CH(CH_2)_4CH_3$$

5.2.29 COFFEE (*COFFEA ARABICA* AND *C. ROBUSTA*)

Coffee seed oil consists mainly of palmitic (32 and 34%), oleic (13 and 8%), and linoleic (41 and 44%) acids for the robusta and arabica oils, respectively[11,65].

5.2.30 COHUNE (*ATTALEA COHUNE*)

Cohune seeds contain a lauric type of oil rich in short- and medium-chain acids: 8:0 (7 to 9%), 10:0 (6 to 8%), 12:0 (44 to 48%), 14:0 (16 to 17%), 16:0 (7 to 10%), 18:0 (3 to 4%), 18:1 (8 to 10%), 18:2 (1%)[11-13].

5.2.31 CORIANDER (*CORIANDRUM SATIVUM*)

See caraway. Attempts are being made to develop coriander with its high level of petroselinic acid as an agricultural crop. Efforts to transfer the necessary Δ-6 desaturase to rape would provide an alternative source of petroselinic if successful[8,11-13,66].

5.2.32 CORN GERM

See maize germ.

5.2.33 CRAMBE (*CRAMBE ABYSSINICA*, *C. HISPANICA*)

Present interest in this oil, particularly in North Dakota and in Holland, depends on the fact that it is a potential source of erucic acid (50 to 55%) which finds several industrial uses. This was once the major acid in rapeseed oil but modern varieties of this seed produce a low-erucic oil (such as canola) suitable for food use. High-erucic rapeseed oil is still grown for industrial purposes and attempts are being made to increase the level of this C_{22} acid from around 50% to over 65% and even to 90% by genetic engineering[11-13,64-71].

TABLE 5.3
Fatty Acid Composition (% of Total) of Oils from Selected *Cuphea* spp.

	8:0	10:0	12:0	14:0	16:0	18:1	18:2
C. pulcherrina	94.4	3.3	0.0	0.0	0.6	0.7	1.0
C. koehneana	0.6	91.6	1.5	0.6	1.3	1.1	3.1
C. calophylla	0.1	5.0	85.0	6.8	1.1	0.5	1.3
C. salvadorensis	25.3	0.9	2.8	64.5	5.2	0.5	0.5
C. denticula	0.0	0.0	0.0	0.0	33.0	9.8	53.2

Adapted from Pandey, V., Banerji, R., Dixit, B.S., Singh, M., Shukla, S., and Singh, S.P., *Eur. J. Lipid Sci. Technol.*, 102, 463–466, 2000.

5.2.34 CUPHEA

Cuphea plants furnish seeds with oils which may be rich in C_8, C_{10}, C_{12}, or C_{14} acids (Table 5.3). They generally contain >30% of oil and are expected to produce a commercial crop in the period 2005–2010. Problems of seed dormancy and seed shattering have already been solved. Since markets for lauric oils already exist there should be no difficulty in substituting cuphea oils. More recently it has been reported that cuphea will be used as a commercial source of lauric acid from 2003 onwards. Pandey et al. (73) have described the oil (17 to 29%) from *Cuphea procumbens* containing 88 to 95% of decanoic acid[11–13,72,73].

5.2.35 CUPUACU BUTTER (*THEOBROMA GRANDIFLORA*, ALSO CALLED CUPU ASSU KERNEL OIL)

This is a solid fat containing palmitic (6 to 12%), stearic (22 to 35%), arachidic (10 to 12%), oleic (39 to 47%), and linoleic (3 to 9%) acids. With this fatty acid composition it will be rich in SOS triacylglycerols and have melting properties similar to cocoa butter[12].

5.2.36 DATE SEED (*PHOENIX DACTYLIFERA* L.)

Two cultivars of this species (Deglit Nour and Allig) have been examined. They contain oil (10.2 and 12.7%) which is rich in oleic acid (41 and 48%) along with a range of C_8 to C_{18} acids. These include lauric (18 and 6%), myristic (10 and 3%), palmitic, (11 and 15%), linoleic (12 and 21%), and other minor acids[11,13,74].

5.2.37 DIMORPHOTHECA (*DIMORPHOTHECA PLUVIALIS*)

The seed of *Dimorphotheca pluvialis* is not very rich in oil (13 to 28%, typically about 20%) but it contains an unusual C_{18} hydroxy fatty acid (~60%) with the hydroxyl group adjacent (allylic) to a conjugated diene system[11,12]. Because of this structural feature this acid is chemically unstable and easily dehydrates to a mixture of conjugated 18:3 acids ($\Delta8,10,12$ and $\Delta9,11,13$). Dimorphecolic acid ($CH_3(CH_2)_4CH=CHCH=CHCH(OH)(CH_2)_7COOH$; 9-OH 10*t*12*c*-18:2) provides a convenient source of 9-hydroxy- and 9-oxostearate and of hydroxy epoxy esters.

5.2.38   ECHIUM (*ECHIUM PLANTAGINEUM*)

A number of seeds are known to contain stearidonic acid (Δ-6,9,12,15-18:4) and attempts are being made to grow *Echium plantagineum* as a source of this acid (see Table 5.2). Most of the oils from these seeds contain linoleic and γ-linolenic acids as well as α-linolenic and stearidonic acids[13,44,75,76]. A typical analysis of refined echium oil is given as palmitic (6.1%), stearic (2.3%), oleic (16.2%), linoleic (14.4%), α-linolenic (37.1%), γ-linolenic (8.7%), and stearidonic (13.6%). With almost 60% of the acids having three or four double bonds the oil is highly unsaturated. Another convenient source of stearidonic acid is the readily available blackcurrant seed oil even though it only contains 2.5 to 3.0% of this acid.

Stearidonic acid is the first metabolite in the conversion of α-linolenic to EPA and DHA and the arguments for dietary supplements containing GLA can also be applied to stearidonic acid.

5.2.39   *EUPHORBIA LATHYRIS* (CAPER SPURGE)

Attempts are being made to develop this plant as a commercial crop yielding an oleic-rich oil (80 to 85%). It is a Mediterranean annual with about 50% oil in its seed but problems associated with seed shattering and the presence of a carcinogenic milky sap have still to be overcome through plant breeding[13].

5.2.40   *EUPHORBIA LAGASCAE*

This euphorbia species is one of a limited number of plants that contain significant proportions of epoxy acids in their seed oils (see vernonia oil). With ~64% of vernolic acid (12,13-epoxyoleic) and minor proportions of palmitic (4%), oleic (19%), and linoleic (9%) acids this oil will be rich in triacylglycerols containing two or three vernolic chains, However, not all reports include vernolic acid and there may be some confusion between the species examined[11-13].

5.2.41   EVENING PRIMROSE (*OENOTHERA BIENNIS*, *O. LAMARCKIANA*, AND *O. PARVIFLORA*)

See Borage.

5.2.42   FLAX (*LINUM USITATISSIMUM*)

Linseed oil is one of the most unsaturated vegetable oils because of its high levels of linoleic and linolenic acids (Table 5.4). The names given to both these acids were based on their occurrence in linseed oil. It is oxidized and polymerized very readily and its industrial use in paints, varnishes, inks, and linoleum production is based on these properties.

With recognition of the importance of dietary n-3 acids there is a growing use of the seed and its oil in food products for humans and for animals. The oil used for human consumption is generally obtained by cold pressing or by extraction with supercritical carbon dioxide and is sold under the name flaxseed oil[77-81].

Using chemical mutation, plant breeders in Australia and later in Canada developed varieties of linseed with much reduced levels of linolenic acid and enhanced levels of linoleic acid, which were called linola or solin. These grow in the same temperate zones as rapeseed/canola and are used in linoleic-rich spreads as an alternative to sunflower oil[82].

TABLE 5.4
Fatty Acid Composition of Linseed Oil and Linola

	Saturated	18:1	18:2	18:3
Linseed	10	16	24	50
Linola	10	16	72	2
Solin	10	14	73	2
High-palmitic	31	11	7	44
High-oleic	19	49	22	9

Adapted from Oomah, B.D. and Mazza, G., in *Functional Foods: Biochemical and Processing Aspects*, Mazza, G., Ed., Technomic, Lancaster, PA, 1998, pp. 91–138.

5.2.43 GOLD OF PLEASURE (*CAMELINA SATIVA*, ALSO CALLED FALSE FLAX)

In addition to its interesting fatty acid composition, this plant attracts attention because it grows well with lower inputs of fertilizers and pesticides than traditional crops. The plant can be grown on poorer soils and is reported to show better gross margins than either rape or linseed. The seed yield is in the range 1.5 to 3.0 tons per hectare and the oil content between 36 and 47%. The oil has an unusual fatty acid composition. It contains significant levels of oleic acid (10 to 20%), linoleic acid (16 to 24%), linolenic acid (30 to 40%), and of C_{20} and C_{22} acids, especially eicosenoic (15 to 23%). Another paper reports 30 to 38% oil containing oleic (14 to 20%), linoleic (19 to 24%), linolenic (27 to 35%), eicosenoic (12 to 15%), and other acids (12 to 20%, of which saturated acids comprise about 10%) along with a range of tocols (5 to 22, mean 17 mg/100 g). Detailed tocopherol analysis shows that over 80% of the total is γ-tocopherol and/or β-tocotrienol[83]. Despite its high level of unsaturation, the oil shows reasonable oxidative stability. Attempts are being made to optimize the agronomy. Its use in paints, varnishes, inks, cosmetics, and even as a food oil is being examined and developed. This vegetable oil is unusual in that it contains cholesterol at a level of 188 ppm which is remarkably high for a vegetable source[13,84–90].

5.2.44 GRAPESEED (*VITIS VINIFERA*)

These seeds produce variable levels of an oil (6 to 20%) now available as a gourmet oil and for which Codex values have been reported[11–13]. The oil is rich in linoleic acid (60 to 76%) and also contains palmitic (6 to 8%), stearic (3 to 6%), and oleic (12 to 25%) acids. In common with other oils rich in linoleic it is reported to have a beneficial effect on the skin. Some samples of grapeseed oil have higher PAH (polycyclic aromatic hydrocarbon) levels than is desired and Moret et al. have described the effect of processing on the PAH content of the oil[11–13,91].

5.2.45 CHILEAN HAZELNUT (*GEVUINA AVELLANA*)

Chilean hazelnuts are native to Argentina and Chile and attempts are being used to produce a commercial crop in Chile and in New Zealand. The fatty acid composition is unusual in that the unsaturated centers occupy unconventional positions and in the range of chain lengths (C_{16} to C_{24}). The oil content of the kernels is 40 to 48% and contains a significant quantity of α-tocotrienol (130 mg/kg). The oil is rich in monounsaturated acids and is often compared with macadamia oil but gevuina seeds do not have the hard shell of macadamia nuts. The fatty acid composition is given below. The double bond positions are unusual and unrelated to each other except that three are n-5 olefininc groups. These unexpected results require confirmation[13,92].

Saturated	Unsaturated	
C_{16}	1.9	22.7 ($\Delta 11$)
C_{18}	0.5	39.4 ($\Delta 9$), 6.2 ($\Delta 12$), 5.6 ($\Delta 9,12$)
C_{20}	1.4	1.4 ($\Delta 11$), 6.6 ($\Delta 15$)
C_{22}	2.2	7.9 ($\Delta 17$), 1.6 ($\Delta 19$)
C_{24}	0.5	

5.2.46 HAZELNUT (*CORYLUS AVELLANA*, FILBERTS)

Hazelnut oil is rich in oleic acid (65 to 75% or even higher) and also contains linoleic acid (16 to 22%). The level of saturated acids is low. Hazelnuts are produced mainly in Turkey and also in New Zealand. The nuts produce 55 to 63% of oil with saturated acids (6 to 8%), monoene acids (74 to 80%), and linoleic acid (6 to 8%). This fatty acid composition is very similar to that of olive oil and hazelnut oil is sometimes added as an adulterant of the more costly olive oil. There have been several reports on methods of detecting this adulteration, many of them related to the presence of filbertone ((*E*)-5-methylhept-2-en-4-one; $H_3CH_2CH(CH_3)COCH=CHCH_3$) in hazelnut oil. Hazelnuts appear in a short list of nuts for which a health claim may be made[93]. Other details can be found in references[92,94–101].

5.2.47 HEMP (*CANNABIS SATIVA*, MARIJUANA)

Hemp seed oil (25 to 34% of whole seed, 42 to 47% of dehulled seed) has an interesting fatty acid composition. One report gives the following values: palmitic (4 to 9%), stearic (2 to 4%), oleic (8 to 15%), linoleic (53 to 60%), α-linolenic (15 to 25%), γ-linolenic (0 to 5%), and stearidonic (0 to 3%) acids. The oil is a rich source of tocopherols — virtually entirely the γ-compound — at 1500 mg/kg and is used in cosmetic formulations. Evidence from a study in Finland indicates that dietary consumption of hemp seed oil leads to increased levels of GLA in blood serum. The growing of hemp is banned in the U.S. and therefore supplies of hemp seed oil, if any, must be imported. Further details are available in references[11–13,38,102–104].

5.2.48 HONESTY (*LUNARIA ANNUA*)

This seed oil contains significant levels of erucic (22:1, 41%) and nervonic (24:1, 22%) acids and is being studied as a new crop because it is a good source of the latter acid which may be useful in the treatment of demyelinating disease[11,13,105].

5.2.49 HYPTIS (*HYPTIS* SPP.)

Hagemann et al. have reported the fatty acid composition of oils from six different hyptis species. Five contain high levels of linolenic (51 to 64%) and linoleic (22 to 31%) acids but the oil from *Hyptis suaveolens* contains less than 1% of linolenic acid with 77 to 80% of linoleic acid and palmitic (8 to 15%) and oleic (6 to 8%) acids[13,106].

5.2.50 ILLIPE

See Borneo Tallow.

5.2.51 JOJOBA (*SIMMONDSIA CHINENSIS*)

Jojoba oil is a valuable source of C_{20} and C_{22} compounds. The oil has already been developed as a marketable product but only in limited supply. It is produced by a drought-resistant plant that resists desert heat. It takes 5 to 7 years to first harvest, 10 to 17 years to full yield, and has a life span of about 100 years. Jojoba plants are being grown in southwestern U.S., Mexico, Latin America, Israel, South Africa, and Australia. Yields are reported to be about 2.5 tons of oil per hectare.

Jojaba oil contains only traces of triacylglycerols and is predominantly a mixture of wax esters based mainly on C_{20} and C_{22} monounsaturated acids and alcohols. The oil contains esters with 40, 42, and 44 carbon atoms with two isolated double bonds, one in the acyl chain and one in the alkyl chain. The oil serves as a replacement for sperm whale oil which is proscribed in many countries because the sperm whale is an endangered species. As a high-priced commodity, jojoba oil is used in cosmetics. As it gets cheaper through increasing supplies it will be used as a superior lubricant and also as a biofuel. The oil is fairly pure as extracted, has a light color, and is resistant to oxidation because its two double bonds are well separated. The oil can be chemically modified by reaction of the double bonds such as hydrogenation, stereomutation, epoxidation, and sulfochlorination[11–13,107].

5.2.52   KAPOK (*BOMBAX MALABARICUM, CEIBA PENTANDRA*)

This name is applied to a number of tropical trees of the bombax family. The oil is a byproduct of kapok fiber production. Its major component acids are palmitic (22%), oleic (21%), and linoleic (37%) but it also contains about 13% of cyclopropene acids (malvalic and sterculic) which make it unsuitable for food use[11–13].

5.2.53   KARANJA (*PONGAMIA GLABRA*)

Karanja seed oil from India is rich in monounsaturated acids (C_{18} and C_{20}). It contains the following acids: palmitic (4 to 8%) stearic (2 to 9%), arachidic (2 to 5%), oleic (44 to 71%), linoleic (2 to 18%), and eicosenoic (9 to 12%)[6,13].

5.2.54   KIWI (*ACTINIDIA CHINENSIS, A. DELICIOSA*)

The seed of this fruit furnishes a linolenic-rich oil (~63%) with lower levels of linoleic (~16), oleic (~13), and saturated (~8%) acids[13].

5.2.55   KOKUM (*GARCINIA INDICA*)

Both kokum and mahua fats are rich in saturated and oleic acids and contain high levels of SOS triacylglycerols (Table 5.5). They can be fractioned separately or as blends of the two oils to produce stearins which can be used as cocoa butter extenders. Kokum butter is one of the six permitted fats (palm oil, illipe butter, kokum butter, sal fat, shea butter, and mango kernel fat) that can partially replace cocoa butter in chocolate in some countries. Kokum butter is a stearic acid-rich fat and Bhattacharyya[6] has reported an approximate composition of palmitic (2.5 to 5.3%), stearic (52.0 to 56.4%), oleic (39.4 to 41.5%), and linoleic (trace to 1.7%) acids[11–13,108].

5.2.56   KUKUI (*ALEURITES MOLUCCANA*)

The oil from this nut is reported to contain palmitic (6 to 8%), stearic (2 to 5%), oleic (24 to 29%), linoleic (33 to 39%), and linolenic (21 to 30%) acids[13].

5.2.57   KUSUM (*SCHLEICHERA TRIJUGA*)

Kusum seed oil is an unusual oleic-rich oil (40 to 67%) in that it also contains significant quantities of arachidic acid (20:0, 20 to 31%) and lower levels of palmitic (5 to 9%), stearic (2 to 6%), and linoleic (2 to 7%) acids[6,11–13].

TABLE 5.5
Fatty Acids and Triacylglycerols of Kokum and Madhua Fats

Oil source	Fatty acids				Triacylglycerols[a]		
	16:0	**18:0**	**18:1**	**18:2**	**StOSt**	**POSt**	**POP**
Kokum	2.0	49.0	49.0	0	72.3	7.4	0.5
Mahua	23.5	20.0	39.0	16.7	10.6	22.2	18.9
Stearin[b]	15.7	37.8	35.5	11.1	46.2	15.0	9.7

[a] P = palmitic, O = oleic, St = stearic, S = saturated.
[b] Obtained by dry fractionation of a 1:1 mixture of the two oils.

5.2.58 LESQUERELLA (*LESQUERELLA FENDLERI*)

The only oil of commercial significance with a hydroxy acid is castor oil, but among the new crops being seriously developed are two containing hydroxy acids. Lesquerella oils have some resemblance to castor oil but *Dimorphotheca pluvialis* seed oil contains a different kind of hydroxy acid. Plants of the lesquerella species are characterized by the presence of the C_{20} bis-homolog of ricinoleic acid — lesquerolic acid — sometimes accompanied by other acids of the same type at lower levels: ricinoleic acid, 12-OH Δ9-18:1; densipolic acid, 12-OH Δ 9,15-18:2; lesquerolic acid, 14-OH Δ 11-20:1; auricolic acid, 14-OH Δ 11,17-20:2.

A typical analysis of *L. fendleri* seed oil showed the presence of palmitic (1%), stearic (2%), oleic (15%), linoleic (7%), linolen (14%), lesquerolic (54%), and auricolic (4%) acids. Since lesquerolic acid is the C_{20} homolog of ricinoleic with the same β-hydroxy alkene unit, it undergoes similar chemical reactions but produces (some) different products. For example, pyrolysis should give heptanal and 13-tridecenoic acid (in place of 11-undecenoic acid from castor oil). This could be converted to 13-aminotridecanoic acid, the monomer required to make nylon-13. Similarly, alkali fusion will give 2-octanol and dodecadienoic acid in place of decadienoic (sebacic) acid from castor oil. This C_{12} dibasic acid is already available from petrochemical products and has a number of applications. The free hydroxyl group in castor and lesquerella oils can be esterified with fatty acids such as oleic acid to give an estolide.

The status and potential of lesquerella as an industrial crop was reviewed in 1997. Lesquerella plants can be grown in saline soils[12,13,109,110].

5.2.59 LINSEED

See Flax.

5.2.60 MACADAMIA (*MACADAMIA INTEGRIFOLIA, M. TETRAPHYLLA*)

The nuts are used as a snack food and it has been claimed that their consumption reduces total and LDL cholesterol[111]. They are rich in oil (60 to 70%) that is used in cosmetics and is available as a gourmet oil. It is characterized by its high level of monoene acids (~80%): 16:1 16 to 23%, 18:1 55 to 65%, and 20:1 1 to 3%. Its high level of monoene acids makes it good for skin care but low levels of tocopherols limit its oxidative stability. Gevuina oil has similar composition and is sometimes used in place of macadamia oil. Certain health benefits have been claimed for hexadecenoic acid with the most convenient sources being sea buckthorn oil and macadamia oil[112–114].

5.2.61 MAHUA (*MADHUCA LATIFOLIA*)

See Kokum fat and mango kernel fat. Bhattacharyya gives the fatty acid composition of mahua fat as palmitic (22.4 to 37.0 %), stearic (18.6 to 24.0 %), and oleic (37.1 to 45.5 %) acids[6,13].

5.2.62 MAIZE GERM (*ZEA MAIS*, CORN GERM)

Maize oil is obtained from seeds that contain only 3 to 5% oil. The crop is grown as a source of starch and a byproduct in the recovery of the starch is the maize germ which contains around 30% of oil. This is the source of a product generally called corn oil but, more correctly, is corn germ oil. The oil is mainly glycerol esters of palmitic (~11%), oleic (~25%), and linoleic (~60%) acids[11,13,115].

5.2.63 MANGO (*MANGIFERA INDICA*)

Mango is consumed in large quantities as a fruit. The kernel contains 7 to 12% of lipid with palmitic (3 to 18%), stearic (24 to 57%), oleic (34 to 56%), and linoleic (1 to 13%) acids. In a typical case these values were 10.3, 35.4, 49.3, and 4.9%, respectively. It is fractionated to give a lower melting olein with excellent emollient properties. The accompanying stearin can serve as a cocoa butter equivalent (POP 1%, POSt 12%, StOSt 56%) and as a component of a *trans*-free bakery shortening along with fractioned mahua fat. It is one of six permitted fats (palm oil, illipe butter, kokum butter, sal fat, shea butter, and mango kernel fat) that can partially replace cocoa butter in chocolate[6,11–13,116,117].

5.2.64 MANKETTI (*RICINODENDRON RAUTTANENNI*)

This oil, used widely in Namibia as an emollient, contains eleostearic and other conjugated octadecatrienoates (total 6 to 28%) as well as oleic (18 to 24%), linoleic (39 to 47%), and a range of minor acids[13].

5.2.65 MARIGOLD (*CALENDULA OFFICINALIS*)

See Calendula.

5.2.66 MARULA (*SCLEROCARYA BIRREA*)

Marula oil is oleic rich (typically 75%) and also contains palmitic (11%), stearic (7%), and linoleic (5%) acids[13].

5.2.67 MEADOWFOAM (*LIMNANTHES ALBA*)

This oil is unusual in that over 95% of its component acids are C_{20} or C_{22} compounds and include Δ5-20:1 (63 to 67%), Δ5-22:1 (2 to 4%), Δ13-22:1 (16 to 18%), and Δ5,13-22:2 (5 to 9%). It is being grown in the U.S. and its potential uses are being thoroughly examined. Winter cultivars now being developed are expected to improve the suitability of the crop to conditions in northern Europe[11–13,118,119]. Potential uses of this oil include cosmetic applications, production of dimer acid, as a lubricant, and via a wide range of novel derivatives based on reaction at the Δ5 double bond[120,121].

5.2.68 MELON (*CITRULLUS COLOCYTHIS* AND *C. VULGARIS*)

This seed oil has been examined in terms of its fatty acids and phospholipids by Akoh and Nwosu. They report the major fatty acids in the total lipids from two samples to be palmitic (11 and 12%), stearic (7 and 11%), oleic (10 and 14%), and linoleic (71 and 63%) acids. In a later paper three cutivars (Hy-mark, Honey Dew, and Orange Flesh) are described in terms of lipid content (25.7 to 28.6%) and fatty acid composition[11,13,122,123].

5.2.69 MORINGA (MORINGA OLEIFERA, M. STENOPETALA)

Dried moringa seeds contain about 35% of an oil rich in oleic acid (palmitic 12.3%, stearic 4.1%, oleic 76.8%, linoleic 2.4%, linolenic 1.6%, and eicosenoic 2.1%). The oil has high oxidative stability resulting in part from its fatty acid composition (low levels of polyunsaturated fatty acids) and from the presence of the flavone myricetin which is a powerful antioxidant. In a recent study cold-pressed oil (36%) is compared with that extracted by chloroform/methanol (45%) and the composition of the fatty acids and sterols is reported. The oils contain 20:0, 20:1, 22:0, 22:1, and 26:0 at around 11% (total)[6,11,124,125]. One specification[19] indicates the presence of palmitoleic acid (8%) in this oil.

5.2.70 MOWRAH (MADHUCA LATIFOLIA, M. LONGIFOLIA, M. INDICA)

This comes mainly from India where the fat is used for edible and industrial purposes. The nuts contain 46% of oil with variable levels of palmitic (15 to 32%), stearic (16 to 26%), oleic (32 to 45%), and linoleic (14 to 18%) acids. The mowrah fat examined by De et al. with levels of 27, 9, 39, and 24% for these four acids differs somewhat, particularly in the levels of stearic and linoleic acids[6,11,13,126].

5.2.71 MURUMURU BUTTER (ASTROCARYUM MURUMURU)

This solid fat is a type of lauric oil with 89% saturated acids (mainly lauric 42% and myristic 37%) and 11% of unsaturated acids (almost entirely oleic)[11].

5.2.72 MUSTARD (BRASSICA ALBA, B. HIRTA, B. NIGRA, B. JUNCEA, B. CARINATA)

Mustard seeds contain 24 to 40% of oil characterized by the presence of erucic acid. Typical values are oleic 23%, linoleic 9%, linolenic 10%, eicosenoic 8%, and erucic 43% acids. The plant is grown extensively in India. Canadian investigators have bred *Brassica juncea* (oriental mustard) from an Australian line with low erucic acid and low glucosinolate so that it has a fatty acid composition (palmitic 3%, stearic 2%, oleic 64%, linoleic 17%, and linolenic 10% acids) similar to that of canola oil from *B. napus* and *B. rapa*. This makes it possible to expand the canola growing area of western Canada[11–13,127].

5.2.73 NEEM (AZADIRACHTA INDICA)

This interesting seed oil contains chemicals used to control 200 species of insects. For example, the oil prevents larval insects from maturing. Bhattacharyya has reported a fatty acid composition of palmitic 16 to 19%, stearic 15 to 18%, arachidic 1 to 3%, oleic 46 to 57%, and linoleic 9 to 14% acids[6,11–13,128,129].

5.2.74 NGALI NUT (CANARIUM SPP.)

Analyses of five different *Canarium* spp. have been reported[13] (Table 5.6). They contain the same major component acids but at differing levels.

5.2.75 NIGELLA (NIGELLA SATIVA, BLACK CUMIN)

Typically nigella oil contains palmitic (10%), oleic (35%), and linoleic (45%) acids. Related species (*N. arvensis* and *N. damascena*) give similar oils with less oleic and more linoleic acid. The

TABLE 5.6
Fatty Acid Composition of the Seed Oils from Five *Canarium* spp.[13]

	Palmitic	Stearic	Oleic	Linoleic
C. commune	30	10	40	19
C. ovatum	33–38	2–9	44–60	0–10
C. patentinervium	33	10	27	28
C. schweinfurthii	1	—	84	15
C. vulgare	29	12	49	10

presence of low levels of 20:1 ($\Delta 11c$, 0.5 to 1.0%) and higher levels of 20:2 ($\Delta 11c14c$, 3.6 to 4.7%) in all these oils may be of taxonomic significance. In one analysis the oil contained the following major triacylglycerols: LLL 25%, LLO 20%, LLP 17%, LOP 13%, and LOO 10%, reflecting the high level of linoleic acid. The seeds appear to contain an active lipase and the oil quickly develops high levels of free acid. The oil is also a good source of thymoquinone and is reported to assist in the treatment of prostate problems[13,130–134].

5.2.76   NIGER (*GUIZOTIA ABYSSINICA*)

This oil comes mainly from Ethiopia. The seeds contain 29 to 39% of oil rich in linoleic acid (71 to 79%) along with palmitic, stearic, and oleic acids, each at levels of 6 to 11%. The major triacylglycerols and sterols (particularly sitosterol, campesterol, stigmasterol, and $\Delta 5$-avenasterol) have been identified. The oil is rich in α-tocopherol (94 to 96% of total values ranging from 657 to 853 mg/kg) and is therefore a good source of vitamin E. It is used for both edible and industrial purposes[12,13,135,136].

5.2.77   NUTMEG (*MYRISTICA MALABARICA* AND OTHER *MYRISTICA* SPP.)

Not surprisingly, considering its botanical name, seeds of the *Myristica* spp. are rich in myristic acid (~40%). Higher levels (60 to 72%) were quoted in earlier work and one report gives 12:0 (3 to 6%), 14:0 (76 to 83%), 16;0 (4 to 10%), 18:1 (5 to 11%), and 18:2 (0 to 2%)[11–13,137].

5.2.78   OATS (*AVENA SATIVA*)

This grain seed contains 4 to 8% of lipid, though somewhat more in certain strains. The major component acids are palmitic (13 to 28%), oleic (19 to 53%), linoleic (24 to 53%), and linolenic (1 to 5%). The oil contains triacylglycerols (51%), di- and mono-acylglycerols (7%), free acids (7%), sterols and sterol ester (each 3%), glycolipids (8%), and phospholipids (20%). The special features of this oil are utilized in various ways. It is reported to show cholesterolemic and antithrombotic activity, is present in "Olibra" used as an appetite suppressant, is used in cosmetics by virtue of its glycolipids, and can be used in baking at levels as low as 0.5% to increase loaf volume. Oat lipids are the subject of several recent reviews[11,12,138–142].

5.2.79   OITICICA (*LICANIA RIGIDA*)

The kernel oil obtained from this Brazilian tree is characterized by its high level (~78%) of licanic acid (4-oxo-9c11t13t-octadecatrienoic acid), a keto derivative of the more familiar eleostearic acid. The oil shows drying properties but does not dry as quickly as tung oil[11–13].

5.2.80   OYSTER NUT (*TELFAIRIA PEDATA*, JICONGER NUT, KOEME NUT)

The kernel contains ~60% of oil. A similar oil in *T. occidentalis* is reported to contain palmitic (16%), stearic (13%), oleic (30%), and linoleic (40%) acids[13,143].

5.2.81   PARSLEY (*PETROSELINIUM SATIVUM*)

See caraway.

5.2.82   PASSIONFRUIT (*PASSIFLORA EDULIS*)

This popular fruit contains about 20% of oil in its seed and is available as a gourmet oil for use in specialty foods and salad dressings. It is linoleic rich (65 to 75%), but also contains palmitic (8 to 12%) and oleic (13 to 20%) acids. Its high level of linoleic acid makes the oil good for skincare[11,13,144].

5.2.83   PEACH (*PRUNUS PERSICA*)

Peach kernels contain 44% of an oleic-rich oil (67%) along with palmitic (9%) and linoleic (21%) acids[11–13,28].

5.2.84   PECAN (*CARYA PECAN, C. ILLINOENSIS*)

Pecan oil contains palmitic (5 to 11%), oleic (49 to 69%), and linoleic (19 to 40%) acids. It is reported to lower blood cholesterol and the USFDA allows such a claim to be made in respect of pecan nuts[11,12,95].

5.2.85   PERILLA (*PERILLA FRUTESCENS*)

Perilla is a linolenic-rich oil (57 to 64%) used as a drying oil. It also contains oleic (13 to 15%) and linoleic (14 to 18%) acids and comes mainly from Korea or India. Recent descriptions of this oil come from these two countries[11–13,145–147].

5.2.86   PHULWARA BUTTER (*MADHUCA BUTYRACEAE* OR *BASSIA BUTYRACEA*)

This solid fat is exceptionally rich in palmitic acid and would be expected to contain high levels of POP among its triacylglycerols. Bhattycharyya reports palmitic 60.8%, stearic 3.2%, oleic 30.9%, and linoleic 4.9% acids[6,11,13].

5.2.87   PISTACHIO (*PISTACHIO VERA*)

Pistachio nuts, produced mainly in Iran, are widely consumed as shelled nuts. They contain about 60% of an oil used for cooking and frying. Mean fatty acid values for five varieties are given as palmitic (10%), stearic (3%), oleic (69%), and linoleic (17%). Triacylglycerol composition has been suggested as a method of determining the country of origin of pistachio nuts. The USFDA allows the following claim with respect to pistachio nuts: "Scientific evidence suggest but does not prove, that eating 1.5 ounces (40 g) a day of most nuts as part of a diet low in saturated fat and cholesterol may reduce the risk of heart disease"[11,12,95,148–150].

5.2.88   PLUM (*PRUNUS DOMESTICA*)

The kernels contain oil (41%) that is rich in oleic acid (71%) along with significant levels of linoleic acid (16%)[11,28].

5.2.89   POPPY (*PAPAVER SOMNIFERIUM*)

Opium is obtained from unripe capsules and from straw of the poppy plant but the narcotic is not present in the seed which is much used for birdseed. It contains 40 to 70% of a semidrying oil used by artists and also as an edible oil. Rich in linoleic acid (72%), it also contains palmitic (10%), oleic (11%), and linolenic (5%) acids[11–13,151].

5.2.90   PUMPKIN (*CUCURBITA PEPO*)

Pumpkin seed oil is a linoleic-rich oil containing palmitic (4 to 14%), stearic (5 to 6%), oleic (21 to 47%), and linoleic (35 to 59%) acids. It has attracted attention because of its reported potential to cure prostate disease. A recent paper reports lipid classes, fatty acids, and triacylglycerols in three pumpkin cultivars emphasizing the marked differences between these[11–13,152–155].

5.2.91   PURSLANE (*PORTULACA OLERACEA*)

The plant (leaves, stem, and whole plant) is reported to be the richest vegetable source of n-3 acids including low levels of the 20:5, 22:5, and 22:6 members. This is a surprising and unlikely result and needs to be confirmed[156]. These have not been identified in the seed oil which contains palmitic (15%), stearic (4%), oleic (18%), linoleic (33%), and linolenic (26%) acids.

5.2.92   RASPBERRY (*RUBUS IDAEUS*)

Raspberry seed oil is highly unsaturated with palmitic (3%), oleic (9%), linoleic (55%) and linolenic (33%) acids. It is reported to be a rich source of tocopherols (3300 mg/l of oil divided between the α- (500), γ- (2400), and δ-compounds (400))[38].

5.2.93   RED PALM OIL

Crude palm oil contains 400 to 1000 ppm of carotenes of which over 90% are the α- and β-isomers. These levels fall to virtually zero after physical refining, but by appropriate modification of the refining process it is possible to obtain a product containing ~540 ppm of carotenes. This is marketed as red palm oil with added nutritional value because of the presence of the carotenes which serve as pro-vitamin A[157,158].

5.2.94   RICEBRAN OIL (*ORYZA SATIVA*)

Rice (*Oryza sativa*) is an important cereal with an annual production of above 500 to 800 MMT. To produce white rice the hull is removed and the bran layer is abraded giving 8 to 10% of the rice grain. The bran contains the testa, cross cells, aleurone cells, part of the aleurone layer, and the germ and includes almost all the oil of the rice coreopsis. Gopala Krishna[162] considers that there is a potential for over 5 MMT of ricebran oil per annum but present production is only about 0.7 MMT and not all this is of food grade. India (0.50 MMT), China (0.12 MMT), and Japan (0.08 MMT) are the major countries producing ricebran oil.

   Lipases liberated from the testa and the cross cells promote rapid hydrolysis of the oil and therefore it should be extracted within hours of milling. Attempts have been made to upgrade oil with 30% free acid by reaction with glycerol and the enzyme Lipozyme (*Mucor miehei* lipase) followed by neutralization. The major acids in ricebran oil are palmitic (12 to 18%, typically 16%) oleic (40 to 50%, typically 42%), and linoleic (29 to 42%, typically 37%). The oil contains phospholipids (~5%), a wax which may be removed for industrial use, and unsaponifiable matter including sterols, 4-methylsterols, triterpene alcohols, tocopherols (~860 ppm), and squalene, among others.

Kochar[163] reports that the major tocols in ricebran oil (total 860 mg/kg of oil) are α-tocopherol (292), γ-tocopherol (144), α-tocotrienol (71), and γ-tocotienol (319). These are mean values from 22 samples of oil.

Refined ricebran oil is an excellent salad oil and frying oil with high oxidative stability resulting from its high level of tocopherols and from the presence of the oryzanols — ferulic acid esters of sterols and triterpene alcohols (ferulic acid is 3-(3'-methoxy-4'-hydroxyphenyl)propenoic acid). Rice bran oil also finds several nonfood uses.

Rice bran oil is reported to lower serum cholesterol by reducing LDL and VLDL without changing the level of HDL. This effect seems not to be related to fatty acid or triacylglycerol composition but to the unsaponifiable fraction (4.2%) and probably to the oryzanols (1.5 to 2.0% of the oil). These can be isolated in concentrated form from ricebran oil soapstock but have not yet been accepted for food use[11–13,159–167].

5.2.95   ROSE HIP (*ROSA CANINA*, HIPBERRY)

Rose hips are best known for the high level of vitamin C in their fleshy parts but the seeds contain a highly unsaturated oil reported to contain only 5.4% of saturated acids and 8.4% of oleic acid with the balance being linoleic (54%) and linolenic (32%) acids[11]. The oil is used in cosmetics.

5.2.96   SACHA INCHI (*PILKENETIA VOLUBILIS*, INCA PEANUT)

This highly unsaturated oil from plants in the tropical jungles of America is rich in linoleic (34 to 39%) and linolenic (47 to 51%) acids with only low levels of oleic (6 to 9%) and saturated (5 to 7%) acids. This makes it comparable but not identical to linseed oil[168].

5.2.97   SAFFLOWER (*CARTHAMUS TINCTORIUS*)

After dehulling, safflower seeds contain about 40% of an oil normally rich in linoleic acid (~75%) along with oleic acid (14%) and saturated acids (10%) and favored as a starting for the preparation of conjugated 18:2 acids. High-oleic varieties (~74%) have also been developed (see Chapter 19)[11–13,169].

5.2.98   SAL (*SHOREA ROBUSTA*)

This tree, which grows in northern India, is felled for timber. Its seed oil is rich in stearic acid and can be used as a cocoa butter equivalent (CBE). The major acids are palmitic (2 to 8%), stearic (35 to 48%), oleic (35 to 42%), linoleic (2 to 3%), and arachidic (6 to 11%) acids. Its major triacylglycerols are of the SUS type required of a CBE. Sal olein is an excellent emollient and sal stearin, with POP 1%, POSt 13%, and StOSt 60%, is a superior CBE . It is one of the six permitted fats (palm oil, illipe butter, kokum butter, sal fat, shea butter, and mango kernel fat) that can partially replace cocoa butter in chocolate[11–13,55,170–173].

5.2.99   *SALICORNIA BIGELOVII*

This annual dicotyledon is of interest because it is a halophyte growing in areas that support only limited vegetation. When growing it can be irrigated with salt water. It produces seed at a level of 0.5 to 1.0 t/acre which furnishes oil (25 to 30%) and meal with 40% protein. The oil is rich in linoleic acid (74%) and also contains oleic (12%), palmitic (8%), and lower levels of stearic and α-linolenic acids. Its tocopherols (720 ppm) are mainly the α- and γ-compounds and its sterol esters (4%) are mainly stigmasterol, β-sitosterol, and spinasterol[174].

TABLE 5.7
Fatty Acid Composition (wt% Mean of 21 Samples) of Seed Oil and Berry Oil from Sea Buckthorn[178]

	16:0	9-16:1	18:0	9-18:1	11-18:1	18:2	18:3
Seed oil	7.7	—	2.5	18.5	2.3	39.7	29.3
Berry oil	23.1	23.0	1.4	17.8	7.0	17.4	10.4

5.2.100 SEA BUCKTHORN (*HIPPOPHAE RHAMNOIDES*)

This is a hardy bush growing wild in several parts of Asia and Europe and now cultivated in Europe, North America, and Japan. It is resistant to cold, drought, salt, and alkali. Different oils are available in the seeds and in the pulp/peel but these are not always kept separate. The seed oil is rich in oleic, linoleic, and linolenic acids but the berry oil contains significant levels of palmitoleic acid (see Table 5.7). Several health benefits are claimed for this oil which is now available in encapsulated form and is being incorporated into functional foods. The oil is rich in sterols, carotenoids (especially β-carotene), and tocopherols (2470 mg/l of oil). Sea buckthorn pulp is rich in α-tocopherol (2000 mg/l) while the seed oil is rich in α- (1000 mg/l) and γ-tocopherols (1000 mg/l)[38,175-181].

It has been claimed that the annual demand in North America could be about 10 tons of oil from 1500 tons of fruit and that this could be a commercial crop in Canada supplying American needs and also exporting to Europe[179].

5.2.101 SESAME (*SESAMUM INDICUM*)

This oil has an annual production of just under 1 MMT. It is grown mainly in India and China but also in Myanmar, Sudan, Egypt, and Mexico. The seed contains 40 to 60% oil with almost equal levels of oleic (35 to 54%, average 40%) and linoleic (39 to 59%, average 46%) acids along with palmitic (8 to 10%) and stearic (5 to 6%) acids. The oil has high oxidative stability because it contains sesamin and sesaminol and is favored as a component of frying oils. This antioxidant is converted to a more powerful antioxidant (sesamol) when heated[11-13,182-184].

5.2.102 SHEA (*BUTYROSPERMUM PARKII*, SHEA BUTTER, KARITE BUTTER)

This fat comes from trees grown mainly in West Africa and contains an unusually high level of unsaponifiable material (~11%) including polyisoprene hydrocarbons. It is rich in stearic acid but its fatty acid composition varies with its geographical source. It contains palmitic (4 to 8%), stearic (23 to 58%), oleic (33 to 68%), and linoleic (4 to 8%) acids. It can be fractionated to give a stearin (POP 1%, POSt 8%, and StOSt 68%) which can be used as a CBE. It is one of the six permitted fats (palm oil, illipe butter, kokum butter, sal fat, shea butter, and mango kernel fat) which, in some countries at least, can partially replace cocoa butter in chocolate[11-13,55,170-172].

5.2.103 SHIKONOIN SEED (*LITHOSPERMUM SPP.*)

Several plants of the lithospermum genus have been examined. They belong to the Boraginaceae family and therefore it is not surprising that many of them contain γ-linolenic and stearidonic acid (see borage)[13,185].

5.2.104   SISYMBRIUM IRIO

Sisymbrium irio is one of several sisymbrium oils reported by Ucciani[13]. These are brassica and therefore it is not surprising that many contain long-chain monoene acids. *Sisymbrium irio* seed oil, for example, is reported to contain 20:0 (0 to 8%) and 22:1 (erucic acid, 6 to 10%) in addition to palmitic (14 to 16%), oleic (17 to 19%), linoleic (13 to 16%), and linolenic (33 to 37%) acids.

5.2.105   TAMANU (*CALOPHYLLUM TACANAHACA*)

Tamanu oil is obtained from nuts which grow on the ati tree and is important in Polynesian culture. It has four major component acids (palmitic 12%, stearic 13%, oleic 34%, and linoleic 38%) and is claimed to be helpful in the treatment of many skin ailments[186].

5.2.106   TEASEED (*THEA SINENSIS, T. SASANGUA*)

The seeds contain 56 to 70% oil with palmitic (5 to 17%), oleic (58 to 87%), and linoleic (7 to 17%) acids as the major fatty acids. Myristic, stearic, eicosenoic, and docosenoic acids may also be present[11,12].

5.2.107   TOBACCO (*NICOTIANA TOBACUM*)

Tobacco seeds contain an oil rich in linoleic acid (> 70%) but with virtually no linolenic acid. After refining it can be used for edible purposes or as a non-yellowing drying oil. In one sample of the oil that was analyzed the major triacylglycerols were LLL (38%), LLO (24%), and LLS (20%)[11–13,15].

5.2.108   TOMATO SEED (*LYCOPERSICUM ESCULENTUM*)

Tomato seed oil is a linoleic-rich vegetable oil with an unusually high level of cholesterol. The fatty acid composition is reported to be palmitic (12 to 16%), oleic (16 to 25%), and linoleic (50 to 60%) with ~2% of linolenic acid[11–13].

5.2.109   TUNG (*ALEURITES FORDII*)

This oil comes mainly from China which explains its alternative name of China wood oil. It is characterized by the presence of a conjugated triene acid (α-eleostearic, Δ9c11t13t-18:3, ~69%). The oil dries more quickly than linseed with its nonconjugated triene acid but oxidized tung oil contains less oxygen (5%) than does oxidized linseed oil (12%). Put another way, tung oil hardens at a lower level of oxygen uptake than linseed oil. This oil is exported mainly from China (30,000 to 40,000 tons) and imported mainly by Japan, South Korea, Taiwan, and the U.S.[11–13]

5.2.110   UCUHUBA (*VIROLA SURINAMENSIS*)

This tree grows in South America. Its seeds provide one of the few oils rich in myristic acid (69%) along with lauric (13%), palmitic (7%), and oleic and linoleic (together 10%) acids. These values are reflected in the triacylglycerol composition: MMM 43%, MML 31%, and LMP 10%, where L, M, and P represent lauric, myristic, and palmitic acid chains[11–13].

5.2.111   VERNONIA OILS

A small number of seed oils contain epoxy acids and sometimes these unusual acids attain high levels. Such oils show a wide range of reactions producing compounds of unusual structure with properties of potential value. The most common acid of this type is vernolic (12,13-epoxyoleic) acid which is one of the two monoepoxides of linoleic acid. First identified in *Vernonia anthelmintica* with 72% of vernolic acid, the acid has also been recognized in the seed oils of *V. galamensis* (73 to 78%), *Cephalocroton cordofanus* (62%), *Stokes aster* (65 to 79), *Euphorbia lagescae* (57 to 62%), *Erlanga tomentosa* (52%), *Crepis aureus* (52 to 54%), and *C. biennis* (68%). With these high levels of venolic acid the triacylglycerols in these seed oils will be rich in esters with two or three vernolic acid groups. *V. galamensis* seed oil, for example, is reported to contain 50 to 60% of trivernolin and 21 to 28% of glycerol esters of the type V_2X where V and X represent vernolic and other acyl groups, respectively. Attempts are being made to domesticate *Vernonia galamensis* and *Euphorbia lagescae*[11–13,187].

5.2.112 WALNUT (*JUGLANS REGIA*)

Walnut oil is an unsaturated oil containing both linoleic (50 to 60%) and linolenic (13 to 15%) acids and rich in tocopherols (~1500 mg/kg of oil). It is used as a gourmet oil in Japan, France, and other countries. A recent paper gives the detailed composition (fatty acids, triacylglycerols, sterols, and tocopherols) of oil extracted with hexane and with supercritical carbon dioxide. The two products contain ~300 and 400 ppm of tocopherols of which 82% is the β/γ-compounds[11–13,95,188,189].

5.2.113 WATERMELON (*CITRULLUS VULGARIS*)

Watermelon seeds yield an oil rich in linoleic acid and in lycopene. It is reported to contain palmitic (9 to 11%), oleic (13 to 19%), and linoleic (62 to 71%) acids[12,13,190].

5.2.114 WHEAT GERM (*TRITICUM AESTIVUM*)

This oil is highly unsaturated with linoleic (~60%) and some linolenic (~5%) acids. It is valued for its high tocopherol levels (~2500 mg/l of oil) and is reported to lower total and LDL cholesterol levels. α-, β-, γ-, δ-Tocopherols are present at levels of 1210, 65, 24, and 25 mg/l of oil in addition to small amounts of tocotrienols[38,191,192].

TABLE 5.8
Fatty acid composition of minor oils rich in C_{16} and C_{18} acids

Oil	Palmitic	Stearic	Oleic	Linoleic	α-Linolenic	Other (a)
Aceituno	12	28	58			
Alfalfa				34	25	
Almond			65-70			
Amaranthus	19	3	22	45		
Argemone	12-15		28-29	55		
Apricot	5	1	66	29		
Argane			42-47	31-37		
Arnebia	7		14	23	45	GLA 3, SA 4
Avocado	10-20		60-70	10-15		
Babassu				30		See Table 5.9
Baobab	22		34	30		
Basil	6-11		9-13	18-31	44-65	
Blackcurrant						See Table 5.1
Borage						See Table 5.1

(Continued)

TABLE 5.8
(Continued)

Oil	Palmitic	Stearic	Oleic	Linoleic	α-Linolenic	Other (a)
Borneo tallow	18	46	35			
Brazil nut						See text
Buffalo gourd	9	2	25	62		
Calendula				28-34		Cal 53-62
Camelina						See Gold of Pleasure
Candlenut	6-8	2-3	12-25	38-45	25-30	
Caraway						Pet 35-43
Carrot						Pet 66-73
Cashew	9-14	6-12	57-65	16-18		
Chaulmoogra						See text
Cherry			30-40	40-50		Elst 6-12
Chestnut	15		54	25		
Chia	9-11		7-8	20-21	52-63	
Chinese vegetable tallow	75		20-25			See text
Coffee	32		13	41		See text
Cohune						See Table 5.9
Coriander						Pet 31-75
Crambe						Er 50-55
Cuphea						See Table 5.3
Cupuacu date seed	6-12	22-35	39-47	3-9		Ar 10-12
Date						See Table 5.9
Dimorphotheca						See text
Echium						See Table 5.1
Euphorbia lathyris			80-85			
Euphorbia lagascae	4		19	9		Ver 64
Evening primrose						See Table 5.1
Flaxseed						See Table 5.4
Gold of pleasure			10-20	19-24	27-33	Er 12-15 See text
Grapeseed	6-8	3-6	12-25	60-76		
Gevuina						See text
Hazelnut			65-75	16-22		See text
Hemp	4-9	2-4	8-15	53-60	15-25	GLA 0-5, SA 0-3
Honesty						Er 41, Ner 22
Hyptis				22-31	51-64	See text
Illipe						See Borneo tallow
Jojoba						See text
Kapok	22		21	37		CP acids 13
Karanja	4-8	2-9	44-71	2-18		Ar 2-5, Eic 9-12
Kiwi			13	16	63	Sat 8
Kokum	3-5	52-56	39-42	0-2		
Kukui	6-8	2-5	24-29	33-39	21-30	
Kusum	5-9	2-6	40-67	2-7		Ar 20-31
Lesquerella						See text
Macadamia			55-65			Hex 16-23, Eic 1-3
Mahua	22-37	19-24	37-46			
Maize germ	11		25	60		
Mango	10	35	49	5		
Manketti			18-24	39-43		See text
Marula	11	7	75	5		

(Continued)

TABLE 5.8
(Continued)

Oil	Palmitic	Stearic	Oleic	Linoleic	α-Linolenic	Other (a)
Meadowfoam						See text
Melon	11	7	10	71		See text
Moringa	12	4	77	2	2	Eic 2
Mowrah	27	9	39	24		See text
Murumuru						See Table 5.9
Mustard (low erucic)	3	2	64	17	10	See text
Neem	16-19	15-18	46-57	9-14		Ar 1-3
Ngali						See Table 5.6
Nigella	10		35	45		See text
Niger				71-79		See text
Nutmeg						See Table 5.9
Oats	13-28		19-53	24-53	1-5	
Oiticica						See text
Oyster nut	16	13	30	40		
Parsley						Pet 80
Passionfruit	8-12		13-20	65-75		
Peach	9		67	21		
Pecan	5-11		49-69	19-40		
Perilla			13-15	14-18	57-64	
Phulwara butter	61	3	31	5		
Pistachio	10	3	69	17		
Plum			71	16		
Poppy	10		11	72	5	
Pumpkin	4-14	5-6	21-47	35-59		
Purslane	15	4	18	33	26	See text
Raspberry	3		9	55	33	
Ricebran	16		42	37		
Rose hip			8	54	32	Sat 5
Sacha inchi			6-9	34-39	47-51	Sat 5-7
Safflower			14	75		Sat 10
Sal	2-8	35-48	35-42	2-3		Ar 6-11
Salicornia bigelovii	8		12	74		
Sea buckthorn						See Table 5.7
Sesame	9	65	40	46		
Shea	4-8	23-58	33-68	4-8		
Shikoin						See text
Sisymbrium irio	14-16		17-19	13-16	33-37	Ar 0-8 Er 6-10
Stillingia						See Chinese vegetable tallow
Tamanu	12	13	34	38		
Teased	5-17		58-87	7-17		
Tobacco				>70		
Tomato	12-16		16-25	50-60	2	
Tung						Elst 69
Ucuhuba						See Table 5.9
Vernonia						See text
Walnut				50-60	13-15	
Watermelon	9-11		13-19	62-71		
Wheatgerm				60	5	

(a) eicosenoic acid, Elst = eleaostearic acid, Er = ericic acid, GLA = γ-linolenic acid, Hex = hexadecenoic acid, Ner = nervonic acid; Pet = petroselinic acid, SA = stearidonic acid, Ver = vernolic acid.

TABLE 5.9
Fatty acid composition of minor oils rich in short and medium chain acids

Oil	8:0	10:0	12:0	14:0	16:0	18:0	18:1	18:2
Babassu	6	4	45	17	9	3	13	3
Cuphea		See Table 5.3						
Cohune	7-9	6-8	44-48	16-17	7-10	3-4	8-10	1
Date			18	10	11		41	12
Murumuru			42	37			11	
Nutmeg			3-6	76-83	4-10		5-11	0-2
Ucuhuba			13	69	7		Unsat 11	

REFERENCES

1. Gunstone, F.D., Ed., *Vegetable Oils in Food Technology: Composition, Properties, and Uses,* Blackwell, Oxford, 2002.
2. Harris, N., Putting speciality and novel oils to good use, *Oils Fats Int.,* 20, 20–21, 2004.
3. Aitzetmuller, K., Matthaus, B., and Friedrich, H., A new database for seed oil fatty acids: the database SOFA, *Eur. J. Lipid Sci. Technol.,* 105, 92–103, 2003. www.bagkf.de/sofa
4. Gunstone, F.D., Ed., *Lipids for Functional Foods and Nutraceuticals,* The Oily Press, Bridgewater, U.K., 2003.
5. Cosmetic, Toiletries, and Fragrances Association, *CTFA International Cosmetic Ingredient Dictionary and Handbook,* 9th ed., Micelle Press, Dorset, U.K, 2002.
6. Bhattacharyya, D.K., Lesser-known Indian plant sources for fats and oils, *Inform,* 13, 151–157, 2002.
7. Gunstone, F.D., Minor oils, specialty oils, and super-refined oils, in *Structured and Modified Lipids,* Gunstone, F.D., Ed., Marcel Dekker, New York, 2001, pp. 185–208.
8. Gunstone, F.D., What else beside commodity oils and fats?, *Fett/Lipid,* 101, 124–131, 1999.
9. Mazza, G., Ed., *Functional Foods: Biochemical and Processing Aspects,* Technomic, Lancaster, PA, 1998.
10. Abbott, T.P., Phillips, B.S., Butterfield, R.O., Isbell, T.A., and Kleiman, R., On-line chemical database for new crop seeds, *J. Am. Oil Chem. Soc.,* 74, 723–726, 1997. www.ncaur.usda.gov/nc/ncdb/search.html-ssi
11. Padley, F.B., Major vegetable fats in *The Lipid Handbook,* 2nd ed., Gunstone F.D., Harwood, J.L. and Padley, F.B., Eds., Chapman and Hall, London, 1994, pp. 53–146.
12. Firestone, D. Ed., *Physical and Chemical Characteristics of Oils, Fats, and Waxes,* AOCS Press, Champaign, IL, 1996.
13. Ucciani, E., *Nouveau Dictionnaire des Huiles Vegetales,* Tec and Doc, New York, 1995.
14. Kamel, B.S. and Kakuda, Y., Eds., *Technological Advances in Improved and Alternative Sources of Lipids,* Blackie Academic and Professional, London, 1994.
15. Pritchard, J.L.R., Analysis and properties of oilseeds, in *Analysis of Oilseeds, Fats, and Fatty Foods,* Rossell, J.B. and Pritchard, J.L.R., Eds., Elsevier Applied Science, London, 1991, pp. 39–102.
16. Robbelen, G., Downey, R.K., and Ashri. A., Eds., *Oil Crops of the World: Their Breeding and Utilization,* McGraw-Hill, New York, 1989.
17. Eckey, E.W., *Vegetable Fats and Oils,* Reinhold, New York, 1954.
18. A and E Connock, www.connock.co.uk/vegetable_oils.htm and www.creative-compounds.co.uk
19. Statfold oils, www.statfold-oils.co.uk/specifications
20. FOSFA Anatomy of seeds, www.fosfa.org
21. Prats-Moya, M.S., Grane-Teruel, N., Berenguer-Navarro, V., and Martin-Carratala, M.L., A chemometric study of genotypic variation in triacylglycerol composition among selected almond cultivars, *J. Am. Oil Chem. Soc.,* 76, 267–272, 1990.
22. Budin, J.T., Breene, W.M., and Putnam, D.H., Some compositional properties of seeds and oils of eight *Amaranthus* species, *J. Am. Oil Chem. Soc.,* 73, 475–481, 1996.
23. Lehmann, J.W., Putnam, D.H., and Qureshi, A.A., Vitamin E isomers in grain amaranthus (*Amaranthus* spp.), *Lipids,* 29, 177–181, 1994.

24. Jahaniaval, F., Kakuda, Y., and Marcone, M.F., Fatty acid and triacylglycerol compositions of seed oils of five *Amaranthus* accessions and their comparison to other oils, *J. Am. Oil Chem. Soc.*, 77, 847–852, 2000.

25. Sun, H., Wiesenborn, D., Tostenson, K., Gillespie, J., and Rayas-Duarte, P., Fractionation of squalene from amaranth seed oil, *J. Am. Oil Chem. Soc.*, 74, 413–418, 1997.

26. Anon, Cherry and apricot oils are safe for food use, *Lipid Technol.*, 5, 53, 1993.

27. Kamel, B.S. and Kakuda, Y., Characterization of the seed oil and meal from apricot, cherry, nectarine, peach and plum, *J. Am. Oil Chem. Soc.*, 69, 492–494, 1992.

28. Zlatanov, M. and Janakieva, I., Phospholipid composition of some fruit-stone oils of Rosaceae species, *Fett/Lipid,* 100, 312–315, 1998.

29. Yaghmur, A., Aserin, A., Mizrahi, Y., Nerd, A., and Garti, N., Argon oil-in-water emulsions; preparation and stabilization, *J. Am. Oil Chem. Soc.*, 76, 15–18, 1999.

30. Werman, M.J. and Neeman, I., Avocado oil production and chemical characteristics, *J. Am. Oil Chem. Soc.*, 64, 229–232, 1987.

31. Swisher, H.O., Avocado oil: from food use to skin care, *J. Am. Oil Chem. Soc.*, 65, 1704–1706, 1993.

32. Lozano, Y.F., Mayer, C.D., Bannon, C., and Gaydou, E.M., Unsaponifiable matter, total sterol and tocopherol contents of avocado oil varieties, *J. Am. Oil Chem. Soc.*, 70, 561–565, 1993.

33. Bruin, N. et al., FR2822068 and EP 1213975, cited in *Lipid Technol.* 15, 71, 2003.

34. Eyres, L., Sherpa, N., and Hendriks, G., Avocado oil: a new edible oil from Australasia, *Lipid Technol.*, 13, 84–88, 2001.

35. Birkbeck, J., Health benefits of avocado oil, *Food New Zealand*, April/May 2002, pp. 40–42.

36. PhytoTrade Africa, www.sanprota.com/products/baobob

37. Angers, P., Morales, M.E., and Simon, J.E., Fatty acid variation in seed oil among O*cimum* species, *J. Am. Oil Chem. Soc.*, 73, 393–395, 1996.

38. Yang, B., Natural vitamin E: activities and sources, *Lipid Technol.*, 15, 125–130, 2003.

39. Horrobin, D.F., Nutritional and medical importance of gamma-linolenic acid, *Prog. Lipid Res.*, 31, 163–194, 1992.

40. Clough, P., Sources and production of speciality oils containing GLA and stearidonic acid, *Lipid Technol.*, 13, 9–12, 2001.

41. Clough, P.M., Specialty vegetable oils containing γ-linolenic acid and stearidonic acid, in *Structured and Modified Lipids*, Gunstone, F.D., Ed., Marcel Dekker, New York, 2001, pp. 75–117.

42. Ju, Y-H. and Chen, T-C., High purity γ-linolenic acid from borage oil fatty acids, *J. Am. Oil Chem. Soc*, 79, 29–32, 2002.

43. Hayes, D.G., van Alstine, J.M., and Setterwall, F., Urea-based fractionation of seed oil samples containing fatty acids and triacylglycerols of polyunsaturated and hydroxy fatty acids, *J. Am. Oil Chem. Soc.*, 77, 207–213, 2000.

44. Chen, T.C. and Ju, Y-H., Polyunsaturated fatty acid concentrates from borage and linseed oil fatty acids, *J. Am. Oil Chem. Soc.*, 78, 485–488, 2001.

45. Traitler, H., Winter, H., Richli, U., and Ingenbleek, Y., Characterisation of γ-linolenic acid in *Ribes* seed, *Lipids*, 19, 923–928, 1984.

46. Liu, J-W., DeMichelle, S., Bergana, M., Bobik, E. Jr., Hastilow, C., Chuang, L-T., Mukerji, P., and Huang, Y-S., Characterization of oil exhibiting high γ-linolenic acid from a genetically transformed canola strain, *J. Am. Oil Chem. Soc.*, 78, 489–493, 2001.

47. Wretensjo, I. and Karlberg, B., Pyrrolizidine alkaloid content in crude and processed borage oil from different processing stages, *J. Am. Oil Chem. Soc.*, 80, 963–970, 2003.

48. Wretensjo, I. and Karlberg, B., Characterization of sterols in refined borage oil by GC-MS. *J. Am. Oil Chem. Soc.*, 79, 1069–1074, 2002.

49. Eskin, N.A.M., Authentication of evening primrose, borage, and fish oils, in *Oils and Fats Authentication*, Jee. M., Ed., Blackwell, Oxford, 2002, pp. 95–114.

50. Huang. Y-S, and Mills, D.E., γ-Linolenic Acid, AOCS Press, Champaign, IL, 1966.

51. Huang, Y-S. and Ziboh, V.A., γ-Linolenic Acid: Recent Advances in Biotechnology and Clinical Applications, AOCS Press, Champaign, IL, 2001.

52. Gunstone, F.D., γ-Linolenic acid (GLA), in *Lipids for Functional Foods and Nutraceuticals*, Gunstone, F.D., Ed., The Oily Press, Bridgewater, U.K., 2003, pp. 272–279.

53. Gunstone, F.D., Gamma-linolenic acid: occurrence and physical and chemical properties, *Prog. Lipid Res.*, 31, 145–161, 1992.

54. Campbell, I., 2002, www.britanniafood.com/invite
55. Stewart, I.M. and Timms, R.E., Fats for chocolate and sugar confectionery, in *Fats in Food Technology,* Rajah, K.K., Ed., Sheffield Academic Press, Sheffield, U.K., 2002, pp. 159–191.
56. Timms, R.E., *Confectionery Fats Handbook: Properties, Production and Application,* The Oily Press, Bridgewater, U.K., 2003.
57. Vasconcellos, J.A., Bemis, W.P., Berry, J.W., and Weber, C.W., The buffalo gourd, *Cucurbita foetidissima* HBK, as a source of edible oil, in *New Sources of Fats and Oils,* Pryde, E.H., Princen, L.H., and Mukherjee, K.D., Eds., AOCS Press, Champaign, IL, 1981, p. 55–68.
58. Janssens, R.J. and Vernooij, W.P., *Calendula officinalis*: a new source for pharmaceutical, oleochemical, and functional compounds, *Inform,* 12, 468–477, 2000.
59. Janssens, R.J., *Calendula* oil: seed classification, oil processing and quality aspects. *Lipid Technol.,* 12, 53–58, 2000.
60. Unilever, U.S. Patent 6,365,175 and Elizabeth Arden, U.S. Patent 6,455,057 and EP 1200058, cited in *Lipid Technol.,* 15, 118 and 119, 2003.
61. Toschi, T.G., Caboni, M.F., Penazzi, G., Lercker, G., and Capella, P., A study on cashew nut oil composition, *J. Am. Oil Chem. Soc.,* 70, 1017–1020, 1993.
62. Comes, F., Farines. M., Aumelas, A., and Soulier, J., Fatty acids and triacylglycerols of cherry seed oil, *J. Am. Oil Chem. Soc.,* 69, 1224–1227, 1992.
63. Ayerza, R., Oil content and fatty acid composition of chia (*Salva hispanica* L) from five Northwestern locations in Argentina, *J. Am. Oil Chem. Soc.,* 72, 1079–1081, 1995.
64. Jeffrey, B.S.J. and Padley, B.F., Chinese vegetable tallow: characterization and contamination by stillingia oil, *J. Am. Oil Chem. Soc.,* 68, 123–127, 1991.
65. Rui Alves, M., Casai, S., Oliveira, M.B.P.P., and Ferreira, M.A., Contribution of FA profile obtained by high-resolution GC/chemometric techniques to the authenticity of green and roasted coffee varieties, *J. Am. Oil Chem. Soc.,* 80, 511–517, 2003.
66. Foglia, T.A., Jones, H.C., and Sonnet, P.E., Selectivity of lipases: isolation of fatty acids from castor, coriander, meadowfoam oils, *Eur. J. Lipid Sci. Technol.,* 102, 612–617, 2000.
67. Lessman, K.J., Anderson, W.P., Pryde, E.H., Princen, L.H., and Mukherjee K.D., Eds., *New Sources of Fats and Oils,* AOCS Press, Champaign, IL, 1981, pp. 223–246.
68. Leonard, C., Sources and commercial applications of high-erucic vegetable oils, *Lipid Technol.,* 6, 79–83, 1994.
69. Steinke, G. and Mukherjee, K.D., Lipase-catalyzed and alkali-catalyzed alcoholysis of crambe oil and camelina oil for the preparation of long-chain esters, *J. Am. Oil Chem. Soc.,* 77, 361–366 and 367–370, 2000.
70. Hebard, A.B., Growing high-erucic rapeseed oil as an example of identity preservation, *Lipid Technol,* 14, 53–55, 2002.
71. Temple-Heald, C., Crambe oil, in *Rapeseed and Canola Oil,* Gunstone, F.D., Ed., Blackwell, Oxford, 2004, pp. 115–116.
72. Isbell, T.A., Current progress in the development of *Cuphea, Lipid Technol.,* 14, 77–80, 2002.
73. Pandey, V., Banerji, R., Dixit, B.S., Singh, M., Shukla, S., and Singh, S.P., *Cuphea* a rich source of medium chain triglycerides: fatty acid composition and oil diversity in *Cuphea procumbens, Eur. J. Lipid Sci. Technol.,* 102, 463–466, 2000.
74. Besbes, S., Blecker, C., Deroanne, C., Drira, N-E., and Attia. H., Date seeds: chemical composition and characteristic profiles of the lipid fraction, *Food Chem.* 84, 577–584, 2004.
75. Anon, Echium oil, *Lipid Technol Newsletter,* 9, 56, 2003.
76. Kallio, H., Stearidonic acid, in *Lipids for Functional Foods and Nutraceuticals*, Gunstone, F.D., Ed., The Oily Press, Bridgewater, U.K., 2003, pp. 280–284.
77. Ackman, R.G. and Daun, J.K., Identity characteristics for edible oils: origin, current usefulness, and possible future directions, *Inform,* 12, 998–1003, 2001.
78. Thompson, L.U. and Cunnane, S.C., *Flaxseed in Human Nutrition,* 2nd ed., AOCS Press, Champaign, IL, 2003.
79. Oomah, B.D. and Mazza, G., Flaxseed products for disease prevention, in *Functional Foods: Biochemical and Processing Aspects,* Mazza, G., Ed., Technomic, Lancaster, PA, 1998, pp. 91–138.
80. Li, D., Bode, O., Drummond, H., and Sinclair A.J., Omega-3 (n-3) fatty acids, in *Lipids for Functional Foods and Neutraceuticals,* Gunstone, F.D., Ed., The Oily Press, Bridgewater, U.K., 2003, pp. 225–262.
81. Williams, C.M. and Roche, H.M., Achieving optimal fatty acid intake through manufactured foods, in *Structured and Modified Lipids,* Gunstone, F.D., Ed., Marcel Dekker, New York, 2001, pp. 465–484.

82. Green, A.G. and Dribnenki, J.C.P., Linola: a new premium polyunsaturated oil, *Lipid Technol.*, 6, 29–33, 1994.

83. Budin, J.T., Breene, W.M., and Putnam, D.H., Some compositional properties of camelina (*Camelina sativa* L. Crantz) seed and oils, *J. Am. Oil Chem. Soc.*, 72, 309–315, 1995.

84. Steinke. G., Kirchoff, R., and Mukherjee, K.D., Lipase catalysed alcoholysis of crambe oil and camelina oil for the preparation of long-chain esters, *J. Am. Oil Chem. Soc.*, 77, 361–366, 2000.

85. Steinke, G., Schonwiese, S., and Mukherjee, K.D., Alkali-catalysed alcoholysis of crambe oil and camelina oil for the preparation of long-chain esters, *J. Am. Oil Chem. Soc.*, 77, 367–371, 2000.

86. Shukla, V.K.S., Dutta, P.E., and Artz, W.E., Camelina oil and its unusual cholesterol content, *J. Am. Oil Chem. Soc.*, 79, 965–969, 2002.

87. Bonjean, A., Monteuuis, B., and Messean, A., Que penser de la cameline en 1995?, *Oleagineux Corps gras Lipides*, 2, 97–100, 1995.

88. Zubr, J., Oil-seed crop: *Camelina sativa*, *Ind. Crops Prod.*, 6, 113–119, 1997.

89. Hebard, A., *Camelina sativa*: a pleasurable experience or another false hope, *Lipid Technol.*, 10, 81–83, 1998.

90. Leonard, E.C., Camelina oil: α-linolenic source. *Inform*, 9, 830–838, 1998.

91. Moret, S., Dudine, A., and Conte, L.S., Processing effects on the polyaromatic hydrocarbon content of grapeseed oil, *J. Am. Oil Chem. Soc.*, 77, 1289–1292, 2000.

92. Bertoli, C., Fay, F.B., Stancanelli, M., Gumy, D., and Lambelet, P., Characterization of Chilean hazelnut (*Gevuina avellana* Mol) seed oil, *J. Am. Oil Chem. Soc.*, 75, 1037–1040, 1998.

93. Anon, New health claim guidelines published, *Inform*, 14, 476, 2003.

94. Bewadt, S. and Aparicio, R., The detection of the adulteration of olive oil with hazelnut oil: a challenge for the chemist, *Inform*, 14, 342–343, 2003.

95. Bernardo-Gil, M.G., Grenha, J., Santos, J., and Cardoso, P., Supercritical fluid extraction and characterisation of oil from hazelnut, *Eur. J. Lipid Sci. Technol.*, 104, 402–409, 2002.

96. Savage, G.P., McNeil, D.L., and Dutta, P.C., Lipid composition and oxidative stability of oils in hazelnuts (*Corylus avellana* L.) grown in New Zealand, *J. Am. Oil Chem. Soc.*, 74, 755–759, 1997.

97. Gordon, M.H., Covell, C., and Kirsch, N., Detection of pressed hazelnut oil in admixtures with virgin olive oil by analysis of polar components, *J. Am. Oil Chem. Soc.*, 78, 621–624, 2001.

98. Zabaras, D. and Gordon, M.H., Detection of pressed hazelnut oil in virgin olive oil by analysis of polar components: improvement and validation of the method, *Food Chem.*, 84, 475–483, 2004.

99. Ruiz del Castillo, M.L., Herraiz, M., Molero, M.D., and Herrera, A., Off-line coupling of high-performance liquid chromatography and ¹H nuclear magnetic resonance for the identification of filbertone in hazelnut oil, *J. Am. Oil Chem. Soc.*, 78, 1261–1265, 2001.

100. Ruiz del Castillo, M.L. and Herraiz, M., Ultrasonically assisted solid-phase extraction and GC analysis of filbertone in hazelnut oil, *J. Am. Oil Chem. Soc.*, 80, 307–310, 2003.

101. Christopoulou, E., Lazaraki, M., Komaitis, M., and Kaselimis, K., Effectiveness of determination of fatty acids and triglycerides for the detection of adulteration of olive oils with vegetable oils, *Food Chem.*, 84, 463–474, 2004.

102. Anon, Hemp oil study finds higher GLA levels, *Inform*, 12, 1213, 2001.

103. Przybylski, R. and Fitzpatrick, K.C., Does hempseed need to be a forbidden crop?, *Inform*, 13, 489–491, 2002.

104. www.hemptrade.ca

105. Sargent, J. and Coupland, K., Applications of specialised oils in the nutritional therapy of demyelinating disease, *Lipid Technol.*, 6, 10–14, 1994.

106. Hagemann, J.M., Earle, F.R., Wolff, I.A., and Barclay, A.S., Search for new industrial oils: XIV. Seed oils of Labiatae, *Lipids*, 2, 371–380, 1967.

107. Wisniak. J., Ed., *The Chemistry and Technology of Jojoba Oil*, AOCS Press, Champaign, IL, 1987.

108. Jeyarani, Y. and Yella Reddy, S., Heat-resistant cocoa butter extenders from mahua (*Madhuca latifolia*) and kokum (*Garcinia indica*) fats, *J. Am. Oil Chem. Soc.*, 76, 1431–1436, 1999.

109. Abbott, T.P., Status of lesquerella as an industrial crop, *Inform*, 8, 1169–1175, 1997.

110. Isbell, T.A. and Cermak, S.C., Synthesis of triglyceride estolides from lesquerella and castor oils, *J. Am. Oil Chem. Soc.*, 79, 1227–1233, 2002.

111. Colquhoun, D.M., Humphries, J.A., Moores, D., and Somerset, S.M., Effects of a macadamia nut enriched diet on serum lipids and lipoproteins compared with a low fat diet, *Food Austral.*, 48, 531–537, 1996.

112. Macfarlane. N. and Harris, R.V., Macadamia nuts as an edible oil source, in *New Sources of Fats and Oils*, Pryde, E.H., Princen, L.H., and Mukherjee, K.D., Eds., AOCS Press, Champaign, IL, 1981, pp. 103–108.

113. Croy, C., Macadamia nut oil, *Inform*, 5, 970–971, 1994.

114. Yang, B., Palmitoleic acid, in *Lipids for Functional Foods and Nutraceuticals*, Gunstone, F.D., Ed., The Oily Press, Bridgewater, U.K., 2003, pp. 266–272.

115. Moreau, R.A., Corn oil, in *Vegetable Oils in Food Technology*, Gunstone, F.D., Ed., Blackwell, Oxford, 2002, pp. 278–296.

116. Osman, S.M., Mango fat, in *New Sources of Fats and Oils*, Pryde, E.H., Princen, L.H., and Mukherjee, K.D., Eds., AOCS Press, Champaign, IL, 1981, pp. 129–140.

117. Yella Reddy. S. and Jeyarani, T., Trans-free bakery shortenings from mango kernel and mahua fats by fractionation and blending, *J. Am. Oil Chem. Soc.*, 78, 635–640, 2001.

118. Jolliff, G.D., Development and production of meadowfoam (*Limnanthes alba*), in *New Sources of Fats and Oils*, Pryde, E.H., Princen, L.H., and Mukherjee, K.D., Eds., AOCS Press, Champaign, IL, 1981, pp. 269–285.

119. Isbell, T.A., Development of meadowfoam as an industrial crop through novel fatty acid derivatives, *Lipid Technol.*, 9, 140–144, 1997.

120. Isbell, T.A., Novel chemistry of Δ5 fatty acids, in *Lipid Synthesis and Manufacture*, Gunstone, F.D., Ed., Sheffield Academic Press, Sheffield, U.K., 1998, pp. 401–421.

121. Isbell, T.A. and Cermak, C.S., Synthesis of δ-eicosanolactone and δ-docosanolactone from meadowfoam oil, *J. Am. Oil Chem. Soc.*, 78, 527–531, 2001.

122. Akoh, C.C. and Nwosu, C.V., Fatty acid composition of melon seed oil lipids and phospholipids, *J. Am. Oil Chem. Soc.*, 69, 314–316, 1992.

123. Bora, P.S., Narain, N., and de Mello, M.L.S., Characterization of the seed oils of some commercial cultivars of melon, *Eur. J. Lipid Sci. Technol.*, 102, 266–269, 2000.

124. Lalas, S. and Tsaknis, J., Extraction and identification of natural antioxidant from the seeds of the *Moringa oleifea* tree variety of Malawi, *J. Am. Oil Chem. Soc.*, 79, 677–683, 2002.

125. Lalas, S., Tsaknis, J., and Stiomos, K., Characterisation of *Moringa stenopetala* seed oil variety "Marigat" from island Kokwa, *Eur. J. Lipid Sci. Technol.*, 105, 23–31, 2003.

126. De. B.K., Bhattacharyya, K., and Bandyopadhyay, K., Bio- and auto-catalytic esterification of high acid mowrah fat and palm kernel oil, *Eur. J. Lipid Sci. Technol.*, 104, 167–173, 2002.

127. Gunstone, F.D., Oilseed crops with modified fatty acid composition, *J. Oleo Science*, 50, 269–279, 2000.

128. Anon, Neem: source of insecticide, fungicide, *Inform*, 5, 713, 1994.

129. Melton, L., Slimming pills for insects, *Chem. Br.*, 39, 24–26, 2003.

130. Atta, M.B., Some characteristics of nigella (*Nigella sativa* L.) seeds cultivated in Egypt and its lipid profile, *Food Chem.*, 83, 63–68, 2003.

131. Ustun, G., Kent, L., Cekin, N., and Civelekoglu, H., Investigation of the technological properties of *Nigella sativa* (black cumin) seed oil, *J. Am. Oil Chem. Soc.*, 67, 958–960, 1990.

132. Zeitoun, M.A.M. and Neff, W.E., Fatty acid, triacylglycerol, tocopherol, sterol phospholipid composition and oxidative stability of Egyptian *Nigella sativa* seed oil, *Oleagineux Corps gras Lipides*, 2, 245–248, 1995.

133. Aitzetmuller, A., Werner, G., and Ivanov, S.A., Seed oils of *Nigella* species and of closely related genera, *Oleagineux Corps gras Lipides*, 4, 385–388, 1997.

134. Takruri, H.R.H. and Dameh, M.A.F., Study of the nutritional value of black cumin seeds (*Nigella sativa* L), *J. Sci. Food Agric.*, 76, 404–410, 1998.

135. Marini, F., Magri, A.L., Marini, D., and Balastrieri, F., Characterization of the lipid fraction of niger seeds (*Guizotia abyssinica* Cass) from different regions of Ethiopia and India and chemometric authentication of their geographical origin, *Eur. J. Lipid Sci. Technol.*, 105, 697–704, 2003.

136. Dutta, P.C., Helmersson. S., Kebedu, E., Alemaw, G., and Appelqvist, L.A., Variation in lipid composition of niger seed (*Guizotia abyssinica*) samples collected from different regions in Ethiopia, *J. Am. Oil Chem. Soc.*, 71, 839–845, 1994.

137. Hilditch, T.P. and Williams, P.N., *The Chemical Constitution of Natural Fats*, 4th ed., Chapman and Hall, London, 1964, p. 336.

138. Anon, Olibra confirmed as appetite suppressant, *Lipid Technol. Newsletter*, 4, 105–106, 1998.

139. Holmback, J.M., Karlsson, A.A., and Arnoldsson, K.C., Characterisation of *N*-acylphosphatidylethanolamine and acylphosphatidylglycerol in oats, *Lipids*, 36, 153–165, 2001.

140. Herslof, B.G., Glycolipids herald a new era for food and drug products, *Lipid Technol.*, 12, 125–128, 2000.
141. Zhou, M., Robards, K., Glennie-Holmes, M., and Helliwell, S., Oat lipids: a review, *J. Am. Oil Chem. Soc.*, 76, 159–169, 1999.
142. Peterson, D.M., Oat lipids: composition, separation, and applications, *Lipid Technol.*, 15, 56–59, 2002.
143. Badifu, G.I.O., Chemical and physical analyses of oils from four species of Cucurbitaceae, *J. Am. Oil Chem. Soc.*, 68, 428–432, 1991.
144. Advisory Committee in Novel Foods and Procedures (U.K.), annual report 1991, pp. 7, 30–31.
145. Longvah, T. and Deosthale, Y.G., Chemical and nutritional studies on hanshi (*Perilla frutescens*), a traditional oilseed from northeast India, *J. Am. Oil Chem. Soc.*, 68, 781–784, 1991.
146. Shin, H-S. and Kim S-W., Lipid composition of perilla seed, *J. Am. Oil Chem. Soc.*, 71, 619–622, 1994.
147. Kim, I-H., Kim, H., Lee, K-T., Chung, S-H., and Ko, S-M., Lipase catalysed acidolysis of perilla oil with caprylic acid to produce structured lipids, *J. Am. Oil Chem. Soc.*, 79, 363–367, 2002.
148. Dyszel, S.M. and Pettit, B.C., Determination of the country of origin of pistachio nuts by DSC and HPLC, *J. Am. Oil Chem. Soc.*, 67, 947–951, 1990.
149. Maskan, M. and Karatas, S., Fatty acid oxidation of pistachio nuts stored under various atmospheric conditions and different temperatures, *J. Sci. Food Agric.*, 77, 334–340, 1998.
150. Yildiz, M., Gurcan, S.T., and Ozdemir, M., Oil composition of pistachio nuts (*Pistachio vera* L) from Turkey, *Fett/Lipid*, 100, 84–86, 1998.
151. Singh, P.S., Shukla, S., Khanna, K.R., Dixit, B.S., and Banerji, R., Variation of major fatty acids in F$_8$ generation of opium poppy (*Papaver somniferum* x *Papaver setigerum*) genotypes, *J. Sci. Food Agric.*, 76, 168–172, 1998.
152. Fruhwirth, G.O., Wennzl, T., El-Toukhy, R., Wagner, F.S., and Hermetter, A., Flourescence screening of antioxidant capacity in pumpkin seed oil and other natural oils, *Eur. J. Lipid Sci. Technol.*, 105, 266–274, 2003.
153. Murkovic, M., Piironen, V., Lampi, A.M., Kraushofer, T., and Sontag, G., Changes in chemical composition of pumpkin seeds during the roasting process for production of pumpkin seed oil: 1. Non volatile compounds. *Food Chem.*, 84, 359–365, 2004.
154. Siegmund, B. and Murkovic, M., Changes in chemical composition of pumpkin seeds during the roasting process for production of pumpkin seed oil: 2. Volatile compounds, *Food Chem.*, 84, 367–374, 2004.
155. Yoshida, H., Shougaki, Y., Hirakawa, Y., Tomiyama, Y., and Mizushina, Y., Lipid classes, fatty acid composition and triacylglycerol molecular species in the kernels of pumpkin (*Cucurbita* spp) seeds, *J. Sci. Food Agric.*, 84, 158–163, 2004.
156. Omara-Alwala, T.R., Mebrahtu, T., Prior, D.E., and Ezekwe, M.O., Omega-three fatty acids in purslane (*Portulaca oleracea*), *J. Am. Oil Chem. Soc.*, 68, 198–199, 1991.
157. Berger, K.G., Palm oil, in *Structured and Modified Lipids*, Gunstone, F.D., Ed., Marcel Dekker, New York, 2001, pp. 119–153.
158. Siew Wai Lin, Palm oil, in *Vegetable Oils in Food Technology: Composition, Properties, and Uses*, Gunstone, F.D., Ed., Blackwell, Oxford, 2002, pp. 59–97.
159. Hall, C., III, Rice bran oil, in *Lipids for Functional Foods and Nutraceuticals*, Gunstone, F.D., Ed., The Oily Press, Bridgewater, U.K., 2003, pp. 73–83.
160. Kochar, S.P., Rice-bran oil, in *Vegetable Oils in Food Technology*, Gunstone, F.D., Ed., Blackwell, Oxford, 2002, pp. 308–318.
161. Moldenhauer, K.A., Champagne, E.T., McCaskill, D.R., and Guraya, H., Functional products from rice, in *Functional Foods: Biochemical and Processing Aspects*, Mazza, G., Ed., Technomic, Lancaster, PA, 1998, p. 71–90.
162. Gopala Krishna, A.G., Nutritional components of rice bran oil in relation to processing, *Lipid Technol.*, 14, 80–84, 2002.
163. Gingras. L., Refining of rice bran oil, *Inform*, 11, 1196–1203, 2000.
164. Goffman, F.D., Pinson, S., and Bergman, C., Genetic diversity for lipid content and fatty acid profile in rice bran, *J. Am. Oil Chem. Soc.*, 80, 485–490, 2003.
165. Xu, Z. and Godber, J.S., Antioxidant activities of major components of γ-oryzanol from rice bran using a linoleic acid model, *J. Am. Oil Chem. Soc.*, 78, 645–649, 2001.
166. Dunford, N.T. and King, J.W., Thermal gradient deacidification of crude rice bran oil utilizing supercritical carbon dioxide, *J. Am. Oil Chem. Soc.*, 78, 121–125, 2001.

167. Gopala Krishna, A.G., Khatoon, S., Shiela, P.M., Sarmandal. C.V., Indira, T.N., and Mishra, A., Effect of refining of crude rice bran oil on the retention of oryzanol in the refined oil, *J. Am. Oil Chem. Soc.,* 78, 127–131, 2001.

168. Guillen, M.D., Ruiz, A., Cabo, N., Chirinos, R., and Pascual, G., Characterisation of sachi inchi (*Plukenetia volubilis* L) oil by FTIR spectroscopy and ^{1}H NMR. Comparison with linseed oil, *J. Am. Oil Chem. Soc.,* 80, 755–762, 2003.

169. Smith, J.R., *Sunflower*, AOCS Press, Champaign, IL, 1996.

170. Shukla, V.K.S., Confectionery fats, In *Developments in Oils and Fats*, Hamilton, R.J., Ed., Blackie, London, 1995, pp. 66–94.

171. Lipp, M., and Anklam, E., Review of cocoa butter and alternative fats for use in chocolate: A. Compositional data. B Analytical approaches for identification and determination, *Food Chem.,* 62, 73–97 and 99–108, 1998.

172. Birkett, J., Manufacture and use of fats in chocolate, in *Oils and Fats Handbook, Vol 1, Vegetable Oils and Fats*, Rossell, B., Ed., Leatherhead Food RA, Leatherhead, U.K., 1999, chap. 1-22.

173. Talbot, G., Manufacture and use of fats in non-chocolate confectionery, in *Oils and Fats Handbook, Vol 1, Vegetable Oils and Fats*, Rossell, B., Ed., Leatherhead Food RA, Leatherhead, U.K., 1999, Co. 1–12.

174. Lu, Z., Glenn. E.P., and John, M.M., Salicornia bigelovii: an overview, *Inform,* 11, 418–423, 2000.

175. Yang, B. and Kallio, H., Fatty acid composition of lipids in sea buckthorn (*Hippophae* rhamnoides L.) berries of different origin, *J. Agric. Food Chem.,* 49, 1939–1947, 2001.

176. Kallio, H., Yang, B., Peippo, P., Tahnonen, R., and Pan, R., Triacylglycerols, glycerophospholipids, tocopherols and tocotrienols in berries and seeds of two subspecies (ssp *sinensis* and ssp. *mongolica)* of sea buckthorn (*Hippophae* rhamnoides), *J. Agric. Food Chem.,* 50, 3004–3009, 2002.

177. Yang, B., Karlsson, R., Oksman, P., and Kallio, H., Phytosterols in sea buckthorn (*Hippophae* rhamnoides L.) berries: identification and effects of different origins and harvesting times, *J. Agric. Food Chem.,* 49, 5620–5629, 2001.

178. Yang, B., Gunstone, F.D., and Kallio, H., Oils containing oleic, palmitoleic, γ-linolenic and stearidonic acids, in *Lipids for Functional Foods and Nutraceuticals,* Gunstone, F.D., Ed., The Oily Press, Bridgewater, U.K., 2003, pp. 263–290.

179. Anon, Good prospects for sea buckthorn in North America, *Lipid Technnol.,* 15, 99–100, 2003.

180. Zadernowski, R., Naczk, M., and Amarowicz, R., Tocopherols in sea buckthorn (*Hipppphae* rhamnoides L) berry oil, *J. Am. Oil Chem. Soc.,* 80, 55–58, 2003.

181. Li., T.S.C. and Wang, L.C.H., Physiological components and health effects of ginseng, *Echinacea* and sea buckthorn, in *Functional Foods: Biochemical and Processing Aspects*, Mazza, G., Ed., Technomic, Lancaster, PA, 1998, pp. 329–356.

182. Hall, C., III, Sesame seed oil, in *Lipids for Functional Foods and Nutraceuticals*, Gunstone, F.D., Ed., The Oily Press, Bridgewater, U.K, 2003, pp. 83–90.

183. Kochar, S.P., Sesame seed oil, in *Vegetable Oils in Food Technology,* Gunstone, F.D., Ed., Blackwell, Oxford, 2002, pp. 297–308.

184. Kochar, S.P., Sesame seed oil: a powerful antioxidant, *Lipid Technol. Newsletter,* 6, 35–39, 2000.

185. Miller, R.W., Earle, F.R., Wolff, I.A., and Barclay, A.S., Search for new seed oils: XV. Oils of Boraginaceae, *Lipids,* 3, 43–45, 1968.

186. http://tamanu.biz

187. Sherringham, J.A., Clark, A.J., and Keene, B.R.T., New chemical feedstocks from unsaturated oils, *Lipid Technol.,* 12, 129–132, 2000.

188. Oliveira, R., Rodrigues, M.F., and Bernardo-Gil, M.G., Characterisation and supercritical carbon dioxide extraction of walnut oil, *J. Am. Oil Chem. Soc.,* 79, 225–230, 2002.

189. Crowe, T.D. and White, P.J., Oxidative stability of walnut oils extracted with supercritical carbon dioxide, *J. Am. Oil Chem. Soc.,* 80, 575–578, 2003.

190. Anon, Watermelon offers as much lycopene as tomatoes, *Lipid Technol. Newsletter,* 8, 56–57, 2002.

191. Barnes, P.J. and Taylor, P.W., The composition of acyl lipids and tocopherols in wheat germ oils from various sources, *J. Sci. Food Chem.,* 31, 997–1006, 1980.

192. Barnes, P.J., Lipid composition of wheat germ and wheat germ oil, *Fat Sci. Technol.,* 84, 256–269, 1982.

6 Sphingolipids

*Fang Fang, Hang Chen, Chi-Tang Ho,
and Robert T. Rosen*
Department of Food Science and Center for Advanced Food Technology,
Rutgers University, New Brunswick, New Jersey

CONTENTS

6.1 INTRODUCTION

Sphingolipids are found in all eucaryotic cells, but are especially abundant in the plasma membrane and related cell membranes, such as endoplasmic reticulum, golgi membranes, and lysosomes. They play an important role in maintaining membrane structure, and participate in intracellular signaling[1]. As receptors and ligands, they are involved in interactions between cells, and cells and matrix; they also serve as a binding site for toxins of bacterial and nonbacterial origin and hormones and viruses, among others[2,3].

Sphingolipids are a major topic of current research for several reasons. Firstly, sphingolipids are mediators of the signaling pathway of growth factors (e.g., platelet-derived growth factor), cytokines (e.g., tumor necrosis factor), and chemotherapeutics, and play an important role in regulation of cell growth, differentiation, and death. The hydrolysis products of sphingolipids, ceramides, sphingosine, and sphingosine-1-phosphate, are highly bioactive compounds that can, as a potent mitogen (sphingosine 1-phosphate, sphinganine 1-phosphate), act as lipid "second messengers" in the signal transduction pathways that either induce apoptosis (sphingosine, sphinganine, ceramides) or inhibit apoptosis[2,4]. Secondly, disruption of sphingolipid metabolism is implicated in several animal diseases and possibly human cancer. Finally, although relatively little is known about the mechanism of sphingolipids as dietary components, it is reasonable to believe that they contribute to disease prevention.

Studies about dietary sphingolipids found that they can suppress colon carcinogenesis. Milk sphingolipids were fed to female CF1 mice, which were previously administered 1,2-dimethylhydrazine. It was found that sphingolipids reduced the number of aberrant colonic crypt foci and aberrant crypts per focus, both of which are early indicators of colon carcinogenesis, by 70 and 30%, respectively. A longer term study found that sphingolipids had no effect on colon tumor incidence, but up to 31% of the tumors of mice fed sphingolipids were adenomas, while all of the tumors of mice fed without sphingolipids were adenocarcinomas[5–7]. Different classes of sphingolipids, containing different headgroups (sphingomyelin, glycosphingolipids and ganglioside), showed similar effects[7].

A model of colon cancer inhibition by dietary sphingolipids was suggested[7]. The digestion, uptake, and subsequent metabolism of these compounds were studied in this model. It has been established that dietary sphingomyelin and glycosphingolipids can be hydrolyzed into ceramides, sphingosine, and other sphingoid bases in the intestinal lumen by intestinal enzymes and microflora. The released metabolites are taken up by colonic cells. These cells resynthesize ceramides and sphingomyelin or degrade sphingoid bases. Both ceramides and sphingoid bases have biological functions involved in regulating cell growth, induction of cell differentiation, and apoptosis. Hence, the digestion of dietary sphingolipids into sphingoid bases and ceramides might help to reduce the risk of colon cancer.

In addition to colon cancer inhibition activity, sphingolipids were found to have other beneficial effects. In short-[8] and long-term[9] animal studies (feeding experiments with rats), sphingolipids were found to reduce plasma cholesterol, a risk factor for atherosclerosis. Also, the sphingolipids in foods may protect humans against bacterial toxins and viruses[10]. Many microorganisms, microbial toxins, and viruses bind to cell membranes through sphingolipids; therefore, sphingolipids in food can compete for cellular binding sites and facilitate the elimination of pathogenic microorganisms or toxins through the intestines.

6.2   STRUCTURE OF SPHINGOLIPIDS

There are over 300 known sphingolipids with considerable structural variation, but they all have in common a sphingoid base backbone, an amide-linked nonpolar aliphatic tail, and a polar head group. There are over 60 different sphingoid base backbones[11] that vary in alkyl chain lengths (from 14 to 22 carbon atoms), degree of saturation and position of double bonds, presence of a hydroxyl group at position 4, and branching of the alkyl chain. The amino group of the sphingoid base is often substituted by a long-chain fatty acid to produce ceramides. The fatty acids vary in chain length (14 to 30 carbon atoms), degree of saturation (but are normally saturated), and presence or absence of a hydroxyl group on the α- (or the ω-, in the case of ceramides of skin) carbon atom. More complex sphingolipids are formed when a polar head group is added at position 1 of a sphingoid base[10]. Figure 6.1 shows the general structure of sphingolipids.

As mentioned earlier, the structure of sphingolipids is of a complex nature; there is considerable variation among different organisms with respect to the type of sphingoid backbone, the polar group, and fatty acids. The sphingoid backbones of most mammalian sphingolipids consist mainly of sphingosine (*trans*-4-sphingenine, d18:1Δ^4), and a lesser amount of sphinganine (d18:0) and 4-hydroxysphinganine (t18:0), whereas plants contain sphinganine, 4-hydroxysphinganine, and *cis* and *trans* isomers of 8-sphingenine (d18:1Δ^8), 4,8-sphingadiene (d18:2$\Delta^{4,}\Delta^8$), and 4-hydroxy-8-sphingenine (t18:1Δ^8). The core structure of the sphingoid base is 2-amino-1,3-dihydroxyoctadecane, named sphinganine or d18:0, where d denotes a dihydroxy base. It can be substituted by an additional hydroxyl group at position 4, named 4-hydroxysphinganine or t18:0, where t denotes a trihydroxy base. If it has double bonds at position 4, 8, or 4 and 8, it is called sphingosine, d18:1Δ^4; 8-sphingenine, d18:1Δ^8; or 4,8-sphingadiene, d18:2$\Delta^{4,}\Delta^8$, respectively. Unlike mammalian sphingolipids, which consist of many different polar head groups (phosphocholine, glucose, galactose, *N*-acetylneuraminic acid, fructose and other carbohydrates), plant sphingolipids have mainly glucose (to a lesser extent oligosaccharides containing glucose and mannose, and inositol) as their head group. Depending on the head groups, sphingolipids are divided into two major classes: phosphosphingolipids, with a phosphoric acid linked to the position 1 of a ceramide through an ester bond (sphingomyelin with phosphocholine as the head group is the major component); and glycosphingolipids, with a glycosidic bond to a sugar moiety. The latter are further divided into neutral and acid glycosphingolipids. The neutral sphingolipids are cerebrosides with glucose, galactose, lactose, or oligosaccharides as the head group. The acid sphingolipids include gangliosides and sulfides. Gangliosides have oligoglycosidic head groups containing one or more sialic acid group

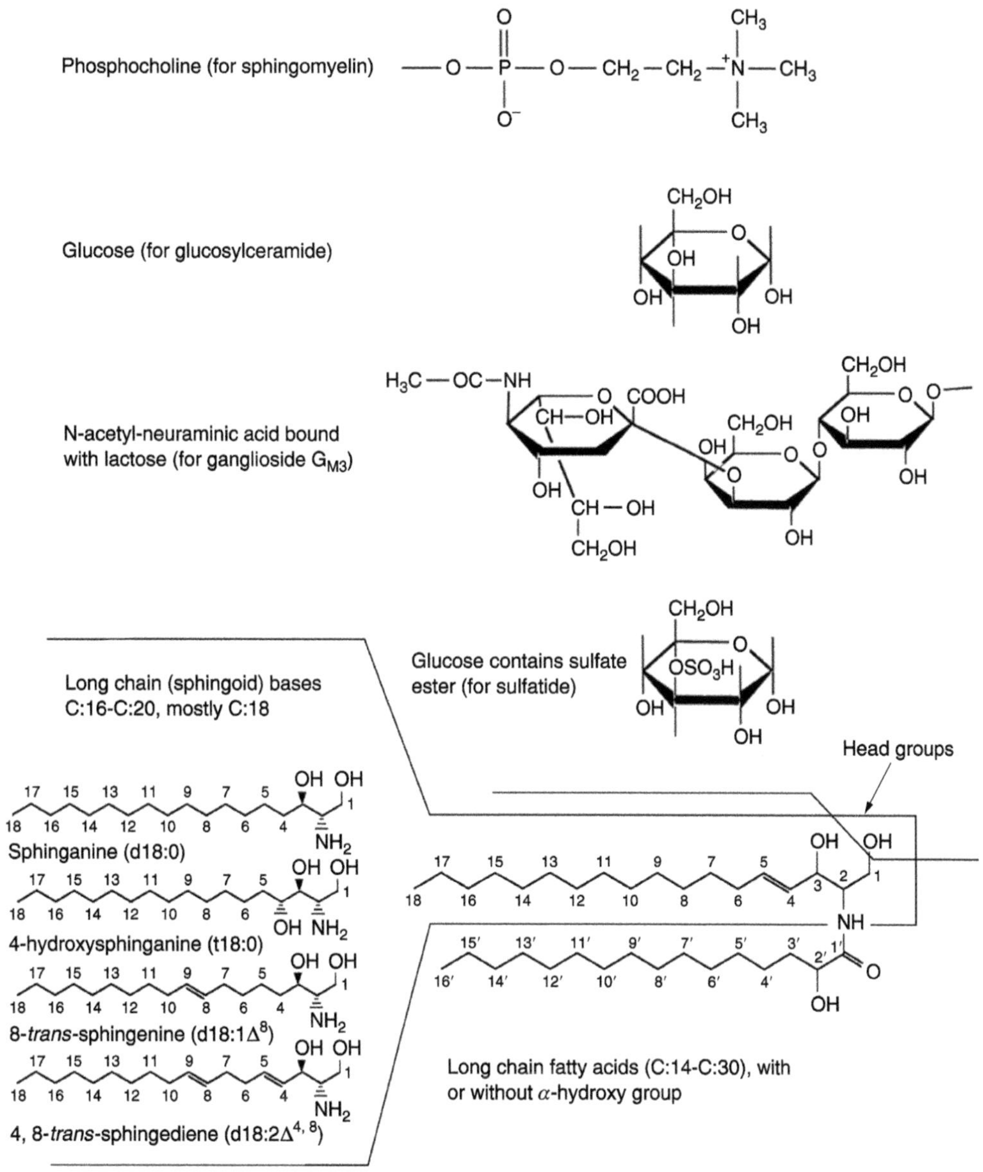

FIGURE 6.1 General structure of sphingolipids.

(N-acyl, especially acetyl derivatives of neuraminic acid). Sulfides have sulfate ester bound with the sugar moiety. In the case of fatty acids, it appears that plants typically contain mostly 2-hydroxy fatty acids, saturated or monoenoic, ranging from C_{14} to C_{26}, while sphingolipids from animals have fatty acids both with and without a 2-hydroxy group.

6.3 SPHINGOLIPIDS IN FOODS

Sphingolipids are components of a variety of foods. The amounts vary considerably. There is no evidence to indicate that sphingolipids are required for growth or survival. Generally, foods of mammalian origin (dairy products, eggs) are rich in sphingolipids. However, a study found high amounts of cerebrosides in soybean[12]. Considerable amounts of sphingolipids were also found in cereals, fruits, and vegetables[13–16]. Most recently, Sang et al.[17] studied the constituents of almond nuts, and one of the cerebrosides was found to be a major component of it.

The structures of the sphingolipids in food vary considerably. The sphingolipids of mammalian tissues, lipoproteins, and milk typically contain ceramides, sphingomyelins, cerebrosides, and gangliosides; plants, fungi, and yeast mainly have cerebrosides and phosphoinositides[10]. For example, sphingomyelin is the major mammalian sphingolipid, and it is rarely present in plants or microorganisms. Milk contains many kinds of sphingolipids, including sphingomyelin, glucosylceramide, lactosylceramide, and gangliosides.

Structural analysis of plant sphingolipids was carried out on a limited number of species and tissue types. Plant cerebrosides have been isolated from seeds and/or leaves of a few species, including rice, oats, wheat, winter rye, soybean, mung bean, spinach, and almond[13,17,18]. In vegetables and fruits, cerebrosides were also isolated from tubers of white yam, sweet potato, and potato, as well as apples, tomatoes, and red bell pepper[15,19–22]. In soybean, only glucose (Glc) ceramides (Cer) were found, with d18:2$\Delta^{4,}\Delta^8$ as the major backbone, and 2-hydroxy palmitic acid (C16:0h) (here h stands for hydroxyl group in fatty acid) as the main fatty acid. Other sphingoid backbones, such as d18:0, d18:1Δ^4, d18:1Δ^8, t18:0, and t18:1Δ^8 were found in minor amounts. Fatty acids such as C22:0h to C26:0h were found in trihydroxy base-containing species[12]. Bell pepper and tomato also contained mainly GlcCer with isomers of sphingadienine (d18:2$\Delta^{4,}\Delta^8$), sphingenine (d18:1), and 4-hydroxysphingenine (t18:1) as the predominant backbones. A major fatty acid, C16:0h, was almost exclusively associated with d18:2$\Delta^{4,}\Delta^8$ and d18:1 sphingoids, and C22:0h to C24:0h were primarily associated with t18:1 sphingoid[15]. Similar specificity of the amide-linked fatty acids was also found in the leaves of spinach[14], pea seeds[23], scarlet runner beans, and kidney beans[24]. The main glucocerebroside of whole rye leaf and plasma membrane was found to be C24:0h-t18:1 with lesser amounts of C24:0h-d18:2$\Delta^{4,}\Delta^8$ and C22:0h-t18:1[25]. The cerebrosides found in rice grain contained mainly Glc, but also Man-Glc, Glc-Glc, [Man]$_2$-Glc, Glc-Man, and [Man]$_3$-Glc as head groups and C24:0h and C22:0h as the major fatty acids. The sphingoid base d18:2$\Delta^{4,}\Delta^8$ was mainly found in glucocerebrosides, while t18:1 was normally connected with di-, or oligoglycosyl-ceramides[26]. Similar patterns were also found in pea seeds and wheat grain[27]. While the major cerebrosides in wheat grain were Glc-ceramide, Man-Glc-ceramide, Man-Man-Glc-ceramide, and Man-Man-Man-Glc-ceramide, the principal fatty acids were hydroxypalmitic and hydroxyarachidic acids, the major sphingoid was *cis*-8-sphingenine[27].

While most sphingolipids found in plants are cerebrosides, free ceramides were also found in some plant species, such as rice grain[26], pea seeds[23], black gram[28], scarlet runner beans, and kidney beans[24]. In plants, the molecular species of ceramides are different from the ceramide residue of cerebrosides, whereas in the case of animal tissue, free ceramide is supposed to be a precursor of cerebroside, due to their component structural similarity. The major sphingoids were mostly t18:1 and t18:0, and C24:0h was found to be the most popular fatty acid in plants.

In spite of the biological activities of these compounds, studies about the sphingolipid content in food are quite sparse. Table 6.1 summarizes the sphingolipid content in food from several available references. The value of the sphingolipid content is measured by the sum of sphingomyelin and glycosphingolipid. Since some amounts of sphingolipids have been published in moles, the conversion to grams was calculated by using 747 g/mol as the average molecular weight for glycosylceramide and 751 g/mol as the average molecular weight for sphingomyelin[10]. Five major classes of food are listed in Table 6.1. In dairy products, the content is calculated with the milk density of 1.03 g/ml. The estimation of sphingolipid in bovine whole milk is based on the sum of 0.019 g/kg of glycosphingolipid[29] and 0.089 g/kg of sphingomyelin[30]. As for the other products in dairy food, the amounts of sphingolipids are estimated from the sphingolipid content in bovine whole milk and the milk fat content of the products. The estimation of sphingolipid content in butter, eggs, meat products, fish, nuts, fruits, and some vegetables is based only on the sphingomyelin content since the glycosphingolipid content has not been published. As for sweet potato, wheat flour, and bell

TABLE 6.1
Sphingolipids in Food

Food	Sphingolipid content (g/kg)	Ref.
Diary products		10
Whole milk (3.5%)	0.11	30
Low-fat milk (<2%)	0.063	29
Cheese (29%)	0.91	29
Frozen dairy (11%)	0.35	29
Cream (37%)	1.16	29
Evaporated and condensed milk (9%)	0.28	29
Butter	0.35	16
Eggs	1.69	16
Meat products		10
Beef	0.29	31
Pork	0.26	31
Turkey	0.29	31
Chicken	0.39	31
Fish		10
Salmon	0.12	31
Catfish	0.075	31
Fruits		10
Apple	0.011	16
Banana	0.015	16
Orange	0.018	16
Nuts		10
Peanut	0.059	16
Cereals		10
Wheat flour	0.288	13
Vegetables		10
Cauliflower	0.14	16
Cucumber	0.02	16
Iceberg lettuce	0.038	16
Potato	0.05	16, 19
Tomato	0.031	15, 16
Sweet potato	1.269	14
Spinach	1.79	14
Soybean	2.23	12
Bell pepper	0.027	15

pepper, the estimation is based on the glycosylceramide content only. Other than the dairy products, the amount of sphingolipid in spinach, soybean, and tomato are calculated by the sum of sphingomyelin and glycosylceramide. Most data in the list were based on studies that were originally designed to elucidate chemical structures rather than to quantify sphingolipid content. Many utilized indirect measurement, such as phosphorous content of sphingomyelin, hexose content of cerebrosides, or nitrogen content to calculate the amount of sphingolipid. Structural variations were not considered. This summary shows that milk[30], egg[16], and soybeans[12] have the highest sphingolipid content, followed by meat (chicken, beef, pork)[31] and cereal (wheat), and fruits and vegetables[10,16] which have relatively low sphingolipid contents.

Recent research stresses the importance of sphingolipids in biological systems and their preventive effect on colon carcinogenesis. Due to the enormous structural diversity described above, it is important to establish structure–function relationships. There are no studies about the biological effect of plant sphingolipids, but studies with human adenocarcinoma cell line (HT29 cells) found that the toxicities of soy and wheat ceramides were comparable to brain ceramide[32]. Studies are underway using dietary sphingolipids as chemopreventive material. Knowledge about the structures and concentrations of sphingolipids in food and their metabolic pathways in the human body are very important. Also, little is known about variations of sphingolipid amounts over season and during food processing. Easy and sensitive methods are needed to determine sphingolipids in fresh and processed foods, both qualitatively and quantitatively.

6.4 SPHINGOLIPIDS ANALYSIS

Sphingolipids pose an enormous challenge to analytical chemists for the following reasons. Firstly, it is difficult quantitatively to isolate them in completely pure form, since the content in biological materials is very low. Secondly, it is difficult to separate and identify the molecular species because of the structural variations mentioned in Section 6.3. Thirdly, the lack of a chromophore makes it impossible to use UV detection. Different techniques were used to analyze this class of compounds, including thin layer chromatography (TLC), gas chromatography mass spectrometry (GC/MS), high-performance liquid chromatography (HPLC) with various detection methods, mass spectrometry (MS), and tandem mass spectrometry (MS/MS).

TLC is a common method used to separate and purify compounds. According to the literature[33–37], lipid extracts are usually subject to alkaline hydrolysis and then separated on silica gel plates. The mobile phase is a mixture of chloroform, methanol, and acetic acid in different proportions. Ceramides were identified by staining with copper sulfate in orthophosphoric acid or 8-anilino-1-naphthalene sulfonic acid. Quantification was carried out by densitometry. In one method[35], various types of ceramides from human stratum corneum cells were identified and quantified using TLC silica plates and chloroform/methanol/glacial acetic acid (190:9:1, v/v/v) as the mobile phase. The method was sensitive; ceramides were separated into five different fractions based on the presence or absence of hydroxyl groups in the fatty acid or long-chain bases. However, the method required previous separation of the ceramide fraction. Also, separation of the ceramide species could not be achieved by a single TLC run, and a second TLC run was required.

GC/MS was used by some researchers to analyze plant glucocerebrosides[15,25]. Fatty acid composition was analyzed after acid hydrolysis. Glucocerebrosides were dissolved in methanolic HCl and refluxed at 70°C for an extended time. The fatty acid and 2-hydroxy fatty acid methyl esters were extracted with petroleum ether, dried under nitrogen, and silylated with tetramethylsilane (TMS). The resulting fatty acid and O-trimethylsilyl 2-hydroxyl fatty acid methyl esters were analyzed by GC/MS. The long-chain base composition was determined after strong alkaline hydrolysis. The glucocerebrosides were refluxed in 10% barium hydroxide in dioxane at 110°C for 24 hours. Hydrolysis products were extracted with ethyl acetate. The long-chain bases were purified by TLC, and converted to N-acetyl derivatives by reaction with methanol:acetic anhydride 4:1 for 16 to 18 hours. After drying under nitrogen, the derivatives were dissolved in a chloroform/methanol/water mixture, and the long-chain bases were recovered in the chloroform phase. Following silylation, the N-acetyl-O-TMS long-chain base derivatives were analyzed by GC/MS. Before GC/MS analysis, the purified glucocerebroside fraction needs to be separated. The intact glucocerebroside molecular species were separated by C18 reverse-phase HPLC. The effluents were collected. Following drying and silylation, they were identified by GC/MS as TMS-ether derivatives.

The most popular chromatographic method used for analyzing this class of compounds is HPLC. However, lack of a chromophore makes it impossible to identify these molecules using UV detection. Even though some methods were developed involving the monitoring of the UV absorption at as low as 206 nm[38,39], they were very insensitive. Higher sample concentrations were required and the peak shape was not good. Various methods have been developed based on the derivatization of sphingolipids with fluorescent or UV-absorbing compounds and subsequent analysis of the derivatized compounds[40–43]. The most often-used derivatization reagents are benzoyl chloride and benzoyl anhydride, producing *N*-acyl derivatives, which have strong UV absorption in the 230 to 280 nm range, depending on the type of compound. Benzoylation was achieved by reaction with benzoyl chloride or benzoyl anhydride in pyridine at 70°C for as long as 4 hours. Benzoyl anhydride was preferred for ceramides containing nonhydroxy fatty acids, because in HPLC analysis *N*-acyl benzoyl derivatives overlap methyl benzoate, a byproduct of the reaction. However, for ceramides containing hydroxyl fatty acids and phytosphingosine, benzoyl chloride and prolonged reaction time were necessary, due to the steric hindrance of the hydroxy group of fatty acids. The method was cumbersome and time-consuming, and the reagents used are toxic; benzoyl chloride is carcinogenic. The chemicals involved must be prepared fresh each time, and the reaction is very sensitive to water. Another major disadvantage is the low stability of the derivatized products, which necessitates that the analysis be done shortly after the reaction. Fluorescent derivatization reagents used to label and identify ceramides were also reported[41,43–45]. Lester et al.[44] used the fluorescent amino group reagent 6-aminoquinolyl-*N*-hydroxysuccinimidyl carbamate to tag the molecular species of dihydrosphingosines and phytosphingosines and their 1-phosphates; Previati et al.[41] determined ceramide after coupling to the fluorescent label (+)-6-methoxy-α-methyl-2-naphthaleneacetic acid. The reaction was achieved after incubation at –20°C for 3 hours using catalytic agents 4-dimethylaminopyridine and *N*,*N*'-dicyclohexylcarbodiimide. Phospholipids should be separated before initiating the reaction, because of their inhibition effect on the reaction. Yano et al.[43] analyzed ceramides after reaction with anthroyl cyanide, a fluorescent reagent. The reaction was run in the presence of acetonitrile/dichloromethane (1:2 v/v) containing 0.4% quinuclidine at 4°C overnight. The methods with fluorescent derivatives are normally quite sensitive, but they have drawbacks, such as being time consuming and cumbersome as well as inhibition of the reactions by the impurities, and toxicity of the reagent chemicals.

The introduction of the evaporative light-scattering detector (ELSD) enables direct analysis without prior derivatizations[46]. ELSD is a mass detector, which is ideal for the analysis of compounds without a UV chromophore. ELSD is designed to separate nonvolatile solute particles from a volatile eluant. Quantification can be achieved since the response is a function of mass, but the relationship is generally not linear. Identification may be carried out by comparison with pure standards. However, ELSD has greatly simplified the analysis of all lipid classes and has become the method used frequently in lipid analysis by HPLC.

Both normal-phase (NP) HPLC and reverse-phase (RP) HPLC have been used in sphingolipid separation. Generally, NP-HPLC separates the sphingolipids (or other lipids) in classes, and RP-HPLC can separate the lipid subclasses into molecular species. Demopoulos et al.[38] separated individual glycolipid classes (gangliosides, sulfatides, *N*-palmitoyl-sphingosine, cardiolipin, digalactosyl-diacylglycerols, galactosyl-cerebrosides, and ceramides) using a silica column; the mobile phase used was a linear gradient from 100% acetonitrile to 100% methanol. Nomikos et al.[39] used a NP aminopropyl-modified silica gel HPLC column to separate several polar lipids, phospholipids, glycolipids, and phenolics. The separation was performed using a gradient elution with acetonitrile/methanol, methanol, and water. Cyano column[41] and diol column[46,47] have also been used. The mobile phase used for CN column was a gradient of 100% hexane, 1% 2-propanol in hexane, and 10% 2-propanol in hexane. For diol-modified silica column, a solvent gradient was made of solvent A, hexane–1-propanol–formic acid–triethylamine (63:35:0.6:0.08 v/v/v/v) and solvent B, 1-propanol–water–formic acid–triethylamine (89:10:0.6:0.08, v/v/v/v). C18 RP HPLC was reported to separate some glycolipid subclasses into molecular species[38]. The linear gradient was from 100% methanol/water (4:1, v/v) to 100% acetonitrile/methanol (7:5, v/v) and then hold for 10 minutes. Yano et al.[43] used C18 to separate ceramide molecular species; acetonitrile–methanol–ethyl acetate (12:1:7 v/v/v) was used as the mobile phase. Long-chain bases were also separated on a C18 column[44]. A novel PR-HPLC analysis employing a C6 (hexyl) column was carried out by Whitaker[15]. Using this column, he was able to separate cerebroside species from bell pepper and tomato fruits. The mobile phase used was a gradient elution with acetonitrile and water.

One of the most powerful techniques used in lipid analysis today is HPLC coupled with mass spectrometry (HPLC/MS). Several mass spectrometric ionization techniques, such as fast atom bombardment[31], electrospray ionization[47,48], ionspray ionization[49], and atmospheric pressure chemical ionization[47,50], have been used. By using HPLC/MS, one can get information on the molecular structure of the intact lipids, which helps differentiate molecular species within different lipid classes. By using MS/MS, identification of molecular species of different sphingolipids can be achieved in an easier and more sensitive way. There are many other advantages to using mass spectrometry, such as small sample size, minimal sample preparation, no need for derivatization, speed, and sensitivity. In the literature, sphingolipids of both animal and plant origin were analyzed by mass spectrometry.

Ceramide profile of a bovine brain extract and a lipid extract of cultured T-cells were analyzed by electrospray ionization mass spectrometry (ESI/MS) and tandem mass spectrometry (ESI/MS/MS)[48]. The sample, either directly or after clean up, was infused into the mass spectrometer. Collision-induced fragmentation results in characteristic product ions, m/z at 264, 282 for sphingosine and at 266, 284 for sphinganine, regardless of the length of the fatty acid chain. By using precursor ion scan analysis, sphingosine- and sphinganine-based ceramide species were detected. The change of ceramide levels in complex biological mixtures was measured quantitatively by comparison with mass intensity of an internal standard. Ceramides were also analyzed using atmospheric pressure chemical ionization-mass spectrometry (APCI/MS)[50]. Ceramide species from the cells were separated by RP HPLC, and detected by APCI/MS. Selected ion monitoring (SIM) was used to detect sphingosine-based ceramides by monitoring the common fragment ion m/z at 264 at high cone voltage. Quantification was carried out by comparing with known amounts of authentic samples. Bovine milk is a good source of sphingolipids, which include glucosyl ceramides, lactosyl ceramides, and sphingomyelins. Molecular species of these sphingolipids were analyzed by LC/MS[51], a method based on NP HPLC on-line with discharge-assisted thermospray (plasmaspray) mass spectrometry. Through tandem mass spectrometry using CID specific long-chain base and fatty acid compositions of the ceramide units can be revealed. In a paper a year later the same authors[47] discussed the analysis of a molecular species of sphingomyelin from bovine milk. Both ESI/MS and APCI/MS were used for structural determination. The sphingomyelin fraction was separated by NP HPLC. Firstly, using ESI, protonated molecules were detected; secondly, using APCI, fragmentation was achieved in the ion source. With the ceramide ions as precursors, ions representing both the long-chain bases and fatty acids were identified via collision-induced decomposition using APCI/MS/MS.

Sphingolipid determination of plant origin by mass spectrometry was also reported. The molecular species of the major sphingolipid, glucosylceramides, of soybean and wheat were analyzed by low- and high-resolution MS/MS using positive ion fast atom bombardment (FAB)[32]. The glucosylceramides were purified from soybean and wheat, and directly introduced into a mass spectrometer using a FAB probe. By analyzing the fragmentation pattern, different glucosylceramides were identified, but the principle used in the analysis was quite complicated.

REFERENCES

1. Hakomori, S.I., Bifunctional role of glycosphingolipids. Modulators for the transmembrane signaling and mediators for cellular interactions, *J. Biol. Chem.*, 265, 18713–18716, 1991.
2. Merrill, A.H., Jr., Schmelz, E.M., Dillehay, D.L., Spiegel, S., Shayman, J.A., and Schroeder, J.J., Sphingolipids. The enigmatic lipid class: biochemistry, physiology, and pathophysiology, *Toxicol. Appl. Pharmacol.*, 142, 208–225, 1997.
3. Riboni, L., Viani, P., Bassi, R., Prinetti, A., and Tettamanti, G., The role of sphingolipids in the process of signal transduction, *Prog. Lipid Res.*, 36, 153–195, 1997.
4. Schmelz, E.M., Dietary sphingomyelin and other sphingolipids in health and disease, *Nutr. Bull.*, 25, 135–139, 2000.

5. Dillehay, D.L., Webb, S.J., Schmelze, E.M., and Merrill, A.H., Jr., Dietary sphingomyelin inhibits 1,2-dimethylhydrazine-induced colon cancer in CF1 mice, *J. Nutr.*, 124, 615–620, 1994.

6. Schmelz, E.M., Dillehay, D.L., Webb, S.K., Reiter, A., Adams, J., and Merrill, A.H., Jr., Sphingomyelin consumption suppresses aberrant colonic crypt foci and increases the proportion of adenomas versus adenocarcinomas in CF1 mice treated with 1,2- dimethylhydrazine: implications for dietary sphingolipids and colon carcinogenesis. *Cancer Res.*, 56, 4936–4941, 1996.

7. Schmelz, E.M., Sullards, M.C., Dillehay, D.L., and Merrill, A.H., Jr., Colonic cell proliferation and aberrant crypt foci formation are inhibited by dairy glycosphingolipids in 1,2-dimethylhydrazine treated CF1 mice, *J. Nutr.*, 130, 522–527, 2000.

8. Imaizumi, K., Tominaga, A., and Sato, M., Effects of dietary sphingolipids on levels of serum and liver lipids in rats. *Nutr. Res.*, 12, 543–548, 1992.

9. Kobayashi, T., Schimizugawa, T., Osakabe, T., Watanabe, S., and Okuyama, H., A long- term feeding of sphingolipids affected the levels of plasma cholesterol and hepatic triacylglycerol but not tissue phospholipids and sphingolipids. *Nutr. Res.*, 17, 111–114, 1997.

10. Vesper, H., Schmelz, E.M., Nilolova-karakashian, M.N., Dillehay, D.L., Lynch, D.V., and Merrill, A.H., Jr., Sphingolipids in food and the emerging importance of sphingolipids to nutrition, *J. Nutr.*, 129, 1239–1250, 1999.

11. Karlsson, K.A., On the chemistry and occurrence of sphingolipid long-chain bases, *Lipids*, 5, 6–43, 1970.

12. Ohnishi, M. and Fujino, Y., Sphingolipids in immature and mature soybeans, *Lipids*, 17, 803–810, 1982.

13. Laine, R.A. and Renkomen, O., Ceramide di- and trihexosides of wheat flour, *Biochem.*, 13, 2837–2843, 1974.

14. Ohnishi, M., Ito, S., and Fujino, Y., Characterization of sphingolipids in spinach leaves, *Biochim. Biophys. Acta*, 752, 416–422, 1983.

15. Whitaker, B.D., Cerebrosides in mature-green and red-ripe bell pepper and tomato fruits, *Phytochemistry*, 42, 627–632, 1996.

16. Zeisel, S.H., Choline, in *Modern Nutrition in Health and Disease*, 8th ed., Vol. 1, Shils, M.E., Olson, J.A., and Shike, M., Eds., Williams & Wilkins, Baltimore, MD, 1994, p. 451.

17. Sang, S.M., Kikuzaki, H., Lapsley, K., Rosen, R.T., Nakatani, N., and Ho, C.-T., Sphingolipid and other constituents from almond nuts (*Prunus amygdalus* Batsch), *J. Agric. Food Chem.*, 50, 4709–4712, 2002.

18. Lynch, D.T., Sphingolipids, in *Lipid Metabolism in Plants*, Moore, T.S., Jr., Ed., CRC Press, Boca Raton, FL, 1993, pp. 285–308.

19. Galliard, T., Aspects of lipid metabolism in higher plants: I. Identification and quantitative determination of the lipids in potato tubers, *Phytochemistry*, 7, 1907–1914, 1968.

20. Galliard, T., Aspects of lipid metabolism in higher plants: II. The identification and quantitative analysis of lipids from the pulp of pre- and post-climacteric apples, *Phytochemisty*, 7, 1915–1922, 1968.

21. Osagie, A.U. and Opute, F.I., Major lipid constituents of Dioscorea rotundata tuber during growth and maturation, *J. Exp. Botany*, 32, 737–740, 1981.

22. Walter, W.M., Jr., Hansen, A.P., and Purcell, A.E., Lipids of cured centennial sweet potatoes, *J. Food Sci.*, 36, 795–797, 1971.

23. Ito, S., Ohnishi, M., and Fujino, Y., Investigation of sphingolipids in pea seeds, *Agric. Biol. Chem.*, 49, 539–540, 1985.

24. Kojima, M., Ohnishi, M., and Ito, S., Composition and molecular species of ceramide and cerebroside in scarlet runner beans (*Phaseolus coccineus* L.) and kidney beans (*Phaseolus vulgaris* L.), *J. Agric. Food Chem.*, 39, 1709–1714, 1991.

25. Cahoon, E.B. and Lynch, D.V., Analysis of glucocerebrosides of rye (*Secale cereale* L. cv Puma) leaf and plasma membrane, *Plant Physiol.*, 95, 58–68, 1991.

26. Fujino, Y., Ohnishi, M., and Ito, S., Molecular species of ceramide and mono-, di-, tri-, and tetraglycosylceramides in bran and endosperm of rice grains, *Agric. Biol. Chem.*, 49, 2753–2762, 1985.

27. Fujino, Y. and Ohnishi, M., Sphingolipids in wheat grain, *J. Cereal Sci.*, 1, 159–168, 1983.

28. Kondo, Y. and Nakano, M., Kinetic changes in free ceramide and cerebroside during germination of black gram, *Agric. Biol. Chem.*, 50, 2553–2559, 1986.

29. Newburg, D.S. and Chaturvedi, P., Neutral glycolipids of human and bovine milk, *Lipids*, 27, 923–927, 1992.

30. Zeisel, S.H., Char, D., and Sheard, N.F., Choline, phosphatidylcholine and sphingomyelin in human and bovine milk and infant formulas, *J. Nutr.*, 116, 50–58, 1986.

31. Blank, M.L., Cress, E.A., Smith, Z.L., and Snyder, F., Meat and fish consumed in the American diet contain substantial amounts of ether-linked phospholipids, *J. Nutr.*, 122, 1656–1661, 1992.

32. Sullards, M.C., Lynch, D.V., Merrill, A.H., Jr., and Adams, J., Structure determination of soybean and wheat glucosylceramide by tandem mass spectrometry, *J. Mass Spect.*, 35, 347–353, 2000.

33. Bose, R., Chen, P., Loconti, A., Abrams, J., and Kolesnick, R., Ceramide generation by the reaper protein is not blocked by the caspase inhibitor p35, *J. Biol. Chem.*, 273, 28852–28859, 1998.

34. Robson, K.J., Stewart, M.E., Michelson, S., Lazao, N.D., and Downing, D.T., 6-Hydroxy-4-sphingenine in human epidermal ceramides, *J. Lipid Res.*, 35, 2060–2068, 1994.

35. Motta, S., Monti, M., Sesana, S., Caputo, R., Carelli, S., and Ghidoni, R., Ceramide composition of the psoriatic scale, *Biochim. Biophys. Acta*, 1182, 147–151, 1993.

36. Haak, D., Gable, K., Beeler, T., and Dunn, T., Hydroxylation of Saccharomyces cerevisiae ceramide requires sur2p and scs7p, *J. Biol. Chem.*, 272, 29704–29710, 1997.

37. Selvam, R. and Radin, N.S., Quantitation of lipids by charring on thin-layer plates and scintillation quenching: application to ceramide determination, *Anal. Biochem.*, 112, 338–345, 1981.

38. Demopoulos, C.A., Kyrili, M., Antonopoulou, S., and Andrikopoulos, N.K., Separation of several main glycolipids into classes and partially into species by HPLC and UV-detection, *J. Liquid Chromatogr. Related Technol.*, 19, 771–781, 1996.

39. Nomikos, T., Karantonis, H.C., Fragopoulou, E., and Demopoulos, C.A., One-step separation system for the main phospholipids, glycolipids, and phenolocs by normal phase HPLC. Application to polar lipid extracts from olive and sunflower oils, *J. Liquid Chromatogr. Related Technol.*, 25, 137–149, 2002.

40. Iwamori, M., Costello, C., and Moser, H.W., Analysis and quantitation of free ceramide containing non-hydroxy and 2-hydroxy fatty acids, and phytosphingosine by high-performance liquid chromatography, *J. Lipid Res.*, 20, 86–96, 1979.

41. Previati, M., Bertolaso, L., Tramarin, M., Bertagnolo, V., and Capitan, S., Low nanogram range quantitation of diglycerides and ceramide by high performance liquid chromatography, *Anal. Biochem.*, 233, 108–114, 1996.

42. Snada, S., Uchida, Y., Anraku, Y., Izawa, A., Iwamori, M., and Nagai, Y., Analysis of ceramide and mono-hexaosyl glycolipid derivatives by high-performance liquid chromatography and its application to the determination of the molecular species in tissues, *J. Chromatogr.*, 400, 223–231, 1987.

43. Yano, M., Kishida, E., Muneyuki, Y., and Masuzawa, Y., Quantitative analysis of ceramide molecular species by high performance liquid chromatography, *J. Lipid Res.*, 39, 2091–2098, 1998.

44. Lester, R.L. and Dickson, R.C., High-performance liquid chromatography analysis of molecular species of sphingolipid-related long chain bases and long chain base phosphates in Saccharomyces cerevisiae after derivatization with 6-aminoquinolyl-N-hydroxysuccinimidyl carbamate, *Anal. Biochem.*, 298, 283–292, 2001.

45. Merrill, A.H., Jr., Wang, E., Mullins, R.E., Jamison, W.C.L., Nimkar, S., and Liotta, D.C., Quantitation of free sphingosine in liver by high-performance liquid chromatography, *Anal. Biochem.*, 171, 373–381, 1988.

46. McNabb, T.J., Cremesti, A.E., Brown, P.R., and Fischl, A.S., The separation and direct detection of ceramides and sphingoid bases by normal-phase high-performance liquid chromatography and evaporative light-scattering detection, *Anal. Biochem.*, 276, 242–250, 1999.

47. Karlsson, A.Å., Michélsen, P., and Odham, G., Molecular species of sphingomyelin: determination by high-performance liquid chromatography/mass spectrometry with electrospray and high-performance liquid chromatography/tandem mass spectrometry with atmospheric pressure chemical ionization, *J. Mass Spect.*, 33, 1192–1198, 1998.

48. Gu, M., Kerwin, J.L., Watts, J.D., and Aebersold, R., Ceramide profiling of complex lipid mixtures by electrospray ionization mass spectrometry, *Anal. Biochem.*, 244, 347–356, 1997.

49. Mano, N., Oda, Y., Yamada, K., Asakawa, N., and Katayama, K., Simultaneous quantitative determination method for sphingolipid metabolites by liquid chromatography/ionspray ionization tandem mass spectrometry, *Anal. Biochem.*, 244, 291–300, 1997.

50. Couch, L.H., Churchwell, M.I., Doerge, D.R., Tolleson, W.H., and Howard, P.C., Identification of ceramides in human cells using liquid chromatography with detection by atmospheric pressure chemical ionization-mass spectrometry, *Rapid Commun. Mass Spect.*, 11, 504–512, 1997.

51. Karlsson, A.Å., Arnoldsson, K.C., Westerdahl, G., and Odham, G., Common molecular species of glucosyl ceramides, lactosyl ceramides and sphingomyelins in bovine milk determined by high performance liquid chromatography–mass spectrometry, *Milchwissenschaft*, 52, 554–559, 1997.

7 Modification and Purification of Sphingolipids and Gangliosides

Scott Bloomer
Land O'Lakes Inc., St. Paul, Minnesota

CONTENTS

7.1 INTRODUCTION

Milk is a safe and convenient source of sphingolipids and gangliosides for nutritional, medical, and structural research and development. Interest has focused on these bioactive lipids due to their impact on cellular metabolism and implication in cancer prevention and apoptosis. Their structures offer several functional groups which lend themselves to modification for enhancement and alteration of activity. This chapter discusses the sources of sphingolipids and gangliosides. Purification and modification are then discussed, with emphasis on patented methods to help the reader deduce where the remaining opportunities are present.

Sphingolipids and gangliosides are biologically active polar lipids present in very small amounts in most biological materials. Their surface activity and low concentration offers worthwhile analytical and purification challenges to chemists, and their unexplored biological activities provide opportunities for biochemical and medical research.

Sphingolipids and gangliosides have long been associated with brain tissues, owing to their early discovery in bovine brain[1]. Several industrial processes rely on gangliosides from bovine brain[2–5], which have become more risky since the spread of bovine spongiform encephalitis (BSE). However, sphingolipids and gangliosides are readily obtained from bovine milk and colostrums[6,7], and one of the purposes of this chapter is to bring this safe source to the attention of the reader.

7.1.1 SPHINGOLIPIDS

Sphingolipids bear certain structural similarities to phospholipids. This is evident by comparing the structures of phosphatidylcholine (PC) **1** and sphingomyelin (SM) **2.** The glycerol backbone of **1** is replaced with 2-amino-1,3-dihydroxypropane in **2** and in all other sphingolipids and gangliosides[8]. The three points of difference in SM are: (1) the hydroxyl group corresponding to C1 of PC is unesterified; (2) one of the C1 hydrogens of PC is replaced with a long-chain acyl group anchored by a double bond; and (3) an amide base instead of a common ester bond anchors a fatty acid chain to the carbon corresponding to the C2 carbon of PC. The backbone of sphingolipids and gangliosides originates from serine[9]. The side-chain serine hydroxyl becomes the link to a polar headgroup; the distal free hydroxyl oxygen originates from palmitoyl CoA reduced by NADPH. Over 300 structurally discrete sphingolipids are known.

7.1.2 GANGLIOSIDES

Gangliosides are a subset of glycosphingolipids; the other subsets, neutral glycolipids and sulfatides, are not reviewed here. Gangliosides are differentiated by the presence of sialic acid **3** (*N*-acyl- or *O*-acylneuraminic acid) linked to sugar moieties. Ganglioside classification is on the basis of carbohydrate structure, which seems strange for lipids[8]. This is because it is the carbohydrate moiety that projects from membranes and determines much of the function of the molecules. Glycosphingolipids are the glycosides of *N*-acylsphingosine **4** (the trivial name is ceramide). Typical milk gangliosides include GM3 **5** (sialosyllactosylceramide; NeuAcα2-3Galβ1-4Glcβ1-1Cer) and GD3 (di-*N*-acetyl-neuraminosyl hematoside; NeuGly2-8NeuGly2-3Galβ1-4Glcβ1-1Cer). Interested readers should consult the definitive review by Hakomori[8].

STRUCTURE 7.1 Phosphatidylcholine (PC).

STRUCTURE 7.2 Sphingomyelin (SM).

STRUCTURE 7.3 Sialic acid (*N*-acyl- or *O*-acyl neuraminic acid).

STRUCTURE 7.4 Ceramide (*N*-acylsphingosine).

STRUCTURE 7.5 GM3 (Sialosyllactosylceramide).

7.2 SOURCES OF SPHINGOMYELIN AND GANGLIOSIDES FOR RESEARCH

7.2.1 SPHINGOLIPIDS

For quantities useful for research, sphingomyelin and gangliosides can be obtained from brain tissue[10], meat and fish[11], egg yolk[10], and dairy products[12]. Bovine milk is a convenient source of sphingomyelin and gangliosides, and is currently considered to be safer than bovine brain tissue.

Sphingolipids from different sources often have significant structural differences in acyl groups, polar head groups, and sphingolipid backbone. Plant sphingolipids contain less variety than mammalian sphingolipids, and are composed mainly of cerebrosides[13]. Bovine brain SM is enriched in C18:0 and C24:1 fatty acids; chicken egg yolk is enriched in C16:0 fatty acids; and bovine milk SM is enriched in long-chain saturated fatty acids C22:0, C23:0, and C24:0[10]. Human milk is somewhat higher in SM than bovine milk[14].

Annual human ingestion of sphingolipids has been estimated to be about 115 to 140 g; about a quarter is from milk products[15]. The sphingolipid contents of several foods are listed in Table 7.1. They are not hydrolyzed in the stomach, but substantial hydrolysis takes place in the small intestine and colon (primarily by microflora). Much of the products are taken up by the intestinal cells. Sphingomyelin levels in the colon can be elevated by feeding; this has implications in deterring colon cancer (see below)[16]. The sphingolipid contents from food have been reviewed[15].

7.2.2 GANGLIOSIDES

Although gangliosides are present in a wide variety of tissues, much of the gangliosides used in research to date have been obtained from brain tissue[8] and milk. The spread of BSE (since 1986)

TABLE 7.1
Sphingolipid Content of Foods and Dairy Process Streams (mg/g)

Chicken	0.4	Butter	0.35
Eggs	1.7	Buttermilk	2.5
Soybeans	1.8	Cheese	1.0
Wheat flour	0.43	Whey	0.4
Milk	0.12	Cream	1.3

Adapted from Vesper, H., Schmelz, E.-M., Nikolova-Karakashian, N., Dillehay, D., Lynch, D., and Merrill A., *J. Nutr.*, 129, 1239–1250, 1999.

TABLE 7.2
Major Milk Gangliosides; State of Lactation Not Specified

Ganglioside	Human (µg/l)	Bovine (µg/l)
GM1	1.2	1.2
GM2	250	700
GM3	8.1×10^3	300

Adapted from Walsh, M. and Wiemer, B., abstracts of papers, 90th Annual Meeting of the American Oil Chemists' Society, Orlando, FL, May 9–12, 1999, American Oil Chemists' Society, Champaign, IL, 1999.

has compromised bovine brain as a source, necessitating development of protocols for purification of gangliosides from bovine brain tissue in a manner that ensures removal of the virus[17]. Gangliosides are often tightly associated with cholesterol, forming rigid microdomains referred to as "rafts"[18]. Dissociation of these rafts may be important in purification.

The gangliosides of human and bovine milk differ somewhat in their composition (Table 7.2). In addition, the relative proportions of gangliosides in human and bovine milk change as lactation proceeds, presumably to meet the changing needs of the infant. In the first three weeks of human lactation, GD3 is the most abundant ganglioside; after three weeks, GM3 becomes dominant[8,19]. Bovine milk gangliosides are also most abundant in GD3 in the beginning of lactation (5 days), decreasing thereafter as the content of GM3 increases[7]. These changing targets have implications for infant formula manufacture. The infant brain and central nervous system contain significant levels of sialic acid, so it is deemed important in development of structure and function. Human milk contains 0.3 to 1.5 mg/ml of sialic acid, most of which is bound to oligosaccharides but a significant fraction is found in gangliosides[20]. Human milk gangliosides are much more effective inhibitors of enterotoxins than bovine gangliosides. In bovine milk, GD3 and GM3 predominate; about 70 mg of GD3 can be recovered per kilogram of bovine cream[21].

Bovine milk is a very attractive source for sphingomyelin and gangliosides for further study. It is widely available and produced according to "good manufacturing practice" to ensure a wholesome starting material. In addition, the materials of interest are enriched in several products of milk processing, especially cream[12], buttermilk[22–24], and whey[25–27]. These processing streams are enriched in milkfat globule membrane material, which is the preferred location for polar molecules in milk[28,29]. The surface area of milkfat globule membrane in milk is 46 m^2/l, resulting in abundant surface area to accommodate sphingolipids and gangliosides[12].

STRUCTURE 7.6 Sphingosine.

7.3 FUNCTIONS AND USES OF SPHINGOLIPIDS AND GANGLIOSIDES

Sphingolipids and gangliosides are not ordinary fats; they are powerful biomolecules which serve structural functions, have profound regulatory functions, and are effective at low concentrations[30].

7.3.1 FUNCTIONS AND USES OF SPHINGOLIPIDS

Sphingolipids help define the structural properties of membranes, lipoproteins, and the water barrier of skin[31]. They participate in cell–cell communication, cell recognition, and anchor membrane proteins. Sphingolipids and metabolites thereof exert an effect on the growth, differentiation, and apoptosis of most cell types that have been studied[15]. Sphingolipids can be hydrolyzed to form potent second messenger substances with a variety of functions[32]. The sphingolipid metabolite sphingosine **6** is one of the most potent inhibitors of protein kinase C found in mammalian cells; it also inhibits or stimulates a host of other enzymes[31].

Sphingolipids are hydrolyzed *in vivo* to ceramide **4** and sphingosine **6**, which are important in transmembrane signal transduction and cell regulation[33]. These molecules serve as second messengers for extracellular agonists, such as cytokines, hormones, and growth factors.

Sphingolipids are considered tumor suppressor lipids due to their participation in cellular pathways associated with the suppression of oncogenesis[33]. Sphingomyelin hydrolysis turns on one or more of three antiproliferative pathways: inhibition of cell growth, induction of differentiation, or apoptosis[34]. These provide useful decelerators of cell growth. For example, extracellular activation of receptors of the tumor necrosis factor superfamily stimulates sphingomyelinases so that sphingomyelin in the cell membrane is hydrolyzed to ceramide and phosphorylcholine within minutes. Ceramide stimulates the induction of apoptosis by cytotoxic humoral factors[35] and apoptotic DNA degradation closely follows.

Anticarcinogenic activity in the colon has been demonstrated by feeding higher levels of sphingomyelin than are present in a normal diet[30]. Sphingomyelinase is present in very small amounts *in vivo*, so sphingomyelin is digested slowly through the small intestine and colon[33]. About 70% inhibition of the development of aberrant crypt foci, an early marker of colon carcinogenesis, has been clearly demonstrated[15] after SM feeding.

Sphingolipids have been implicated in atherogenesis by mediating cellular events believed to be crucial in the formation of vascular lesions[13]. Several components of atherogenic lesions (oxidized low-density lipoprotein (LDL), growth factors, and cytokines) stimulate the hydrolysis of sphingomyelin to generate ceramide[36]. A direct correspondence between the amount of sphingolipids and cholesterol in tissues has been observed[31]. Sphingolipids are usually found in close association with cholesterol[37], and the metabolism of sphingolipids and cholesterol is related.

Sphingomyelins are critical in the maintenance of membrane microdomains and thus are ubiquitous in eukaryotic cell membranes[34], primarily on the outer leaflet of the plasma membrane[14,38,39]. Sphingomyelins are enriched in epidermis cells[40] and serum lipoproteins[41] (especially LDL[15]).

Sphingomyelin from milk readily forms lamellar phases, liposomes, and oil-in-water emulsions. The high gel-to-liquid crystal transition temperature of sphingomyelin suggests unique stability advantages for use in pharmaceutical compositions and cosmetics[41]. They can combine with phospholipids to form an aqueous dispersion of vesicles useful for cosmetic emulsions; milk sphingolipids are preferred to animal brain sphingolipids for application to human skin[42]. Sphingolipids have proven bactericidal activity[43].

7.3.2 FUNCTIONS AND USES OF GANGLIOSIDES

Gangliosides exert significant biological activity. GM3 and GM1 can slow cell reproduction[31]. Ganglioside modulation of receptors may be due to their rapid synthesis and turnover, allowing cells to respond to changing external stimuli; ganglioside aggregation on cell walls produces high local concentrations, facilitating binding of low-affinity receptors. GM3 is present in large quantities in the plasma membrane of CD4 super (+) lymphocytes and macrophages; when several HIV-1 and HIV-2 glycoproteins were tested, all of them interacted with GM3[44]. GM3 and GD3 are enriched in some cancer cells; levels are significantly elevated in human gastric tumors and some mammary tumors[45]. In melanomas GD3 is derived from GM3. The ratio of GM3 to GD3 is apparently useful for monitoring the progress of Stage II melanoma[46].

Gangliosides, administered alone or in combination, are effective in relief of pain resulting from peripheral neuropathies[47]. This effect was noted with naturally occurring gangliosides in different combinations. Ganglioside GM1, either in the native state, or modified to form salts or so-called "inner esters," can prevent the development of tolerance to the analgesic effect of morphine and related opiates[48]. This allows low morphine doses to be used for prevention of pain and obviates the problems of adverse reactions to increasing doses of morphine to compensate for human tolerance.

Both human and bovine gangliosides have enterotoxin inhibitory activity, inhibiting bacterial adhesion by *Escherichia coli* and enterotoxin binding of *Vibreo cholera*. Thus, they exert a protective effect against enterotoxin-induced diarrhea in infants[49]. A necessary feature for this protection is resistance to acid hydrolysis in the infant stomach; 80% of the sialic acid of GD3 and GM3 remains intact under infant stomach conditions, remaining able to exert biological activity in the infant intestine. Milk gangliosides are as effective as sphingolipids in the inhibition of the development of aberrant crypt foci, an early marker of colon carcinogenesis[15]. Dietary gangliosides are useful in modifying colon bacterial populations of preterm newborns (decreasing *E. coli* and increasing bifidobacteria counts)[50].

An improved application for controlling diarrhea involves combining GM1 or derivatives thereof with activated carbon or cellulose to form a matrix which binds cholera toxin and facilitates removal from the site of activity[51].

Immobilized gangliosides for simultaneous capture and detection of several pathogens have been thoroughly investigated[24]. A biosensor that takes advantage of human ganglioside interactions for detection of a large number of toxins, protozoa, viruses, and bacteria has been developed. Capture of pathogens from 16 discrete organisms has been demonstrated. This approach is much more selective than the use of antibodies; false positives were obtained with antibody capture of dead *E. coli* 0157 H7 cells, but ganglioside interactions responded only to live cells.

Lactoferrin, a milk protein with significant antibacterial activity, is a very minor component of human and bovine milk (20 to 200 mg/l)[52]. Walsh et al. have taken advantage of ganglioside–lactoferrin interactions to produce concentrates in which lactoferrin constituted 40% of the protein eluted. A similar approach was used to enrich transferrin, which was only present at 20 to 200 μg/l, to 14% concentration[53].

7.4 PURIFICATION STRATEGIES

Sphingolipids and gangliosides have been subjected to a variety of purification schemes on many scales. Most of the published and patented methods start from animal sources, especially bovine or porcine brain, milk, or epidermis[54]. Sphingolipids and gangliosides are enriched in buttermilk and the lipid fraction of whey, making them safe, attractive starting materials[55].

7.4.1 SOLVENT EXTRACTION AND PURIFICATION METHODS

Hakomori's method, based on Folch washing, is a common starting point for ganglioside purification[8]. In this method, brain tissue is homogenized with 20 volumes of chloroform/methanol 2/1 (v/v). For large amounts of brain tissue (kilograms) prehomogenization in 5 to 10 volumes of acetone followed by filtration to remove solids, evaporation, and reextraction of recovered residue yields an acetone powder. This powder is subjected to the chloroform/methanol extraction with only small losses of gangliosides[8]. The review by Hakomori[8] is the definitive work on identification of purified gangliosides; the field has also been reviewed more recently[56].

Classic chloroform/methanol extraction (Folch extraction) is useful for both sphingolipid and ganglioside purification for analysis[22] and experimentation[25]. When materials rich in neutral lipids are extracted, the Folch method is preceded by extraction with cold acetone to remove triacylglycerols and other neutral lipids[7]. Ganglioside GM1 can be extracted from bovine brain in a straightforward manner[57].

Ganglioside purification from natural sources can be effected by extraction with isopropanol/hexane/water, chloroform/methanol, or 90% ethanol. Further purification steps such as Folch partitioning, ion exchange chromatography, and HPLC are required to produce a material suitable for further modification[8].

Ganglioside extraction for immobilized pathogen capture beds was very straightforward[24]. Fresh buttermilk (30% solids) was ultrafiltered through a membrane with a 1 kDa cut-off to remove lactose and extracted with chloroform/methanol/water (40/80/30 v/v/v). Fresh buttermilk is a preferred starting material, as some ganglioside degradation takes place in production of dairy powders in commercial spray dryers.

A patent teaches the purification of gangliosides from milk using a single solvent[23]. This method starts from butter from which the aqueous phase has been removed (butter serum, 100 g); to this 1000 g of an 80% ethanol solution were added. This mixture was stirred for an hour at 60°C and filtered to remove undissolved solids. After cooling the filtrate to −20°C overnight, a precipitate was formed. This precipitate (6.3 g) contained 4.7% ganglioside, a 16.7-fold enrichment. As this procedure avoids the use of chloroform, food-grade concentrates are obtained.

Due to the presence of large amounts of protein and lactose in commercial spray-dried whey powders, there are special difficulties in obtaining efficient extraction of lipids from an otherwise very useful starting material. A patent teaches a method that solves this problem and enables downstream processing of lipids extracted from whey[58]. Normally these components are, through the drying process and other processes employed in manufacturing the powder, so well embedded in an impenetrable matrix of lactose, calcium phosphate, and protein that they do not allow themselves to be extracted. Through the proper choice of solvent and extraction conditions this matrix is opened up and rendered permeable for the dissolved substances. The method relies on the proper ratios of polar short-chain aliphatic alcohol, water, and dried whey powder at the proper temperature to prevent the formation of an intractable, swollen cake from which extraction is impossible. In our hands, sphingolipids and gangliosides from buttermilk powder and procream powder (a lipid-enriched byproduct of the production of whey protein isolate) were easily extracted according to the method by adding a fourfold excess (v/w) of a solution of isopropanol containing 10% water and stirring at 50°C. In addition to good recoveries of lipids, the resulting powders were lighter in color and had lost some of their characteristic dairy odors.

Whey fat concentrate obtained by the Swedish method[58] was the foundation for US and European patents, which teach a method of extracting sphingomyelin from a phospholipid-containing fat concentrate[59,60]. In this method, a fat concentrate originating from extraction of dried whey or buttermilk with a polar organic solvent is the starting material. To 100 kg of this whey fat concentrate[58] were added 200 l of 75% ethanol and 100 l of n-heptane. A biphasic mixture was formed, the sphingolipids and phospholipids being dissolved in the n-heptane phase, so the ethanol phase was removed. To the resulting heptane phase was added a solvent of intermediate polarity, such as acetone, in which sphingolipids are not soluble. A precipitate of sphingolipids (60%) was formed by keeping the temperature of the mixture at 20°C. The precipitate was removed and phospholipids were recovered from the *n*-heptane/acetone mixture by cooling to 0 to 5°C. If desired, the sphingolipid-enriched precipitate can be subjected to a second intermediate-polarity solvent wash to yield a precipitate containing 70% sphingomyelin. With a subsequent column purification step, sphingomyelin of 95% purity may be obtained[59].

7.4.2 COLUMN PURIFICATION METHODS

Svennerholm purified gangliosides on a powdered cellulose column in 1956[8]. More recently, he and co-workers reported quantitative separation of ganglioside into mono-, di-, tri-, tetra, and pentasialo-ganglioside fractions with a new anion exchange resin[57]. Ganglioside GD3 (Glac2) was recently prepared from bovine cream by liquid-phase extraction with methanol or ethanol followed by anion exchange chromatography. This method affords 70 mg of pure GD3 from 1 kg of bovine cream[21].

For the preparation of small amounts of sphingolipids, solid-phase extraction columns readily produced high-purity substances[61]. They are routinely used for preparation of sphingolipids and derivatives for analysis[62]. Sphingolipids could even be fractionated from complex lipid mixtures into different classes using this approach[63].

HPLC remains a popular technique for effecting purifications where high-purity materials are essential, such as in cell culture research. Equine erythrocyte GM3 was purchased and purified using a Zorbax-NH2 column in less than 15 minutes and by employing an aqueous polar solvent eluent. The approach was also effective for isolation of rat liver GM3[64]. Separation of gangliosides from mouse hybrodoma cells on a larger scale (batches of up to 500 mg of gangliosides) was carried out using a strong anion exchange gel eluted with a buffer-methanol gradient.

7.4.3 MISCELLANEOUS METHODS

Chitosan is useful for selective precipitation of milkfat globule membrane fragments in cheese whey[65], followed by their extraction from the complex. This would serve as a useful starting point for sphingolipid and ganglioside purification.

Gangliosides have also been purified by cloud point extraction in which use of a surfactant improves recovery of the less polar gangliosides[66]. A novel embodiment of this method uses carbonated water.

Sphingolipids in concentrations greater than 6% have been obtained in a food-grade process which avoids the use of solvents. When whey is microfiltered to make whey protein concentrate, the fat, including sphingolipids, and some of the protein are retained. This retained material is a product of commerce dubbed "ProCream." When we subjected ProCream to proteolysis to make peptide mixtures, a phospho-lipoprotein fraction further enriched in sphingolipids and phospholipids was obtained. It is widely known that application of phospholipases to phospholipoprotein complexes provides a mixture with improved emulsifying properties, i.e., separation of components from the mixture becomes more difficult[67–71]. However, a process that overcomes the difficulties encountered with phospholipase hydrolysis enhancement of emulsions to yield a mixture depleted in phospholipids and enriched in sphingolipids has been developed[72].

7.4.4 IDEAS FOR THE FUTURE

When a hyperpolarizing electric field is imposed on a monolayer lipid film, glycosphingolipids and gangliosides modify the polarizability of the films[73]. Although this phenomenon has not been taken advantage of for purification of these target molecules from complex mixtures, it appears to offer a fruitful field of inquiry. Buttermilk sphingolipids can be concentrated from bulk solution by addition of cholesterol[22], which caused aggregation of buttermilk sphingomyelin and formation of vesicles; these could be separated by ultrafiltration with a 300 kDa cut-off membrane[22].

7.5 MODIFICATION STRATEGIES

Sphingolipids and gangliosides have several functional moieties and thus provide fertile ground for modification by chemical and biological catalysts.

7.5.1 MODIFICATION OF SPHINGOLIPIDS

Sphingolipids **2** contain a free hydroxyl group, an amide bond, and a phosphodiester group. The *N*-acyl linkage holding a carboxyl moiety bound to the sphingosine base of sphingomyelin can be cleaved by reacting with hydrazine and alkaline hydrolysis in alcohols[8] or solvents to yield a free fatty acid and lysosphingomyelin **7**[74] (also called sphingosylphosphocholine), or can be prepared by acid methanolysis of sphingomyelin[51]. However, these approaches suffer the usual lack of specificity inherent in chemical hydrolysis processes. Specific deacylation can be effected by incubation of sphingomyelin with a sea mud bacterium, *Shewanella alga* NS-589[75] and this is easily carried out enzymatically by sphingolipid ceramide N-deacylase (SCDase, EC 3.5.1.69)[6]. SCDase is now commercially available from PanVera Corporation (Madison, WI). It was recently demonstrated that 100% yields in lysosphingolipid production could be obtained by extractive bioconversion in a two-phase system with SCDase[76].

Like so many hydrolases, SCDase is capable of carrying out synthetic (reverse hydrolysis) reactions analogous to the well-known production of specific-structured triacylglycerols[77]. Synthesis of specific-structured sphingolipids (SSS) has been brought to maturity[78]. Thus, labeled fatty acid has been transferred to sphingosine to make a labeled substrate for ceramidase activity assays; the reaction has been carried out with a range of fatty acid/lysoglycosphingolipid combinations[62,79]. The SSS which inhibit the activation of protein kinase C with a resulting utility in treating nerve pathologies have been synthesized[80,81]. In a similar approach, a commercial lipase preparation was used to carry out the amidation of lysosphingolipid to make "hybrid ceramides" containing new acyl groups not found in the parent sphingolipid[54]. However, this patent relied heavily on the use of commercial bacterial lipase and porcine pancreas lipase preparations, which are usually crude mixtures of many enzyme activities[82,83], and have recently been shown to contain high levels of protease activity[84]. As the amidation reaction falls into the range of reactions catalyzed by proteases, there is a possibility that the synthesis of hybrid ceramides was actually catalyzed by protease impurities in the commercial lipase preparation.

Sphingomyelinase (sphingomyelin cholinephosphohydrolase, EC 3.1.4.12) cleaves the phosphodiester of sphingomyelin, releasing phosphorocholine and a ceramide base **4**. Sphingomyelinase cleavage is analogous to the action of phospholipase C action on glycerophospholipids. This enzyme is available commercially from Asahi Chemical Enzymes (*Streptomyces* spp.; Tokyo, Japan), Higeta Shoyu Co. Ltd (*Bacillus cereus*; Tokyo, Japan)[75], and Sigma-Aldrich (*Bacillus cereus*; St. Louis, MO)[85]. This enzyme is useful in studies of sphingomyelin metabolism, as an acid sphingomyelenase has been isolated from human milk[14], and alkaline sphingomyelinases have been found in human bile[86].

STRUCTURE 7.7 Lysosphingomyelin (sphingosylphosphocholine).

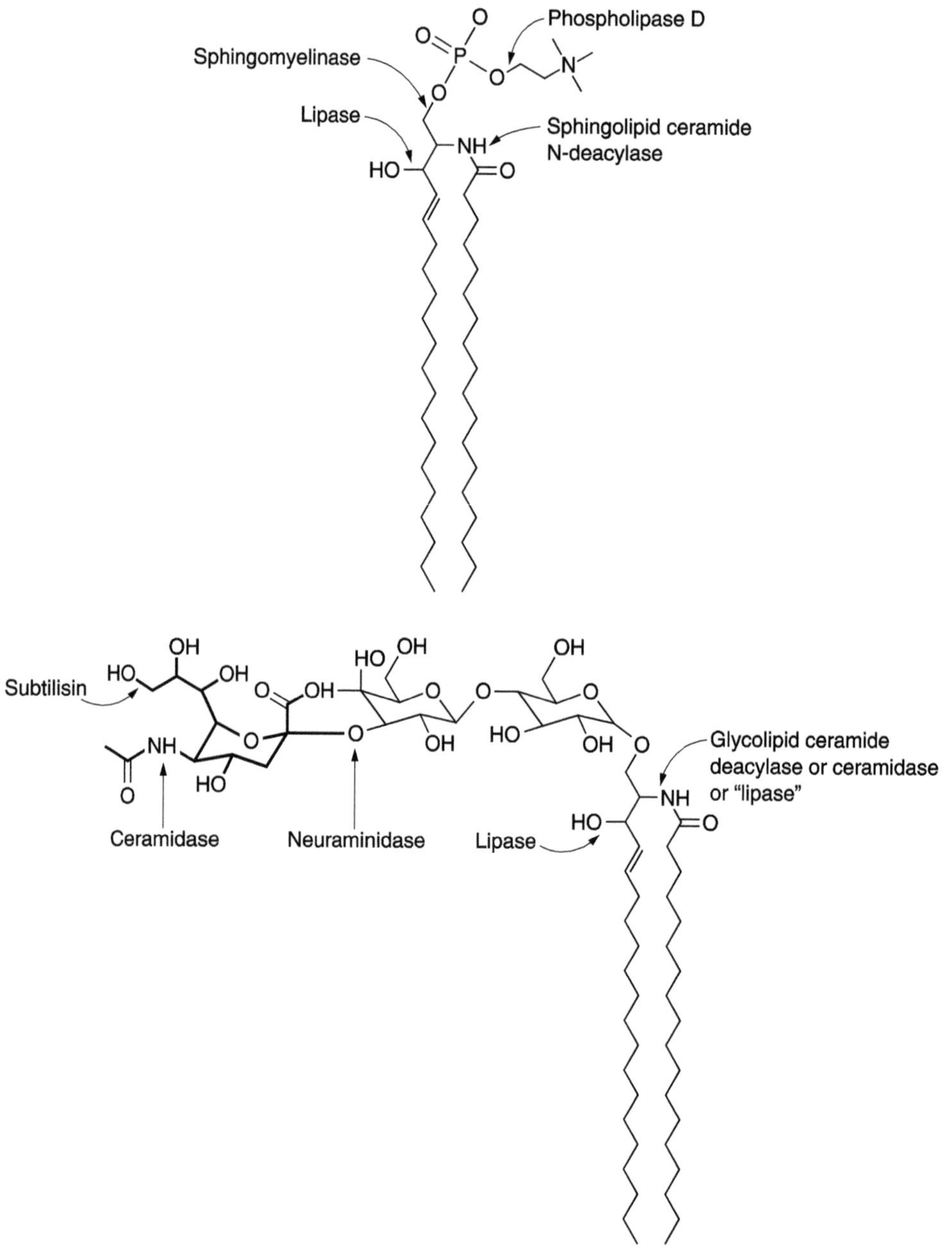

FIGURE 7.1 Sites of enzymatic modification of sphingomyelin (top) and gangliosides (bottom).

Phospholipase D has been used to modify sphingomyelin (Figure 7.1). The choline group is removed and a serine group is substituted[87] to provide a compound useful for treatment of nervous system autoimmune reactions and nervous system disorders including Alzheimer's, Parkinson's, and Huntington's syndromes, senile dementia, multiple sclerosis, rheumatoid arthritis, and insulin-dependent diabetes.

The pioneering efforts of Merrill's group at Emory University have resulted in a large number of sphingolipid structural derivatives[88]. These structures are potent biomolecules useful for treatment of abnormal cell proliferation (including benign and malignant tumors), the promotion of cell differentiation, the induction of apoptosis, inhibition of protein kinase C, the treatment of inflammatory conditions, and treating intestinal bacterial infections.

7.5.2 MODIFICATION OF GANGLIOSIDES

In addition to the free hydroxyl group and amide-bonded acyl group of sphingolipids, gangliosides also contain one or more sugar moieties. Although total ganglioside synthesis is very difficult, some research groups have worked toward this goal[8]. Sialic acid derivatives can be built and linked to a ceramide base to form the ganglioside iso-GM3[89] or GM3[90]. These approaches are very sophisticated and not for the faint of heart; fortunately, synthetic organic chemists are a courageous group.

In a manner analogous to sphingolipids, the N-acyl moiety of gangliosides can be cleaved to yield free fatty acids and sphingosine. This reaction can be carried out chemically[8] or enzymatically. SCDase (see above) is able to hydrolyze the N-acyl group from gangliosides, giving rise to an alternative name (glycolipid ceramide deacylase)[51]. Ceramidase (EC 3.5.1.23) carries out the same reaction[62]; however, it is less specific in that it also removes acetyl groups located on ganglioside sialic acids (see below).

Further removal of acyl groups of sialic acid amides results in a material to which new acyl groups can be introduced to make structured gangliosides[91]. One form, an acyl-di-lysoganglioside, has been shown to inhibit the activation of protein kinase C, and is thus a potent agent for nerve system therapy[92]. This is especially effective when short-chain fatty acids are added back to the nitrogens through an amide bond[93]. Sulfation of lysogangliosides and dilysogangliosides imparts neuroprotective (anticytoneurotoxic) effects to gangliosides, and modulates the expression of CD4 receptors on lymphocyte membranes[94,95]. Modulation of CD4 expression has been associated with inhibition of proliferation of HIV.

Simple removal of the acetate moiety from ganglioside sialic acids forms compounds useful for stimulation of growth of human and animal cells both *in vitro* and *in vivo*[3]. In another approach, the carboxyl and hydroxyl groups of the sialic acid and oligosaccharide moieties are derivatized by esterification or amidation of carboxyls, or peracylation of hydroxyl groups by a broad range of molecules[96,97]. Treatment of nervous system pathologies by the modified gangliosides is taught. Ganglioside modifications also include the conversion of a sialic acid carboxyl moiety into the carboxylamide of an aliphatic amino acid or amino sulfonic acid[98].

GM3 contains an acetyl moiety on the polar sugar headgroup. This acetyl can be selectively hydrolyzed by base hydrolysis to make deacetyl-GM3; alternatively, the N-acyl group can be selectively removed by base hydrolysis to make lyso-GM3. Of course, the two can be combined to make a deacyl-lyso-GM3[99]. Deacylation by base hydrolysis in a polar solvent and production of substantially pure de-N-acetyl GM3 is taught in a patent[100]. The target compound stimulates the growth of human cells and is especially directed toward acceleration of wound healing.

One research group has made strong advances in the modification of gangliosides, including the synthesis of so-called "inner esters"[101,102]. These compounds stimulate nerve sprouting and activate membrane enzymes involved in conducting nervous stimuli, with a result that nerve regeneration is promoted. Thus, these inner esters are useful for treating disorders of the nervous system[103]. They are formed by the esterification reaction between the carboxylic acid moiety of a sialic acid with a hydroxyl group from the same ganglioside molecule, either from a carbohydrate or an adjoining sialic acid.

Modification of the free hydroxyl group on the sphingosine base of gangliosides by classic organic synthesis has been thoroughly investigated[88]. A large number of ganglioside modifications produced molecules useful as membrane receptors, tumor markers, cell growth controlling substances, and cancer immunotherapy agents.

Nonspecific alkaline hydrolysis of ganglioside GM1 produces a deacetylated and deacylated derivative useful in binding cholera toxin and controlling diarrhea[2]. However, gentler modification of GM1 removes only the N-acyl moiety. The free amino groups are then reacted with aldehydes bound to beads to form imino groups, which are reduced to form stable secondary amines. The resulting matrix is useful in purifying cholera toxins from crude culture filtrates[104]. This technology has been adapted to a useful biosensor for determining cholera toxin concentrations in drinking water[24].

GM3 can be converted to ceramide lactoside by treatment with Neuraminidase (Sigma, St. Louis, MO). *Arthrobacter ureafaciens* bacteria are capable of synthesizing two neuraminidase isoenzymes. Neuraminidase isoenzyme L catalyzes general removal of sialic acids from mixed gangliosides to yield Asialo GM1[4]. Neuraminidase isoenzyme S catalyzes specific removal of *N*-acetylneuraminic acid only from mixed gangliosides to form the monosialoganglioside GM1[105]. The exact same reaction is catalyzed by immobilized *Clostridium perfringens* neuraminidase[5].

9-*O*-Acetyl GD3 is a malignant melanoma cell-specific antigen, making it an attractive target for immunotherapy. GD3 can be converted to 9-*O*-acetyl GD3 by ganglioside *O*-acetyl transferase; however, this requires acetyl CoA as the acyl donor. Recently, a serine protease (Subtilisin BPN) was used to accomplish the same reaction, using the much cheaper vinyl acetate as the acyl donor[106].

A structured lipid (GM5) not found in nature combines two ceramide bases onto a single sialic acid-containing sugar headgroup[107]. The resulting novel compound is reportedly useful as a tumor marker. Another group has succeeded in oxidizing the sphingoid base double bond to an aldehyde by ozonolysis, which is easily reductively amidated to form a unique reactive aldehyde useful for a broad range of further reactions[108,109]. For example, the reactive aldehyde can be used as a starting point for synthesis of vaccines for neuroectodermal tumors, and the study of mammalian receptors of glycosphingolipids.

7.6   CONCLUSIONS

Sphingolipids and gangliosides are potent biological molecules which offer a broad platform for modification. Their extraction and purification present significant room for research and improvements. The analysis of these complicated molecules, especially at the low concentrations found in most biological matrices, is a formidable challenge. Although there is a significant body of knowledge about their modification and resulting biological activities, there remain many fascinating questions and vast unexplored areas. Advances and breakthroughs in sphingolipid and ganglioside research are likely for years to come.

REFERENCES

1. Thudichum, J.L., *A Treatise on the Chemical Constitution of the Brain*, Baillaire, Tindall & Cox, London, 1884.
2. Mynard, M.-C. and Tayot, J.-C., Ganglioside Containing Compositions for Treating Diarrhea, U.S. Patent 4, 347, 244, August 31, 1982.
3. Nores, G., Hanai, N., Dohl, T., Nojiri, H., and Hakomori, S.-I., Gangliosides Containing De-*N*-acetyl-sialic Acid and Their Applications as Modifiers of Cell Physiology, European patent application EP 328, 420 A2, August 16, 1989.
4. Sugimori, T., Tsukada, Y., and Ohta, Y., Process for Preparing Asialo GM1, European patent application EP 451, 270 A1, October 16, 1991.
5. Mammarella, C., Cumar, F., and Rodriguez, P. Method of Obtaining Monosialogangliosides, European patent application EP 540, 790 A1, May 12, 1993.
6. Ito, M., Kurita, T., and Kita K., A novel enzyme that cleaves the *N*-acyl linkage of ceramides in various glycosphingolipids as well as sphingomyelin to produce their lyso forms, *J. Biol. Chem.*, 270, 24370–24374, 1994.
7. Puente, R., Garcia-Pardo, L.A., and Hueso, P., Gangliosides in bovine milk, *Biol. Chem. Hoppe-Seyler*, 373, 283–288, 1992.
8. Hakomori, S., Chemistry of glycosphingolipids, in *Sphingolipid Biochemistry*, Hakamori, J.K., Ed., Plenum Press, New York, 1983, pp. 1–165.
9. Merrill, A. and Wang, E., Enzymes of ceramide biosynthesis, *Methods Enzymol.*, 209, 427–437, 1992.

10. Valeur, A., Olsson, N.U., Kaufmann, P., Wada, S., Kroon, C.G., Westerdahl, G., and Odham, G., Quantification and comparison of some natural sphingomyelins by on-line high-performance liquid chromatography/discharge-assisted thermospray mass spectrometry, *Biol. Mass Spectrom.*, 23, 313–319, 1994.

11. Hellgren, L., Occurrence of bioactive sphingolipids in meat and fish products, *Eur. J. Lipid Sci. Technol.*, 103, 661–667, 2001.

12. Jensen, R., The composition of bovine milk lipids, January 1995 to December 2000, *J. Dairy Sci.*, 95, 295–350, 2002.

13. Pfeuffer, M. and Schrezenmeir, J., Dietary sphingolipids, metabolism and potential health implications, *Kieler Milchwirtschaftliche Forschungsberichte.*, 53, 31–42, 2001.

14. Nyberg, L., Farooqi, A., Blackberg, L., Duan, R.-D., Nilsson, A., and Hernell, O., Digestion of ceramide by human milk bile salt stimulated lipase, *J. Pediatric Gastroent. Nutr.*, 27, 560–567, 1998.

15. Vesper, H., Schmelz, E.-M., Nikolova-Karakashian, N., Dillehay, D., Lynch, D., and Merrill A., Sphingolipids in food and the emerging importance of sphingolipids to nutrition, *J. Nutr.*, 129, 1239–1250, 1999.

16. Nyberg, L., Nilsson, A., Lundgren, P., and Duan, R.-D., Localization and capacity of sphingomyelin digestion in the rat intestinal tract, *J. Nutr. Biochem.*, 8, 112–118, 1997.

17. della Valle, F., Callegaro, L., and Lorenzi, S., Method for the Preparation of a Mixture of Glycosphingolipids Free from Contamination by Non-conventional Viruses, U.S. Patent 5, 521, 164, May 28, 1996.

18. Muthing, J., Analyses of glycosphingolipids by high-performance liquid chromatography, *Methods Enzymol.*, 312, 45–63, 2000.

19. Rueda, R., Puente, R., Hueso, P., Maldonado, J., and Gil, A., New data on content and distribution of gangliosides in human milk, *Biol. Chem. Hoppe-Seyler*, 376, 723–727, 1995.

20. Nakano, T., Sugawara, M., and Kawakami, H., Sialic acid in human milk, composition and functions, *Acta Pediatrica Taiwanica*, 42, 11–17, 2001.

21. Jennemann, R. and Wiegandt, H., A rapid method for the preparation of ganglioside Glac2 (GD3), *Lipids*, 29, 365–368, 1994.

22. Eckhardt, E., Moscette, A., Renooij, W., Goerdayal, S., van Berge-Henegouwen, G., and van Erpecum, K., Asymmetric distribution of phosphatidylcholine and sphingomyelin between micellar and vesicular phases, potential implications for canicular bile formation, *J. Lipid Res.*, 11, 2022–2033, 1999.

23. Morita, M., Sugawara, M., Eto, M., Miura, S., and Kotani, M., Method of Manufacturing Compositions With High Ganglioside Content, U.S. Patent 6,265,555, July 24, 2001.

24. Walsh, M. and Wiemer, B., Pathogen Capture with Immobilized Glycolipids, abstracts of papers, 90th Annual Meeting of the American Oil Chemists' Society, Orlando, FL, May 9–12, 1999, American Oil Chemists' Society, Champaign, IL, 1999.

25. Dufour, E., Subirade, M., Loupil, F., and Riaublanc, A., Whey proteins modify the phase transition of milk fat globule phospholipids, *Lait*, 79, 217–228, 1999.

26. Attebury, J., Removing Lipid Material from Whey, U.S. Patent 3,560,219, February 2, 1971.

27. Baumy, J., Gestin, L., Fauquant, J., Boyaval, E., and Maubois, J., Technologies de purification des phospholipides du lactoserum, *Process* (Rennes), 1047, 29–33, 1990.

28. Kitchen, B., Fractionation and characterization of the membranes from bovine milk fat globules, *J. Dairy Res.*, 44, 469–482, 1977.

29. Dapper, C., Valivullah, M., and Keenan, T., Use of polar aprotic solvents to release membranes from milk lipid globules, *J. Dairy Sci.*, 70, 760–765, 1987.

30. Schmelz, E., Dietary sphingomyelin and other sphingolipids in health and disease, *Nutr. Bull.*, 25, 135–139, 2000.

31. Merrill, A., Hannun, Y., and Bell, R., Introduction, Sphingolipids and their metabolites in cell regulation, *Adv. Lipid Res.*, 25, 1–24, 1993.

32. Carrer, D. and Maggio, B., Phase behavior and molecular interactions of ceramide with dipalmitoylphosphatidylcholine, *Lipids*, 40, 1978–1989, 1999.

33. Parodi, P., Conjugated linoleic acid and other anticarcinogenic agents of bovine milk fat, *J. Dairy Sci.*, 82, 1339–1349, 1999.

34. Hannun, Y. and Linardic, C., Sphingolipid breakdown products, anti-proliferative and tumor-suppressor lipids, *Biochem. Biophys. Acta*, 1154, 223–236, 1993.

35. Bonini, C., Cazzato, C., Cernia, E., Palocci, C., Soro S., and Viggiani, L., Lipase enhanced catalytic efficiency in lactonisation reactions, *J. Mol. Catal. B. Enzym.*, 16, 1–5, 2001.

36. Auge, N., Negre-Salvayre, A., Salvayre, R., and Levade, T., Sphingomyelin metabolites in vascular cell signalling and atherogenesis, *Prog. Lipid Res.*, 39, 207–229, 2000.

37. Balcao, V., Vieira, M., and Malcata, F., Adsorption of protein from several commercial lipase preparations onto a hollow-fiber membrane module, *Biotechnol. Prog.*, 12, 164–172, 1996.

38. Parodi, P., Cows' milk fat components as potential anticarcenogenic agents, *J. Nutr.*, 127, 1055–1060, 1997.

39. Yu, R. and Ariga, T., Ganglioside analysis by high-performance thin-layer chromatography, *Methods Enzymol.*, 312, 115–134, 2000.

40. Higuchi, K., Kawashima, M., Takagi, Y., Kondo, H., Yada, Y., Ichikawa, Y., and Imokawa, G., Sphingosylphosphorylcholine is an activator of transglutaminase activity in human keratinocytes, *J. Lipid Res.*, 42, 1562–1570, 2001.

41. Malmsten, M., Bergenstahl, B., Nyberg, L., and Odham, G., Sphingomyelin from milk: characterization of liquid crystalline, liposome and emulsion properties, *J. Am. Oil Chem. Soc.*, 71, 1021–1026, 2001.

42. Handjani, R.-M., Ribier, A., and Colarow, L., Composition Cosmetique ou Dermopharmaceutique Contenant des Vesicules Formees par un Melange Phospholipides/Glycolipides, European Patent EP 455, 528 B1, November 6, 1991.

43. Sprong, R., Hulstein, M., and Van der Meer, R., Bactericidal activities of milk lipids, *Antimicrob. Agents Chemother.*, 45, 1298–1301, 2001.

44. Hammache, D., Pieroni, G., Yahi, N., Delezay, O., Koch, N., Lafont, H., Tamalet, C., and Fantini, J., Specific interaction of HIV-1 and HIV-2 surface envelope glycoproteins with monolayers of galactosyl-ceramide and ganglioside GM3, *J. Biol. Chem.*, 273, 7967–7971, 1998.

45. Dyatlovitskaya, E., Tekieva, E., Lemenovskaya, A., Somova, O., and Berberlson, L., Gangliosides GM3 and GD3 in human gastric and mammary tumors, *Biochem.-USSR*, 56, 374–377, 1991.

46. Ravindranath, M., Tsuchida, T., Morton, D., and Irie, R., Ganglioside GM3, GD3 ratio as an index for the management of melanoma, *Cancer*, 67, 3029–3035, 1991.

47. della Valle, F., Romeo, A., and Lorenzi, S., Gangliosides Mixtures, Useful as a Therapeutic Tool for Eliminating Painful Effects of Peripheral Neuropathies, U.S. Patent 4, 707, 469, November 27, 1987.

48. della Valle, F. and Toffano, G., Use of Monosialoganglioside GM1 to Prevent the Development of Tolerance to the Analgesic Effect of Morphine and Related Drugs, U.S. Patent 5, 183, 807, February 2, 1993.

49. Kawakami, H., Ishiyama, Y., and Idota, T., Stability of milk gangliosides and formation of GD3 lactone under acidic conditions, *Biosci. Biotech. Biochem.*, 58, 1314–1315, 1994.

50. Rueda, R., Maldonaldo, J., Narbona, E., and Gil, A., Neonatal dietary gangliosides, *Early Human Devel.*, 53, S135–S147, 1998.

51. Ito, M., Kurita, T., and Kita, K., Glycolipid Ceramide Deacylase, European patent application EP 707, 063 A1, August 16, 1989.

52. Walsh, M. and Nam, S., Affinity enrichment of bovine lactoferrin in whey, *Prep. Biochem. Biotechnol.*, 31, 229–240, 2001.

53. Walsh, M. and Nam, S., Rapid fractionation of bovine transferrin using immobilized gangliosides, *Prep. Biochem. Biotechnol.*, 31, 89–102, 2001.

54. Smeets, J., De Pater, R.J., and Lambers, J., Enzymatic Synthesis of Ceramides and Hybrid Ceramides, U.S. Patent 5, 610, 040, March 11, 1997.

55. Huang, R., Isolation and characterization of the gangliosides of buttermilk, *Biochim. Biophys. Acta*, 306, 82–84, 1973.

56. Schnaar, R., Isolation of glycosphingolipids, *Methods Enzymol.*, 230, 348–370, 1994.

57. Fredman, P., Nilsson, O., Tayot, J.-L., and Svennerholm, L., Separation of gangliosides on a new type of anion-exchange resin, *Biochim. Biophys. Acta*, 618, 42–52, 1980.

58. Lindqvist, B., Förfarande för att Foerädla Torkade Vassleprodukter, Swedish Patent SE7801821-5, August 3, 1983.

59. Nyberg, L. and Burling, H., Method for Extracting Sphingomyelin, U.S. Patent 5, 677, 472, October 17, 1997.

60. Nyberg, L. and Burling, H., Method for Extracting Sphingomyelin, European Patent EP 689, 579 B1, Sept 15, 1999.

61. Hamilton, J. and Comai, K., Rapid separation of neutral lipids, free fatty acids and polar lipids using prepacked silica sep-pac columns, *Lipids*, 23, 1146–1149, 1998.

62. Mitsutake, S., Kita, K., Okino, N., and Ito, M.,[14C] ceramide synthesis by sphingolipid ceramide N-deacylase, new assay for ceramidase activity detection, *Anal. Biochem.*, 247, 52–57, 1997.

63. Bodennec, J., Koul, O., Aguado, I., Brichon, G., Zwingelstein, G., and Portoukalian, J., A procedure for fractionation of sphingolipid classes by solid-phase extraction on aminopropyl cartridges, *J. Lipid Res.*, 41, 1524–1531, 2000.

64. Menzeleev, R., Krasnopolsky, Y., Zvonkova, E., and Shvets., V., Preparative separation of ganglioside GM3 by high-performance liquid chromatography, *J. Chromatogr. A.*, 678, 183–187, 1994.

65. Damodaran, S., Removing Lipids from Cheese Whey Using Chitosan, U.S. Patent 5,436,014, July 25, 1995.

66. Terstappen, G., Futerman, A., and Schwarz, A., Cloud-point extraction of gangliosides using nonionic detergent C14EO6, *Methods Enzymol.*, 312, 187–196, 2000.

67. Buikstra, F., van der Kruis, A., and van der Heijden, P., Heat-Stable Oil-in-Water Emulsions Stabilized by Hydrolyzates, U.S. Patent 5,650,190, July 22, 1997.

68. Andreae, C., Dazo, P., Kuil, G., Matthijsses, G., and Mulder, J., Process of Making a Sterile, Packed Food Emulsion, U.S. Patent 5,945,149, August 31, 1999.

69. van Dam, A., Emulsions, U.S. Patent 4,034,124, July 5, 1977.

70. Schenk, B., Process for the Preparation of a Water and Oil Emulsion, U.S. Patent 5,028,447, July 2, 1991.

71. van Dam, A., Oil-in-Water Emulsion and Process for the Preparation Thereof, U.S. Patent 4,119,564, October 10, 1978.

72. Bloomer, S. and Brody, E., Method of Preparing a Milk Polar Lipid Enriched Concentrate and a Sphingolipid Enriched Concentrate, U.S. patent application 20040047947, March 11, 2004.

73. Maggio, B., Modulation of phospholipase A2 by electrostatic fields and dipole potential of glycosphingolipids in monolayers, *J. Lipid Res.*, 40, 930–939, 1999.

74. Gaver, S. and Sweeley, C., Methods for methanolysis of sphingolipids and direct determination of long-chain bases by gas chromatography, *J. Am. Oil Chem. Soc.*, 42, 294–298, 1965.

75. Sueyoshi, N., Izu, H., and Ito, M., Preparation of a naturally occurring D-erythro-(S2,3R)-sphingosylphosphocholine using *Shewanella* alga NS-589, *J. Lipid Res.*, 38, 1923–1927, 1997.

76. Kurita, T., Izu, H., Sano, M., and Kato, I., Producing Lysosphingolipids by Hydrolyzing the Acid Amide Bond Between Sphingoid and Fatty Acid in Sphingolipid–Lysosphingolipid Production Using Enzyme in Two-Phase System, WO patent application WO 99050433 A1, October 7, 1999.

77. Xu, X., Production of specific-structured triacylglycerols by lipase-catalyzed reactions, a review, *Eur. J. Lipid Sci. Technol.*, 4, 287–303, 102.

78. Ito, M., Kurita, T., Mitsutake, S., and Kita, K., Process for the Preparation of Sphingolipids and Sphingolipid Derivatives, European Patent EP 940,409 B1, May 12, 2004.

79. Tani, M., Kita, K., Komori, H., Nakagawa, T., and Ito, M., Enzymatic synthesis of omega-amino-ceramide, preparation of a synthetic fluorescent substance for ceraminidase, *Anal. Biochem.*, 263, 183–188, 1998.

80. della Valle, F. and Romeo, A., Lysosphingolipid Derivatives, U.S. Patent 5,519,007, May 21, 1996.

81. della Valle, F. and Romeo, A., Lysosphingolipid Derivatives, U.S. Patent 5,792,858, August 11, 1998.

82. Plou, F., Sogo, P., Calvo, M., Burguillo, F., and Ballesteros, A., Kinetic and enantioselective behavior of isoenzymes A and B from *Candida rugosa* lipase in the hydrolysis of lipids and esters, *Biocatal. Biotransform.*, 15, 75–89, 1997.

83. Bjurlin, M., Bloomer, S., and Haas, M., Composition and activity of commercial triacylglycerol hydrolase preparations, *J. Am. Oil Chem. Soc.*, 78, 153–160, 2001.

84. Bjurlin, M. and Bloomer, S., Proteolytic activity in commercial triacylglycerol hydrolase preparations, *Biocatal. Biotransform.*, 20, 179–188, 2001.

85. Fanani, M. and Maggio, B., Kinetic steps for the hydrolysis of sphingomyelin by *Bacillus cereus* sphingomyelinase in lipid monolayers, *J. Lipid Res.*, 41, 1832–1840, 2000.

86. Nyberg, L., Duan, R.-D., Axelsson, J., and Nilsson, Å., Identification of an alkaline sphingomyelinase activity in human bile, *Biochim. Biophys. Acta*, 1300, 42–48, 1996.

87. Romeo, A., Kirshner, G., and Menon, G., Therapeutic Use of Phosphoryl-L-serine-N-acyl-sphingosine, U.S. Patent 5,556,843, September 17, 1996.

88. Liotta, D., Merrill, A., Keane, T., Schmelz, E., and Bhalla, K., Sphingolipid Derivatives and Their Methods of Use, WO patent application WO99041266, August 19, 1999.

89. Ogawa, T., Sugimoto, M., Numata, M., Yoshimura, S., Ito, M., and Shitori, Y., Sialic Acid Derivatives and Process Thereof, European patent application 479,769 A2, April 8, 1992.

90. Ogawa, T., Sugimoto, M., Shitori, Y., and Ito, M., Sialic Acid Derivatives, Galactose Derivatives; Method for Producing the Same, European patent application 166,442 A3, January 2, 1986.

91. della Valle, F. and Romeo, A., Modified Gangliosides and the Functional Derivatives Thereof, U.S. Patent 5, 264, 424, November 23, 1993.

92. della Valle, F. and Romeo, A., Di-lysoganglioside Derivatives, U.S. Patent 5,523,294, June 4, 1996.

93. della Valle, F. and Romeo, A., Semisynthetic Ganglioside Analogues, U.S. Patent 5,484,775, January 16, 1996.

94. della Valle, F. and Romeo, A., O-Sulfated Gangliosides and Lysoganglioside Derivatives, U.S. Patent 5, 849, 717, December 15, 1998.

95. Romeo, A., Kirchner, G., Chizzolini, C., Manev, H., and Facci, L., Sulfated Lysoganglioside Derivatives, U.S. Patent 5, 795, 869, August 18, 1998.

96. della Valle, F. and Romeo, A., Ganglioside Derivatives, U.S. Patent 4, 713, 374, December 15, 1989.

97. della Valle, F. and Romeo, A., Ganglioside Derivatives, U.S. Patent 4, 849, 413, July 18, 1989.

98. Romeo, A., Toffano, G., and Loen, A., Ganglioside Derivatives, U.S. Patent 5, 350, 841, September 27, 1994.

99. Valiente, O., Mauri, L., Casellato, R., Fernandez, L., and Sonnino, S., Preparation of deacetyl-, lyso-, and deacetyl-lyso-GM3 by selective alkaline hydrolysis of GM3 ganglioside, *J. Lipid Res.*, 42, 1318–1324, 2001.

100. Hakomori, S.-I., Nores, G., Hanai, N., Levery, D.T.S., Salyan, M., and Nojiri, H., Naturally Occurring Gangliosides Containing De-N-acetyl-sialic Acid and Their Applications as Modifiers of Cell Physiology, U.S. Patent 5, 272, 138, December 21, 1993.

101. della Valle, F. and Romeo, A., Method for Preparing Ganglioside Derivatives and Use Thereof in Pharmaceutical Compositions, U.S. Patent 4, 476, 119, October 9, 1984.

102. della Valle, F. and Romeo, A., Method for Preparing Ganglioside Derivatives and Use Thereof in Pharmaceutical Compositions, U.S. Patent 4, 716, 223, December 29, 1987.

103. della Valle, F. and Romeo, A., Inner Esters of Gangliosides with Analgesic–Anti-inflammatory Activity, U.S. Patent 5, 045, 532, September 3, 1991.

104. Tayot, J.-L., Holmgren, J., Svennerholm, L., Lindblad, M., and Tardy, M., Receptor-specific large-scale purification of cholera toxin on silica beads derivatized with LysoGM1 ganglioside, *Eur. J. Biochem.*, 113, 249–258, 1981.

105. Sugimori, T., Tsukada, Y., and Ohta, Y., Process for Preparing Ganglioside GM1, European patent application 451,267 B1, August 9, 1995.

106. Takayama, S., Livingston, P., and Wong, C.-H., Synthesis of the melanoma-associated ganglioside 9-*O*-acetyl GD3 through regioselective enzymatic acetylation of GD3 using *Subtilisin*, *Tetrahedron Lett.*, 37, 9271–9274, 1996.

107. Ogawa, T., Sugimoto, M., Numata, M., Shitori, Y., and Ito, M., Sialocylceramides and Production Method Thereof, U.S. Patent 4, 730, 058, March 8, 1988.

108. Wiegand, H. and Bosslet, S., Glycosphingolipids with a Group Capable of Coupling in the Sphingoid Portion, the Preparation and Use Thereof, U.S. Patent 5, 571, 900, November 5, 1996.

109. Wiegand, H. and Bosslet, S., Glycosphingolipids with a Group Capable of Coupling in the Sphingoid Portion, the Preparation and Use Thereof, U.S. Patent 5, 599, 914, February 14, 1997.

8 Hydroxy Fatty Acids

Thomas A. McKeon, Charlotta Turner, Xiaohua He, Grace Chen, and Jiann-Tsyh Lin
USDA-ARS WRRC, Albany, California

CONTENTS

8.1 INTRODUCTION

Production of hydroxy fatty acids in plants is of current interest principally due to the novel physical and chemical properties that are characteristic of hydroxy fatty acids. Castor oil is currently the only major source of hydroxy fatty acids. It has a long history in medicinal applications, serving as a laxative and during labor to promote the birthing process. Its compatibility as an emollient has promoted its use as a massage oil and in cosmetics. This chapter provides a brief description of hydroxy fatty acids, principally focused on castor oil, and discusses pharmacological and physiological applications.

8.2 SOURCES OF HYDROXY FATTY ACIDS

The castor plant (*Ricinus communis* L.) provides the only current commercially available source of hydroxy fatty acid, ricinoleate, 12-hydroxyoleic acid (Figure 8.1). Castor oil comprises up to 60% of the seed weight, and ricinoleate represents up to 90% of the fatty acid content. We have recently confirmed the high content of triricinolein, 71% of the triacylglycerol[1]. A recent report describes a castor variety that contains considerably less ricinoleate, 14%, with 78% oleate content[2]. This variety is of biochemical interest, as, presumably, the biosynthetic pathway is blocked in the hydroxylation of oleate.

There are other plant sources of hydroxy fatty acids. *Lesquerella fenderlii* is a "new crop" that has been developed as an alternative source of hydroxy fatty acid for U.S. production. Although this plant grows in desert climates and other marginal agricultural land, it has not yet reached the volume needed to be considered a commercial success or a commodity. *Lesquerella* contains up to 55% lesquerolic acid, the 20-carbon analog of ricinoleate. It is derived from ricinoleate by a two-carbon elongation to 14-hydroxy-11-eicosenoate. Its lubricant properties and cosmetic properties are similar to those of ricinoleate[3], but it has no apparent physiological effect, and is not

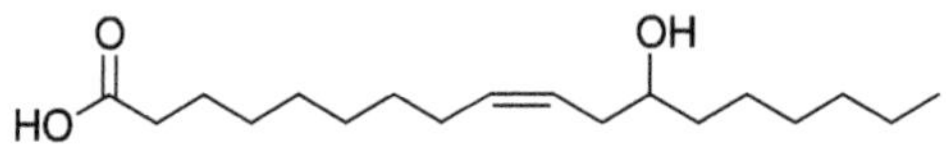

Ricinoleic acid

FIGURE 8.1 Structure of ricinoleic acid.

considered in this chapter. *Dimorphotheca pluvialis* produces an oil containing up to 54% of dimor-phecolic acid, 9-hydroxy-10-*trans*-12-*trans*-octadecadienoate, and this oil can be dehydrated to produce a fatty acid with conjugated double bonds and nonyellowing durable drying quality, simi-lar to eleostearate in tung oil. Because of this desirable property, it is currently a crop of interest but has not yet achieved commercial availability.

Until the late 1960s, castor was grown in the U.S. and supplied about half of the U.S. demand for castor oil. The rest of the need was filled by importation of castor beans for processing. As a result of the loss of crop parity and the rising cost of energy needed to detoxify and dealleregenize castor meal, castor oil production in the U.S. ceased. The issuance of the Presidential Executive Order 13134 in August 1999 supporting a drive to generate more products from biological sources provided consid-erable impetus to reintroduce castor as a U.S. crop. As a result, efforts to detoxify and deallergenize castor seed using genetic engineering have arisen in order to meet expanding needs for a safe source of castor oil[4].

8.3   USES OF CASTOR OIL AND RICINOLEATE

The mid-chain hydroxyl group in ricinoleate has a dramatic effect on the physical and chemical properties of castor oil. The viscosity of castor oil is greatly enhanced in comparison to a common vegetable oil due to interchain hydrogen bonding of the hydroxyl groups. As a result of the mid-chain polarity, castor oil and ricinoleate derivatives have greater interaction with metals and are generally more effective greases and lubricants than other seed oils. Chemically, castor oil has the capability of forming interpenetrating polymer networks depending on the presence of additional polymerizing groups on the fatty acyl chain[5], and of forming a potentially rich source of monomers for novel biobased plastics (Figure 8.2). The chemical bonds next to the hydroxyl group are sus-ceptible to quantitative cleavage to an array of products depending on the conditions of the reaction. These reactions produce derivatives that currently serve as monomers for producing plastics and other types of polymers listed in Table 8.1[6,7]. Current uses for hydroxy fatty acids, oils containing them, and derivatives obtained from them include greases and other lubricants, surfactants, drying oils, emollients, engineering plastics, thermoplastics, and insulators[6]. Industrial uses for hydroxy fatty acids are limited by their availability. For example, hydroxy fatty acid methyl esters are excel-lent lubricity additives for diesel fuel, eliminating the need to add sulfur compounds that impart lubricity and reducing soot output, thus providing a biobased additive that significantly reduces air pollution. The volume needed to supply just U.S. needs is over 15 times the current U.S. import of castor oil.

The presence of the hydroxyl group allows sulfonation, leading to what is known as Turkey Red Oil, the first synthetic surfactant[6]. Dehydration of the oil results in the formation of conjugated linoleic acid which is used as a nonyellowing drying oil. The epoxidized (blown) oil is useful as a plasticizer and as a replacement for volatile petroleum-derived compounds, resulting in a low VOC (volatile organic carbon) coating. There are numerous other uses that have been described, some of which have not been implemented due to the limited supply of castor oil.

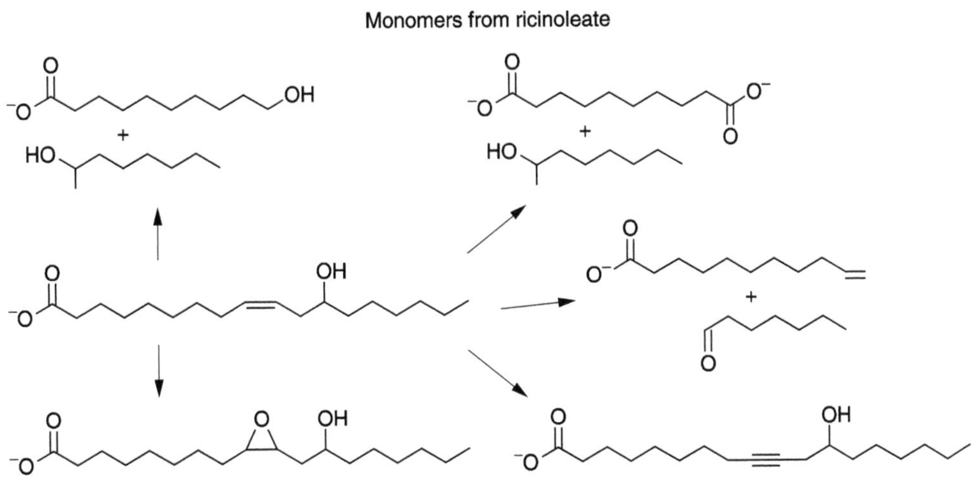

Monomers from ricinoleate

FIGURE 8.2 Monomeric compounds derived from ricinoleate.

TABLE 8.1
Products Derived from Castor Oil

Product	Application, area
Lubricants	Lithium grease; heptanoate esters for jet engines
Coatings	Nonyellowing drying oil; low VOC oil-based paints
Surfactants	Turkey Red Oil
Plasticizers	Blown oil, for polyamides, rubber; heptanoates, low-temperature uses
Cosmetics	Lipstick
Pharmaceuticals	Laxative
Polymers	Polyesters, from sebacic acid; polyamides, nylon 11, nylon 6,10; polyurethanes
Perfumes	Odorants include 2-octanol, heptanal, and undecenal
Fungicides	Undecenoic acid and derivatives

8.4 BIOSYNTHESIS OF RICINOLEATE

The biosynthesis of castor oil has long been a matter of interest, due to the hydroxylation reaction. Research that led to the cloning of the gene for oleoyl-12-hydroxylase has yielded interesting insights into lipid biosynthesis and control of the fatty acid composition of oil. While castor seed contains up to 60% oil with 90% ricinoleate, other plants expressing the hydroxylase gene do not seem to produce oil of more than 20% hydroxy fatty acid[8].

Since the hydroxylase gene alone was not sufficient to elicit high levels of hydroxy fatty acid production, it seemed that there must be other enzymes that are required in order to achieve high ricinoleate levels in the oil[9]. Based on intermediates that accumulated during *in vitro* castor oil biosynthesis carried out by castor seed microsomes, several enzymatic steps that appear to be important for high ricinoleate levels have been identified[10–12]. The pathway derived from this research is shown in Figure 8.3. The recent cloning of the cDNA for diacylglycerol acyltransferase (DGAT) from castor has led to the determination that castor DGAT displays at least a two-fold preference for ricinoleoyl substrates[13], providing support for the contention that high ricinoleate results from the combined effects of several enzymes in the castor oil biosynthetic pathway.

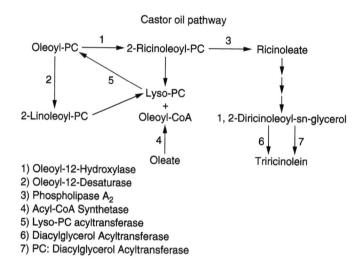

FIGURE 8.3 Abbreviated pathway for castor oil biosynthesis. Numbered enzymes are those identified[9] as being components that appear to be important in the incorporation of ricinoleate into oil while maintaining oleate available for conversion to ricinoleate.

8.5   SEPARATION AND ANALYSIS OF HYDROXY FATTY ACIDS

The presence of a hydroxyl group on acyl chains of an oil simplifies the development of chromatographic systems for separation and identification, as their mobility is usually very different from that of fatty acids with hydrocarbon chains. Lin et al.[14] devised a simple separation in the course of carrying out castor oil metabolic studies[12]. This system enabled the separation and identification of radiolabeled triacylglycerols formed during incubation of radiolabeled fatty acids with castor microsomes. The chromatographic system uses a C_{18} reversed phase column (250 × 4.6 mm) and triacylglycerols are eluted with a linear gradient of 100% methanol to 100% 2-propanol in 40 minutes. This approach was expanded to a preparative HPLC column that allowed separation of the principal triacylglycerols of castor oil, and collection of triricinolein in pure form, approximately 0.3 g per run on a 250 × 15 mm C_{18} column[15].

Turner et al.[16] devised an automated method for analyzing castor oil fatty acids by enzymatic conversion of castor oil to the methyl ester using supercritical carbon dioxide as both extraction and reaction solvent, followed by GC analysis. The method appears to be as reliable as transesterification in methanolic HCl and GC analysis, and considerably reduces the amount of organic solvent required.

Additional methods for analysis of seed oils containing hydroxy fatty acid include HPLC coupled to atmospheric pressure chemical ionization (APCI) mass spectrometry[17] and a combination of LC-APCI MS and LC-UV-MALDI[18] which allow identification and quantitation of individual triacylglycerol components.

8.6   PHYSIOLOGICAL AND PHARMACOLOGICAL EFFECTS

Ricinoleate represents the simplest "model" for an oxidized fatty acid. As described above, efforts to produce this fatty acid in transgenic plants have not approached the levels of production seen in castor. It has been hypothesized that ricinoleate accumulates as phospholipid in the membrane, and this is eliminated from the membrane by the action of a phospholipase A2[19] which serves an editing function by removing oxidized fatty acids from the membrane, to prevent them interfering with proper membrane function. The free fatty acid is susceptible to further oxidation by lipoxygenase or by β-oxidation, thus generating a futile cycle. Although ricinoleate represents a model of an oxidized fatty acid, fatty acids with similar structures, including the naturally occurring lesquerolate[3], 12-hydroxy stearate, 10-hydroxy stearate, ricinelaidate (12-hydroxy-*trans*-octadec-9-enoate), and methyl ricinoleate, were not physiologically active in their effect on intestinal smooth muscle contractions of several small animal models[20].

The effect of castor oil is directly related to the ricinoleate content, as the free fatty acid duplicates the effect of castor oil, so that ricinoleate and castor oil have both been used in experiments to identify the pharmacological effects of castor oil[21]. The laxative action of castor oil and ricinoleate is brought about by direct physiological effects that appear to involve several signaling pathways including NO, platelet-activating factor, and eicosanoids. Experiments by Izzo et al.[22] indicate that the contractile effect of ricinoleate results from direct action on the smooth muscle of rat ileum. This action is enhanced by the inclusion of NO inhibitors, and the effect is blocked by NO synthase substrates, suggesting that the ricinoleate-induced contractions are modulated by endogenous NO. However, pretreatment of rats with 7-nitroindazole, a selective inhibitor of nerve NO synthase, caused inhibition of the laxative effect of castor oil[23], as did intraperitoneal injection of N(G)-nitro-L-arginine methyl ester, another NO synthase inhibitor[24]. This NO inhibition did not prevent the damage done by castor oil- and ricinoleate-induced colitis in these studies.

Castor oil has long been used in birthing, usually to induce or stimulate contractions. However, recent studies indicate that castor oil has no apparent effect on cervical ripening and induction of labor in human subjects[25,26].

Although castor oil is generally safe for external applications to skin, there are reports of ricinoleate derivatives causing an allergic reaction resulting in contact dermatitis. These compounds include propylene glycol ricinoleate, glyceryl ricinoleate, and zinc ricinoleate. The esters are found in some lipsticks and the zinc compound in some deodorants[27–29]. Viera et al.[29] found that topical application of ricinoleic acid could reduce the inflammatory response in several models of the inflammation response. Repeated topical applications of ricinoleic acid significantly reduced inflammation, both acute and subchronic.

8.7   CONCLUSIONS

Castor oil and ricinoleate have potent physiological effects when taken internally. Given their potential for causing intestinal damage resulting in an inflammatory response, and the availability of safer, milder effective laxatives, castor oil should not be the laxative of choice. Nonetheless, castor oil appears to have some beneficial effects in topical applications.

REFERENCES

1. Lin, J.T., Turner, C., Liao, L.P., and McKeon, T.A., Identification and quantification of the molecular species in castor oil by HPLC using ELSD, *J. Liq. Chromatogr. Rel. Technol.*, 26, 1051–1058, 2003.
2. Rojas-Barros, P., DeHaro, A., Munoz, J., and Fernandez-Martinez, J.M., Isolation of a natural mutant in castor with high oleic/low ricinoleic acid content in the oil, *Crop Sci.*, 44, 76–80, 2004.
3. Arquette, J.G. and Brown, J.H., Development of a cosmetic grade oil from *Lesquerella fendleri* seed, in *New Crops*, Janick, J. and Simon, J.E., Eds., Wiley, New York, 1993, pp. 367–371.
4. Auld, D.L., Rolfe, R.D., and McKeon, T.A., Development of castor with reduced toxicity, *J. New Seeds*, 3, 61–69, 2001.
5. Barrett, L.W., Sperling, L.H., and Murphy, C.J., Naturally functionalized triglyceride oils in interpenetrating polymer networks, *J. Am. Oil Chem. Soc.*, 70, 523–534, 1993.
6. Caupin, H.-J., Products from castor oil: past, present, and future, in *Lipid Technologies and Applications*, Gunstone, F.D. and Padley, F.B., Eds., Marcel Dekker, New York, 1997, pp. 787–795.
7. Pryde, E.H. and Rothfus, J.A., Industrial and nonfood uses of vegetable oils, in *Oil Crops of the World*, Robbelen, G., Downey, R.K., and Ashri, A., Eds., McGraw-Hill, New York, 1989, pp. 87–117.

8. Broun, P. and Somerville, C., Accumulation of ricinoleic, lesquerolic, and densipolic acids in seeds of transgenic *Arabidopsis* plants that express a fatty acyl hydroxylase cDNA from castor bean, *Plant Physiol.*, 113, 933–942, 1997.

9. McKeon, T.A. and Lin, J.T., Biosynthesis of ricinoleic acid for castor oil production, in *Lipid Biotechnology*, Kuo, T.M. and Gardner, H.W., Eds., Marcel Dekker, New York, 2002, pp. 129–139.

10. Bafor, M., Smith, M.A., Jonsson, L., Stobart, K., and Stymne, S., Ricinoleic acid biosynthesis and triacylglycerol assembly in microsomal preparations from developing castor-bean (*Ricinus communis*) endosperm, *Biochem. J.*, 280, 507–514, 1991.

11. Lin, J.T., Woodruff, C.L., Lagouche, O.J., McKeon, T.A., Stafford, A.E., Goodrich-Tanrikulu, M., Singleton, J.A., and Haney, C.A., Biosynthesis of triacylglycerols containing ricinoleate in castor microsomes using 1-acyl-2-oleoyl-*sn*-glycerol-3-phosphocholine as the substrate of oleoyl-12-hydroxylase, *Lipids*, 33, 59–69, 1998.

12. Lin, J.T., Chen, J.M., Liao, L.P., and McKeon, T.A., Molecular species of acylglycerols incorporating radiolabeled fatty acids from castor (*Ricinus communis* L.) microsomal incubations. *J. Agric. Food Chem.*, 50, 5077–5081, 2002.

13. He, X., Turner, C., Chen, G.Q., Lin, J.T., and McKeon, T.A., Cloning and characterization of a cDNA encoding diacylglycerol acyltransferase from castor bean, *Lipids*, 39, 311–318, 2004.

14. Lin, J.T., Woodruff, C.L., and McKeon, T.A., Non-aqueous reversed-phase high performance liquid chromatography of synthetic triacylglycerols and diacylglycerols, *J. Chromatogr. A*, 782, 41–48, 1997.

15. Turner, C., He, X., Nguyen, T., Lin, J.T., Wong, R., Lundin, R., Harden, L., and McKeon, T., Lipase-catalyzed methanolysis of triricinolein in organic solvent to produce 1,2 (2,3)-diricinolein, *Lipids*, 38, 1197–1206, 2003.

16. Turner, C., Whitehand, L.C., Nguyen, T., and McKeon, T., Optimization of a supercritical fluid extraction/reaction methodology for the analysis of castor oil using experimental design, *J. Agric. Food Chem.*, 52, 26–32, 2004.

17. Byrdwell, W.C. and Neff, W.E., Analysis of hydroxy-containing seed oils using atmospheric pressure chemical ionization mass spectrometry, *J. Liq. Chromarogr. Rel. Technol.*, 21, 1485–1501, 1998.

18. Stubiger, G., Pittenauer, E., and Allmaier, G., Characterisation of castor oil by on-line and off-line non-aqueous reverse-phase high-performance liquid chromatography-mass spectrometry (APCI and UV/MALDI), *Phytochem. Anal.*, 14, 337–346, 2003.

19. Stahl. U., Banas, A., and Stymne, S., Plant microsomal phospholipid hydrolases have selectivities for uncommon fatty acids. *Plant Physiol.*, 107, 953–962, 1995.

20. Stewart, J.J., Gaginella, T.S., and Bass, P., Actions of ricinoleic acid and structurally related fatty acids of the gastrointestinal tract: I. Effects on smooth muscle contractility in vitro, *J. Pharmacol. Exp. Ther.*, 195, 347–354, 1975.

21. Gaginella, T.S., Capasso, F., Mascolo, N., and Perilli, S., Castor oil: new lessons from an ancient oil, *Phytotherapy Res.*, 12 (Suppl. 1), S128–S130, 1998.

22. Izzo, A.A., Mascolo, N., Viola, P., and Capasso, F., Inhibitors of nitric oxide synthase enhance rat ileum contractions induced by ricinoleic acid in vitro, *Eur. J. Pharmacol.*, 243, 87–90, 1993.

23. Uchida, M., Kato, Y., Matsueda, K., Shoda, R., Muraoka, A., and Yamato, S., Involvement of nitric oxide from nerves on diarrhea induced by castor oil in rats, *Jpn. J. Pharmacol.*, 82, 168–170, 2000.

24. Capasso, F., Mascolo, N., Izzo, A.A., and Gaginella, T.S., Dissociation of castor oil induced diarrhea and intestinal mucosal injury in rat: effect of N(G)-nitro-L-arginine methyl ester, *Br. J. Pharmacol.*, 113, 1127–1130, 1994.

25. Kelly, A.J., Kavanagh, J., and Thomas, J., Castor oil, bath and/or enema for cervical priming and induction of labour. *Cochrane Database Syst. Rev.*, 2, CD003099, 2001.

26. Tenore, J.L., Methods for cervical ripening and induction of labor, *Am. Fam. Physician*, 67, 2076–2084, 2003.

27. Sowa, J., Suzuki, K., Tsurata, K., Akamatsu, H., and Matsunaga, K., Allergic contact dermatitis from propylene glycol ricinoleate in a lipstick, *Contact Dermatitis*, 48, 228–229, 2003.

28. Magerl, A., Heiss, R., and Frosch, P.J., Allergic contact dermatitis from zinc ricinoleate in a deodorant and glyceryl ricinoleate in a lipstick, *Contact Dermatitis*, 44, 119–121, 2001.

29. Vieira, C., Evangelista, S., Cirillo, R., Lippi, A., Maggi, C.A., and Manzini, S., Effect of ricinoleic acid in acute and subchronic experimental models of inflammation, *Mediators Inflammation*, 9, 223–228, 2000.

9 Tree Nut Oils and Byproducts: Compositional Characteristics and Nutraceutical Applications

Fereidoon Shahidi and H. Miraliakbari
Department of Biochemistry, Memorial University of Newfoundland,
St. John's, Newfoundland, Canada

CONTENTS

9.1 INTRODUCTION

Tree nuts, their oils, and byproducts (defatted meals and hulls) contain several bioactive and health-promoting components. Epidemiological evidence indicates that the consumption of tree nuts may exert a number of cardioprotective effects which are speculated to arise from their lipid and nonlipid components, including unsaturated fatty acids, phytosterols, and phenolic antioxidants[1]. Recent investigations have also shown that dietary consumption of tree nut oils may provide even more beneficial effects than consumption of whole tree nuts, possibly due to the replacement of dietary carbohydrate with unsaturated lipids and/or other components present in the oil extracts[2]. Tree nut byproducts are utilized as sources of dietary protein and as health-promoting phytochemicals such as natural antioxidants.

Generally, tree nuts are rich in fat and contain high amounts of monounsaturated fatty acids (MUFAs; predominantly oleic acid), but also contain lower amounts of polyunsaturated fatty acids (PUFAs; predominantly linoleic acid) and small amounts of saturated lipids[3]. In many parts of the world, such as the Middle East and Asia, tree nuts are cultivated for use as oil crops and snack foods and are important sources of energy and essential dietary nutrients as well as phytochemicals[4]. Tree nut oils also constitute components of some skin moisturizers and cosmetic products[5].

This chapter summarizes the chemical characteristics and potential health effects of almonds, hazelnuts, and walnuts as well as their oils and byproducts, including antioxidant extracts. The protein compositions of tree nut byproducts are also discussed.

9.2 ALMOND

The almond tree (*Prunus delcis* and *Prunus amara*) and its fruit (containing the almond kernel or "almond") have long been recognized as being commercially valuable and nutritionally important. California and Italy are the major almond-producing regions of the world; however, other parts of Europe, Asia, and Australia also contribute a lower level of production[6]. The only other economically important product of almond trees is the almond hull, which is traditionally used in animal feed preparations, but it has gained some recent recognition in terms of its health-promoting components. Several studies have reported that almond consumption may improve blood lipid profile by lowering low-density lipoprotein (LDL) cholesterol and raising plasma high-density lipoprotein (HDL) cholesterol levels. Thus, there is much current interest in almond oil as a health-promoting edible oil[7]. The proximate composition of almond is 50.6% lipid, 21.3% protein, 19.7% carbohydrate, 5.3% moisture, and 3.1% ash (w/w)[3].

The defatted meals and hulls of almonds contain several antioxidative compounds as well as other health-promoting substances. Senter et al.[8] performed a comparative analysis of phenolic acids in selected tree nut meals including pine nut, almond, hazelnut, chestnut, and walnut, among others. The results of this study showed that gallic acid was the predominant phenolic compound in all tree nut meals except pine nut (caffeic acid), almond, and hazelnut (protocatechuic acid). Other phenolic compounds identified included *p*-hydroxybenzoic, *p*-hydroxyphenylacetic, vanillic, syringic, and ferulic acids (Table 9.1)[8]. The antiradical activity of ethanolic extracts of almond and almond byproducts including brown skins and hulls has been reported[9]. The Trolox equivalent antioxidant activities of brown skins and hulls were 13 and 10 times greater than that of the whole almond extracts. At 200ppm, ethanolic extracts of almond skins and hulls had strong scavenging activities against superoxide radical (95 and 99%, respectively), hydrogen peroxide (91%), hydroxyl radical (100 and 56%, respectively), and 2,2-diphenyl-1-picrylhydrazyl (DPPH) radical (100%)[9]. Sang et al.[10] isolated nine phenolic compounds from almond skins and assessed their DPPH scavenging activity; catechin and protocatechuic acid exhibited the best antioxidant activity, followed by 3′-*O*-methylquercetin 3-*O*-β-D-galactopyranoside, then 3′-*O*-methylquercetin 3-*O*-β-D-glucopyranoside and 3′-*O*-methylquercetin 3-*O*-α-L-rhamnopyranosyl-(1→6)-β-D-glucopyranoside and vanillic and *p*-hydroxybenzoic acids, naringenin 7-*O*-β-D-glucopyranoside, and finally kaempferol 3-*O*-α-L-rhamnopyranosyl-(1→6)-β-D-glucopyranoside[10]. Frison-Norrie and Sporns[11] quantitatively assessed the flavonol glycoside composition of blanched almond skins using matrix-assisted laser desorption/ionization time-of-flight mass spectrometry (MALDI-TOF MS), showing the presence of isorhamnetin rutinoside (51 µg/g), isorhamnetin glucoside (18 µg/g), kaempferol rutinoside (18 µg/g), and kaempferol glucoside (6 µg/g). More recently, Pinelo et al.[12] reported the total phenolics content and DPPH scavenging activity of almond hull ethanolic extracts at 3.74 mg/g and 58%, respectively. Sang et al.[13] also isolated potentially health-promoting sterols, nucleotides, and one sphingolipid, 1-*O*-β-D-glucopyranosyl-(2S,3R,4E,8Z)-2-[(2R)-2-hydroxyhexadecanoylamino]-4,8-octadecadiene-1,3-diol, from defatted almond meals. In light of data showing that tree nuts, tree nut oils, and tree nut byproducts contain heath-promoting phytochemicals, Davis and Iwashi[14] examined the effects of dietary consumption of whole almonds, almond oil, and almond meal on aberrant crypt foci development in a rat model of colon carcinogenesis. This landmark study showed that both almond oil and almond meal reduced aberrant crypt foci development, but whole almonds showed a significantly stronger anticancer effect in this model, implying a synergistic anticancer activity between the lipidic and nonlipidic constituents of almonds[14].

Shi et al.[15] assessed the fatty acid composition of almond oil; oleic acid was the major fatty acid present (68%), followed by linoleic acid (25%), palmitic acid (4.7%), and small amounts (<2.3%) of palmitoleic, stearic, and arachidic acids. Almond oil is also a rich source of α-tocopherol (around 390 mg/kg) and contains trace amounts of other tocopherol isomers as well as phylloquinone (70µg/kg)[3]. Almond oil contains 2.6 g/kg phytosterols, mainly β-sitosterol, with trace amounts of stigmasterol and campesterol[3]. The compositional characteristics of almond oil show that it is rich

TABLE 9.1
Phenolic Acid Constituents (μg/g) of Selected Tree Nut Meals

Phenolic acid	Almond	Hazelnut	Walnut
p-Hydroxybenzoic	0.30	<0.01	0.06
Phenyl acetic	0	0	0.02
Vanillic	0.07	<0.01	0.09
Proto-catechuric	0.70	0.36	0.02
Syringic	0	0	0.02
Gallic	<0.01	<0.01	0.02
Caffeic	0	<0.01	0.10
Ferulic	0	0	<0.01
Total	1.08	0.36	0.51

Adapted from Senter, S.D., Horvat, R.J., and Forbus, W.R.J., *J. Food Sci.*, 48, 798–799, 1983.

in several health-promoting nutrients, many of which may be responsible for the observed beneficial effects of dietary almond consumption in cardiovascular diseases[16] and in weight management[17]. However, few investigations have explored this topic. Hyson et al.[18] conducted a dietary intervention study to determine whether the consumption of whole almonds or almond oil for six weeks would result in similar or different effects on plasma lipids and *ex-vivo* LDL oxidation. Both groups consumed diets with identical almond oil and total fat levels. This study showed that both whole almond and almond oil consumption caused similar reductions in plasma cholesterol and LDL (4 and 6%, respectively) as well as a 14% decrease in fasting plasma triacylglycerols. These findings indicate that the lipid component of almond is responsible for its cardioprotective effects, but may warrant further investigation[18].

Several lines of evidence suggest that regular consumption of whole almonds as part of a healthy diet can help improve several parameters related to cardiovascular health which include lowering of LDL cholesterol and total plasma lipids[19]. Sabaté et al.[20] compared the effects of almond intake with those of a National Cholesterol Education Program (NCEP) Step I diet on serum lipids, lipoproteins, apolipoproteins, and glucose in healthy and mildly hypercholesterolemic adults. The NCEP Step I diet is known to reduce LDL cholesterol by 3 to 10%. The experimental diets included a Step I diet, a low-almond diet, and a high-almond diet, in which almonds contributed 0, 10, and 20% of total energy, respectively[20]. An inverse relationship was observed between the percentage of energy in the diet from almonds and the subjects' total cholesterol, LDL cholesterol, and apolipoprotein B concentrations and the ratios of LDL to HDL cholesterol and apolipoprotein B to apolipoprotein A. Compared with the Step I diet, the high-almond diet significantly reduced ($p < 0.01$) total cholesterol by 0.24 mmol/l or 4.4%, LDL cholesterol by 0.26 mmol/l or 7.0%, and apolipoprotein B by 6.6 mg/dl or 6.6%, increased HDL cholesterol by 0.02 mmol/l or 1.7%, and decreased the ratio of LDL to HDL cholesterol by 8.8%. Results of this study showed that incorporation of 68 g of almonds (20% of energy) into a 2000 kcal Step I diet markedly improved serum lipid profile of healthy and mildly hypercholesterolemic adults[20]. Similar findings have been reported by other researchers using roasted almonds[21]. Furthermore, animal model studies have confirmed the cardioprotective effects of almond consumption[22].

9.3 HAZELNUTS

Hazelnuts or filberts (*Corylus* spp.) are a rich source of energy with a 61 to 63% lipid content (w/w)[3,23]. Other components of hazelnuts are carbohydrate (15.3%), protein (13.0%), moisture (5.4%), and ash $(3.6\%)^3$. Turkey is the world's largest producer of hazelnuts, accounting for approximately 75% of total hazelnut production, followed by Italy which accounts for 10% of total global production. In the U.S., the state of Oregon is the largest producer and in Canada, southwestern British Columbia produces a small amount of hazelnuts; North America contributes less than 5% to the total world hazelnut production which is about 850,000 metric tons (unshelled basis)[24].

Few researchers have investigated the potential of hazelnuts as a source of natural antioxidants. Yurttas et al.[25] assessed the phenolic composition of methanolic extracts of defatted hazelnuts (hazelnut meal), showing that gallic acid, p-hydroxybenzoic acid, caffeic acid, epicatechin, sinapic acid, and quercetin were the predominant phenolics present. The composition of phenolic acid constituents in hazelnut meal has also been assessed by Senter et al.[8] using GC-MS (Table 9.1). Proto-catechuric acid was shown to be the main phenolic compound present in hazelnut meal ($0.36\mu g/g$), but trace amounts ($<0.1\mu g/g$) of p-hydroxybenzoic, vanillic, gallic, and caffeic acids were also present[8]. Moure et al.[26] examined the antioxidant activity of ethanolic hazelnut hull extracts, showing DPPH bleaching activities ranging from 86.2 to 94.4%. Similar values have been reported by Krings and Berger[27] using ethanolic extracts of both roasted and unroasted hazelnut meals. The extracts of roasted and unroasted hazelnut meals exhibited comparable antioxidant activities in both the DPPH bleaching assay and stripped corn oil model system[27]. Wu et al.[28] recently examined the antioxidant capacities of both lipophilic and hydrophilic extracts of hazelnuts using the oxygen radical absorbance capacity (ORAC) assay with fluorescein as the fluorescent probe. Grated hazelnuts were packed into extraction cells with sand and extracted with two solvent systems using a Dionex ASE 200 accelerated solvent extractor. During the first treatment, lipophilic extracts were obtained with hexane:dichloromethane (1:1 v/v), followed by a second treatment with acetone/water/acetic acid (70:29.5:0.5 v/v/v) to obtain the hydrophilic extracts. Results of this study[28] showed that lipophilic hazelnut extracts had ORAC values of $3.7\mu mol$ Trolox equivalents/g hazelnut, whereas the hydrophilic extracts had ORAC values of $92.8\mu mol$ Trolox equivalents/g hazelnut.

The fatty acids of hazelnut oil included 78 to 83% oleic acid, 9 to 10% linoleic acid, 4 to 5% palmitic acid, and 2 to 3% stearic acid as well as other minor fatty acids[3,24]. Parcerisa et al.[29] examined the lipid class composition of hazelnut oil and demonstrated that triacylglycerols constituted 98.4% of total lipids; glucolipids comprised 1.4% of total lipids, while trace amounts ($<0.2\%$) of phosphatidylcholine and phosphatidylinositol were also present. Hazelnut oil contains 1.2 to 1.4 g/kg of phytosterols primarily in the form of β-sitosterol and is a very good source of α-tocopherol (382 to 472 mg/kg)[3,24]. The main odorant in hazelnut oil responsible for its characteristic flavor is 5-methyl-(E)-2-hepen-4-one or filbertone, which can produce intense hazelnut oil-like aroma at the very low odor threshold of 5 ng/kg oil[30]. The oil from unroasted hazelnuts typically contains about $6\mu g$ filbertone/kg oil whereas the oil from roasted hazelnuts contains over $315\mu g$ filbertone/kg oil[30].

Several reports have shown that hazelnut is a health-promoting food and a contributing factor for the beneficial health effects of the Mediterranean-style diet[31]; however, few studies have investigated the health effects of hazelnut oil. Balkan et al.[32] examined the effects of hazelnut oil administration on plasma peroxide levels, plasma lipid profiles, plasma LDL and VLDL levels, and atherosclerotic plaque development in male New Zealand white rabbits. In this study, animals were fed control diets, control diets rich in cholesterol (0.5% w/w), control diets rich in cholesterol (0.5% w/w) together with hazelnut oil supplementation (5% w/w), or a control diet with hazelnut oil supplementation (5% w/w) for 14 weeks. The results showed that when supplemented in control diets, hazelnut oil reduced plasma cholesterol and apoB-100-containing lipoprotein levels by an insignificant level. No differences were observed in the high cholesterol diet group supplemented with hazelnut oil which implies that hazelnut oil may be an effective health-promoting agent in diets with normal lipid intake, but cannot reverse the effects of high cholesterol intake[32].

9.4   WALNUTS

Walnuts (nux juglandes) are harvested from the walnut tree (*Juglans regia*) and are the most popular nut ingredient in North American cooking. Over 30 varieties of walnut trees are currently harvested that have been developed for various characteristics including pest tolerance, early/late harvest, and shell thickness. The major walnut-producing nations are the U.S. (California), China, Turkey, India, France, Italy, and Chile[33].

Walnuts contain about 65% lipids; however, considerable differences exist among varieties (range: 52 to 70% w/w)[3,34]. Walnuts also contain 15.8% protein, 13.7% carbohydrate, 4.1% moisture, and 1.8% ash (w/w)[3]. The defatted meals of walnuts are a good source of natural antioxidants, containing predominantly caffeic, vanillic, and *p*-hydroxybenzoic acids[8] (Table 9.1). Wu et al.[28] showed that lipophilic walnut extracts had ORAC values of 4.8 μmol Trolox equivalents/g walnut and hydrophilic walnut extracts had ORAC values of 130.6 μmol Trolox equivalents/g walnut. Gunduc and El[35] have assessed the total phenolics contents of ethanolic extracts of several Turkish foods including walnuts using the Folin-Ciocalteu colorimetric method and reported a total phenolics content of 7.1 mg/g (as gallic acid equivalents) for whole walnuts. This group also compared the ability of food extracts to inhibit the *in vitro* oxidation of LDL, showing that both walnut and red wine extracts inhibited LDL oxidation to the greatest degree among the food samples tested[35]. Fukuda et al.[36] studied the composition and antioxidant activity of walnut polyphenol extracts in butanol. Using semipreparative liquid chromatography and one- and two-dimensional NMR analyses, Fukuda et al.[36] isolated 14 polyphenolic constituents from walnut extracts including three new hydrolyzable tannins, glansrins A, B, and C (ellagitannins with a tergalloyl or related polyphenolic acyl group), along with pendunculagin, tellimagrandin I and II, casuarinin, rugosin C, casuarictin, and ellagic acid. Adenosine and adenine were also identified in the walnut extracts[36]. The 14 walnut polyphenols had superoxide dismutase-like activities and strong DPPH bleaching activities, indicating that ellagitannin polyphenols act as strong antioxidants[36]. Similar findings were reported by Anderson et al.[37] who studied the composition of methanolic extracts of walnut and their ability to inhibit both azo-mediated and Cu^{2+}-mediated LDL oxidation. Anderson et al.[37] reported walnut total phenolics contents of 20 mg/g (as gallic acid equivalents), and LC-MS analysis confirmed the presence of ellagic acid and other related ellagitannins; no tocopherols were reported in the walnut extracts. Walnut extracts inhibited both azo- and Cu^{2+}-mediated LDL oxidation in a dose-dependant manner, but the extent of inhibition was significantly greater in the Cu^{2+}-mediated oxidation system[37]. Sze-Tao et al.[38] reported the hydrolyzable tannin content of several walnut batches using two modified vanillin assays, with values ranging from 363 to 1095 mg catechin equivalents/100 g of sample. Differences in total hydrolyzable tannin contents of the various walnut samples were attributed mainly to the different processing and storage conditions employed for each walnut batch[38]. Recently, walnut phenolic extracts have been shown to inhibit fibrillar amyloid beta-protein (A) production which may exert beneficial effects in Alzheimer's disease suffers since fibrillar amyloid beta-protein (A) is the principal component of amyloid plaques commonly seen in Alzheimer's disease[39]. Fukuda et al.[40] have studied the effects walnut polyphenols on blood lipid profiles and oxidative stress in type II diabetic mice (nine-week-old C57/BL/KsJ-db/db male mice). In this study[40] seven mice were supplemented orally with purified ethanolic walnut extracts at a daily level of 200 mg/kg body weight for four weeks, while eight mice were used as controls. Results of this study showed that supplementation of walnut polyphenolics significantly reduced serum triacylglycerols and urinary 8-hydroxy-2′-deoxyguanosine (an *in vivo* marker of oxidative stress) after four weeks. No significant differences were observed in body weight, blood glucose, or total serum cholesterol between the experimental and control groups[40].

The fatty acid composition of walnut oil is unique compared to other tree nut oils for two reasons: walnut oil contains predominantly linoleic acid (49 to 63%) and also a considerable amount of α-linolenic acid (8 to 15.5%). Other fatty acids present include oleic acid (13.8 to 26.1%), palmitic acid (6.7 to 8.7%), and stearic acid (1.4 to 2.5%)[34]. The tocopherol content of walnut oil varies among different cultivars and extraction procedures and ranges between 268 and 436 mg/kg. The predominant tocol isomer is γ-tocopherol (>90%), followed by α-tocopherol (6%), and then β- and Δ-tocopherols[41]. Nonpolar lipids have been shown to constitute 96.9% of total lipids in walnut oil, while polar lipids account for 3.1%. The polar lipid fraction consisted of 73.4% sphingolipids (ceramides and galactosylceramides) and 26.6% phospholipids (predominantly phosphatidylethanolamine)[42]. Walnut oil contains approximately 1.8 g/kg phytosterols[1], primarily β-sitosterol (85%), followed by Δ-5-avenasterol (7.3%), campesterol (4.6%), and finally cholesterol (1.1%)[42].

Evidence from epidemiological studies, intervention studies, and clinical trials show that walnut consumption has favorable effects on serum lipid levels in humans such as lowering LDL, raising HDL, and reducing total serum triacylglycerol levels, all of which reduce the likelihood of suffering from a cardiovascular event[19,43,44]. Many of the beneficial effects associated with walnut consumption have previously been attributed to the polyunsaturated fatty acid intake and have prompted health researchers to investigate which of these effects, if any, can be attributed to the lipid component of walnuts. Lavedrine et al.[45] conducted a cross-sectional study to assess the association between whole walnut and walnut oil consumption and blood lipid levels. This study included 933 men and women aged 18 to 65 years living in Dauphine, France (a major walnut-producing area). Factors used to assess cardiovascular disease risk included a one-year dietary recall questionnaire and serum levels of HDL, LDL, total cholesterol, and levels of the apolipoproteins apoA1 and apoB. Results from this study showed that higher levels of HDL cholesterol and apoA1 were associated with higher amounts of walnut oil and kernel consumption, with a positive trend existing between the various degrees of walnut oil/kernel consumption in this cohort. Other blood lipids did not show any significant association with walnut consumption; the nature of the cohort group made it impossible to separate the effects of whole walnut and walnut oil consumption[45]. More recently, Zibaeenezhad et al.[46] examined the effects of walnut oil consumption on plasma triacylglycerol levels in hyperlipidemic men and women. In this trial, 29 patients were given 3 g/day of walnut oil (six 500 mg capsules per day) for 45 days; 31 patients were given placebo and were used as controls. Supplementation of walnut oil reduced serum levels of LDL, triacylglycerol, and total cholesterol while increasing serum HDL levels; however, only the decrease in serum triacylglycerol reached significance[46]. The fatty acid composition of walnut oil has been suggested as being responsible for its cardioprotective feature, but results from studies such as that of Espin et al.[47] show that the antioxidative components of walnut oil have significant antiradical properties that may exert a protective effect against the oxidation of biomacromolecules such as LDL, a known risk factor for atheroma development and thus heart disease.

9.5 UTILIZATION OF DEFATTED TREE NUT MEALS AND OTHER BYPRODUCTS AS PROTEIN SOURCES

Defatted tree nut meals and hulls are traditionally utilized as animal feed due to their low cost and high nutritional value of their proteins and other constituents such as vitamins and phytochemicals[48]. Tree nut byproducts have many food[49] and biochemical applications[50]. Tree nut meals are rich in several antioxidative compounds and other health-promoting substances, which have led some research groups to investigate their potential as functional food ingredients and as possible sources of nutraceuticals[49]. The predominant nutritional component of tree nut meals is protein, constituting around 40% of total weight[49] and the protein component is of high quality compared to other defatted meals, containing at least some amount of all essential amino acids[3]. As an example, cashew nut meal contains 42% crude protein and compared to soybean meal, cashew nut meal enhances livestock weight gain curves and has a higher protein score (93 versus 97, respectively)[51].

TABLE 9.2
Amino Acid Profiles (%) of Tree Nut Proteins

Amino acid	Almond	Hazelnut	Pecan	Walnut	Pistachio	Brazil	Pine	Macadamia	Cashew
Tryptophan	0.87	1.42	1.04	1.07	1.36	0.90	1.74	0.64	1.43
Threonine	3.08	2.95	3.44	3.78	3.35	2.33	4.39	3.56	3.44
Isoleucine	3.14	3.75	3.77	3.96	4.49	3.32	5.38	3.02	3.94
Leucine	6.68	7.26	6.72	7.42	7.75	7.44	9.98	5.79	7.35
Lysine	2.73	2.63	3.22	2.68	5.74	3.17	5.19	0.17	4.63
Methionine	0.85	1.07	2.05	1.49	1.68	6.49	2.47	0.22	1.80
Cystine	1.28	1.51	1.70	1.31	1.78	2.36	2.51	0.05	1.96
Phenylalanine	5.22	4.53	4.78	4.50	5.29	4.06	5.30	6.40	4.75
Tyrosine	2.41	2.99	2.41	2.57	2.07	2.70	5.07	4.92	2.53
Valine	3.63	4.37	4.61	4.77	6.18	4.87	7.15	3.49	5.46
Arginine	11.2	14.2	13.2	14.4	10.1	13.8	26.9	13.5	10.6
Histidine	2.69	2.16	2.94	2.48	2.52	2.48	3.31	1.87	2.27
Alanine	4.54	4.67	4.46	4.41	4.59	3.71	7.24	3.73	4.18
Aspartic acid	12.4	10.5	10.4	11.6	9.06	8.67	12.6	10.5	8.96
Glutamic acid	23.5	23.3	20.5	17.8	19.1	20.3	23.5	21.8	22.5
Glycine	6.67	4.65	5.09	5.17	4.75	4.62	7.05	4.37	4.68
Proline	4.40	3.36	4.08	4.47	4.05	4.23	7.44	4.50	4.05
Serine	4.57	4.41	5.32	5.92	6.11	4.40	5.87	4.03	5.38

Values adapted from United States Department of Agriculture (USDA) Nutrient Database Version 17, www.nal.usda.gov/fnic/foodcomp/search (accessed May 28, 2005).

Similar findings have also been reported for walnut meal[52]. The amino acid compositions of proteins from several tree nut meals are given in Table 9.2 and show that in most cases glutamic acid, arginine, and aspartic acid account for about 40% of the amino acids in these proteins, whereas tryptophan is a limiting amino acid in all tree nut proteins except macadamia nut protein which contains only trace amounts of cystine. Thus, the defatted meals of tree nuts serve as high-quality protein sources.

9.6   CONCLUSIONS

Several tree nut varieties serve as valuable food crops with several food applications due to their unique flavor, texture, and healthful lipid composition. Byproducts of tree nuts also have several uses including functional food ingredients and as sources of nutraceutical extracts and dietary protein. Compared to most other vegetable oils, tree nut oils show high oxidative stability which is due to high levels of monounsaturated fatty acids rather than polyunsaturated fatty acids and high concentrations of antioxidative minor components[53]. The use of tree nut oils and byproducts in everyday cooking is very common in some parts of the world and is becoming more widespread due to increased consumer demand for alternative and health-promoting foods. The consumption of high fat tree nuts and their oils has been shown to have antiatherogenic effects, which may be related to the known positive cardiovascular health effects of unsaturated fatty acids, phytosterols, and tocol isomers. Other minor phytochemicals present in tree nut oils may also contribute to their observed health effects. However, little information is available regarding the health effects of tree nut byproducts.

REFERENCES

1. Hu, F.B., Stampfer, M.J., Manson, J.E., Rimm, E.B., Colditz, G.A., Rosner, B.A., Speizer, F.E., Hennekens, C.H., and Willett, W.C., Frequent nut consumption and risk of coronary heart disease in women: prospective cohort study, *Br. Med. J.*, 317, 1341–1345, 1998.

2. Grundy, S.M., Florentin, L., Nix, D., and Whelan, M.F., Comparison of monounsaturated fatty acids and carbohydrates for reducing raised levels of plasma cholesterol in man, *Am. J. Clin. Nutr.*, 47, 965–969, 1988.

3. United States Department of Agriculture (USDA) Nutrient Database Version 17, www.nal.usda.gov/fnic/foodcomp/search (accessed May 28, 2005).

4. Bonvehi, J.S., Coll, F.V., and Rius, I.V., Liquid chromatographic determination of tocopherols and tocotrienols in vegetable oils, formulated preparations, and biscuits, *J. AOAC Int.*, 83, 627–634, 2000.

5. Madhaven, N., Final report on the safety assessment of Corylus avellana (hazel) seed oil, Corylus americana (hazel) seed oil, Corylus avellana (hazel) seed extract, Corylus americana (hazel) seed extract, Corylus avellana (hazel) leaf extract, Corylus americana (hazel) leaf extract, and Corylus rostrata (hazel) leaf extract, *Int. J. Toxicol.*, 20, 15–20, 2001.

6. Martins, M., Tenreiro, R., and Oliveira, M.M., Genetic relatedness of Portuguese almond cultivars assessed by RAPD and ISSR markers, *Plant Cell Rep.*, 22, 71–78, 2003.

7. Spiller, G.A., Jenkins, D.A.J., Bosello, O., Gates, J.E., Cragen, L.N., and Bruce, B.J., Nuts and plasma lipids: an almond-based diet lowers LDL-C while preserving HDL-C, *J. Am. College Nutr.*, 17, 285–290, 1998.

8. Senter, S.D., Horvat, R.J., and Forbus, W.R.J., Comparative GLC-MS analysis of phenolic acid of selected tree nuts, *J. Food Sci.*, 48, 798–799, 1983.

9. Siriwardhana, S. and Shahidi, F., Antiradical activity of extracts of almond and its by-products, *J. Am. Oil Chem. Soc.*, 79, 903–906, 2002.

10. Sang, S., Lapsley, K., Jeong, W.S., Lachance, P.A., Ho, C.T., and Rosen, R.T., Antioxidative phenolic compounds isolated from almond skins (Prunus amygdalus Batsch), *J. Agric. Food Chem.*, 50, 2459–2463, 2002.

11. Frison-Norrie, S. and Sporns, P.J., Identification and quantification of flavonol glycosides in almond seedcoats using MALDI-TOF MS, *J. Agric. Food Chem.*, 50, 2782–2787, 2002.

12. Pinelo, M., Rubilar, M., Sineiro, J., and Nu'nez, M.J., Extraction of antioxidant phenolics from almond hulls (Prunus amygdalus) and pine sawdust (Pinus pinaster), *Food Chem.*, 85, 267–273, 2004.

13. Sang, S., Kikuzaki, H., Lapsley, K., Rosen, R.T., Nakatani, N., and Ho, C.T., Sphingolipid and other constituents from almond nuts (Prunus amygdalus Batsch), *J. Agric. Food Chem.*, 50, 4709–4712, 2002.

14. Davis, P.A. and Iwahashi, C.K., Whole almonds and almond fractions reduce aberrant crypt foci in a rat model of colon carcinogenesis, *Cancer Lett.*, 165, 27–33, 2001.

15. Shi, Z., Fu, Q., Chen, B., and Xu, S., Analysis of physicochemical property and composition of fatty acid of almond oil, *Chinese J. Chromatogr.*, 17, 506–507, 1999.

16. Sabaté, J. and Fraser, G.E., Nuts: a new protective food against coronary heart disease. *Curr. Opin. Lipidol.*, 5, 11–16, 1999.

17. Fraser, G.E., Bennett, H.W., Jaceldo, K.B., and Sabaté, J., Effect on body weight of a free 76 Kilojoule (320 calorie) daily supplement of almonds for six months, *J. Am. College Nutr.*, 21, 275–283, 2002.

18. Hyson, D.A., Schneeman, B.O., and Davis, P.A., Almonds and almond oil have similar effects on plasma lipids and LDL oxidation in healthy men and women, *J. Nutr.*, 132, 703–707, 2002.

19. Abbey, M., Noakes, M., Belling, G.B., and Nestel, P.J., Partial replacement of saturated fatty acids with almonds or walnuts lowers total serum cholesterol and low-density-lipoprotein cholesterol, *Am. J. Clin. Nutr.*, 59, 995–999, 1994.

20. Sabaté, J., Haddad, E., Tanzman, J.S., Jambazian, P., and Rajaram, S., Serum lipid response to the graduated enrichment of a Step I diet with almonds: a randomized feeding trial, *Am. J. Clin. Nutr.*, 77, 1379–1384, 2003.

21. Spiller, G.A., Miller, A., Olivera, K., Reynolds, J., Miller, B., Morse, S.J., Dewell, A., and Farquhar, J.W., Effects of plant-based diets high in raw or roasted almonds, or roasted almond butter on serum lipoproteins in humans, *J. Am. College Nutr.*, 22, 195–200, 2003.

22. Yan, X.S., Wang, J., and Liang, S., Effects of nuts rich in monounsaturated fatty acids on serum lipids of hyperlipidemia rats, *Wei Sheng Yan Jiu*, 32, 120–122, 2003.

23. Alasalvar, C., Shahidi, F., Liyanapathirana, C.M., and Ohshima, T., Turkish Tombul hazelnut (Corylus avellana L.): 1. Compositional characteristics, *J. Agric. Food Chem.*, 51, 3790–3796, 2003.

24. Alasalvar, C., Shahidi, F., Ohshima, T., Wanasundara, U., Yurttas, H.C., Liyanapathirana, C.M., and Rodrigues, F.B., Turkish Tombul hazelnut (Corylus avellana L.): 2. Lipid characteristics and oxidative stability, *J. Agric. Food Chem.*, 51, 3797–3805, 2003.

25. Yurttas, H.C., Shafer, H.W., and Warthesen, J.J., Antioxidant activity of nontocopherol hazelnut (Corylus spp.) phenolics, *J. Food Sci.*, 65, 276–280, 2000.

26. Moure, A., Franco, D., Sineiro, J., Dominguez, H., and Nunez, M.J., Simulation of multistage extraction of antioxidants from Chilean hazelnut (Gevuina avellana) hulls, *J. Am. Oil Chem. Soc.*, 80, 389–397, 2003.

27. Krings, U. and Berger, R.G., Antioxidant activity of some roasted foods, *Food Chem.*, 72, 223–231, 2001.

28. Wu, X., Beecher, G.R., Holden, J.M., Haytowitz, D.B., Gebhart, S.B., and Prior, R.L., Lipophilic and hydrophilic antioxidant capacities of common foods in the United States, *J. Agric. Food Chem.*, 52, 4026–4037, 2004.

29. Parcerisa, J., Richardson, D.G., Rafecas, M., Codony, R., and Boatella, J., Fatty acid distribution in polar and nonpolar lipid classes of hazelnut oil (*Corylus avellana* L.), *J. Agric. Food Chem.*, 45, 3887–3890, 1997.

30. Pfnuer, P., Matsui, T., Grosch, W., Guth, H., Hofmann, T., and Schieberle, P., Development of a stable isotope dilution assay for the quantification of 5-methyl-(E)-2-hepten-4-one: application to hazelnut oils and hazelnuts, *J. Agric. Food Chem.*, 47, 2044–2047, 1999.

31. Kris-Eterton, P.M., A new role for diet in reducing the incidence of cardiovascular disease: evidence from recent studies, *Curr. Atherosclerosis Rep.*, 3, 185–187, 1999.

32. Balkan, J., Hatipoğlu, A., Gülcin, A., and Uysal, M., Influence on hazelnut oil administration on peroxidation status of erythrocytes and apolipoprotein B 100-containing lipoproteins in rabbits fed on a high cholesterol diet, *J. Agric. Food Chem.*, 51, 3905–3909, 2003.

33. Walnut industry fact sheet, www.walnut.org/pdfs/walnuts_factsheet.pdf (accessed May19, 2005).

34. Zwarts, L., Savage, G.P., and McNeil, D.L., Fatty acid content of New Zealand-grown walnuts (Juglans regia L.), *Int. J. Food Sci. Nutr.*, 50, 189–194, 1999.

35. Gunduc, H. and El, S.N., Assessing antioxidant activities of phenolic compounds of common Turkish food and drinks on in vitro low-density lipoprotein oxidation, *J. Food Sci.*, 68, 2591–2595, 2003.

36. Fukuda, T., Ito, H., and Yoshida, T., Antioxidative polyphenols from walnuts, *Phytochemistry*, 63, 795–801, 2003.

37. Anderson, K.J., Teuber, S.S., Gobeille, A., Cremin, P., Waterhouse, A.L., and Steinberg, F.M., Walnut polyphenolics inhibit in vitro human plasma and LDL oxidation, *J. Nutr.*, 131, 2837–2842, 2001.

38. Sze-Tao, K.W.C., Shrimpf, J.E., Teuber, S.S., Roux, K.H., and Sath, S.K., Effects of processing and storage on walnut (*Juglans regia* L) tannins, *J. Sci. Food Agric.*, 81, 1215–1225, 2001.

39. Chauhan, N., Wang, K.C., Wegiel, J., and Malik, M.N., Walnut extract inhibits the fibrillization of amyloid beta-protein, and also defibrillizes its preformed fibrils. *Curr. Alzheimer Res.*, 1,183–188, 2004.

40. Fukuda, T., Ito, H., and Yoshida, T., Effect of the walnut polyphenol fraction on oxidative stress in type 2 diabetes mice, *Biofactors*, 21, 251–253, 2004.

41. Savage, G.P., Dutta, P.C., and McNeil, D.L., Fatty acid and tocopherol contents and oxidative stability of walnut oils, *J. Am. Oil Chem. Soc.*, 76, 1059–1065, 1999.

42. Tsamouris, G., Hatzinantoniou, S., and Demetzos, C., Lipid analysis of Greek walnut oil (*Juglans regia* L.), *J. Biosci.*, 57, 51–56, 2002.

43. Zambon, D., Sabate, J., Munoz, S., Campero, B., Casals, E., Merlos, M., Laguna, J.C., and Ros, E., Substituting walnuts for monounsaturated fat improves the serum lipid profile of hypercholesterolemic men and women. A randomized crossover trial, *Ann. Intern. Med.*, 132, 538–546, 2000.

44. Sabate, J., Fraser, G.E., Burke, K., Knutsen, S.F., Bennet, H., and Lindsted, K.D., Effects of walnuts on serum lipid levels and blood pressure in normal men, *N. Engl. J. Med.*, 328, 603–607, 1993.

45. Lavedrine, F., Zmirou, D., Ravel, A., Balducci, F., and Alary, J., Blood cholesterol and walnut consumption: a cross-sectional survey in France, *Preventive Med.*, 28, 333–341, 1999.

46. Zibaeenezhad, M.J., Rezaiezadeh, M., Mawla, A., Ayatollahi, S.M., and Panjahshahin, M.R., Antihypertriglyceridemic effect of walnut oil, *Angiology*, 54, 411–414, 2003.

47. Espin, J.C., Soler-Rivas, C., and Wichers, H.J., Characterization of the total free radical scavenger capacity of vegetable oils and oil fractions using 2,2-diphenyl-1-picrylhydrazyl radical, *J. Agric. Food Chem.*, 48, 648–657, 2000.

48. Harvey, D., Commonwealth Bureau of Animal Nutrition, 19, 105–113, 1970.
49. Souza, M.L. and Menezes, H.C., Processing of Brazil nut and meal and cassava flour: quality parameters, *Ciencia e Tecnologia de Alimentos*, 24, 120–128, 2004.
50. Wolf, W.J. and Sathe, S.K., Ultracentrifugal and polyacrylamide gel electrophoretic studies of extractability and stability of almond meal proteins, *J. Sci. Food Agric.*, 78, 511–521, 1998.
51. Piva, G., Santi, E., and Ekpenyong, T., Nutritive value of cashewnut extraction meal, *J. Sci. Food Agric.*, 22, 22–23, 1971.
52. Savage, G.P., Cardioprotective nutrients in walnuts, *Proc. Nutr. Soc. New Zealand*, 25, 19–25, 2000.
53. Romero, N., Robert, P., Masson, L., Ortiz, J., Pavez, J., Garrido, C., Foster, M., and Dobarganes, C., Effect of alpha-tocopherol and alpha-tocotrienol on the performance of Chilean hazelnut oil (Gevuina avellana Mol) at high temperature, *J. Sci. Food Agric.*, 84, 943–948, 2004.

10 Gamma-Linolenic Acid (GLA)

Yao-wen Huang
Department of Food Science and Technology, University of Georgia,
Athens, Georgia, USA

Chung-yi Huang
Department of Food Science, National I-Lan University, I-Lan, Taiwan, ROC

CONTENTS

10.1   INTRODUCTION

Gamma-linolenic acid (GLA), an all-*cis* omega-6 (n-6) long-chain polyunsaturated fatty acid (PUFA), is an important essential fatty acid (EFA). EFAs are required for human health but cannot be made in the body and must be obtained from food. GLA is comprised of 18 carbon atoms with three double bonds. It is known as 18:3n-6, 6,9,12-octadecatrienoic acid, (Z,Z,Z)-6,9,12-octadecatrienoic acid, *cis*-6, *cis*-9, *cis*-12-octadecatrienoic acid, and gamolenic acid. On January 27, 1993 the U.S. Court of Appeals for the Seventh Circuit ruled that GLA is a single food ingredient and not subject to food additive regulation[1], and thus many commercial products as dietary supplements are currently available on the market.

169

10.2 BIOCHEMISTRY

Two EFAs for the human body are linoleic acid (LA, 18:2n-6) in the n-6 family and alpha-linolenic acid (ALA, 18:3n-3) in the n-3 family. In the n-6 family, LA can be oxidized by enzyme, delta-6 desaturase, of the endoplasmic reticulum of mammalian liver cells to GLA. GLA is then rapidly elongated by the addition of two carbons from acetyl-CoA to form eicosatrienoic acid (dihomo-gamma-linolenic acid, DHGLA, 20:3n-6). DHGLA can be further oxidized to form arachidonic acid (AA, 20:4n-6) by delta-5 desaturase at a small percentage[2,3].

AA is then esterified and made a component of membrane phospholipids. AA also serves as a pool of immediate precursor for the prostaglandins, a variety of hormone-like molecules, 2-series of prostaglandins (PEG$_2$) by cyclooxygenase and/or 4-series leukotrienes, RAF, by 5-lipoxygenase. Both products have proinflammatory properties[1]. Delta-6 and delta-5 desaturases are rate-limiting enzymes which determine the concentrations of DHGLA and AA in the cells. They will affect the rate of biosyntheses of their metabolites, the prostaglandins. Prostaglandins are important for the regulation of inflammation, pain, and swelling; blood pressure; heart function; gastrointestinal function and secretions; kidney function and fluid balance; blood clotting and platelet aggregation; allergic response; nerve transmission; and steroid production and hormone synthesis[4].

In addition to forming AA, however, DHGLA can be metabolized by cyclooxygenase into 1-series of prostaglandins (PEG$_1$), vasodilators, which have antiinflammatory properties to inhibit thrombus formation. In several cell types such as neutrophils, macrophage/monocytes, and epidermal cells, DHGLA can also be metabolized by 15-lipoxygenase into 15-(S)-hydroxyl-8, 11, 13-eicosatrienoic acid (15-HETrE)[5]. 15-HETrE is capable of inhibiting the formation of AA-derived 5-lipoxygenase metabolites, the 4-series leukotrienes, e.g., LTC$_4$ and LTB$_4$[6,7]. These products are associated with several pathogenic inflammatory, hyperproliferative disorders[8]. Cho and Ziboh[9] reported 15-HETrE can be incorporated into the membrane phospholipids, phosphatidylinositol 4,5-bisphosphate, and released as 15-HETrE-containing diacylglycerol which is capable of inhibiting protein kinase C beta, a mediator of the cell cycle in select cell types. These findings along with inhibition of leukotriene biosynthesis can further shed light on the mechanism of GLA on antiinflammatory and hyperproliferative responses[1].

In the n-3 family, the synthesis of eicosapentanoic acid (EPA, 20:5n-3) is dependent on the ALA precursor. ALA that is oxidized by the same endoplasmic reticulum enzymes is converted to an 18:4n-3 fatty acid. This fatty acid is further elongated by two carbon atoms from acetyl-CoA to form 20:4n-3 fatty acid. This fatty acid is oxidized by delta-5 desaturase to form EPA. EPA is further metabolized by cyclooxygenase into 3-series of prostaglandins (PEG$_3$). Since delta-5 desaturase prefers the n-3 to n-6 fatty acids, the amounts of AA and EPA formed depend on the amounts of their respective LA and ALA precursors.

In cells, prostaglandins act differently: PEG$_1$ and PEG$_3$ prevent platelet stickiness, improve blood flow, and reduce inflammation, while PEG$_2$ promotes platelets sticking together leading to hardening of the arteries, heart disease, and stroke. Therefore, manipulating prostaglandin metabolism may help treat certain health conditions. From the pathway of syntheses of PUFA and prostaglandins from n-6 and n-3 fatty acids, the type of dietary oils consumed and stored in cell membranes may play an important role for the production of metabolites. Reducing the level of AA and increasing the level of DHGLA and EPA may be achieved by reducing intake of animal fat (except fish oil) and supplementing with diets rich in GLA[4].

Although GLA formation is dependent on the activity of delta-6 desaturase, some factors such as aging, nutrient deficiency, smoking, *trans* fatty acids, and excessive alcohol consumption will reduce the capability of this enzyme[5]. Dietary supplementation of GLA bypasses the rate-limited delta-6 desaturation step and is quickly elongated to DHGLA by elongase, with only a very limited amount being desaturated to AA by delta-5 desaturase[1]. The kinetics of dietary GLA supplementation in humans has been explained[11]; triacylglycerols (TAGs) of GLA following ingestion undergo hydrolysis to form monoacylglycerols (MAGs) and free fatty acids by lipases. The MAGs and free fatty acids are immediately absorbed by enterocytes. In the enterocytes, a reacylation takes place reforming TAGs that are then assembled with phospholipids, cholesterol, and apoproteins into chylomicrons. The chylomicrons are released into the lymphatics from where they are transported to the systemic circulation. In the circulation, the chylomicrons are degraded by lipoprotein lipase. The fatty acids including GLA are finally distributed to various tissues in the body. GLA is normally found in the free state in the cell at a small level but occurs as components of phospholipids, neutral lipids, and cholesterol esters, mainly in cell membranes.

The balance of n-3 to n-6 lipids is critical to proper prostaglandin metabolism. The n-6 fatty acids in excess can contribute to inflammatory processes and impede absorption of n-3 fatty acids. A combination of ALA (or EPA and docosahexaenoic acid; DHA) with GLA may antagonize conversion to AA[12]. A good ratio of n-6 to n-3 fatty acids is reported as 4:1[13]; however, Americans consume 10 to 20 times the amount of n-6 fatty acids that is needed. Taking GLA should also supplement the intake of EPA-rich fish oil[4]. EPA is a precursor of the series-3 prostaglandins (in a higher level), the series-5 leukotriens, and the series-3 thromboxanes (in a lower level).

10.3   THERAPEUTIC APPLICATIONS

GLA appears to be of benefit in some conditions due to the production of various prostaglandins and leukotrienes. Some of these substances may increase symptoms, while others decrease them. A diet rich in GLA may affect the balance to more favorable prostaglandins and leukotrienes and make it helpful for some diseases. Some of the diseases listed in the following have been studied in the past:

1. Attention deficit hyperactivity disorder (ADHD)[14]: a chronic behavioral disorder characterized by hyperactiveness and/or inattentiveness.
2. Cyclic mastalgia[15]: a condition whereby a woman's breast becomes painful during the week before menstrual period. The discomfort is accompanied by swelling, inflammation, and sometimes actual cysts that form in the breasts. The symptom is also called fibrocystic breast disease, cyclic mastitis, and mastodynia. The study indicated the symptoms seem to be associated with an imbalance of fatty acids in the body.
3. Diabetic neuropathy[16,17]: a condition of gradual deterioration of nerves caused by diabetes.
4. Eczema[18–21]: a condition whereby a person has superficial inflammation of the skin, and may be characterized by vesicles with acute, redness, edema, oozing, crusting, scaling, and usually itching. Scratching or rubbing may lead to lichenification. The term is often synonymously used as dermatitis.
5. Obesity[22]: overweight caused by family history.
6. Osteoporosis[23–25]: a generalized, progressive diminution in bone tissue mass causing weakness of skeletal strength, even though the ratio of mineral to organic elements is unchanged in the remaining morphologically normal bone. A person with uncomplicated osteoporosis may remain asymptomatic or have aching pain in the bones, particularly the back.
7. Menopausal symptoms[26]: caused by the physiologic cessation of menses as a result of decreasing ovarian function. Hot flushes and sweating secondary to vasomotor instability affect 75% of women. Psychological and emotional symptoms of fatigue, irritability, insomnia, and nervousness may relate to both estrogen deprivation and the stress of aging and changing roles.
8. Raynaud's phenomenon[27]: a condition whereby a person has spasm of arterioles, usually in the digits and occasionally other acral parts including the nose and tongue, with intermittent pallor or cyanosis of the skin. Intermittent attacks of blanching or cyanosis of the digits are precipitated by exposure to cold or by emotional upsets.

9. Rheumatoid arthritis[28,29]: an autoimmune disease whereby a person has usually symmetric inflammation of the peripheral joints, potentially resulting in progressive destruction of articular and pariarticular structures; generalized manifestations may also be present. Onset may be abrupt, with simultaneous inflammation in multiple joints, or more frequently insidious, with progressive joint involvement.

10. Premenstrual syndrome[30,31]: a condition characterized by nervousness, irritability, emotional instability, depression, and possibly headaches, edema, and mastalgia. It occurs during the 7 to 10 days before menstruation and disappears a few hours after onset of menstrual flow. The syndrome is also called premenstrual tension. Most women experience some symptoms referable to the menstrual cycle; in many cases the symptoms are significant but of short duration and are not disabling.

10.4   SOURCES

GLA is found naturally in some plant seed oils such as borage (18 to 26 g/100 g GLA), black currant (15 to 20 g/100 g GLA), and evening primrose (7 to 10 g/100 g GLA), and in fungal oil of *Mucor javanicus* (23 to 26 g/100 g GLA)[32]. It is also found in hemp seed oil at 1 to 6%[11]. Although borage seed oil has the highest level of GLA, it may contain low levels of hepatotoxic unsaturated pyrrolizidine alkaloids[33]. GLA is also found in human milk and in small amounts in organ meats. GLA is produced naturally in the body as the delta 6-desaturase metabolite of the essential fatty acid linoleic acid. Under certain conditions such as decreased activity of the delta-6 desaturase, GLA may become a conditionally EFA. GLA is present in the form of TAGs and the stereospecificity is varied among different oil sources.

10.4.1   BORAGE

Borage (*Borago officinalis* L.) has common names of beebread, starflower, and ox's tongue. It belongs to the family of Boraginaceae. It is related to forget-me-not. The large plant, a hardy annual, grows to 50 cm tall. The whole plant is rough with white, stiff, prickly hairs in the lower stem. It tends to grow on rough ground and in ditches. Borage, native to western Mediterranean areas including Spain and North Africa, spreads all over Europe and is naturalized to North America up to an altitude of 1800 m.

The leaves are alternate, large, wrinkled, deep green, oval and pointed, of 7 cm long and 3 cm broad (Figure 10.1). The star-shaped bright-blue flowers, much like forget-me-not, have black anthers. The flowers bloom from May to September. The fruit consists of four brownish-black mutlets. The fresh herb is rich in honey-producing juices. It has cucumber-like odor and taste.

10.4.1.1   Parts Used

The parts of the plant used are leaves, flowers, and seeds.

10.4.1.2   Constituents

The plant contains mucilage, malic acid, potassium nitrate, and tannins. Oil extracted from seeds contains 30 to 40% linoleic acid, 18 to 25% GLA, 15 to 20% oleic acid, 9 to 12% palmitic acid, and 3 to 4% stearic acid[34]. GLA present in borage oil is in the form of TAGs. It is concentrated in the *sn*-2 position in the TAGs. Although borage oil contains 2 to 3 times more GLA than evening primrose oil, the latter appears to provide the most benefit in nutritional and clinical studies[35].

FIGURE 10.1 Borage (*Borago officinalis* L.)

10.4.1.3　Utilization

1. Food use. In the early 1990s the young tops of plants were used as a kitchen flavoring herb. The flowers were used to garnish salads, while stems and leaves were eaten raw or cooked[36]. Germans commonly added it to soups, omelets, and doughnuts.
2. Traditional medicinal use. The leaves and flowers were seeped in wine to dispel melancholy. Today, the plant is still largely used in claret cup. The leaf was used as an infusion to treat fever, coughs, and sore throats. It has also been used as a diuretic, as a poultice for inflammation and swelling, as an expectorant, and for depression[37]. The herb is used by European herbalists as a mild amphoteric to the HPA axis.
3. Current use. Borage oil has been used to lower blood pressure[38], inhibit platelet aggregation, improve psoriasis, relieve premenstrual symptoms, improve atopic eczema, and improve infantile seborrhea. It was also found useful in controlling arthritic pain at a recommended dose of 3 g GLA per day[39].

10.4.1.4　Toxicity and Interactions

Seed oil is not associated with toxicity, while leaf contains pyrrolizidine alkaloids (supinin and lycopsamin) that are known to be hepatotoxic[36,37]. This herb should not be used at the same time with other hepatotoxic drugs or herbs[37]. Due to the effect of GLA on prostaglandin synthesis, there is a potential for interaction with oral anticoagulants and platelet-inhibiting drugs[40]. Due to lowering the seizure threshold, borage oil should not be used with anticonvulsants[37].

10.4.2 BLACKCURRANT

Blackcurrant (*Ribes nigrum* L.) has common names of European blackcurrant, grosellero negro, quinsy berries, and squinancy berries. It belongs to the family of Grossulariaceae. It has brownish black berries called quinsy berries ripening from July to August. The flowers are reddish inside and greenish outside and bloom in April and May. The plant grows wild in Central and Eastern Europe, but mostly cultivated in temperate regions. The wild blackcurrant can be found in wet forests, hedges, alder swamps, and the bottom of valleys up to an altitude of 2000 m. At present, blackcurrant also grows in New York.

Blackcurrant is a thick 1.5 m tall perennial shrub with 3 to 5 lobed, doubly serrate leaves that have a large number of yellow, glandular dots on the underside (Figure 10.2). This makes the blackcurrant easily recognized and distinguished from other currants such as redcurrant, which is not used as a medicine.

10.4.2.1 Parts Used

The parts of the plant that have been used are the fruits, leaves, roots, and seeds.

10.4.2.2 Constituents

The plant contains substantial quantities of vitamin C (500 to 2000 mg/l), potassium, rutin, tannic acid, and black pigment, anthocyanins[42]. The blackcurrant oil (BCO) extracted from seeds contains 47 to 48% linoleic acid, 16 to 17% GLA, 12 to 14% ALA, 9 to 11% oleic acid, 6% palmitic acid, 2.5 to 3.5% stearidonic acid (SDA), and 1.5% stearic acid[43]. The GLA is concentrated in the *sn*-3 position and contains ALA, GLA, and the unique SDA.

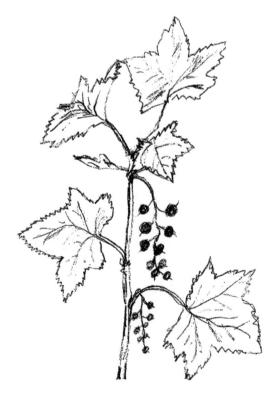

FIGURE 10.2 Blackcurrant (*Ribes nigrum* L.)

10.4.2.3   Utilization

1. Food use. As redcurrant (*R. rubrum*), the redcurrant berries have been processed into juice, jam, jelly, syrup, tea, wine, and other beverages.
2. Traditional medicinal use. The juice has been drunk to ward off the beginning of colds and flu. It is recommended that a glass of the juice be taken at noon and in the evening during the convalescent period. The leaf has diuretic and diaphoretic actions and is used as a gargle for sore throat. It has also been used to treat mild, unspecific diarrhea[42]. BCO is also used as an ingredient for cosmetic cream for damaged and baby hair, body, hand, and facial care.
3. Current use. As a GLA-rich source, BCO has been used for indications including cardiovascular disease and rheumatoid arthritis[11] and symposiums including high blood pressure, arthritis, inflammation, pain, candida, alcoholism, and respiratory and skin problems.

10.4.2.4   Toxicity and Interactions

There are no reports of toxic effects.

10.4.3   EVENING PRIMROSE

Evening primrose (*Oenothera biennis* L.) has common names of king's cure-all, donkeys' herb, gardeners' ham, and evening star. It belongs to the family of Onagaceae. The plant, native to North America and Europe, is a sturdy yellow-flowered biennial herb growing up to 1 to 2 m tall (Figure 10.3). The plant can be seen growing along roadsides in North America.

FIGURE 10.3 Evening primrose (*Oenothera biennis* L.)

Evening primrose got its name from the fact that clusters of flowers first open in the evening, then remain wide open during the next day[44]. The yellow flowers have four petals in a cross-like form. It grows in gardens, waste ground, fallow lands, ditches, rubble on embankments, and sand dunes.

10.4.3.1 Parts Used

Leaves, bark, roots, and seeds have been utilized.

10.4.3.2 Constituents

Evening primrose oil (EPO) extracted from seeds contains 8 to 14% GLA, 65 to 80% linoleic acid (LA), 6 to 11% oleic acid, 7% palmitic acid, and 2% stearic acid[44]. The GLA is concentrated in sn-3 position. In addition to GLA and LA, EPO also contains 1 to 2% unsaponifiables rich in beta-sitosterol and citrostadienol.

10.4.3.3 Utilization

1. Food use. Seeds have been used for food by Native Americans. Roots and young leaves collected in the fall, with a peculiar, bitter flavor were consumed as a vegetable. In Germany, the seeds, with an aromatic taste reminiscent of poppy seed oil, were used as a coffee substitute during wartime.
2. Traditional medicinal use. Whole plant was used as a poultice for bruises, while root was used to treat hemorrhoids. Tea made from seed was used for sore throats and gastrointestinal irritation. Leaf and root bark were used to treat spastic coughs and irritable bowel syndrome[36]. It has also been used for treatment on other conditions including menstrual discomfort, menopausal symptoms, hypertension, rheumatoid arthritis, multiple sclerosis (MS), cardiovascular disease, and skin conditions[45]. The beneficial effect of the oil in MS may be of considerable value[23].
3. Current use. EPO has been used for the following symptoms or conditions: asthma, migraines, allergic-induced eczema[36], cardiovascular disease[46], diabetic neuropathy, dermatitis and atopic eczema, schizophrenia, and tardive dyskinesia[47]. EPO is also an important ingredient in many cosmetic creams for moisturizing dry skin and in anti-wrinkle and nail products[44].

10.4.3.4 Toxicity and Interactions

No toxic effect is known. A few people taking large doses have reported mild side effects of abdominal discomfort, nausea, or headache[36]. An interaction between EPO and anesthetics might result in seizures[48]. As with other oil sources for GLA, some interactions with oral anticoagulants and platelet-inhibiting drugs[40], and anticonvulsant[37] may occur. The use of EPO in pregnant women is unwise[41].

10.5 COMMERCIAL PRODUCTS

GLA on the market has several forms including concentrate and capsules. The concentrations of GLA may vary in different oil preparations. EPO, borage oil, and BCO are available for alleviating different conditions. Doses for rheumatoid arthritis and other conditions may range from 360 mg to 2.9 g daily and are usually taken with meals[11]. GLA is also available in capsule form of 200, 300, and 1000 mg.

10.6   INTERACTIONS WITH MEDICATIONS AND NUTRIENTS

Positive or negative effects may occur when those who are being treated with some medications or are taking other nutrients[49]:

1. Ceftazidine: GLA may increase the effectiveness of ceftazidine, an antibiotic, against bacterial infections[49].
2. Chemotherapy drugs: GLA may increase the effects of anticancer treatment drugs including doxorubicin, cisplatin, carboplatin, idarubicin, mitoxantrone, tamoxifen, vincristine, and vinblastine[49].
3. Corticosteroids: GLA can decrease dosage or even lead to discontinuation of medications completely[50].
4. Cyclosporine: GLA may increase the immunosuppressive effects of cyclosporine, a drug used to suppress the immune system after an organ transplant, and may protect against kidney damage[49].
5. Phenothiazines for schizophrenia: GLA may interact with phenothiazine (i.e., chlorpromazine, fluphenazine, perphenazine, promazine, and thioridazine), which is used to treat schizophrenia, to increase the risk of seizures[49].
6. Warfarin and hemophiliacs: with possible antithrombotic activity, those on such medications should stop the use of GLA before surgery[11].
7. Zinc, ascorbic acid, and vitamin B_6: aid in conversion of GLA to PGE_1[5].

10.7   CONCLUSIONS

GLA converts to PGE_1 which has antiinflammatory, antithrombotic, antiproliferative, vasodilation, and lipid-lowering potential. GLA is also an important component of phospholipids in membranes where it enhances their fluidity and integrity[51,52]. Due to potential side effects and interactions with drugs and nutrients, supplementation of GLA should be under the supervision of healthcare providers.

REFERENCES

1. Fan, Y. and Chapkin, R.S., Importance of dietary gamma-linolenic acid in human health and nutrition, *J. Nutr.*, 128, 1411–1414, 1998.
2. Johnson, M.M., Swan, D.D., Surette, M.E., Stegner, J., Chilton, T., Fontech, A.N., and Chilton, F.H., Dietary supplementation with gammaplinolenic acid alters fatty acid content and eicosanoid production in healthy humans, *J. Nutr.*, 127, 1435–1444, 1997.
3. Zurier, R.B., Rossetti, R.G., Jacobson, E.W., DeMarco, D.M., Liu, N.Y., Terming, J.E., White, R.B., and Laposata, M., Gamma-linolenic acid treatment of rheumatoid arthritis. A randomized, placebo-controlled trial, *Arthritis Rheum.*, 39, 1808–1817, 1996.
4. Murray, M.T., Essential fatty acid (part four), *Encyclopedia of Nutritional Supplements: The Essential Guide for Improving Your Health Naturally*, Prima, Rocklin, CA, 1996, pp. 235–278.
5. Horrobin, D.F., Loss of delta-6 desaturase activity as a key factor in aging, *Med. Hypotheses*, 7, 1211–1220, 1981.
6. Borgeat, P., Hamberg, M., and Samuelsson, S., Transformation of arachidonic acid and homo-gamma-linolenic acid by rabbit polymorphonuclear leukocytes. Monohydroxy acids from novel lipoxygenase, *J. Biol. Chem.*, 251, 7816–7820, 1976.
7. Chapkin, R.S., Miller, C.C., Somers, S.D., and Erickson, K.L., Ability of monohydro-eicosatrienoic acid (15-OH-20:3) to modulate macrophage arachidonic acid metabolism, *Biochem. Biophys. Res. Commun.*, 153, 799–804, 1988.
8. Miller, C.C., Tang, W., Ziboh, V.A., and Fletcher, M.P., Dietary supplementation with ethyl ester concentrates of fish oil (n-3) and borage oil (n-6) polyunsaturated fatty acids induces epidermal generation of local putative anti-inflammatory metabolites, *J. Investig. Dermatol.*, 96, 98–103, 1991.

9. Goulet, J.L., Snouwaert, J.N., Latour, A.M., Coffman, T.M., and Koller, B.H., Altered inflammatory responses in leukotrien-deficient mice, *Proc. Natl. Acad. Sci. USA*, 91, 12852–12856, 1994.

10. Cho, Y. and Ziboh, V.A., A novel 15-hydroxyeicosatrienoic acid-substituted diacylglycerol (15-HETrE-DAG) selectively inhibits epidermal protein kinase C-beta, *Biochim. Biophys. Acta*, 1349, 67–71, 1997.

11. Anon, Gamma-linolenic acid (GLA), in *PDR for Nutritional Supplements*, Hendler, S.S. and Rarvik, D., Eds., Medical Economics, Montvale, NJ, 2001, pp, 171–174.

12. Cleland, L.G., James, M.J., and Proudman, S.M., The role of fish oils in the treatment of rheumatoid arthritis, *Drugs*, 63, 845–853, 2003.

13. Schlomo, Y. and Carasso, R.L., Modulation of learning, pain thresholds, and thermoregulation in the rat by preparations of free purified alpha-linolenic and linoleic acids: determination of the optimal w3-to-w6 ratio, *Proc. Natl. Acad. Sci. USA*, 10, 10345–10347, 1993.

14. Richardson, A.J. and Puri, B.K., The potential role of fatty acids in attention-deficit/hyperactivity disorder, *Prostaglandins Leukot. Essent. Fatty Acids*, 63, 79–87, 2000.

15. Pye, J.K., Mansel, R.E., and Hughes, L.E., Clinical experience of drug treatments for mastalgia, *Lancet*, 2, 373–377, 1985.

16. Horrobin, D.F., The use of gamma-linolenic acid in diabetic neuropathy, *Agents Actions Suppl.*, 37, 120–144, 1992.

17. Jamal, G.A. and Carmichael, H., The effect of gamma-linolenic acid on human diabetic peripheral neuropathy: a double-blind placebo-controlled trial, *Diabet. Med.*, 7, 319–323, 1990.

18. Berth-Jones, J. and Graham-Brown, R.A., Placebo-controlled trial of essential fatty acid supplementation in atopic dermatitis, *Lancet*, 341, 1557–1560, 1993.

19. Hederos, C.A. and Berg, A., Epogam evening primrose oil treatment in atopic dermatitis and asthma, *Arch. Dis. Child*, 75, 494–497, 1996.

20. Horrobin, D.F., Gamma-linolenic acid, *Rev. Contemp. Physiol.*, 1, 1–41, 1990.

21. Worm, M. and Henz, B.M., Novel unconventional therapeutic approaches to atopic eczema, *Dermatology*, 201, 191–195, 2000.

22. Garcia, C.M., Carter, J., and Chou, A., Gamma-linolenic acid causes weight loss and lower blood pressure in overweight patients with family history of obesity, *Swed. J. Biol. Med.*, 4, 8–11, 1986.

23. Horrobin, D.F., Multiple sclerosis: the rational basis for treatment with colchicine and evening primrose oil, *Med. Hypotheses*, 5, 365–378, 1979.

24. Kruger, M.C. and Horrobin, D.F., Calcium metabolism, osteoprosis and essential fatty acids: a review, *Prog. Lipid Res.*, 36, 131–151, 1997.

25. Kruger, M.C., Coetzer, H., de Winter, R., Gericke, G., and van Papendorp, D.H., Calcium, gamma-linolenic acid and eicosapentaenoic acid supplementation in senile osteoprosis, *Aging Clin. Exp. Res.*, 10, 385–394, 1998.

26. Chenoly, R., Hussain, S., Tayob, Y., O'Brien, P.M., Moss, M.Y., and Morse, P.F., Effect of oral gamolenic acid from evening primrose oil on menopausal flushing, *BMJ*, 19, 501–503, 1994.

27. Belch, J.J.F., Shaw, B., O'Dowd, A., Saniabadi, P., Lieberman, P., and Sturrock R.D., Evening primrose oil (efamol) in the treatment of Raynaud's phenomenon: a double blind study, *Thrombosis Haemostasis*, 54, 490–494, 1985.

28. Leventhal, L.J., Boyce, E.G., and Zurier, R.B., Treatment of rheumatoid arthritis with blackcurrant seed oil, *Br. J. Rheumatol.*, 33, 847–852, 1994.

29. Rothman, D., Deluca, P., and Zurier, R.B., Botanical lipids: effects on inflammation, immune responses, and rheumatoid arthritis. *Semin. Arthritis Rheum.*, 25, 87–96, 1995.

30. Budeiri, D., Li Wan Po, A., and Dornan, J.C., Is evening primrose oil of value in the treatment of premenstrual syndrome, *Control Clin. Trials*, 17, 60–68, 1996.

31. Horrobin, D.F., The role of essential fatty acids and prostalandins in the premenstrual syndrome, *J. Reprod. Med.*, 28, 465–468, 1983.

32. Lawson, L.D. and Huges, B.G., Triacylglycerol structure of plant and fugal oils containing gamma-linolenic acid, *Lipids*, 23, 313–317, 1988.

33. Tyler, V.E., *Herbs of Choices: The Therapeutic Use of Phytomedicinals*, Haworth Press, Binghamton, NY, 1994, pp. 137–139.

34. http://www.greencottage.com/oils/borage.html (accessed October 2004).

35. Redden, P.R., Lin, X., and Horrobin, D.F., Dilinoleoyl-mono-gamma-linolenin (DLGM) and Di-gamma-linolenoyl-monolinolein (DGML): naturally occurring structured triacylglycerols in evening

primrose oil, in *Structural Modified Food Fats: Synthesis, Biochemistry, and Use*, Christophe, A.B., Ed., AOCS Press, Champaign, IL, 1998, pp. 121–128.

36. Kuhn, M.A. and Winston, D., *Herbal Therapy & Supplements: A Scientific & Traditional Approach*, Lippinocott, Philadelphia, PA, 2000, pp. 70–73, 123–126.

37. Miller, L.G., Herbal medicinals: selected clinical considerations focusing on known or potential drug-herb interactions, *Arch. Intern. Med.*, 58, 2200–2211, 1998.

38. Mills, D.E., Dietary fatty acid supplementation alters stress reactivity and performance in man, *J. Hum. Hypertens.*, 3, 111–112, 1989.

39. Chrubasik, S. and Pollak, S., Pain management with herbal antirheumatic drugs, *Wien. Med. Wochenscher.*, 152, 198–203, 2002.

40. Heck, A.M., Dewitt, B.A., and Luke, A.L., Potential interactions between alternative therapies and warfarin, *Am. J. Health Syst. Pharm.*, 57, 1221–1227, 2000.

41. Philp, R.B., *Herbal–Drug Interactions and Adverse Effects: An Evidence-based Quick Reference Guide*, McGraw-Hill, New York, 2004, pp. 50–51, 107–109.

42. Weiss, R.F. and Fintelmann, V., *Herb Medicine*, Thieme: Stuttgart, Germany, 2000, pp. 218–219.

43. http://www.greencottage.com/oils/blackcurrant.html (accessed October 2004).

44. http://www.greencottage.com/oils/primrose.html (accessed October 2004).

45. Coleman, E. and Taylor, M., Evening Primrose Oil, HCRC FAQ sheet, Healthcare Reality Check website.

46. Philp, R.B., *Methods of Testing Proposed Antithrombotic Drugs*, CRC Press, Boca Raton, FL, 1981, pp. 44–46.

47. Vaddadi, K., Dyskinesias and their treatment with essential fatty acid: a review, *Prostaglandins Leukot. Essent. Fatty Acids*, 55, 89–94, 1996.

48. Fugh-Berman, A. and Ernst, E., Herb–drug interactions: review and assessment of report reliability, *Br. J. Clin. Pharmacol.*, 52, 587–595, 2001.

49. http://www.umm.edu/altmed/ConsSupplements/Interactions/GammaLinolenicAcidGLAcs.htm (accessed January 2005).

50. Belch, J.J., Ansell, D., Madhok, R., O'Doud, A., and Sturrock, R.D., Effects of altering dietary essential fatty acids on requirements for non-steroidal anti-inflammatory drugs in patients with rheumatoid arthritis: a double-blind placebo-controlled study, *Ann. Rheum. Dis.*, 47, 96–104, 1988.

51. Mayes, P., Metabolism of unsaturated fatty acids and eicosanoids, In *Harper's Biochemistry*, 24th ed., Murray, R., Granner, D., Mayes, P., and Rodwell, V., Eds., Appleton & Lange, Stamford, CT, 1996, pp. 236–244.

52. Voet, D. and Voet, J., Lipid metabolism, in *Biochemistry*, 2nd ed., John Wiley, New York, 1995, pp. 704–713.

11 Diacylglycerols (DAGs) and their Mode of Action

Brent D. Flickinger
Archer Daniels Midland Company, Decatur, Illinois

CONTENTS

The role of dietary fat has become a renewed focus in understanding the influence of nutrition in health and disease. Guidelines and programs by public health agencies and professional health organizations encourage certain dietary choices and adequate exercise to improve overall health[1-3]. Dietary levels of fat, especially saturated fat, and balancing the types of fat in the diet are common recommendations from these groups.

Considerable research continues in order to shed light on the understanding of the link between dietary fat and its metabolism with lifestyle diseases, particularly cardiovascular disease and obesity[4]. The native structure of a fatty acid and its position on the glycerol backbone potentially influence nutritional aspects of dietary fats. With regard to native structure, the most obvious consideration is the degree of unsaturation. Omega-6 and omega-3 polyunsaturated fatty acids, linoleic and alpha-linolenic acids, respectively, are nutritionally essential[5]. Elongation and desaturation products of linoleic and alpha-linolenic acids (arachidonic (ARA), eicosapentaenoic (EPA), and docosahexaenoic (DHA) acids) appear to be as important. EPA and DHA and the bioactive molecules created from their metabolism appear to be more efficient and effective in their actions when EPA and DHA are consumed directly from dietary sources. Clear differences on risk factors for various health conditions have been observed between saturated, monounsaturated, and polyunsaturated fatty acids[6]. Also, omega-6 (including gamma-linolenic) and omega-3 polyunsaturated fatty acids may exert different impacts on these risk factors[6]. The number of carbons comprising the fatty acid chain length is the next obvious consideration. Specific types of saturated (stearic versus lauric, myristic, palmitic) and polyunsaturated (18-carbon vs. 20-carbon vs. 22-carbon) fatty acids may have different effects *in vivo*[6]. Medium-chain (8- and 10-carbon) fatty acids have different physical properties which result in their metabolism being partially independent of transport and oxidation pathways needed by long-chain fatty acids.

Both nutritional and pharmaceutical products commonly have focused on limiting fat digestion and/or absorption (i.e., structured lipids, nondigestible fats, lipase inhibitors, fat absorbers) or enhancing fat catabolism (i.e., caffeine, ephedra). Newer products have focused not on limiting fat digestion and absorption but on edible oils that have an impact on fat metabolism through inherent differences in their natural digestion and absorption. Using the nature of fat digestion and absorption, new types of fats are being developed to deliver differing nutritional properties. Such products include diacylglycerol (DAG)-rich oils and medium-chain triacylglycerol (MCT)-rich oils (MCTs have been utilized extensively in special clinical settings such as fat malabsorption and treatment for burn patients for several decades). The position of a fatty acid may influence its biological impact, particularly saturated fatty acids, due to the nature of fat digestion and absorption. Fat digestion from triacylglycerols (TAGs) creates free fatty acids and 2-monoacylglycerols (2-MAGs). These digestion products are passively absorbed into intestinal lumen cells then reassembled to TAGs.

DAGs occur naturally in edible oils to varying degrees (Table 11.1)[7,8]. Cottonseed and olive oils contain greater amounts of DAGs than other commonly used edible oils. DAGs have been utilized by the food industry as emulsifiers for some time. However, in 1999 the Kao Corporation of Japan introduced a DAG oil as a cooking oil which contains greater than 80% DAGs that looks and tastes like conventional edible oils (being pale yellow with a light, bland flavor) and performs to the standards of conventional oil for many home-use and ingredient applications. DAG oil, unlike conventional edible oils, contains predominantly DAGs while having a current fatty acid composition primarily of linoleic and oleic fatty acids (Table 11.2). DAG oil is prepared using a process involving an *sn*-1,3-specific lipase which has been described elsewhere[9]. The majority of naturally occurring and DAG oil DAGs have fatty acids in the *sn*-1,3 configuration due to the thermodynamic equilibrium with the 1,2 configuration (Table 11.2). This enriched composition of DAGs in DAG oil makes an important difference in their absorption and subsequent metabolism, particularly the difference due to 1,3-DAGs.

TABLE 11.1
Relative Contribution of Mono-, Di-, and Triacylglycerols in Selected Edible Oils

Oil	Monoacylglycerol[a]	Diacylglycerol	Triacylglycerol	Others
Soybean	—	1.0	97.9	1.1
Cottonseed	0.2	9.5	87.0	3.3
Palm	—	5.8	93.1	1.1
Corn	—	2.8	95.8	1.4
Safflower	—	2.1	96.0	1.9
Olive	0.2	5.5	93.3	2.3
Rapeseed	0.1	0.8	96.8	2.3
Lard	—	1.3	97.9	0.8

[a] wt% of total acylglycerol content.

TABLE 11.2
Typical Relative Acylglycerol Portion and Fatty Acid Composition of DAG Oil

	Amount (wt%)
Acylglycerols	
Diacylglycerol[a]	82
sn-1,3	57
sn-1,3	25
Triacylglycerol	17
Monoacylglycerol	1
Fatty acids	
16:0	2
18:0	1
18:1	38
18:2	54
18:3	5

[a] Minimum content.

The body digests DAG oil as it would a conventional oil yielding monoacylglycerol and free fatty acids. However, the 1-monoacylglycerol (1-MG) produced during digestion of 1,3-DAG is different from 2-MG resulting from TAG hydrolysis[10]. The gut normally reassembles TAGs beginning with 2-MG. Previous literature indicates that providing 1-MG results in lower amounts of fat-rich particles appearing in serum following consumption. This difference in fat metabolism is apparent with DAG oil as reflected in fewer fat-rich particles appearing in the blood following a meal containing DAG oil. This difference in fat metabolism leaves fatty acids that must be handled by the gut and/or liver. This difference in fat metabolism appears to result in changes in serum TAGs (both postprandial and fasting levels) and fat oxidation with the latter being important for body weight and body fat regulation.

When substituted for a conventional TAG oil with closely matched fatty acid composition, the DAG oil was absorbed in a manner that resulted in lower secretion of lymph, lower serum TAGs, and decreased levels of postmeal TAG-rich particles in blood[11,12]. This first observation of a metabolic difference between DAG oil and TAG oil was observed in animals with regard to plasma TAG metabolism. In human studies, similar differences in postmeal TAG levels in the serum were observed. DAG oil decreased the elevation in serum TAGs with most pronounced effects above a single dose of 20 g[13]. Using a single oil dose of 55 g, DAG oil significantly reduced the increase in postprandial serum TAGs at 2 and 4 hours after consumption compared to a conventional TAG oil with a matched fatty acid profile[14]. Remnant lipoprotein cholesterol levels were decreased significantly also at 2 and 4 hours postconsumption.

A significant decrease in fasting serum TAGs was observed in Japanese type II diabetics with elevated fasting serum TAGs (>150 mg/dl) when DAG oil was substituted for conventional oil at 10 g per day for 12 weeks as part of a low-fat diet[15]. Additionally, reductions in glycosylated hemoglobin over the course of the study indicated improved blood sugar control in the diabetics consuming DAG oil compared to conventional oil.

The maintenance of lower body weight and body fat by DAG consumption has been observed in animals and humans. Several well-controlled studies have been conducted in humans examining the impact of DAG oil on body weight and body fat. When consuming approximately 5% of total calories from DAG oil for 16 weeks, significantly greater losses in body weight ($p < 0.01$) and body fat area ($p < 0.05$) were observed when compared to subjects consuming conventional oil[16]. In a subsequent study using considerably more subjects, overweight individuals consuming 15% of total energy from DAG oil for 6 months as part of a diet with mild caloric restriction (500 to 800 kcal/d) demonstrated a greater extent of body weight ($p < 0.025$) and body fat loss ($p < 0.037$) over the period of dietary intervention when compared to subjects consuming a conventional TAG oil[17]. Both studies show similar apparent changes between Japanese individuals and overweight adult Americans. As a result, DAG oil appears to enable greater degrees of body fat and body weight loss compared to conventional oil when used as a dietary aid as part of a healthy diet or caloric management plan. More importantly, DAG oil was utilized as the oil ingredient for food items including mayonnaise, crackers, muffins, and instant soups with no apparent decrease in subject compliance due to product quality or flavor.

With the apparent energy value of DAG oil being nearly identical to conventional oil[18], a return to differences in the process of fat digestion and absorption is necessary in order to begin to understand why DAG oil has an impact on fat and energy metabolism. 2-MAG and free fatty acids (FFAs) are digestion products of conventional TAG oils following action by an *sn*-1,3-specific pancreatic lipase. These digestion components migrate via passive absorption into cells that line the gut and then are reassembled into TAGs. Beginning with 2-MAG, triacylglycerol resynthesis follows sequential addition of FFAs by monoacylglycerol acyltransferase (MAGAT), a process which is most efficient using 2-MAG[19], then by diacylglycerol acyltransferase (DAGAT). TAGs are packaged into fat-rich particles known as chylomicrons then secreted into the lymph that carries fat-rich particles away from the gut then into the bloodstream for circulation throughout the body. The level of TAG-rich particles temporarily increases in the blood following a meal containing dietary fat then decreases as tissues take up fatty acids before residual dietary fatty acids eventually reach the liver.

Using animal models, experimental evidence shows increased oxygen consumption indicating increased energy expenditure, increased portal vein FFAs, increased beta-oxidation in the liver and small intestine, and increased mRNA expression of acyl CoA oxidase (ACO) and uncoupling protein-2 (UCP-2) in the small intestine following 1,3-DAG or DAG oil consumption[10,20–22]. Activities of enzymes related to fatty acid oxidation in the liver were observed to increase while activities of enzymes related to fatty acid synthesis correspondingly decreased following 14 days of DAG oil consumption[20]. These differences in liver enzyme activities were accompanied by lower hepatic TAG content and lower serum cholesterol. Further studies in animals demonstrated different fat digestion products in the gut as well as a tendency to utilize greater amounts of oxygen after DAG oil consumption[10]. Using mice (C57BL/6J) prone to diet-induced obesity, lower body weight during their lifespan was observed during *ad libitum* DAG oil consumption in place of conventional oil[21,22]. In the liver of mice fed DAG oil, ACO mRNA levels were also significantly increased, consistent with increased ACO activity, compared to a conventional TAG oil[21]. DAG oil enhanced beta-oxidation in the small intestine in mice fed DAG oil compared to TAG oil[22]. Greater beta-oxidation in the small intestine was associated with increased expression of genes involved in beta-oxidation and lipid metabolism including ACO, medium-chain acyl-CoA dehydrogenase (MCAD), liver fatty acid binding protein (L-FABP), fatty acid transporter (FAT), and UCP-2. However, compared to rat where changes occurred in the liver, these changes in beta-oxidation and mRNA expression in mice occurred solely in the small intestine.

The different physical effects from animal and human results observed following DAG oil consumption may be explained, in part, by this shift in fat metabolism. An increased use of fat for energy utilization through increased energy expenditure rather than fat storage appears to occur following the use of DAG through differential handling of 1,3-DAG. However, the literature suggests that increased fat oxidation may have an impact on appetite. Numerous scientific observations indicate that enhanced beta-oxidation or inhibition of fatty acid synthase may enhance suppression of appetite[23–26].

In Japan, DAG cooking oil is approved as a food for specific health use (FOSHU) pertaining to postmeal blood lipids and body weight by the Japanese Ministry of Health, Labour and Welfare[27]. The professional association of Japanese physicians which administers annual physicals in Japan (known as the Japanese Society of Human Dry Dock and affiliated with the Japan Hospital Association) officially recommends DAG cooking oil as part of a healthy diet. These recommendations provide the Japanese public reliable support that the use of DAG oil can impact postmeal blood lipids and body weight and body fat in a healthy manner when used as part of a healthy diet. In the marketplace, DAG oil (known as "Healthy Econa oil") continues to be a leading selling cooking oil in Japan with over 160 million bottles (600 g oil/bottle) sold since being commercially launched in February 1999.

In the U.S., DAG oil has been self-affirmed by expert panel review as notification for generally recognized as safe (GRAS) use[28]. DAG oil can be used as an edible cooking and salad oil (home-use) as well as in products using oil as an ingredient such as spreads, dressings for salads, baked goods, healthy bars, and numerous additional food product categories.

New products promoted as "interfering with fat digestion and/or absorption" often receive considerable scrutiny as a consequence of undigested fat reaching the lower bowel and decreased serum levels of certain fat-soluble nutrients. Examples of a food ingredient and prescription drug that block fat digestion and absorption are Olestra, a nondigestible fat substitute, and Orlistat, a pancreatic lipase inhibitor, respectively. With normal digestion and absorption like conventional oil, DAG oil consumption does not result in fatty stools resulting in fat-soluble vitamins appearing to be absorbed normally and to remain unchanged in their serum levels[18,29].

With further understanding on the impact of dietary fat on body weight and lipid metabolism, products such as DAG oil offer the potential for incorporating dietary fat into the American diet with the aim of promoting decreased obesity as well as both fasting and postprandial serum triglycerols. Following the Japanese marketplace which has been and continues to be an environment for providing consumers with a variety of unique dietary fats with healthy characteristics, DAG oil may provide consumers the opportunity to affect positively their health through informed choices on incorporating prudent decisions in their choices of dietary fats.

REFERENCES

1. Trumbo, P., Schlicker, S., Yates, A.A., and Poos, M., Dietary reference intakes for energy, carbohydrate, fiber, fat, fatty acids, cholesterol, protein and amino acids, *J. Am. Diet. Assoc.*, 102, 1621–1630, 2002.
2. Krauss, R.M., Eckel, R.H., Howard, B., Appel, L.J., Daniels, S.R., Deckelbaum, R.J., Erdman, J.W., Jr., Kris-Etherton, P., Goldberg, I.J., Kotchen, T.A., Lichtenstein, A.H., Mitch, W.E., Mullis, R., Robinson, K., Wylie-Rosett, J., St Jeor, S., Suttie, J., Tribble, D.L., and Bazzarre, T.L., AHA Dietary Guidelines: revision 2000: a statement for healthcare professionals from the Nutrition Committee of the American Heart Association, *Circulation* 102, 2284–2299, 2000.
3. General, U.S.S., *Healthy People 2010: Understanding and Improving Health*, U.S. Department of Health and Human Services, Washington, DC, 2000.
4. Astrup, A., Ryan, L., Grunwald, G.K., Storgaard, M., Saris, W., Melanson, E., and Hill, J.O., The role of dietary fat in body fatness: evidence from a preliminary meta-analysis of ad libitum low-fat dietary intervention studies, *Br. J. Nutr.*, 83 (Suppl. 1), S25–S32, 2000.
5. Dietary fats: total fat and fatty acids, in *Dietary Reference Intakes for Energy, Carbohydrate, Fiber, Fat, Fatty Acids, Cholesterol, Protein, and Amino Acids*, Institute of Medicine, Washington, DC, 2002, pp. 335–432.
6. Population nutrient intake goals for preventing diet-related chronic diseases, in *Diet, Nutrition And The Prevention Of Chronic Diseases*, report of a joint WHO/FAO expert consultation, Geneva, 2003, pp. 54–133.
7. Abdel-Nabey, A.A., Shehata, A.A.Y., Ragab, M.H., and Rossell, J.B., Glycerides of cottonseed oils from Egyptian and other varieties, *Riv. Ital. Sostanze Grasse*, 69, 443–447, 1992.
8. D'Alonzo, R.P., Kozarek, W.J., and Wade, R.L., Glyceride composition of processed fats and oils as determined by glass capillary gas chromatography, *J. Am. Oil Chem. Soc.*, 59, 292–295, 1982.
9. Watanabe, T., Yamaguchi, H., Yamada, N., and Lee, I., Manufacturing process of diacylglycerol oil, in *Diacylglycerol Oil*, Katsuragi, Y., Yasukawa, T., Matsuo, N., Flickinger, B., Tokimitsu, I., and Matlock, M., Eds., AOCS Press, Champaign, IL, 2004, pp. 253–261.
10. Watanabe, H., Onizawa, K., Taguchi, H., Kobori, M., Chiba, H., Naito, S., Matsuo, N., Yasukawa, T., Hattori, M., and Shimasaki, H., Nutritional characterization of diacylglycerol in rats, *J. Japan Oil Chem. Soc.*, 46, 301–308, 1997.
11. Hara, K., Onizawa, K., Honda, H., Otsuji, K., Ide, T., and Murata, M., Dietary diacylglycerol-dependent reduction in serum triacylglycerol concentration in rats, *Ann. Nutr. Metab.*, 37, 185–191, 1993.
12. Murata, M., Hara, K., and Ide, T., Alteration by diacylglycerols of the transport and fatty acid composition of lymph chylomicrons in rats, *Biosci. Biotech. Biochem.*, 58, 1416–1419, 1994.
13. Taguchi, H., Watanabe, H., Onizawa, K., Nagao, T., Gotoh, N., Yasukawa, T., Tsushima, R., Shimasaki, H., and Itakura, H., Double-blind controlled study on the effects of dietary diacylglycerol on postprandial serum and chylomicron triacylglycerol responses in healthy humans, *J. Am. Coll. Nutr.*, 19, 789–796, 2000.
14. Tada, N., Watanabe, H., Matsuo, N., Tokimitsu, I., and Okazaki, M., Dynamics of postprandial remnant-like lipoprotein particles (RLP) in serum after loading of diacylglycerols, *Clin. Chem. Acta*, 311, 109–117, 2001.
15. Yamamoto, K., Asakawa, H., Tokunaga, K., Watanabe, H., Matsuo, N., Tokimutsu, I., and Yagi, N., Long-term ingestion of dietary diacylglycerol lowers serum triacylglycerol in Type II diabetic patients with hypertriglyceridemia, *J. Nutr.*, 131, 3204–3207, 2001.
16. Nagao, T., Watanabe, H., Goto, N., Onizawa, K., Taguchi, H., Matsuo, N., Yasukawa, T., Tsushima, R., Shimasaki, H., and Itakura, H., Dietary diacylglycerol suppresses accumulation of body fat compared to triacylglycerol in men in a double-blind controlled trial, *J. Nutr.*, 130, 792–797, 2000.
17. Maki, K.C., Davidson, M.H., Tsushima, R., Matsuo, N., Tokimitsu, I., Umporowicz, D.M., Dicklin, M.R., Foster, G.S., Ingram, K.A., Anderson, B.D., Frost, S.D., and Bell, M., Consumption of diacylglycerol oil as part of a reduced-energy diet enhances loss of body weight and fat in comparison with consumption of a triacylglycerol control oil, *Am. J. Clin. Nutr.*, 76, 1230–1236, 2002.

18. Taguchi, H., Nagao, T., Watanabe, H., Onizawa, K., Matsuo, N., Tokimitsu, I., and Itakura, H., Energy value and digestibility of dietary oil containing mainly 1,3-diacylglycerol are similar to those of triacyl-glycerol, *Lipids*, 36, 379–382, 2001.

19. Bierbach, H., Triacylglycerol biosynthesis in human small intestinal mucosa. Acyl-CoA: monoglyceride acyltransferase, *Digestion*, 28, 138–147, 1983.

20. Murata, M., Ide, T., and Hara, K., Reciprocal responses to dietary diacylglycerol of hepatic enzymes of fatty acid synthesis and oxidation in the rat, *Br. J. Nutr.*, 77, 107–121, 1997.

21. Murase, T., Mizuno, T., Omachi, T., Onizawa, K., Komine, Y., Kondo, H., Hase, T., and Tokimitsu, I., Dietary diacylglycerol suppresses high fat and high sucrose diet-induced body fat accumulation in C57BL/6J mice, *J. Lipid Res.*, 42, 372–378, 2001.

22. Murase, T., Aoki, M., Wakisaka, T., Hase, T., and Tokimitsu, I., Anti-obesity effect of dietary diacyl-glycerol in C57BL/6J mice: dietary diacylglycerol stimulates intestinal lipid metabolism, *J. Lipid Res.*, 43, 1312–1319, 2002.

23. Strauss, R.S., Fatty acid synthase inhibitors reduce food intake and body weight, *Pediatr. Res.*, 48, 422, 2000.

24. Krotkiewski, M., Value of VLCD supplementation with medium chain triglycerides, *Int. J. Obes. Relat. Metab. Disord.*, 25, 1393–1400, 2001.

25. Van Wymelbeke, V., Louis-Sylvestre, J., and Fantino, M., Substrate oxidation and control of food intake in men after a fat-substitute meal compared with meals supplemented with an isoenergetic load of carbohydrate, long-chain triacylglycerols, or medium-chain triacylglycerols, *Am. J. Clin. Nutr.*, 74, 620–630, 2001.

26. Stubbs, R.J., and Harbron, C.G., Covert manipulation of the ratio of medium- to long-chain triglycerides in isoenergetically dense diets: effect on food intake in ad libitum feeding men, *Int. J. Obes. Relat. Metab. Disord.*, 20, 435–444, 1996.

27. Current FOSHU list, in Vol. 2003, Japanese Ministry of Health, Labour and Welfare, 2003.

28. GRAS Notification 115, in Vol. 2003, U.S. Food and Drug Administration, 2003.

29. Watanabe, H., Onizawa, K., Naito, S., Taguchi, H., Goto, N., Nagao, T., Matsuo, N., Tokimitsu, I., Yasukawa, T., Tsushima, R., Shimasaki, H., and Itakura, H., Fat-soluble vitamin status is not affected by diacylglycerol consumption, *Ann. Nutr. Metab.*, 45, 259–264, 2001.

12 Conjugated Linoleic Acids (CLAs): Food, Nutrition, and Health

Bruce A. Watkins and Yong Li
Center for Enhancing Foods to Protect Health, Lipid Chemistry and
Molecular Biology Laboratory, Purdue University, West Lafayette, Indiana

CONTENTS

12.1 INTRODUCTION

The group of fatty acids often called conjugated linoleic acids (CLAs) are positional and geometric isomers of octadecadienoic acid (18:2). The double bonds in CLAs are conjugated which means that they are not separated by a methylene group ($-CH_2-$) that is typical of the double bonds in polyunsaturated fatty acids (PUFAs), such as linoleic acid (LA or 18:2n-6), an omega-6 essential fatty acid. Many foods contain some CLA isomers[1] but the chief sources are found in dairy and beef products (or those derived from ruminant sources) since the process of bacterial biohydrogenation of PUFAs in the rumen leads to their formation[2-4].

Octadecadienoic acids have been reported to contain conjugated double bonds at positions 7,9; 8,10; 9,11; 10,12; 11,13; and 12,14 along the alkyl chain (counting from the carboxyl end of the molecule) in chemically prepared CLA mixtures or natural products[5-8]. The positional conjugated diene isomers of CLAs can occur in one or more of the following four geometric configurations: *cis,trans*; *trans,cis*; *cis,cis*; or *trans,trans* ("*c*" and "*t*" will be used to represent "*cis*" and "*trans*" respectively in the following), which would potentially yield up to 24 possible isomers of CLAs[7]. Many of the isomers were reportedly identified in commercially available preparations of CLAs which are produced under alkali conditions from vegetable oils containing a high concentration of LA[9]. In contrast to natural CLAs, which are formed through bacterial biohydrogenation and contained mainly the *c*9,*t*11 isomer (about 80% of total CLA isomers present in foodstuff)[10], chemically synthesized CLA sources for commercial purposes are roughly comprised of equal amounts of two major CLA isomers ($c9,t11$ and $t10,c12$) and smaller amounts of others with varying degree of total CLA in the preparation that is affected by the raw material and manufacturing process[11]. The composition of commercial CLA products should be carefully checked before they are used in research work since the isomeric distribution varies greatly between manufacturers and even between batches within the same manufacturer[12].

The most common CLA isomer found in meat from ruminant species and bovine dairy food products is octadeca-$c9,t11$-dienoic acid[13], even though minor components, such as the $t7,c9$, $t8,c10$, $t10,c12$, $t11,c13$, $c11,t13$, and $t12,t14$ isomers, and their c,c and t,t isomers, were also reported in these products[1]. The CLA in ruminant meat and dairy products is believed to be formed by bacterial isomerization of LA and possibly α-linolenic acid (18:3n-3) from grains and forages to the $c9,t11$-18:2 in the rumen of these animals[2,3,14]. CLAs may also be formed during cooking and processing of foods[14].

12.2   BIOCHEMICAL AND MOLECULAR ASPECTS OF CLA ACTIONS

The reported mechanisms of CLA actions include antioxidant properties, inhibition of carcinogen–DNA adduct formation, inducing apoptosis and cytotoxic activity, altering tissue fatty acid composition and prostanoid formation, and affecting the expression and action of cytokines and growth factors[1]. Though CLA isomers are purported to possess numerous biological actions, the evidence consistently observed includes anticancer effects in rodents and cancer cell cultures and reduction of body fat in growing animals and certain human studies. In some cases the biological responses from CLA isomers were influenced by the amounts of dietary n-6 and n-3 PUFAs[15–17].

The cytotoxic effects of CLA isomers on growth of various human and animal-derived cancer cells appear to be mediated by decreasing the expression of the gene transcription factor Bcl-2 family whose members inhibit apoptotic cell death and/or induce caspase-dependent apoptosis[18–22]. CLAs also prevented basic fibroblast growth factor-induced angiogenesis[23], a critical process for growth and metastasis of cancers. There is further evidence that isomers of CLAs alter the action of peroxisome proliferator-activated receptors (PPARs), especially PPARα, to reduce carcinogenesis[24]. In addition, the effects of CLA isomers on fat and energy metabolism may, in part, be directed through changes in both PPARα and PPARγ[25,26].

Several studies have demonstrated that CLAs reduce body fat accumulation in growing animals[27–29] but not all CLA isomers contributed to this effect equally as found in mice[30,31]. Hargrave et al.[27] reported that the $t10,c12$ isomer, but not the $c9,t11$, induced body fat loss and adipocyte apoptosis in mice. Other investigations indicated a fat loss or redistribution when CLA isomers were given to chickens[32], rats[33], and pigs[34], but when tested in human subjects the results were not consistent[35–37]. In most investigations in human subjects the use of supplements containing both the $c9,t11$ and $t10,c12$ isomers of CLA reduced body fat in men and women[38]. Based on the data from growing animals and human subjects, one approach for advancing the understanding of CLA isomers in fat reduction is to examine carefully its effect on related endocrine factors during different physiological states on energy metabolism and expenditure for applications in weight loss in humans.

One target for understanding weight loss is the protein called leptin. The hormone leptin has important effects in regulating body weight, metabolism, and reproductive function. Recent experiments have revealed that CLA isomers reduced leptin concentrations or expression in animal and cell culture studies[39–42]. As a regulator of appetite and lipid metabolism and a proposed neuroendocrine regulator of bone mass, leptin is a potential target of CLA action in mediating its effects on body fat as well as bone growth and bone mass[43,44].

Specific effects of CLA isomers on the activity and expression of enzymes associated with anabolic pathways of fatty acid formation have been reported[45]. For example, CLA was observed to decrease the mRNA level of the $\Delta 9$-desaturase enzyme in both liver tissue and hepatocyte cultures[46].

In immune function, CLA diminished the production of an array of proinflammatory products in macrophages through activation of PPARγ[25] and it lowered basal and lipopolysaccharide (LPS) stimulated interleukin (IL)-6 and basal tumor necrosis factor (TNF) production by rat resident peritoneal macrophages[15]. Yu et al.[25] reported that by activation of PPARγ, CLA decreased interferon-γ-induced mRNA expression of cyclooxygenase (COX)-2, inducible NOS (iNOS), TNFα, and proinflammatory cytokines (IL-1β and IL-6) in RAW macrophage cell cultures.

Specific PUFAs including LA, arachidonic acid (AA, 20:4n-6), eicosapentaenoic acid (EPA, 20:5n-3), and docosahexaenoic acid (DHA, 22:6n-3) are known agonists or antagonists of COX-2 expression through the activation of PPARs[47]. It has been reported that dietary CLA isomers reduced *ex vivo* prostaglandin E$_2$ (PGE$_2$) production in rat bone organ cultures[16]. Other investigators have also observed that CLA reduced PGE$_2$ production in various biological systems[48–50].

A unifying aspect of action for CLA isomers, both biochemical and molecular, include leptin and enzymes of lipid metabolism, PPARs, and the gene targets of PPARs including COX-2. These proposed mechanisms where CLA isomers have an impact on biology are likely areas of investigation to advance the understanding of CLAs in fat metabolism and for improving health.

12.3 SALIENT BIOLOGICAL ACTIONS OF CLA

CLA is the only known antioxidant and anticarcinogen nutrient associated with foods originating from animal sources. An early report on CLA from beef suggested that the extracts protected against chemically induced cancer[51]. In this study, CLAs isolated from extracts of grilled ground beef were found to reduce skin tumors in mice treated with 7,12-dimethylbenz[α]anthracene (DMBA), a known carcinogen[51].

The broadening research on CLA isomeric mixtures has relied almost entirely on animal models and various cell culture systems. The properties of CLAs include anticarcinogenic[52–57], antioxidative[57,58], and immunomodulative[15,59,60]. In recent years, preliminary data suggest that CLAs may have a role in controlling obesity[61–63], reducing the risk of diabetes[64], and modulating bone metabolism[16,17,42].

The specific investigations into the anticarcinogenic properties of CLA isomers included mixtures (mainly $c9,t11$ and $t10,c12$) of variable purity which demonstrated reduced chemically induced tumorigenesis in rat mammary gland and colon[52,53,56,65–67] and modulated chemically induced skin carcinogenesis[51,68]. The CLA isomers also inhibited the growth of human tumor cell lines in culture[55,69,70] and in SCID (severe combined immunodeficient) mice[54,71].

The apparent weight-control effect of CLAs found in animal studies was not easily repeated in clinical trials using human subjects[35,35–37]. Among the published human study reports, some indicate a decline in body fat when CLA was administered, while none of them observed any effect on whole body weight reduction[38]. The dose of CLA supplementation used in these trials ranged from 0.4 to 6.8 g/d and the duration of treatments was from 4 weeks to 6 months. Subjects involved in these investigations were either healthy or having various disease conditions, such as overweight/obese or with type 2 diabetes mellitus. In all cases, a mixture of CLA isomers was used for the dietary treatment and most of these preparations had equal amounts of $c9,t11$ and $t10,c12$ isomers. The CLA effects on body fat reduction were observed in both genders.

12.4 CLA IN FOOD AND FOOD ENRICHMENT

12.4.1 FOOD SOURCES AND ESTIMATED INTAKES OF CLA ISOMERS

The highest concentrations of CLA in food are in dairy products[72,73] and fat in the meats of lamb, veal calves, and cattle[74]. The $c9,t11$ isomer is the chief isomer of CLA present in food although many different isomers can be found and in varying concentrations[1]. A brief summary of the sources

TABLE 12.1

Food Content and Isomeric Distribution of CLA in Commercial and Natural Food Products

Products	CLA content (mg/g fat)[a]	c9,t11 (%)	Identified CLA isomers[b]	Ref.[c]
Natural cheeses				
Blue	0.55–7.96	15–100	1/2, 4, 5, 8, 9, 10, 12/14	113, 83, 86, 114, 115, 116
Cheddar	1.36–5.86	18–100	1/2, 4, 5, 8, 9, 10, 12/14	
Cottage	4.5–5.9	83–100	1	
Cougar Gold	3.20–5.17	85–100	1/2, 4, 5, 8, 10, 12/14	
Monterey Jack	4.80	100	1	
Mozzarella	4.31–4.96	84–100	1	
Swiss	5.45–14.2	90–100	1	
Processed cheeses				
Processed	1.81–6.2	18–100	1/2, 4, 5, 8, 9, 10, 12/14	113
Cheese whiz	4.9–8.81	21–100	1/2, 4, 5, 8, 9, 10, 12/14	
Kraft American singles	3.19	58	1	
Butter and milk				
Cow milk	0.7–10.1	59–100	1/2, 4, 5, 6, 8, 9, 10, 12/14	83, 82, 86, 117, 115, 118
Butter	4.7–8.11	78–90	1	
Fermented dairy products				
Buttermilk	4.66–5.4	89–100	1	83, 86, 115
Sour cream	4.14–7.49	78–100	1	
Yogurt	1.7–9.01	71–100	1	
Ice cream				
Ice cream	3.6–4.95	76–86	1	83
Beef				
Beef products, uncooked	1.2–8.5	21–61	1/2, 4, 5, 8, 9, 10, 12/14	119
Cooked beef	3.3–9.9	19–84	1/2, 4, 5, 8, 9, 10, 12/14	

[a] Some values were originally expressed as percentage of total FAME or mg/g fatty acids. These values are converted to mg/g fat by using a multiplication factor of 9.5 or 0.95, respectively[120].
[b] Each number represents a CLA isomer designated as follows: 1 = c9,t11; 2 = t9,c11; 3 = c8,t10, 4 = c10,t12, 5 = t10,c12; 6 = t7,c9; 7 = c8,c10; 8 = c9,c11; 9 = c10,c12; 10 = c11,c13; 11 = t11,t13, 12 = t9,t11; 13 = t8,t10; 14 = t10,t12. Isomers separated by "/" co-elute during chromatographic analysis either by GC or HPLC analysis.
[c] General references for all food sources:[3,14,75]; and sources for dairy:[76]. Also see text.

of CLA in commonly consumed food products is presented in Table 12.1. The CLA content has been extensively examined in many foods[3,14,75] and well characterized in dairy products[76].

One of the earliest estimates of CLA intake calculated a range from 0.3 to 1.5 g/person/d which was dependent on gender and the intake of food from animal and vegetable origins[72]. Habitual dietary intake of total CLA for humans has been estimated to be in the range of 0.1 to 0.4 g/d, and the intake of the c9,t11 CLA isomer was determined to be 94.9 ± 40.6 mg/d for 22 free-living Canadians by analyzing two 7 d diet records taken 6 months apart[77]. Total CLA (c9,t11 and t10,c12 isomers, at a ratio of about 1.32:1) intake was estimated to be 0.21 ± 0.01 g/d for men and 0.15 ± 0.01 g/d for women (in the western U.S.) in a survey using food duplicate methodology[78]. In another study of young female college students in Germany, CLA (c9,t11) intake was calculated to be 0.25 g/d by food frequency questionnaires or 0.32 g/d by a 7 d estimated diet record[79]. These values for CLA intakes are close to those reported earlier (0.36 g/d for women and 0.44 g/d for men) in German subjects[80].

Some investigators suggest that the human estimated intake of CLA by dietary sources is not enough to exert the potential beneficial biochemical, molecular, and physiological effects against cancer, atherosclerosis, and obesity based on studies in animal models. For example, Ip et al.[65] estimated that a 70 kg human should consume 3.0 g CLA/d to achieve the beneficial effects in inhibiting mammary carcinogenesis. This calculation was determined from rats given a diet supplemented with CLA at 0.1% of the total diet and would reflect about a three-fold higher intake of 1 g CLA/d by an average person in the U.S.[14]. We calculated the human equivalent CLA intake, based on 0.1% dietary CLA given to rats, as 0.72 g/d for a 70 kg person adjusting for the difference in metabolic rate of a human compared to the rat. This calculation assumes a rat intake of 15 g of diet per day at a body weight of 350 g and using allometric scaling to adjust intake to metabolic body weight with a coefficient of 0.73[81]. This lower estimate of human intake to achieve the biological and physiological effects of CLA is feasible by enriching CLA in food products. Therefore, even with the recent estimate at only 0.2 g CLA/person/d[78], it is possible to achieve the potential health-promoting benefits of CLA without taking supplements, but by enriching existing food sources.

12.4.2   FOOD ENRICHMENT WITH CLA ISOMERS

There are two approaches to increase the dietary intake of CLA isomers from food. The first way is to encourage more consumption of CLA-rich foods including dairy and beef products. This is not a reasonable method since dietary guidelines limit the intake of conventional CLA-containing foods. In addition, many CLA-rich foods are also significant sources of saturated fat and cholesterol that should be limited to reduce the risk of cardiovascular disease and cancer. The second way is to increase the CLA content in eggs, milk, and meat leading to the development of animal-derived designed foods. The latter approach is more practical since it would not depend on changing dietary practices or elevating the daily intake of nutrients that contribute to chronic diseases. Increasing the CLA content of food products like milk and meat also has the potential of increasing their nutritional and health value, and could favorably influence the marketing of value-added designed foods.

Several factors during every stage from the field to table, including raw material production, processing, packaging, storage, and food preparation before serving, can influence the CLA content in food. Generally, the intrinsic CLA content is determined in the raw food or after minimal processing. The exception is the analysis of the CLA content in various cheeses[1]. Subsequent processing, storage, and food preparation, however, will modify the CLA content to some extent, although the variation is fairly small compared to the large natural variation found in dairy products as an example[82,83].

Although CLA is naturally present in dairy foods, experiments have evaluated methods to enhance the CLA content in milk to further increase its levels in products made from milk. The CLA concentrations in various dairy products (cheeses, milk, butter, buttermilk, sour cream, ice cream, and yogurt) ranged from 0.55 to 24 mg/g fat[1]. The average CLA content in milk is about 10 mg/g milk fat[72] but natural cheeses contain the greatest variation in the amount of CLA isomers[1]. Seven CLA peaks that could represent nine isomers were present in dairy products; among these $c9,t11$, $t10,c12$, $t9,t11$, and $t10,t12$ accounted for more than 89%[14]. The CLA content in cheeses is primarily dependent on the CLA content of the milk, which varies in CLA concentration due to seasonal changes, geography, nutrition of the cow, and management practices. In addition, CLA content of cheese, to a limited extent, is affected by the production process and maturation[84].

Type of feed (nutritional factors), season, genetic variation, and management factors can influence the concentration of CLA in dairy and in ruminant fats and meat products[82,85,86]. Dhiman et al.[82] showed that the CLA content in bovine milk could be increased linearly as the amount of pasture was increased in the ration. Besides modifying the ration composition with varying amounts of grass and grains, oilseeds such as soybeans and fishmeal and the oils in these products have been used to supplement the rations of cows to increase the CLA content in milk. As a result, the CLA concentrations in milk of cows given these oil supplements were much higher (50% or more) compared to cows fed the CLA-enhancing rations of grasses and forages[85,87–91]. A recent study by

Reklewska et al.[92] showed that feeding Friesian cows 21 g/d of linseed and 21 g/d trace element–mineral mixture (Mg, Fe, Cu, Co, Mn, Zn, Se, Cr, and Ca) not only elevated the CLA content in milk, but also significantly reduced the cholesterol level in the milk by as much as 32% compared to the milk from the control cows given a total mixed ration. Species of cows and geographical and seasonal influences were also examined in affecting CLA content in milk[93,94]. Besides modifying the diets of cows, feeding exogenous CLA directly to cows has been observed to increase the CLA content of milk[95,96]. The exogenous CLA isomers (CLA-60, Natural Lipids, Hovdebygda, Norway) altered milk fatty acid composition, reduced bovine milk fat content and yield by as much as 55%, presumably by inhibiting *de novo* fatty acid synthesis in the mammary tissue[96]. Factors that affect the CLA content in milk should have the same effect on the meat of similar species.

CLA is also present in small amounts in other food animal products. Turkey meat has the highest CLA content of 2.5 mg/g fat for nonruminant species[14]. Chicken contained CLA (0.9 mg/g fat) as did pork (0.6 mg/g fat) with the *c*9,*t*11 being the major isomer at 84 and 82% of the total CLA, respectively[3]. The amount of CLA in chicken egg yolk lipids ranged from 0 to 0.6 mg/g fat[3,97–100]. Though these animal sources usually contain only a trace amount of CLA, they can be enriched by dietary manipulations. CLA supplements have been administered to various animals to enhance the concentrations of these isomers in pork[101–106], chicken meat and egg[98–100], and fish fillet[107,108].

The efficiency of CLA enrichment differs among animal species. Based on the CLA amount per gram of fat, egg yolk had the highest CLA content at 11% of total fatty acids when hens were fed 5% CLA in the diet. When given a diet containing 1% CLA, the amount of CLA in fish fillets (hybrid striped bass) reached 8% of the total fatty acids. In pigs, the highest CLA content was 6% in adipose tissue when these animals were fed a diet supplemented with 2% CLA. Reported values for CLA in milk and beef were around 2 to 6% of total fatty acids, which is significantly lower than that in nonruminant species. However, the methods of enrichment in ruminant animals (dairy cows and cattle) compared with other domestic food animal species (pigs, chicken, and fish) are quite different. Enrichment of CLA in milk and beef is primarily achieved by supplementing animals with precursor fatty acids of CLA compared to the direct feeding of CLA isomers to fish, pigs, and chickens.

Besides elevating CLA isomers in pork, supplementation to pigs led to a decease in carcass fat and improvements in pork quality. In contrast, high levels of CLA supplementation (0.9 g/d) to laying hens resulted in undesirable quality changes in fresh and hard-cooked eggs[100] which included detrimental effects on the vitelline membrane of egg yolk[109]. Based on these studies, further research is needed to understand the effects of CLA isomers on the fatty acid composition and quality aspects of animal-derived foods. New research should examine how specific CLA isomers and other nutraceutical lipids affect the quality and nutritional value of animal-derived food products.

12.5   IMPORTANCE OF CLA IN HUMAN NUTRITION

The biological and physiological effects of CLA isomers investigated on various health conditions in humans include body weight control, lipid metabolism, diabetes and insulin resistance, immune function, and epidemiological findings (summarized in Table 12.2). The evidence for any benefit to human health is inconclusive at this time and some investigators suggest that the consumption of CLA isomers by humans should not be recommended until further research is conducted[110].

In the recent literature, several studies have been published to investigate the actions of CLA on weight loss to suggest some benefit; however, it is difficult to make a simple recommendation for the use of any supplement at present. This is due to the differences in CLA isomeric compositions of the supplements, age, gender, health status of the subjects, and duration of the treatments used in the design of these trials. Yet, the consistent evidence of the positive effects of CLA on various biological targets in animal and cell culture studies justifies the need to continue the work in human

TABLE 12.2
Health Benefits of CLA and Applicable References from Human Clinical Trials

Effect	CLA dose[a] (g/d)	Duration	Subjects	Ref.
Reduced body fat but not body weight	1.7–4.2	4 weeks–6 months	Healthy or overweight/ obese subjects	121–124
Lowered leptin, no effect on fat mass	3	94 d (64 d CLA treatment)	Healthy women subjects	125
Reduced blood triacylglycerols and VLDL/HDL cholesterol	0.7–3.0	8 weeks	Healthy or normolipidaemic subjects	126, 127
The $t10,c12$ enhanced response to hepatitis B vaccination	1.7 or 1.6	12 weeks	Healthy Caucasian males	128

[a] Whenever possible, CLA dose was given as the amount of pure CLA when the CLA source was a mixture of CLA and other fatty acids. CLA sources used in the studies mentioned above included roughly equal amount of $c9,t11$ and $t10,c12$ isomers with varying percentages based on total fatty acid analysis.

subjects. New research should limit the scope of biological endpoints such that specific actions and biological targets can be rigorously tested in the human. An integrative approach to understanding fat accumulation, insulin actions, immune function, and bone biology would involve COX-2, PPARs, and other specific transcription factors. Therefore, recognizing the potential diverse actions of these nutrients provides an opportunity for collaborative scientific inquiry to study systematically the functions of CLA isomers for improving health. Since CLA isomers, as well as other PUFAs, are recognized as natural PPAR ligands, investigating how these isomers modulate transcription factors and their target genes could lead to the elucidation of the benefits of CLA in humans. Future investigations with CLAs should take into account the recognized actions of these isomers in specific physiological states and candidate targets for these nutrients.

The marketing of commercial CLA products has been largely directed towards body weight reduction to address the obesity problem in the U.S. However, it is premature to recommend CLA as a means to control obesity based on the limited published research. The future application of CLAs and other nutraceutical fatty acids as supplements and in functional food formulations continues to be an important area of public health research[111] and a major effort of product development for the food industry[112].

12.6 SUMMARY

CLA isomers are a group of unusual unsaturated fatty acids that may be involved in a variety of biological functions related to health. The metabolic and physiological effects of CLA isomers described in this chapter focused on body weight and cancer; however, obesity, diabetes, immune function, and osteoporosis are other areas that have been studied. Research with CLAs on the regulation of transcription factors and genes common to many chronic diseases provide an opportunity for cooperative scientific investigation. Since CLA isomers, as well as other PUFAs, are recognized as natural PPAR ligands, investigating how these isomers modulate transcription factors and their target genes could lead to the elucidation of the potential benefits of CLAs in humans. Future investigations with CLAs should take into account the known actions and potential applications of these isomers to specific physiological states and chronic diseases.

ACKNOWLEDGMENT

The authors acknowledge support by a grant from the 21st Century Research and Technology Fund and the Center for Enhancing Foods to Protect Health (www.efph.purdue.edu).

REFERENCES

1. Watkins, B.A. and Li, Y., Conjugated linoleic acid: the present state of knowledge, in *Handbook of Nutraceuticals and Functional Foods*, Wildman, R.E.C., Ed., CRC Press, Boca Raton, FL, 2001, pp. 445–476.
2. Bartlet, J.C. and Chapman, D.G., Detection of hydrogenated fats in butter fat by measurement of *cis–trans* conjugated unsaturation, *J. Agric. Food Chem.*, 9, 50–53, 1961.
3. Chin, S.F., Liu, W., Storkson, J.M., Ha, Y.L., and Pariza, M.W., Dietary sources of conjugated dienoic isomers of linoleic acid, a newly recognized class of anticarcinogens, *J. Food Comp. Anal.*, 5, 185–197, 1992.
4. Parodi, P.W., Conjugated octadecadienoic acids of milk fat, *J. Dairy Sci.*, 60, 1551–1553, 1977.
5. Davis, A.L., McNeill, G.P., and Caswell, D.C., Analysis of conjugated linoleic acid isomers by C-13 NMR spectroscopy, *Chem. Phys. Lipids*, 97, 155–165, 1999.
6. Sehat, N., Kramer, J.K., Mossoba, M.M., Yurawecz, M.P., Roach, J.A., Eulitz, K., Morehouse, K.M., and Ku, Y., Identification of conjugated linoleic acid isomers in cheese by gas chromatography, silver ion high performance liquid chromatography and mass spectral reconstructed ion profiles. Comparison of chromatographic elution sequences, *Lipids*, 33, 963–971, 1998.
7. Sehat, N., Yurawecz, M.P., Roach, J.A., Mossoba, M.M., Kramer, J.K., and Ku, Y., Silver-ion high-performance liquid chromatographic separation and identification of conjugated linoleic acid isomers, *Lipids*, 33, 217–221, 1998.
8. Sehat, N., Rickert, R., Mossoba, M.M., Kramer, K., Yurawecz, M.P., Roach, J.A., Adlof, R.O., Morehouse, K.M., Fritsche, J., Eulitz, K.D., Steinhart, H., and Ku, Y., Improved separation of conjugated fatty acid methyl esters by silver ion-high-performance liquid chromatography, *Lipids*, 34, 407–413, 1999.
9. Ackman, R.G., Laboratory preparation of conjugated linoleic acids, *J. Am. Oil Chem. Soc.*, 75, 1227, 1998.
10. Martin, J.C. and Valeille, K., Conjugated linoleic acids: all the same or to everyone its own function?, *Reproduc. Nutr. Dev.*, 42, 525–536, 2002.
11. Reaney, M.J.T., Liu, Y.D., and Westcott, N.D., Commercial production of conjugated linoleic acid, in *Advances in Conjugated Linoleic Acid Research*, Yurawecz, M.P., Mossoba, M.M., Kramer, J.K.G., Pariza, M.W., and Nelson, G.J., Eds., AOCS Press, Champaign, IL, 1999, pp. 39–54.
12. Adlof, R.O., Copes, L.C., and Walter, E.L., Changes in conjugated linoleic acid composition within samples obtained from a single source, *Lipids*, 36, 315–317, 2001.
13. Parodi, P.W., Cows' milk fat components as potential anticarcinogenic agents, *J. Nutr.*, 127, 1055–1060, 1997.
14. Ha, Y.L., Grimm, N.K., and Pariza, M.W., Newly recognized anticarcinogenic fatty acids: identification and quantification in natural and processed cheeses, *J. Agric. Food Chem.*, 37, 75–81, 1989.
15. Turek, J.J., Li, Y., Schoenlein, I.A., Allen, K.G.D., and Watkins, B.A., Modulation of macrophage cytokine production by conjugated linoleic acids is influenced by the dietary n-6:n-3 fatty acid ratio, *J. Nutr. Biochem.*, 9, 258–266, 1998.
16. Li, Y. and Watkins, B.A., Conjugated linoleic acids alter bone fatty acid composition and reduce ex vivo prostaglandin E2 biosynthesis in rats fed n-6 or n-3 fatty acids, *Lipids*, 33, 417–425, 1998.
17. Li, Y., Seifert, M.F., Ney, D.M., Grahn, M., Grant, A.L., Allen, K.G., and Watkins, B.A., Dietary conjugated linoleic acids alter serum IGF-I and IGF binding protein concentrations and reduce bone formation in rats fed (n-6) or (n-3) fatty acids, *J. Bone Miner. Res.*, 14, 1153–1162, 1999.
18. Liu, J.R., Chen, R.Q., Yang, Y.M., Wang, X.L., Xue, Y.B., Zheng, Y.M., and Liu, R.H., Effect of apoptosis on gastric adenocarcinoma cell line SGC-7901 induced by cis-9, trans-11-conjugated linoleic acid, *World J. Gastroenterol.*, 8, 999–1004, 2002.
19. Miller, A., Stanton, C., and Devery, R., Cis 9, trans 11- and trans 10, cis 12-conjugated linoleic acid isomers induce apoptosis in cultured SW480 cells, *Anticancer Res.*, 22, 3879–3887, 2002.

20. Kim, E.J., Jun, J.G., Park, H.S., Kim, S.M., Ha, Y.L., and Park, J.H., Conjugated linoleic acid (CLA) inhibits growth of Caco-2 colon cancer cells: possible mediation by oleamide, *Anticancer Res.*, 22, 2193–2197, 2002.

21. Palombo, J.D., Ganguly, A., Bistrian, B.R., and Menard, M.P., The antiproliferative effects of biologically active isomers of conjugated linoleic acid on human colorectal and prostatic cancer cells, *Cancer Lett.*, 177, 163–172, 2002.

22. Yamasaki, M., Chujo, H., Koga, Y., Oishi, A., Rikimaru, T., Shimada, M., Sugimachi, K., Tachibana, H., and Yamada, K., Potent cytotoxic effect of the trans10, cis12 isomer of conjugated linoleic acid on rat hepatoma dRLh-84 cells, *Cancer Lett.*, 188, 171–180, 2002.

23. Moon, E.J., Lee, Y.M., and Kim, K.W., Anti-angiogenic activity of conjugated linoleic acid on basic fibroblast growth factor-induced angiogenesis, *Oncol. Rep.*, 10, 617–621, 2003.

24. Thuillier, P., Anchiraico, G.J., Nickel, K.P., Maldve, R.E., Gimenez-Conti, I., Muga, S.J., Liu, K.L., Fischer, S.M., and Belury, M.A., Activators of peroxisome proliferator-activated receptor-alpha partially inhibit mouse skin tumor promotion, *Mol. Carcinog.*, 29, 134–142, 2000.

25. Yu, Y., Correll, P.H., and Vanden Heuvel, J.P., Conjugated linoleic acid decreases production of proinflammatory products in macrophages: evidence for a PPAR gamma-dependent mechanism, *Biochim. Biophys. Acta*, 1581, 89–99, 2002.

26. Peters, J.M., Park, Y., Gonzalez, F.J., and Pariza, M.W., Influence of conjugated linoleic acid on body composition and target gene expression in peroxisome proliferator-activated receptor alpha-null mice, *Biochim. Biophys. Acta*, 1533, 233–242, 2001.

27. Hargrave, K.M., Li, C.L., Meyer, B.J., Kachman, S.D., Hartzell, D.L., Della-Fera, M.A., Miner, J.L., and Baile, C.A., Adipose depletion and apoptosis induced by trans-10, cis-12 conjugated linoleic acid in mice, *Obes. Res.*, 10, 1284–1290, 2002.

28. Tsuboyama-Kasaoka, N., Takahashi, M., Tanemura, K., Kim, H.J., Tange, T., Okuyama, H., Kasai, M., Ikemoto, S., and Ezaki, O., Conjugated linoleic acid supplementation reduces adipose tissue by apoptosis and develops lipodystrophy in mice, *Diabetes*, 49, 1534–1542, 2000.

29. Akahoshi, A., Goto, Y., Murao, K., Miyazaki, T., Yamasaki, M., Nonaka, M., Yamada, K., and Sugano, M., Conjugated linoleic acid reduces body fats and cytokine levels of mice, *Biosci. Biotechnol. Biochem.*, 66, 916–920, 2002.

30. Takahashi, Y., Kushiro, M., Shinohara, K., and Ide, T., Activity and mRNA levels of enzymes involved in hepatic fatty acid synthesis and oxidation in mice fed conjugated linoleic acid, *Biochim. Biophys. Acta*, 1631, 265–273, 2003.

31. Warren, J.M., Simon, V.A., Bartolini, G., Erickson, K.L., Mackey, B.E., and Kelley, D.S., Trans-10, cis-12 CLA increases liver and decreases adipose tissue lipids in mice: possible roles of specific lipid metabolism genes, *Lipids*, 38, 497–504, 2003.

32. Badinga, L., Selberg, K.T., Dinges, A.C., Corner, C.W., and Miles, R.D., Dietary conjugated linoleic acid alters hepatic lipid content and fatty acid composition in broiler chickens, *Poult. Sci.*, 82, 111–116, 2003.

33. Yamasaki, M., Ikeda, A., Oji, M., Tanaka, Y., Hirao, A., Kasai, M., Iwata, T., Tachibana, H., and Yamada, K., Modulation of body fat and serum leptin levels by dietary conjugated linoleic acid in Sprague Dawley rats fed various fat-level diets, *Nutrition*, 19, 30–35, 2003.

34. Meadus, W.J., MacInnis, R., and Dugan, M.E., Prolonged dietary treatment with conjugated linoleic acid stimulates porcine muscle peroxisome proliferator activated receptor gamma and glutamine-fructose aminotransferase gene expression in vivo, *J. Mol. Endocrinol.*, 28, 79–86, 2002.

35. Kamphuis, M.M.J.W., Lejeune, M.P.G.M., Saris, W.H.M., and Westerterp-Plantenga, M.S., The effect of conjugated linoleic acid supplementation after weight loss on body weight regain, body composition, and resting metabolic rate in overweight subjects, *Int. J. Obes.*, 27, 840–847, 2003.

36. von Loeffelholz, C., Kratzsch, J., and Jahreis, G., Influence of conjugated linoleic acids on body composition and selected serum and endocrine parameters in resistance-trained athletes, *Eur. J. Lipid Sci. Technol.*, 105, 251–259, 2003.

37. Riserus, U., Smedman, A., Basu, S., and Vessby, B., CLA and body weight regulation in humans, *Lipids*, 38, 133–137, 2003.

38. Li, Y., and Watkins, B.A., CLA in human nutrition and health, in *Handbook of Functional Lipids*, Akoh, C.C., Ed., CRC Press, Boca Raton, FL, 2005.

39. Rodriguez, E., Ribot, J., and Palou, A., Trans-10, cis-12, but not cis-9, trans-11 CLA isomer, inhibits brown adipocyte thermogenic capacity, *Am. J. Physiol. Regul. Integr. Comp. Physiol.*, 282, R1789–R1797, 2002.

40. Rahman, S.M., Wang, Y., Yotsumoto, H., Cha, J., Han, S., Inoue, S., and Yanagita, T., Effects of conjugated linoleic acid on serum leptin concentration, body-fat accumulation, and beta-oxidation of fatty acid in OLETF rats, *Nutrition*, 17, 385–390, 2001.

41. Kang, K. and Pariza, M.W., trans-10, cis-12-Conjugated linoleic acid reduces leptin secretion from 3T3-L1 adipocytes, *Biochem. Biophys. Res. Commun.*, 287, 377–382, 2001.

42. Watkins, B.A., Li, Y., Romsos, D.R., and Seifert, M.F., CLA and bone modeling in rats, in *Advances in Conjugated Linoleic Acid Research*, Sebedio, J.L., Christie, W.W., and Adlof, R., Eds., AOCS Press, Champaign, IL, 2003, pp. 218–250.

43. Reseland, J.E. and Gordeladze, J.O., Role of leptin in bone growth: central player or peripheral supporter?, *FEBS Lett.*, 528, 40–42, 2002.

44. Thomas, T. and Burguera, B., Is leptin the link between fat and bone mass?, *J. Bone Miner. Res.*, 17, 1563–1569, 2002.

45. Belury, M.A., Moya-Camarena, S.Y., Liu, K.L., and Vanden Heuvel, J.P., Dietary conjugated linoleic acid induces peroxisome-specific enzyme accumulation and ornithine decarboxylase activity in mouse liver, *J. Nutr. Biochem.*, 8, 579–584, 1997.

46. Lee, K.N., Pariza, M.W., and Ntambi, J.M., Conjugated linoleic acid decreases hepatic stearoyl-CoA desaturase mRNA expression, *Biochem. Biophys. Res. Commun.*, 248, 817–821, 1998.

47. Meade, E.A., McIntyre, T.M., Zimmerman, G.A., and Prescott, S.M., Peroxisome proliferators enhance cyclooxygenase-2 expression in epithelial cells, *J. Biol. Chem.*, 274, 8328–8334, 1999.

48. Sugano, M., Tsujita, A., Yamasaki, M., Yamada, K., Ikeda, I., and Kritchevsky, D., Lymphatic recovery, tissue distribution, and metabolic effects of conjugated linoleic acid in rats, *J. Nutr. Biochem.*, 8, 38–43, 1997.

49. Bulgarella, J.A., Patton, D., and Bull, A.W., Modulation of prostaglandin H synthase activity by conjugated linoleic acid (CLA) and specific CLA isomers, *Lipids*, 36, 407–412, 2001.

50. Liu, K.L. and Belury, M.A., Conjugated linoleic acid reduces arachidonic acid content and PGE_2 synthesis in murine keratinocytes, *Cancer Lett.*, 127, 15–22, 1998.

51. Ha, Y.L., Grimm, N.K., and Pariza, M.W., Anticarcinogens from fried ground beef: heat-altered derivatives of linoleic acid, *Carcinogenesis*, 8, 1881–1887, 1987.

52. Ip, C., Jiang, C., Thompson, H.J., and Scimeca, J.A., Retention of conjugated linoleic acid in the mammary gland is associated with tumor inhibition during the post-initiation phase of carcinogenesis, *Carcinogenesis*, 18, 755–759, 1997.

53. Banni, S., Angioni, E., Casu, V., Melis, M.P., Carta, G., Corongiu, F.P., Thompson, H., and Ip, C., Decrease in linoleic acid metabolites as a potential mechanism in cancer risk reduction by conjugated linoleic acid, *Carcinogenesis*, 20, 1019–1024, 1999.

54. Cesano, A., Visonneau, S., Scimeca, J.A., Kritchevsky, D., and Santoli, D., Opposite effects of linoleic acid and conjugated linoleic acid on human prostatic cancer in SCID mice, *Anticancer Res.*, 18, 1429–1434, 1998.

55. Cunningham, D.C., Harrison, L.Y., and Shultz, T.D., Proliferative responses of normal human mammary and MCF-7 breast cancer cells to linoleic acid, conjugated linoleic acid and eicosanoid synthesis inhibitors in culture, *Anticancer Res.*, 17, 197–203, 1997.

56. Ip, M.M., Masso-Welch, P.A., Shoemaker, S.F., Shea-Eaton, W.K., and Ip, C., Conjugated linoleic acid inhibits proliferation and induces apoptosis of normal rat mammary epithelial cells in primary culture, *Exp. Cell Res.*, 250, 22–34, 1999.

57. Nicolosi, R.J., Rogers, E.J., Kritchevsky, D., Scimeca, J.A., and Huth, P.J., Dietary conjugated linoleic acid reduces plasma lipoproteins and early aortic atherosclerosis in hypercholesterolemic hamsters, *Artery*, 22, 266–277, 1997.

58. Lee, K.N., Kritchevsky, D., and Pariza, M.W., Conjugated linoleic acid and atherosclerosis in rabbits, *Atherosclerosis*, 108, 19–25, 1994.

59. Chin, S.F., Storkson, J.M., Liu, W., Albright, W.J., and Pariza, M.W., Conjugated linoleic acid (9,11- and 10,12-octadecadienoic acid) is produced in conventional but not germ-free rats fed linoleic acid, *J. Nutr.*, 124, 694–701, 1994.

60. Sugano, M., Tsujita, A., Yamasaki, M., Noguchi, M., and Yamada, K., Conjugated linoleic acid modulates tissue levels of chemical mediators and immunoglobulins in rats, *Lipids*, 33, 521–527, 1998.

61. DeLany, J.P., Blohm, F., Truett, A.A., Scimeca, J.A., and West, D.B., Conjugated linoleic acid rapidly reduces body fat content in mice without affecting energy intake, *Am. J. Physiol. Regul. Integr. Comp. Physiol.*, 45, R1172–R1179, 1999.

62. Park, Y., Storkson, J.M., Albright, K.J., Liu, W., and Pariza, M.W., Evidence that the *trans*-10, *cis*-12 isomer of conjugated linoleic acid induces body composition changes in mice, *Lipids*, 34, 235–241, 1999.

63. Yamasaki, M., Mansho, K., Mishima, H., Kasai, M., Sugano, M., Tachibana, H., and Yamada, K., Dietary effect of conjugated linoleic acid on lipid levels in white adipose tissue of Sprague-Dawley rats, *Biosci. Biotechnol. Biochem.*, 63, 1104–1106, 1999.

64. Houseknecht, K.L., Heuvel, J.P.V., Moya-Camerena, S.Y., Portocarrero, C.P., Nickel, K.P., and Belury, M.A., Dietary conjugated linoleic acid normalizes impaired glucose tolerance in the Zucker diabetic fatty fa/fa rat, *Biochem. Biophys. Res. Commun.*, 244, 678–682, 1998.

65. Ip, C., Singh, M., Thompson, H.J., and Scimeca, J.A., Conjugated linoleic acid suppresses mammary carcinogenesis and proliferative activity of the mammary gland in the rat, *Cancer Res.*, 54, 1212–1215, 1994.

66. Ip, C., Briggs, S.P., Haegele, A.D., Thompson, H.J., Storkson, J., and Scimeca, J.A., The efficacy of conjugated linoleic acid in mammary cancer prevention is independent of the level or type of fat in the diet, *Carcinogenesis*, 17, 1045–1050, 1996.

67. Thompson, H., Zhu, Z., Banni, S., Darcy, K., Loftus, T., and Ip, C., Morphological and biochemical status of the mammary gland as influenced by conjugated linoleic acid: implication for a reduction in mammary cancer risk, *Cancer Res.*, 57, 5067–5072, 1997.

68. Belury, M.A., Nickel, K.P., Bird, C.E., and Wu, Y., Dietary conjugated linoleic acid modulation of phorbol ester skin tumor promotion, *Nutr. Cancer*, 26, 149–157, 1996.

69. Schonberg, S. and Krokan, H.E., The inhibitory effect of conjugated dienoic derivatives (CLA) of linoleic acid on the growth of human tumor cell lines is in part due to increased lipid peroxidation, *Anticancer Res.*, 15, 1241–1246, 1995.

70. Shultz, T.D., Chew, B.P., and Seaman, W.R., Differential stimulatory and inhibitory responses of human MCF-7 breast cancer cells to linoleic acid and conjugated linoleic acid in culture, *Anticancer Res.*, 12, 2143–2145, 1992.

71. Visonneau, S., Cesano, A., Tepper, S.A., Scimeca, J.A., Santoli, D., and Kritchevsky, D., Conjugated linoleic acid suppresses the growth of human breast adenocarcinoma cells in SCID mice, *Anticancer Res.*, 17, 969–973, 1997.

72. Fritsche, J. and Steinhart, H., Analysis, occurrence, and physiological properties of *trans* fatty acids (TFA) with particular emphasis on conjugated linoleic acid isomers (CLA): a review, *Fett-Lipid*, 100, 190–210, 1998.

73. Molkentin, J., Bioactive lipids naturally occurring in bovine milk, *Nahrung*, 43,185–189, 1999.

74. Haumann, B.F., Conjugated linoleic acid, *INFORM*, 7, 152–159, 1996.

75. Ma, D.W.L., Wierzbicki, A.A., Field, C.J., and Clandinin, M.T., Conjugated linoleic acid in Canadian dairy and beef products, *J. Agric. Food Chem.*, 47, 1956–1960, 1999.

76. Lin, H., Boylston, T.D., Chang, M.J., Luedecke, L.O., and Shultz, T.D., Survey of the conjugated linoleic acid contents of dairy products, *J. Dairy Sci.*, 78, 2358–2365, 1995.

77. Ens, J.G., Ma, D.W.L., Cole, K.S., Field, C.J., and Clandinin, M.T., An assessment of c9,t11 linoleic acid intake in a small group of young Canadians, *Nutr. Res.*, 21, 955–960, 2001.

78. Ritzenthaler, K.L., McGuire, M.K., Falen, R., Shultz, T.D., Dasgupta, N., and McGuire, M.A., Estimation of conjugated linoleic acid intake by written dietary assessment methodologies underestimates actual intake evaluated by food duplicate methodology, *J. Nutr.*, 131, 1548–1554, 2001.

79. Fremann, D., Linseisen, J., and Wolfram, G., Dietary conjugated linoleic acid (CLA) intake assessment and possible biomarkers of CLA intake in young women, *Public Health Nutr.*, 5, 73–80, 2002.

80. Fritsche, J., Rickert, R., Steinhart, H., Yurawecz, M.P., Mossoba, M.M., Sehat, N., Roach, J.A.G., Kramer, J.K.G., and Ku, Y., Conjugated linoleic acid (CLA) isomers: formation, analysis, amounts in foods, and dietary intake, *Fett-Lipid*, 101, 272–276, 1999.

81. Brody, S., *Bioenergetics and Growth with Special Reference to the Efficiency Complex in Domestic Animals*, Hafner, New York, 1964.

82. Dhiman, T.R., Anand, G.R., Satter, L.D., and Pariza, M.W., Conjugated linoleic acid content of milk from cows fed different diets, *J. Dairy Sci.*, 82, 2146–2156, 1999.

83. Shantha, N.C., Ram, L.N., O'Leary, J., Hicks, C.L., and Decker, E.A., Conjugated linoleic acid concentrations in dairy products as affected by processing and storage, *J. Food Sci.*, 60, 695–697, 1995.

84. Lavillonniere, F., Martin, J.C., Bougnoux, P., and Sebedio, J.L., Analysis of conjugated linoleic acid isomers and content in French cheeses, *J. Am. Oil Chem. Soc.*, 75, 343–352, 1998.

85. Kelly, M.L., Berry, J.R., Dwyer, D.A., Griinari, J.M., Chouinard, P.Y., Van, A.ME, and Bauman, D.E., Dietary fatty acid sources affect conjugated linoleic acid concentrations in milk from lactating dairy cows, *J. Nutr.*, 128, 881–885, 1998.

86. Jiang, J., Bjorck, L., and Fonden, R., Conjugated linoleic acid in Swedish dairy products with special reference to the manufacture of hard cheeses, *Int. Dairy J.*, 7, 863–867, 1997.

87. Dhiman, T.R., Satter, L.D., Pariza, M.W., Galli, M.P., Albright, K., and Tolosa, M.X., Conjugated linoleic acid (CLA) content of milk from cows offered diets rich in linoleic and linolenic acid, *J. Dairy Sci.*, 83, 1016–1027, 2000.

88. Solomon, R., Chase, L.E., Ben Ghedalia, D., and Bauman, D.E., The effect of nonstructural carbohydrate and addition of full fat extruded soybeans on the concentration of conjugated linoleic acid in the milk fat of dairy cows, *J. Dairy Sci.*, 83, 1322–1329, 2000.

89. Chouinard, P.Y., Corneau, L., Butler, W.R., Chilliard, Y., Drackley, J.K., and Bauman, D.E., Effect of dietary lipid source on conjugated linoleic acid concentrations in milk fat, *J. Dairy Sci.*, 84, 680–690, 2001.

90. Bauman, D.E., Barbano, D.M., Dwyer, D.A., and Griinari, J.M., Technical note: production of butter with enhanced conjugated linoleic acid for use in biomedical studies with animal models, *J. Dairy Sci.*, 83, 2422–2425, 2000.

91. Baer, R.J., Ryali, J., Schingoethe, D.J., Kasperson, K.M., Donovan, D.C., Hippen, A.R and Franklin, S.T., Composition and properties of milk and butter from cows fed fish oil, *J. Dairy Sci.*, 84, 345–353, 2001.

92. Reklewska, B., Oprzadek, A., Reklewski, Z., Panicke, L., Kuczynska, B., and Oprzadek, J., Alternative for modifying the fatty acid composition and decreasing the cholesterol level in the milk of cows, *Livest. Prod. Sci.*, 76, 235–243, 2002.

93. White, S.L., Bertrand, J.A., Wade, M.R., Washburn, S.P., Green, J.T. Jr., and Jenkins, T.C., Comparison of fatty acid content of milk from Jersey and Holstein cows consuming pasture or a total mixed ration, *J. Dairy Sci.*, 84, 2295–2301, 2001.

94. Collomb, M., Butikofer, U., Sieber, R., Bosset, O., and Jeangros, B., Conjugated linoleic acid and trans fatty acid composition of cows' milk fat produced in lowlands and highlands, *J. Dairy Res.*, 68, 519–523, 2001.

95. Baumgard, L.H., Corl, B.A., Dwyer, D.A., Saebo, A., and Bauman, D.E., Identification of the conjugated linoleic acid isomer that inhibits milk fat synthesis, *Am. J. Physiol. Regul. Integr. Comp. Physiol.*, 278, R179–R184, 2000.

96. Chouinard, P.Y., Corneau, L., Barbano, D.M., Metzger, L.E., and Bauman, D.E., Conjugated linoleic acids alter milk fatty acid composition and inhibit milk fat secretion in dairy cows, *J. Nutr.*, 129, 1579–1584, 1999.

97. Ahn, D.U., Sell, J.L., Jo, C., Chamruspollert, M., and Jeffrey, M., Effect of dietary conjugated linoleic acid on the quality characteristics of chicken eggs during refrigerated storage, *Poult. Sci.*, 78, 922–928, 1999.

98. Chamruspollert, M. and Sell, J.L., Transfer of dietary conjugated linoleic acid to egg yolks of chickens, *Poult. Sci.*, 78, 1138–1150, 1999.

99. Du, M., Ahn, D.U. and Sell, J.L., Effect of dietary conjugated linoleic acid on the composition of egg yolk lipids, *Poult. Sci.*, 78, 1639–1645, 1999.

100. Watkins, B.A., Devitt, A.A., Yu, L., and Latour, M.A., Biological activities of conjugated linoleic acids and designer eggs, in *Egg Nutrition and Biotechnology*, Sim, J.S., Nakai, S., and Guenter, W., Eds., CABI, Wallingford, U.K., 1999, pp. 181–195.

101. Dugan, M.E.R., Aalhus, J.L., Schaefer, A.L., and Kramer, J.K.G., The effect of conjugated linoleic acid on fat to lean repartitioning and feed conversion in pigs, *Can. J. Anim. Sci.*, 77, 723–725, 1997.

102. Ostrowska, E., Muralitharan, M., Cross, R.F., Bauman, D.E., and Dunshea, F.R., Dietary conjugated linoleic acids increase lean tissue and decrease fat deposition in growing pigs, *J. Nutr.*, 129, 2037–2042, 1999.

103. Kramer, J.K., Sehat, N., Dugan, M.E., Mossoba, M.M., Yurawecz, M.P., Roach, J.A., Eulitz, K., Aalhus, J.L., Schaefer, A.L., and Ku, Y., Distributions of conjugated linoleic acid (CLA) isomers in tissue lipid classes of pigs fed a commercial CLA mixture determined by gas chromatography and silver ion-high-performance liquid chromatography, *Lipids*, 33, 549–558, 1998.

104. Thiel-Cooper, R.L., Parrish, Jr., F.C., Sparks, J.C., Wiegand, B.R., and Ewan, R.C., Conjugated linoleic acid changes swine performance and carcass composition, *J. Anim. Sci.*, 79, 1821–1828, 2001.

105. O'Quinn, P.R., Nelssen, J.L., Goodband, R.D., Unruh, J.A., Woodworth, J.C., and Tokach, M.D., Effects of modified tall oil versus a commercial source of conjugated linoleic acid and increasing levels of modified tall oil on growth performance and carcass characteristics of growing-finishing pigs, *J. Anim. Sci.*, 78, 2359–2368, 2000.

106. Eggert, J.M., Belury, M.A., Kempa-Steczko, A., Mills, S.E., and Schinckel, A.P., Effects of conjugated linoleic acid on the belly firmness and fatty acid composition of genetically lean pigs, *J. Anim. Sci.*, 79, 2866–2872, 2001.

107. Twibell, R.G., Watkins, B.A., Rogers, L., and Brown, P.B., Effects of dietary conjugated linoleic acids on hepatic and muscle lipids in hybrid striped bass, *Lipids*, 35, 155–161, 2000.

108. Twibell, R.G., Watkins, B.A., and Brown, P.B., Dietary conjugated linoleic acids and lipid source alter fatty acid composition of juvenile yellow perch, Perca flavescens, *J. Nutr.*, 131, 2322–2328, 2001.

109. Watkins, B.A., Feng, S., Strom, A.K., Devitt, A.A., Yu, I., and Li, Y., Conjugated linoleic acids alter the fatty acid composition and physical properties of egg yolk and albumen, *J. Agric. Food Chem.*, 51, 6870–6876, 2003.

110. Kelley, D.S. and Erickson, K.L., Modulation of body composition and immune cell functions by conjugated linoleic acid in humans and animal models: benefits vs. risks, *Lipids*, 38, 377–386, 2003.

111. *Exploring a Vision: Integrating Knowledge for Food and Health*, a workshop summary, National Research Council of the National Academies, Rouse, T.I. and Davis, D.P., Eds., National Academy Press, Washington, DC, 2004.

112. Camire, M.E., Childs, N., Hasler, C.M., Pike, L.M., Shahidi, F., and Watkins, B.A., Nutraceuticals for health promotion and disease prevention, Issue Paper number 24, 1–16, 2003.

113. Shantha, N.C., Decker, E.A., and Ustunol, Z., Conjugated linoleic acid concentration in processed cheese, *J. Am. Oil Chem. Soc.*, 69, 425–428, 1992.

114. Lin, H., Boylston, T.D., Luedecke, L.O., and Shultz, T.D., Conjugated linoleic acid content of cheddar-type cheeses as affected by processing, *J. Food Sci.*, 64, 874–878, 1999.

115. Banni, S., Carta, G., Contini, M.S., Angioni, E., Deiana, M., Dessi, M.A., Melis, M.P., and Corongiu, F.P., Characterization of conjugated diene fatty acids in milk, dairy products, and lamb tissues, *J. Nutr. Biochem.*, 7, 150–155, 1996.

116. Werner, S.A., Luedecke, L.O., and Shultz, T.D., Determination of conjugated linoleic acid content and isomer distribution in three Cheddar-type cheeses: effects of cheese culture, processing, and aging, *J. Agric. Food Chem.*, 40, 1817–1821, 1992.

117. Yurawecz, M.P., Roach, J.A., Sehat, N., Mossoba, M.M., Kramer, J.K., Fritsche, J., Steinhart, H., and Ku, Y., A new conjugated linoleic acid isomer, 7 trans, 9 cis-octadecadienoic acid, in cow milk, cheese, beef and human milk and adipose tissue, *Lipids*, 33, 803–809, 1998.

118. Jahreis, G., Fritsche, J., Mockel, P., Schone, F., Moller, U., and Steinhart, H., The potential anticarcinogenic conjugated linoleic acid, cis-9,trans-11 C18:2, in milk of different species: cow, goat, ewe, sow, mare, woman, *Nutr. Res.*, 19, 1541–1549, 1999.

119. Shantha, N.C., Crum, A.D., and Decker, E.A., Evaluation of conjugated linoleic acid concentrations in cooked beef, *J. Agric. Food Chem.*, 42, 1757–1760, 1994.

120. Precht, D. and Molkentin, J., C18:1, C18:2 and C18:3 *trans* and *cis* fatty acid isomers including conjugated *cis* delta 9, *trans* delta 11 linoleic acid (CLA) as well as total fat composition of German human milk lipids, *Nahrung*, 43, 233–244, 1999.

121. Thom, E., Wadstein, J., and Gudmundsen, O., Conjugated linoleic acid reduces body fat in healthy exercising humans, *J. Int. Med. Res.*, 29, 392–396, 2001.

122. Smedman, A. and Vessby, B., Conjugated linoleic acid supplementation in humans: metabolic effects, *Lipids*, 36, 773–781, 2001.

123. Blankson, H., Stakkestad, J.A., Fagertun, H., Thom, E., Wadstein, J., and Gudmundsen, O., Conjugated linoleic acid reduces body fat mass in overweight and obese humans, *J. Nutr.*, 130, 2943–2948, 2000.

124. Riserus, U., Berglund, L., and Vessby, B., Conjugated linoleic acid (CLA) reduced abdominal adipose tissue in obese middle-aged men with signs of the metabolic syndrome: a randomised controlled trial, *Int. J. Obes. Relat. Metab. Disord.*, 25, 1129–1135, 2001.

125. Medina, E.A., Horn, W.F., Keim, N.L., Havel, P.J., Benito, P., Kelley, D.S., Nelson, G.J., and Erickson, K.L., Conjugated linoleic acid supplementation in humans: effects on circulating leptin concentrations and appetite, *Lipids*, 35, 783–788, 2000.

126. Mougios, V., Matsakas, A., Petridou, A., Ring, S., Sagredos, A., Melissopoulou, A., Tsigilis, N., and Nikolaidis, M., Effect of supplementation with conjugated linoleic acid on human serum lipids and body fat, *J. Nutr. Biochem.*, 12, 585–594, 2001.

127. Noone, E.J., Roche, H.M., Nugent, A.P., and Gibney, M.J., The effect of dietary supplementation using isomeric blends of conjugated linoleic acid on lipid metabolism in healthy human subjects, *Br. J. Nutr.*, 88, 243–251, 2002.

128. Albers, R., van der Wielen, R.P.J., Brink, E.J., Hendriks, H.F.J., Dorovska-Taran, V.N., and Mohede, I.C.M., Effects of cis-9, trans-11 and trans-10, cis-12 conjugated linoleic acid (CLA) isomers on immune function in healthy men, *Eur. J. Clin. Nutr.*, 57, 595–603, 2003.

13 Occurrence of Conjugated Fatty Acids in Aquatic and Terrestrial Plants and their Physiological Effects

Bhaskar Narayan, Masashi Hosokawa, and Kazuo Miyashita
Laboratory of Bio-functional Material Chemistry, Hokkaido University, Hakodate, Japan

CONTENTS

13.1 INTRODUCTION

Conjugated fatty acids are attracting increased interest due to their beneficial effects in terms of human health. Among them conjugated linoleic acid (CLA) has been researched and reviewed extensively in relation to its occurrence[1], metabolism, and physiological effects[2,3]. Several researchers have reported the occurrence of various conjugated fatty acids including trienes, tetraenes, and pentaenes in different plant sources including those from terrestrial and aquatic origins. Attempts to use fatty acid composition as an aid in taxonomical conclusions have also been reviewed thoroughly[4,5]. The occurrence, health effects, and industrial uses of conjugated dienes, especially CLA/CLA isomers, have also been reviewed extensively and thoroughly by several researchers. Hence, the present chapter covers the conjugated fatty acids other than conjugated dienes including CLA. In the context of this chapter, fatty acids with three or more conjugated double bond systems that occur naturally in plants of terrestrial and aquatic origin are only considered along with their reported/documented physiological effects. For the purpose of this chapter, fatty acids with conjugated unsaturation that also contain acetylenic bonds or oxygen functions are excluded. Also, only macroalgae (sea-grasses) are considered as aquatic plants and the microalgae are excluded. Thus, a comprehensive picture of conjugated fatty acids that occur in plants of terrestrial and aquatic origin with emphasis on their physiological and health effects is provided. It is worth noting that some plant seeds contain conjugated linolenic acid (CLN) at high level (30 to 70 wt% lipid), although other kinds of conjugated fatty acids including CLA are only found in natural products at concentrations less than 1%. Thus, CLN isomers are major conjugated fatty acids of natural origin. Hence, in this chapter, we describe in detail the physiological effects of CLN isomers that occur in some plant seeds. It is hoped that this overview serves as a reference to multidisciplinary researchers working on biochemistry, physiology, and nutrition, especially with reference to conjugated fatty acids.

13.2   OCCURRENCE OF CONJUGATED FATTY ACIDS

Fatty acids having three or four conjugated double bonds occur in various seed oils and in aquatic plants, especially those of marine origin. For convenience, the occurrence can be divided based on the origin, i.e., from a terrestrial or aquatic environment.

Several conjugated fatty acids have been reported to occur in terrestrial plant lipids, especially seed oils; and most of these are 18-carbon compounds originating from oleic, linoleic, linolenic, and stearidonic acids. They include dienes, trienes, and tertraenes. The conjugated trienoic fatty acids from plant sources mainly include α-eleostearic acid (*cis*(c),11*trans*(t),13t-18:3), catalpic acid (9t,11t,13c-18:3), punicic acid (9c,11t,13c-18:3), calendic acid (8t,10t,12c-18:3), and jacaric acid (8c,10t,12c-18:3)[5]. A high content of calendic acid in pot marigold seed oil, punicic acid in pomegranate seed oil, and α-eleostearic acid in tung and bitter gourd seed oils has been reported[6]. α-Eleostearic acid has been shown to be the principal component, contributing to more than 50%, of bitter gourd seed oil. However, the flesh of the bitter gourd is reported to contain catalpic acid[7]. The only well-known conjugated diene and tetraene of plant origin is 10t, 12t-18:2[8] and α-parinaric acid (9c,11t,13t,15c-18:4)[9,10]. To date, there are no reports on the occurrence of conjugated fatty acids with more than 18 carbon atoms in lipids of plant origin, and seed oils seem to be the major source of all these conjugated trienoic fatty acids. Most of the conjugated C18 acids found in seed oils are positional and geometrical isomers of α-linolenic acid (18:3n-3). The naturally occurring trienoic and tetraenoic conjugated fatty acids are summarized in Table 13.1 along with their typical sources (mainly plant seed oils).

Apart from the plant sources mentioned above, conjugated trienes have also been reported in other plants/seed oils. A conjugated trienoic fatty acid with 10 carbon atoms has been found in the latex of the poinsettia, *Euphorbia pulcherrima*. This conjugated fatty acid was identified as 2t,4t,6c-decatrieonic acid (10:3) along with four other isomers which were present as minor components[11].

Unlike their terrestrial counterparts, the aquatic plants have been found to possess conjugated fatty acids, with carbon chain length varying from 16 to 22 carbon atoms, as natural constituents in their lipids. Both trienes and tetraenes occur in aquatic plant lipids (see Table 13.2). Although many research groups have reported the fatty acid composition of seaweeds from different regions of the world, little information is available on the occurrence of conjugated polyunsaturated fatty acids (PUFAs) in seaweeds. These include conjugated trienes in *Ptilota*[12,13] and *Acanthophora spicifera*[14], and tetraenes in *Bosiella orbigniana*[15], *Lithothamnion corallioides*[16], and *Anadyomene stellata*[17]. The work on investigation of conjugated polyenes from the seaweed *Ptilota filicina* resulted in the definition of a polyenoic fatty acid isomerase (PFI)[13] while research on an enzyme from *L. coralliodes* explained the mechanism of formation of tetraene by that enzyme[16]. Recently, PFI was characterized and functionally expressed by DNA cloning[18].

TABLE 13.1
Natural Conjugated Fatty Acids of Terrestrial Plant Origin[5,6,9]

Configuration	Trivial name	Typical source
Trienes		
8c,10t,12c-18:3	Jacaric	*Jacaranda mimoifolia*
8t,10t,12c-18:3	Calendic	*Calendula officianalis* (pot marigold)
9c,11t,13c-18:3	Punicic	*Punicia grannatum* (pomegranate)
		Trichosanthes anguina (snake gourd)
9c,11t,13t-18:3	α-Eleostearic	*Aleurites fordit* (tung)
		Momordica charantia (bitter gourd)
9t,11t,13c-18:3	Catalpic	*Catalpa ovata*
9t,11t,13t-18:3	β-Eleostearic	*Aleurites fordit* (tung)
Tetraenes		
9c,11t,13t,15c-18:4	α-Parinaric	*Impatiens edgeworthii*
9t,11t,13t,15t-18:4	β-Parinaric	*Impatiens edgeworthii*

TABLE 13.2
Natural Conjugated Fatty Acids of Aquatic Plant Origin[12,14,15,17]

Configuration	Trivial name	Typical source
Trienes		
5c,7t,9t,14c-20:4		*Acanthophora spicifera* (red algae)
5t,7t,9t,14c-20:4		*Acanthophora spicifera* (red algae)
5c,7t,9t,14c,17c-20:5		*Ptilota filcina* (red algae)
		Acanthophora spicifera (red algae)
5t,7t,9t,14c,17c-20:5		*Ptilota filcina* (red algae)
		Acanthophora spicifera (red algae)
Tetraenes		
5c,8c,10t,12t,14c-20:5	Basseopentaenoic	*Bassiella orbingnian* (red algae)
4c,7c,9t,11t,13c,16c,19c-20:7	Stellaheptaenoic	*Anadyomene stellata* (green algae)

13.3　FACTORS RESPONSIBLE FOR THE FORMATION OF CONJUGATED POLYENES

Various enzymes in both terrestrial and aquatic plants are thought to be responsible for the formation of conjugated trienes/tetraenes endogenously. The major factors that result in the formation of these trienes or tetraenes and recent experiments involving biosynthesis as well as isotope studies for deciphering information on these substances are discussed here, but hypothetical pathways purely based on structural analysis are not entertained. Readers may refer to several publications, including reviews[19–27] that provide extensive information on this subject. The diversity of fatty acids in nature has been attributed to the combinations of the numbers and locations of double and triple bonds, with conjugated fatty acids being no exception. A family of structurally related enzymes including desaturases and their diverged forms such as hydroxylases, epoxygenases, acetylenases,

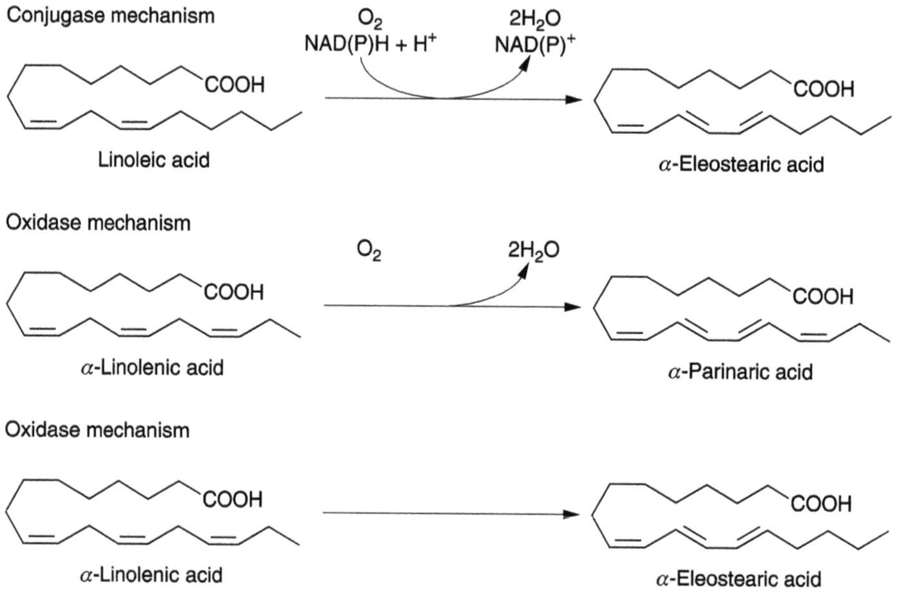

FIGURE 13.1 Summary of mechanisms involved in the formation of conjugated fatty acids.

and fatty acid conjugases are responsible for this diversity[28,29]. Several enzymes have been identified in a variety of higher plants, including those from the aquatic environment, that are capable of synthesizing conjugated fatty acids such as calendic, α-eleostearic, α-parinaric, and basseopentaenoic acids[13,25,29–32]. Some of these enzymes have also been successfully characterized for their protein make up and also expressed functionally in cDNA[18,29] including the genes responsible for such enzymes. The enzymes responsible for the formation of conjugated fatty acids can be grouped into three main categories of conjugases, oxidases, and isomerases. At least three different mechanisms have been documented for the biosynthesis of conjugated fatty acids in aquatic and terrestrial plants; these are summarized in Figure 13.1.

Crombie and Holloway[24] were the first to provide an isotopic evidence for conversion of linoleic acid into calendic acid in the developing seeds of *Calendula officianalis*. Further, the biosynthesis of α-eleostearic acid from linoleate in the developing seeds of *Momordica charantia* (bitter gourd) has also been demonstrated[30]. It was observed that linoleate esterified into phosphotidylcholine served as the precursor in this process. Further, they concluded that in spite of lacking the ability to synthesize α-linolenic acid, the lipid metabolic machinery of bitter gourd seeds has the ability to incorporate this fatty acid into lipids. However, this fatty acid was never converted to α-eleostearic acid. It was also demonstrated that the conjugated double bonds arise from the modification of an existing *cis* double bond[29].

13.4 PHYSIOLOGICAL EFFECTS OF CONJUGATED POLYENES ON CANCER CELL LINES

There are only a few reports on the inhibitory effect of conjugated polyenes on the growth of cancer cell lines. The effects and the underlying mechanism responsible for such effects are reviewed in this section. For the purpose of this section, documented reports related to both natural and chemically prepared conjugated polyenes are considered.

13.4.1 CONJUGATED POLYENES OTHER THAN CLN

Cornelius et al.[33] reported the strong cytotoxic effect of conjugated octadecatrienoic acid (9c,11t,13t,15c-18:4; α-parinaric acid) from garden balsam seed oil on human cancer cells. α-Parinaric acid exhibited toxicity to human monocytic leukemia (HL-60) and retinoblastoma (Y-79) cells at concentrations as low as 5 μM or less. Cytotoxicity exhibited by α-parinaric acid was attributed to the cell death via cellular lipid peroxidation. This was proved by the fact that the cytotoxic action was blocked by the addition of antioxidants such as butylated hydroxytoluene (BHT). Similarly, conjugated eicosapentaenoic acid (CEPA) and conjugated docosahexaenoic acid (CDHA) showed marked cytotoxicity against colorectal adenocarcinoma (DLD1) cells at concentration as low as 12 μM[34]. They also exhibited selective cytotoxicity against other cancer cell lines including HepG2, A549, and MCF-7 cells. The underlying mechanism of the toxicity was similar to that of α-parinaric acid. The amounts of phospholipid hyderoperoxides and thiobarbituric acid reactive substances (TBARS) were significantly increased in cells supplemented with CEPA and CDHA and the toxic action of CEPA and CDHA was blocked by the addition of α-tocopherol[34]. It has been reported that the cytotoxic activity of fatty acids involves lipid peroxidation and alteration in fatty acid composition of membrane phospholipids, and changes in eicosanoid synthesis and membrane fluidity[35–38]. In particular, several studies show the evidence that lipid hydroperoxides and aldehydes can mediate cell death and cell growth arrest including apoptosis[39–44].

Bégin et al.[35,36] reported the toxic effect of eicosapentaenoic acid (EPA) and docosahexaenoic acid (DHA) on several kinds of tumor cells. Moreover, other polyunsaturated fatty acids, i.e., arachidonic acid (20:4n-6), α-linolenic acid (18:3n-3), and γ-linolenic acid (18:3n-6) have cytotoxic effect on several tumor cell lines at concentrations above 50 μM. The cytotoxic effects of CEPA and CDHA were clearly observed at concentrations above 10 to 50 μM[34]. The cytotoxic effects of CEPA and CDHA were stronger than those of their corresponding nonconjugated fatty acids, EPA and DHA.

CEPA and CDHA were prepared by alkaline isomerization of corresponding EPA and DHA. The resultant CEPA and CDHA were composed of mixtures of conjugated dienoic, trienoic, and tetraenoic acids. Among these, conjugated trienoic acids played the most important role in the cytotoxic effects of CEPA and CDHA on tumor cells. In a study comparing CLA and CLN, Igarashi and Miyazawa[45] found that CLN and not CLA was highly cytotoxic to various cultured human tumor cells that included DLD-1, HepG2, A549, MCF-7, and MKN-7 cells. These authors concluded that the conjugated trienoic structure was more responsible for the cytotoxicity than other types of isomers. This was confirmed by the comparison of cytotoxic effects of two kinds of alkaline isomerized CLN mixture (CLN-1 and CLN-2) on human monocytic leukemia cells (U-937), which contained the same overall concentration of conjugated fatty acids (48 to 49%) with different diene (45.3% for CLN-1 and 28.6% for CLN-2) and triene (3.7% for CLN-1 and 19.7% for CLN-2) contents[46] (Figure 13.2). Although linoleic acid (LA) and α-linolenic acid (LN) had no cytotoxic effect on the cells, conjugated fatty acids (CLA, CLN-1, CLN-2) reduced cell viability at concentrations up to 324 μM for CLA, 327 μM for CLN-1, and 41 μM for CLN-2, indicating higher cytotoxic effect of conjugated trienes than conjugated dienes.

13.4.2 CONJUGATED LINOLENIC ACID (CLN)

13.4.2.1 Effects of CLN: *In Vivo* Studies

Although CLA has many kinds of biological activities and CEPA shows inhibitory effects on the growth of cancer cells, the content of these conjugated fatty acids in natural products is usually less than 1%. In contrast, CLN is present in high amounts in some seed oils. Takagi and Itabashi[6] have reported the occurrence of calendic acid (8t,10t,12c-18:3; 62.2%) in pot marigold seed oil, punicic

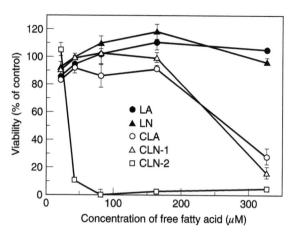

FIGURE 13.2 Cytotoxic effect of LA, CLA, LN, and isomerized linolenic acid (CLN-1 and CLN-2) on U-937 cells[46]. Viability was assessed spectrophotometrically by using WST-1 reagent. Each point is the mean + SD of three values obtained from separate cultures.

acid (c9c,11t,13c-18:3; 83.0%) in pomegranate seed oil, α-eleostearic acid (9c,11t,13t-18:3) in tung (67.7%) and bitter gourd (56.2%) seed oils, as well as catalpic acid (9t,11t,13c-18:3; 42.3%) in catalpa seed oil. Among plant seeds containing CLN, bitter gourd and pomegranate are edible plants and catalpa is used in Chinese medicine. Bitter gourd in particular is an important cultivated food crop in Asia. Therefore, it is interesting to study the physiological effects of CLN from these seed oils.

Suzuki et al.[46] reported the cytotoxic effect of conjugated trienoic fatty acids of natural origin on mouse tumor (SV-T2) cells. Fatty acid from pot marigold (8t,10t,12c-18:3; 33.4%) had no effect on either cell line, but other kinds of fatty acids from seed oils were cytotoxic to SV-T2 cells. The same effect was observed in the case of human monocytic leukemia cells (U-937). The fatty acids from seed oils of pomegranate, tung, and catalpa showed cytotoxity at concentrations exceeding 10 μM for pomegranate and tung, and 20 μM for catalpa. However, pot marigold fatty acids were cytotoxic to U-937 cells at concentrations exceeding 50 μM. The study on the effect of each CLN isomer confirmed isomeric specificity[46] and that 9,11,13-18:3 were more cytotoxic than 8,10,12-18:3 isomers (Figure 13.3). In addition, it was revealed that the difference in *cis/trans* configuration among 9,11,13-CLN isomers did not affect their cytotoxicity.

The cytotoxicity of each 9,11,13-CLN isomer was completely inhibited by the addition of BHT as an antioxidant to fatty acid at a 1:4 mole ratio. As described earlier, the mechanism of the cytotoxity of CLN is presumed to involve lipid peroxidation. Furthermore, superior cell toxicity exerted by 9,11,13-18:3 compared to 8,10,12-18:3 could be attributed to a difference in oxidative stability of these CLN isomers. The oxidative stability of four kinds of CLN isomers and LN in an aqueous phase is depicted in Figure 13.4. The stabilities of three kinds of 9,11,13-18:3 isomers were the same, but lower than that of the 8t,10t,12c-18:3 isomer. Hence, the higher cytotoxity of 9,11,13-18:3 as compared to 8,10,12-18:3 may partly be due to their different susceptibilities to peroxidation[46].

Involvement of lipid peroxidation was also noticed in the cytotoxic effect of CLN on the growth of human colon cancer cells (DLD-1, HepG2, A549, HL-60)[47]. A fatty acid mixture rich in CLN (α-eleostearic acid; 9c,11t,13t-18:3) showed a stronger dose-dependent inhibitory effect than two kinds of CLA isomers (9c,11t-18:2 and 10t,12c-18:2) on cultured human colon cancer cells by activating the apoptotic pathway. In addition, caspases, which are apoptosis-promoting factors, were activated by the addition of an α-eleostearic acid-rich fatty acid mixture to DLD-1 cells. The treatment of DLD-1 cells with the fatty acid mixture increased the amounts of membrane phospholipid peroxidation as reflected in TBARS values. In contrast, the addition of α-tocopherol suppressed the

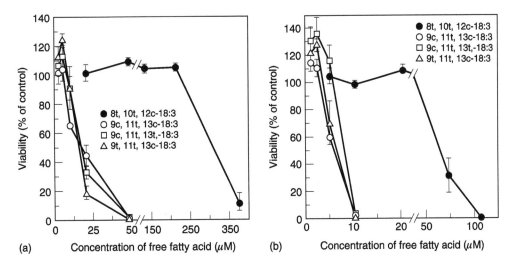

FIGURE 13.3 Cytotoxic effect of each CLN isomer on SV-T2 cells (A) and on U-937 cells (B)[46]. Viability was assessed spectrophotometrically by using WST-1 reagent. Each point is the mean + SD of three values obtained from separate cultures.

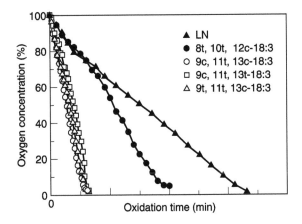

FIGURE 13.4 Comparison of oxidative stabilities of CLN isomers and LN in an aqueous dispersion. Fatty acids (1.08 mg/3 mL) were incubated with 2,2′-azobis(2-amidinopropane)dihydrochloride (AAPH) (1.0 mM) in 3 mL of phosphate buffer containing 0.1 wt% of Triton X-100. Oxidation was monitored by measuring oxygen uptake during oxidation.

oxidative stress and induction of apoptosis by α-eleostearic acid-rich fatty acid mixture. Hence, it can be concluded that CLN induces apoptosis in tumor cells via lipid peroxidation.

The higher anticarcinogenic effect of CLN compared to CLA has further been confirmed by another study[48]. A fatty acid mixture from bitter gourd seed oil (BGO-FFA), which contained more than 60% α-eleostearic acid (9c,11t,13t-18:3), exhibited stronger growth inhibition and apoptosis induction in colon cancer cells (DLD-1, HT-29, Caco-2) than CLA (9c,11t-18:2). This study also showed that the inhibitory effect of CLN on the growth of colon cancel cells was related to the regulation of peroxisome proliferator-activated receptor (PPAR)γ. PPARγ has been focused on as

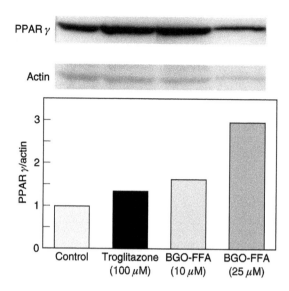

FIGURE 13.5 PPARγ expression in Caco-2 cells treated with BGO-FFA and troglitazone[48]. Cells were preincubated for 24 h and then BGO-FFA and troglitazone was added into the cultured medium. Cells were incubated for an additional 24 h. Cellular protein was extracted, and levels of PPARγ were detected using Western blot analysis. The results were densitometrically analyzed using Scion Image (Scion Corporation, USA) and normalized against the Acthin signal. Relative PPARγ protein was assigned the control ratio to a value of 1.0.

one of the target molecules to prevent cancer[49–51]. McCarty[52] has suggested that PPARγ activation by CLA is also associated with its cancer-retardant activity[52]. PPARγ is predominantly found in adipose tissue[53], and is also expressed in colon[54,55], breast[56], and prostate cancer cells[57]. Recent research has shown that PPARγ activation induces growth arrest and apoptosis in colon[58,59] and breast cancer[60]. It has been reported that troglitazone, a specific PPARγ ligand, effectively suppresses the development of aberrant crypt foci (ACF), which are putative precursor lesions for colonic adenocarcinoma, induced by the treatment with azoxymethane (AOM) and dextran sodium sulfate in rats[61,62]. PPARγ ligands such as troglitazone and 15-d-prostaglandin (PG) J$_2$ were reported to cause growth inhibition and induce apoptosis in cancer cells[58,63,64]. The effect of troglitazone was also found in the growth inhibition of human colon cancer cells with apoptosis, in which PPARγ expression in Caco-2 cells increased by 1.5-fold (Figure 13.5) compared with untreated cells[48]. BGO-FFA enhanced the expression of PPARγ protein in a dose-dependent manner. The expression level of PPARγ protein in Caco-2 cells increased by approximately three-fold compared to control cells after incubation with 25 μM of BGO-FFA for 24 h, indicating a higher ligand activity on PPARγ than troglitazone. Two possible mechanisms of the anticarcinogenic activity of CLN can be hypothesized: these are induction of apoptosis via lipid peroxidation and upregulation of some gene expressions by PPARγ. Despite all this, the mechanism of action of CLN still remains unknown. Further studies are required to investigate the interaction of PPARγ signaling, gene expressions, and lipid peroxidation in the cytotoxic effect of CLN on the growth of cancer cells.

13.4.2.2 CLN as Anticancer Nutrient: *In Vivo* Studies

One third of human cancers might be associated with dietary habits and lifestyle and the amount and type of dietary fat consumed are of particular importance[65–68]. Studies in humans and experimental animals have indicated the protective effect of fish oils which are rich in n-3 polyunsaturated fatty acids such as EPA and DHA; the mechanism of protection is thought to be mainly related to their interference with biosynthesis of two-series of prostaglandins from arachidonic acid[69]. Other

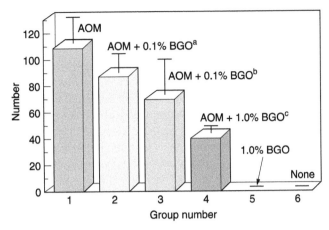

FIGURE 13.6 Incidence of ACF in rats treated with AOM and/or CLN[75]. a–c: significantly different from group 1 (a: $P < 0.05$; b $P < 0.01$; c $P < 0.001$).

fatty acids may also have antitumorigenic properties. CLN from bitter gourd seed oil (BGO) showed a significant reduction in the frequency of colonic ACF in rat (Figure 13.6) and ACF is the precursor of colon carcinogenesis[75]. All rats belonging to groups 1 to 4 were initiated with AOM and developed ACF. The dietary administration of BGO caused significant inhibition of ACF formation. A significant effect was found in the 0.01% BGO diet, which contained 0.006% CLN (9c,11t,13t-18:3). In addition, there were significant decreases in the total number of aberrant crypts (ACs) per colon and the number of ACs per focus. In groups 5 and 6, which were given a CLN diet alone and unreacted, respectively, there were no microscopically observable changes, including ACF, in colonic morphology. This study also showed that diets containing BGO support normal growth in rats without any adverse effects and histology of liver revealed no morphological alterations, such as fatty liver.

The study on the inhibitory effect of CLN from BGO on rat colonic ACF formation[75] also revealed that the proliferating cell nuclear antigen (PCNA)-labeling indices in ACF and normal-appearing crypts were decreased by dietary feeding of CLN. Similar findings were reported for retinoids and some other natural compounds[76–79]. Cell proliferation plays an important role in multistage carcinogenesis with multiple genetic changes. Furthermore, feeding of CLN enhanced apoptotic cells in ACF without affecting the surrounding normal-appearing crypts. These results indicate that the inhibitory effect of CLN could, in part, be due to modification of cell proliferation and apoptosis induction, which may be derived from gene regulation by the CLN molecule.

The chemopreventive ability of BGO on rat colon cancer was also confirmed in a long-term *in vivo* assay[80]. Dietary administration of BGO rich in CLN (9c,11t,13t-18:3) significantly inhibited the development of colonic adenocarcinoma induced by AOM in male F344 rats without causing any adverse effects. In addition, a significant reduction in the multiplicities of colorectal carcinoma (number of carcinomas/rats) in rats given BGO-containing diets at all dose levels (0.01, 0.1, or 1%) was found when compared to the AOM-administered group alone. The other CLN isomer (9c,11t,13c-18:3) from pomegranate seed oil (PGO) also showed a chemopreventive effect on rat colon cancer induced by AOM[81]. Dietary feeding of PGO suppressed progression of adenoma to malignant neoplasm in the postinitiation phase of colon cancer. In these studies[80,81], the protective effect of BGO and PGO against colon carcinogenesis was not dose-dependent.

Dietary feeding of BGO and PGO also enhanced PPARγ expression in nonlesional colonic mucosa[80,81]. This result is consistent with that of an *in vitro* study[48] in which CLN (9c,11t,13t-18:3) induced apoptosis in human colon cancer cells and enhanced PPARγ expression in the cells.

Synthetic ligands for PPARγ and PPARγ effectively inhibit AOM-induced ACF in rats[61,62]. Thus, it may be possible that BGO and PGO suppress colon carcinogenesis by means of altering PPARγ expression in colonic mucosa.

The antioxidant activity of CLN may be another possible explanation for the inhibitory effect of colon carcinogenesis by feeding of BGO and PGO diet. Dhar et al.[87] reported that CLN (9c,11t,13t-18:3) from BGO acts as an antioxidant. Rat plasma lipid peroxidation was significantly lower in the group supplemented with BGO as compared with the control group (sunflower oil supplementation). Feeding of BGO significantly reduced lipoprotein peroxidation and erythrocyte ghost membrane lipid peroxidation. In compounds with more than two conjugated double bonds, conjugation increases the rate of lipid peroxidation[88] and polyunsaturated conjugated fatty acids are more susceptible to lipid oxidation than those of the corresponding nonconjugated fatty acids containing the same number of double bonds. Thus, in the *in vivo* study conjugated trienoic fatty acids are also likely to be more rapidly oxidized than linoleates by picking up more free radicals, thereby eliminating or reducing the formation of hydroperoxides.

In another study, Tsuzuki et al.[47] demonstrated that the anticarcinogenic effect of CLN was directly associated with lipid peroxidation. They transplanted human colon cancer cells (DLD-1) into nude mice and CLA (9c,11t- and 10t,12c-18:2) and CLN (9c,11t,13t-18:3) were administered to animals. Tumor growth was suppressed by supplementation of CLA and CLN, and the extent of growth suppression was in the order CLN > 9c,11t-CLA > 10t,12c-CLA. Furthermore, DNA fragmentation was enhanced and lipid peroxidation was increased in tumor cells of the CLN-fed mouse. The same result was also observed in a study on the effect of CEPA[89]. This study indicated that CEPA had an extremely strong antitumor effect on tumor cells that were transplanted into nude mice, as compared to EPA and CLA. In the tumor cells of the mice, the membrane phospholipid hydroperoxide and TBARS levels showed an increase when mice were fed with CEPA, suggesting the involvement of lipid peroxidation in the anticarcinogenic effect of CEPA.

13.4.2.3 Does the Anticancer Effect of CLN Come from Bioconversion of CLN to CLA?

Each CLN isomer separated from plant seed oils showed inhibitory effects on some kinds of cancer cells[46]. The cytotoxity of CLN could be attributed to its conjugated triene structure[45] and this effect was affected by the position of the conjugated double bonds[46]. Furthermore, dietary administration of CLN (9c,11t,13t-18:3) from BGO caused a significant and dose-dependent reduction in the frequency of azoxymethane-induced colonic ACF in F344 rats, which are precursor lesions for rat colon cancer[75]. CLN administration lowered the proliferating cell nuclear antigen index and induced apoptosis in ACF[75]. Long-time bioassay on the CLN (9c,11t,13t-18:3 and 9c,11t,13c-18:3) from BGO and PGO also confirmed the anticarcinogenic effect of CLN from seed oils[80,81]. These findings suggested possible chemopreventive activity of CLN in the early phase of colon tumorigenesis through modulation of cryptal cell proliferation activity and/or apoptosis.

In a study on the inhibitory effect of CLN from BGO and PGO, CLA isomer (9c,11t-18:2) was found in the liver lipids of rats fed with CLN. Preliminary studies indicated that CLA was a powerful anticancer agent in the rat mammary tumor model with an effective range of 0.1 to 1.0% in the diet[70,71,90]. Therefore, CLA found in rat liver lipids may be correlated with the anticancer effect of CLN (9c,11t,13t-18:3 and 9c,11t,13c-18:3). A possible pathway for the formation of CLA in liver lipids of rats fed BGO and PGO is the bioconversion of CLN (9c,11t,13t-18:3 and 9c,11t,13c-18:3) to CLA (9c,11t-18:2)[91,92].

Table 13.3 and Table 13.4 show the fatty acid composition of total lipids (TL) and phosphatidylcholine from the liver of the rats fed diets containing CLA and four kinds of CLN originating from BGO, PGO, catalpa seed oil (CTO) (9t,11t,13t-18:3), and pot marigold seed oil (PMO) (8t,10t,12c-18:3), respectively. When the dietary fat containing CLN or CLA was supplemented to animals, a part of the basal dietary oil (soybean oil) was substituted by triacylglycerol containing

TABLE 13.3
Fatty Acid Composition (%) of Rat Liver TL

	Group					
Fatty acid	Control	CLA	PGO	BGO	CTO	PMO
16:0	19.6 ± 1.0[a]	22.3 ± 0.8[a]	21.1 ± 1.1[a]	20.2 ± 1.0[a]	20.8 ± 1.5[a]	19.8 ± 1.0[a]
18:0	17.9 ± 1.0[a,b]	19.0 ± 1.4[b]	16.2 ± 1.9[a,c]	16.8 ± 0.6[a,c]	16.0 ± 0.6[c]	15.9 ± 1.7[c]
18:1n-7	3.4 ± 0.2[a]	3.2 ± 0.5[a]	4.8 ± 0.7[b,c]	5.2 ± 0.6[b]	4.4 ± 0.5[c]	5.1 ± 0.6[b,c]
18:1n-9	7.7 ± 0.7[a]	7.5 ± 2.0[a]	8.2 ± 0.8[a]	7.9 ± 0.8[a]	7.3 ± 1.1[a]	8.9 ± 3.3[a]
18:2n-6	18.0 ± 1.6[a]	15.0 ± 1.0[b]	13.3 ± 1.1[b,c]	12.4 ± 1.5[c]	14.5 ± 1.7[b]	13.3 ± 0.6[c]
20:4n-6	16.8 ± 2.2[a,b]	15.5 ± 2.8[b,c]	17.5 ± 1.6[a,b]	20.0 ± 1.0[a]	18.8 ± 1.6[a]	19.4 ± 3.0[a]
22:6n-3	7.5 ± 1.5[a,b]	8.0 ± 1.3[a]	6.0 ± 0.8[b]	6.9 ± 0.9[a,b]	6.9 ± 0.8[a,b]	7.2 ± 0.8[a,b]
CLA(9c,11t)	ND	0.9 ± 0.2[b]	1.1 ± 0.3[c]	0.7 ± 0.1[b]	ND	ND
CLA(10t,12c)	ND	0.4 ± 0.1[b]	ND	ND	ND	ND
CLA(9t,11t)	ND	ND	ND	ND	1.0 ± 0.2[b]	ND
CLA(8t,10t)	ND	ND	ND	ND	ND	1.2 ± 0.1[b]
CLN	ND	ND	ND	ND	ND	ND

Note: Data are represented as means ± SE of seven rats. ND, not detected. Different letters indicate a significant difference (*P* < 0.01) within the same fatty acid.

TABLE 13.4
Fatty Acid Composition (%) of Rat Liver Phosphatidylcholine

	Group					
Fatty acid	Control	CLA	PGO	BGO	CTO	PMO
16:0	19.3 ± 0.6[a]	19.3 ± 0.7[a]	19.3 ± 0.8[a]	18.8 ± 0.9[a]	18.8 ± 1.0[a]	18.3 ± 0.8[a]
18:0	23.1 ± 1.3[a]	22.3 ± 1.0[a]	20.3 ± 1.6[b]	20.2 ± 1.1[b]	19.0 ± 1.0[b,c]	17.8 ± 1.2[c]
18:1n-7	3.1 ± 0.3[a]	2.7 ± 0.4[a]	4.2 ± 0.5[b,c]	4.9 ± 0.7[b,d]	3.9 ± 0.6[c]	5.0 ± 0.6[d]
18:1n-9	2.8 ± 0.2[a]	2.7 ± 0.2[a]	3.5 ± 0.4[b]	3.8 ± 0.3[b]	3.6 ± 0.3[b]	4.4 ± 0.6[c]
18:2n-6	13.4 ± 1.0[a,b]	12.5 ± 0.7[a]	12.4 ± 0.7[a]	12.6 ± 1.5[a]	14.6 ± 1.3[b]	12.6 ± 0.4[a]
20:4n-6	1.0 ± 1.2[a]	1.2 ± 2.0[a,b]	1.8 ± 1.6[c]	1.6 ± 1.4[b]	1.9 ± 1.4[c]	1.6 ± 1.2[b]
22:6n-3	10.6 ± 1.8[a,b]	12.2 ± 1.5[b]	9.1 ± 0.9[a,c]	7.0 ± 1.7[d]	6.4 ± 1.3[d]	8.5 ± 0.4[a,d]
CLA(9c,11t)	ND	0.2 ± 0.0[b]	0.4 ± 0.1[c]	0.3 ± 0.0[b]	ND	ND
CLA(10t,12c)	ND	0.3 ± 0.1[b]	ND	ND	ND	ND
CLA(9t,11t)	ND	ND	ND	ND	1.1 ± 0.2[b]	ND
CLA(8t,10t)	ND	ND	ND	ND	ND	1.1 ± 0.1[b]
CLN	ND	ND	ND	ND	ND	ND

Note: Data are represented as means ± SE of seven rats. ND, not detected. Different letters indicate a significant difference (*P* < 0.01) within the same fatty acid.

each CLN or CLA isomer to give about 20% CLA or CLN concentration in the dietary fat, which corresponded to about 1.4% (w/w) of the diet. As shown in Table 13.3 and Table 13.4, no CLN isomer was detected in the liver TL and phosphatidylcholine of the rats fed CLN diets, although dietary fat in PGO, BGO, CTO, and PMO diets contained 19.1 to 21.3% CLN. CLA was found in these lipids. GC and GC-MS analysis corroborated the presence of 9c,11t-CLA isomer in the liver lipids of the PGO and BGO diet-fed rats, whereas the occurrence of 9t,11t-CLA isomer and 8t,10t-CLA isomer

TABLE 13.5
Changes in Fatty Acid Composition (%) of TL from Caco-2 Cells After Incubation with or without 9c,11t,13t-CLN

| | Incubation time | | | | | |
| | 6 h | | 24 h | | 48 h | |
Fatty acid	CLN(–)	CLN(+)	CLN(–)	CLN(+)	CLN(–)	CLN(+)
14:0	2.5	2.1	2.9	2.4	2.1	2.6
16:0	21.7	18.9	22.8	19.3	20.7	20.5
18:0	7.3	7.6	6.6	8.3	5.7	9.3
16:1n-7	8.8	6.6	11.3	7.4	11.9	5.9
18:1n-7	12.7	11.0	14.2	11.0	14.8	10.4
18:1n-9	23.3	23.7	25.4	28.9	25.7	33.3
18:2n-6	1.9	4.0	1.5	4.0	1.0	6.3
20:4n-6	5.9	4.9	4.0	3.0	2.7	3.0
22:6n-3	3.3	2.8	2.2	1.7	2.0	ND
CLA(9t,11t)	ND	0.7	ND	2.4	0.1	3.6
CLA(9t,11t,13t)	ND	3.9	ND	2.3	ND	ND

Note: ND, not detected.

was found in the liver lipids of rats fed with CTO diet and PMO diet, respectively. CLA (9c,11t-18:2) was also found in the liver and adipose tissue TL of BGO-fed mice[92]. Therefore, the formation of the CLA isomer in the tissue lipids of the CLN diet-fed animals could be explained by the enzymatic conversion of CLN to CLA, namely the biohydrogenation at carbon 13 or carbon 12 double bonds. This metabolic pathway was confirmed by the fact that no CLN was detected in the liver lipids of the CLN diet-fed animals and the CLA concentration in the liver and adipose tissue lipids of the mice increased with increasing CLN content in the diet.

As shown in Table 13.3, there was little difference in the content of each CLA isomer in the liver TL of the rats fed with CLN diets, suggesting that the bioconversion rate of each CLN isomer was the same as that of the biohydrogenation rate at 13c double bond of 9c,11t,13t-18:3 and 9t,11t,13c-18:3, at 13t double bond of 9t,11t,13t-18:3, and at 12c double bond of 8t,10t,12c-18:3. The contents of 9t,11t-18:2 and 8t,10t-18:2 in the liver phosphatidylcholine of rats fed CTO and PMO diets were higher than those of 9c,11t-18:2 of rats fed PMO and BGO diets. This suggests a higher incorporation rate of t/t CLA isomer into the cell membrane lipids.

Significant differences in tissue fatty acid composition, other than CLA and CLN, were also observed between the dietary treatment groups. In particular, dietary CLA and CLN affected the composition of n-6 fatty acids in the tissue lipids of animals. Rats fed CLA and CLN had a significantly lower concentration of 18:2n-6 in liver TL than those kept on control diets (Table 13.4). The same effect was observed in the mouse liver TL and the mouse adipose tissue TL[92]. This effect of CLN may be due to the corresponding CLA isomer originating from CLN. Li and Watkins[93] reported that dietary CLA decreased the concentration of 18:2n-6. However, the effects of dietary CLN on other fatty acid compositions were not always consistent with those of CLA. For example, the effects of CLN on 18:0, 18:1, 20:4n-6, 22:6n-3 in rat liver TL were different from those of CLA (Table 13.4). It has been reported that dietary CLA decreased the concentration of 18:1, but increased the concentration of 18:0 and 22:6n-3 in rat liver TL[93]. In contrast, in the liver phosphatidylcholine of rats fed CLA and CLN, no significant decrease in 18:2n-6 was observed, while CLN treatment induced significantly higher levels of 20:4n-6 in rat liver phosphatidylcholine than control and CLA treatments[92].

Preliminary studies indicated that CLA was an anticancer agent in the rat mammary tumor model[70,71,90]; 9c,11t-CLA isomer is considered to be one of the active constituents. As shown in Table 13.3, only 9c,11t-CLA isomer was accumulated in the liver lipids of rats and mice fed BGO and PGO diets. The amount of 9c,11t-CLA in the liver lipids was comparable to that of the rats fed a CLA diet; therefore, the potential anticancer effects of 9c,11t,13t-18:3 in BGO[80] and 9c,11t,13c-18:3 in PGO[81] are expected to be partly due to the presence of 9c,11t-18:2 isomer derived from both 9c,11t,13t-18:3 in BGO and 9c,11t,13c-18:3 in PGO. Furthermore, as shown in Table 13.3, the contents of linoleic acid (18:2n-6) in the liver lipids of rats fed CLN-containing diets were significantly lower than those of rats fed diets without CLN. This reduction in the content of linoleic acid may also contribute to the inhibitory effect of CLN on colon carcinogenesis. However, judging from the powerful inhibitory activity of CLN at lower dose levels found in animal studies[75,80,81], other factors such as direct action of CLN as PPARγ ligand[80,81] and acceleration of lipid peroxidation followed by apoptosis[47] should also be considered.

13.5   CONCLUSIONS

The fact that CLA has potent beneficial health and biological effects is indisputable[2,3]. Also, CLA has potential for use as a functional food component and in nutraceuticals. CLA is known to occur naturally in dairy products, but its concentration is usually less than 1%[1]; thus, alkaline isomerized linoleic acid is generally used as a CLA source. However, the CLA obtained by isomerization of linoleic acid is a mixture of many kinds of positional and geometrical isomers. Seed oils such as BGO, PGO, CTO, and PTO contain one specific CLN isomer. Particularly, pomegranate and bitter gourd are widely cultivated in the world and the lipid content of pomegranate seed and bitter seed are more than 20 and 40% on a dry weight basis, respectively. As CLN gives the same benefits as CLA, seed oils rich in CLN would be very interesting sources for use in functional foods and neutraceuticals.

With increasing interest in CLN, more information on the oxidative stability of CLN is required for application of CLN or CLN-containing seed oil. Oxidation of CLN not only produces rancid flavors in foods but can decrease their nutritional quality and safety. Therefore, it is important to establish effective control methods against oxidation of conjugated CLN. CLN is more susceptible to oxidation than its corresponding nonconjugated fatty acid, α-linolenic acid[46,88]. Comparative studies on the oxygen consumption, peroxide formation, and polymer formation[88] clearly showed that main oxidation products of CLN were dimers, but less peroxide was accumulated during oxidation. Although CLN is oxidatively less stable than α-linolenic acid and linoleic acid, the addition of antioxidants, tocopherol and Trolox, effectively inhibited the oxidation. The hydrophilic antioxidant Trolox was more effective than tocopherol in this case[88]. Especially, in the early stage of oxidation, the oxidation of CLN was better inhibited than that of α-linolenic acid. These results suggest that antioxidants are very important in the application of oils containing CLN for use in functional foods and neutraceutical formulations.

Finally, as can be seen from various works covered in this review, conjugated polyenes have a great potential as ingredients in functional/health food formulations. Since information on the behavior of these fatty acids in physiological system is needed to enhance the credibility of these conjugated polyenes for health-related uses, further research is needed in this direction. Aquatic plants offer themselves as suitable candidates for such further research as terrestrial plant seed oils are already being explored for such possibilities.

REFERENCES

1. Fritsche, J. and Steinhart, H., Analysis, occurrence, and physiological properties of trans fatty acids (TFA) with particular emphasis on conjugated linoleic acid isomers (CLA): a review, *Fett/Lipid*, 100, 190–21, 1998.

2. Yurawecz, M.P., Mossoba, M.M., Kramer, J.K.G., Pariza, M.W., and Nelson, G.J., Eds., *Advances in Conjugated Linoleic Acid Research*, Vol. 1, AOCS Press, Champaign, IL, 1999.

3. Sébédio, J.-L., Christie, W.W., and Adlof, R., Eds., *Advances in Conjugated Linoleic Acid Research*, Vol. 2, AOCS Press, Champaign, IL, 2003.

4. Badami, R.C. and Patil, K.B., Structure and occurrence of unusual fatty acids in minor seed oils, *Prog. Lipid Res.*, 19, 119–153, 1981.

5. Smith, Jr., C.R., Occurrence of unusual fatty acids in plants, *Prog. Chem. Plants Other Lipids*, 11, 137–177, 1971.

6. Takagi, T. and Itabashi, Y., Occurrence of mixtures of geometrical isomers of conjugated octadecatrienoic acids in some seed oils: analysis by open tubular gas liquid chromatography and high performance liquid chromatography, *Lipids*, 16, 546–551, 1981.

7. Suzuki, R., Arato, S., Noguchi, R., Miyashita, K., and Tachikawa, O., Occurrence of conjugated linolenic acid in flesh and seed of bitter gourd, *J. Oleo Sci.*, 50, 753–758, 2001.

8. Hopkins, C.Y. and Chisholm, M.J., Isolation of natural isomer of linoleic acid from seed oil, *J. Am. Oil Chem. Soc.*, 41, 42–44, 1964.

9. Bagby, M.O., Smith, Jr., C.R., and Wolff, I.A., Stereochemistry of α-parinaric acid from *Impatiens edgeworthi* seed oil, *Lipids*, 1, 263–267, 1966.

10. Spitzer, V., Marx, F., and Pfeilsticker, K., Electron impact mass spectra of the oxazoline derivatives of some conjugated diene and triene C_{18} fatty acids, *J. Am. Oil Chem. Soc.*, 71, 873–876, 1994.

11. Warnaar, F., Deca-2,4,6-trienoic acid, a new conjugated fatty acid isolated from the latex of *Euphorbia pulcherrima* willd, *Lipids*, 12, 707–710, 1977.

12. Lopez, A. and Gerwick, W.H., Two new icosapentaenoic acids from the temperate red seaweed *Ptilota filicina* J. Agardh, *Lipids*, 22, 190–194, 1987.

13. Wise, M.L., Hamberg, M., and Gerwick, W.H., Biosynthesis of conjugated triene containing fatty acids by novel isomerase from the red marine algae *Ptilota filicina*, *Biochemistry*, 33, 15223–15232, 1994.

14. Bhaskar, N., Kinami, T., Miyashita, K., Park, S., Endo, Y., and Fujimoto, K., Occurrence of conjugated polyenoic fatty acids in seaweeds from the Indian ocean, *Z. Naturforsch.*, 59c, 310–314, 2004.

15. Burgess, J.R., De la Rosa, R.I., Jacobs, R.S., and Butler, A., A new eicosapentaenoic acid formed from arachdonic acid in the coralline red algae *Bassiella orbigniana*, *Lipids*, 26, 162–165, 1991.

16. Hamberg, M., Metabolism of 6,9,12-octadecatrienoic acid in the red alga *Lithothamnion corrallioides*: mechanism of formation of a conjugated tetraene fatty acid, *Biochem. Biophys. Res. Commun.*, 188, 1220–1227, 1992.

17. Mikhailova, M.V., Bemis, D.L., Wise, M.L., Gerwick, W.H., Norris, J.N., and Jacobs, R.S., Structure and biosynthesis of novel conjugated polyene fatty acids from the marine green alga *Anadyomene stellata*, *Lipids*, 30, 583–589, 1995.

18. Zheng, W., Wise, M.L., Wyrick, A., Metz, J.G., Yuan, L., and Gerwick, W.H., Polyenoic fatty acid isomerase from the marine algae *Ptilota filicina*: protein characterization and functional expression of the cloned cDNA, *Arch. Biochem. Biophysics*, 401, 11–20, 2002.

19. Gunstone, F.D., Hamilton, R.J., Padley, F.B., and Qureshi, M. I., Glyceride studies: V. The distribution of unsaturated acyl groups in vegetable triglycerides, *J. Am. Oil Chem. Soc.*, 42, 965–970, 1965.

20. Morris, L.J. and Marshall, M.O., Occurrence of stearoilic acid in *Santalaceae* seed oils, *Chem. Ind.*, 460–461, 1966.

21. Morris, L.J. and Marshall, M.O., Occurrence of *cis,trans*-linoleic acid in seed oils, *Chem. Ind.*, 1493–1494, 1966.

22. Hopkins, C.Y. and Chisholm, M.J., A survey of the conjugated fatty acids of seed oils, *J. Am. Oil Chem. Soc.*, 45, 176–182, 1968.

23. Crombie, L. and Holloway, S.J., Origins of conjugated triene fatty acids. The biosynthesis of calendic acid by *Calendula offcinalis*, *J. Chem. Soc. Chem. Commun.*, 15, 953–955, 1984.

24. Crombie, L. and Holloway, S.J., The biosynthesis of calendic acid, octadeca-(8E,10E,12Z)-trienoic acid by developing marigold seeds: origins of (E,E,Z) and (Z,E,Z) conjugated triene acids in higher plants, *J. Chem. Soc. Perkin Trans.* 1, 2425–2434, 1985.

25. Gerwick, W.H. and Bernart, M.W., Eicosanoids and related compounds from marine algae, in *Marine Biotechnology*, Vol. I, Attaway, D.H. and Zabrosky, O.R., Eds., Plenum Press, New York, 1993, pp. 101–152.

26. Gerwick, W.H., Structure and biosynthesis of marine algal oxylipins, *Biochim. Biophys. Acta*, 1211, 243–255, 1994.
27. Moore, B.S., Biosynthesis of marine natural products: microorganisms and macroalgae, *Nat. Prod. Rep.*, 16, 653–674, 1999.
28. Lee, M., Lenman, M., Banaś, A., Bafor, M., Singh, S., Schweizer, M., Nilsson, R., Liljenberg, C., Dahlqvist, A., Gummeson, P.-O., Sjödahl, S., Green, A. and Stymne, S., Identification of non-heme diiron proteins that catalyze triple bond and epoxy group formation, *Science*, 280, 915–918, 1998.
29. Cahoon, E.B., Carlson, T.J., Ripp, K.G., Shweiger, B.J., Cook, G.A., Hall, S.E., and Kinney, A.J., Biosynthetic origin of conjugated double bonds: production of fatty acid components of high-value drying oils in transgenic soybean embryos, *Proc. Natl. Acad. Sci. USA*, 96, 12935–12940, 1999.
30. Liu, L., Hammond, E.G., and Nikalou, B.J., In vivo studies of the biosynthesis of α-eleostearic acid in the seed of *Momordica charantia* L., *Plant Physiol.*, 113, 1343–1349, 1997.
31. Cahoon, E.B., Ripp, K.G., Hall, S.E., and Kinney, A.J., Formation of conjugated delta8, delta10-double bonds by delta12-oleic-acid desaturase-related enzymes: biosynthetic origin of calendic acid, *J. Biol. Chem.*, 276, 2637–2643, 2001.
32. Dyer, J.M., Chapital, D.C., Kuan, J.-C.W., Mullen, R.T., Turner, C., McKeon, T.A., and Pepperman, A.B., Molecular analysis of a bifunctional fatty acid conjugase/desaturase from tung. Implications for the evolution of plant fatty acid diversity, *Plant Physiol.*, 130, 2027–2038, 2002.
33. Cornelius, A.S., Yerram, N.R., Kratz, D.A., and Spector, A.A., Cytotoxic effects of cis-parinaric acid in cultured malignant cells. *Cancer Res.*, 51, 6025–6030, 1991.
34. Igarashi, M. and Miyazawa, T., Do conjugated eicosapentaenoic acid and conjugated docosahexaenoic acid induce apoptosis via lipid peroxidation in cultured human tumor cells?, *Biochem. Biophys. Res. Comm.*, 270, 649–656, 2000.
35. Bégin, M.E., Ells, G., Das, U.N., and Horrobin, D.F., Different killing of human carcinoma cells supplemented with n-3 and n-6 polyunsaturated fatty acids, *J. Natl. Cancer Inst.*, 77, 1053–1062, 1986.
36. Bégin, M.E., Ells, G., and Horrobin, D.F., Polyunsaturated fatty acid-induced cytotoxicity against tumor cells and its relationship to lipid peroxidation, *J. Natl. Cancer Inst.*, 80, 188–194, 1988.
37. Das, U.N., Gamma-linolenic acid, arachidonic acid, and eicosapentaenoic acid as potential anticancer drugs, *Nutrition*, 6, 429–434, 1990.
38. Cave, Jr., W.T., Dietary n-3 (ω-3) polyunsaturated fatty acid effects on animal tumorigenesis, *FASEB J.*, 5, 2160–2166, 1991.
39. Esterbauer, H., Cytotoxicity and genotoxicity of lipid-oxidation products, *Am. J. Clin. Nutr.*, 57, 779S–786S, 1993.
40. Grune, T., Siems, W.G., Zollner, H., and Esterbauer, H., Metabolism of 4-hydroxynonenal, a cytotoxic lipid peroxidation product, in Ehrlich mouse ascites cells at different proliferation stages, *Cancer Res.*, 54, 5231–5235, 1994.
41. Sandstrom, P.A., Tebbey, P.W., Van Cleave, S., and Buttke, T.M., lipid hydroperoxides induce apoptosis in T cells displaying a HIV-associated glutathione peroxidase deficiency, *J. Biol. Chem.*, 269, 798–801, 1994.
42. Sandstrom, P.A., Pardi, D., Tebbey, P.W., Dudek, R.W., Terrian, D.M., Folks, T.M., and Buttke, T.M., Lipid hydroperoxide-induced apoptosis: lack of inhibition by Bcl-2 over-expression. *FEBS Lett.*, 365, 66–70, 1995.
43. Aoshima, H., Satoh, T., Sakai, N., Yamada, M., Enokiko, Y., Ikeuchi, T., and Hatanaka, H., Generation of free radicals during lipid hydroperoxide-triggered apoptosis in PC12h cells, *Biochem. Biophys. Acta*, 1345, 35–42, 1997.
44. Ji, C., Rouzer, C.A., Marnett, L.J., and Pietenpol, J.A., Induction of cell cycle arrest by the endogenous product of lipid peroxidation, malondialdehyde, *Carcinogenesis*, 19, 1275–1283, 1998.
45. Igarashi, M. and Miyazawa, T., Newly recognized cytotoxic effects of conjugated trienoic fatty acids on the cultured human tumor cells, *Cancer Lett.*, 148, 173–179, 2000.
46. Suzuki, R., Noguchi, R., Ota, T., Abe, M. Miyashita, K., and Kawada, T., Cytotoxic effect of conjugated trienoic fatty acids on mouse tumor and human monocytic leukemia cells, *Lipids*, 36, 477–482, 2001.
47. Tsuzuki, T., Tokuyama, Y., Igarashi, M., and Miyazawa, T., Tumor growth suppression by α-eleostearic acid, a linolenic acid isomer with a conjugated triene system, via lipid peroxidation, *Carcinogenesis*, 25, 1417–1425, 2004.

48. Yasui, Y., Kudo, M., Suzuki, R., Hosokawa, M., Kohno, H., Tanaka, T., and Miyashita, K., Conjugated linolenic acid inhibits cell growth an induce apoptosis and PPARγ expression in human colon cancer cell lines, *Cancer Lett.*, submitted.

49. Sporn, M.B., Suh, N., and Mangelsdorf, D.J., Prospects for prevention and treatment of cancer with selective PPARγ modulators (SPARMs), *Trends Mol. Med.*, 7, 395–400, 2001.

50. Gupta, R.A. and Dubois, R.N., Controversy: PPARγ as a target for treatment of colorectal cancer, *Am. J. Physiol.*, 283, G266–G269, 2002.

51. Girnum, G.D., Smith, W.M., Drori, S., Sarraf, P., Mueller, E., Eng, C., Nambiar, P., Rosenberg, D.W., Bronson, R.T., Edelmann, W., Kucherlapati, R., Gonzalez, F.J., and Spiegelman, B.M., APC-dependent suppression of colon carcinogenesis by PPARγ. *Proc. Natl. Acad. Sci. USA*, 99, 13771–13776, 2002.

52. McCarty, M.F., Activation of PPARgamma may mediate a portion of the anticancer activity of conjugated linoleic acid, *Med Hypotheses*, 55, 187–188, 2000.

53. Tontonoz, P., Hu, E., and Speigelman, B.M., Stimulation of adipogenesis in fibroblasts by PPAR gamma 2, a lipid-activated transcription factor, *Cell*, 79, 1147–1156, 1994.

54. Dubois, R.N., Gupta, R., Brockman, J., Reddy, B.S., Krakow, S.L., and Lazar, M.A., The nuclear eicosanoid receptor, PPARγ, is aberrantly expressed in colonic cancers, *Carcinogenesis* 19, 49–53, 1998.

55. Kitamura, S., Miyazaki, Y., Shimomura, Y., Kondo, S., Kanayama, S., and Matsuzawa, Y., Peroxisome proliferator-activated receptor γ induces growth arrest and differentiation markers of human colon cancer cells, *Jpn. J. Cancer Res.*, 90, 75–80, 1999.

56. Elstner, E., Muller, C., Koshizuka, K., Williamson, E.A., Park, D., Asou, H., Ahintaku, P., Said, J.W., Heber, D., and Koeffler, H.P., Ligands for peroxisome proliferator-activated receptorgamma and retinoic acid receptor inhibit growth and induce apoptosis of human breast cancer cells in vitro and in BNX mice, *Proc. Natl. Acad. Sci. USA*, 95, 8806–8811, 1998.

57. Kubota, T., Koshizuka, K., Williamson, E.A., Asou, H., Said, J.W., Holden, S., Miyoshi, I., and Koeffler, H.P., Ligand for peroxisome proliferator-activated receptor gamma (troglitazone) has potent antitumor effect against human prostate cancer both in vitro and in vivo, *Cancer Res.*, 58, 3344–3352, 1998.

58. Yang, W.L. and Frucht, H., Activation of the PPAR pathway induces apoptosis and COX-2 inhibition in HT-29 human colon cancer cells, *Carcinogenesis* 22, 1379–1383, 2001.

59. Shimada, T., Kojima, K., Yoshimura, K., Hiraishi, H., and Terano, A., Characteristic of the peroxisome proliferators activated receptor γ (PPARγ) ligand induced apoptosis in colon cancer cells, *Gut*, 50, 658–664, 2002.

60. Clay, C.E., Namen, A.M., Atsumi, G., Willingham, M.C., High, K.P., Kute, T.E., Trimboli, A.J., Fonteh, A.N., Dawson, P.A., and Chilton, F.H., Influence of J series prostaglandins on apoptosis and tumorigenesis of breast cancer cells, *Carcinogenesis*, 20, 1905–1911, 1999.

61. Tanaka, T., Kohno, K., Yoshitani, S., Takashima, S., Okumura, A., Murakami, A., and Hosokawa, M., Ligands for peroxisome proliferator-activated receptors alpha and gamma inhibit chemically induced colitis and formation of aberrant crypt foci in rats, *Cancer Res.*, 61, 2424–2428, 2001.

62. Kohno, H., Yoshitani, S., Takashima, S., Okumura, A., Hosokawa, M., Yamaguchi, N., and Tanaka, T., Troglitazone, a ligand for peroxisome proliferator-activated receptor gamma, inhibits chemically-induced aberrant crypt foci in rats, *Jpn J. Cancer Res.*, 92, 396–403, 2001.

63. Tsubouchi, Y., Sano, H., Kawahito, Y., Mukai, S., Yamada, R., Kohno, M., Inoue, K., Hla, T., and Kondo, M., Inhibition of human lung cancer cell growth by the peroxisome proliferator-activated receptor-gamma agonists through induction of apoptosis, *Biochem. Biophys. Res. Commun.*, 270, 400–405, 2000.

64. Clay, C.E., Namen, A.M., Atumi, G., Willingham, M.C., High, K.P., Kute, T.E., Trimboli, A.J., Fonteh, A.N., Dawson, P.A., and Chilton, F.H., Influence of J series prostaglandins on apoptosis and tumorigenesis of breast cancer cells, *Carcinogenesis*, 20, 1905–1911, 1999.

65. Wynder, E.L., Kajitani, T., Ishikawa, S., Dodo, H., and Takano, A., Environmental factors of cancer of colon and rectum, *Cancer*, 23, 1210–1220, 1969.

66. Reddy, B.S., Tanaka, T., and Simi, B., Effect of different levels of dietary *trans* fat or corn oil on azoxymethane-induced colon carcinogenesis in F344 rats, *J. Natl. Cancer Inst.*, 75, 791–798, 1985.

67. Bartsch, H., Nair, J., and Owen, R.W., Dietary polyunsaturated fatty acids and cancers of the breast and colorectum: emerging evidence for their role as risk modifiers, *Carcinogenesis*, 20, 2209–2218, 1999.

68. Rao, C.V., Hirose, Y., Indranie, C., and Reddy, B.S., Modulation of experimental colon tumorigenesis by types and amounts of dietary fatty acids, *Cancer Res.*, 61, 1927–1933, 2001.

69. Reddy, B.S., Dietary fat and colon cancer: animal model studies, *Lipids*, 27, 807–813, 1992.

70. Ip, C., Chin, S.F., Scimeca, J.A., and Pariza, M.W., Mammary cancer prevention by conjugated dienoic derivative of linoleic acid, *Cancer Res.*, 51, 6118–6124, 1991.

71. Ip, C., Singh, M., Thompson, H.J., and Scimeca, J.A., Conjugated linoleic acid suppresses mammary carcinogenesis and proliferative activity of the mammary gland in the rat, *Cancer Res.*, 54, 1212–1215, 1994.

72. Liew, C., Schut, H.A.J., Chin, S.F., Pariza, M.W., and Dsshwood, R.H., Protection of conjugated linoleic acids against 2-amino-3-methylimidazo[4,5-*f*]quinoline-induced colon carcinogenesis in the F344 rat: a study of inhibitory mechanisms, *Carcinogenesis*, 16, 3037–3043, 1995.

73. Xu, M. and Dashwood, R.H., Chemoprevention studies of heterocyclic amine-induced colon carcinogenesis, *Cancer Lett.*, 143, 179–183, 1999.

74. Belury, M.A., Nickel, K.P., Bird, C.E., and Wu, Y., Dietary conjugated linoleic acid modulation of phorbol ester skin tumor promotion, *Nutr. Cancer*, 26, 149–157, 1996.

75. Kohno, H., Suzuki, R., Noguchi, R., Hosokawa, M., Miyashita, K., and Tanaka, T., Dietary conjugated linolenic acid inhibits azoxymethane-induced colonic aberrant crypt foci in rats, *Jpn. J. Cancer Res.*, 93, 133–142, 2002.

76. Kohno, H., Maeda, M., Honjo, S., Murakami, M., Shimada, R., Masuda, S., Simuda, T., Azuma, Y., Ogawa, H., and Tanaka, T., Prevention of colonic preneoplastic lesions by the β-cryptxanthin and hesperidin rich powder prepared from *Citrus unshiu* Marc. juice in male F344 rats, *J. Toxicol. Pathol.*, 12, 209–215, 1999.

77. Zheng, Y., Kramer, P.M., Olson, G., Lubet, R.A., Steele, V.E., Kelloff, G.J., and Pereira, M.A., Prevention by retinoids of azoxymethane-induced tumors and aberrant crypt foci and their modulation of cell proliferation in the of rats, *Carcinogenesis*, 18, 2119–2125, 1997.

78. Tanaka, T., Shimizu, M., Kohno, H., Yoshitani, H., Tsukio, Y., Murakami, A., Safitri, R., Takahashi, D., Yamamoto, K., Koshimizu, K., Ohigashi, H., and Mori, H., Chemoprevention of azoxymethnae-induced rat abberant crypt foci by dietary zerumbone isolated from *Zingiber zerumbet*, *Life Sci.*, 69, 1935–1945, 2001.

79. Tanaka, T., Kohno, H., Shimada, R., Kagami, S., Yamaguchi, F., Kataoka, S., Ariga, T., Murakami, A., Koshimizu, K., and Ohigashi, H., Prevention of colonic aberrant crypt foci by dietary feeding of garcinol in male F344 rats, *Carcinogenesis*, 21, 1183–1189, 2000.

80. Kohno, H., Yasui, Y., Suzuki, R., Hosokawa, M., Miyashita, K., and Tanaka, T., Dietary seed oil rich in conjugated linolenic acid from bitter melon inhibits azoxymethane-induced rat colon carcinogenesis through elevation of colonic PPAR gamma expression and alteration of lipid composition, *Int. J. Cancer*, 110, 896–901, 2004.

81. Kohno, H., Suzuki, R., Yasui, Y., Hosokawa, M., Miyashita, K., and Tanaka, T., Pomegranate seed oil rich in conjugated linolenic acid suppresses chemically induced colon carcinogenesis in rats, *Cancer Sci.*, 95, 481–486, 2004.

82. Kimoto, N., Hirose, M., Futakuch,i M., Iwata, T., Kasai, M., and Shirai, T. Site-dependent modulating effects of conjugated fatty acids from safflower oil in a rat two-stage carcinogenesis model in femal Sprague-Dawley rats, *Cancer Lett.*, 168, **15**–21, 2001.

83. Ip, C. and Scimeca, J.A., Conjugated linoleic acid and linoleic acid are distinctive modulators of mammary carcinogenesis, *Nutr. Cancer*, 27, 131–135, 1997.

84. Moya-Camarena, S.Y., Vanden Heuvel, J.P., and Belury, M.A., Conjugated linoleic acid activates peroxisome proliferator-activated receptor a and b subtypes but does not induce hepatic peroxisome proliferation in Sprague-Dawley rats, *Biochim. Biophys. Acta*, 1436, 331–342, 1999.

85. Houseknecht, K.L., Vanden Heuvel, J.P., Moya-Camarena, S.Y., Portocarrero, C.P., Peck, L.W., Nickel, K.P., and Belury, M.A., Dietary conjugated linoleic acid normalizes impaired glucose tolerance in the Zucker diabetic fatty *fa/fa* rat, *Biochem. Biophys. Res. Commun.*, 244, 678–682, 1989.

86. Ha, Y.L., Storkson, J., and Pariza, M.W., Inhibition of benzo(*a*)pyrene-induced mouse forestomach neoplasia by conjugated dienoic derivatives of linoleic acids, *Cancer Res.*, 50, 1097–1101, 1990.

87. Dhar, P., Ghosh, S., and Bhattacharyya, D.K., Dietary effects of conjugated octadecatrienoic fatty acid (9*cis*, 11*trans*, 13*trans*) levels on blood lipids and nonenzymatic *in vitro* lipid peroxidation in rats, *Lipids*, 34, 109–114, 1999.

88. Suzuki, R., Abe, M., and Miyashita, K., Comparative study on the autoxidation of TAG containing conjugated and non-conjugated C18 polyunsaturated fatty acids, *J. Am. Oil Chem. Soc.*, in press.

89. Tsuzuki, T., Igarashi, M., and Miyazawa, T., Conjugated eicosapentaenoic acid (EPA) inhibits transplanted tumor growth via membrane lipid peroxidation in nude mice, *J. Nutr.*, 134, 1162–1166, 2004.

90. Ha, Y.L., Grimm, N.K., and Pariza, M.W., Anticarcinogens from fried ground beef: heat-altered derivatives of linoleic acid, *Carcinogenesis*, 8, 1881–1887, 1987.

91. Noguchi, R., Yasui, Y., Suzuki, R., Hosokawa, M., Fukunaga, K., and Miyashita, K., Dietary effects of bitter gourd oil on blood and liver lipids of rats, *Arch. Biochem. Biophys.*, 396, 207–212, 2001.

92. Noguchi, R., Yasui, Y., Hosokawa, M., Fukunaga, K., and Miyashita, K., Bioconversion of conjugated linolenic acid to conjugated linoleic acid, in *Essential Fatty Acids and Eicosanoids*, Huang, Y.-S., Lin, S.-J., and Huang, P.-C., Eds., AOCS Press, Champaign, IL, 2003, pp. 353–359.

93. Li, Y. and Watkins, B.A., Conjugated linoleic acids alter bone fatty acid composition and reduce *ex vivo* prostaglandin E2 biosynthesis in rats fed n-6 or n-3 fatty acids, *Lipids*, 33, 417–425, 1998.

14 Marine Conjugated Polyunsaturated Fatty Acids

Yasushi Endo, Si-Bum Park, and Kenshiro Fujimoto
Graduate School of Agricultural Science, Tohoku University, Sendai, Japan

CONTENTS

14.1 DISTRIBUTION

Certain plants contain conjugated polyunsaturated fatty acids (CPUFAs) as major fatty acids in their seed lipids[1,2]. *Aleurites fordii* (Euphorbiaceae) has 9Z,11E,13E-octadecatrienoic acid known as α-eleostearic acid and β-eleostearic acid (9E,11E,13E-octadecatrienoic acid) at a level of more than 70% of total fatty acids. *A. fordii* produces tung oil which is used in the paints and coating industries. Karela (*Momordica charantia*) also contains α-eleostearic acid at a level of 50% in its seed oil. *Catalpa ovata* or *C. bignonioides* (Bignoniaceae) has catalpic acid (9E,11E,13Z-octadecatrienoic acid), while *Punica granatum* (Punicaceae), *Cucurbita digitata* or *C. palmata* (Cucurbitaceae) have punicic acid (9Z,11E,13Z-octadecatrienoic acid). *Calendida officinalis* (Compositae) and *Jacaranda mimosifolia* (Bignoniaceae) have calendic acid (8E,10E,12Z-octadecatrienoic acid) and jarcaric acid (8Z,10E,12Z-octadecatrienoic acid), respectively. These CPUFAs are generally called conjugated linolenic acid (CLN), because they are positional and geometrical isomers of linolenic acid. Most CPUFAs present in plants contain three conjugated double bonds and 18 carbons, but α-parinaric acid (9Z,11E,13E,15E-octadecatrienoic acid), which contains four conjugated double bonds, exists in *Impatiens edgworthii*[1].

Recently, some CPUFAs have been found in marine algae (Table 14.1 and Figure 14.1), although they are minor fatty acids (usually below 1% of total fatty acids). Lopez and Gerwick[3] found the conjugated triene-containing fatty acids 5Z,7E,9E,14Z,17Z- and 5E,7E,9E,14Z,17Z-eicosapentaenoic acids (1, 2 in Figure 14.1) in red alga, *Ptilota filicina*, on the Oregon coast. Park et al.[4] also found 5Z,7E,9E,14Z,17Z- and 5E,7E,9E,14Z,17Z-eicosapentaenoic acids in red alga, *Ptilota pectinata*, collected at the Hokkaido coast in Japan. They also observed the presence of 5Z,7E,9E,14Z-and 5E,7E,9E,14Z-eicosatetraenoic acids (3, 4 in Figure 14.1) in it. Narayan et al.[5] found 5Z,7E,9E,14Z,17Z- and 5E,7E,9E,14Z,17Z-eicosapentaenoic acids and 5Z,7E,9E,14Z- and 5E,7E,9E,14Z-eicosatetraenoic acids in red alga, *Acanthophora spicifera*, collected in the Indian Ocean. Burgess et al.[6] reported the presence of a conjugated tetraene-containing fatty acid in red alga,

Bosseiella orbigiana. The CPUFA was identified as 5Z,8Z,10E,12E,14Z-eicosapentaenoic acid (5 in Figure 14.1) and named bossepentaenoic acid. Bossepentaenoic acid was also found in red alga, *Lithothamnion coralloides*[7]. It is interesting that most CPUFAs found in red algae consisted of 20 carbons but not 18 carbons. Probably, these CPUFAs may be derived from eicosapentaenoic acid (5Z,8Z,11Z,14Z,17Z-eicosapentaenoic acid, EPA) and arachidonic acid (5Z,8Z,11Z,14Z-eicosatetraenoic acid, AA) which are the major fatty acids of lipids in marine algae.

CPUFAs were also found in green and brown algae. Mikhailova et al.[8] found bossepentaenoic acid in green alga, *Anadyomene stellata*. They also found 4Z,7Z,9E,11E,13E,16Z,19Z-docosaheptaenoic acid (6 in Figure 14.1) in the alga and named it stellaheptaenoic acid. Both of these CPUFAs have four conjugated double bonds. Our group found 5Z,7E,9E,14Z,17Z- and 5E,7E,9E,14Z,17Z-eicosapentaenoic acids in brown alga, *Dicyopteris divaricata* (unpublished). The presence of conjugated tetraene-containing fatty acids in brown alga was also found, but the CPUFAs have not been identified. Conjugated diene-containing fatty acids have not been found yet, although they may also exist in marine algae.

14.2   BIOSYNTHESIS

Two types of mechanisms are considered for biosynthesis of CPUFAs in marine algae. As for conjugated tetraene-containing fatty acids such as bossepentaenoic acid and stellaheptaenoic acid, the biosynthesis by the fatty acid oxidase related to D-amino acid oxidase has been hypothesized[9]. Table 14.2 shows conjugated tetraene fatty acids produced after incubation of PUFAs with enzymes from marine algae. Burgess et al.[6] observed the formation of bossepentaenoic acid after incubation of AA with aqueous extracts from red alga, *Bossiella orbigniana*. That is, AA was enzymatically

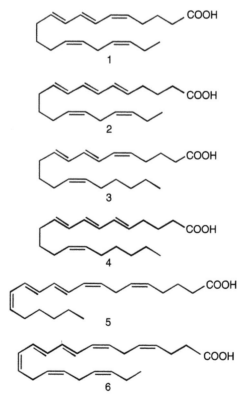

FIGURE 14.1 Structure of marine conjugated and polyunsaturated fatty acids.

TABLE 14.1
Conjugated and Polyunsaturated Fatty Acids in Marine Algae

Alga	Conjugated and polyunsaturated fatty acid	Origin	Ref.
Rhodophyta			
Ptilota filicina	5Z,7E,9E,14Z,17Z-eicosapentaenoic acid	EPA	3
	5E,7E,9E,14Z,17Z-eicosapentaenoic acid	EPA	
	5Z,7E,9E,14Z-eicosatetraenoic acid	AA	
Bossiella orbigiana	5Z,8Z,10E,12E,14Z-eicosapentaenoic acid	AA	6
Lithothamnion coralloides	5Z,8Z,10E,12E,14Z-eicosapentaenoic acid	AA	7
Ptilota pectinata	5Z,7E,9E,14Z,17Z-eicosapentaenoic acid	EPA	4
	5E,7E,9E,14Z,17Z-eicosapentaenoic acid	EPA	
	5Z,7E,9E,14Z-eicosatetraenoic acid	AA	
	5E,7E,9E,14Z-eicosatetraenoic acid	AA	
Acanthophora spicifera	5Z,7E,9E,14Z,17Z-eicosapentaenoic acid	EPA	5
	5E,7E,9E,14Z,17Z-eicosapentaenoic acid	EPA	
	5Z,7E,9E,14Z-eicosatetraenoic acid	AA	
	5E,7E,9E,14Z-eicosatetraenoic acid	AA	
Chlorophyta			
Anadyomene stellata	4Z,7Z,9E,11E,13Z,16Z,19Z-docosaheptaenoic acid	DHA	8
	5Z,8Z,10E,12E,14Z-eicosapentaenoic acid	AA	
Phaeophyta			
Dictyopteris divaricata	5Z,7E,9E,14Z,17Z-eicosapentaenoic acid	EPA	Park et al. (unpublished)
	5E,7E,9E,14Z,17Z-eicosapentaenoic acid	EPA	

Note: AA, arachidonic acid; (20:5n-6); EPA, eicosapentaenoic acid (20:5n-3); DHA, docosahexaenoic acid (22:6n-3).

TABLE 14.2
Enzymatic Production of Conjugated Tetraene Fatty Acids by Marine Algae

Alga	Origin	Products	Ref.
Bossiella orbigiana	5Z,8Z,11Z,14Z-eicosatetraenoic acid (AA)	20:5	6
Lithothamnion coralloides	6Z,9Z,12Z-octadecatrienoic acid	18:4	10
Anadyomene stellata	6Z,9Z,12Z,15Z-octadecatetraenoic acid	16:5, 18:4, 20:5, 20:6	8
	5Z,8Z,11Z,14Z-eicosatetraenoic acid (AA)	20:5, 22:6	
	5Z,8Z,11Z,14Z,17Z-eicosapentaenoic acid (EPA)	16:5, 18:4, 20:5, 20:6	
	7Z,10Z,13Z,16Z-docosatetraenoic acid	16:5, 18:4, 20:5, 20:6, 22:7	
	4Z,7Z,10Z,13Z,16Z,19Z-docosahexaenoic acid (DHA)	16:5, 18:4, 20:5, 20:6, 22:7	

Note: 16:5 = 4Z,7Z,9E,11E,13Z-hexadecapentaenoic acid; 18:4 = 6Z,8E,10E,12Z-octadecatetraenoic acid; 20:5 = 5Z,8Z,10E,12E,14Z-eicosapentaenoic acid; 20:6 = 5Z,8Z,10E,12E,14E,17Z-eicosahexaenoic acid; 22:7 = 4Z,7Z,9E,11E,13Z,16Z,19Z-docosaheptaenoic acid.

desaturated and isomerized to bossepentaenoic acid in red alga. Hamberg[10] also observed that the crude enzyme from red alga, *Lithothamnion corallioides*, could produce 6Z,8E,10E,12Z-octadecatrienoic acid from γ-linolenic acid (6Z,9Z,12Z-octadecatrienoic acid). Mikhailova et al.[8] also reported that octadecatetraenoic acid, AA, EPA, and docosahexaenoic acid (DHA) could be transformed to the corresponding tetraene-containing fatty acids by the crude enzymes from *Anadyomene stellata*.

Biosynthesis of conjugated triene-containing fatty acids such as 5Z (or 5E),7E,9E,14Z,17Z-eicosapentaenoic acid and 5Z (or 5E),7E,9E,14Z-eicosatetraenoic acid has been considered to

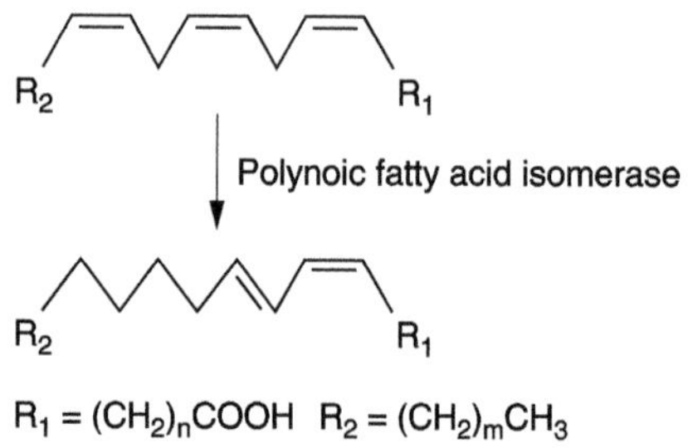

$$R_1 = (CH_2)_nCOOH \quad R_2 = (CH_2)_mCH_3$$

FIGURE 14.2 Isomerization of methylene-interrupted PUFA by polyenoic fatty acid isomerase.

involve polyenoic fatty acid isomerase (PFI). Wise et al.[11] prepared PFI from *Ptilota filicina* and investigated its enzymatic characteristics. They demonstrated that PFI could produce 5Z,7E,9E,14Z,17Z-eicosapentaenoic acid and 5Z,7E,9E,14Z-eicosatetraenoic acid from EPA and AA, respectively. Especially, PFI was more specific to EPA. Park et al.[4] also observed that EPA and AA were transformed to 5Z,7E,9E,14Z,17Z-eicosapentaenoic acid and 5Z,7E,9E, 14Z-eicosatetraenoic acid, respectively, by the PFI extracted from *Ptilota pectinata*. Probably, 5E,7E,9E,14Z,17Z-eicosapentaenoic acid and 5E,7E,9E,14Z-eicosatetraenoic acid could be automatically isomerized from 5Z,7E,9E,14Z,17Z-eicosapentaenoic acid and 5Z,7E,9E,14Z-eicosatetraenoic acid, respectively.

14.3 POLYENOIC FATTY ACID ISOMERASE (PFI)

Bioconversion of conjugated fatty acids from the methylene-interrupted PUFAs has been reported in bacteria[12,13] and higher plants[14], which accumulate considerable levels of conjugated fatty acids. In marine algae, biosynthesis of conjugated fatty acids has been reported in red algae, *Bossiella orbigniana*[6], *Ptilota filicina*[3], and *Lithothamnion coralllioides*[10] and in a green alga *Anadyomene stellata*[8]. In *Lithothamnion coralllioides* and *Bossiella orbigniana* oxidative pathways have been suggested for the production of some of these CPUFAs[6,10].

In the studies on the biosynthesis of the conjugated fatty acid 5Z,7E,9E,14Z,17Z-eicosapentaenoic acid from EPA (5Z,8Z,11Z 14Z,17Z-eicosapentaenoic acid) by *Ptilota filicina*, a new fatty acid isomerase was discovered[11]. This enzyme, as noted earlier, was termed PFI and it was able to catalyze the isomerization of substrates containing three or more methylene-interrupted double bonds into a Z,E,E-conjugated triene functionality (Figure 14.2)[11]. The *P. filicina* PFI shows a catalytic activity over a wide pH range, with activity being optimal below pH 6.0. EPA and AA showed the highest V_{max}[11].

The *P. filicina* PFI was purified to electrophoretic homogeneity and the cloning was reported for the first time as the conjugase from an algal species[15]. A single band on sodium dodecyl sulfate–polyacrylamide gel electrophoresis (SDS-PAGE) showed a mass of 61 kDa, while the native enzyme had a mass of approximately 125 kDa. This means the protein exists as a dimer. Two very similar cDNA clones encoding novel 500-amino acid proteins, both with calculated molecular weights of 55.9 kDa, were isolated. The native *P. filicina* PFI is a glycoprotein and chromophoric with a flavin-like UV spectrum.

A very similar PFI in a red alga, *Ptilota pectinata*, which grows in northern Japan, was also discussed[4]. This enzyme catalyzes the formation of conjugated trienes from various methylene-interrupted PUFAs with three or more double bonds. The structures of conjugated fatty acids produced by the *P. pectinata* PFI are similar to those biosynthesized by the *P. filicina* PFI. Eicosapentaenoate was the best substrate followed by arachidonate, docosapentaenoate (n-3), and octadecatetraenoate (n-3). The purified preparation of PFI exhibited a single band on SDS-PAGE and had a molecular weight of approximately 65 kDa.

14.4 BIOLOGICAL FUNCTION

The biological effects of CPUFAs present in marine algae have been unknown because they are trace components. However, several biological functions have been reported for CPUFAs such as α-eleostearic acid and α-parinaric acid contained in plants[16–24].

14.4.1 LIPID MODIFICATION

Dhar et al.[16] reported that lipid peroxidation in liver and the erythrocyte membrane was reduced in rats fed α-eleostearic acid ($9Z,11E,13E$-octadecatrienoic acid)-rich karela seed oil, although the serum cholesterol level was not affected by dietary karela oil. Noguchi et al.[17] reported that the serum free cholesterol and liver weight were reduced in rats fed bitter gourd oil containing α-eleostearic acid. The linoleic acid level was lower in liver lipids of rats fed bitter gourd oil than that in the control group, while the DHA level was higher. However, the level of hydroperoxides and α-tocopherol in plasma was not affected by dietary bitter gourd oil. Koba et al.[18] reported that dietary CLN prepared from perilla oil by alkaline isomerization reduced adipose tissue weight and serum cholesterol level in rats, but it increased serum triacylglycerol and free fatty acid levels.

Lee et al.[19] observed that dietary tung oil containing α- and β-eleostearic acids reduced adipose tissue weight and plasma cholesterol level, but not the weight of the liver and heart in laying hens. They also observed a reduced level of linolenic acid and an increased proportion of conjugated linoleic acid in all tissues of laying hens. From these results, it was concluded that α-eleosteraic acid had the ability to modify fatty acid composition and lipid level and reduce body fat mass. Probably, conjugated EPA and AA present in marine algae may also have similar lipid modification effects.

14.4.2 ANTICANCER PROPERTIES

Cornelius et al.[20] reported that α-parinaric acid was cytotoxic to cultured human malignant cells. However, human fibroblast, bovine aortic endothelial cells, and Caco-2 colonic mucosal cells were not sensitive to α-parinaric acid. Matsumoto et al.[21] reported that α-parinaric acid of balsam seed oil was not cytotoxic to human breast and colorectal cancer cells.

However, several researchers observed anticancer effects of CLN. Igarashi and Miyazawa[22] reported that conjugated triene-containing linolenic acid prepared by alkaline isomerization was cytotoxic to human tumor cells and that conjugated diene-containing octadecatrienoic acid was not toxic. Matsumoto et al.[21] also observed that α-eleostearic acid of tung oil inhibited the growth of human breast and colorectal cancer cells. Suzuki et al.[23] investigated the effects of the position of double bonds on the anticancer function of CLN, and observed that 9,11,13-octadecatrienoic acid was more cytotoxic to mouse tumor cells and human monocytic leukemia cells than 8,10,12-octadecatrienoic acid. Kohno et al.[24] reported the *in vivo* anticancer effect of α-eleostearic acid. They observed that dietary α-eleostearic acid of bitter gourd oil inhibited azoxymethane-induced colonic aberrant crypt foci in rats.

Igarashi and Miyazawa[25] observed that conjugated EPA and DHA prepared by alkaline isomerization showed intensive cytotoxicity to human tumor cells. They suggested that the cytotoxicity of conjugated EPA and DHA was due to the induction of apoptosis via lipid peroxidation of cell membranes. Matsumoto et al.[21] also observed that supplementation with α-tocopherol as a natural antioxidant reduced the cytotoxicity of conjugated EPA and DHA to cultured human cancer cells.

Park et al.[26] demonstrated the anticancer effect of naturally occurring CPUFAs such as $5Z,7E,9E,14Z,17Z$-eicosapentaenoic acid and $5Z,7E,9E,14Z$-eicosatetraenoic acid. They enzymatically prepared $5Z,7E,9E,14Z,17Z$-eicosapentaenoic acid and $5Z,7E,9E,14Z$-eicosatetraenoic acid from EPA and AA, respectively, using crude extracts of red alga, *Ptilota pectinata*, and investigated their cytotoxicity to human cancer cells. Figure 14.3 shows the cytotoxicity of enzymatically conjugated EPA and AA to human cancer cell lines. Enzymatically conjugated EPA and AA exhibited

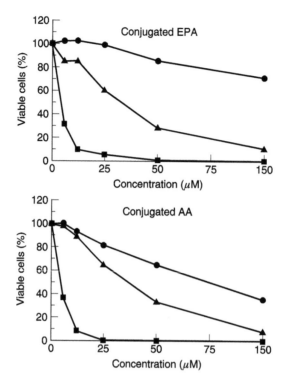

FIGURE 14.3 Cytotoxicity of enzymatically conjugated EPA and AA to human cancer cell lines: ●, HepG2; ▲, A-549; ■, DLD-1.

strong cytotoxicity to human lung (A-549) and colorectal (DLD-1) cancer cells, but had a weak effect to human liver (HepG2) cancer cells. Conjugated EPA and AA prepared by alkaline isomerization showed an intensive cytotoxicity to all tested human cancer cells. They completely inhibited the growth of all cancer cell lines including human breast (MCF-7) and colorectal (HT-29) cancer cells at a concentration of 100 μM.

From these results, it is evident that the anticancer effect of CPUFAs depends on numbers and positions of conjugated double bonds in them. Especially, conjugated triene-containing fatty acid has a strong anticancer effect. Therefore, CPUFAs present in marine algae are expected to possess anticancer effects.

REFERENCES

1. Murase, Y., Conjugated aliphatic unsaturated fatty acids in natural fats, *J. Jpn. Oil Chem. Soc.*, 15, 602–607, 1966.
2. Takagi, T. and Itabashi, Y., Occurrence of mixtures of geometrical isomers of conjugated octadecatrienoic acids in some seed oils: analysis by open-tubular gas liquid chromatography and high performance liquid chromatography, *Lipids*, 16, 546–551, 1981.
3. Lopez, A. and Gerwick, W.H., Two new icosapentaenoic acids from the temperate red seaweed *Ptilota filicina* J. Agardh, *Lipids*, 22, 190–194, 1987.
4. Park, S.-B., Matsuda, H., Endo, Y., Fujimoto, K., and Taniguchi, K., Biosynthesis of conjugated trienoic fatty acids by red alga *Ptilota pectinata*, *Biosci., Biotechnol. Biochem.*, submitted.
5. Narayan, B., Kinami, T., Miyashita, K., Park, S.-B., Endo, Y., and Fujimoto, K., Occurrence of conjugated polyenoic fatty acids in seaweeds from the Indian Ocean, *Z. Natuforsch. C*, 59c, 310–314, 2004.

6. Burgess, J.R., de la Rosa, R.I., Jacobs, R.S., and Butier, A., A new eicosapentaenoic acid formed from arachidonic acid in the coralline red algae *Bossiella orbigniana*, *Lipids*, 26, 162–165, 1991.

7. Gerwick, W.H., Asen, P., and Hambergs, M., Biosynthesis of 13R-hydroxyarachidonic acid, an usual oxylipin from the red alga *Lithothamnion coralloides*, *Phytochemistry*, 34, 1029–1033, 1993.

8. Mikhailova, M.V., Bemis, D.L., Wise, M.L., Gerwick, W.H., Norris, J.N., and Jacobs, R.S., Structure and biosynthesis of novel conjugated polyene fatty acids from the marine green alga *Anadyomene stellata*, *Lipids*, 30, 583–589, 1995.

9. Gerwick, W.H., Structure and biosynthesis of marine algal oxylipins, *Biochim. Biophys. Acta*, 1211, 243–255, 1994.

10. Hamberg, M.m Metabolism of 6, 9, 12-octadecatrienoic acid in the red alga *Lithothamnion coralloides*: mechanism of formation of a conjugated tetraene fatty acid, *Bichem. Biophys. Res. Commun.*, 188, 1220–1227, 1992.

11. Wise, M.L., Hamberg, M., and Gerwick, W.H., Biosynthesis of conjugated triene-containing fatty acids by a novel isomerase from the red marine alga *Ptilota filicina*, *Biochemistry*, 33, 15223–15232, 1994.

12. Jiang, J., Bjorck, L., and Fonden, R., Production of conjugated linoleic acid by dairy starter cultures, *J. Appl. Microbiol.*, 85, 95–102, 1998.

13. Ogawa, J., Matsumura, K., Kishino, S., Omura, Y., and Shimizu, S., Conjugated linoleic acid accumulation via 10-hydroxy-12-octadecaenoic acid during microaerobic transformation of linoleic acid by Lactobacillus acidophilus, *Appl. Environ. Microbiol.*, 67, 1246–1252, 2001.

14. Conacher, H.B.S., Gunstone, F.D., Hornby, G.M., and Padley, F.B., Glyceride structures. IX: Intraglyceride distribution of vernolic acid and of five conjugated octadecatrienoic acids in seed glycerides, *Lipids*, 5, 434–441, 1970.

15. Zheng, W., Wise, M.L., Wyrick, A., Mets, J.G., Yuan, L., and Gerwick, W.H., Polyenoic fatty acid isomerase from marine alga Ptilota filicina: protein characterization and functional expression of the cloned cDNA, *Arch. Biochem. Biophys.*, 401, 11–20, 2002.

16. Dhar, P., Ghosh, S., and Bhattacharyya, D.K., Dietary effects of conjugated octadecatrienoic fatty acid (9cis, 11trans, 13trans) levels on blood lipids and nonenzymatic in vitro lipid peroxidation in rats, *Lipids*, 34, 109–114, 1999.

17. Noguchi, R., Yasui, Y., Suzuki, R., Hosokawa, M., Fukunaga, K., and Miyashita, K., Dietary effects of bitter gourd oil on blood and liver lipids of rats, *Arch. Biochem. Biophys.*, 396, 207–212, 2001.

18. Koba, K., Akahoshi, A., Yamasaki, M., Tanaka, K., Yamada, K., Iwata, T., Kamegai, T., Tsutsumi, K., and Sugano, M., Dietary conjugated linolenic acid in relation to CLA differently modifies body fat mass and serum and liver lipid levels in rats, *Lipids*, 37, 343–350, 2002.

19. Lee, J.-S., Takai, J., Takahashi, K., Endo, Y., Fujimoto, K., Koike, S., and Matsumoto, W., Effect of dietary tung oil on the growth and lipid metabolism of laying hens, *J. Nutr. Sci. Vitaminol.*, 48, 142–148, 2002.

20. Cornelius, A.S., Yerram, N.R., Kratz, D.A., and Spector, A.A., Cytotoxic effect of cis-parinaric acid in cultured malignant cells, *Cancer Res.*, 51, 6025–6030, 1991.

21. Matsumoto, N., Endo, Y., Fujimoto, K., Koike, S., and Matsumoto, W., The inhibitory effect of conjugated and polyunsaturated fatty acids on the growth of human cancer cell lines, *Tohoku J. Agric. Res.*, 52, 1–12, 2001.

22. Igarashi, M. and Miyazawa, T., Newly recognized cytotoxic effect of conjugated trienoic fatty acids on cultured human tumor cells, *Cancer Lett.*, 148, 173–179, 2000.

23. Suzuki, R., Noguchi, R., Ota, T., Abe, M., Miyashita, K., and Kawada, T., Cytotoxic effect of conjugated trienoic fatty acids on mouse tumor and human monocytic leukemia cells, *Lipids*, 36, 477–482, 2001.

24. Kohono, H., Suzuki, R., Noguchi, R., Hosokawa, M., Miyashita, K., and Tanaka, T., Dietary conjugated linolenic acid inhibits azoxymethane-induced colonic aberrant crypt foci in rats, *Jpn. J. Cancer Res.*, 93, 133–142, 2002.

25. Igarashi, M. and Miyazawa, T., Do conjugated eicosapentaenoic acid and conjugated docosahexaenoic acid induce apotosis via lipid peroxidation in cultured human tumor cells?, *Biochem. Biophys. Res. Commun.*, 270, 649–656, 2000.

26. Park, S.-B., Matsuda, H., Endo, Y., Fujimoto, K., and Taniguchi, K., Cytotoxicity of conjugated trienoic eicosapentaenoic and arachidonic acids produced by crude enzyme from red alga, *Ptilota pectinata*, on the human cancer cells, *Biosci. Biotechnol. Biochem.*, submitted.

15 Marine Oils: Compositional Characteristics and Health Effects

Fereidoon Shahidi and H. Miraliakbari
Department of Biochemistry, Memorial University of Newfoundland,
St. John's, Newfoundland, Canada

CONTENTS

15.1   INTRODUCTION

There is a considerable body of evidence suggesting the beneficial health effects of seafood and marine oil consumption. These findings have long been attributed to their long-chain polyunsaturated omega-3 (n-3) fatty acids comprised mainly of eicosapentaenoic acid (EPA) and docosahexaenoic acid (DHA)[1]. The earliest reports of these findings were cross-cultural epidemiological studies involving Greenland Inuits and Danish settlers, which showed that the traditional Greenlandic diet, rich in marine mammals and fish, significantly reduced the incidence of cardiovascular diseases[2]. Although the mechanisms by which marine n-3 fatty acids exert their cardioprotective effects are not fully understood, it appears that EPA and DHA can reduce the likelihood of fatal arrhythmias[3] by altering the myocyte membrane fatty acid composition[4], among others. Larger doses of EPA and DHA have been shown to lower serum triacylglycerol levels and reduce platelet aggregation[6]. Long-chain omega-3 fatty acids have also been shown to act as disease-modifying agents in several human diseases including arthritis, cancer, diabetes, inflammatory bowel disease, and mental diseases[6].

Marine lipids originate from the flesh of fatty fish, the liver of white lean fish, and the blubber of marine mammals. They contain triacylglycerols, that is, glycerol esterified to primarily long-chain fatty acids, with small amounts of long-chain alcohols esterified to fatty acids (wax esters). Most marine animal oils contain small amounts of unsaponifiable matter, such as hydrocarbons, fatty alcohols, and waxes, among others. Although EPA and DHA occur mainly in marine organisms, their total amounts and relative proportions vary widely depending on their source, the composition of the plankton/feedstuff, and the time of harvest. Some sources can easily provide the required daily amount of long-chain omega-3 fatty acids (~800 mg[6]) in one 2 g oil capsule, whereas other sources require higher consumption levels in order to reach these amounts. Improvements in the production of fish oil concentrates have led to marine oil products with EPA and DHA levels exceeding 85% (w/w)[7]. The main component of marine lipids is triacylglycerols that are rich in monounsaturated as well as polyunsaturated n-3 fatty acids. Both fish and marine mammal oils are rich in EPA and DHA, but marine mammal oils also contain a relatively large amount of docosapentaenoic acid (DPA, n-3). The spatial distribution of triacylglycerol fatty acids in fish and marine mammal oils differs in that fish oils contain long-chain polyunsaturated fatty acids in the sn-2 position of triacylglycerols, whereas marine mammal lipids contain long-chain polyunsaturated fatty acids predominantly in the sn-1 and sn-3 positions. These factors greatly influence the metabolism and potential health effects of marine lipids[6], and have also been shown to influence their oxidative stability.

Whale oil, one of the earliest widely produced marine oils, was used primarily in the cosmetic, wax, and paint industries. Currently menhaden and sardine oils are the most widely produced marine oils; however, the livers of cod, haddock, halibut, shark, whales, and tuna are commonly used as sources of marine liver oils that are rich in the fat-soluble vitamins, particularly vitamins A and D[8]. It is estimated that over 70 million metric tons of fats and oils are produced each year and marine oils account for about 2% of this, with whole fish body oils comprising 97% of total marine oil production[8]. Table 15.1 shows the distribution of world whole fish body oil production[9].

This chapter provides an overview of the lipid composition of marine animal oils including those of various fish, marine mammals, cephalopods, and crustaceans. The health effects of selected marine animal lipids are also covered.

TABLE 15.1
World Production of Fish Body Oils (in 1000's of metric tons[a])[9]

Country/region	1990	1994	1996	1998	2000
Scandinavia	200	320	350	350	350
Japan	400	65	50	50	50
USA	170	150	160	150	160
Chile	200	300	300	100	170
Peru	195	500	450	200	600
Russian Federation	25	20	10	7	10
Others	100	140	130	150	170
World	1264	1505	1381	865	1417

[a] Estimated values.

15.2 FISH OILS

15.2.1 MENHADEN OIL

Menhaden (*Brevoortia tyrannus*) is a fish species that grows rapidly as a filter feeder on an abundant supply of plankton in estuaries, with most reaching maturity at one year of age. The availability of menhaden is high in near shore waters of the Atlantic coast of the United States and in the shores of the Gulf of Mexico. They form large schools, usually of the same size and age group[10].

The fatty acid composition of menhaden oil shows that it contains 30% saturated fatty acids of which palmitic acid is the most abundant, 22% monounsaturated fatty acids, 18% EPA, and 9.6% DHA (Table 15.2)[10]. The fatty acid composition of menhaden oil has also been reported by Nichols and Davies[11] using both gas chromatography–mass spectrometry (GC-MS) and high-performance liquid chromatography–mass spectrometry (LC-MS) combined with ultraviolet (UV) detection, showing good agreement between the percentage composition of the fatty acid components determined by GC-MS and LC-UV analyses.

Torres and Hill[12] recently incorporated conjugated linoleic acid into menhaden oil using lipase-catalyzed acidolysis. Under optimal conditions, this group was able to obtain 9% incorporation of conjugated linoleic acid and less than 10% diacylglycerols in the final oil product[12]. Rice et al.[13] were able to produce a marine n-3 fatty acid concentrate from menhaden oil using a bioreactor. Lipase from *Candida cylindracea* was immobilized by adsorption on microporous polypropylene fibers and then used to hydrolyze selectively the saturated and monounsaturated fatty acid residues of menhaden oil at 40°C and pH 7.0. In 3.5 h, the shell and tube reactor containing the hollow fibers gave a fractional release of the saturated and monounsaturated fatty acid residues in the order of C14:0, C16:0, C16:1, C18:0, C18:1. After one winterization step, the percentages of saturated fatty acids present were reduced by 18.2, 81.8, and 60% for myristic, palmitic, and stearic acids, respectively[13]. The remainder of the saturated and monounsaturated fatty acids may then be removed using a urea complexation process[13]. Porsgaard and Hoy[14] studied the spatial distribution of menhaden oil triacylglycerols by employing Grignard degradation with allyl magnesium bromide followed by isolation and analysis of the resultant fatty acids from positions *sn*-1 and *sn*-3 as well as the *sn*-2 monoacylglycerol fraction (Table 15.3)[14]. The cholesterol content of menhaden oil has been reported to be 1.4 g/kg[15].

Cell culture and animal model studies have shown that dietary menhaden oil is cardioprotective[16,17] and in some cases anticarcinogenic[18]; however, few reports showing the health effects in humans have been published. Goodie et al.[22] performed a randomized, double-blind, placebo-controlled trial to assess whether supplementation with a menhaden oil concentrate could improve the vascular function of peripheral small arteries in hypercholesterolemic patients. The results of this report show that menhaden oil improved endothelial function in peripheral small arteries in hypercholesterolemic patients, which may provide a mechanism for the beneficial effects of EPA and DHA in coronary heart disease[22]. Yuan et al.[15] studied the effects of dietary menhaden oil on plasma cholesterol, systolic blood pressure, and antioxidant parameters of plasma and livers in spontaneously hypertensive as well as Wistar Kyoto rats. In both types of rats, inclusion of menhaden oil reduced plasma cholesterol compared to diets containing butter, beef tallow, or soybean oil. In this study, the dietary fat source did not significantly influence systolic pressure or tissue antioxidant status[15]. Menhaden oil supplementation has also been shown to reduce the development of rodent mammary gland tumors[19].

15.2.2 HERRING

Herring is one of the most widely processed fish species. The two most common herring species are the Atlantic herring (*Clupea harengus harengus*) which contains 9% fat, and the Pacific herring (*Clupea harengus pallasi Valenciennes*) containing 14% fat[18]. Herring is also fished in the North

TABLE 15.2
Fatty Acid Composition (%) of Menhaden Oil[10]

Fatty acid	Content (%)
12:0	0.15
14:0	7.30
15:0	0.65
16:0	19.45
16:1	9.05
16:2 n-7	0.50
16:2 n-4	1.55
16:3 n-4	1.70
16:3 n-3	0.15
16:4 n-4	0.20
16:4 n-1	2.60
17:0	1.05
18:0	4.45
18:1	10.40
18:2 n-9	0.20
18:2 n-6	1.30
18:2 n-4	0.50
18:3 n-6	0.30
18-3 n-3	0.65
18:4 n-3	2.65
19:0	0.10
20:0	0.30
20:1	1.45
20:2 n-9	0.15
20:2 n-6	0.30
20:3 n-6	0.20
20:3 n-3	1.00
20:4 n-6	0.15
20:4 n-3	0.80
20:5 n-3	18.30
21:0	0.10
21:5 n-3	0.90
22:0	0.15
22:1	1.55
22:2	0.10
22:3 n-3	0.20
22:4 n-3	0.60
22:5 n-3	1.80
22:6 n-3	9.60
23:0	0.15
24:0	0.10
24:1	0.70
Iodine value (Wijs)	162.1 g I_2/100 g oil

Sea[10]. One- to three-year-old juvenile herring are steamed, seasoned, smoked, and packed or canned for human consumption. Herring is an important bait species for other fisheries such as the lobster fishery, and is processed to produce herring fishmeal which is commonly used in the animal, livestock, and aquaculture industries[9].

The fatty acid composition of herring oil includes 18% saturated fatty acids, 50% monounsaturated fatty acids (primarily cetoleic acid, C22:1 n-9), 7.5% EPA, and 6.8% DHA (Table 15.4)[10]. The seasonal variation in herring oil fatty acid composition was reported by Adios et al.[22], showing that

TABLE 15.3
Spatial Distribution (%) of Fatty Acids of Menhaden Oil[14]

Fatty acid	sn-1 and 3	sn-2
14:0	10.5	14.3
16:0	18.6	22.5
16:1	13.1	13.4
18:0	3.2	4.1
18:1 n-9	7.2	2.6
18:1 n-7	3.5	1.6
18:2 n-6	1.3	1.0
18:3 n-3	0.9	0.8
18:4 n-3	2.8	—
20:1 n-9	0.9	2.7
20:5 n-3	18.2	14.6
22:5 n-3	2.7	5.1
22:6 n-3	12.6	14.8

the percentage of saturated fatty acids remains relatively constant, whereas C22:1 levels changed by 15% during a four-month period[22]. More recently, the α-tocopherol content of three menhaden oils was analyzed using HPLC with fluorescence detection, with values of 28 to 98 mg/kg oil[23]. This report also illustrates that the freshness and processing techniques applied to the herring samples markedly affect the tocopherol content as well as the overall quality of the extracted oil, as determined by conjugated diene levels, peroxide values, and free fatty acid contents[23]. Neutral lipids (NL), phospholipids (PL), and free fatty acids (FFA) in crude herring oil were separated and quantified using solid-phase extraction[22]. Menhaden oil contained 97% NL, 2.3% FFA, and 0.7% PL[24].

Studies investigating the health effects of herring oil consumption have shown that herring oil may help reduce some symptoms associated with rheumatoid arthritic and cardiovascular disease[25].

15.2.3 COD

There are two species of cod: Atlantic cod (*Gadus morhua* (L.)) and Pacific cod (*Gadus macrocephalus Tilesius*). The production of whole body cod oil is not common due to the low fat content of cod, which is less than 1% for both cod types[21]. However, the byproducts of codfish processing (livers and in some cases heads) are commonly extracted for their oils. Cod liver oil gained prominence in the 17th century for its ability to cure night blindness, and later to prevent rickets. With the development of vitamin chemistry, these beneficial effects were related to the presence of vitamins A and D in cod liver oil[8].

Cod liver oil is a rich source of both vitamins A and D, containing 1000 IU/g and 10 IU/g, respectively[21]. Cod liver oil contains 5.7 g/kg cholesterol[21], and also contains 120 mg/kg α-tocopherol[26]. The fatty acid composition of cod liver oil has been reported[18], showing 23% saturated fat, 47% monounsaturated, 7% EPA, and 11% DHA. Copeman and Parrish[27] examined lipid class compositions of crude Atlantic cod flesh oil and cod liver oil using Iatroscan (Table 15.5), showing that cod flesh oil contained predominantly muscle phospholipids at 54.9%, whereas cod liver oil contained mostly triacylglycerols at 66.9%[27]. This group also examined the fatty acid compositions of both whole fish cod oil and cod liver oil (Table 15.6)[27].

The potential health benefits of cod liver oil have been investigated in several human disease states. Brox et al.[28] examined the effects of dietary cod liver oil on total serum cholesterol, high-density lipoprotein cholesterol, postprandial triacylglycerol, apolipoproteins Al and B100, lipoprotein (a), monocyte function expressed as monocyte-derived tissue factor expression, and tumor necrosis

TABLE 15.4
Fatty Acid Composition (%) of Herring Oil[10]

Fatty Acid	Content (%)
12:0	0.10
14:0	6.10
15:0	0.40
16:0	10.70
16:1	7.30
16:2 n-7	0.20
16:2 n-4	0.40
16:3 n-4	6.70
16:4 n-4	0.10
16:4 n:1	1.20
17:0	0.35
18:0	1.40
18:1	10.30
18:2 n-9	<0.01
18:2 n-6	0.95
18:2 n-4	0.10
18:3 n-6	0.05
18-3 n-3	2.00
18:4 n-3	3.15
19:0	0.20
20:0	0.10
20:1	13.40
20:2 n-6	0.15
20:3 n-6	0.10
20:3 n-3	0.30
20:4 n-6	<0.01
20:4 n-3	0.75
20:5 n-3	7.45
21:0	0.10
21:5 n-2	0.25
22:0	0.05
22:1	21.25
22:2	0.20
22:4 n-3	0.25
22:5 n-3	0.75
22:6 n-3	6.75
23:0	0.10
24:0	0.15
24:1	0.75
Iodine value (Wijs)	135.7 g I_2/100 g oil

factor, showing that none of these parameters were significantly altered. However, cod liver oil did significantly alter plasma fatty acid composition[28].

15.2.4 CAPELIN AND SARDINE

Capelin (*Mytilus edulis*) is a small marine finfish and is widely processed for its whole body oil. It is also one of the major prey species for larger finfish and marine mammals and is commonly used as a bait species in some fisheries. Whole capelin contains 9.9% fat (wet weight) and its oil is dominated by triacylglycerols (85.8%) (Table 15.7)[27]. The fatty acid composition of capelin oil has been reported by Copeman and Parrish[27], and was shown to contain 17% saturated, 63%

TABLE 15.5
Lipid Class Composition of Cod Flesh Oil and Cod Liver Oil[27]

	Cod flesh	Cod liver
Lipid content (%)		
Wet weight	0.5	24.0
Dry weight	2.7	49.8
Lipid classes (% total lipids)		
Hydrocarbons	0.8	3.8
Sterol esters	2.5	0.8
Triacylglycerols	11.4	66.9
Diacylglycerols	3.7	4.9
Free fatty acids	6.6	2.4
Sterols	13.8	2.3
Phospholipids	54.9	12.3

TABLE 15.6
Fatty Acid Compositions (%) of Cod Flesh Oil and Cod Liver Oil[27]

Fatty acid	Cod flesh oil	Cod liver oil
14:0	1.4	3.6
15:0	0.2	0.3
16:0	18.0	11.0
17:0	0.1	0.1
18:0	3.4	0.3
22:0	0.1	—
23:0	0.1	—
15:1	—	0.1
16:1 n-7	3.3	11.7
16:1 n-5	0.0	0.2
18:1 n-11	0.8	—
18:1 n-9	6.4	15.7
18:1 n-7	3.0	5.7
18:1 n-5	0.5	0.6
20:1 n-9	2.4	8.9
20:1 n-7	0.2	0.2
22:1 n-11	0.5	3.0
22:1 n-9	0.2	0.2
24:1	0.4	0.3
16:3 n-4	0.1	0.3
16:4 n-3	0.1	—
16:4 n-1	0.1	0.4
18:2 n-6	1.0	1.5
18:2 n-4	0.1	0.3
18:3 n-3	0.1	0.1
18:4 n-3	0.2	1.0
18:4 n-1	0.3	0.2
20:2 n-6	0.3	0.3
20:3 n-6	0.1	—
20:3 n-3	0.1	0.1
20:4 n-6	2.0	—
20:5 n-3	19.1	12.2
22:4 n-6	<0.1	—
22:5 n-3	2.0	1.7
22:6 n-3	32.6	12.7

TABLE 15.7
Lipid Class Composition of Capelin Oil[27]

Lipid class	% Total lipids
Hydrocarbons	0.6
Triacylglycerols	85.8
Free Fatty Acids	0.7
Sterols	0.9
Phospholipids	7.2

TABLE 15.8
Fatty Acid Composition (%) of Capelin Oil[27]

Fatty acid	Content (%)
14:0	5.9
15:0	0.2
16:0	8.7
17:0	1.0
18:0	0.6
20:0	0.1
15:1	0.1
16:1 n-7	10.3
16:1 n-5	0.2
18:1 n-9	3.8
18:1 n-7	1.6
18:1 n-5	0.6
20:1 n-9	16.6
20:1 n-7	1.0
22:1 n-11	9.3
22:1 n-9	18.5
24:1	0.5
16:3 n-4	0.8
16:4 n-3	0.1
16:4 n-1	2.0
18:2 n-6	0.5
18:2 n-4	0.1
18:3 n-3	0.2
18:4 n-3	1.2
18:4 n-1	0.3
20:2 n-6	0.1
20:4 n-6	0.2
20:5 n-3	9.3
22:5 n-3	0.9
22:6 n-3	4.1

monounsaturated, and 20% polyunsaturated fatty acids (Table 15.8). Similar fatty acid compositions have been reported by Budge et al.[29].

Sardine (*Sardina pilchardus*) is another small finfish that is commonly processed for its body oil (4.8% of wet weight)[30]. The oil contains 40.8% triacylglycerols, 18.4% phospholipids, and 121 μg/g (ppm) α-tocopherol[30]. Passi et al.[30] examined the fatty acid composition of the triacylglycerol and phospholipid fractions of sardine oil, showing significant differences in total monounsaturated and polyunsaturated fatty acids between the two fractions (Table 15.9).

TABLE 15.9
Fatty Acid Composition (%) of Sardine Oil Triacylglycerols and Phospholipids[30]

Fatty acid	Triacylglycerol	Phospholipid
14:0	6.5	1.8
15:0	1.3	0.7
16:0	18.1	20.9
17:0	1.5	0.9
18:0	5.3	9.9
20:0	0.5	0.4
24:0	1.0	1.0
15:1	—	0.1
16:1	7.4	0.9
17:1	1.2	1.1
18:1	14.3	10.0
20:1	2.6	1.8
22:1	0.9	0.6
18:2 n-6	2.2	1.5
18:3 n-3	1.0	0.7
18:4 n-3	1.2	0.9
20:2 n-6	0.9	0.6
20:3 n-3	0.0	1.9
20:4 n-6	1.7	1.4
20:5 n-3	10.7	5.0
22:4 n-6	0.0	0.2
22:5 n-3	1.6	0.1
22:5 n-6	0.7	0.8
22:6 n-3	19.2	33.8

15.2.5 OTHER FISH OILS

The fatty acid compositions of oils from other popular fish species have been studied and are reported in Table 15.10; these include skipjack tuna (*Katsuwonus pelamis*), butterfish (*Peprilus triacanthus*), flounder species (*Pseudopleuronectes* spp. and *Limanda* spp.), haddock (*Melanogrammus aeglefinus*), halibut (*Hippoglossus hippoglossus*), mackerel (*Scomber scombrus*), and Atlantic salmon (*Salmo salar*).

Among these oils, halibut and skipjack tuna contain the highest amounts of DHA comprising around 30% of total fatty acids, while flounder species and haddock contain the highest amounts of EPA with nearly 15% of total fatty acids.

15.3 MARINE MAMMAL OILS

Traditional uses of marine mammal oils were as lubricants and as lamp oil; currently there is much interest in the use of marine mammal oils as nutritional supplements for their beneficial health effects[33]. Marine mammals contain fat-rich deposits known as blubber which serves as an energy deposit, thermal insulation, and to alter their body's buoyancy. As mentioned in the introduction, marine mammal triacylglycerols have different spatial distributions of long-chain polyunsaturated fatty acids from fish oil triacylglycerols, and also contain significant amounts of docosapentaenoic acid (DPA).

15.3.1 SEAL

Seal blubber comprises 29% of the total seal body weight and is the major lipid deposit in seals; however, lipids are also present in other tissues including skeletal muscles, kidneys, livers, hearts,

TABLE 15.10
Fatty Acid Compositions (%) of Selected Fish Oils

Fatty acid	Skipjack tuna[a]	Butterfish[b]	Yellowtail flounder[b]	Winter flounder[b]	Haddock[b]	Halibut[b]	Mackerel[b]	Salmon[c]
14:0	8.2	5.0	2.7	2.0	2.0	1.0	3.6	5.5
16:0	26.5	16.5	14.0	15.0	14.4	17.6	16.4	10.2
18:0	2.7	4.8	3.6	4.5	4.1	5.7	4.6	2.7
16:1	8.2	3.1	6.5	5.2	3.1	3.3	3.0	8.1
18:1	6.8	24.8	14.1	10.9	12.8	11.7	14.5	16.9
20:1	0.5	6.0	3.8	4.4	4.6	1.7	6.1	15.1
22:1	—	5.8	0.9	0.6	2.2	0.4	7.1	14.4
24:1	—	0.7	0.7	0.6	1.2	1.18	1.3	—
18:2 n-6	1.1	0.8	1.0	0.6	0.8	0.6	1.5	4.5
18:3 n-3	—	—	—	—	—	—	—	0.9
18:4 n-3	1.3	0.8	1.0	0.5	0.8	0.2	1.4	1.7
20:4 n-6	3.1	1.6	2.6	3.6	2.5	5.4	1.5	0.6
20:5 n-3	11.1	5.1	15.0	14.4	14.8	9.6	8.0	6.2
22:5 n-3	—	2.4	3.3	3.8	1.9	2.6	1.6	1.8
22:6 n-3	29.1	10.8	18.7	20.1	24.8	30.6	19.3	9.1

[a] Values from[31].
[b] Values from[29].
[c] Values from[32].

TABLE 15.11
Lipid Content (% Wet Weight) of Selected Tissues of Harp, Grey, Ringed, and Hooded Seals[34]

Tissue	Harp	Grey	Ringed	Hooded
Blubber	93.8	91.9	93.6	89.4
Muscle	1.9	1.8	1.9	2.4
Brain	8.1	10.3	6.9	7.4
Kidney	3.0	3.4	3.6	3.1
Heart	2.2	1.8	2.3	2.0
Liver	3.8	5.6	3.7	3.7
Lung	2.2	2.0	2.1	1.8

lungs, and brains. Seal milk is very rich in lipids[34]. Seal blubber contains between 90 and 94% lipids, of which 98.9% is neutral lipid[34]. The lipid contents of other seal tissues include 1.2% for muscle tissues, 7.8% in seal pelts, and 43% in seal milk[34] (Table 15.11). The most common seal species in North Atlantic Ocean waters are the harp seal (*Phoca groenlandica*), bearded seal (Erignathus barbatus), grey seal (*Halichoerus gyrus*), harbor seal (*Phoca vitulina*), hooded seal (*Cystophora cristata*), and ringed seal (*Phoca hispida*). The fatty acid compositions of seal species have been reviewed[35] and are reported in Table 15.12, showing that oleic acid is a main fatty acid in most seal blubber oils. The α-tocopherol content of crude and processed seal blubber oil has been reported to be 28 mg/kg (ppm) for crude seal blubber oil, 32 mg/kg (ppm) for alkali-refined

TABLE 15.12
Fatty Acid Compositions of Selected Seal Blubber Oils[35]

Fatty acid	Bearded	Grey	Harbor	Harp	Hooded	Ringed
12:0	0.05	0.14	—	—	0.21	0.02
14:0	2.76	3.52	4.52	3.73	6.10	1.68
15:0	0.40	0.29	—	0.23	0.29	0.22
16:0	9.96	9.16	8.03	5.98	8.61	3.23
17:0	0.80	0.62	—	0.92	0.47	0.41
18:0	2.30	0.52	0.85	0.88	0.86	0.73
19:0	0.52	0.15	—	—	0.06	0.35
20:0	0.23	0.03	1.19	0.11	0.08	0.07
14:1	0.26	0.14	—	1.09	2.29	0.56
16:1	14.05	19.84	19.26	18.02	22.49	9.52
17:1	0.33	0.50	—	0.55	0.40	0.27
18:1	21.69	32.85	23.76	20.83	25.34	16.21
19:1	0.33	0.14	—	—	0.03	0.08
20:1	8.99	8.06	9.06	12.16	8.05	8.17
22:1	3.26	0.73	0.31	2.01	2.00	2.01
24:1	0.56	—	—	—	—	0.76
16:2 n-6	0.44	0.07	—	—	0.06	0.32
18:2 n-9	0.82	—	—	—	0.15	0.17
18:2 n-6	0.16	0.97	1.27	1.51	1.89	0.59
20:2 n-9	0.24	—	—	—	—	0.13
20:2 n-6	0.12	0.10	—	0.16	0.09	0.10
22:2 n-6	—	—	—	—	0.01	0.20
16:3 n-6	<0.01	0.24	—	—	0.13	0.04
18:3 n-6	0.11	0.08	—	0.19	0.06	0.07
18:3 n-3	0.21	0.27	—	0.40	0.78	0.22
20:3 n-9	0.11	0.03	—	—	0.11	0.39
20:3 n-6	0.11	0.06	—	0.09	0.03	0.06
20:3 n-3	<0.01	0.03	—	0.05	0.03	0.06
18:4 n-3	0.73	0.69	—	1.00	1.75	0.75
20:4 n-6	0.76	0.66	0.44	0.46	0.51	0.37
22:4 n-6	0.51	0.10	—	0.11	—	1.16
20:5 n-3	9.27	5.23	9.31	6.41	4.29	10.57
21:5 n-3	0.51	0.16	—	—	0.15	0.21
22:5 n-3	4.76	4.94	4.22	4.66	3.48	14.55
22:6 n-3	13.38	7.12	7.76	7.58	7.47	26.19

seal blubber oil, 31 mg/kg (ppm) for refined-bleached seal blubber oil, and 24 mg/kg (ppm) for refined-bleached and deodorized seal blubber oil[36].

15.3.2 Whale

Large-scale whaling processes date back to the early 17th century, when the blubber and body oils of whales were a valuable commodity due to their use as fuels and as components of high-quality soaps and cosmetics. Like other marine mammals, the blubber of whale is a rich source of lipids, and these lipids are predominantly neutral lipids. The composition of whale oils differs from fish and seal oils in that whale oil contains wax esters (long-chain fatty alcohols esterified to fatty acids) in addition to triacylglycerols; in some cases wax esters comprise the entire neutral lipid fraction[33]. The purpose of wax esters in the blubber of whales may be to act as thermal insulation, biosensor, or buoyancy modifier. The blubber of the *Physeteridae* spp. and of the beaked whales contains predominantly wax esters (93 to 99%), sperm whale blubber contains 79% wax esters, and the dwarf sperm whale contains 42% wax esters[37,38]. Wax esters of 16:0, 16:1 occur in the bottlenose whale (*Physeter catodon*), as does pristine (a C49 long-chain unsaturated hydrocarbon)[39]. The blubbers of dolphins have not been extensively studied but have been shown to resemble whale blubber oil, containing moderate to high amounts of wax esters and long-chain hydrocarbons[40,41].

15.4 CRUSTACEAN AND CEPHALOPOD OILS

Crustacean and cephalopod species are popular seafoods but are not widely processed to extract their oils; however, reports of the fatty acid compositions and lipid class compositions of their oils have been reported. Crustaceans are members of the class Crustaces and are easily identified by their hard outer shells (exoskeletons), which are usually composed of polymers of chitin and calcium-based inorganics. Crustaceans include several shrimp species (order Stomatopoda), lobsters, crabs, and shellfish species (order Gasteropodes) including mussels, oysters, and clams[30]. The cephalopods are popular class of marine animals that characteristically do not have a hard endoskeleton or exoskeleton, and include octopuses, squids, and scallops[30].

15.4.1 SHRIMP

Shrimp has been reported to contain 1.8 to 2.6% lipids (wet weight), and the fatty acid composition of whole shrimp lipids have been reported[29,30]. *Pandalus borealis* oil contains 16% saturated fatty acids, 44% monounsaturated fatty acids, and 30.8% polyunsaturated fatty acids (Table 15.13)[29]. The lipid class composition of *Squilla mantis* oil has been reported by Passi et al.[30], showing 47.8% triacylglycerols and 17.8% phospholipids. The α-tocopherol content of *Squilla mantis* oil has been estimated at about 635 mg/kg (ppm), making it a rich source of this fat-soluble vitamin[30]. The fatty acid compositions of the triacylglycerol and phospholipid fractions of *Squilla mantis* oil have been studied and are reported in Table 15.14[30].

15.4.2 LOBSTER, CRAB, MUSSELS, OYSTER, AND CLAM

The lipid contents of lobsters and crabs are generally low (0.8 to 2.0%, wet weight), and as such the oils of these species are not produced commercially. The fatty acids of Nova Scotia crab and lobster reflect a high content of long-chain n-3 fatty acids as shown in Table 15.15[29].

Mussels (*Mytilus* spp.), oysters (*Ostrea* spp.), clams (*Spisula* spp.), and scallops (*Chlamys* spp.) are known as bivalves; they are filter feeding animals that occur on ocean floors up to near shore areas. Bivalves are commonly caught by sweeping the ocean floor with a type of net known as a dredge[27]. In shallow waters near shores, bivalves are simply collected during low tide. Although the lipid content of bivalves are generally below 1%, their lipid component contains several interesting and potentially health-promoting compounds including several types of sterols[27] and α-tocopherol[30]. Copeman and Parrish[27] have studied the lipid class (Table 15.16), fatty acid (Table 15.17), and sterol (Table 15.16) compositions of lipids of bivalve species from southern Labrador. Results from this work[27] show that bivalves have low amounts of lipids (0.5 to 0.8%, wet weight) that contain predominantly phospholipids. Bivalve lipids are rich in polyunsaturated fatty acids (50 to 64%) and EPA (~30% of all fatty acids), and contain a wide array of sterols but predominantly cholesterol[27]. Passi et al.[30] have also reported the lipid class and fatty acid compositions of lipid classes from several bivalve species, reporting lower amounts of phospholipids and higher amounts of triacylglycerols than Copeman and Parrish[27]. This group[30] has also shown that bivalve triacylglycerols contain higher amounts of saturated and monounsaturated fatty acids and lower amounts of polyunsaturated fatty acids than bivalve phospholipids. Passi et al.[30] have also shown that in the common mussel (*Mytilus edulis*), EPA is found mainly in triacylglycerols whereas DHA is found mainly in phospholipids.

TABLE 15.13
Fatty Acid Composition (%) of Shrimp (*Pandalus borealis*) Oil Fatty Acids[29]

Lipid content	2.6
Fatty acid	90.7
14:0	2.89
16:0	11.42
18:0	1.93
16:1 n-7	8.74
18:1 n-9	11.76
18:1 n-7	6.83
20:1 n-11	1.40
20:1 n-9	4.85
20:1 n-7	1.53
22:1 n-11	6.74
22:1 n-9	1.56
24:1	0.30
18:2 n-6	1.00
18:4 n-3	0.71
20:4 n-6	1.66
20:5 n-3	15.26
22:5 n-3	0.74
22:6 n-3	11.37

TABLE 15.14
Fatty Acid Compositions (%) of Triacylglycerols and Phospholipids of Shrimp (*Squilla mantis*) Oil[30]

Fatty acid	Triacylglycerol	Phospholipid
14:0	5.4	1.2
15:0	2.7	1.2
16:0	22.2	14.6
17:0	1.7	0.9
18:0	7.1	9.7
20:0	0.6	0
24:0	1.6	1.9
16:1	9.6	6.4
17:1	1.3	1.6
18:1	12.4	15.4
20:1	4.8	3.3
22:1	1.7	0.5
18:2 n-6	2.1	1.7
20:2 n-6	1.1	1.5
20:4 n-6	1.3	1.9
22:4 n.6	1.8	1.9
22:5 n-6	0.9	1.8
18:3 n-3	0.6	1.4
18:4 n-3	0.2	0.1
20:3 n-3	0.2	0
20:4 n-3	0.2	0.2
20:5 n-3	8.9	14.8
22:3 n-3	0.3	0.6
22:5 n-3	2.2	2.5
22:6 n-3	9.1	14.9

TABLE 15.15
Fatty Acid Composition (%) of Red Crab (*Geryon quinquedens*), Rock Crab (*Cancer irroratus*), and Lobster (*Homarus americanus*)[29]

Fatty acid	Red crab	Rock crab	Lobster
Lipid content	1.8	0.8	2.0
14:0	1.95	1.69	2.65
16:0	10.15	10.44	11.41
18:0	2.40	3.18	3.16
16:1 n-7	6.17	5.93	6.52
18:1 n-9	15.28	8.16	10.40
18:1 n-7	4.84	7.45	6.53
20:1 n-7	0.92	1.84	1.69
20:1 n-9	6.65	3.85	4.53
20:1 n-11	1.78	1.37	1.68
22:1 n-11	6.69	3.63	3.51
22:1 n-9	1.00	0.61	0.71
24:1	0.44	0.30	0.18
18:2 n-6	0.83	1.00	0.84
18:4 n-3	0.25	0.42	0.83
20:4 n-6	3.04	4.05	6.33
20:5 n-3	12.13	20.74	17.04
22:5 n-3	2.25	2.06	1.29
22:6 n-3	11.93	10.35	7.69

15.4.3 OCTOPUS AND SQUID

Octopus and squid species are generally low in fat but have lipid-rich livers from which edible oil is produced. The largest world market for octopus and squid is in Japan, where they are widely consumed[42]. The fatty acid composition of oils of octopus and squid resemble fish oils and are used in the aquaculture applications; however, oils from deep-water species have been shown to contain high amounts of wax esters and diacylglyceryl ethers which may make them unsuitable for human consumption[43]. The fatty acid compositions of oils from octopus (*Octopus vulgaris*) and two squid species (*Loligo vulgaris* and *Illex illecebrosus*) are presented in Table 15.18, showing that these animals are rich in EPA and DHA.

15.5 HEALTH EFFECTS OF MARINE OILS

15.5.1 CARDIOVASULAR DISEASE

It is becoming apparent that regular consumption of seafoods and marine oils containing long-chain polyunsaturated n-3 fatty acids lowers the rate of incidence and death from cardiovascular heart disease[44,45]. The cardioprotective effects of fish oils were first postulated in the 1950s based on cross-cultural studies done on Greenland Inuits and Danish settlers of Greenland[46]. These studies revealed that the Greenland Inuits had a significantly lower incidence of heart disease compared to the Danish settlers, despite comparable fat intakes (40% of caloric intake). This anomaly was referred to as the "Eskimo paradox"[47]. Epidemiological studies done in the 1970s suggested a strong correlation between the low incidence of coronary heart disease in Greenland Inuits and their high consumption of fish and marine mammals, both being rich in long-chain n-3 fatty acids[47]. Other cross-cultural epidemiological studies among coastal Japanese and Alaskan populations have resulted in similar findings, showing inverse relationships between long-chain polyunsaturated n-3 fatty acid intake and cardiovascular disease[48,49].

TABLE 15.16
Lipid Class and Sterol Compositions (%) of Bivalves from Southern Labrador[27]

	Surf clam	Greenland cockle	Blue mussel	Icelandic scallop
Total lipids	0.8	0.6	0.6	0.5
Lipid class				
Hydrocarbons	0.5	0.1	0.1	0.0
Sterol esters	0.4	1.0	1.8	0.0
Triacylglycerols	0.0	14.3	34.7	0.5
Diacylglycerols	0.7	0.7	0.8	0.0
Phospholipids	63.3	49.9	37.8	74.8
Free fatty acids	6.2	12.4	9.0	0.6
Sterols	18.1	10.9	8.6	21.3
Sterols (% total sterols)				
24-Nordehydrocholesterol	2.8	9.5	5.6	9.1
24-Nordehydrocholestanol	0.4	0.3	0.1	0.1
Occelasterol	1.9	2.6	2.8	3.7
trans-22-Dehydrocholesterol	10.2	10.1	8.0	11.5
Cholesterol	35.2	19.7	39.4	25.2
Cholestanol	1.4	0.5	0.8	0.2
Brassicasterol	9.2	5.8	8.8	11.7
C27 Steraldienol	0.0	3.9	7.5	0.0
Brassicastanol	0.3	0.2	0.1	0.0
Stellasterol	2.2	2.5	2.8	1.8
24-Methylenecholesterol	7.9	16.8	7.6	12.1
Campesterol	13.4	2.9	1.3	2.7
Stigmasterol	0.9	1.3	1.1	0.9
C28 Steradienol	2	8.2	5.4	6.2
Sitosterol	5.5	6.3	2.9	5.7
Fucosterol	0.0	3.8	1.5	3.9
C29 Stanol	2	0.9	1.9	0.9
Isofucosterol	0.8	0.8	0.6	0.7
Dinosterol	0.4	0.6	0.2	1.2
Dinostanol	2.9	3.5	1.8	2.5

The biochemical basis for cardioprotective effects of n-3 fatty acids are unknown but are probably multifactorial and may collectively result in increased heart rate variability (antiarrhythmic), reduced atheroma development (antiatherogenic), and decreased platelet reactivity/aggregation (antithrombotic). Investigations on the link between fish oils and cardiovascular disease in both animal and human models have concluded that this effect may be mediated by substrate competition between n-3 fatty acids and arachidonic acid (20:4, n-6) for cyclooxygenase (COX) enzymes that produce prostaglandins and thromboxanes. Competition between n-3 fatty acids and arachidonic acid could result in positive health benefits for the following reasons: (1) n-3 fatty acids inhibit the production of arachibonic acid through substrate competition for the Δ^6 desaturase[50]; (2) long-chain n-3 fatty acids compete with arachidonic acid for incorporation into the sn-2 position of membrane phospholipids thereby reducing membrane acachidonic acid levels[51]; and (3) eicosanoids produced from EPA have antiinflammatory and antiaggregatory effects, for example increasing the membrane EPA/arachidonic acid ratio shifts eicosanoid production from the proaggregatory eicosanoids PGI_2 and TXA_2 towards the antiaggregatory TXA_3 in platelets[51] and PGI_3 in endothelial cells[52]. These actions would result in vasodilation and decreased platelet aggregation, both having antithrombotic effects.

The oils of seal and whale blubber have been shown to exert beneficial cardiovascular health effects[2]. The spatial distribution of long-chain polyunsaturated fatty acids in marine mammal oils allows for faster and more thorough metabolism of these fatty acids, which may explain the

TABLE 15.17
Fatty Acid Composition (%) of Bivalves from Southern Labrador[27]

Fatty acid	Surf clam	Greenland cockle	Blue mussel	Icelandic scallop
14:0	2.6	6.5	4.1	2.0
i-15:0	1.1	0.1	0.0	0.0
ai-15:0	Tr	0.1	Tr	Tr
15:0	Tr	0.6	0.5	0.6
i-16:0	0.8	0.4	0.1	0.1
ai-16:0	Tr	0.5	0.4	0.1
16:0	10.5	10.2	12.1	16.6
i-17:0	1.2	1.0	0.4	0.5
ai-17:0	0.8	0.7	0.7	0.5
17:0	1.1	0.3	0.5	0.5
18:0	3.0	5.2	2.7	3.7
20:0	5.7	0.1	0.2	0.1
15:0	0.0	Tr	0.1	Tr
16:1n-5	0.7	0.4	0.3	0.5
16:1n-7	5.6	7.1	10.4	3.9
17:1	0.0	0.0	0.1	0.0
18:1n-5	2.0	0.3	0.2	0.2
18:1n-7	3.4	3.2	3	3.4
18:1n-9	1.5	1.2	1.1	1.9
18:1n-11	0.0	0.2	0.2	0.3
20:1n-7	4.3	0.2	1.9	0.7
20:1n-9	4.1	5.2	1.7	0.8
22:1	0.0	0.1	0.1	0.0
24:1	Tr	Tr	0.2	Tr
16:2n-4	Tr	0.8	0.9	0.2
16:3n-4	0.9	0.8	0.7	0.2
16:4n-1	Tr	0.2	0.7	0.2
16:4n-3	Tr	0.6	5.3	0.3
18:2n-4	0.9	0.4	0.4	0.3
18:2n-6	0.9	0.8	1.4	1.0
18:3n-3	Tr	0.5	0.9	0.4
18:3n-4	Tr	0.2	0.2	Tr
18:3n-6	Tr	0.1	0.2	0.1
18:4n-1	0.6	0.9	0.3	0.0
18:4n-3	1.0	1.1	3	3.1
20:2n-6	2.0	0.6	0.5	0.5
20:3n-3	Tr	0.2	0.3	0.1
20:3n-6	0.5	0.1	Tr	Tr
20:4n-3	Tr	0.4	0.1	0.1
20:4n-6	3.4	1.3	2	1.2
20:5n-3	22.9	22.6	19.6	26.9
21:5n-3	1.1	1.6	1.2	0.9
22:4n-6	Tr	0.2	0.1	0.3
22:5n-3	1.9	1.6	1	0.7
22:5n-6	Tr	0.1	0.0	0.0
22:6n-3	14.3	16.5	13.2	25.9

Tr: < 0.1%.

TABLE 15.18
Fatty Acid Composition (%) of Octopus and Squid Species[42]

Fatty acid	Octopus vulgaris	Loligo vulgaris	Illex illecebrosus
14:0	2.7	2.5	2.2
16:0	23.8	24.6	27.6
18:0	10.0	2.8	4.4
16:1	0.7	1.5	0.4
18:1 n-9	7.2	4.1	4.9
18:1 n-7	0.0	1.4	0.0
20:1	3.8	4.6	4.9
22:1	1.7	0.4	0.5
18:2 n-6	0.0	0.0	0.1
18:3 n-3	0.0	0.9	0.1
18:4 n-3	6.3	1.3	0.0
20:4 n-3	0.0	0.2	0.4
20:5 n-3	16.1	14.3	13.9
22:5 n-3	1.8	0.4	1.3
22:6 n-2	20.6	31.6	16.9

enhanced observed health effects of these oils compared to fish oils or vegetable sources of n-3 fatty acids. Also, marine mammal lipids contain DPA (up to ten times the amount present in fish oils), which has been implicated as being important in maintaining the soft, plaque-free integrity of arteries and as to stimulate endothelial cell migration, both of which having antiatherosclerotic effects[36]. Seal oil supplementation in humans has been shown to reduce cardiovascular disease risk factors including reduced risk of thrombosis and reduced plasma cholesterol[36].

The Diet and Reinfarction Trial (DART) was the earliest controlled trial to examine the effects of dietary intervention in the secondary prevention of myocardial infarction[53]. The study included 2033 Welsh men who had recovered from a previous heart attack and were allocated to one of two groups that received or did not receive advice on each of three dietary factors: reduced fat intake and an increased ratio of polyunsaturated to saturated fat, increased fatty fish intake, and increased cereal fiber intake. The results of this landmark trial revealed that men receiving at least two fatty fish meals per week (200 to 400g of fish meat or 1.5 g of fish oil through supplement per day) had a 29% reduction in fatal heart attack risk which was apparent after 4 months of intervention (RR 0.71, 95% CI: 0.54–0.93) when compared to the nonfish consuming groups, but there was no significant reduction in the incidence of nonfatal heart attacks. The results of this study strongly suggest that marine n-3 fatty acids have a specific antiarrhythmic effect rather than antiatherogenic or antithombotic effects[53].

The Gruppo Italiano per lo Studio della Sopravvienenza nell'Infarto Miocardio (GISSI) Prevenzione study was initiated in 1993 and was carried out for 3.5 years[54]. It was a multicentered trial conducted in Italy (172 centers) and included 11,324 patients who had suffered a heart attack less than three months prior to recruitment. Subjects were randomly assigned to one of the four groups in non-blinded fashion, one group receiving 0.85 g of purified eicosapentaenoic and docosahexaenoic acid ethyl esters as capsules, another receiving 300 mg of vitamin E through supplement, a third group receiving both marine n-3 fatty acid ethyl esters and vitamin E supplements at the stated amounts, and finally a control group receiving no supplement. Most of the participants were already on some type of drug regime to reduce their cardiovascular risk; drugs used included aspirin, beta-blockers, statins, and angiotensin converting enzyme inhibitors. The primary endpoints of the GISSI-Prevenzione study were the cumulative rates of the following: (1) total death, nonfatal heart attack, and nonfatal stroke, and (2) total cardiovascular death, nonfatal heart attack, and nonfatal stroke. Secondary analysis was carried out for each component in the primary endpoint. By the end of the

3.5-year study period there were 1017 (9.0%) coronary deaths and 1500 (13.3%) total coronary events. Results published in 1999[54] concluded that in the n-3 group there was a relative risk reduction of 15% in the primary endpoint (total death, nonfatal heart attack, and nonfatal stroke) and 20% in the other primary endpoint (cardiovascular death, nonfatal heart attack, and nonfatal stroke). Secondary analyses showed risk reductions of 30% for cardiovascular death and 45% for sudden cardiovascular death among the n-3 group[54]. These results for the n-3 group are more impressive than those of the DART study[53] considering that this intervention trial included only Italian heart disease patients whose diet habits were typical Mediterranean, implying better results may be seen in typical western diets high in saturated fats and low in polyunsaturates. The vitamin E intervention group showed significant, but less impressive results compared to the n-3 group, relative risk reductions compared to the control group showed a significant 14% reduction in risk for one primary endpoint (total death, nonfatal heart attack, and nonfatal stroke) and a nonsignificant 11% risk reduction for the other primary endpoint (cardiovascular death, nonfatal heart attack, and nonfatal stroke). The time-course data reanalysis of GISSI-Prevenzione study results for the n-3 group revealed that survival curves for n-3 fatty ethyl ester treatment diverged early after randomization, and total mortality was significantly lowered after 3 months of treatment (RR 0.59, 95% CI: 0.36–0.97, $P = 0.037$). The reduction in risk of sudden cardiac death was statistically significant after 4 months (RR 0.47, 95% CI; 0.23–0.99, $P = 0.048$). Just as in the DART study[53], the GISSI-Prevenzione study revealed that marine n-3 fatty acid intake conferred early and progressive risk reductions for cardiovascular disease.

The few dietary intervention trials assessing the cardioprotective effects of n-3 fatty acids do give promising results but are limited by the numerous dietary changes in these trials and the possibility that the cardioprotective effects observed in these trials are due to other components in the diets of these subjects. Further trials conducted at the multinational level are needed, not only to confirm the effects of n-3 fatty acid supplementation but also to explore their effects in societies with different dietary habits and risks for cardiovascular disease.

15.5.2 CHRONIC INFLAMMATION

Chronic inflammation associated with diseases such as inflammatory bowel disease, psoriasis, arteriosclerosis, and rheumatoid arthritis may be caused or attenuated by alterations of normal cytokine pathways resulting in overproduction of inflammatory cytokines. Many antiinflammatory pharmacotherapies have been developed to inhibit the production of proinflammatory cytokines. Omega-3 fatty acids may be of use in the treatment and/or management of inflammatory diseases because they have been shown to alter the pattern of cytokine biosynthesis both *in vitro* and *in vivo*[55], and currently there is much research investigating their therapeutic potential for inflammatory diseases.

Many animal models have been used to study the effects of n-3 fatty acid supplementation in inflammatory bowel disease. Shoda et al.[56] studied the effects of fish oil and safflower oil (n-6 fatty acid rich) supplementation on ulcer formation and proinflammatory cytokine production in rats. Their results showed that the fish oil group had less plasma LTB_4 levels and ulcer formation when compared to the safflower oil group. These researchers also showed a significant inverse association between LTB_4 formation and degree of ulceration ($p < 0.05$)[56]. Later, Shoda et al.[57] examined the incidence of Crohn's disease and dietary habits among Japanese men and women over a 19-year period, showing that individuals with lower dietary n-6/n-3 ratios were 21% less likely to suffer from Crohn's disease (RR 0.79)[57]. A similar study by Alsan and Triadafilopoulos[58] reported clinical improvements in 11 patients with moderate ulcerative colitis following fish oil supplementation. Not all studies have supported the therapeutic effects of n-3 fatty acids in inflammatory bowel disease sufferers. For example, Lorenz-Meyer et al.[59] performed a double-blind, placebo-based trial on 204 Crohn's disease patients in remission to study the effects of highly concentrated polyunsaturated n-3 fatty acids on the maintenance of remission over a 12-month period. The subjects of this trial were randomly distributed into a control group or an n-3 fatty acid group (experimental group). The experimental group was supplemented with 5.1 g of n-3 fatty acid ethyl esters per day while the control group was given a placebo. At the end of this trial there was no difference in the maintenance of remission between the n-3 and control groups; specifically, 30% of patients in both groups remained in remission[59]. However, at the end of this study[59] it was noticed that the n-3 group required less drug therapy (prednisolone) to manage the disease compared to the control group.

Several intervention trials have been conducted to study the effects of n-3 fatty acid supplementation in patients with arthritis, particularly rheumatoid arthritis[60]. Kremer et al.[61] examined the effects of dietary fat intake manipulation on clinical measures in patients with rheumatoid arthritis in a prospective, double-blind, placebo-based study that lasted 12 weeks. Seventeen patients consumed an experimental diet with a high ratio of polyunsaturated fat to saturated fat, and a 1.8g supplement of eicosapentaenoic acid daily. Twenty patients took a control diet with a lower polyunsaturated fat to saturated fat ratio and one placebo supplement daily. This research group[61] reported favorable results for the n-3 group after 12 weeks (reduced morning stiffness and number of tender joints), but their results do not directly implicate eicosapentaenoic acid as therapeutic because the observed benefits could have been caused by dietary modifications such as low saturated fat intake. Responses on a follow-up questionnaire revealed that the improvements observed in the n-3 group were totally abolished 4 to 8 weeks after the fish oil supplementation period, which allowed the authors to further attribute the beneficial results to the intervention regimen conducted on the n-3 group[61]. Volker et al.[62] performed a randomized, placebo-based, double-blind clinical study to determine the effects of fish oil supplementation (40 mg/kg/day, 60% n-3 fatty acids) on clinical variables in 50 rheumatoid arthritis patients who consumed less than 10 g of n-6 fatty acids per day. After 15 weeks of supplementation, there was a significant improvement ($p < 0.02$) in the clinical status of patients in the n-3 group compared to the placebo group. Five subjects in the n-3 group and three in the control group met the American College of Rheumatology 20% improvement criteria[62]. This research group[62] also reported significant incorporation of n-3 fatty acids into monocyte lipids and increased levels of plasma eicosapentaenoic acid in the n-3 group but not in the placebo group.

15.5.3 CANCER

Experimental and epidemiological studies have demonstrated that the composition of dietary fat affects the incidence and progression of some cancers; n-3 fatty acids have been shown to have anticarcinogenic effects while saturated and n-6 fatty acids may promote cancer development[63].

Early evidence from epidemiological studies indicated that n-3 fatty acids might be protective against prostate cancer. Cross-cultural studies among the Inuit and non-Inuit peoples of Canada, Alaska, and Greenland from 1969 to 1988 showed that the incidence rate of prostate cancer among the Inuit populations were 70 to 80% less than the non-Inuit populations[63]. This observation was attributed to dietary differences between the two populations, in particular the traditional seafood diet of Inuit peoples that are exceptionally rich in n-3 fatty acids were speculated as having anticarcinogenic effects[63]. Terry et al.[64] studied the association between fatty fish consumption and prostate cancer in a long-term prospective cohort of 6272 Swedish men. In 1967, 107-item questionnaires investigating dietary and lifestyle habits were sent to all participants. Participants were followed up until diagnosis of prostate cancer, death, or end of the follow-up period (December 31, 1997). Follow-up data for each participant were obtained from the Swedish National Cancer Register and Swedish National Death Register, no follow-up questionnaires were sent to participants. After 30 years of follow-up, an inverse association between fatty fish consumption and prostate cancer was observed. Interestingly, fish consumption was positively associated with healthy diet and lifestyle habits.

Western populations exhibit significantly higher colon cancer incidence and mortality rates compared to Asian populations, which experts have long associated with high dietary fat and animal fat consumption by Western populations[65]. Caygill et al.[66] examined colon cancer mortality data from 24 European countries, showing a significant inverse correlation between colon cancer mortality and fish meat/fish oil consumption. This inverse correlation was significant for both men and women who consumed fish or fish oil for 1 year, 10 years, or 23 years before cancer mortality.

This study strongly suggests that fish oil consumption can significantly reduce colorectal cancer mortality. Unfortunately dietary amounts of fish or fish fatty acids were not adequately assessed in this study; making it impossible to assess critically these findings[66].

Experimental, animal, and human studies have shown that high dietary fat intake increases breast cancer risk[67]. Similar studies show n-6 fatty acids promote while n-3 fatty acids inhibit breast cancer development[66]. Holmes et al.[68] analyzed data from 1982 breast cancer patients (mean age 54 years) registered in the 18-year Nurses Health Study. The results of this assessment showed that n-3 fatty acid intake significantly reduced breast cancer mortality by 48% (RR = 0.52; 95% CI: 0.30–0.93). Recently, Holmes et al.[69] re-examined the data of the 121,700 female nurses registered in the Nurses Health Study to find associations between breast cancer and dietary intake of meat, fish meat, and eggs. Data from food frequency questionnaires sent in 1976 and every two years thereafter until 1994 were compiled and organized into quintiles for highest to lowest intakes of meat fish and eggs. After the 18-year follow-up period 4107 cases of breast cancer were diagnosed. Women in the highest quintile for meat, egg, and fish intake showed no difference in breast cancer risk; secondary analyses did not affect these results[69]. Similar results have been observed in the 8-cohort international pooling project involving 350,000 women who were followed up for 15 years[70]. The proposed antitumorigenic effects of n-3 fatty acids in breast cancer have been studied using *in vitro* and animal models of this disease; several studies show n-3 fatty acids are able to modulate second messenger systems and cell signaling cascades in cancerous breast cells. Protein kinase-A (PKA) and protein kinase-C (PKC) are overexpressed in tumors of most poor-prognosis breast cancer patients, and inhibition of these second messenger system components are reported to arrest the development of cancerous breast cells[71]. Several nonhuman studies support the premise that n-3 fatty acids inhibit breast carcinoma development by influencing the biochemical events that follow tumor initiation. Unfortunately, these findings do not correlate well with human breast cancer studies. This may imply that n-3 fatty acids at attainable human dietary levels (1 to 3% of total calories) do not affect breast cancer development.

15.5.4 Mental Health and Development

Docosahexaenoic acid is especially important during prenatal human brain development; incorporation of docosahexaenoic acid into growing neurons is a prerequisite for synaptogenesis (formation of synapses)[72]. The importance of n-3 fatty acids during prenatal development are best indicated by the observation that deficiency of these fatty acids during development greatly increases the likelihood of diminished visual acuity, cerebellar dysfunction, and several cognitive impairments and neurological disorders[73]. The importance of n-3 fatty acids during human development is also evident by the fact that both the placenta and mammary tissues supply large amounts of docosahexaenoic acid to developing young[74].

The effects of n-3 fatty acids on the clinical symptoms of depression and schizophrenia have received considerable attention. Recent studies among Inuit populations show an overall decline in mental health characterized by increased rates of depression as well as other mental illnesses, which may be linked to the rapid alteration of their traditional culture to a more westernized one that has led to dietary changes from traditional seafoods to processed foods[75]. The fact that eicosapentaenoic acid-derived eicosanoids are the least proinflammatory eicosanoids provides a possible explanation for the beneficial effects of n-3 fatty acid supplementation in depression, since depression has been linked to proinflammatory cytokine production. Because the eicosanoid products of n-3 fatty acids do not activate macrophages to any extent compared to those derived from n-6 fatty acids, replacement of membrane n-6 fatty acids with n-3 fatty acids would reduce proinflammatory cytokine production, especially if cyclooxygenase activity is enhanced in depressive patients. Several lines of evidence support the beneficial effects of n-3 fatty acids on depressive disorders, but this evidence is far from conclusive and currently no mental health agency warrants n-3 fatty acid supplementation for the treatment of depressive symptoms[76].

Many research groups have evaluated the effects of n-3 fatty acid supplementation in schizophrenia. Previous family history of schizophrenia is the major risk factor for this disease; however, oxidative injury to neuronal cells and abnormal neuronal membrane phospholipid composition have been observed in schizophrenic patients post mortem[77]. Reduced docosahexaenoic acid levels have been observed in neurons of schizophrenic patients that may be the result of phospholipase A_2 overexpression[78]. Many schizophrenic patients show signs of excessive *in vivo* lipid peroxidation; these include increased plasma thiobarbituric acid reactive substances (TBARS)[79] and breath pentane[80], which suggests that reduced docosahexaenoic acid in schizophrenics may be due to increased oxidative stress. Hibbeln et al.[81] recently quantified the erythrocyte fatty acid compositions of 76 medicated schizophrenic patients before and after 16 weeks of eicosapentaenoic acid (3 g/day) or placebo supplementation. Several schizophrenic indices were performed on each patient before and after the supplementation period. Although plasma eicosapentaenoic acid levels were increased in the n-3 fatty acid group ($P < 0.05$, Mann–Whitney tests), these differences did not correlate with reduced schizophrenia symptoms[81].

15.6   CONCLUSIONS

Marine oils differ widely with respect to their lipid compositions, but are generally characterized by moderate to large amounts of both saturated and polyunsaturated fatty acids, the latter being rich in long-chain polyunsaturated n-3 fatty acids.

Marine oils are derived from the whole body of fish or from certain parts including muscle, blubber, and liver, among others. The livers of marine animals are known to serve as rich sources of lipids and especially lipid-soluble vitamins. All marine oils are abundant in long-chain polyunsaturated fatty acids such as EPA and DHA; however, only marine mammal oils contain a significant proportion of DPA. The positional distribution of triacylglycerols of marine lipids exhibit some marked differences depending on their specific source: fish oils have long-chain polyunsaturated fatty acids occupying the *sn*-2 position of triacylglycerols, whereas marine mammals have long-chain polyunsaturated fatty acids predominantly in the *sn*-1 and *sn*-3 positions of the triacylglycerols molecules.

Regular consumption of marine animals and/or their oils are known to promote health, attributed mainly to their long-chain polyunsaturated n-3 fatty acids. However, more studies are needed in order to establish optimal doses and ideal sources of marine animal oils for the many conditions they have been speculated to influence.

REFERENCES

1. Harris, W.S., Fish oil supplementation: evidence for health benefits, *Cleveland Clin. J. Med.*, 71, 208–221, 2004.
2. Dyerberg, J.B., Bang, H.O., Stoffersen, E., Moncada, S., and Vane, J.R., Eicosapentaenoic acid and prevention of thrombosis and atherosclerosis?, *Lancet*, 2, 117–119, 1978.
3. Leaf, A.K., Xiao, Y.F., and Billman, G.E., Clinical prevention of sudden cardiac death by n-3 polyunsaturated fatty acids and mechanism of prevention of arrhythmias by n-3 fish oils, *Circulation*, 107, 2646–2652, 2003.
4. Nair, S.S.L. and Garg, M.L., Specific modifications of phosphatidylinositol and nonesterified fatty acid fractions in cultured porcine cardiomyocytes supplemented with n-3 polyunsaturated fatty acids, *Lipids*, 34, 697–704, 1999.
5. Kris-Etherton, P.M.H., W.S., and Appel, L.J., Fish consumption, fish oil, omega-3 fatty acids, and cardiovascular disease, *Circulation*, 106, 2747–2757, 2002.
6. Shahidi, F. and Miraliakbari., H., Omega-3 (n-3) fatty acids in health and disease: 1. Cardiovascular disease and cancer, *J. Med. Food*, 7, 387–401, 2004.
7. Jonzo, M.D.H., Zagol, L., Druet, D., and Comeau, L., Concentrates of DHA from fish oil by selective esterification of cholesterol by immobilized isoforms of lipase from Candida *rugosa*, *Enzyme Microbial Technol.*, 27, 443–450, 2000.

8. Bimbo, A.P., Production of fish oil, in *Fish Oils in Nutrition*, Stansby, M.E., Ed., Van Nostrand Reinhold, New York, 1990, pp. 141–180.

9. *Fishmeal and Oil: Supplies and Markets*, Barlow, S.M., Ed., International Fishmeal and Fish Oil Organization, UK, 2002. Published online at: http://www.iffo.org.uk/tech/alaska.htm

10. Ackman, R.G., Fish oils, in *Bailey's Industrial Oil and Fat Products*, Shahidi, F., Ed., John Wiley, NJ, 2005, pp. 279–317.

11. Nichols, D.S. and Davies, N.W., Improved detection of polyunsaturated fatty acids as phenacyl esters using liquid chromatography-ion trap mass spectrometry, *J. Microbiol. Methods*, 50, 103–113, 2002.

12. Torres, C.F. and Hill, C.G., Lipase-catalyzed acidolysis of menhaden oil with conjugated linoleic acid: effect of water content, *Biotechnol. Bioeng.*, 78, 509–516, 2002.

13. Rice, K.E., Watkins, J., and Hill, C.G., Hydrolysis of menhaden oil by a *Candida cylindracea* lipase immobilized in a hollow-fiber reactor, *Biotechnol. Bioeng.*, 63, 33–45, 1999.

14. Porsgaard, T. and Hoy, C.E., Lymphatic transport in rats of several dietary fats differing in fatty acid profile and triacylglycerol structure, *J. Nutr.*, 130, 1619–1624, 2000.

15. Yuan, Y.V., Kitts, D.D., and Godin, D.V., Variations in dietary fat and cholesterol intakes modify antioxidant status of SHR and WKY rats, *J. Nutr.*, 128, 1620–1630, 1999.

16. Pakala, R., Pakala, R., and Benedict, C., Eicosapentaenoic acid and docosahexaenoic acid selectively attenuate U46619-induced smooth muscle cell proliferation, *Lipids*, 34, 915–920, 1999.

17. Lopez, D., Orta, X., Casos, K., Saiz, M.P., Puig-Parellada, P., Farriol, M., and Mitjavila, M.T., Upregulation of endothelial nitric oxide synthase in rat aorta after ingestion of fish oil-rich diet, *Am. J. Physiol. Heart Circ. Phys.*, 287, 567–572, 2004.

18. Hilakivi-Clarke, L., Cho, E., Cabanes, A., DeAssis, S., Olivo, S., Helferich, W., Lippman, M.E., and Clarke, R., Dietary modulation of pregnancy estrogen levels and breast cancer risk among female rat offspring, *Clin. Cancer Res.*, 8, 3601–3610, 2002.

19. Welsch, C.W., Dietary fat, calories, and mammary gland tumorigenesis, *Adv. Exp. Med. Biol.*, 322, 203–222, 1992.

20. Goodie, G.K., Garcia, S., and Heagerty, A.M., Dietary supplementation with marine fish oil improves in vitro small artery endothelial function in hypercholesterolemic patients: a double-blind placebo-controlled study, *Circulation*, 96, 2802–2807, 1997.

21. United States Department of Agriculture, USDA National Nutrient Database for Standard Reference Version 17, http://www.nal.usda.gov/fnic/foodcomp/ search/ (accessed May 6, 2005).

22. Adios, I., Van Der Padt, A., Luten, J.B., and Boom, R.M., Seasonal changes in crude and lipid composition of herring fillets, byproducts, and respective produced oils, *J. Agric. Food Chem.*, 50, 4589–4599, 2002.

23. Adios, I., Schelvis-Smit, R., Veldman, M., Luten, J.B., Van Der Padt, A., and Boom, R.M., Chemical and sensory evaluation of crude oil extracted from herring byproducts from different processing operations, *J. Agric. Food Chem.*, 51, 1897–1903, 2003.

24. Adios, I., Van Der Padt, A., Boom, R.M., and Luten, J.B., Upgrading of maatjes herring byproducts: production of crude fish oil, *J. Agric. Food Chem.*, 49, 3697–3701, 2001.

25. Sidhu, K.S., Health benefits and potential risks related to consumption of fish or fish oil, Regulatory Toxicol. Pharmacol., 38, 336–344, 2003.

26. Slover, H.T. and Thompson, R.H., Determination of tocopherols and sterols by capillary gas chromatography,. *J. Am. Oil Chem. Soc.*, 60, 1524–1528, 1983.

27. Copeman, L. and Parrish, C.C., Lipids classes, fatty acids, and sterols in seafood from Gilbert Bay, Southern Labrador, *J. Agric. Food Chem.*, 52, 4872–4881, 2004.

28. Brox, J., Olaussen, K., Osterud, B., Elvevoll, E.O., Bjornstad, E., Brattebog, G., and Iversen, H., A long-term seal- and cod-liver-oil supplementation in hypercholesterolemic subjects, *Lipids*, 36, 7–13, 2001.

29. Budge, S.M., Iverson, S.J., Bowen, W.B., and Ackman, R.G., Among- and within-species variability in fatty acid signatures of marine fish and invertebrates on the Scotian Shelf, Georges Banks and southern Gulf of St. Lawrence, *Can. J. Fish. Aqua. Sci.*, 59, 886–898, 2002.

30. Passi, S., Cataudella, S., Di Marco, P., De Simone, F., and Rastrelli, L., Fatty acid composition and antioxidant levels in muscle tissue of different Mediterranean marine species of fish and shellfish, *J. Agric. Food Chem.*, 50, 7314–7322, 2002.

31. Tanabe, T., Suzuki, T., Ogura, M., and Watanabe, Y., High proportion of docosahexanenoic acid in the lipid of juvenile and young skipjack tuna, *Katsuwonus pelamis* from the tropical western pacific, *Fish. Sci.*, 65, 806–807, 1999.

32. Aursand, M., Bleivik, B., Rianuzzo, J.R., Jorgensen, L., and Mohr, V., Lipid distribution and composition of commercially farmed Atlantic salmon (*Salmo salar*), *J. Sci. Food Agric.*, 64, 239–248, 1994.

33. Ackman, R.G., Ed., *Marine Biogenic Lipids, Fats and Oils*, CRC Press, Boca Raton, FL, 1989.

34. Durnford, E. and Shahidi, F., Analytical and physical chemistry: comparison of FA compositions of selected tissues of phocid seals of eastern Canada using one-way and multivariate techniques, *J. Am. Oil Chem. Soc.* 79, 1095–1102, 2002.

35. Shahidi, F., Ed., *Seal Fishery and Product Development*, Sciencetech, St. John's, 1998.

36. Shahidi, F. and Zhong, Y., Marine mammal oils, in *Bailey's Industrial Oil and Fat Products*, Shahidi, F., Ed., John Wiley, NJ, 2005, pp. 259–278.

37. Litchfield, C., Greenberg, A.J., Ackman, R.G., and Eaton, C.A., Distinctive medium chain wax esters, triglycerides, and diacyl glyceryl ethers in the head fats of the Pacific beaked whale, Berardius bairdi, *Lipids*, 13, 860–866, 1978.

38. Bottino, N.R., Lipids of the antarctic sei whale, Balaenoptera borealis, *Lipids*, 13, 18–23, 1978.

39. Litchfield, C., Greenberg, A.J., Caldwell, D.K., Caldwell, M.C., Sipos, J.C., and Ackman, R.G., Comparative lipid patterns in acoustical and nonacoustical fatty tissues of dolphins, porpoises and toothed whales, *J. Comp. Physio.* (B), 50, 591–597, 1975.

40. Koopman, H.N., Iverson, S.J., and Read, A.J., High concentrations of isovaleric acid in the fats of odontocetes: variation and patterns of accumulation in blubber vs. stability in the melon. *J. Comp. Physio.* (B), 173, 247–261, 2003.

41. Hooker, S.K., Iverson, S.J., Ostrom, P., and Smith, S.C., Diet of northern bottlenose whales inferred from fatty-acid and stable-isotope analyses of biopsysamples, *Can. J. Zool.*, 79, 1442–1454, 2001.

42. Arts, M.T., Ackman, R.G., and Holub, B.J., "Essential fatty acids" in aquatic ecosystems: a crucial link between diet and human health and evolution, *Can. J. Fish. Aqua. Sci.*, 58, 122–137, 2001.

43. Hayashi, K., Wax esters in the stomach content lipids of gonatid squid Gonatopsis borealis, *Nippon Suisan Gakkaishi*, 55, 1463, 1989.

44. Albert, C.M., Campos, H., Stampfer, M.J., Ridker, P.M., Manson, J.E., Willett, W.C., and Ma, J., Blood levels of long-chain n-3 fatty acids and the risk of sudden death, *N. Engl. J. Med.*, 346, 1113–1118, 2002.

45. Hu, F.B., Bronner, L., Willet, W.C., Stampfer, M.J., Rexrode, K.M., Albert, C.M., Hunter, D., and Manson, J.E., Fish and omega-3 fatty acid intake and risk of coronary heart disease in women, *J. Am. Med. Assoc.*, 287, 1815–1821, 2002.

46. Sinclair, H.M., Deficiency of essential fatty acids and atherosclerosis, etcetera, *Lancet*, 267, 381–383, 1956.

47. Bang, H.O., Dyerberg, J., and Sinclair, H.M., The composition of the Eskimo food in north western Greenland, *Am. J. Clin. Nutr.*, 33, 2657–2661, 1980.

48. Hirai, A., Terano,T., Tamura, Y., and Yoshida, S., Eicosapentaenoic acid and adult diseases in Japan: epidemiological and clinical aspects, *J. Int. Med.*, 225 (Suppl.), 69–75, 1989.

49. Davidson. M., Bulkow, L.R., and Gellin, B.G., Cardiac mortality in Alaska's indigenous and non-native residents, *Int. J. Epidemiol.*, 22, 62–71, 1993.

50. Garg, M.L., Sebokova, E., and Thompson, A.B.R., Δ^6-Desaturase activity in liver microsomes of rats fed diets enriched with cholesterol and/or n-3 fatty acids, *Biochem. J.*, 249, 351–356, 1988.

51. Coker, S.J. and Parratt, J.R., AH23848, a thromboxane antagonist, suppresses ischemia and reperfusion induced in anaesthetized greyhounds, *Br. J. Pharm.*, 86, 259–264, 1985.

52. Fischer, S. and Webber, P.C., Prostaglandin I_3 is formed in-vivo in man after dietary eicosapentaenoic acid, *Nature*, 307, 165–168, 1984.

53. Burr, M.L., Fehily, A.M., Gilbert, J.M., Rodgers, S., Holliday, R.M., Sweetnam, P.M., Elwood, P.C., and Deadman, N.M., Effects of changes in fat, fish, and fibre intakes on death and myocardial reinfarction: Diet and Reinfarction Trial (DART), *Lancet*, 2, 757–761, 1989.

54. GISSI-Prevenzione study investigators, Dietary supplementation with n-3 polyunsaturated fatty acids and vitamin E in 11,324 patients with myocardial infraction: results of the GISSI-Prevenzione trial, *Lancet*, 354, 447–455, 1999.

55. Meydani, S.N., Endres, S., Woods, M.M., Goldin, B.R., Soo, C., Morrill-Labrode, A., Dinarello, C.A., and Gorbach, S.L., Oral n-3 fatty acid supplementation suppresses cytokine production and lymphocyte proliferation: comparison between young and older women, *J. Nutr.*, 121, 547–55, 1991.

56. Shoda, R., Matsueda, K., Yamato, S., and Umeda, N., Therapeutic efficacy of N-3 polyunsaturated fatty acid in experimental Crohn's disease, *J. Gastro.*, 30, 98–101, 1995.

57. Shoda, R., Matsueda, K., Yamato, S., and Umeda, N., Epidemiologic analysis of Crohn's disease in Japan: increased dietary intake of n-6 polyunsaturated fatty acids and animal protein relates to the increased incidence of Crohn's disease in Japan, *Am. J. Clin. Nutr.*, 63, 741–745, 1996.

58. Alsan, A. and Triadafilopoulos, G., Fish oil fatty acid supplementation in active ulcerative colitis: a double blind, placebo controlled, crossover study, *Gut*, 35, 345–357, 1993.

59. Lorenz-Meyer, H., Bauer, P., Nicolay, C., Schultz, B., Purrmann, J., Fleig, W.E., Scheurlen, C., Koop, I., Pudel, V., and Carr, L., Omega-3 fatty acids and low carbohydrate diet for the maintainance of remission in Crohn's disease. a randomized controlled multicenter trial (German Crohn's Disease Study Group), *Scand. J. Gastro.*, 31, 778–785, 1996.

60. Cleland, L.G., James, M.J., and Proudman, S.M., The role of fish oils in the treatment of rheumatoid arthritis, *Drugs*, 69, 845–853, 2003.

61. Kremer, J.M., Bigauoette, J., Michalek, A.V., Timchalk, M.A., Lininger, L., Rynes, R.I., Huyck, C., and Zieminski, J., Effects of manipulation of dietary fatty acids on clinical manifestations of rheumatoid arthritis, *Lancet*, 1, 184–187, 1985.

62. Volker, D., Fitzgerald, P., Major, G., and Garg, M., Efficacy of fish oil concentrate in the treatment of rheumatoid arthritis, *J. Rheum.*, 10, 2343–2346, 2000.

63. Prener, A., Storm, H.H., and Nielsen, N.H., Cancer of the male genital tract in circumpolar Inuit, Acta Oncologica (Stockholm), 35, 589–593, 1996.

64. Terry, P., Lichtenstein, P., Feychting, M., Ahlbom, A., and Wolk, A., Fatty fish consumption and risk of prostate cancer, *Lancet*, 357, 1764–1766, 2001.

65. Wynder, E.L., Kajitani, T., Ishikawa, S., Dodo, H., and Takano, A., Environmental factors of cancer of the colon and rectum: II. Japanese epidemiological data, *Cancer*, 12, 1210–1220, 1969.

66. Caygill, C.P., Charlett, A., and Hill, M.J., Fat, fish oil and cancer, *Br. J. Cancer*, 74, 159–164, 1996.

67. Wu, A., Pike, M., and Stram, D., Meta analysis: dietary fat intake, serum estrogen levels, and the risk of breast cancer, *J. Natl. Cancer Inst.*, 91, 529–534, 1999.

68. Holmes, M.D., Stampfer, M.J., Colditz, G.A., Rosner, B., Hunter, D.J., and Willet, W.C., Dietary factors and the survival of women with breast carcinoma, *Cancer*, 86, 826–835, 1999.

69. Holmes, M.D., Colditz, G.A., Hunter, D.J., Hakinson, S.E., Rosner, B., Speizer, F.E., and Willet, W.C., Meat, fish and egg intake and risk of breast cancer, *Int. J. Cancer Res.*, 104, 221–227, 2003.

70. Bougnoux, P., n-3 Fatty acids and cancer, *Curr. Opin. Nutr. Metabolic Care*, 2, 121–126, 1999.

71. Rose, D.P. and Connolly, J.M., Omega-3 fatty acids as cancer chemopreventative agents, *J. Natl. Cancer Inst.*, 83, 217–244, 1999.

72. Martin, R.E. and Bazan, N.G., Changing fatty acid content of growth cone lipids prior to synaptogenesis, *J. Neurochem.*, 59, 318–325, 1992.

73. Chamberlain, J.G., Fatty acids in human brain phylogeny, *Perspect. Biol. Med.*, 436–445, 1996.

74. Neuringer, M., Cerebral cortex docosahexaenoic acid is lower in formula fed than in breast fed infants, *Nutr. Rev.*, 51, 238–241, 1993.

75. McGrath-Hanna, N.K., Greene, D.M., Tavernier, R.J., and Bult-Ito, A., Diet and mental health in the arctic: is diet an important risk factor for mental health in circumpolar peoples? A review, *Int. J. Circumpolar Health*, 62, 228–241, 2003.

76. Keller, J.R., Omega-3 fatty acids may be effective in the treatment of depression, *Topics Clin. Nutr.*, 17, 21–27, 2002.

77. Mahadik, S.P., Evans, D., and Lal, H., Oxidative stress and role of antioxidant and n-3 essential fatty acid supplementation in schizophrenia, *Prog. Neuropsychol. Pharmacol. Biol. Psychiatry*, 25, 463–493, 2001.

78. Horrobin, D.F., The relationship between schizophrenia and essential fatty acids and eicosanoid metabolism, *Prostaglandins, Leukotrienes Essen. Fatty Acids*, 46, 71–77, 1992.

79. McCreadie, R.G., Macdonald, E., Wiles, D., Campell, G., and Patterson, J.R., Plasma lipid peroxide and serum vitamin E levels in patients with and without tardive dyskinesia and normal subjects. *Br. J. Psych.*, 167, 1–8, 1995.

80. Phillips, M., Erickson, G.A., Sabas, N., Smith, J.P., and Greenberg, J., Volatile organic compounds in the breath of schizophrenic patients, *J. Clin. Pathol.*, 48, 466–469, 1995.

81. Hibbeln, J.R., Makino, K.K., Martin, C.E., Dickerson, F., Boronow, J., and Fenton, W.S., Smoking, gender, and dietary influences on erythrocyte essential fatty acid composition among patients with schizophrenia or schizoaffective disorder, *Biol. Psychiatry*, 53, 431–441, 2003.

16 Single-Cell Oils as Sources of Nutraceutical and Specialty Lipids: Processing Technologies and Applications

S.P.J.N. Senanayake and Jaouad Fichtali
Martek Biosciences Corporation, Winchester, Kentucky

CONTENTS

16.1 INTRODUCTION

16.1.1 BIOLOGICAL SIGNIFICANCE OF POLYUNSATURATED FATTY ACIDS

The therapeutic significance of n-3 polyunsaturated fatty acids (PUFAs), especially docosa-hexaenoic acid (DHA; 22:6n-3) and eicosapentaenoic acid (EPA; 20:5n-3), has been demonstrated by numerous clinical and epidemiological studies[1-3]. The traditional source of n-3 fatty acids is fish oil. However, the use of fish oil as a food component is limited due to problems associated with its typical fishy smell, unpleasant taste, and poor oxidative stability. Furthermore, the presence of EPA in fish oil is considered undesirable for application in infant food[4]. Thus, alternative sources are of interest. Considerable evidence has indicated that n-3 fatty acids in fish oils actually derive from zooplankton that consumes algae[5]. Therefore, microalgae are considered as the most promising source of DHA.

DHA is a long-chain fatty acid with 22 carbon atoms and 6 methylene-interrupted *cis* double bonds (Figure 16.1). DHA, along with EPA, is believed to provide the health benefits associated with the consumption of certain marine fish and their oils. DHA accumulates in the membranes of human nervous, visual, and reproductive tissues and is also the most abundant fatty acid in the gray matter of the brain. In addition, DHA is considered to be particularly important in infant nutrition and brain development. In the fetus and infant, DHA is required for the development and maturing of the eye, where it constitutes about 25% of the fatty acids in the retina, and of the nervous system, where it makes up about 12 to 15% of total fatty acids in gray matter of the cortex and brain stem[6]. The gray matter of the brain and cells of the retina have the highest concentration of DHA of any tissue in the body[7]. The demand for DHA is highest during the latter part of pregnancy and in the first few months of infancy[8]. EPA does not appear to play a particular role in these processes, other than serving as a precursor of DHA.

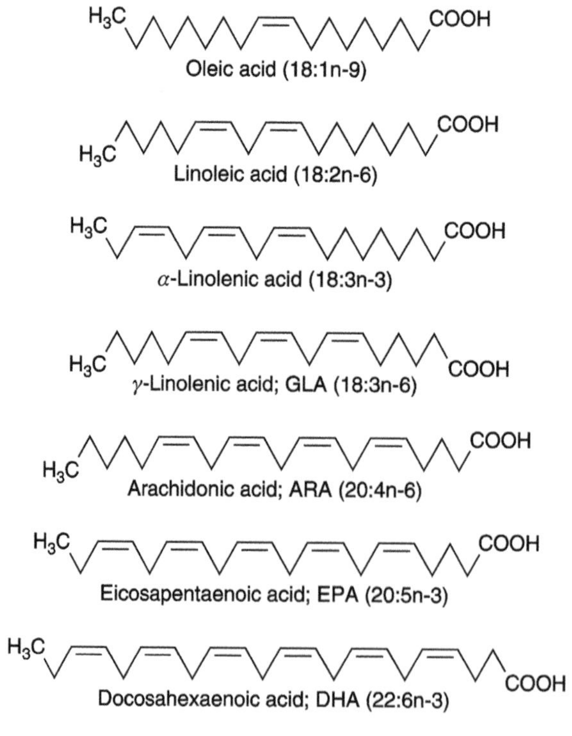

FIGURE 16.1 Chemical structures of selected long-chain PUFAs.

The n-6 fatty acids are another important group of PUFAs. There are several different types of n-6 fatty acids. Examples of n-6 fatty acids include linoleic acid (18:2n-6), γ-linolenic acid (GLA; 18:3n-6), dihomo-γ-linolenic acid (DGLA; 20:3n-6), and arachidonic acid (ARA; 20:4n-6). The n-6 PUFAs have at least four roles: (1) modulation of membrane structure; (2) formation of short-lived, local regulating, biologically active molecules; (3) control of water impermeability of the skin; and (4) regulation of cholesterol transport and cholesterol synthesis. An essential fatty acid, linoleic acid is the chief n-6 PUFA in the North American diet. Most n-6 fatty acids in the diet come from vegetable oils. Good dietary sources of n-6 fatty acids include cereals, eggs, poultry, most vegetable oils, whole-grain breads, baked goods, and margarine. There has been increasing interest in the microbiological production of ARA. Owing to its high physiological activity, ARA has found wide application in medicine, pharmacology, cosmetics, the food industry, agriculture, and other fields.

16.1.2   Nutritional Significance of DHA and ARA

Although common plant sources and vegetable oils lack DHA, some do contain significant amounts of the n-3 fatty acid α-linolenic acid (ALA; 18:3n-3), which in mammalian organisms can be metabolically converted to DHA via desaturation and elongation reactions. However, the conversion efficiency of ALA to DHA is very limited in human adults with conversion efficiencies estimated to be approximately 5%[9] and even more limited in infants where the conversion efficiency of ALA to DHA appears to be less than 1%[10]. Thus, the most direct means of maintaining optimum levels of DHA in the human body is by direct consumption of preformed dietary DHA such as that from microalgae.

As noted earlier, DHA and EPA are present in fish oils as a result of the bioaccumulation through the marine food chain. DHA is naturally synthesized only by some algae and some algae-like microorganisms. As outlined in Figure 16.2, DHA is synthesized from ALA by a series of alternating desaturation (in which a carbon–carbon double bond is introduced) and elongation (in which two carbon atoms are added) reactions. DHA can be metabolically converted back into EPA via retroconversion. Although this reaction is believed to be a minor pathway in humans[11], there is evidence for retroconversion of DHA to docosapentaenoic acid (DPA; 22:5n-3) and EPA in the human body[12]. After ingestion of ethyl esters of DHA, the DHA and EPA in plasma phospholipids were increased, but DPA remained essentially unchanged[13]. However, ingestion of DHA increased the levels of DHA, DPA, and EPA in the phosphatidylcholine (PC) and phosphatidylethanolamine (PE) fractions of the platelets. Blood platelet aggregation was significantly decreased by ingestion of DHA, thus supporting the view that dietary n-3 PUFAs may alleviate certain forms of cardiovascular dysfunction.

The essential fatty acids linoleic acid and ALA cannot be synthesized *in vivo* in mammals and must be ingested as part of the diet. Although in most western diets linoleic acid levels are adequate or even high, the conversion of linoleic acid to GLA is rate limiting and may not be adequate due to the impaired Δ6-desaturase activity. Elongation of GLA to DGLA is, however, rapid. ARA is produced directly from DGLA by a Δ5-desaturase. Many of the physiological effects attributed to ARA relate to its role in eicosanoid production. Eicosanoids are short-lived, hormone-like substances and exert diverse actions on the cardiovascular, reproductive, respiratory, renal, endocrine, skin, nervous, and immune systems. Eicosanoids include the prostanoids (prostaglandins, prostacyclins, and thromboxanes), leukotrienes, and hydroxy fatty acids. ARA produces prostanoids of 2-series and leukotrienes of 4-series. Prostanoids are produced in most tissues, whereas leukotrienes are generally formed in different blood cells. In platelets, ARA forms thromboxane A_2 (TXA$_2$) and in endothelial cells of blood vessels, the major product of ARA is prostacyclin I_2 (PGI$_2$). Eicosanoids derived from ARA promote the aggregation of blood platelets, the clotting of blood within blood vessels (thrombosis), and inflammatory reactions. ARA is the most tightly regulated fatty acid in cell membrane phospholipids because it affects the way cells behave, and its actions have far-ranging effects[14].

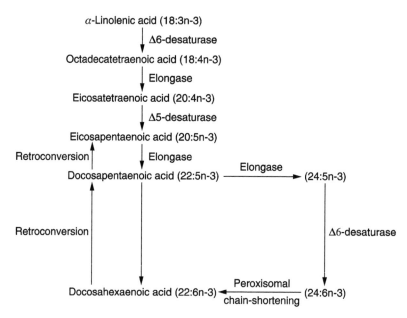

FIGURE 16.2 Metabolism of docosahexaenoic acid.

16.2   DHA AND ARA IN INFANT NUTRITION

Infant formula manufacturers have introduced new versions of infant formula, supplemented with two nutritionally important fatty acids, DHA and ARA. Trade names under which DHA- and ARA-supplemented infant milk-based formulas are currently marketed are Enfamil Lipil® (Mead Johnson Nutritionals, Evansville, IN), Similac Advance® (Ross Products, Columbus, OH), Parents Choice® (Wyeth Nutritionals International, Collegeville, PA), and Good Start® Supreme DHA & ARA (Nestlé, Switzerland). DHA and ARA are present in breast milk and are thought to be associated with visual and mental development in infants. The rationale is that formulas enhanced with DHA and ARA may promote improved visual and mental development outcomes in formula-fed babies, more similar to breast-fed babies. The U.S. Food and Drug Administration (FDA) has stated it has no objections to the addition of DHA and ARA to formula for term infants[15]. The British Nutrition Foundation, the Food and Agriculture Organization of the United Nations/World Health Organization (FAO/WHO), and the International Society for the Study of Fatty Acids and Lipids (ISSFAL) all recommend supplementation of premature infant formulas with both DHA and ARA[16–19]. FAO/WHO[20] reviewed all the available literature on DHA and ARA supplementation and recommended that all full-term infant formulas should deliver 20 mg DHA/kg/day and 40 mg ARA/kg/day (0.3% DHA and 0.6% ARA assuming consumption of 6.5 g fat/kg/day) to babies[20]. The European Commission (EU) Scientific Committee on Foods[21] has also published some recommendations regarding ratios of DHA and ARA in infant formulas. Having reviewed the available literature the committee sees the evidence as insufficient to set an obligatory minimum level of long-chain PUFAs. The committee considered that a statement relating to the presence of DHA in a formula should only be made if the content of DHA is not less than 0.2% of the total fatty acids. The proposed level was based on the available studies on effects of DHA in infants and the fact that the level is at the lower end of the range of human milk DHA content worldwide. To avoid relative deficiency of ARA, which may have negative effects on growth, and also to keep a proper balance between n-6 and n-3 PUFAs for other reasons, the concentration of n-6 PUFAs should not be lower than that of DHA in all types of formulas with added long-chain PUFAs. It was proposed for both infant formula and follow-on formula that n-6 PUFAs should not exceed 2% of total fatty acids and that of n-3 PUFAs 1% of total fatty acids. ARA should not contribute more than 1% of total fatty acids. The ratio between EPA and DHA should be < 1[21]. While there is no agreement on DHA and ARA levels, studies that have shown benefits for visual acuity and cognitive development have used higher DHA and ARA levels. A meta-analysis[22] of 14 controlled trials using a DHA equivalents index (estimated bioavailable DHA based upon intake of preformed DHA and precursors) found indeed that across studies, higher DHA levels translated to better efficacy, and this could explain the differences in findings between studies.

The optimal fatty acid composition of infant formula has not been determined. Experts agree that infant formulas should be designed to approximate the fatty acid composition of breast milk and include n-3 fatty acids[23]. Infants may have a unique need for essential fatty acids, particularly DHA and ARA, which are important PUFAs in human milk. Worldwide, DHA concentrations in breast milk range from 0.07% to greater than 1.0% of total fatty acids, with a mean value of about 0.34%. ARA concentrations are greater in magnitude and lower in variability, averaging 0.53%. Given the importance of all n-3 PUFAs in the diets of infants, the U.S. Institute of Medicine (Washington, DC) set an adequate intake of 0.5 g of n-3 fatty acids per day during the first 12 months of life[24]. The n-6/n-3 ratio of infant formula may be especially important, as the relative amounts of these fatty acids influence the amounts of DHA and ARA in tissues[25].

Studies have been done to examine the effect of dietary DHA and ARA on visual function. A one-year study measured the red blood cell fatty acid composition and visual function of 108 infants over 52 weeks. Twenty-nine of the infants were breast-fed and the rest were fed either plain infant formula or formula supplemented with DHA and/or ARA. It was found that the fatty acid make-up and visual function were similar between the breast-fed and fatty acid-supplemented groups but were inferior in the unsupplemented formula group[26]. Another similar study measured visual acuity in breast-fed, formula-fed, and DHA supplemented formula-fed infants at 16 and 30 weeks of age. It was found that the infants in the unsupplemented formula group had poorer visual function than the other two groups. Red blood cell fatty acid levels of DHA correlated well with visual acuity in all infants at both ages tested. Interestingly, infants breast-fed for less than 16 weeks had poorer visual acuity than infants receiving a steady supply of DHA did, whether from breast milk or supplemented formula[27].

In a series of studies, infants were fed formula containing DHASCO® and ARASCO® (with DHA at 0.36% of total fat and ARA at 0.72% of total fat) for their first four months and compared with infants fed formula without DHA and ARA. The authors reported no adverse effects that were attributed to the study formula[28,29]. A published abstract report on a study conducted with infants weaned from human milk and fed either DHA- and ARA-supplemented formula or control formula (i.e., formula without DHA or ARA) until one year of age. The authors did not report any negative effects from the DHA- and ARA-supplemented formula[30].

In another study, Birch et al.[31] found that infants who were breast-fed for six weeks and then weaned to DHA- and ARA-supplemented infant formula had significantly better visual acuity at 17, 26, and 52 weeks of age and significantly better stereoacuity at 17 weeks of age than infants who were weaned to nonsupplemented formula. In a similar study, Hoffman et al.[32] concluded that infants who were breast-fed for four to six months and then weaned to DHA- and ARA-supplemented infant formula demonstrated more mature visual acuity than those infants who were weaned to nonsupplemented formula.

Compared to formula feeding, breastfeeding has been found to produce long-lasting improvements in cognitive ability and educational achievement[33]. Data analysis has determined that DHA and ARA content of breast milk is at least partly responsible for this difference. One study found an IQ advantage of preterm infants fed breast milk *by tube* compared to a nonsupplemented formula-fed group[34], indicating an effect beyond the actual act of breastfeeding. There have also been several studies that have compared regular infant formula to formula supplemented with DHA and ARA as to their effect on mental development. These studies have found infants receiving DHA- and ARA-enriched formula to have greater cognitive ability than infants fed unsupplemented formula[35,36].

Carlson et al.[37] found that premature infants fed DHA had better visual acuity at two and four months of age. Werkman and Carlson[38] also showed a positive difference in premature infants fed DHA and ALA at ages up to twelve months after term. In another study, Uauy et al.[39] documented positive effects of DHA and DHA plus ARA on retinal function and visual acuity in premature and term infants, respectively.

A clinical study conducted by Birch et al.[28] concluded that infants who were fed formula supplemented with DHA and ARA scored 7 points higher on the mental development index (MDI) than infants who were fed formula without DHA and ARA. Both the cognitive and motor subscales of the MDI showed a significant developmental age advantage for DHA- and DHA+ARA-supplemented groups over the control group. The formula-supplemented infants also exhibited better visual acuity (equivalent to one line on the eye chart) than the nonsupplemented infants and similar visual acuity to that of the breast-fed infants.

Smuts et al.[40] found that expectant mothers who increased their dietary intake of DHA during the last trimester of pregnancy through DHA-enriched eggs increased their length of gestation by 6 days. Studies with stable isotopes indicate that preterm and term infants can convert linoleic acid to ARA and ALA to DHA[10]. However, sufficient synthesis to enable tissue accretion is limited at birth because of immature enzyme systems.

Clandinin et al.[41] have shown that the accretion of DHA and ARA by the fetal brain during the last trimester of gestation is essential; therefore, infants born prematurely are at an increased risk of having a decreased level of these two fatty acids. Autopsy studies of term infants who died during the first year of life have also shown such PUFA differences in the brains and retinas of infants fed formulas not supplemented with DHA[42,43]. Studies of fatty acids in autopsied sudden-infant-death syndrome individuals showed reductions in DHA around 10% in those fed conventional formula compared to those fed breast milk[42]. All of the above studies support the importance of DHA and ARA for normal neural development.

Willatts and colleagues[35] at the University of Dundee in Scotland obtained parental cooperation on behalf of 44 healthy, full-term newborns. Starting shortly after birth, half the babies received a standard infant formula while the others received the same formula supplemented with DHA and ARA. The fat supplement was derived from milk fat, vegetable oils, and egg lipids. When tested at 10 months, both groups had normal physical development and were equally able to solve simple mental problems. However, faced with a more complex mental challenge, those taking DHA-supplemented formula did better and their advantage was statistically significant.

Numerous studies have compared the red blood cell fatty acid profiles of breast-fed and formula-fed infants and have found significantly lower levels of DHA and ARA in the formula-fed infants, regardless of varying the concentration of ALA and linoleic acids[25,44,45]. Supplementation of infant formulas with DHA and ARA results in red blood cell fatty acid profiles extremely similar to those of breast-fed infants[45]. Red blood cell fatty acid levels are a good reflection of levels within the brain and, indeed, breast-fed infants have been found to have significantly more DHA within brain tissue than formula-fed infants[42]. Newborns start off with a supply of DHA within their tissues, provided that the mother had adequate DHA status during pregnancy. If supplied with a dietary source, the brain continues to incorporate DHA and ARA for several years after birth. Stored DHA is quickly depleted in the newborn. This situation is especially problematic for premature infants because the brain and the retina do a lot of developing during the last trimester of pregnancy. Premature birth interferes with this development. A premature infant fed a diet devoid of preformed DHA and ARA will become deficient more quickly than will a full-term infant.

16.3   DHA IN CARDIOVASCULAR HEALTH

Cardiovascular disease includes all diseases of the blood vessels and heart, such as coronary heart disease and stroke. Cardiovascular disease is the leading cause of death in Canada and the United

States[46]. Research studies have indicated that the increased consumption of fish containing n-3 fatty acids can protect against death from cardiovascular disease[47]. Research indicates that the long-chain n-3 PUFAs, especially DHA and EPA, may be effective in reducing the clinical risk of cardiovascular disease by favorably altering lipid and hemostatic factors such as bleeding time and platelet aggregation[48]. Dietary supplementation of DHA, EPA, and other n-3 PUFAs has also been recommended to lower the risk of cardiovascular disease and to improve the overall health of humans. Controlling blood lipids (total cholesterol, HDL, LDL, and triacylglycerols) is important for reducing the risk of developing heart disease. One of the cardioprotective effects of DHA appears to be to help normalize certain blood lipids. Simon et al.[49] examined the relation between serum fatty acids and coronary heart disease (CHD) in men. Interestingly, the levels of DHA in serum phospholipid were inversely correlated with CHD risk in a multivariate model that controlled for the effects of the HDL-cholesterol to LDL-cholesterol ratio. DHA has been found to exhibit antiplatelet aggregatory potential in some studies[50], triacylglycerol (TAG)-lowering effects in the majority of human studies using supplemental DHA, and important antiarrhythmatic effects in experimental studies[51].

Grimsgaard et al.[52] used highly purified preparations of DHA and EPA in human studies and reported similar serum TAG-lowering effects of each but divergent effects on serum fatty acids. Ethyl ester preparations of DHA and EPA over a seven-week period provided a decrease in serum TAG levels by 26% in the DHA group, which was accompanied by a 4% rise in HDL-cholesterol, and a 4% decrease in the total cholesterol to HDL-cholesterol ratio. In a double-blind, placebo-controlled trial of subjects with abnormal lipids, supplementation with more than 1.2 g DHA/day for six weeks resulted in a clinically significant drop of approximately 20% in TAGs and a statistically significant increase in HDL levels[53].

Recently, Engler et al[54] studied the effects of DHA supplementation on endothelial function in children with inherited high cholesterol levels. The study concludes that restoration of normal blood vessel function has the potential for preventing the progression of early coronary heart disease in high-risk children. Their findings are the first to suggest that supplementation with DHA may improve vascular health in hyperlipidemic children at high risk for early heart disease. This study used Martek DHA™, which is produced from microalgae.

Several scientific bodies have made recommendations regarding the intake of n-3 fatty acids and cardiovascular health. A 1999 workshop sponsored by the National Institute of Health (NIH) and ISSFAL recommended a daily DHA intake of 220 mg as an adequate intake (AI) for healthy adults[55]. In 2000 the American Heart Association (AHA) advised the U.S. population to eat at least two servings of fatty fish per week for their omega-3 cardiovascular benefits. In 2002 the AHA expanded upon its recommendation to include, specifically, people with documented CHD and people with high TAG levels. The AHA recommended that CHD patients consume 1 g of DHA and EPA daily and patients with high TAG levels consume 2 to 4 g DHA and EPA daily[56]. In January 2005 the U.S. Department of Health and Human Services and U.S. Department of Agriculture released updated Dietary Guidelines mentioning for the first time DHA, an omega-3 fatty acid[57]. The 2005 Dietary Guidelines recognize that "limited evidence suggests an association between consumption of fatty acids in fish and reduced risks of mortality from cardiovascular disease for the general population." In addition, the 2005 Dietary Guidelines recognize that there are other sources of DHA and EPA that may provide similar cardiovascular benefits, but state that more research is needed.

16.4   DHA AND NEUROLOGICAL/BEHAVIORAL DISORDERS

Neurological diseases associated with low levels of DHA such as Alzheimer's disease, depression, schizophrenia, as well as various peroxisomal disorders have been reported[58]. Researchers at the National Institute of Alcohol Abuse and Alcoholism (Rockville, MD) have linked the increasing rate of depression to an imbalance in the ratio of n-6 to n-3 fatty acids[59]. Studies of Paleolithic nutrition and modern hunter-gatherer populations suggest that humans evolved on a diet different from today's typical Western diet. The diets of the Paleolithic era in which our genetic patterns were established consisted of less total and saturated fats and small but roughly equal amounts of n-6 and n-3 fatty acids, giving an n-6/n-3 ratio of about 1:1. Typical U.S. diets are high in n-6 and low in n-3 fatty acids compared with the Paleolithic diet on which humans evolved, resulting in an n-6/n-3 ratio of about 9:1[60]. The n-6/n-3 ratio may be as high as 17:1 in the overall Western diet[61].

An interesting finding concerning PUFAs may shed some light on children with attention deficit hyperactivity disorder (ADHD). Stevens et al.[62] found that boys diagnosed with ADHD had significantly lower levels of n-3 and n-6 fatty acids in their blood compared to children without ADHD. It was also noted that characteristic symptoms of fatty acid deficiency, such as thirst, frequent urination, and dry hair and skin were also prevalent in the boys with ADHD.

DHA-containing formula increased neurobehavioral scores in 1-month-old rhesus monkeys compared to controls[63]. This study used DHA at 1.0% of fatty acids (0.5% energy), compared to the highest amount in U.S. formulas of about 0.35% of fatty acids (0.17% energy) and is among the first studies to raise the possibility that higher amounts of DHA may be required to confer benefits that can be detected with behavioral testing.

Animal models of DHA deficiency can be used to help define the nature and extent of the functional deficits in the nervous system that are associated with this dietary treatment. In one study, a diet low in n-3 fatty acids was given to rats for three generations to induce a significant decline in levels of brain DHA. It has been observed in this study that rats on a diet low in n-3 fatty acids performed more poorly in learning and cognitive-related behavioral performances[64].

Epidemiological studies suggest that increased intake of the DHA is associated with reduced risk of Alzheimer's disease[65,66]. Recently, a group of neuroscientists showed for the first time that a diet high in the DHA helps protect the brain against the memory loss and cell damage caused by Alzheimer's disease[65]. The new research suggests that a DHA-rich diet may lower one's risk of Alzheimer's disease and help slow progression of the disorder in its later stages. A more recent study, which observed aged mice bred with genetic mutations that cause brain pathology linked to Alzheimer's disease (the APPsw transgenic mouse model), demonstrated that the pathogens linked to Alzheimer's disease (total Aβ and plaque burden) were significantly reduced in mice fed a diet enriched with DHA[66]. This study is significant because it shows that DHA added to the diet altered the processing of the amyloid precursor protein and the accumulation of its toxic amyloid protein metabolite that is widely believed to cause Alzheimer's.

Scientists are also exploring the influence of DHA on such varied disorders as schizophrenia, multiple sclerosis, and aggression. Such research will help further to clarify the function of DHA and possibly identify its beneficial uses for both children and adults. In the meantime, the role of DHA in infant development proves it is a highly important component in our nutritional arsenal.

16.5   DHA AND PREGNANCY/LACTATION

DHA is important throughout pregnancy and lactation for the health of both the mother and the fetus/infant. The functional role of DHA for the structural and functional development of fetus/infants seems to be at least partly established. During pregnancy DHA is preferentially transferred from the mother to the fetus[67], particularly in the last trimester of pregnancy, and impairments in this passage are associated with intrauterine growth retardation[68]. In infancy, brain DHA levels are higher if the infant has some dietary sources of DHA (i.e., human milk) and neurodevelopmental scores are associated with DHA levels in body pools represented by blood indices[69]. The DHA intake in early infancy has also been associated with lower blood pressure values at 5 years of age[70]. A NIH workshop of fatty acid experts, recognizing the importance of maternal DHA intake, recommended 300 mg/day of DHA as AI for pregnant and lactating women[55].

Reported studies have consistently shown that women's diets in the U.S. have one of the lowest levels of DHA in the world[26,71]. Clinical studies have demonstrated that women who supplement their diets with DHA during pregnancy experience elevated DHA levels in their blood[72]. One study, in a population at risk for preterm delivery, demonstrated that increasing DHA consumption during pregnancy was shown to improve outcomes by extending the length of pregnancy[73]. The same study also exhibited a trend toward higher birth weight, length, and head circumference of infants. Smuts et al.[40] have also demonstrated that expectant mothers who increased their dietary intake of DHA during the last trimester of pregnancy through DHA-enriched eggs had an increased length of gestation. Another study reported a 4 IQ point advantage at 4 years of age for children of mothers who had supplemented their diet with DHA during pregnancy and lactation versus those children whose mothers did not[74]. Studies have also suggested an association between lower DHA status after pregnancy and the occurrence of postpartum depression[75].

A recent study[76] has shown that supplementation with Martek's DHA™ (200 mg) during the first four months of nursing leads to improved attention skills in children at 5 years of age. This same group[77] showed that there was an improvement (8 points) on the Bayley's psychomotor development index (PDI) in the same children whose mothers received the DHA supplement during nursing.

Recently, Mead Johnson Nutritionals (Evansville, IN) has launched a DHA supplement for pregnant and nursing women containing Martek DHA. The product, Expecta™ LIPIL® DHA supplement, is a nonprescription prenatal supplement that provides 200 mg of DHA for pregnant and nursing women. The typical American diet is lower in DHA relative to many other countries and, consequently, U.S. women have among the lowest breast milk DHA levels in the world[26,71]. The introduction of Expecta™LIPIL® DHA supplement extends Mead Johnson's use of Martek DHA beyond infant formula and responds to the growing body of evidence suggesting that both mothers and their babies benefit from an adequate maternal intake of DHA.

16.6　SOURCES OF DHA AND ARA

There are several sources of DHA and ARA used around the world, including various marine and fish oils, egg-derived lipids, and lipids from microorganisms. The n-3 PUFAs are found almost exclusively in aquatic organisms where they exist in varying amounts and ratios. The principal source of DHA in the biosphere is from algae. Numerous algal species produce DHA. Two main species in particular (*Crypthecodinium cohnii* and *Schizochytrium* spp.) have been manipulated to produce elevated levels of DHA. These organisms have been cultivated on the industrial scale and the biomass has been used for the production of high DHA-containing oils. Although fish and fish oil contain varying levels and mixtures of DHA and EPA, it is the DHA that accumulates at high levels in neuronal and retinal tissues.

DHA is found primarily in esterified form associated with phospholipids in the cellular membrane as well as within storage lipid, primarily in the form of TAGs. DHA is one of the main components of the gray matter of the brain, the phospholipids of the retina, the testes, and sperm. DHA is lacking from all common plant food sources including the commercial vegetable oils. However, microalgae and other sources of DHA are available for nutritional and food applications.

ARA has been reported to accumulate in algae, yeast, and fungi. The fungi of the genus mortierella, belonging to the class Phycomycetes, have been considered as promising producers of ARA. Mortierella fungi are known to accumulate high contents of ARA. The soil fungus from *Mortierella alpina* was selected for the development of an industrial process because it meets two important criteria: it has a high ARA content, which can be more than 50% of total fatty acids, and it is oleaginous, accumulating intracellular TAGs[78]. ARA is found in both the polar and the nonpolar lipids; however, it is the TAG fraction that is currently used as the main commercial product. Companies actively involved in commercial production of ARA include Martek Biosciences Corporation (USA), DSM Food Specialties (The Netherlands), and Suntory (Japan).

16.7   SINGLE-CELL OILS (SCO)

Lipophilic compounds derived from microbial sources, referred to as single-cell oils (SCO), are of potential industrial interest due to their specific characteristics[79,80]. For instance, oleaginous bacteria accumulate mostly certain lipids such as polyhydroxyalkanoates[80]. In contrast, oleaginous yeasts and molds accumulate TAGs rich in PUFAs or having specific structure[79,81,82]. Oleaginous microorganisms may provide an economical source of PUFAs, provided that most of the PUFAs occur in TAGs, which are the preferred forms for dietary intake. Furthermore, microorganisms preferably contain one specific PUFA rather than a mixture of various PUFAs. This gives the microbial oils an additional value as compared to fish oils, which contain mixtures of PUFAs. In addition, PUFAs can be purified more easily and economically from oils that contain one PUFA instead of a mixture of PUFAs. It should also be noted that microbial oils are intrinsically more pure than plant or fish oils in that there should be no residue arising from the use of pesticides, herbicides, or fungicides and contamination of heavy metals.

Microorganisms capable of producing PUFAs containing 20 or more carbon atoms include lower fungi (sometimes referred to as the Phycomycetes or the Zygomycetes), bacteria, and marine microalgae[83]. With few exceptions, bacteria are probably not suitable as SCO producers, as they do not accumulate high amounts of TAGs[84]. Considerable diversity in long-chain PUFA profiles occurs across the classes of phototrophic algae (Table 16.1). For instance, green algae either do not contain or have only very low levels of EPA and DHA; diatoms contain elevated EPA, as do eustigmatophytes; dinoflagellates contain DHA and other C_{18} PUFAs; red algae may contain elevated proportions of ARA (Table 16.1). This diversity in long-chain PUFAs is potentially valuable, and there is interest in the cultivation of these phototrophic algae in photobioreactors for production of high-purity long-chain PUFAs. Currently, at least two commercial processes exist for the production of n-3 PUFAs by heterotrophic cultivation of marine microalgae.

In order to be able to develop economically feasible oil production processes by cultivation of oleaginous microorganisms in bioreactors, it is important that the oils or fatty acids are produced at a high productivity and are of a sufficiently high value. The factors that determine the overall volumetric productivity are the final dry biomass, oil content of the biomass, desired fatty acid of the oil, and the process duration. The value of the oils and fatty acids are dependent on market mechanisms. Lipid accumulation is a dynamic process, which depends on the microorganism, the

TABLE 16.1
Distribution of PUFAs in Microalgae, Fungi, and Other Groups

Group	Genus/species	PUFAs
Dinoflagellates	*Crypthecodinium cohnii*	DHA
Thraustochytrids	*Schizochytrium* spp.	DHA
Dinoflagellates	*Amphidinium* spp.	DHA
Fungi	*Thraustochytrium aureum*	DHA
Fungi	*Mortierella alpina*	ARA
Red algae	*Porphyridium*	ARA
Eustigmatophytes	*Nannochloropsis*	EPA
Diatoms	*Chaetoceros*	EPA
Diatoms	*Skeletonema costatum*	EPA
Phytoflagellates	*Isochrysis galbana*	EPA
Cyanobacteria (blue-green algae)	*Spirulina* spp.	GLA
Fungi	*Mucor javanicus*	GLA

PUFAs, polyunsaturated fatty acids; DHA, docosahexaenoic acid; ARA, arachidonic acid; EPA, eicosapentaenoic acid; GLA, γ-linolenic acid.

growth conditions (pH, temperature, aeration, nutrients, etc.), and the growth phase. Therefore, the proper selection of the microorganism, process optimization, and timing of harvest are essential for efficient SCO production. Most oleaginous microorganisms start to accumulate oil whenever an excess carbon source is present while at the same time the growth is limited by another nutrient, such as nitrogen.

16.8   SOURCES OF MICROALGAE

Microalgae are a very diverse group of organisms that consist of both prokaryotic and eukaryotic forms. Most microalgae are phototrophic; they require light as the free-energy source for growth. However, some species of microalgae are capable of heterotrophic growth, as they do not need light as a free-energy source, but derive metabolic energy from the dissimilation of organic carbon compounds. Dinoflagellates are early eukaryotic marine microalgae that can be phototrophic and/or heterotrophic and form an important part of the marine plankton. Dinoflagellates in general grow slowly and have been found to be quite shear-sensitive[85]. DHA is the characteristic PUFA of the marine dinoflagellates[86].

Crypthecodinium cohnii is a unicellular, nonphotosynthetic marine dinoflagellate and is found naturally in association with decaying seaweed. It is evident that *Crypthecodinium cohnii* is probably the major source of a TAG oil rich in DHA and free of all other PUFAs, making it not only unique in such a fatty acid profile but also highly desirable commercially. This is remarkable as most marine microalgae rich in PUFAs contain intermediate fatty acids in the cascade of desaturation and elongation. The predominant fatty acids present in *Crypthecodinium cohnii* are myristic acid (14:0), palmitic acid (16:0), oleic acid (18:1), and DHA. Nearly 30 to 55% of its constituent fatty acids are DHA and no other PUFAs are present in excess of 1% (Table 16.2). Thus, it is easy to separate DHA from the fatty acid mixtures. Since it grows easily in bioreactors, it is an ideal source of DHA. For these reasons, *Crypthecodinium cohnii* represents a promising microalga for the commercial production of DHA. *Crypthecodinium cohnii* has been specifically selected for the commercial product because it has been in culture over the last 90 years and in all the numerous studies using this species there has never been any indications of pathogenicity or toxigenicity.

Schizochytrium spp. is a thraustochytrid and is a member of the Chromista kingdom. This is not the same as the kingdom to which the bluegreen or dinoflagellate microalgae belong. The kingdom Chromista includes golden algae, diatoms, yellow-green algae, haptophyte and cryptophyte algae, oomycetes, and thraustochytrids[87]. Thraustochytrids are single-cell organisms that produce a high oil and long-chain PUFA content. Although they were initially thought to be primitive fungi, they

TABLE 16.2
Comparison of DHA Sources

	Fish liver	Egg yolk	Algal oil from *Crypthecodinium cohnii*	Algal oil from *Schizochytrium* spp.
22:6n-3	8.6	3.4	55.0	39.4
22:5n-6	Not detected	0.3	Not detected	14.9
20:5n-3	9.3	0.3	Not detected	1.5
14:0	0.1	Not detected	14.0	9.0
16:0	11.6	28.1	12.0	22.5
18:1	25.2	36.1	10.0	2.0
20:1	13.1	Not detected	Not detected	Not detected
Form	TAG	Phospholipid	TAG	TAG

have been reassigned to the subclass Thraustochytridae (Chromista, Heterokonta), which aligns them more closely with diatoms and brown algae. Under cultivation conditions, thraustochytrids can achieve a considerably higher biomass yield (> 20 g/L) than other microalgae. In addition, thraustochytrids can be grown in fermenters on an organic carbon source and therefore are a highly attractive, renewable, and contaminant-free source of n-3 fatty acid-rich oils. This class of microalga is primarily saprotrophic (obtain food by absorbing dissolved organic matter) and is found throughout the world in estuarine and marine habitats. They are generally found associated with organic detritus, decomposing algal and plant material, and sediments. Microalgae and other microscopic organisms are primarily consumed by a wide range of filter feeding invertebrates (e.g., clams and mussels) in the marine ecosystem. There are no reports of toxicity or pathogenicity associated with schizochytrium in the literature[88] and it contains no algal toxins as determined by analytical methods. Successful cultivation of *Crypthecodinium cohnii* and *Schizochytrium* spp. to produce commercial edible oils containing DHA has been achieved only by former OmegaTech Inc. (now Martek Biosciences Boulder Corporation) and Martek Biosciences Corporation in the United States.

In algae, similar to other organisms, lipids and fatty acids are basic cellular constituents that can serve both as structural components of the cell and as a storage product for the cell. Algae contain many of the lipid classes found in other organisms. The major nonpolar lipids present in algae are the TAGs and hydrocarbons, and the major polar lipid classes include phosphatidylcholine (PC), phosphatidylethanolamine (PE), phosphatidylinositol (PI), phosphatidylserine (PS), phosphatidylglycerol (PG), cardiolipin, and glycolipids[89]. Algae contain a wide variety of saturated fatty acids from C_{12} to C_{18}, but the major fatty acids are 12:0, 14:0, 16:0, and 18:0. A number of unsaturated fatty acids are found in algae, ranging from C_{16} to C_{22} with 1 to 6 carbon–carbon double bonds in the *cis* configuration. It is important to note that algae are the predominant producers of long-chain n-3 PUFAs (e.g., DHA) in the biosphere.

16.9   SOURCES OF FUNGI

The development of a microbial PUFA production process requires the selection of the proper microorganism and optimized cultivation techniques. Several oleaginous fungi are capable of the production of high amounts of PUFAs in their lipid. Several species of mortierella and some other genera of fungi were identified as producers of ARA. Their oils do not contain n-3 PUFAs.

The genus mortierella belongs to the Phycomycetes. This group of lower fungi has some properties that are quite different from the more archetypal ascomycetes, such as penicillium or aspergillus. The mycelium of Phycomycetes does not contain "real" septate cells; instead, filaments contain sections filled with cytoplasm, interspersed by sections that are empty. The cytoplasm harbors multiple nuclei, even in spores (single-celled reproductive bodies). Thus, the life cycle does not include any uninucleate stage. This complicates the selection and maintenance of strains, since one always deals with populations rather than with individuals.

Filamentous and pellet morphologies are well known in submerged cultures of mycelial microorganisms. Growth of many fungi in liquid culture can be either dispersed or in pellets. The morphology is influenced by culture conditions, such as the concentration of the substrates, dissolved oxygen concentration, mineral addition, and the natural nitrogen source. Many researchers indicate that filamentous morphology in culture increases the viscosity in the culture broth, leading to an increase in mass transfer resistance in the reactor[90]. In contrast, pellet cultures allow easier mixing and better mass transfer to the culture broth. In *Mortierella alpina*, it is interesting to note that the morphological properties are quite variable between isolates. This applies both to sporulation (the process of spore development) on solid media and to the tendency to grow dispersed or as pellets in liquid culture[78]. Like all fungi, *Mortierella alpina* is a psychrotrophic, nonphotosynthetic organism, which requires a reduced carbon source for growth and, like most soil fungi, it may be found associated with common root crops and directly in the food chain of humans.

16.10   MANUFACTURING PROCESS OF SCO

16.10.1   FERMENTATION AND RECOVERY OF SCO FROM MICROALGAE

Crypthecodinium cohnii and *Schizochytrium* spp. are grown by heterotropic fermentation. As noted earlier, heterotrophic fermentation is independent of light. The fermentation occurs in bioreactors, which can be operated axenically and under controlled optimum conditions. Further benefits include higher biomass concentrations, increased reproducibility, and straightforward scale-up of the fermentation processes.

Crypthecodinium cohnii is a unique heterotropic marine dinoflagellate in that DHA is almost exclusively the only PUFA present in its lipid and can be as high as 65% of the total fatty acids[91]. *Schizochytrium* spp. is also a heterotropic microalgae belonging to the order Thrausochytriales within the phylum Heterokonta, which can yield about 40% of DHA from its total fatty acid production. Strains of the genus schizochytrium are ideal candidates for fermentation development for a variety of reasons: (1) heterotrophic nature; (2) small size; (3) ability to use a wide range of carbon and nitrogen sources; (4) ability to grow at low salinities and exhibit enhanced lipid production at low salinities; and (5) the high DHA content in their lipids even at elevated temperatures[92].

The industrial DHA production potential of microalgae mainly depends on the primary strain selection for the fatty acid composition, yield, and adaptability to the fermenter. As microalgae are highly environmentally dependent and the synthesis of fatty acids, especially PUFAs, is influenced by many parameters, such as culture age, salinity, medium composition, temperature, and aeration, a cost-effective fermentation process should be established through systematic investigations into the effects of various nutrient and environmental conditions and the use of the various high cell-density strategies on the growth and DHA production of the selected microalga, *Crypthecodinium cohnii*. The production strains can be selected for rapid growth and high levels of production of the specific oils. Master seed banks of all strains are maintained under liquid nitrogen conditions and working seed stocks, prepared from this master seed bank, are also maintained cryogenically. On initiation of a production run, an individual ampoule from a working seed is used to inoculate a shake flask. The medium used to grow *Crypthecodinium cohnii* and *Schizochytrium* spp. from shake flask to production scale contains carbon source, nitrogen source, salts, and a number of other micronutrients. The cultures are transferred successively to large fermenters based on specific growth parameters. Throughout the process the concentration of carbon substrate, pH, temperature, pressure, airflow, agitation, and dissolved oxygen are regularly monitored and controlled. All fermentations involved in such high-value products require axenic culture (pure culture of one species only). Therefore, at each transfer stage in the inoculum sequence, broth samples are plated to establish the microbial purity. In addition, the purity of the cultures are also monitored every 24 hours by manual observation of a sample under a microscope and the plating of culture broth samples under several conditions to confirm the presence or absence of any microbial contamination. In the final fermentation vessel the cultures are allowed to go into nitrogen limitation at which time they begin producing their storage products, TAGs which are enriched in DHA.

DHA is contained entirely within the algal cells and is distributed in both structural lipids and storage lipids. The viscosity of the broth may increase drastically due to production of extracellular polysaccharides. In addition, the pH tends to change rapidly during the final holding time. All these factors tend to increase the difficulties of product recovery. To ensure good recovery of the oil at optimum quality, speed of operation is important because of the sensitive nature of the product to contamination, cell lysis, and lipid oxidation. At the conclusion of the fermentation, DHA-rich biomass is concentrated via centrifugation or ultrafiltration in order to make it compatible with the subsequent drying operation and to reduce the energy cost associated with drying.

16.10.2 FERMENTATION AND RECOVERY OF SCO FROM FUNGI

A large-scale fermentation process for the production of ARA-rich oil has been developed by DSM in The Netherlands in conjunction with Martek Biosciences Corporation in the United States. A selected nontoxigenic strain of the soil fungus *Mortierella alpina* is currently used as a commercial source of oil rich in ARA. Starting from a working cell bank (either as spores or as vegetative mycelia), shake flasks and inoculum fermenters are used for the initial phases of biomass production. The goal is to generate sufficient biomass of the desired macroscopic morphology[93]. The growth medium of *Mortierella alpina* in the fermentation process is based on glucose and either yeast extract or hydrolyzed vegetable protein as a nitrogen source. During fermentation, the temperature, pH, airflow, agitation, pressure, glucose concentration, and dissolved oxygen concentration are monitored and controlled and the cultures are transferred to successively larger vessels based on predefined criteria.

Lipid synthesis in oleaginous fungi is known to occur mainly in the second phase of batch culture growth under conditions of limitation or inhibition by some component (or factor), when excessive carbon substrate is still available in the medium. For instance, the TAGs are generally best produced under limited amounts of nitrogen substrate, and this is the regime of choice for the main fermentation. Use of a dual-feed system, in which the nitrogen and the carbon substrate(s) can be dosed independently, is advisable[78]. The dosage of the nitrogen source may determine the formation of productive (lipid-free) biomass. Once the maximal biomass concentration has been reached the nitrogen feed can be stopped, and the carbon feed can be used to keep lipid accumulation going. Carbon is best dosed as nonlimiting feed, adjusted to the consumption rate to ensure a continuous, but moderate, excess of the carbon substrate. This allows maximum lipid accumulation while avoiding inhibitory effects of high concentrations of the carbon substrate. Intense lipid synthesis in microorganisms may be considered as a combination of growth-associated processes (formation of functional lipids, mainly phospholipids) and processes unessential for cell growth (accumulation of storage lipids, mainly TAGs)[94]. It seems quite reasonable to assume that ARA, which exhibits a high physiological activity, is a component of functional lipids and, therefore, its synthesis should proceed concurrently with mycelial growth.

As noted earlier, the organism can metabolize the lipid that forms. This is problematic if the lipid is broken down after the fermentation has been stopped, possibly due to two metabolic processes: (1) lipase activity, which lowers the TAG content of the oil, and (2) subsequent oxidation of the fatty acids. To avoid these processes, the biomass should be inactivated as quickly as possible. It has been found that pasteurization at 63°C is sufficient to attain a fully stable situation, which could be maintained for more than 24 h without observable changes in the oil content or the fatty acid composition[78]. Typically, production of single-cell organisms containing long-chain PUFAs is carried out in fermenters having a capacity of 16,000 to 53,000 gallons[78]. In the industrial-scale fermentation, cells are harvested at maximum volumetric ARA productivity. The fungus *Mortierella alpina* synthesizes ARA in an array of reactions involving various desaturases and elongases (Figure 16.3). The main components of the fatty acid spectrum are the intermediates in the biosynthetic pathway leading to ARA.

Different techniques may be used for harvesting ARA fermentation broth. One of the protocols reported by Streekstra[78] includes dehydration of biomass via extrusion drying. The dry biomass is a rather stable production intermediate, but cool storage under a nitrogen atmosphere is necessary for optimal quality, because the granules are susceptible to oxidative damage. When kept for some weeks at room temperature in the presence of air, the levels of ARA decreased. This also applied to samples stored for analytical evaluation. The dry biomass particles are extracted with hexane to yield a crude oil that is relatively stable, probably due to the presence of endogenous antioxidants. These minor components, however, are lost in the subsequent steps of refining, bleaching, and deodorization. The refined oil is protected by an added antioxidant system, usually containing mixed natural tocopherols and ascorbyl palmitate. These are all standard operations used for production of high-quality edible oils.

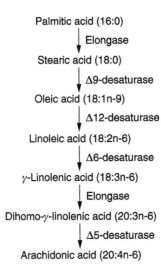

Palmitic acid (16:0)

↓ Elongase

Stearic acid (18:0)

↓ Δ9-desaturase

Oleic acid (18:1n-9)

↓ Δ12-desaturase

Linoleic acid (18:2n-6)

↓ Δ6-desaturase

γ-Linolenic acid (18:3n-6)

↓ Elongase

Dihomo-γ-linolenic acid (20:3n-6)

↓ Δ5-desaturase

Arachidonic acid (20:4n-6)

FIGURE 16.3 Biosynthetic pathway of arachidonic acid in fungi.

16.10.3 CELL DISRUPTION AND PRETREATMENT

In the recovery of oil from algal or fungal biomass, cell disruption is the most critical step. This is because the efficiency at this step of the operation directly affects the subsequent downstream processing and losses occurring at this initial stage cannot be regained. All microorganisms being considered in this section are protected by strong cell walls. In order to release their cellular contents a number of techniques for cell disruption have been employed. Although many methods are available, only a limited number have been shown to be suitable for large-scale applications. All methods shown in Figure 16.4 are able to liberate products from microbial cells more or less effectively and can be classified into physical, chemical, and biological methods, depending on the basic effects that cause disintegration. The knowledge of cell wall structure and composition can aid in developing more efficient methods and is especially important for chemical and biological methods of cell digestion. In general, mechanical disruption is favored at the process scale. Bead mill and high-pressure homogenizers are most widely used.

Bead mills consist of either a horizontal grinding chamber containing impellers or rotating discs mounted concentrically or off-centered on a motor-driven shaft. The grinding action is due to glass or plastic beads typically occupying 80 to 85% of the working volume of the chamber. Bead mills are usually operated in batch or continuous mode. The beads are retained in the grinding chamber either by a sieve plate or a similar device to retain the beads. The units require high-capacity cooling systems. Horizontal units are preferred for cell disruption to reduce the fluidizing effect in vertical units. The parameters involved in the process are numerous and include bead type and size, configuration of disc/impeller, speed of disc/impeller, loading and density, cell properties, cell concentration, process feed flow rate, and residence time, among others. Given the large number of variables that need to be optimized for disruption of cells in the bead mill, it is not surprising that this technique lags behind high-pressure homogenization. Although the cell disruption can be batchwise, continuous operation is more practical with large fluid volumes. The bead mill has been used in the disruption of yeast cells, bacteria, algae, and filamentous organisms.

High-pressure homogenizers lend themselves to a relatively user-friendly operation. Cell disruption is essentially achieved by passing the cells at high pressures through a small valve or orifice. Several types of equipment including APV high-pressure homogenizers (APV Gaulin, Germany), Rannie hyperhomogenizers (APV Rannie AS, Denmark), ultrahigh-pressure cell disrupters

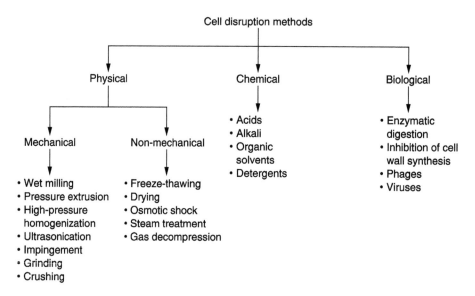

FIGURE 16.4 Cell disruption methods for single-cell oil microorganisms.

(Constant Systems Ltd, U.K.), and microfluidizers (Microfluidics, U.S.) are currently available in the market. Some of the high-pressure homogenizers can achieve pressures of up to 1500 bar. Relatively new valve design and material have improved the cell breakage efficiency and reduced wear and tear. The Microfluidics design, which is based on two impinging process fluid jets, can operate at high pressures even though requiring a supply of pressurized air to activate the equipment. Ultrahigh-pressure cell disrupters are relatively new to the field. The equipment has been tested for containment and can reach very high pressures of up to 2700 bar depending on the prototype.

There are various ways of assessing cell disruption, namely microscopic inspection and counting of cells, measurement of soluble proteins and enzymes, and particle measurement. With microscopy, it may not be easy to differentiate between intact and damaged cells even if staining techniques are used. However, it is advisable to perform visual inspection of samples to better understand the process. Soluble protein measurements are easy and results are reliable. Use of particle measurement techniques requires a special type of equipment and expertise.

The influence of pressure and number of passes on cell disruption is well documented[95,96]. Cell disruption follows first-order kinetics with respect to the number of passes, which means that, at a given pressure, cell disintegration is gradual and is dependent on the number of passes through the equipment until all cells are disrupted. As the pressure increases the amount of cell breakage increases accordingly. With high pressures, the general intention is to reduce the number of passages through the equipment to a minimum, preferably keeping it to a single pass. This may save considerable time and possibly energy. Use of high pressures with a single pass may be considered a reasonable alternative, as cells are not micronized as a result of multiple passes. An increase in fluid process temperature results in a higher disruption rate; however, this is not desirable due to oxidative degradation of the product if lipid is produced.

After cell disruption, the first major step is to separate the cell debris from the homogenate. Typically, this involves centrifugation or filtration. In both cases, the performance of the process is dependent on the characteristics (cell debris size, density, viscosity, etc.) of the homogenate. The reduction in viscosity of the homogenate is desirable to improve solid–liquid separation. Small particles and fluids with high viscosity are detrimental to the separation process. In centrifugation, the settling velocity is decreased. In the case of filtration, clogging and reduction in flow rate may occur.

When cells are particularly resilient to cell disruption, it may be necessary to pretreat the cells via chemical or enzymatic means. This is assuming that these additions will not be harmful to the product. If cells are hexane washed, the weak pretreated cells may break in the centrifuge prior to oil extraction. Once the cells are pretreated, better cell disruption can be achieved at lower pressures and with fewer passes through the homogenizer.

16.10.4 EXTRACTION AND REFINING

The processes for the extraction and refining of algal and fungal oils are the same as those used in edible oil processing. The oil is extracted from the algal or fungal biomass by blending the biomass in hexane in a continuous extraction process. The hexane/biomass slurry may be homogenized to ensure that all cells are broken and that maximum oil can be extracted. The solid–liquid separation technique can be used to recover miscella. The hexane associated with the oil is removed by evaporation to produce hexane-free oil. A minimum of a two-stage evaporation system using different vacuum/temperature condition may be needed to remove hexane to parts per million levels and make crude oil suitable for refining operations. The oil can also be winterized to remove a high-melting TAG fraction.

DHA- or ARA-rich crude oils are not suitable for human consumption because of the presence of impurities, odor, taste, and lack of clarity. Hence, crude oils need to be refined. This is achieved via conventional vegetable oil refining procedures including degumming, caustic-refining, bleaching, and deodorization. Impurities and minor components that are removed or reduced in the refining step include free fatty acids, water, phospholipids, minerals, carotenoids, sterols, tocopherols, waxes, and residual cell debris. Because of the sensitivity of DHA- or ARA-rich oil to oxidative damage, the best processing operations use short reaction times at reduced temperatures and with a constant blanket of nitrogen throughout the process. The deodorized oil is diluted with high-oleic sunflower oil to bring the DHA or ARA level to an industry standard of 40%. The oil is then stabilized by adding antioxidants, mainly ascorbyl palmitate and tocopherols. The oil is finally packaged in nitrogen-purged containers and stored frozen until shipment. Oil samples are collected at each step of the process for analyses of peroxide value, free fatty acids, residual soaps, phosphorus content, and fatty acid composition. All unit operations are carried out according to current good manufacturing practice (cGMP) regulations as designated by the FDA.

16.11   QUALITY ASPECTS OF SCO

Unlike crude oils from oilseeds and fish, crude oils from *Crypthecodinium cohnii* and *Schizochytrium* spp., produced by fermentation, are free from pesticides, aflatoxins, insecticides, heavy metals, and other pollutants often found in fish and plant oils. This simplifies the final refining operations aimed at removing impurities that have been coextracted from the cells along with the TAGs and therefore does not compromise the quality of the final oil. Shim et al.[97] analyzed several fish oil and algal oil dietary supplements for polychlorinated biphenyl (PCB) residues. They reported that fish oil supplements contained elevated levels of PCB residues. However, PCBs were not detected in algal oil supplements because these were derived from *Crypthecodinium cohnii* and PCB contamination was avoided.

A rigorous quality control is used by the industry to ensure consistency and product quality. At each step of the operation, exposure to heat, air, light, and heavy metals is minimized. State-of-the-art analytical techniques and a trained sensory panel are employed to improve and maintain the highest quality standards of final products for food and therapeutic applications. The SCO from microalgae exhibit improved oxidative stability. The oxidative stability of various food/pharmaceutical grade DHA-rich oils is measured using Metrohm Rancimat. The induction time at 60°C (average of two determinations) was as follows[98]: fish oil containing 12% DHA, 89 h; tuna head oil

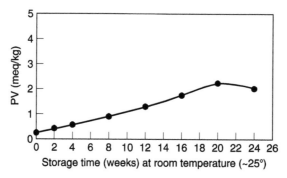

FIGURE 16.5 Oxidative stability of single-cell oil derived from *Crypthecodinium cohnii*.

containing 25% DHA, 119 h; algal oil containing 25% DHA, 152 h; and algal oil containing 42% DHA, 118 h. The typical storage stability of algal oil derived from *Crypthecodinium cohnii* is shown in Figure 16.5.

16.12   COMPOSITION OF BIOMASS

Results of proximate analysis conducted on dried microalgae from *Crypthecodinium cohnii* and *Schizochytrium* spp. are reported in Table 16.3. Crude fat (40 to 66%) and protein (20 to 23%) were the predominant components of dried microalgae with minor amounts of ash (~4.0%), crude fiber (2.0 to 5.8%), and moisture (~3.0%) accounting for the balance. DHA content, calculated on a dry weight basis, in the whole cell microalgae from *Crypthecodinium cohnii* and *Schizochytrium* spp. comprised 20.4 to 21.0% of the cells. Lipid class composition of the dried microalgae from *Schizochytrium* spp. was determined by extracting crude lipid from the cells and fractionating into various lipid class components. TAGs represented the major lipid class fraction in the crude lipids, accounting for 90 to 92% of the total crude lipid. Minor amounts of fatty acid sterol esters (0.4%), diacylglycerols (1%), free sterols (1%), and free fatty acids (0.1%) were present in the crude lipid fraction isolated from *Schizochytrium* spp. β-Carotene was identified as the primary carotenoid present in crude lipid. Fatty acid profiles of crude lipids have been determined. In *Schizochytrium* spp., DHA (C22:6n-3) and DPA (C22:5, n-6) were shown as the major PUFAs in the crude lipid fraction accounting for 42 and 12.5%, respectively. In addition, myristic (9%) and palmitic (26%) were the other major fatty acids. Cholesterol, brassicasterol, and stigmasterol were identified as the major sterol components in the fatty acid sterol ester and free sterol fraction in the crude lipid isolated from *Schizochytrium* spp. Regiospecific fatty acyl position was obtained on the TAG fraction isolated from the crude oil from *Schizochytrium* spp. ^{13}C NMR analysis was conducted[99] with results demonstrating approximately 23 mol% combined DHA and DPA esterified to the *sn*-2 position and 27 mol% esterified to the *sn*-1 and *sn*-3 positions of the glycerol molecules. DHA (55.0%), myristic (14.0%), palmitic (12.0%), and oleic (10.0%) were the predominant fatty acids present in crude oil derived from *Crypthecodinium cohnii*.

16.13   CHARACTERISTICS OF DHASCO®

DHASCO®, produced by Martek Biosciences Corporation, is extracted from the marine microalgal species *Crypthecodinium cohnii*. The final product contained approximately 40% (w/w) DHA. It is a free-flowing liquid oil, which is yellow-orange in color due to the coextraction of carotene pigments. The final product contains about 95% TAGs, with some diacylglycerols and unsaponifiable

TABLE 16.3

Proximate Compositions (%) of *Schizochytrium* spp. and *Crypthecodinium cohnii* Whole Cell Microalgae Produced via Fermentation Process

	Schizochytrium spp.	*Crypthecodinium cohnii*
Moisture	3.0	2.7
Ash	4.0	3.9
Crude protein	20.0	23.1
Total lipids	66	40.0
Crude fiber content	2.0	5.8
DHA (dry weight basis)	21.0	20.4

material, as is typical for all food-grade vegetable oils. Because of the controlled manufacturing process of SCO, as discussed earlier, the potential for contamination with environmental pollutants (pesticide residues, dioxin, etc.) or heavy metals (e.g., Pb, As, and Hg) is eliminated. The fatty acid composition of DHASCO® oil is given in Table 16.4. The fatty acid profile of this algal oil is unique in that it contains no PUFAs other than DHA, except a small quantity of linoleic acid (~0.5%) from high-oleic sunflower oil diluent[91]. The DHASCO® has been used for the supplementation of infant formulas.

The unsaponifiable matter of DHASCO® oil is generally about 1.5% and is made up mainly of sterols[100]. The main sterol has been identified as the 4-methylsterol, dinosterol. The principal components of the sterol fraction in DHASCO® (i.e., dinosterol) are found in the normal metabolic pathway of cholesterol biosynthesis and have been identified in several common food sources including fish and shellfish. A study providing large amounts of the isolated unsaponifiable fraction of crude DHASCO® to rats concluded that these sterols had no adverse effects on growth or lipid metabolism[101].

The oil exhibits a remarkable oxidative stability. This is a result of the absence of other long-chain PUFAs, the relatively low levels of prooxidant heavy metals as compared to fish oil, and the favorable distribution of the DHA in the TAG molecules. Under typical storage conditions of –20°C or 4°C, the DHASCO® is stable for months. As a consequence, this oil has organoleptic properties that are greatly accepted. When encapsulated in soft gelatin capsules of 250 to 500 mg each, the oil can go on for several years at room temperature before showing any change in the peroxide values.

Approximately 45% of the DHA found in algal oil is located at the *sn*-2 position of TAG molecules[102]. The TAG structure of algal oil is nearly identical to that of human milk, with respect to the positional distribution of DHA in TAGs. Martin et al.[103] reported that in human milk about 50 to 60% of the DHA is preferentially esterified at the *sn*-2 position of TAGs. Thus, digestion and absorption of DHA in algal oil is expected to be similar to that of DHA in human milk fat.

16.14   CHARACTERISTICS OF DHASCO®-S

DHASCO®-S is TAG oil, extracted from *Schizochytrium* spp., which is enriched to about 40% (w/w) in DHA. It is described as a yellow to light orange-colored oil and contains greater than 90% (w/w) of TAGs, with some diacylglycerols, free fatty acids, carotenoids, squalene, and phytosterols. β-Carotene was identified as the primary carotenoid component of the lipid fraction. The oil contains a range of fatty acids, including EPA and DPA, as well as DHA. However, DHA is the most abundant PUFA component of the oil. Compositional analyses of other components of the oils compare favorably with typical commercial edible oils. In general, the residual extraction solvent is undetectable, and there are no detectable *trans* fatty acids, pesticide residues, or heavy metals such as arsenic, mercury, and lead.

TABLE 16.4
Typical Analyses of DHASCO®, DHASCO®-S, and ARASCO®

	DHASCO®	DHASCO®-S	ARASCO®
Docosahexaenoic acid (g/kg)	min. 400	min. 350	—
Docosahexaenoic acid (%)	40–45	34–39	—
Arachidonic acid (g/kg)	—	—	min. 350
Arachidonic acid (%)	—	—	min. 38
Peroxide value (meq/kg)	0–0.5	max. 5.0	<5
Free fatty acids (%)	0.03–0.1	max. 0.5	<0.4
Moisture and volatiles (%)	0.0–0.02	max. 0.05	max. 0.05
Unsaponifiable matter (%)	1–2	max. 4.5	max. 3
Insoluble impurities (%)	below detection	below detection	below detection
Trans fatty acids (%)	below detection	below detection	below detection
Heavy metals (ppm)	below detection	below detection	below detection
Major fatty acids (%):			
10:0	0–0.5	—	—
12:0	2–5	0–0.5	—
14:0	10–15	9–15	1–3
16:0	10–14	24–28	12–18
16:1	1–3	0.2–0.5	0–2
18:0	0–2	0.5–0.7	10–14
18:1n-9	10–30	0.5–3.0	5–8
18:2n-6	0.4	0.5–1.3	0.5–2
20:0	<0.1	0.2–0.3	2–5
20:3n-6	—	0–0.5	1–5
20:4n-6	<0.1	0.5–0.8	35–43
22:0	0.1	0.1–0.2	0–3
22:5n-3	0.25	12–16	—
22:6n-3	40–45	36–41	—

DHASCO® from *Crypthecodinium cohnii*, DHASCO®-S from *Schizochytrium* spp., ARASCO® from *Mortierella alpina*.

The nonsaponifiable fraction of DHASCO®-S is generally about 1.5% by weight and made up primarily of squalene, sterols, and carotenoids. These components are all present in the food supply. Cholesterol, brassicasterol, and stigmasterol were identified as the major sterol components of the oil.

16.15   CHARACTERISTICS OF ARASCO®

ARASCO® is derived from the soil fungus *Mortierella alpina*. The final commercial product is a clear yellow, free-flowing liquid TAG oil for which there are specified limits for unsaponifiables and free fatty acids. The final ARA content is adjusted to about 40% of total fatty acids by addition of vegetable oil. Compositional analyses of other components of the oil are comparable to typical commercial vegetable oils. In general, there are no detectable *trans* fatty acids, insoluble impurities, pesticide residues, or heavy metals such as arsenic, mercury, or lead. The ARASCO® oil has been fully characterized and the fatty acid composition is uniquely rich in ARA (Table 16.4). Other n-6 fatty acids, such as DGLA and GLA, are also present but to a much lesser extent. In this respect it represents the purest and richest form of ARA that is commercially available to date. Unlike egg yolk ARA, which is predominantly in the form of phospholipids (i.e., phosphatidylcholine), the ARASCO® is a pure TAG oil (Table 16.2).

Like other vegetable oils, the unsaponifiable fraction of the ARASCO® oil contains primarily sterols. The principal component of the sterol fraction in ARASCO® (i.e., desmosterol) is found in the normal metabolic pathway of cholesterol biosynthesis and is commonly found in several common food sources including animal fat, vegetable oils, and human milk. The DHASCO®, DHASCO®-S, and ARASCO® manufactured by Martek comply with the FDA's proposed regulation 21 CFR 170.36 and are classified as generally recognized as safe (GRAS).

16.16   GOOD MANUFACTURING PRACTICE (GMP) AND QUALITY CONTROL

The primary goal is to ensure that the quality, safety, and efficacy of products are maintained batch after batch; therefore, controls have been placed to ensure that this happens. Good manufacturing practice (GMP) is the part of quality assurance that ensures that products are consistently produced and controlled to the quality standards related to their intended use.

Quality control is the part of GMP that is concerned with sampling and testing of raw materials, intermediate products, and final products to defined specifications. The sampling regimes for routine in-process testing during oil processing are structured to address key critical parameters. Ongoing stability programs are also to be put in place, as done by Martek. Analytical methods are selected and validated specifically for stability study purposes. The appropriate documentation and release procedures are available to ensure that products are not released for use until their quality has been judged to meet the required standards. Therefore, GMP and quality control are part of an overall concept of quality assurance, which covers all activities that influence the quality of Martek products.

The primary objective of quality assurance program is to obtain adequate information on all of factors or characteristics of a product affecting its quality. The scholarly interpretation of this information by quality assurance personnel provides management with an index of the entire operation.

A good documentation system is an essential part of GMP as clear and precise documents enable unambiguous communication of specifications, procedures, manufacturing directions, and overall training of product manufacture. Specifications must be available for all raw materials used in the manufacture of bulk products and final product form. They form a basis for quality evaluations and formal release requirements. Written procedures (standard operating procedures, SOPs) are required to define production, engineering, and quality control activities which include sampling, testing, unit operation, cleaning, maintenance, and environmental monitoring, among others.

All analytical procedures have been validated to demonstrate a defined level of performance. Many analytical methods have been evaluated for performance parameters such as accuracy, precision, specificity, linearity of response, limit of detection, limit of quantitation, and robustness. All these performance parameters can be investigated during validation exercises conducted under good laboratory practice (GLP) conditions.

16.17   SAFETY STUDIES OF DHASCO®, DHASCO®- S AND ARASCO®

The safety of lipids is based on a consideration of each compound's toxicological significance, nutritional profile, effect on overall diet, and expected level of use by various age groups of the population. The FDA makes the decisions about the safety of lipids via two major routes[104], with each having its own set of regulatory requirements. The first approach involves a food manufacturer either claiming that a substance qualifies as GRAS or petitioning the FDA to grant the ingredient such a status. Ingredients that the FDA determines are derived from common food components and are generally recognized by scientific experts to be safe for specific applications based on a long history of safe use in foods or extensive scientific evidence can be approved for inclusion in foods as GRAS status. The second approach entails a manufacturer's request for a new ingredient's approval as a food additive. The 1958 Food Additive Amendment to the Federal Food, Drug and Cosmetic (FD&C) Act established a premarket approval process for food additives for their intended use in food. Guidelines for the safety testing of new food ingredients are outlined in the FDA monograph *Toxicological Principles for the Safety Assessment of Direct Food Additives and Color Additives used in Food*, referred to as the Red Book[105]. A food additive is defined as an ingredient not previously found in food whose intended use results, or may reasonably expected to result, directly or indirectly, in its becoming a component or otherwise affecting the characteristics of any food[106]. A food manufacturer's request requires submission of extensive results on the ingredient safety and intended use, and it must await FDA approval before using them in food. Once approved, the FDA establishes recommended limits on consumption and may require monitoring of use and safety over a period of time. This first route was followed for DHASCO®.

A large number of safety studies have been conducted with DHA-rich algal oils from *Crypthecodinium cohnii* and *Schizochytrium* spp. These studies were done according to the FDA's safety guidelines and generally made in compliance with FDA GLP regulations. DHASCO® from *Crypthecodinium cohnii* had undergone extensive safety studies in animals and has been shown to have no acute toxicity when given at the maximum possible dose (20 g/kg body weight) to rats[107]. The safety of DHASCO®-S from *Schizochytrium* spp. is based on the safety of the source organism, the safety of the oil-soluble components of the source organism, i.e., the fatty acid and sterol components, and a battery of classic toxicity studies conducted on the algae. Results of toxicology studies have been published, conducted by dietary administration or gavage of the source algae in laboratory animals and target species of food-producing animals[88]. Safety was further supported by the historical safe use of the algae as a commercial dietary ingredient in several commercial animal species. The extensive battery of safety studies indicated that ARASCO® is a safe dietary source of ARA. Even though ARASCO® is not a food additive, the safety studies were selected to follow the FDA recommendations for food additives.

In 1995, the Regulatory Office of the Ministry of Health in The Netherlands independently evaluated and approved DHASCO® and ARASCO® as safe for use in preterm and term infant formula. Also in 1995 a U.S. select and independent panel of preeminent toxicologists and nutritionists concluded that DHASCO® and ARASCO® are GRAS for use in preterm and term infant formula. Later in 1996, the same panel concluded that Neuromins® DHA is GRAS for use by adults, including pregnant and lactating women. In 1996, the Committee on Toxicology of the U.K. independently evaluated and approved DHASCO® and ARASCO® as safe for use in preterm and term infant formula. In 1996, the Ministry of Health in France also approved DHASCO® and ARASCO® for use in infant formula. In May of 2001 the FDA completed a favorable review of Martek's GRAS notification regarding the use of its DHASCO® and ARASCO® oil blend in infant formula. In 2002, Health Canada completed a favorable review of Martek's submission supporting the use of its proprietary DHASCO® and ARASCO® oils in infant formulas in Canada. As infant formula manufacturers satisfied Health Canada's premarket notification procedures, products containing Martek's oils became available in Canada.

In March 2004, the FDA completed a favorable review of the Martek's GRAS notification for use of the DHA-rich oil derived from *Schizochytrium* spp. in food applications. The agency has issued a letter informing the company that the FDA has no questions regarding the company's notification that the Martek DHA under review is safe for use in food products. This FDA review was not required but was undertaken by Martek because it believes the extra review provides further indication of future food safety. The acknowledgment from the FDA was made after review of the extensive battery of safety and clinical trials data on the Martek DHA oil, and supported the affirmation of an independent panel of expert physicians, scientists, and toxicologists that the Martek oil is GRAS for use in food products. A list of the food products in which the DHA would be included and the maximum level of inclusion were part of the review.

16.18 SCO APPLICATIONS

16.18.1 NUTRITIONAL SUPPLEMENTS AND FOOD INGREDIENTS

Although scientists and government bodies generally recommend a reduction in total fat intake, specific recommendations are appearing that advise increased consumption of n-3 fatty acids. Different bodies, such as FAO/WHO, the U.S. Institute of Medicine, and Health Canada have issued guidelines regarding the intake of n-3 fatty acids as well as the ratio of n-6 to n-3 fatty acids. A joint FAO/WHO committee recommends an n-6/n-3 ratio of between 5:1 and 10:1 and advises individuals consuming diets with a better ratio to consume more foods containing n-3 fatty acids[20]. The U.S. Institute of Medicine supports a ratio of 5:1[24]. Health Canada recommends an n-6/n-3 ratio of 4:1 to 10:1, particularly for infants and pregnant and lactating women[108]. Today, the U.S. diet only supplies about 10% of the recommended daily intake of n-3 fatty acids. The challenge for the fats and oils industry is to close this nutritional gap by increasing the consumption of n-3 fatty acids either by incorporating n-3 fatty acids into food products or by dietary supplements.

Food processors and producers can help consumers increase their n-3 fatty acid intake by developing DHA-fortified and other n-3 fatty acid-enriched foods. Nutritional supplements and food ingredient products with high levels of DHA can be produced from dried microalgae, refined algal oil, DHA concentrates, algal phospholipids, and DHA-enriched eggs (Table 16.5). The market for DHA-rich oil for human consumption can be divided into four categories: food ingredients, nutraceutical/functional foods, health foods, and pharmaceuticals. One of the applications of DHA-rich oil in food products can be microencapsulated oil. Microencapsulation provides protection against oil oxidation and imparts oxidative stability. Hence, algal oils enriched in DHA can be microencapsulated into a powdered product that is relatively stable for storage at ambient temperatures. Algal oil may be concentrated in the form of TAGs, as free fatty acids, or as simple alkyl esters. Most of the algal oil products sold are in the TAG form. Potential food candidates for incorporation of DHA-rich algal oils include yogurt, cheese, nutrition bars, baked goods, and several other applications.

Martek's DHA oil, derived from *Crypthecodinium cohnii*, has been approved for sale in the United States, where it is widely sold through health food stores under the brand name Neuromins®. Martek also supplies the bulk TAG oil from *Crypthecodinium cohnii* to infant formula manufacturers in a number of countries. Martek's DHA oil derived from *Schizochytrium* spp. is designed specifically for use as a food ingredient. It is manufactured using technology that was obtained by Martek

TABLE 16.5
DHA Ingredient Products Derived from Microalgae and Areas of Application

DHA-enriched ingredient	Application
Refined oil	Infant formula, dietary supplements, weaning and toddler foods, broad-based food and dairy products
Microencapsulated oil	Nutritional supplements, broad-based food and dairy products
Dried microalgae	Nutritional supplements, foods, animal feed supplements, pet foods, aquaculture
Ethyl esters	Nutritional supplements, pharmaceuticals
High-purity DHA	Nutritional supplements, pharmaceuticals
Egg-based products (egg yolk powder, defatted egg yolk powder, liquid egg whites)	Baked goods, pasta, sport/nutritional powders, ready-to-eat foods, salad dressings
Phospholipids	Nutritional supplements, foods, pharmaceuticals

in its 2002 acquisition of OmegaTech Inc. (Boulder, CO). PBM Products Inc. (Gordonsville, VA), producers of Bright Beginnings™ infant formula with DHA, has recently launched the first diabetic nutritional drink containing DHA. This patent-pending product is the only diabetic beverage offering the unique combination of n-3 polyunsaturated fatty acids, amino acids, and other desirable nutrients essential for health. GlaxoSmithKline (Middlesex, U.K.) has recently launched a powdered drink mix containing Martek's DHA in India. The product, Junior Horlicks, is specially formulated to meet the needs of a child's developing brain and nervous system and is available in selected outlets in south and east India. Junior Horlicks is the second product offered with DHA in India. In June 2004, GlaxoSmithKline launched DHA-enriched Mother's Horlicks, which is formulated to meet the needs of pregnant and breast-feeding mothers. Mead Johnson Nutritionals (Evansville, IN) has recently launched a DHA supplement for pregnant and nursing women containing DHA. The product, Expecta™ LIPIL® DHA supplement, is a nonprescription prenatal supplement that provides 200 mg of DHA for pregnant and nursing women. It is clear that DHA products from both *Crypthecodinium cohnii* and *Schizochytrium* spp. will be part of the expanding heath food market for adults, toddlers, and for infant foods. It is anticipated that these markets will continue to grow as the direct benefits of increased DHA consumption for both adults and infants appear to be based on strong scientific evidence.

Edible fats containing highly unsaturated fatty acids and suitable for food applications need to be stable. The main problem with the use of unsaturated oils in foods is their poor oxidative stability. The unpleasant odor of oxidized oils poses an entry barrier into mass markets. Good organoleptic quality is essential for lasting market success. In a typical food application study, loaves of bread were baked with a regular pharmaceutical-grade fish oil made from sand eel, a specialty tuna oil, or an algal oil. The breads were standardized to 100 mg of DHA per slice of bread, and a taste panel was asked to evaluate the bread every three days. The taste panel data indicated that off-flavors were less likely to be detected in the DHA bread made with algal oils compared to those made with fish oils[98]. This is likely due to reduced accumulation of off-flavor oxidation products, since the algal source has better oxidative stability.

16.18.1.1 Eggs Enriched with DHA

The wide use of eggs in restaurants, bakeries, and cafeterias offers the opportunity for n-3 fatty acid-enriched eggs to influence the diet of the general population in a healthy way. There are currently eggs on the market enriched with PUFAs that function as dietary supplements in food form. For cost, convenience, and versatility, these eggs provide most consumers with an unsurpassed choice if they want to increase the heart-healthy n-3 fatty acids in their diet. Rather than adding PUFAs directly to the eggs, the nutritional benefit is obtained by feeding hens a diet enriched with fish oils, vegetable oils, or an algal source of DHA. This provides a more cost-effective way for consumers to obtain n-3 fatty acids in the diet over servings of cold-water fish. Martek's DHA rich oil from *Schizochytrium* spp. has been used as a poultry feed additive to give DHA-enriched eggs and meat. These eggs, known as SeaGold eggs, are now available in Europe, and it is the intention of Martek to introduce similar schizochytrium-derived products in the United States. In March 2005 new liquid eggs containing Martek DHA™ were launched by Gold Circle Farms in the United States. Gold Circle Farms DHA omega-3 eggs, produced by Hidden Villa Ranch (Fullerton, CA), contain 150 mg DHA per egg, by utilizing an exclusive process for cultivating marine algae into animal feed for the hens. Like fish, the hens eat the naturally enriched algae that contain DHA resulting in a naturally fortified egg.

In a study conducted at the University of New England, Australia, significant increases in blood n-3 PUFAs and HDL cholesterol were shown in subjects consuming seven enriched eggs per week as compared to controls[109]. The subjects showed a significant gain in body weight and HDL levels, but merely adding one egg per day to the diet should not result in weight or HDL changes. Smuts et al.[40] reported that eggs containing high levels of DHA could be a good alternative source of dietary DHA. The study concluded that relatively modest amounts of dietary DHA during pregnancy appear to extend gestational age and may lead to enhanced fetal growth.

16.18.1.2   Importance of DHA and ARA in Aquaculture

The growth of the aquaculture industry reflects an increasing consumer demand for fish in the face of dwindling marine and freshwater fish stocks. A successful aquaculture industry requires a steady supply of feed with high-quality protein and oil. One of the most significant obstacles to establishing and maintaining an economically feasible aquaculture operation is the difficulty of supplying nutritionally balanced feeds. Larval fish, bivalves, and crustaceans grown in the wild consume a mixed population of feed organisms that collectively furnish balanced nutrition. However, fish larvae, bivalves, and crustaceans raised in aquaculture farms can be difficult to raise and require live feeds (algae, algae-fed rotifers, and Artemia, etc.) for their nutrition. These live feeds are difficult to produce and maintain, need high labor inputs, and specialized facilities, and consequently larval feeds constitute a significant cost to the aquaculture industry.

It is important that aquaculture feeds be nutritionally balanced so that the larvae receive proper nutrition. Microalgae have an important role in aquaculture as a means of enriching zooplankton for on-feeding to fish and other larvae. Microalgae can be utilized in aquaculture as live feeds for all growth stages of bivalve mollusks (e.g., oysters, scallops, clams, and mussels), for the larval/early juvenile stages of abalone, crustaceans, and some fish species, and for zooplankton used in aquaculture food chains. In addition to providing protein (essential amino acids) and energy, they provide other key nutrients such as vitamins, essential PUFAs, pigments, and sterols, which are transferred through the food chain. PUFAs derived from microalgae, i.e., DHA, EPA, and ARA, have been identified as important nutrients that contribute significantly to larval growth and survival[110]. Larvae ultimately acquire these fatty acids from algae, either by directly feeding on algae with high levels of PUFA or by feeding on rotifers and artemia that have been reared on algae high in PUFA. Unfortunately, the algae, artemia, and rotifers used at aquaculture farms are low in DHA and ARA, reducing the survival rates for the larvae below their maximal rate and increasing the cost of the final aquaculture farm product. If sufficient DHA and ARA could be produced to the larvae it is expected that the survival rate for larvae would increase, thus reducing the cost of farm-raised seafood.

A number of processed forms of microalgae have also been assessed as alternatives to live microalgae. Several products based on thraustochytrids from the genus *schizochytrium* have been marketed through Aquafauna Biomarine Inc. (e.g., algaMac 2000) and Sanders Brine Shrimp Co. (e.g., Docosa Gold). These products have high concentrations of DHA[111], and so are being applied as alternatives to commercial oil enrichments (e.g., Selco) for zooplankton fed to larvae.

REFERENCES

1. Bang, H.O., Dyerberg, J., and Hjorne, N., The composition of food consumed by Greenland Eskimos, *Acta Med. Scand.*, 200, 59–73, 1976.
2. Bang, H.O. and Dyerberg, J., Plasma lipids and lipoproteins in Greenlandic West-coast Eskimos, *Acta Med. Scand.*, 192, 85–94, 1972.
3. Bang, H.O. and Dyerberg, J., Lipid metabolism and ischemic heart disease in Greenland Eskimos, *Adv. Nutr. Res.*, 3, 1–21, 1986.
4. Carlson, S.E., Arachidonic acid status of human infants: influence of gestational age at birth and diets with very long chain n-3 and n-6 fatty acids, *J. Nutr.*, 126, 1092–1098, 1996.
5. Groom, H., Oil-rich fish, *Nutr. Food Sci.*, Nov.–Dec., 4–8, 1993.
6. Greiner, R.C.S., Winter, J., Nathanielsz, P.W., and Brenna, J.T., Brain docosahexaenoate accretion in fetal baboons: bioequivalence of dietary α-linolenic and docosahexaenoic acids, *Pediatr. Res.* 42, 826–834, 1997.
7. Duque, A.G., The role of lipids in fetal brain development, *Perinat. Nutr. Rep.*, Spring, 4–5, 1997.

8. Ghebremeskel, K., Min, Y., Crawford, M.A., Nam, J., Kim, A., Koo, J., and Suzuki, H., Blood fatty acid composition of pregnant and nonpregnant Korean women: red cells may act as a reservoir of arachidonic acid and docosahexaenoic acid for utilization by the developing fetus, *Lipids*, 35, 567–574, 2000.

9. Conquer, J.A. and Holub, B.J., Supplementation with an algae source of docosahexaenoic acid increases (n-3) fatty acid status and alters selected risk factors for heart disease in vegetarian subjects, *J. Nutr.*, 126, 3032–3039, 1996.

10. Salem, N., Jr., Wegher, B., Mena, P., and Uauy, R., Arachidonic and docosahexaenoic acids are biosynthesized from their 18-carbon precursors in human infants, *Proc. Natl. Acad. Sci. USA*, 93, 49–54, 1996.

11. Brossard, N., Croset, M., Pachiaudi, C. et al., Retroconversion and metabolism of[13C] 22:6n-3 in humans and rats after intake of a single dose of[13C] 22:6n-3 triacylglycerols, *Am. J. Clin. Nutr.*, 64, 577–586, 1996.

12. Ackman, R.G. and Ratnayake, W.M.N., Fish oils, seal oils, esters and acids: are all forms of omega-3 intake equal?, in *Health Effects of Fish and Fish Oils*, Chandra, R.K., Ed., ARTS Biomedical Publishers and Distributors, St. John's, Newfoundland, 1989, pp. 373–393.

13. Kinsella, J.E., Sources of omega-3 fatty acids in human diets, in *Omega-3 Fatty Acids in Health and Disease*, Lees, R.S. and Karel, M., Eds., Marcel Dekker, New York, 1990, pp. 157–200.

14. Seeds, M.C. and Bass, D.A., Regulation and metabolism of arachidonic acid, *Clin. Rev. Allergy Immunol.*, 17, 5–26, 1999.

15. Krawczyk, T., Do infants need extra DHA, AA?, *Inform*, 12, 1064–1074, 2001.

16. British Nutrition Foundation, Recommendation for intakes of unsaturated fatty acids, in: *Unsaturated Fatty Acids: Nutritional and Physiological Significance*, Chapman and Hall, London, 1992, pp. 152–63.

17. Food and Agriculture Organization of the United Nations/World Health Organization (FAO/WHO) Expert Committee: Fats and Oil in Human Nutrition, Food and Nutrition Paper, Rome, 1994, p. 57.

18. FAO/WHO Joint Expert Consultation, Report of a joint expert consultation, FAO Food and Nutrition Paper 57, Rome, 1995, pp. 49–55.

19. International Society for the Study of Fatty Acids (ISSFAL) Board of Directors, Recommendations for the essential fatty acid requirements for infant formulae, *ISSFAL Newsletter*, 1, 4–5, 1994.

20. WHO and FAO Joint Consultation, Fats and oils in human nutrition, *Nutr. Rev.*, 53, 202–205, 1995.

21. European Commission Scientific Committee on Food, Report of the Scientific Committee on Food on the Revision of Essential Requirements of Infant Formulae and Follow-on Formulae, Brussels, Belgium, May 18, 2003.

22. Uauy, R., Hoffman, D.R., Mena, P., Llanos, A., and Birch, E.E., Term infant studies of DHA and ARA supplementation on neurodevelopment: results of randomized controlled trials, *J. Pediatr.*, 143, S17–S25, 2003.

23. Carver, J.D., Advances in nutritional modifications of infant formulas, *Am. J. Clin. Nutr.*, 77, 1550S–1554S, 2003.

24. Institute of Medicine, *Dietary Reference Intakes for Energy, Carbohydrate, Fiber, Fat, Fatty Acids, Cholesterol, Protein, and Amino Acids*, National Academies Press, Washington, DC, 2002, pp. 7-1–7-69 (dietary fiber), 8-1–8-97 (fat and fatty acids).

25. Jensen, C.L., Chen, H., Fraley, J.K., Anderson, R.E., and Heird, W.C., Biochemical effects of dietary linoleic/alpha-linolenic acid ratio in term infants, *Lipids*, 1, 107–13, 1996.

26. Birch, E.E., Hoffman, D.R., Uauy, R., Birch, D.G., and Prestidge, C., Visual acuity and the essentiality of docosahexaenoic acid and arachidonic acid in the diet of term infants, *Pediatr. Res.*, 44, 201–209, 1998.

27. Makrides, M., Neumann, M., Simmer, K., Pater, J., and Gibson, R., Are long-chain polyunsaturated fatty acids essential nutrients in infancy?, *Lancet*, 345, 1463–1468, 1995.

28. Birch, E.E., Garfield, S., Hoffman, D.R., Uauy, R., and Birch, D.G., A randomized controlled trial of early dietary supply of long-chain polyunsaturated fatty acids and mental development in term infants, *Dev. Med. Child. Neurol.*, 42, 174–181, 2000.

29. Hoffman, D.R., Birch, E.E., Birch, D.G., Uauy, R., Castaneda, Y.S., Lapus, M.G., and Wheaton, D.H., Impact of early dietary intake and blood lipid composition of long-chain polyunsaturated fatty acids on later visual development, *J. Pediatr. Gastroenterol. Nutr.*, 31, 540–553, 2000.

30. Hoffman, D.R., Birch, E.E., Castaneda, Y.S., Fawcett, S.L., Birch, D.G., and Uauy, R., Dietary docosahexaenoic acid (DHA) and visual maturation in the post-weaning term infant, *Invest. Opthalmol. Vis. Sci.*, 42, S122 (abstract), 2001.

31. Birch, E.E., Hoffman, D.R., Castaneda, Y.S., Fawcett, S.L., Birch, D.G., and Uauy, R., A randomized controlled trial of long-chain polyunsaturated fatty acid supplementation of formula in term infants after weaning at 6 wk of age, *Am. J. Clin. Nutr.*, 75, 570–80, 2002.

32. Hoffman, D.R., Birch, E.E., Castaneda, Y.S., Fawcett, S.L., Wheaton, D.H., Birch, D.G., and Uauy, R., Visual function in breast-fed term infants weaned to formula with or without long-chain polyunsaturates at 4 to 6 months: a randomized clinical trial, *J. Pediat.*, 142, 669–677, 2003.

33. Horwood, L.J. and Fergusson, D.M., Breastfeeding and later cognitive and academic outcomes, *Pediatrics*, 101, E9, 1998.

34. Lucas, A., Morley, R., Cole, T.J., Lister, G., and Leeson-Payne, C., Breast milk and subsequent intelligence quotient in children born preterm, *Lancet*, 339, 261–264, 1992.

35. Willatts, P., Forsyth, J.S., DiModugno, M.K, Varma, S., and Colvin, M., Effect of long-chain polyunsaturated fatty acids in infant formula on problem solving at 10 months of age, *Lancet*, 352, 688–691, 1998.

36. Carlson, S.E., Werkman, S.H., Peeples, J.M., and Wilson, W.M., Long-chain fatty acids and early visual and cognitive development of preterm infants, *Eur. J. Clin. Nutr. Suppl.*, 2, S27–S30, 1994.

37. Carlson, S.E., Werkman, S.H., Rhodes, P.G., and Tolley, E.A., Visual-acuity developments in healthy preterm infants: effect of marine-oil supplementation, *Am. J. Clin. Nutr.*, 58, 35–42, 1993.

38. Werkman, S.H. and Carlson, S.E., A randomized trial of visual attention of preterm infants fed docosahexaenoic acid until nine months, *Lipids*, 31, 91–7, 1996.

39. Uauy, R.D., Birch, D.G., Birch, E.E., Tyson, J.E., and Hoffamn, D.R., Effect of dietary omega-3 fatty acids on retinal function of very-low-birth-weight neonates, *Pediatr. Res.*, 28, 485–92, 1990.

40. Smuts, C.M., Huang, M., Mundy, D., Plasse, T., Major, S., and Carlson, S.E., A randomized trial of docosahexaenoic acid supplementation during the third trimester of pregnancy, *Obstet. Gynocol.*, 101, 469–479, 2003.

41. Clandinin, M.T., Chappell, J.E., Leong, S., Heim, T., Swyer, P.R., and Chance, G.W., Intrauterine fatty acid accretion rates in human brain: implications for fatty acid requirements, *Early Hum. Dev.*, 4, 121–129, 1980.

42. Farquharson, J., Cockburn, T., Patrick, W.A., Jamieson, E.C., and Logan, R.W., Infant cerebral cortex phospholipid fatty-acid composition and diet, *Lancet*, 340, 810–813, 1992.

43. Makrides, M., Neumann, M.A., Byard, R.W., Simmer, K., and Gibson, R.A., Fatty acid composition of brain, retina, and erythrocytes in breast- and formula-fed infants, *Am. J. Clin. Nutr.*, 60, 189–94, 1994.

44. Clark, K.J., Makrides, M., Neumann, M.A., and Gibson, R.A., Determination of the optimal ratio of linoleic acid to alpha-linolenic acid in infant formulas, *J. Pediatr.*, 4, S151–S158, 1992.

45. Clandinin, M.T., Van Aerde, J.E., Parrott, A., Field, C.J., Euler, A.R., and Lien, E., Assessment of feeding different amounts of arachidonic and docosahexaenoic acids in preterm infant formulas on the fatty acid content of lipoprotein lipids, *Acta Pediatr.*, 88, 890–896, 1999.

46. Nair, S.S.D., Leitch, J.W., Falconer, J., and Garg, M.L., Prevention of cardiac arrhythmia by dietary (n-3) polyunsaturated fatty acids and their mechanism of action, *J. Nutr.*, 127, 383–393, 1997.

47. Simopoulos, A.P., ω-3 Fatty acids in the prevention-management of cardiovascular disease, *Can. J. Physiol. Pharmacol.*, 75, 234–239, 1997.

48. Uauy-Dagach, R. and Valenzuela, A., Marine oils: the health benefits of n-3 fatty acids, *Nutr. Rev.*, 54, S102–S108, 1996.

49. Simon, J.A., Hodgkins, M.L., Browner, W.S., Neuhaus, J.M., Bernert, J.T., Hulley, S.B., Serum fatty acids and the risk of coronary heart disease, *Am. J. Epidemiol.*, 142, 469–476, 1995.

50. Gaudette, D.C. and Holub, B.J., Docosahexaenoic acid (DHA) and human platelet reactivity. *J. Nutr. Biochem.*, 2, 116–121, 1991.

51. Kang, J.X. and Leaf, A., The cardiac antiarrhythmic effects of polyunsaturated fatty acid, *Lipids*, 31, S41–S44, 1996.

52. Grimsgaard, S., Bonaa, K.H., Hansen, J.B., and Nordoy, A., Highly purified eicosapentaenoic acid and docosahexaenoic acid in humans have similar triacylglycerol-lowering effects but divergent effects on serum fatty acids, *Am. J. Clin. Nutr.*, 66, 649–659, 1997.

53. Mori, T.A., Burke, V., Puddey, I.B., Watts, G.F., O'Neal, D.N., Best, J.D., and Beilin, L.J., Purified eicosapentaenoic and docosahexaenoic acids have different effects on serum lipids and lipoproteins, LDL particle size, glucose, and insulin in mildly hyperlipidemic men, *Am. J. Clin. Nutr.*, 71, 1085–1094, 2000.

54. Engler, M.M., Engler, M.B., Malloy, M., Chiu, E., Besio, D., Paul, S., Stuehlinger, M., Morrow, J., Ridker, P., Rifai, N., and Mietus-Snyder, M., Docosahexaenoic acid restores endothelial function in children with hyperlipidemia: results from the EARLY study, *Int. J. Clin. Pharm. Therap.*, 42, 672–679, 2004.

55. Simopoulos, A.P., Leaf, A., and Salem, N., Jr., Workshop on the essentiality of and recommended dietary intakes for omega-6 and omega-3 fatty acids. *J. Am. Coll. Nutr.*, 18, 487–489, 1999.

56. Kris-Etherton, P.M. et al., AHA Scientific Statement. Fish consumption, fish oil, omega-3 fatty acids and cardiovascular disease, *Circulation*, 106, 2747, 2002.

57. Report of the Dietary Guidelines Advisory Committee on the Dietary Guidelines for Americans, Department of Health and Human Services (HHS) and U.S. Department of Agriculture (USDA), January 2005.

58. Holub, B.J., Docsahexaenoic acid in human health, in *Omega-3 Fatty Acids: Chemistry, Nutrition and Health Effects*, Shahidi, F. and Finley, J.W., Eds., American Chemical Society, Washington, DC, 2001, pp. 54–65.

59. Hibbeln, J. and Salem, N., Dietary polyunsaturated fatty acids and depression: when cholestrol does not satisfy, *Am. J. Clin. Nutr.*, 62, 1–9, 1995.

60. Simopoulos, A.P., New products from the agri-food industry: the return of n-3 fatty acids into the food supply, *Lipids*, 34, S297–S301, 1999.

61. Simopoulos, A.P., n-3 Fatty acids and human health: defining strategies for public policy, *Lipids*, 36, S83–S89, 2001.

62. Stevens, L.J., Zentall, S.S., Deck, J.L., Abate, M.L., Watkins, B.A., Lipp, S.R., and Burgess, J.R., Essential fatty acid metabolism in boys with attention-deficit hyperactivity disorder, *Am. J. Clin. Nutr.*, 62, 761–768, 1995.

63. Champoux, M., Hibbeln, J.R., Shannon, C., Majchrzak, S., Suomi, S.J., Salem, N., Jr., and Higley, J.D., Fatty acid formula supplementation and neuromotor development in rhesus monkey neonates, *Pediatr. Res.*, 51, 273–281, 2002.

64. Moriguchi, T., Greiner, R.S., and Salem, N., Jr., Behavioral deficits associated with dietary induction of decreased brain docosahexaenoic acid concentration, *J. Neurochem.*, 75, 2563–2573, 2000.

65. Calon, F., Lim, G.P., Yang, F., Morihara, T., Teter, B., Ubeda, O., Rostaing, O., Triller, A., Salem, N., Jr., Ashe, K.H., Frautschy, S.A., and Cole, G.M., Docosahexaenioc acid protects from dendritic pathology in an Alzheimer's disease mouse model, *Neuron*, 43, 633–645, 2004.

66. Lim, G.P., Calon, F., Morihara, T., Yang, F., Teter, B., Ubeda, O., Salem, N., Jr., Frautschy, S.A., and Cole, G.M., A diet enriched with the omega-3 fatty acid docosahexaenoic acid reduces amyloid burden in an aged Alzheimer mouse model, *J Neurosci.*, 25, 3032–3040, 2005.

67. Ruyle, M., Connor, W.E., Anderson, G.J., and Lowensohn, R.I., Placental transfer of essential fatty acids in humans: venous-arterial difference for docosahexaenoic acid in fetal umbilical erythrocytes, *Proc. Natl. Acad. Sci. USA*, 87, 7902–7906, 1990.

68. Cetin, I., Giovannini, N., Alvino, G., Agostoni, C., Riva, E., Giovannini, M., and Pardi, G., Intrauterine growth restriction is associated with changes in polyunsaturated fatty acid fetal-maternal relationships, *Pediatr. Res.*, 52, 750–755, 2002.

69. Koletzko, B., Agostoni, C., Carlson, S.E., Clandinin, T., Hornstra, G., Neuringer, M., Uauy, R., Yamashiro, Y., and Willatts, P., Long-chain polyunsaturated fatty acids and perinatal development, *Acta Paediatr.*, 90, 460–464, 2001.

70. Forsyth, J.S., Willatts, P., Agostoni, C., Bissenden, J., Casaer, P., and Boehm, G., Long chain polyunsaturated fatty acid supplementation in infant formula and blood pressure in later childhood: follow up of a randomized controlled trial, *Br. Med. J.*, 326, 953–957, 2003.

71. Auestad, N., Halter, R., Hall, R.T., Blatter, M., Bogle, M.L., Burks, W., Erickson, J.R., Fitzgerald, K.M., Dobson, V., Innis, S.M., Singer, L.T., Montalto, M.B., Jacobs, J.R., Qiu, W., and Bornstein, M.H., Growth and development in term infants fed long-chain polyunsaturated fatty acids: a double-masked, randomized, parallel, prospective, multivariate study, *Pediatrics*, 108, 372–381, 2001.

72. Otto, S.J., Houwelingen, A.C.V., and Hornstra, G., Search for an LCP supplement for pregnant women, *Prostaglandins, Leukotrienes Essen. Fatty Acids*, 57, 190, 1997.

73. Olsen, S.F., Sorensen, J.D., Secher, N.J. et al., Randomised controlled trial of effect of fish oil supplementation on pregnancy duration, *Lancet*, 339, 1003–1007, 1992.

74. Helland, I.B., Smith, L., Saarem, K., Saugstad, O.D., and Drevon, C.D., Maternal supplementation with very-long-chain fatty acids during pregnancy and lactation augments children's IQ at 4 years of age, *Pediatrics*, 111, e39–e44, 2003.

75. Makrides, M. and Gibson, R.A., Long-chain polyunsaturated fatty acid requirements during pregnancy and lactation, *Am. J. Clin. Nutr.*, 71, 307–311, 2000.

76. Jensen, C., Voigt, R., Llorente, A., Peters, S., Prager, T., Zou, Y., Fraley, K., and Heird, W., Effect of maternal docosahexaenoic acid (DHA) supplementation on neuropsychological and visual status of former breast-fed infants, *Pediatr. Res.*, 55, 181A, 2004.

77. Jensen, C., Voigt, R., Prager, T., Zou, Y., Fraley, K., Rozelle, J., Turich, M., Llorente, W., and Heird, W., Effect of maternal docosahexaenoic acid (DHA) supplementation on visual function and neurodevelopment of breast-fed infants, *Pediatr. Res.*, 49, 2572, 2004.

78. Streekstra, H., Fungal production of arachidonic acid-containing oil on an industrial scale, *Inform*, 15, 20, 21, 2004.

79. Ratledge, C., Yeasts, moulds, algae and bacteria as sources of lipids, in *Technological Advances in Improved and Alternative Sources of Lipids*, Kamel, B.S and Kakuda, Y., Eds., Blackie, London, 1994, pp. 235–291.

80. Steinbüchel, A., Polyxydroxyalkanoic acids, in *Biomaterials*, Byrom, D., Ed., McMillan, London, 1991, pp. 123–213.

81. Kavadia, A., Komaitis, M., Chevalot, I., Blanchard, F., Marc, I., and Aggelis, G., Lipid and gamma-linolenic acid accumulation in strains of Zygomycetes growing on glucose, *J. Am. Oil. Chem. Soc.*, 78, 341–346, 2001.

82. Matsuo, T., Terashima, M., Hasimoto, Y., and Hasida, W., Method for Producing Cacao Butter Substitute, U.S. Patent 4,308, 350, 1981.

83. Kendrick, A. and Ratledge, C., Lipids of selected molds grown for production of n-3 and n-6 polyunsaturated fatty acids, *Lipids*, 27, 15–20, 1992.

84. Barclay, W.R., Meager, K.M., and Abril, J.R., Heterotrophic production of long chain omega-3 fatty acids utilizing algae and algae-like microorganisms, *J. Appl. Phycol.*, 6, 123–129, 1994.

85. Spector, D.L., *Dinoflagellates*, Academic Press, New York, 1984.

86. Harrington, G.W., Beach, D.H., Dunham, J.E., and Holz, G.G., Jr., The polyunsaturated fatty acids of marine dinoflagellates, *J. Protozool.*, 17, 213–9, 1970.

87. Zeller, S., Barclay, W., and Abril, R., Production of docsahexaenoic acid from microalgae, in *Omega-3 Fatty Acids: Chemistry, Nutrition and Health Effects*, hahidi, F. and Finley, J.W., Eds., American Chemical Society, Washington, DC, 2001, pp. 108–124.

88. Hammond, B.G., Mayhew, D.A., Kier, L.D., Mast, R.W., and Sander, W.J., Safety assessment of DHA-rich microalgae from Schizochytrium sp., *Regulatory Toxicol. Pharmacol.*, 35, 255–265, 2002.

89. Dembitsky, V.M., Pechenkina-Shubina, E.E., and Rozentsvet, O.A., Glycolipids and fatty acids of some seaweeds and marine grasses from the Black Sea, *Phytochemistry*, 30, 2279–2283, 1991.

90. Hamanaka, T., Higashiyama, K., Fujikawa, S., and Park, E.Y., Mycelial pellet intrastructure and visualization of mycelia and intracellular lipid in a culture of Mortierella alpine, *Appl. Microbiol. Biotechnol.*, 56, 233–238, 2001.

91. Kyle, D.J., Production and use of a single cell oil which is highly enriched in docosahexaenoic acid, *Lipid Technol.*, 9, 107–110, 1996.

92. Barclay, W.R., Process for Growing Thraustochytrium and Schizochytrium Using Non-Chloride Salts to Produce a Microfloral Biomass Having Omega-3-Highly Unsaturated Fatty Acids, U.S. Patent 5, 340, 742, 1994.

93. Park, E.Y., Koike, Y., Higashiyama, K., Fujikawa, S., and Okabe, M., Effect of nitrogen source on mycelial morphology and arachidonic acid production in cultures of *Mortierella alpina*, *J. Biosci. Bioeng.*, 88, 61–67, 1999.

94. Eroshin, V.K., Dedyukhina, E.G., Satroutdinov, A.D., and Chistyakova, T.I., Growth-coupled lipid synthesis in *Mortierella alpina* LPM 301, a producer of arachidonic acid, *Microbiology*, 71, 169–172, 2002.

95. Schutte, H. and Kula, M.R., Pilot- and process-scale techniques for cell disruption, *Biotechnol. Appl. Biochem.*, 12, 599–620, 1990.

96. Chisti, Y. and Moo-Young, M., Disruption of microbial cells for intracellular products, *Enzyme Microbial Technol.*, 8, 194–204, 1986.

97. Shim, S.M., Santerre, C.R., Burgess, J.R., and Deardorff, D.C., Omega-3 fatty acids and total poly-chlorinated biphenyls in 26 dietary supplements, *J. Food Sci.*, 68, 2436–2440, 2003.

98. Becker, C.C. and Kyle, D.J. Developing functional foods. *Food Technol.*, 52, 68–71, 1998.

99. Aursand, M., Rainuzzo, J.R., and Grasdalen, H., Quantitative high-resolution C-13 and H-1 nuclear magnetic resonance of omega-3 fatty acids from white muscle of Atlantic salmon (*Salmo salar*), *J. Am. Oil Chem. Soc.*, 70, 971–981, 1993.

100. Kyle, D.J., The large scale production and use of a single cell oil highly enriched in docosahexaenoic acid, in *Omega-3 Fatty Acids. Chemistry, Nutrition and Health Effects*, Shahidi, F. and Finely, J.W., Eds., ACS Symposium Series 788, American Chemical Society, Washington, DC, 2001, pp. 92–107.

101. Kritchevsky, D., Tepper, S.A., Czarnecki, S.K., and Kyle, D.J., Effects of 4-methylsterols from algae and of β-sitosterol on cholesterol metabolism in rats, *Nutr. Res.*, 19, 1649–1654, 1999.

102. Myher, J.J., Kuksis, A., Geher, K., Park, P.W., and Diersen-Schade, D.A., Stereospecific analysis of triacylglycerols rich in long-chain polyunsaturated fatty acids, *Lipids*, 31, 207–215, 1996.

103. Martin, J.C., Bougnoux, P., Antoine, J.M., Lanson, M., and Couet, C., Triacylglycerol structure of human colostrum and mature milk, *Lipids*, 28, 637–643, 1993.

104. Kurtzweil, P., Taking the fat out of food, *FDA Consumer*, 30, 7, 1996.

105. U.S. Food and Drug Administration, *Toxicological Principles for the Safety Assessment of Direct Food Additives and Color Additives used in Food*, Bureau of Foods, Washington, DC, 1982.

106. U.S. Food and Drug Administration, Department of Health and Human Services, Code of Federal Regulations, Title 21: Food and Drugs, Part 170: Food Additives, U.S. Government Printing Office, Vol. 3, revised as of April 1, 2001, p. 5.

107. Boswell, K., Kosketo, E.K., Carl, L., Glaza, S., Hensen, D.J., Williams, K.D., and Kyle, D.J., Preclinical evaluation of single-cell oils that are highly enriched with arachidonic acid and docosahexaenoic acid, *Food Chem. Toxicol.*, 34, 585–593, 1996.

108. Health and Welfare Canada, Nutrition Recommendations, report of the Scientific Review Committee, Department of Supply and Services, H49-42/1990E, Ottawa, ON, 1990.

109. Farrell, D.J., Enrichment of hen eggs with n-3 long-chain fatty acids and evaluation of enriched eggs in humans, *Am. J. Clin. Nutr.*, 68, 538–44, 1998.

110. Sargent, J.R., McEvoy, L.A., and Bell, J.G., Requirements, presentation and sources of polyunsaturated fatty acids in marine fish larval feeds, *Aquaculture*, 155, 117–127, 1997.

111. Barclay, W. and Zeller, S., Nutritional enhancement of n-3 and n-6 fatty acids in rotifers and *Artemia* by feeding spray-dried *Schizochytrium* sp., *J. World Aqua. Soc.*, 27, 314–322, 1996.

17 Emulsions for the Delivery of Nutraceutical Lipids

Karlene S.T. Mootoosingh and Dérick Rousseau
School of Nutrition, Ryerson University, Toronto, Ontario, Canada

CONTENTS

17.1   INTRODUCTION

The purpose of this chapter is to introduce the reader to aspects of emulsion systems and to their use as vehicles for the delivery of nutraceutical lipids. The areas discussed include an overview of emulsion properties (Section 17.2), emulsion stabilization (Section 17.3), and characterization of emulsion droplet sizes (Section 17.4). The final section (Section 17.5) is devoted to the applications of emulsions as delivery systems for nutraceutical lipids.

17.2   OVERVIEW OF EMULSION PROPERTIES

17.2.1   DEFINITIONS

A simple emulsion consists of two immiscible liquids, typically oil and water, where one forms small spherical droplets within the other. The liquid that forms the droplets is known as the dispersed or internal phase and the liquid that surrounds the droplets is called the continuous or external phase.

Given that the contact between oil and water molecules is thermodynamically unfavorable, emulsions must be kinetically stabilized. Stabilizers in the form of small-molecule surfactants, proteins, and/or thickeners are often used. A lesser known class of stabilizing agents consists of colloidal particles, which act via physical hindrance to promote the kinetic stability of a dispersed droplet phase. Emulsifiers generally play two roles in emulsion kinetic stability. They lower interfacial tension between the oil and water phases and/or form a mechanically, cohesive interfacial film around the droplets thereby minimizing droplet–droplet contact. Surfactants may also impart dynamic properties to the interface, which allow it to resist tangential stresses[1]. Typically, surface-active agents accumulate at the interface when their free energy in the adsorbed state is lower than that in the unadsorbed state. Depending on conditions, emulsions may be more stable at lower temperature due to increased phase viscosity. Stability against coalescence can also be achieved by mechanical means, such as reducing the average droplet size an emulsion contains via homogenization. Generally, surfactants minimize most destabilization processes, which include creaming, sedimentation, flocculation, and coalescence (Figure 17.1).

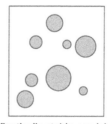

Kinetically stable emulsion

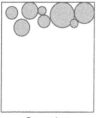 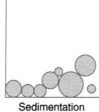 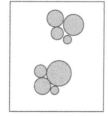 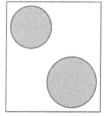

Creaming   Sedimentation   Flocculation   Coalescence

FIGURE 17.1 Mechanisms of emulsion destabilization (creaming, sedimentation, flocculation, and coalescence).

17.2.2 Types of Emulsions

In a two-phase simple emulsion, an oil phase may be dispersed in the form of droplets in an aqueous phase forming an oil-in-water (O/W) emulsion (e.g., milk, cream, and mayonnaise) or an aqueous phase may form droplets in an oil phase, yielding a water-in-oil (W/O) emulsion (e.g., margarine, butter, and spreads). Another case is when water is dispersed in oil droplets, which in turn are dispersed in a continuous aqueous phase. This is known as a water-in-oil-in-water (W/O/W) emulsion. An emulsion of this nature is called a double emulsion or multiple emulsion, and may also exist in the opposite manner, where oil droplets are located within water droplets that are dispersed in a continuous lipid phase (O/W/O) (Figure 17.2). W/O/W emulsions are more commonly used than O/W/O systems and some have been developed for food, pharmaceutical, and cosmetic use.

The equipment most commonly used to generate emulsions can be divided into several categories. The easiest and most basic device used is the high-speed blender. In this method, a stirrer, attached to a rotating shaft, blends the two phases together at a high speed, creating horizontal and vertical flow profiles that enhance mixing. This method produces fairly large droplets in the range of 2 to ~25 μm, although droplet size distributions are not always unimodal.

Colloid mills may also be employed to produce medium- and high-viscosity emulsions in the food industry. In this case, the mixture of oil, water, and surfactant in the form of a coarse emulsion is passed through a narrow gap between a rotor and stator. The phases are disrupted because of the induced stress and the droplets formed are stabilized by a layer of surfactant molecules. The diameter of the droplets produced in this method is typically ~2 μm or greater[1].

High-pressure homogenization is used to produce emulsions with submicrometer droplet size distributions. In this scenario, a coarse emulsion created by a blender is passed through a homogenizer. Pressures between ~5 and ~100 MPa are used in many applications. Several parameters such as the pressure used, the number of passes, and the concentration of surfactant will affect the size of the droplets formed. This process is suitable for the production of emulsions in the food industry using low- and intermediate-viscosity materials.

In general, double emulsions are more difficult to prepare and control than simple emulsions. A two-stage method is necessary where, initially, a W/O emulsion is formed, which is followed by its dispersion in an aqueous phase[2]. Difficulties associated with double emulsions include: (1) they may not be able to withstand high shear stresses during mixing since the structure of their droplets may be distorted or destroyed, (2) they may not withstand heat treatments such as pasteurization without undergoing some phase separation, and (3) commercially sterile double emulsions may be difficult to manufacture, particularly when many ingredients are present[3].

17.2.3 Applications

Emulsions are used in many products such as food, cosmetics, pharmaceuticals, etc. Simple emulsions are the basis for many food products including dairy products, butter and margarine, and many

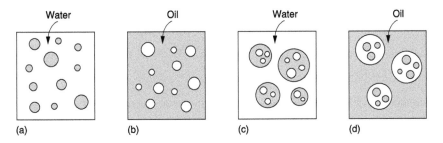

FIGURE 17.2 Two-phase emulsions: (a) oil-in-water (O/W), (b) water-in-oil (W/O), (c) oil-in-water-in-oil (O/W/O), and (d) water-in-oil-in-water (W/O/W).

desserts. Double emulsions have seen use in the pharmaceutical industry, for many applications including the controlled release of drugs and active materials to specific areas of the human body[4-6]. Others applications include the removal of toxic materials via entrapment, and the enhancement in solubility of insoluble materials. In this case, the material is solubilized in an internal phase and partly at the interface. Double emulsions have also shown great potential for use in applications requiring the delivery of specialty and nutraceutical lipids.

17.3   EMULSION STABILITY

17.3.1   Mechanisms of Destabilization

Sedimentation, creaming, flocculation, coalescence, partial coalescence, Ostwald ripening, and phase inversion are the key means of emulsion destabilization[7]. These are now discussed.

17.3.1.1   Creaming and Sedimentation

Sedimentation occurs when a net gravitational force acts upon emulsified droplets that have a density greater than that of the continuous phase, causing them to settle. The opposite is true for droplets with a lower density than the surrounding liquid, which will cause them to rise or cream. Since oils are typically less dense than water, the dispersed phase of O/W emulsions tends to cream, whereas in W/O emulsions, it tends to sediment.

The effects of phase separation by these processes are mostly regarded as being adverse since, if left to progress for any period of time, the visual and textural attributes of the product may be altered. Furthermore, once phase separation begins and is allowed to proceed for any length of time, the rates of flocculation and coalescence may increase.

The rate at which a droplet creams/sediments in an ideal liquid depends upon the net gravitational force. This force is given by:

$$F_g = -\frac{4}{3}\pi r^3 (\rho_2 - \rho_1)g \qquad (17.1)$$

where r is the radius of the particle, g is the acceleration due to gravity, ρ_1 is the density of the continuous phase and ρ_2 is the density of the dispersed phase. A hydrodynamic frictional force also acts against the droplet as it moves through the system:

$$F_f = 6\pi \eta_1 r v \qquad (17.2)$$

where η_1 is the shear viscosity and v is the creaming viscosity. Once a constant velocity is achieved, the upward gravitational force balances out the downward frictional force, therefore giving $F_g = F_f$. Equating these two forces and solving for v, results in the development of Stokes' law equation:

$$v_{Stokes} = -\frac{2gr^2(\rho_2 - \rho_1)}{9\eta_1} \qquad (17.3)$$

This equation is applicable to both creaming and sedimentation as the sign of v indicates whether the droplet moves upwards (+) or downwards (−).

There are notable limitations to the application of Stokes' law. The concentration of the droplets in the emulsion is assumed to be low, since the creaming/sedimentation velocity of droplets in a concentrated emulsion is lower than that of a dilute emulsion. This is due to the hydrodynamic interactions between the droplets. The droplets are assumed to have a uniform size distribution, as larger droplets tend to cream/sediment at a faster rate than smaller droplets. This results in varied rates within an emulsion if it is polydisperse. Lastly, Stokes' law does not take into account the effect that Brownian motion has on the velocity of droplets within an emulsion.

It is desirable to control the rate of droplet separation in emulsions so that the shelf-life of a product may be extended. One means of achieving this is to minimize the difference in density between the droplets and the surrounding liquid, which may reduce gravitational separation.

17.3.1.2 Flocculation

Within an emulsion, there are frequent collisions between droplets. These collisions occur given the continual motion of the droplets resulting from the effects of gravity, thermal energy, applied mechanical forces, etc. Once a collision occurs, one of two events may happen, depending on the relative magnitude of the attractive and repulsive interactions: the droplets may either move apart or form aggregates. During flocculation, two or more droplets contact each other and form an aggregate, yet each droplet remains as an individual entity.

Depending on the desired characteristics of the final product, the effects of flocculation may or may not be beneficial. The negative effects may be seen in systems such as dilute emulsions, where the shelf-life is reduced due to increased gravitational separation. Other unfavorable effects include a significant increase in emulsion viscosity and/or gel formation. On the other hand, the formation of a desired texture may be enhanced by the controlled flocculation of droplets.

The rate of flocculation is described as the number of original droplets that disappear per unit time due to aggregation. As flocculation proceeds, therefore, the total number of individual droplets decreases with time. Equation (17.4) describes the flocculation rate, dn_T/dt, showing that it is dependent on the frequency of collisions between droplets and the fraction of the collisions, which result in aggregate formation:

$$\frac{dn_T}{dt} = -\frac{1}{2}FE \tag{17.4}$$

Here, n_T is the total number of droplets per unit volume, t is the time, F is the collision frequency, and E is the collision efficiency. The factor of one-half is present in the equation since, for every collision between two droplets, the number of individual droplets in the system is reduced by one. The collision frequency is defined as the number of droplet encounters per unit time per unit volume of emulsion.

In the case of small emulsion droplets (<2–3 μm), Brownian motion may increase collision frequency and efficiency. In fact, Brownian motion is the main cause of collisions in quiescent systems. For dilute systems, the collision frequency is given by:

$$F_B = 16\pi D_0 r n^2 \tag{17.5}$$

where F_B is the collision frequency due to Brownian motion, D_0 is the diffusion coefficient of a single droplet, n is the number of droplets per unit volume, and r is the droplet radius. This equation shows that the collision frequency, and therefore the flocculation rate, is proportional to the diffusion coefficient of the droplet, to the droplet radius and to the square of the number of droplets per volume.

Upon applying the Einstein relation $Df = kT$, where f is the friction factor for a droplet, together with Stokes' law, $f = 6\pi\eta a$, a second expression for collision frequency due to Brownian motion is obtained:

$$F_B = \frac{8kT n^2}{3\eta} \tag{17.6}$$

where k is Boltzmann's constant, T is the absolute temperature, and η is the viscosity of the continuous phase. From this equation, it can be deduced that the frequency of collisions between droplets may be reduced by increasing the viscosity of the continuous medium.

Another expression that can be used to understand the rate of destabilization is based on the time required to reduce the original number of droplets by one half, given by:

$$t_{1/2} = \frac{3\eta}{4kTn} \tag{17.7}$$

The rate of droplet flocculation is determined by the collision frequency and the parameters that control the collision efficiency. The collision frequency can be reduced by decreasing the difference in density between the dispersed and the continuous phases, by narrowing the droplet size distribution, by decreasing the droplet concentration, and by increasing the viscosity of the continuous phase. In practice, one mechanism is usually the most influential, and it is best to first determine which dominates before an effective control method can be applied. In emulsions where there is a gel formed or the presence of solid crystalline network, flocculation may be completely inhibited.

17.3.1.3   Coalescence

Droplets in emulsions tend to move towards thermodynamic equilibrium by reducing the contact area between the two phases. In the process of coalescence, two or more droplets merge to form one single larger droplet. Knowledge of the physical mechanism by which coalescence occurs is still incomplete since it is based on the short-range forces that are present at a molecular level. Several factors may induce coalescence of droplets in an emulsion. As mentioned earlier, droplet–droplet collisions are a necessary precursor to coalescence. Due to Brownian motion, emulsion droplets are in continual motion, and the rate at which coalescence occurs depends on the collision frequency and efficiency, which depends upon whether or not the interfacial membranes surrounding the droplets actually rupture or not upon collision. When an emulsifier is not present, the collision efficiency approaches its highest value ($E_c \rightarrow 1$) since there is nothing present to prevent the droplets from merging. In the presence of an emulsifier, however, the collision frequency tends to, but does not necessarily reach, zero ($E_c \rightarrow 0$).

17.3.1.4   Partial Coalescence

Partial coalescence is the process whereby two or more partially crystallized droplets merge to form a single irregularly shaped aggregate (Figure 17.3). The droplets retain some degree of their original identity, while forming an intricate crystalline network with one another[8]. During a cooling regime, lipids present inside droplets may crystallize, and under perikinetic or orthokinetic conditions, ensuing collisions between droplets may lead to flocculation. With fat crystals present at the interface, such collisions may disrupt the dispersed phase given that protruding crystals, being preferentially wetted by the available liquid oil, form bridges between the two droplets. Once penetration occurs, the droplets remain aggregated and, through subsequent collisions, eventually form an aggregated network that restricts the movement of the dispersed phase.

17.3.1.5   Phase Inversion

In the process of phase inversion, an O/W emulsion transforms into a W/O emulsion or vice versa. The production of some foods such as butter or margarine is based on this process, while in other products the effects of phase inversion may be undesirable. A number of factors may trigger phase inversion. These may be related to alterations in the composition or the environmental conditions, such as the dispersed-phase volume fraction, the type and concentration of the emulsifier, the solvent conditions, presence of electrolytes, temperature, or mechanical agitation[9].

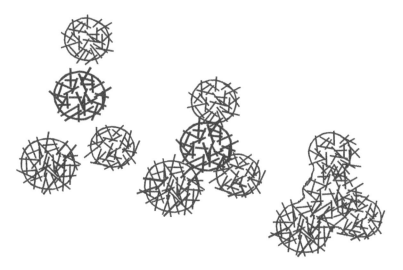

FIGURE 17.3 The partial coalescence of emulsion droplets. The network of crystallized fat is represented by the straight lines.

17.3.1.6 Ostwald Ripening

In Ostwald ripening, mass transfer of the dispersed phase occurs from smaller droplets to larger droplets via passage through the continuous phase. Large droplets therefore grow in size whereas small droplets become smaller and may disappear. The basis of Ostwald ripening is that the solubility of the material within a spherical droplet increases as the droplet decreases in size. The effect of this process is negligible when the mutual solubilities of the oil and water phases are low enough to render the rate of mass transfer also negligible, as in most food applications. Ostwald ripening can be represented by the following:

$$S(r) = S(\infty) \exp\left(\frac{2\gamma V_{\mathrm{m}}}{RTr}\right) \tag{17.8}$$

where V_{m} is the molar volume of the solute, γ is the interfacial tension, $S(\infty)$ is the solubility of the solute in the continuous phase for a droplet with infinite curvature (a planar interface), $S(r)$ is the solubility of the solute when contained in a spherical droplet of radius r, and T is the absolute temperature. This equation shows that the solubility of the material in the spherical droplet increases as the size of the droplet decreases. The material moves from the smaller droplets to the larger ones until steady state has been achieved. Equation (17.9) gives the rate of Ostwald ripening at steady state:

$$\frac{\mathrm{d}\langle r \rangle^3}{\mathrm{d}t} = \frac{8\gamma V_{\mathrm{m}} S(\infty) D}{9RT} \tag{17.9}$$

where D is the translation diffusion coefficient of the solute through the continuous phase. This indicates that the rate of size reduction is directly related to the solubility of the droplet material.

Several factors can be adjusted to control the process of Oswald ripening. The droplet size distribution should be fairly narrow and the average size of the droplets should be relatively large. This minimizes the number of small droplets within the emulsion, thereby preventing an increase in the solubility of the material within the droplets. Having too large an average droplet size may, however, result in an increase in the rates of flocculation, creaming/sedimentation and/or coalescence. Using lipids that are sparingly soluble in water slow downs the rate of Ostwald ripening. Some substances such as alcohols or surfactant micelles, if added to an emulsion, may increase the solubility of lipids in water, hence accelerating the rate of Ostwald ripening.

17.3.2 METHODS TO STABILIZE EMULSIONS

17.3.2.1 Surfactants

Surfactants are amphiphilic molecules that consist of hydrophilic and hydrophobic moieties. In the case of small-molecule surfactants, the surfactant head may be anionic, cationic, zwitterionic, or nonionic in nature whereas the tail portion normally consists of a hydrocarbon chain typically 10 to 20 carbons atoms in length. The capacity of a surfactant to promote emulsification and/or emulsion stability will strongly depend on the characteristics of the head and tail groups.

Surfactant adsorption onto the droplet surface occurs when the hydrophilic head interacts with the aqueous phase and the hydrophobic tail interacts with the nonaqueous phase. The result is a reduction in the interfacial tension between the oil and water, which results in lower energy required for breaking up emulsion droplets during homogenization.

Once emulsified, surfactants may provide a repulsive force between droplets, hence reducing flocculation. For example, the addition of ionic surfactants may yield the same electric charge on all droplets leading to electrostatic repulsion. Short-range repulsive forces such as steric, hydration, and thermal fluctuation interactions are produced by nonionic surfactants, and also aid in preventing droplet aggregation. Other surfactants are capable of reducing droplet aggregation, hence enhancing emulsion stability, by forming multilayers around the emulsion droplets rather than a monolayer.

There are three main characteristics of surfactants that enable them to enhance the formation and stability of emulsions: (1) rapid adsorption of the surfactant molecules onto the droplets during homogenization; (2) monolayer coverage and reduction of the interfacial tension, and (3) reduction of destabilization phenomena[9].

Surfactants can be readily characterized according to Bancroft's rule and their ratio of its hydrophilic to lipophilic groups (HLB number).

17.3.2.1.1 Bancroft's Rule

Bancroft's rule states that the continuous phase formed in an emulsion is the one in which the surfactant is most soluble. A water-soluble surfactant will therefore promote an O/W emulsion, while an oil-soluble surfactant will promote a W/O emulsion. Although widely applicable to systems, there are important exceptions. Some amphiphilic molecules may be highly soluble in both the oil and aqueous phases, but may not be capable of forming stable emulsions due to their low surface activity.

17.3.2.1.2 Hydrophile–Lipophile Balance

The hydrophile–lipophile balance (HLB) of a surfactant describes its affinity towards the oil and aqueous phases of an emulsion. It is based on the chemical structure of the surfactant, according to the ratio of its hydrophilic to lipophilic groups. A high HLB number indicates that the surfactant contains a higher number of hydrophilic groups than lipophilic groups. In order to calculate the HLB number of a surfactant, the number and type of hydrophilic groups and lipophilic groups must be known. This information can also be estimated experimentally from cloud point analysis.

$$HLB = 7 + \sum(\text{hydrophilic group numbers}) - \sum(\text{lipophilic group numbers}) \qquad (17.10)$$

TABLE 17.1
Some Small-Molecule Surfactants and Their Hydrophilic–Lipophilic Balance (HLB) Values.

Surfactant	HLB value
Decaglycerol monooleate	14
Decaglycerol monostearate	13
Glycerol monolaurate	7
Glycerol monostearate	3.8
Hexadecanol	1
Lactoyl monopalmitate	8
Lecithin	~9
Potassium oleate	20
Sodium dodecyl sulfate	40
Sodium lauryl sulfate	40
Sodium oleate	18
Sodium stearoyl lactylate	22
Sodium stearoyl-2-lactoyl lactate	21

After Stauffer, C., *Eagan Press Handbook Series, Emulsifiers*, Eagan Press, MN, 1999.

In general, low HLB numbers in the range of 3 to 6 indicate that the surfactant has a larger amount of lipophilic groups, is oil-soluble and can stabilize W/O emulsions. Longer or more saturated aliphatic chains produce smaller HLB numbers. With a higher HLB number (10 to 18), a surfactant contains more hydrophilic groups, is soluble in water, and can stabilize O/W emulsions. Surfactants with intermediate HLB numbers between 7 and 9 do not exhibit selectivity between the oil and aqueous phases[10]. Surfactants with HLB numbers lower than 3 and higher than 18 are not very surface active and therefore tend to gather in either the oil or aqueous phase. Some examples of common small-molecule surfactants and their HLB values are given in Table 17.1.

Although the use of HLB numbers to classify surfactants is common, it does have one major drawback. The changes induced in the functional properties of surfactant molecules by temperature changes or solution conditions are not taken into account.

17.3.2.2 Addition of Colloidal Particles

Emulsions may also be stabilized by the addition of solid particles. Common emulsifiers used in the food industry, such as saturated monoacylglycerols, may form crystalline structures that adsorb at the oil–water interface, where they serve as a mechanical barrier to prevent droplets from flocculating and coalescing.

For effective stabilization of emulsions, colloidal particles must: (1) be located at the interface before any stabilization can take place; (2) remain at the interface, which will be a function of their size, shape, composition, and wetting behavior; (3) form a monolayer film covering the entirety of the droplet; and (4) some interaction between particles appears necessary for effective stabilization.

As Pickering species, the role that interfacial colloidal particles play on emulsion stability will strongly depend on how they are wetted by the continuous or dispersed phases[11]. During or after emulsification, particles will be adsorbed at the interface if it is energetically favorable. Adsorbed particles will be preferentially wetted by either the aqueous or oil phase depending on the composition of the aqueous and oil phase (i.e., the presence and type of emulsifier(s)) as well as the composition and surface properties of the particles. This behavior is described by the contact angle formed at the boundary of the three phases where the contact angle is the angle the liquid–liquid

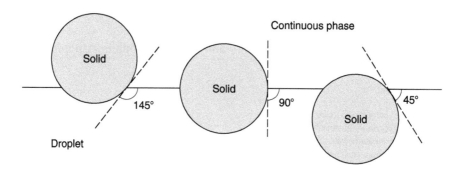

FIGURE 17.4 The stabilization of an emulsion by the adsorption of solid particles at the interface. Contact angles of 90° provide the best stability.

interface makes to the solid phase (Figure 17.4). Particles with contact angles smaller than 90° will stabilize O/W emulsions. With contact angles greater than 90°, the particles will stabilize W/O emulsions[12]. If the particles are completely wetted by either the oil or water phase they become fully dispersed in that phase and will not behave as a Pickering species.

Wetting energy is required to expel the particle from the interface of the continuous medium and into the droplet. The higher the wetting energy, the greater is the stability of the emulsion against coalescence. To determine the amount of energy that is required for a spherical particle to move into the most wetting phase, Equation (17.11) can be applied:

$$\Delta E = \pi r^2 \gamma_{o/w} \, (1 - \cos \theta)^2 \tag{17.11}$$

where ΔE is the energy to expel a spherical particle of radius r from the interface into the phase toward which it has a contact angle of θ. The term $\gamma_{o/w}$ represents the interfacial tension between the two liquids. From this equation, it is clear that a contact angle of 90° between a solid particle and the interface provides the most stability. As the contact angle is reduced, the energy required to completely immerse the particle into one phase is significantly reduced as well. When the angle between the particle and the droplet is greater than 90°, the particle is then forced into the continuous phase and flocculation may be promoted[11].

17.3.2.3 Liquid Crystals

The liquid crystalline (LC) phase demonstrates characteristics of both liquids and crystalline solid, in that the molecules have some of the orientational order present in solids but have lost positional order, similar to molecules in a liquid. The LC phase is formed in an emulsion when the emulsifier concentration is increased beyond a critical concentration, at which point the stability of the emulsion is notably increased. While their stabilization mechanism has not been elucidated, it has been proposed that LCs can form a multilayer film around the droplets that protects droplets from coalescence, likely due to enhanced viscosity[11].

17.3.2.4 Proteins

Proteins are naturally occurring biopolymers that consist of covalently linked monomers. The building blocks of protein are amino acids which, through covalent bonds, form proteins. Each amino acid contains at least one primary amino (–NH$_2$) group, one carboxyl (–COOH) group, and R-group that strongly influences a protein's capacity to stabilize an emulsion. As a result of their amphipathic property, proteins are often the emulsifier of choice for food processors. From a fundamental perspective, the relationship between protein structure and emulsifying activity depends on a combination of factors such as molecular flexibility, molecular size, surface hydrophobicity, net charge, and amino acid composition. The molecular weight of proteins ranges from ~5000 to upwards of 500,000 (e.g., myosin).

17.3.2.5   Gums and Thickeners

Most thickening agents are long-chain, straight or branched polysaccharide-based biopolymers that contain hydroxyl groups capable of hydrogen bonding. Polysaccharides are polymers of mono-saccharides: their degree of polymerization (DP) (or chain length) is typically between 1000 and 10,000[10]. The main protein-based material that falls into this category is gelatin. Starches and cellulose are examples of homopolymers as they are made up of *n* repeating units of the same monomers. The sugar monomers can contain linked side units, or substituent groups, such as sulfates, methyl ethers, esters, and acetals. They can be neutral or anionic.

17.3.2.6   Molecular Interactions and Biopolymers

Several important molecular interactions and entropy effects influence the conformation of biopolymers in aqueous solutions: hydrophobic and electrostatic interactions, hydrogen bonding, disulfide bonds, and configurational entropy[13]. Some of these are now discussed.

17.3.2.6.1   Electrostatic Interactions

The molecular structure and aggregation of biopolymers are strongly influenced by electrostatic interactions. Proteins may contain amino acids that can form positively charged ions (such as arginine, lysine, and proline) or negatively charged ions (such as glutamic and aspartic acids). Polysaccharides may also have ionizable groups along their backbone structures (such as sulfate or phosphate groups). The electrostatic interactions may be affected by the pH of the surrounding aqueous phase and the pK of the ionizable groups[13]. The effects due to electrostatic screening, dependent on the concentration and type of counterions that may be present in the aqueous phase, may also be important. The structure of biopolymers is largely dependent on the type of charged groups it contains. If the groups have similar charges, then the molecule tends to stretch out in order to produce more space between them and hence reduce electrostatic repulsions. If the groups are oppositely charged, the molecule will take on the globular structure, since folding up would maximize the electrostatic attractions.

Aggregation of biopolymer molecules in solution is also dependent on electrostatic interactions. Biopolymers with similar charges tend to remain as individual molecules due to repulsion, whereas aggregation occurs between molecules that are oppositely charged. Electrostatic interactions are also responsible for interactions with low-molecular-weight ions (e.g., Na^+ and Ca^+).

17.3.2.6.2   Hydrogen Bonding

Hydrogen bonding is a relatively weak bonding that exists between molecules, but is strong overall if present in large numbers. Systems attempt to maximize the number and strength of these bonds within their structures. Proteins and polysaccharides both contain monomers that exhibit this type of bonding, which consequently affect their structures. Some may form very ordered structures (such as helices), while others may favor less ordered structures and form hydrogen bonds with the surrounding water molecules. In general, the more highly ordered the conformation is, the less entropically favorable is the structure. Therefore, extensive intramolecular hydrogen bonding is less entropically favorable than extensive intermolecular hydrogen bonding (random-coil conformation). The given environmental conditions also play a key role in the biopolymer's structure formation since the strength of the hydrogen bond is relative to other forces or interactions such as hydrophobic and electrostatic, and configurational entropy that may be involved. Hydrogen bonds may also be formed between biopolymers, causing them to aggregate. High temperatures result in dissociation, but lower temperatures enhance bond stability.

17.3.2.6.3 Hydrophobic Interactions

Hydrophobic interactions exist as a result of the hydrophobicity of certain substances such as fatty acids, alcohols, and some amino acids. The mixing between water and these apolar substances is thermodynamically unfavorable ($\Delta G > 0$). The antagonistic behavior between water and these substances is therefore the driving force for molecules with these groups to restructure themselves into a conformation that reduces the contact of the hydrophobic groups with water. In an aqueous environment, therefore, any apolar groups attached to a substance will associate with each other rather than with the aqueous phase. A reduction in the interfacial surface area between the water and the apolar groups will result in an increase in the overall thermodynamic favorability of the system ($\Delta G < 0$).

Hydrophobic interactions are significant in proteins that contain a large number of nonpolar groups (benzyl group in phenylalanine). Hydrophobic interactions are responsible for the folding of protein molecules into globular structures, so that the contact between the nonpolar groups and water is minimized as these groups move into the interior of the structure. The tertiary structure of proteins is strongly influenced by hydrophobic interactions. Polysaccharides are mainly hydrophilic, with a few exceptions (e.g., gum arabic). The effect of hydrophobic interactions on their structures is therefore not as pronounced as in proteins.

17.3.2.6.4 Disulfide Bonds

Many proteins also have the ability to form disulfide bonds (–S–S–). Disulfide bonds, also known as disulfide bridges, are strong covalent bonds occurring naturally in proteins that are formed by side chains[12]. These bonds may be intramolecular bonds, formed by the bonding of two thiol groups (–SH) in the amino acid cysteine (Cys) that may be present within one protein molecule. The disulfide bond is formed by the oxidation of the thiol group (–SH) according to Figure 17.5 (where R is the remainder of the amino acid attached to the functional group). Two cysteine residues may undergo oxidation to form one residue, with a name change to cystine. The function and stability of the tertiary folded structure of proteins against unfolding are enhanced by these bonds. Disulfide bonds may also be intermolecular bonds that are formed between two different protein molecules. These intermolecular bonds are responsible for aggregation of proteins at interfaces and in gels.

17.3.2.6.5 Configurational Entropy

The configurational entropy of a structure is the amount of free energy within a structure that is as a result of its conformation. The conformation and aggregation of biopolymers depends largely on their configurational entropies, of which there are local and nonlocal contributions. The number of conformations that can be formed by individual monomers in a chain is referred to as the local

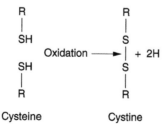

FIGURE 17.5 A schematic of the formation of a disulfide bond between two cysteine residues to form cystine.

entropy. The nonlocal entropy refers to the number of conformations that the whole chain of biopolymer molecules can adopt. The random-coil structure is highly flexible and can take on many conformations, and therefore has a high configurational entropy. More rigid structures such as compact globular proteins have much lower configurational entropies.

17.3.2.7 Some Functional Properties of Biopolymers

This following section outlines the main functional properties of biopolymers, namely emulsification, thickening and stabilization, and gelation.

17.3.2.7.1 *Emulsification*

Biopolymers that are surface-active will accumulate at oil–water interfaces. In the adsorption process, the biopolymers align themselves appropriately so that the nonpolar groups are in contact with the oil phase and the polar groups are in contact with the aqueous phase. The surface area of the contact between the oil and aqueous phases is reduced by the formation of a membrane of the adsorbed biopolymer molecules (Figure 17.6).

The conformation adopted by the molecules depends on their molecular structure and interactions. The physicochemical properties of the membrane that is formed are also dependent on these factors. Consider a random-coil structure that has three predominant regions: polar, nonpolar, and neutral. Because of the flexibility, the polar regions extend into the aqueous phase, the hydrophobic regions extend into the oil phase, and the neutral regions form lines along the interface. Random-coil structures form membranes that are fairly open and thick, and also have low viscoelasticity. Globular proteins, on the other hand, have a definite orientation at an oil–water interface, since the adsorbed nonpolar regions face the oil phase, while the polar regions face the aqueous phase. At an interface, flexible random-coiled biopolymers can rearrange their structural formations at a much faster rate than can rigid globular proteins. Globular proteins are able to unfold at an interface, however, which may result in amino acids that were previously on the interior of the molecule being exposed. This can lead to enhanced hydrophobic interactions and disulfide bond formation. Globular proteins form membranes that are relatively thin and compact, and also have high viscoelasticities. This enhances the membrane's resistance to rupture.

The effectiveness of biopolymers as emulsifiers is based on the speed of adsorption at an interface, so that droplet aggregation can be inhibited. Several factors help prevent droplet aggregation when using biopolymers. Short-range steric repulsive forces exist for sufficiently thick membranes that help prevent flocculation. If the membrane is electrically charged, electrostatic repulsion may prevent flocculation. Ionic strength and pH also affect the magnitude of the electrostatic repulsions between droplets stabilized by charged biopolymers.

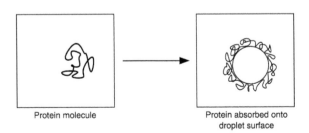

Protein molecule   Protein absorbed onto droplet surface

FIGURE 17.6 A schematic of the structure of the membrane formed by proteins and polysaccharides, illustrating their ability to stabilize emulsions.

17.3.2.7.2 *Thickening and Stabilization*

Increasing the viscosity of the continuous aqueous phase of an emulsion using biopolymers results in both thickening or gelling and stabilization. Thickening enhances the texture and organoleptic properties of the final product, and improves stabilization against separation processes such as creaming or sedimentation. The ability of a biopolymer to increase the viscosity of an emulsion is based on molecular weight, conformation, degree of branching, and flexibility[13]. The biopolymers selected for this purpose are generally highly hydrated molecules having extended structures, or they may be aggregates of molecules.

Equation (17.13) illustrates the linear relationship between the viscosity of an emulsion with a low concentration of droplets and the actual concentration of the droplets in the emulsion:

$$\eta = \eta_0(1 + 2.5\phi) \tag{17.13}$$

where η is the viscosity of the emulsion, η_0 is the viscosity of the pure solvent (water), and ϕ is the volume fraction of droplets in solution. Biopolymers with extended structures have a higher effective volume fraction than their actual volume fraction. Due to the thermal energy of the biopolymer, it always rotates when in solution; it thus entrains a spherical volume of water that is equal in diameter to its longest end-to-end length. The actual volume of the biopolymer within this spherical volume is small; hence its actual volume fraction is smaller than its effectual volume fraction. The smaller the ratio of its actual volume fraction to its effectual volume fraction, the more effective it is considered to be. Molecules with extended structures are preferentially selected over globular structures for this reason.

17.3.2.7.3 *Gelation*

Some biopolymers in emulsion have the ability to cause the aqueous phase to gel. In gel formation, the biopolymers generally entrap water globules within a three-dimensional network, forming a structure that has properties similar to those of a solid (Figure 17.7). The gel properties depend largely on the biopolymer molecules themselves (type, structure, and interactions). Gels may be hard, soft, brittle, or rubbery, have a homogeneous or heterogeneous composition, and may also appear opaque.

The structural organization of the molecules within a gel is either particulate or filamentous. In particulate gels, the biopolymer molecules are in the form of aggregates, which are large enough to scatter light and make the gel optically opaque. The particles are roughly spherical in nature, with diameters ranging in size from 0.1 to 4 μm[14]. In filamentous gels, the biopolymers form a network of thin chain-like structures or filaments. The small pore size of the network allows water to be held

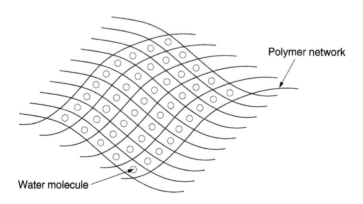

Polymer network

Water molecule

FIGURE 17.7 A schematic of the structure formed by the gelation process in proteins and polysaccharides.

tightly within the network by capillary action. The filaments, with an average thickness of between 20 and 50 nm[14], are thin enough to render the gel transparent.

Gels may be further classified into heat-setting (formed on heating) or cold-setting (formed on cooling) groups, with most biopolymers gelling upon cooling. A notable exception is chitosan, which is a heat-set gelling agent.

The gelation process may be reversible, therefore making gels thermoreversible or thermoirreversible. The reversibility of gels mainly depends on the type of bonding between the biopolymer molecules. Changes in the molecular structure and organization of the molecules during the gelation process also affect the reversibility of the gel. Reversible gels are characterized by the presence of noncovalent bonds and molecules that have not significantly changed prior to gelation. The opposite is true for irreversible gels, where covalent bonds hold the gel together, and where significant changes in the molecules are required for gelation to occur. The interactions responsible depend on the nature of the biopolymers used. Some proteins and polysaccharides may favor hydrogen bond formation (e.g., gelatin and gums). Other biopolymers with many nonpolar groups favor gel formation by hydrophobic interactions (e.g., caseins). Electrostatic interactions are also vital to gel formation.

Of importance for lipid nutraceuticals is a class of gels known as emulsion gels. Emulsion gels are soft solid systems that contain dispersed particles or "fillers." These fillers are emulsion droplets found that can significantly increase the rigidity of gels. Emulsion gels have practical applications in the pharmaceutical, food, and cosmetic industries. The ability of the dispersed droplets to be incorporated into the gel network depends on the composition of the surface-active components.

The use of proteins is very popular in the formation of these three-dimensional networks. Whey proteins are one of the most commonly used building blocks in the formation of viscoelastic gels[15]. Several factors, such as the protein concentration, the thermal processing conditions, and the conditions of the aqueous environments (e.g., pH and ionic strength) all influence the structure and rheological properties of the whey protein gels. It has also been found that the presence, and properties, of the dispersed oil droplet phase incorporated into the gel network also has a strong influence on the rheology and microstructure of the protein gel.

The gel fillers may be described as being active or inactive. Active fillers are those that have a strong interaction between their individual particles and the gel matrix. Inactive fillers generally have the opposite effect, where there is little or no interaction between their individual particles. One way of determining whether or not the filler is active is by examining the composition and properties of the adsorbed monolayer on the droplets. Gel strength is greatly enhanced when protein-coated emulsion droplets become incorporated into the gel network[15].

17.4 METHODS OF EMULSION CHARACTERIZATION

17.4.1 LIGHT SCATTERING

17.4.1.1 Static Light Scattering

Droplet sizes can be determined by static light scattering whereby a coherent light beam is passed through an emulsion[16,17]. When an electromagnetic wave is passed through an emulsion, a scattering pattern, which is used to study the droplet size distribution and concentration, is produced. This pattern is characterized by measuring the intensity of the scattered light that emerges from the emulsion. Mathematical analysis relates light scattering data to droplet size distributions and varies according to the complexity of the emulsion system. Three regimes are defined according to the relationship between the radius of the droplet, r, and the wavelength, λ. The characteristic scattering patterns are: the long-wavelength regime ($r < \lambda/20$), the intermediate-wavelength regime ($\lambda/20 < r < 20\lambda$), and the short-wavelength regime ($r > 20\lambda$).

Knowledge of the droplet size distribution of an emulsion can aid in predicting the long-term stability of the emulsion to destabilizing mechanisms such as creaming, flocculation, coalescence, and Ostwald ripening. Droplet sizes ranging from about 0.1 to 1000 μm can be analyzed using this technique[18]. These instruments are generally used in the food industry in research and development to analyze physicochemical properties of emulsions and for quality control purposes to ensure that the droplet size specifications are met. They are fully automated and easy to use, but are fairly expensive, limiting their use.

There are some limitations associated with the use of light scattering to determine the droplet size distribution of emulsions. If flocculation has occurred, the droplets may no longer be spherical in shape, as the theory to perform the calculation assumes. Only an approximation of the droplet size distribution may be determined using this method. The emulsions may also have to be diluted to prevent multiple scattering and/or stirred to ensure homogeneity. Optically opaque or semisolid emulsions, such as butter or ice cream, cannot be analyzed using this method.

17.4.1.2 Dynamic Light Scattering

Dynamic light scattering techniques have seen use in food applications[19,20]. When droplets are below the lower limit of detection of the static light scattering technique, dynamic light scattering is used. This technique can be used to analyze droplets with diameters ranging from 3 nm to about 3 μm[18]. It is based on the continual movement of the droplets due to their Brownian motion. The variable D, which is the translational diffusion coefficient of the droplets, is determined based on the interaction between the emulsion droplets and the beam. The Stokes–Einstein equation is used to then calculate the radius, r, of the droplets:

$$r = \frac{kT}{6\pi \eta_1 D} \qquad (17.14)$$

Two common measurement techniques used to determine the translational diffusion coefficient of the droplets are photon correlation spectroscopy (PCS) and Doppler shift spectroscopy (DSS).

17.4.2 NUCLEAR MAGNETIC RESONANCE

Nuclear magnetic resonance (NMR) is also used to study the droplet size distributions of emulsion systems[16,21,22]. Droplet sizes ranging between 0.2 and 100 μm can be detected by this technique[18]. Pulsed field gradient NMR (pfg-NMR) allows characterization of emulsion droplet size distributions via measurement of the restricted diffusion of molecules within emulsion droplets. Briefly, pfg-NMR self-diffusion measurements of hydrogen nuclei take place when two equal field gradient pulses are applied within a standard spin-echo pulse sequence, where the intensity of the observed echo is reduced by the effect of molecular diffusion. With emulsion droplets, diffusion is restricted by the presence of the interface. Once this boundary is reached, the echo intensity is no longer reduced. Thus, by monitoring signal attenuation as a function of gradient strength, one may determine droplet sizes within an emulsion and thus generate a droplet size distribution.

NMR is a nondestructive technique that can be used to analyze O/W and W/O emulsions. It can also be used to study droplet crystallization in emulsions, thereby extending its application to emulsions that are optically opaque. The solid content of an emulsion cannot be accurately determined by this technique when the degree of crystallization is less than 1%. The use of NMR is very simple and rapid, although the initial investment required to purchase an instrument is somewhat high.

17.4.3 ULTRASONIC SPECTROMETRY

The droplet size distribution in an emulsion can also be determined by ultrasonic spectrometry, where the interactions between ultrasonic waves and particles are studied[23,24]. One major advantage is that optically opaque emulsions can be analyzed *in situ* without any sample preparation. This technique can be used to measure droplets ranging in size from about 10 nm to 1000 μm[18]. The interactions between the ultrasonic waves and the particles may take three different forms. Upon passing through the emulsion, some of the waves may be scattered in directions that are different from that of the incident wave. Adsorption mechanisms such as thermal conduction and viscous drag may be responsible for the conversion of ultrasonic energy into heat. Finally, there is interference between the waves that travel through the dispersed phase and the continuous medium, and also with the waves that are scattered. These interactions depend on parameters such as the frequency of the ultrasonic wave, the thermophysical properties of the components, and the droplet size and concentration.

17.5 FORMULATION OF EMULSIONS FOR SPECIALTY AND NUTRACEUTICAL LIPID DELIVERY

17.5.1 LIPID EMULSIONS

As discussed earlier, two types of emulsions can be formed: W/O and O/W emulsions. Oil-in-water emulsions, or lipid emulsions, can potentially be used for the delivery of lipophilic materials, for the slow and sustained release of active materials within the bloodstream, and for delivery to targeted tissues in the body. Water-in-oil emulsions, on the other hand, are not used for the delivery of lipophilic substances as the dispersed phase is aqueous in nature, and thus unsuitable for this purpose[25,26].

Lipid emulsions, which normally consist of oil droplets typically 0.1 to 10 μm in diameter, have been used extensively as carriers to deliver active materials to targeted parts of the body[27]. Such emulsions are usually stabilized with phospholipids (lecithins) as the emulsifying agent, which forms a monolayer membrane around the oil droplets. As a result, only oil-soluble active substances can be incorporated within the droplets. These include lipophilic nutraceutical substances, such as fat-soluble vitamins (A, D, E, K), fatty acids (omega-3, medium-chain, PUFAs, etc.), and phytosterols. The key properties that emulsions must offer as delivery systems include: (1) tailored controlled release properties, (2) the protection of emulsified material (e.g., against oxidation), and (3) a lengthy (> 1 year) kinetic stability of the emulsion. Other advantages include high entrapment rates of active substances and simple preparation using a single dispersion step, which can easily be scaled-up for industrial production. Lastly, all emulsions destined to enter the bloodstream must be fully biocompatible.

This section describes recent advances in the use of O/W emulsions as delivery systems. Liu et al.[28] developed castor oil emulsions of castor oil stabilized by phosphatidylcholine (PC) and coemulsifiers (a series of polyethylene glycol (PEG) derivatives). The effect of the PEG on reducing the uptake of the emulsion particles by the reticuloendothelial system, and hence the increase of their blood circulation time, was then tested *in vivo* using an animal model. They found that the addition of PEG derivatives as coemulsifiers prolonged the residence time of the emulsions in the blood. Furthermore, there was a notable effect of droplet diameter on *in vivo* circulation.

In another study on lipid emulsion stability, the relationship between different types of natural oils and emulsion droplet size and stability was investigated[29]. The emulsifier used was egg PC in emulsion systems using 18 different natural oils. The premise of this study was that emulsions made with different oils could have varying degrees of stability, based on differences in the droplet sizes produced by the various oils. Two important physical properties of the system, namely the interfacial tension and the viscosity of the oils, were studied so that a relationship between them and the droplet size could be ascertained. Of the oils tested, squalene, light mineral oil and jojoba bean oil formed emulsions with the smallest droplet sizes where cottonseed, linseed, and evening primrose oils formed emulsions with the droplet sizes. The release of active materials from the stable emulsions was more sustained than from the unstable ones. The relationship between the droplet size and the oil–water interfacial tension was inversely proportional, whereas the effect of the viscosity of the oils on the droplet size was not significant.

Trotta et al.[30] studied the stability of PC-stabilized emulsions as lipid delivery systems, with the aim of enhancing the stability of the emulsion by adjusting the packing geometry of the surfactant molecules. The packing geometry of the surfactant molecule at the interface is based on the ratio between the hydrocarbon volume, the optimum head group area, and the tail length. By reducing its packing geometry, the surfactant molecule can better aid in the formation of the small emulsion droplets required for the delivery of active materials. The packing parameter was reduced in this case by partially substituting the lecithin with a hexanoyl derivative (6-PC), which is a more hydrophilic emulsifier. The mixture of the 6-PC and lecithin formed a flexible monolayer at the oil–water interface since the close packing of the molecules at the interface was prevented by the different structures of the molecules of the two emulsifiers. Emulsions formed using the emulsifier mixture had the smallest mean droplet diameter as compared to emulsions containing only 6-PC or lecithin. This is important since small droplet sizes promote enhanced physical stability.

Upon incorporating an active substance into the different emulsions, adverse effects on their stability were observed in all cases, except for those that contained the emulsifier mixture. The results clearly indicated that there was a marked increase in stability in the emulsions containing the emulsifier mixture produced by high-pressure homogenization rather than high-shear homogenization over the emulsions containing lecithin or 6-PC using the same methods.

Djordjevic et al.[31] determined the optimum conditions for stabilizing O/W emulsions using whey protein isolate (WPI) with high oil contents and low viscosities. They incorporated polyunsaturated fatty acids (PUFAs) within these emulsions and were able to enhance their oxidative stability as a result. The significance of these results is that these WPI-stabilized emulsions have the potential of being used to deliver polyunsaturated lipids (e.g., omega-3 fatty acids), provided that they have fairly low viscosities, are relatively oil-rich, and are stable to thermal processing.

17.5.2 Double Emulsions

As described earlier, in double emulsions, droplets of one liquid are dispersed in another liquid, which in turn is dispersed as droplets in another liquid. The inner droplets are therefore separated from the outer liquid phase by another liquid phase. Most common applications use water-in-oil-in-water (W/O/W) double emulsions, but several specific applications require oil-in-water-in-oil (O/W/O) emulsions. Double emulsions must be stabilized by two types of emulsifiers, one hydrophilic and one hydrophobic, but in most cases a blend of the two gives better stabilization results.

The principal function of the use of double emulsions as delivery systems is for the slow and sustained release of bioactive materials from the droplets dispersed in the interior of the emulsion to the continuous phase. They can, however, also be used in the opposite manner, which is for the removal of toxic matter, by entrapping toxic material from the continuous phase in the inner dispersed droplets[2].

Multiple emulsions are normally stabilized by a combination proteins and small-molecule surfactants. In this regard, it has been found that there is more control over the release of markers when the proteins are used in combination with the lipophilic monomeric emulsifiers, even with very low protein concentrations. As an example, Garti et al.[32] found that the combination of bovine serum albumin (BSA) and monomeric emulsifiers improved the stability of double emulsions and slowed down the release rate of the markers.

Double emulsions have also been stabilized using protein–polysaccharide hybrids, as described by Benichou et al.[33]. In this study, the inner aqueous phase of the double emulsion was stabilized with polyglycerol polyricinoleate, while the outer interface was stabilized by a protein–polysaccharide hybrid. This hybrid was prepared by interacting WPI with different charged and uncharged polysaccharides (xanthan gum and galactomannans). It was found that the double emulsions prepared showed improved stability and reduced polydispersity. The improved stability was obtained by increasing the protein-to-polysaccharide ratio, which resulted in a reduction of the size of the droplets.

The controlled release of thiamin hydrochloride, or vitamin B_1, was also studied using double emulsions. It was found that, because the hybrids gave better coverage of the interface, the release of the entrapped material (vitamin B_1) was slowed down. These protein–polysaccharide biopolymer hybrids were capable of forming thick and effective barriers stabilized by both electrostatic and steric mechanisms, which aided in slowing down the release of active matter entrapped within the inner droplets of the double emulsion.

In conclusion, the development of emulsions as delivery systems for nutraceutical lipids should continue to grow with new applications exploiting the wide array of emulsion types (simple, multiple, emulsion gels, etc.) available. Furthermore, an accrued understanding of the fundamental properties that govern emulsion formation and stability should further improve our ability to tailor emulsions for specific uses.

REFERENCES

1. Dalgleish, D., Food emulsions, in *Emulsions and Emulsion Stability*, Sjoblom, J., Ed., Marcel Dekker, New York, 1996, pp. 287–325.
2. Garti, N. and Benichou, A., Double emulsions for controlled-release applications: progress and trends, in *Encyclopedic Handbook of Emulsion Technology*, Sjoblom, J., Ed., Marcel Dekker, New York, 2001, pp. 377–407.
3. Dalgleish, D., Food emulsions, in *Encyclopedic Handbook of Emulsion Technology*, Sjoblom, J., Ed., Marcel Dekker, New York, 2001, pp. 207–232.
4. Benichou, A., Aserin, A., and Garti, N., Double emulsions stabilized with hybrids of natural polymers for entrapment and slow release of active matters, *Adv. Colloid Interf. Sci.*, 108–109, 29–41, 2004.
5. Garti, N., A new approach to improved stability and controlled release in double emulsions, by the use of graft-comb polymeric amphiphiles, *Acta Polym.*, 49, 606–616, 1998.
6. Cortesi, R., Esposito, E., Luca, G., and Nastruzzi, C., Production of lipospheres as carriers for bioactive compounds, *Biomaterials*, 23, 2283–2294, 2002.
7. Feldman, Y., Skodvin, and Sjoblom, J., Dielectric spectroscopy on emulsion and related collidal system: a review, *Encyclopedic Handbook of Emulsion Technology*, Sjoblom, J., Ed., Marcel Dekker, New York, 2001, pp. 109–168.
8. Goff, H., Emulsion partial coalescence and structure formation in dairy systems, in *Crystallization and Solidification Properties of Lipids*, Widlak, N., Hartel, R., and Narine, S., Eds., AOCS Press, Champaign, IL, 2001, pp. 200–214.
9. McClements, J., Emulsion stability, in *Food Emulsions: Principles, Practice, and Technique*, CRC Press, Washington, DC, 1999, pp. 185–233.
10. Stauffer, C., *Eagan Press Handbook Series, Emulsifiers*, Eagan Press, MN, 1999.
11. Friberg, S., Emulsion stability. in *Food Emulsions*, Friberg, S. and Larsson, K., Eds., Marcel Dekker, New York, 1997, pp. 1–55.
12. Walstra, P., Dispersed systems: basic considerations, in *Food Chemistry*, 3rd ed., Fennema, O., Ed., Marcel Dekker, New York, 1996, pp. 95–155.
13. McClements, J., Emulsion ingredients, in *Food Emulsions: Principles, Practice, and Technique*, CRC Press, Washington, DC, 1999, pp. 83–125.
14. Walstra, P., Soft solids, in *Physical Chemistry of Foods*, Marcel Dekker, New York, 2003, pp. 683–771.
15. Chen, J. and Dickinson, E., Effect of monoglycerides and diglycerol-esters on viscoelasticity of heat-set whey protein emulsion gels, *Int. J. Food Sci. Technol.*, 34, 493–501, 1999.
16. Kiokias, S., Reszka, A., and Bot, A., The use of static light scattering and pulsed-field gradient NMR to measure droplet sizes in heat-treated acidified protein-stabilised oil-in-water emulsion gels, *Int. Dairy J.*, 14, 287–295, 2004.
17. Lindner, H., Fritz, G., and Glatter, O., Measurements on concentrated oil in water emulsions using static light scattering, *J. Colloid Interf. Sci.*, 242, 239–246, 2001.
18. McClements, J., Chracterization of emulsion properties, in *Food Emulsions: Principles, Practice, and Technique*, CRC Press, Washington, DC, 1999, pp. 295–337.

19. Dai, S., Tam, K., and Jenkins, R., Dynamic light scattering of semi-dilute hydrophobically modified alkali-soluble emulsion solutions with varying length of hydrophobic alkyl chains, *Macromolec. Chem. Phys.*, 203, 2312–2321, 2002.

20. Dalgleish, D. and Hallet, F., Dynamic light scattering: applications to food systems, *Food Res. Int.*, 28,181–193, 1995.

21. Hollinsworth, K. and Johns, M., Measurement of emulsion droplet sizes using PFG NMR and regularization methods, *J. Colloid Interf. Sci.*, 258, 383–389, 2003.

22. Malmborg, C., Topgaard, D., and Soderman, O., Diffusion in an inhomogeneous system: NMR studies of diffusion in highly concentrated emulsions, *J. Colloid Interf. Sci.*, 263, 270–276, 2003.

23. Chanamai, R., Herrman, N., and McClements, D., Ultrasonic spectroscopy study of flocculation and shear-induced floc disruption in oil-in-water emulsions, *J. Colloid Interf. Sci.*, 204, 268–276, 1998.

24. Froysa, K. and Nesse, O., Ultrasonic characterization of emulsions, in *Emulsions and Emulsion Stability*, Sjoblom, J., Ed., Marcel Dekker, New York, 1996, pp. 437–468.

25. Washington, C., Stability of lipid emulsions for drug delivery, *Adv. Drug Delivery Rev.*, 20, 131–145, 1996.

26. Garti, N., Progress in stabilization and transport phenomena of double emulsions in food applications, *Lebensm.-Wiss. u.-Technol.* 30, 222–235, 1997.

27. Mizushima, Y., Lipid microspheres (lipid emulsions) as a drug carrier: an overview, *Adv. Drug Delivery Rev.*, 20, 113–115, 1996.

28. Liu, F. and Liu, D., Long-circulating emulsions (oil in water) as carriers for lipophilic drugs, *Pharmaceut. Res.*, 12, 1060–1064, 1995.

29. Chung, H., Kim, T.K., Kwon, M., Kwon, I.C., and Jeong, S.Y., Oil components modulate physical characteristics and function of the natural oil emulsions as drug or gene delivery system, *J. Controlled Rel.*, 71, 339–350, 2001.

30. Trotta, M., Pattarinob, F., and Ignonia, T., Stability of drug-carrier emulsions containing phosphatidylcholine mixtures, *Eur. J. Pharma. Biopharma.*, 53, 203–208, 2002.

31. Djordjevic, D., Kim, H.-J., McClements, D.J., and Decker, E.A., Physical stability of whey protein-stabilized oil-in-water emulsions at pH 3: potential ω-3 fatty acid delivery systems (part A), *J. Food Sci.*, 69, 351–355, 2004.

32. Garti, N., Aserin, A., and Cohen, Y., Mechanistic considerations on the release of electrolytes from multiple emulsions stabilized by BSA and nonionic surfactants, *J. Controlled Rel.*, 29, 41, 1994.

33. Benichou, A., Aserin, A., and Garti, N., Double emulsions stabilised by new molecular recognition hybrids of natural polymers, *Polym. Adv. Technol.*, 13, 1019–1031, 2002.

18 Lipid Emulsions for Total Parenteral Nutrition (TPN) Use and as Carriers for Lipid-Soluble Drugs

Hu Liu and Lili Wang
School of Pharmacy, Memorial University of Newfoundland,
St. John's, Newfoundland, Canada

CONTENTS

18.1 INTRODUCTION AND BACKGROUND

Total parenteral nutrition (TPN) intravenous infusion liquids are used in hospitals and other clinical settings to provide all the components of a balanced nutrition including liquid, carbohydrate, protein, salt, fat, vitamins, and trace elements[1]. Nutrients provided via the vein are broken down into their molecular elements after entering the bloodstream. Living cells transform these nutritional components into energy and nutrients. Parenteral nutrition becomes necessary when the stomach and bowel, due to an illness or operation, cannot do their required tasks. Also, physicians can decide whether a patient should receive parenteral nutrition before an operation, during certain tests, or when the patient is not allowed to eat. The composition of TPNs can be adjusted to the patient's specific needs.

Emulsions are dispersions of one immiscible phase within another, with added surfactants to stabilize the dispersed droplets. The intravenous lipid emulsions are "oil-in-water" emulsions. The impetus to develop such a product was the need for a calorically dense source of energy that could be given daily by way of a peripheral vein. The early work was carried out between 1945 and 1960 at the Harvard School of Public Health by Drs. Frederic Stare and Robert Geyer, using a cottonseed oil-based lipid emulsion with the trade name of Lipomul IV®[1,2]. The product was removed from the market in the mid-1960s because of many clinical complications. In the late 1960s, a soybean oil-based product was developed in Sweden by Dr. Arvid Wretlind to overcome many earlier problems and was marketed in Europe[3]. After its successful application in Europe, the product with a trade name of Intralipid® was approved by the U.S. Food and Drug Administration (FDA) followed by many other countries. Many clinical studies have shown that lipid emulsions can be used as a safe source of dense calories given on a daily basis. Substituting a portion of glucose calories with lipids to produce a well-balanced fuel system avoids an excess of energy source and its possible complications. The intravenous fat emulsions should be compatible with the environment of the circulatory system and should be nontoxic. In addition, the materials contained in the fat emulsions should be chemically and physically compatible with one another, and should be stable enough to withstand the necessary procedures involved in manufacturing, transportation, and storage. At present most of the commercially available intravenous lipid emulsions are derived from vegetable oils such as those of soybean, safflower, and olive. Fish oil products have been introduced to the European market only recently. Some of the commercially available lipid emulsions are summarized in Table 18.1.

In addition to being given intravenously as a source of energy, lipid emulsions can be used as carriers for lipid-soluble drugs for formulation. This would definitely dictate the type of formulation desired.

The lipid composition and particle size distribution of lipid emulsions for TPN use are similar to those of chylomicrons, the endogenous lipoprotein carriers of dietary triacyglycerols in the bloodstream. Intralipid® was developed to mimic the lipoprotein metabolism of chylomicrons for patients who cannot receive nutrients orally. It is generally agreed that parenteral lipid emulsions are taken up along similar routes as natural chylomicrons. In humans, triacylglycerols of chylomicron are hydrolyzed by lipoprotein lipase. Free fatty acids are bound to albumin. After losing the bulk triacyglycerol core, smaller chylomicron remnant particles are taken up by hepatocytes.

18.2   OMEGA-6 (ω-6) VERSUS OMEGA-3 (ω-3) FATTY ACIDS

Fatty acids are characterized by the number of carbon atoms, the number of double bonds, and the position of the first double bond from the methyl end of the molecule. Thus, the designation C18:2, ω-6 stands for linoleic acid, an essential polyunsaturated fatty acid (PUFA) with a chain length of 18 carbon atoms, 2 double bonds, and with the first double bond being at the sixth carbon atom from the methyl end. Meanwhile, ω-3 PUFAs are those with their first double bond being at the

TABLE 18.1
Some Commercially Available Lipid Emulsions[4]

Products	Company (country)	Lipid composition (proportion by weight)	Concentration (%)
Liposyn II	Hospira (U.S.)	Soybean–safflower 50:50	10, 20
Liposyn III		Soybean oil 100%	10, 20
Intralipid	Fresenius Kabi	Soybean oil 100%	10, 20, 30
Structolipid	(Germany/Sweden)	Soybean–MCT 64:36	10, 20, 30
Lipofundin N	B. Braun (Germany)	Soybean oil 100%	10, 20
Lipofundin MCT/LCT		Soybean–MCT 50:50	10, 20
Lipoplus		MCT–soybean–fish oil (50:40:10)	10, 20
Lipovenous	Fresenius Kabi	Soybean oil 100%	10, 20, 30
Lipovenous-MCT	(Germany/Sweden)	Soybean–MCT 50:50	10, 20
Omegaven		Fish oil 100%	10
Clinoleic	Baxter (France)	Olive oil–soybean 80:20	20
Critilip	Baxter (U.S.)	MCT–soybean oil 75:25	20

Note: LCT, long-chain triacylglycerols; MCT, medium-chain triacylglycerols.
Modified from Rombeau, J.L. and Rolandelli, R.H., Eds., *Clinical Nutrition: Parenteral Nutrition*, 3rd ed., WB Saunders, Philadelphia, PA, 2001.

third carbon atom from the methyl group. The structure and biosynthesis of ω-3 and ω-6 PUFAs are provided in other chapters of this book.

Both linoleic acid and α-linolenic acid are essential fatty acids. The most important ω-6 fatty acids derived from linoleic acid are γ-linolenic acid and arachidonic acid (AA), while α-linolenic acid is the parent substance for the long-chain ω-3 fatty acids eicosapentaenoic acid (EPA), docosapentaenoic acid (DPA), and docosahexaenoic acid (DHA). PUFAs are known to have important functions as membrane building blocks and modulators of biochemical processes. As precursors for the synthesis of biologically active eicosanoids, AA and EPA affect inflammatory reactions[5], immunoresponse[6], cardiovascular diseases[7,8], lipid metabolism disorders, thrombosis, and cancer[9–12].

Humans, however, can only synthesize a small quantity of long-chain ω-3 fatty acids from α-linolenic acid by desaturation and chain elongation while marine animals such as fish and seals contain large quantities of EPA, DPA, and DHA. It has been shown that oral, enteral, and parenteral intake of ω-3 fatty acids results in increased proportion of ω-3 to ω-6 fatty acids in membranes of many cell populations such as erythrocytes[13], granulocytes[14,15], thrombocytes[16], endothelial cells, moncytes[17,18], and lymphocytes[19]. Although both ω-6 and ω-3 fatty acids are present in membrane phospholipids and are released by the same enzyme, phospholipase A_2, their metabolites and biological functions are different. AA is an example of ω-6 fatty acids while EPA is an example of ω-3 fatty acids. The two fatty acids are in competition for metabolism by the same enzyme systems (cyclooxygenase and lipoxygenase) and can displace one another through their respective properties[20–22] as shown in Figure 18.1. Depending on the cells (e.g., thrombocytes, endothelial cells, leucocytes, etc.) these fatty acids turn into endoperoxides through the enzyme cyclooxygenase, from which prostaglandins (PG), prostacyclins, and thromboxanes (TX) are produced. The enzyme lipoxygenase results in the formation of leukotrienes (LT). The derivatives produced from EPA are different in their structures and biological activities from the products of AA (Table 18.2)[23,24].

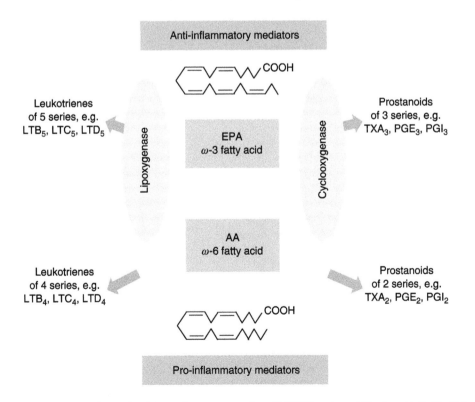

FIGURE 18.1 Eicosanoids derived from eicosapentaenoic acid (EPA) and arachidonic acid (AA). (Adapted from the Information Booklet of Omegaven® provided by Fresnius-Kabi of Germany.)

TABLE 18.2
Biological Effects of AA- and EPA-Derived Eicosanoids in Various Cells

Cell	AA	EPA
Mastocytes	PGD_2: Vasoconstriction PMN activation LTC_4-D_4: Vasoconstriction Bronchoconstriction Permeability increase PAF: Vasoconstriction Bronchoconstriction Edema formation PMN activation Thrombocyte aggregation	PGD_3: Reduced biological effects LTC_5-D_5: Inflammation ↓
Neutrophiles	LTB_4: Chemotaxis PMN activation Permeability increase PAF: Same as above	LTB_5: Inflammation ↓ Immune reaction ↓ PMN adherence ↓
Eosinophiles	LTC_4-D_4: Vasoconstriction Bronchoconstriction Permeability increase PAF: Same as above	LTC_5-D_5: Inflammation ↓
Macrophages	TXA_2: Vasoconstriction Bronchoconstriction PMN activation Thrombocyte activation PGE_2: Vasodilution Bronchorelaxation PMN activation PAF: Same as above	TXA_3: Reduced biological effects PGE_3: Vasodilution Bronchorelaxation PMN activation ↓
Thrombocytes	TXA_2: Vasoconstriction Bronchoconstriction PMN activation Thrombocyte activation PAF: Same as above	TXA_3: Reduced biological effects
Endothelium	PGI_2: Vasodilution Bronchorelaxation PGE2: Vasodilution Bronchorelaxation PMN activation PAF: Same as above	PGI_3: Vasodilution Bronchorelaxation PGE_3: Vasodilution Bronchorelaxation PMN activation ↓

Note: PMN, polymorphonuclear neutrophilic cell; PAF, platelet activation factor.
(Adapted from the Information Booklet of Omegavan® provided by Fresnius-Kabi of Germany.)

The cycloxygenase product of the 3 series, TXA_3, derived from EPA, shows a considerable reduction in proaggregatory and vasoconstrictive properties in comparison with the TXA_2 derived from AA, while PGI_3 is comparable to PGI_2 in antiaggregatory and vasodilatory effects. This means that the prostanoid metabolites of ω-3 fatty acids result in a reduced proaggregatory activity and exhibit a vasodilatory effect. In cells of granulopoiesis and monocyte-macrophage system, AA is metabolized into leukotrienes of the 4 series (LTB_4, C_4, D_4, E_4) that work as potent mediators of leucocyte activation, chemotaxis, and degranulation. EPA is a better substrate of 5-lipoxygenase[25,26] compared to AA. Leukotrienes of the 5 series (LTB_5, C_5, D_5, E_5) derived from EPA have a reduced proinflammatory effect in comparison with AA derivatives[27]. Thus LTB_5 possesses a considerably reduced vasoconstrictive and chemotactic potency in comparison to LTB_4[28,29]. The formation of the platelet activation factor (PAF), which has a strong proinflammatory and platelet-aggregating effect, is reduced by EPA. EPA, therefore, interferes with the PAF precursor pool[30,31]. In addition, it is known that EPA inhibits the formation of proinflammatory cytokines such as interleukin (IL)-1 and tumor necrosis factor α ($TNF\alpha$)[32,33].

As listed in Table 18.2, the overall net biological effects of AA and EPA derivatives are opposite. AA derivatives cause vasoconstriction, bronchoconstriction, PMN activation, and permeability increase while EPA derivatives alleviate inflammation, vasoconstriction, and bronchoconstriction.

As mentioned earlier (Table 18.1), most commercial lipid emulsions are made of vegetable oils, the drawbacks of which are two-fold: a lack of ω-3 PUFAs and the presence of phylloquinone (vitamin K)[34–36]. In addition to their antiinflammatory effects, other benefits of ω-3 unsaturated fatty acids for various conditions such as cardiovascular diseases have been reported. Vitamin K present in these intravenous lipid emulsions may have an undesirable clinical implication on patients who are on anticoagulants such as Warfarin, since vitamin K is one of the key factors in the blood coagulation process. Unpredictable amounts of vitamin K infused to patients may complicate the management of disease conditions. However, oils derived from marine animals do not have such concerns.

The industrial age has led to far-reaching changes in the intake of fats in the diet. As a result, the intake of ω-3 PUFAs has significantly decreased. With this change, incidence of cardiovascular diseases has increased which has led to the recognition that the intake of ω-3 fatty acids may be effective in preventing the development of cardiovascular diseases. It is known that oils from marine animals including fish and seal are rich in ω-3 PUFAs. In recent years, there has been increased understanding of the physiology and the mode of action of the ω-3 fatty acids. In parenteral nutrition, there has been increased interest in the nonenergetic, pharmacological effect of ω-3 fatty acids[37–41].

Based on the known beneficial effects of ω-3 fatty acids, Fresenius Kabi, the world's leading manufacturer of vegetable oil-based lipid TPN emulsions such as Intralipid®, has recently marketed a fish oil-based lipid TPN product, Omegaven®, in Europe and in India. Clinical studies of Omegaven® in Europe and other countries have demonstrated the benefits from the administration of ω-3 fatty acids to the following patient groups[42,43]:

- Posttraumatic and postsurgical patients
- Patients experiencing early stages of sepsis
- Patients with weak immune functions
- Patients with inflammatory bowel diseases (Crohn's disease, ulcerative colitis)
- Patients with inflammatory skin diseases (psoriasis, atopic eczema)

Although fish oil is superior to vegetable oil (presence of ω-3 fatty acids and absence of vitamin K in fish oil), fish oil does have other problems. Fish oil is extracted by a steam-heat process (100°C), which results in possible oxidation of PUFAs. In addition, high temperatures accelerate the conversion of *cis*-fatty acids to *trans*-fatty acids. Both oxidized and *trans*-fatty acids have been implicated in the development of cardiovascular diseases, the inflammatory process, cancer[44–50], and other medical conditions[51]. Some studies have also shown that lipid emulsions containing fish oil are poor substrates for lipoprotein lipase[52] and they tend to accumulate in the circulatory system. In addition, depending on the source of fish oil, the concentration of EPA and DHA, the main ingredients in fish oil, can vary greatly. As a result, the concentrations of EPA and DHA in Omegaven® are 12.5 to 28.2% and 14.4 to 30.9%, respectively[53]. Such a broad concentration range is less than desirable for pharmaceutical use.

To overcome the problems associated with fish oil, potential use of seal oil emulsion for TPN use was investigated. The advantages of seal oil over fish oil include the following:

- Seal oil is processed at a lower temperature (20 to 50°C) and contains a much reduced level of oxidized fatty acids[54]
- Seal oil has a considerably longer shelf life than fish oils (see Table 18.3)
- There is a sustainable, uniform (26 ± 3% of total ω-3 fatty acids), and high-quality supply of seal oil in the North Atlantic Ocean[55,56]

A 10% seal oil-based lipid emulsion has been produced using a high-pressure homogenization method (also known as microfluidization[57]). The particle size of the emulsion was found to be in the range of 300 ± 20 nm. Physical stability studies have been conducted using high-temperature and/or high-gravity accelerated experiments. It was found that the stability of the seal oil emulsion was equal to or better than that of commercial products (Omegaven® and Intralipid®). Chemical stability studies revealed that peroxide value of seal oil emulsion was much lower than that of commercial products indicating a better chemical stability (Figure 18.2)[54]. It has been demonstrated that the 10% seal oil emulsion can be sterilized by autoclave. Endotoxin (pyrogen) test using LAL gel clot[58] indicated that the seal oil emulsion formulation was nonpyrogenic. A more detailed comparison of seal oil emulsion and Omegaven® is summarized in Table 18.3.

The acute toxicity of seal oil emulsion formulation was studied. Sprague-Dawley rats were used and were divided into two groups, one given the fish oil emulsion (Omegaven® from Fresenius-Kabi) and the other given the seal oil emulsion. There were three animals in each group. For a period of ten days the animals were infused for one hour each day via the tail vein with ten times the recommended dose (1 g/day/kg body weight) of fish oil or seal oil emulsion. After the ten-day infusion animals were sacrificed, and tissues including kidney, heart, and liver were collected and fixed for pathological examination which showed no abnormality.

18.3 LIPID EMULSIONS AS CARRIERS FOR WATER-INSOLUBLE DRUGS

With the introduction of new automated organic syntheses, combinatorial chemistry, and high-throughput screening technologies, the capability for screening of drug molecules has soared. However, the scientific community is challenged with finding appropriate vehicles to deliver promising compounds to where and/or when they are needed. As a consequence, many blockbusters

TABLE 18.3
Comparison of Characteristics of Fish Oil and Seal Oil Emulsion Products

	Omegaven® (10% fish oil emulsion)	10% seal oil emulsion
Particle size (nm)	330 ± 24	300 ± 40
Zeta potential (mV)	48.3 ± 17.2	44.7 ± 15.3
pH	7.5–8.7	7.0–8.0
Endotoxin test (LAL)	negative	negative
Sterility	negative	negative
Osmolality (mosm/kg)	308–376	320 ± 20

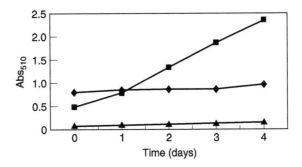

FIGURE 18.2 Lipid peroxidation values of 10% Intralipid® (♦); 10% Omegaven® (■); and 10% seal oil emulsion (▲). Samples (2 mL) were placed in test tubes and incubated at 37°C for 4 days. An aliquot of 20 μL of each sample was removed every 24 h and peroxidation values were determined according to the USP method[59].

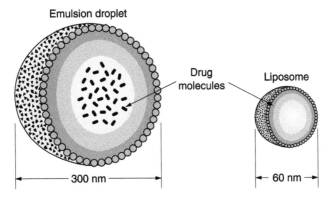

FIGURE 18.3 An illustration of lipid emulsion and liposome as drug carriers. (Adapted from Reference 60.)

have been lost because of the lack of adequate formulations. In many cases, the loss could have simply been due to the water-insolubility of the molecules involved.

To maximize the potential of all drug molecules, efforts have been made to find suitable formulations for their delivery, especially for lipid-soluble drugs. Although chemical modifications may make a lipid-soluble drug water soluble, lipophilic drugs are often more effectively taken up by target cells. In the pharmaceutical industry, both lipid emulsion and liposomes have been used for the formulation of lipid-soluble drugs. Figure 18.3 shows a lipid emulsion and a liposome as drug carriers[60]. An emulsion is a liquid system in which one liquid is dispersed in a second immiscible liquid in droplets. A parenteral lipid emulsion, such as Intralipid®, consists of a water phase with droplets composed of an oily triacylglycerol core (diameter 250 to 400 nm) stabilized with a phospholipid monolayer. The phospholipid monolayer stabilizes the emulsion by long-range repulsive electrostatic forces and short-range repulsive hydration forces. Without core lipids such as triacylglycerols, phospholipid exists as dispersed liposomes, unilamellar closed aggregates with a water core surrounded by a phospholipid bilayer (diameter 60 to 90 nm). Detailed structures of emulsions and liposomes can be examined by electron microscopy and indirect techniques such as NMR. Submicrometer droplet emulsions are typically characterized using laser light-scattering and light-diffraction methods.

Since the introduction of Intralipid® over 30 years ago, triacylglycerol-enriched lipid emulsion formulations as drug carriers for lipid-soluble drugs have been studied. In addition, to overcome the solubility problem of some of the lipid-soluble drugs, emulsions may stabilize otherwise chemically labile drugs against hydrolysis and/or oxidation during processing, sterilization, and storage[60] since the drug would be incorporated into the lipid core of the emulsion. In most cases, other excipients to stabilize drugs are also used, such as pH buffering agents, antioxidants, and chelating agents. Although most nutritional lipid emulsion formulations are highly stable, for up to 18 months at room temperature, a lipid–drug emulsion may require controlled storage at lower temperatures, often 4 to 8°C. The shelf-life of a drug–lipid emulsion is more often limited by the degradation of the incorporated drug molecules than by deterioration of the emulsion media.

Incorporation of lipid-soluble drugs into lipid emulsions could reduce the toxicity of drugs such as the antifungal agent amphotericin B[61,62] and the antineoplastic drug penclomedine[63]. The incorporation of lipid-soluble drugs into lipid emulsion formulations also eliminates the need for organic solvents such as propylene glycol, *tert*-butanol, or dimethyl formamide. These organic solvents are often associated with pain upon injection, phlebitis, and other acute toxicities. Propofol is an anesthetic agent and is highly lipophilic[64,65]. Organic solvents were used for its first generation of formulation and these were associated with the irritation of veins. After being formulated as 10% soybean oil emulsion (marketed as Diprivan®), vein irritation and the pain at the injection site were greatly reduced. Prostaglandin E$_1$ (PGE$_1$) is a potent vasodilator as well as an inhibitor of platelet aggregation (see Table 18.2). Incorporation of PGE$_1$ into a 10% soybean oil emulsion has been marketed successfully in Asian countries and shown an increased biological activity and reduced side effects[66]. Another advantage of using lipid emulsions for lipid-soluble drugs is their reduced adsorption onto plastic containers as well as infusion sets[60], especially for fat-soluble vitamins A, D, E, and K$_1$.

Incorporation of drugs in lipid emulsions may improve the targeting potential of some diagnostic agents. Drug–lipid emulsion formulations have been reported to deliver diagnostic agents to specific organs such as liver, lymph node, and blood pool[67–69]. More recently, it has been shown that lipid emulsions could deliver radioimaging agents to atherosclerotic plaques in animal models[70].

Although considerable effort with liposomes has focused on water-soluble drugs entrapped in the aqueous core, there are a few successful examples of incorporation of lipid-soluble drugs such as Amphitericin B into the hydrophobic regions of phospholipid bilayers. Amphitericin B liposomal formulation has demonstrated reduced nephrotoxicity and improved response rates[71]. Compared to lipid emulsions, the capacity of liposomes in carrying drug molecules is normally much lower. This would result in substantial clinical disadvantages. Furthermore, lipid emulsions are normally ready-to-use formulations stored at room temperature, unless the stability of the incorporated drug itself limits this type of formulation. However, liposomes are generally unstable and come as the lyophilized form which requires reconstitution prior to use. Elimination of a lyophilization step during manufacture is another advantage of lipid emulsions.

At present, most drug–lipid emulsions are vegetable oil-based. With the success of fish oil-based emulsion products for TPN already on the market and the increased understanding of the benefits of ω-3 fatty acids, it is probably reasonable to anticipate the appearance of marine oil-based emulsions as drug carriers in the future.

It should be noted that most of the currently available lipid emulsions are essentially modified versions of Intralipid®. For the controlled and targeted delivery of lipid-soluble drugs, it may be necessary to control and modify the surface properties of the emulsion droplets. Adding block copolymers of ethylene oxide and propylene oxide to stabilize an emulsion may prolong circulation time. Another strategy for avoiding rapid clearance may use phospholipids modified with polyethylene glycol (PEG) or surfactants containing PEG.

18.4 SUMMARY

TPN lipid emulsions commonly used in clinical settings are primarily derived from vegetable oils. However, they lack ω-3 fatty acids, in addition to the presence of unpredictable quantities of vitamin K. As a result, infusion of large quantities of vegetable oil-derived lipid emulsions may complicate the conditions of those patients who also receive anticoagulant therapy. Long-term parenteral use of vegetable oil-based TPN lipid emulsions may result in an unbalanced eicosanoid formation, which may lead to inflammation. It is, therefore, highly undesirable for many hyperinflammatory conditions such as sepsis, Crohn's disease, posttrauma, postsurgery, burn patients, organ-transplant patients who take immunosuppressants, cancer, patients with cardiovascular diseases, and immune-compromised patients to use vegetable oil-based TPN.

The emerging of parenteral lipid emulsions rich in ω-3 fatty acids such as fish oil provides a promising avenue for clinicians to counterbalance the shortcomings of vegetable oil-derived lipid emulsions. The proinflammatory mediators produced from ω-6 fatty acids can be attenuated by partial or complete substitution of the vegetable oil lipid emulsions with marine oil-based products. Such antiinflammatory effects of ω-3 fatty acids are achieved by competing for the same enzyme systems with ω-6 fatty acids, and by producing many eicosanoid modulators which have opposite or much reduced physiological effects in comparison with the metabolites of ω-6 fatty acids.

Fish oils are highly prone to oxidation both in manufacture and storage, which limits their shelf life. The oxidized lipids may counterbalance their beneficial effects[58]. In addition, the contents of the main ω-3 fatty acids in the currently used fish oil lipid emulsion products are very broad. Seal oil, on the other hand, appears to be very stable against oxidation and its ω-3 fatty acid contents are found to be very consistent. Therefore, the lipid emulsion made of seal oil may become an excellent choice for TPN use in the future should the safety and effectiveness of this product be established in animals (preclinical studies) and humans (clinical trials).

In addition, many lipid-soluble drugs have been formulated into emulsions and demonstrated unique advantages over organic solvent-based lipid drug formulations. A new generation of lipid emulsions may contain chemically modified phospholipids and other surfactants which would expand the clinical potentials of many water-insoluble drugs.

REFERENCES

1. Grant, J.P., *Handbook of Total Parenteral Nutrition*, WB Saunders, Philadelphia, PA, 1992, p. 1.
2. Geyer, R., Parenetral nutrition, *Physiol. Rev.*, 40, 150–186, 1960.
3. Wretlind, A., Development of fat emulsions, *J. Parenter. Enteral Nutr.*, 5, 230–235, 1981.
4. Rombeau, J.L. and Rolandelli, R.H., Eds., *Clinical Nutrition: Parenteral Nutrition*, 3rd ed., WB Saunders, Philadelphia, PA, 2001.
5. Burton, J.L., Dietary fatty acids and inflammatory skin disease, *Lancet*, 1, 27–31, 1989.
6. Endrers, S., DeCaterina, R., Schmidt, E.B., Kristensen, S.D., N-3 polyunstaturated fatty acids: update 1995, *Eur. J. Clin. Invest.*, 25, 629–638, 1995.
7. Burr, M.L., Fehily, A.M., and Gilbert, J.F., Effect of changes in fat, fish, and fibre intakes on death and myocardial reinfarction: diet and reinfarction trial (DART), *Lancet*, 2, 757–761, 1989.
8. Hu, F.B., Bronner, L., Willett, W.C., Stampfer, M.J., Rexrode, K.M., Albert, C.M., Hunter, D., and Manson, J.E., Fish and omega-3 fatty acid intake and risk of coronary heart disease in women, *JAMA*, 287, 1815–1821, 2002.
9. Jeeski, L.I., Zerouga, M., and Stillwell, W., Omega-3 fatty acid containing liposomes in cancer therapy, *Proc. Soc. Exp. Biol. Med.*, 210, 227–233, 1995.
10. Tevar, R., Jho, D.H., Babcock, T., Helton, W.S., and Espat, N.J., Omega-3 fatty acid supplementation reduces tumor growth and vascular endothelial growth factor expression in a model of progressive non-metastasizing malignancy, *J. Parenter. Enteral Nutr.*, 26, 285–289, 2002.
11. Barber, M.D., Fearon, K.C., Tisdale, M.J., McMillan, D.C., and Ross J.A., Effect of a fish oil-enriched nutritional supplement on metabolic mediators in patients with pancreatic cancer cachexia, *Nutr. Cancer*, 40, 118–124, 2001.
12. Senzaki, H., Tsubura, A., and Takada, H., Effect of eicosapentaenoic acid on the suppression of growth and metastasis of human breast cancer cells *in vivo* and *in vitro*, *World Rev. Nutr. Diet.*, 88, 117–25, 2001.
13. Brown, A.J., Pang, E., and Roberts, D.C.K., Persistent changes in fatty acid composition of erythrocyte membranes after moderate intake of n-3 polyunsaturated fatty acids: study design implication, *Am. J. Clin. Nutr.*, 54, 668–73, 1991.

14. Chilton, F.H., Patel, M., Fontech, A.N., Hubbard, W.C., and Triggiani, M., Dietary n-3 fatty acid effects on neutrophil lipid composition and mediator production. Influence of duration and dosage, *J. Clin. Invest.*, 91, 115–22, 1993.

15. Morlion, B.J., Torwesten, E., Lessire, H., Sturm, G., Peskar, B.M., Fürst, P., and Puchstein, C., The effect of parenteral fish oil on leukocyte membrane fatty acid composition and leukotriene synthesizing capacity in postoperative trauma, *Metabolism*, 45, 1208–13, 1996.

16. Roulet, M., Frascarolo, P., Pilet, M., and Chapius, G., Effect of intravenously infused fish oil on platelet fatty acid phospholipid composition and on platelet function in postoperative trauma, *J. Parenter. Enteral Nutr.*, 21, 296–301, 1997.

17. Urskaze, M., Hamazaki, T., Makuta, M., Ibuki, F., Kobayashi, S., Yano, S., and Kumagai, A., Infusion of fish oil emulsion: effects on platelet aggregationand fatty acid composition in phospholipids of plasma, platelets, and red blood cell membranes in rabbits, *Am. J. Clin. Nutr.*, 46, 936–940, 1987.

18. Croset, M., Bayon, Y., and Lagarge, M., Incorporation and turnover of eicosapentaenoic and docosahexaenoic acids in human blood platelets in vitro, *Biochem. J.*, 281, 309–316, 1992.

19. Schauder, P., Rohn, U., Schafer, G., Korff, G., and Schenk, H.D., Impact of fish oil enriched total parenteral nutrition on DNA synthesis, cytokine release and receptor expression by lymphocytes in the postoperative period, *Br. J. Nutr.*, 87 (Suppl. 1) S103–S110, 2002.

20. Fischer, S. and Weber, P.C., Prostaglandin I is formed in vivo in man after dietary eicosapentaenoic acid, *Nature*, 307, 165–168, 1984.

21. Strasser, T., Fischer, S., and Weber, P.C., Leukotriene B is formed in human neutrophils after dietary eicosapentaenoic acid, *Proc. Natl. Acad. Sci. USA*, 82, 1540–1543, 1985.

22. Weber, P.C., Fischer, S., von Schacky, C., Lorenz, R., and Strasser, T., Dietary omega-3 polyunsaturated fatty acids and eicosanoid formation in man, in *Health Effects of Polyunsaturated Fatty Acids in Seafoods*, Simopoulos, A.P., Ed., Academic Press, Orlando, FL, 1986, pp. 49–60.

23. Fletcher, J.R., Eicosanoids. Critical agents in the physiological process and cellular injury, *Arch. Surg.* 128, 1192–1196, 1993.

24. Calder, P.C., Polyunsaturated fatty acids, inflammation, and immunity, *Lipids*, 36, 1007–1024, 2001.

25. Lee, T.H., Hoover, R.L., Williams, J.D., Sperling, R.I., Ravalese, J., 3rd, Spur, B.W., Robinson, D.R., Corey, E.J., Lewis, R.A., and Austen, K.F., Effect of dietary enrichment with eicosapentaenoic and docosahexaenoic acids on in vitro neutrophil and monocyte leukotriene generation and neutrophil function, *N. Engl. J. Med.*, 312, 1217–1224, 1985.

26. James, M.J., Gibson, R.A., and Cleland, L.G., Dietary polyunsaturated fatty acids and inflammatory mediator production, *Am. J. Clin. Nutr.*, 71 (Suppl. 1), 343S–348S, 2000.

27. Lewis, R.A., Lee, T.H., and Austen, K.F., Effects of omega-3 fatty acids on the generation of products of the 5-lipoxygenase pathway, in *Health Effects of Polyunsaturated Fatty Acids in Seafoods*, Simopoulos, A.P., Ed., Academic Press, Orlando, FL, 1986, pp. 227–238.

28. Lee, T.H., Sethi, T., Crea, A.E., Peters, W., Arm, J.P., Horton, C.E., Walport, M.J., and Spur, B.W., Characterization of leukotriene B3: comparison of its biological activities with leukotriene B4 and leukotriene B5 in complement receptor enhancement, lysozyme release and chemotaxis of human neutrophils, *Clin. Sci.* (London), 74, 467–475, 1988.

29. Goldman, D.W., Pickett, W.C., and Goetzl, E.J., Human neutrophil chemotactic and degranulating activities of leukotriene B5 (LTB5) derived from eicosapentaenoic acid, *Biochem. Biophys. Res. Commun.*, 117, 282–288, 1983.

30. Mayer, K., Merfels, M., Muhly-Reinholz, M., Gokorsch, S., Rosseau, S., Lohmeyer, J., Schwarzer, N., Krull, M., Suttorp, N., Grimminger, F., and Seeger, W., Omega-3 fatty acids suppress monocyte adhesion to human endothelial cells: role of endothelial PAF generation, *Am. J. Physiol. Heart Circ. Physiol.*, 283, H811–H818, 2002.

31. Weber, C., Aepfelbacher, M., Lux, I., Zimmer, B., and Weber, P.C., Docosahexaenoic acid inhibits PAF and LTD4 stimulated [Ca2+]i-increase in differentiated monocytic U937 cells, *Biochim. Biophys. Acta*, 1133, 38–45, 1991.

32. Pupe, A., Moison, R., De Haes, P., van Henegouwen, GB., Rhodes, L., Degreef, H., and Garmyn, M., Eicosapentaenoic acid, a n-3 polyunsaturated fatty acid differentially modulates TNF-alpha, IL-1alpha, IL-6 and PGE2 expression in UVB-irradiated human keratinocytes. *J. Invest. Dermatol.*, 118, 692–698, 2002.

33. Shimizu, T., Iwamoto, T., Itou, S., Iwata, N., Endo, T., and Takasaki, M., Effect of ethyl icosapentaenoate (EPA) on the concentration of tumor necrosis factor (TNF) and interleukin-1 (IL-1) in the carotid artery of cuff-sheathed rabbit models, *J. Atheroscler. Thromb.*, 8, 45–49, 2001.

34. MacLaren, R., Wachsman, B.A., Swift, D.K., and Kuhl, D.A., Warfarin resistance associated with intravenous lipid administration: discussion of propofol and review of the literature, *Pharmacotherapy*, 17, 1331–1337, 1997.

35. Pettei, M.J., Israel, D., and Levine, J., Serum vitamin K concentration in pediatric patients receiving total parental nutrition, *J. Parenter. Enteral Nutr.*, 17, 465–467, 1993.

36. Carlin, A. and Walker, W.A., Rapid development of vitamin K deficiency in an adolescent boy receiving total parenteral nutrition following bone marrow transplantation, *Nutr. Rev.*, 49, 179–183, 1991.

37. Dyerberg, J., Bang, H.O., and Hjorne, N., Plasma cholesterol concentration in Caucasian Danes and Greenland West-coast Eskimos, *Dan. Med. Bull.*, 24, 52–55, 1977.

38. Dyerberg, J., Bang, H.O., and Hjorne, N., Fatty acid composition of the plasma lipids in Greenland Eskimos, *Am. J. Clin. Nutr.*, 28, 958–966, 1975.

39. Dyerberg, J. and Bang, H.O., Haemostatic function and platelet polyunsaturated fatty acids in Eskimos. 1979, *Nutrition*, 11, 484, 1995.

40. Dyerberg, J., Bang, H.O., Stoffersen, E., Moncada, S., and Vane, J.R., Eicosapentaenoic acid and prevention of thrombosis and atherosclerosis?, *Lancet*, 2, 117–119, 1978.

41. Dyerberg, J. and Bang, H.O., Haemostatic function and platelet polyunsaturated fatty acids in Eskimos, *Lancet*, 2, 433–435, 1979.

42. Heller, A.R., Fischer, S., Rossel, T., Geiger, S., Siegert, G., Ragaller, M., Zimmermann, T., and Koch, T., Impact of n-3 fatty acid supplemented parenteral nutrition on haemostasis after major abdominal surgery, *Br. J. Nutr.*, 87 (Suppl. 1), S95–S101, 2002.

43. Morlion, B.J., Torwesten, E., Lessire, H., Sturm, G., Peskar, B.M., Furst, P., and Puchstein, C., The effect of parenteral fish oil on leukocyte membrane fatty acid composition and leukotriene-synthesizing capacity in patients with postoperative trauma, *Metabolism*, 45, 1208–1213, 1996.

44. Lovejoy, J.C., Smith, S.R., Champagne, C.M., Most, M.M., Lefevre, M., DeLany, J.P., Denkins, Y.M., Rood, J.C., Veldhuis, J., and Bray, G.A., Effects of diets enriched in saturated (palmitic), monounsaturated (oleic), or trans (elaidic) fatty acids on insulin sensitivity and substrate oxidation in healthy adults, *Diabetes Care*, 25, 1283–1288, 2002.

45. Wahle, K.W. and Rotondo, D., Fatty acids and endothelial cell function: regulation of adhesion molecule and redox enzyme expression, *Curr. Opin. Clin. Nutr. Metab. Care*, 2, 109–115, 1999.

46. Sethi, S., Ziouzenkova, O., Ni, H., Wagner, D.D., Plutzky, J., and Mayadas, T.N., Oxidized omega-3 fatty acids in fish oil inhibit leukocyte-endothelial interactions through activation of PPARalpha, *Blood*, 100, 1340–1346, 2002.

47. Slattery, M.L., Curtin, K., Ma, K., Edwards, S., Schaffer, D., Anderson, K., and Samowitz, W., Dietary activity and lifestyle associations with p63 mutations in colon tumors, *Cancer Epidemiol. Biomarkers Prev.*, 11, 541–548, 2002.

48. Luo, J., Zha, S., Gage, W.R., Dunn, T.A., Hicks, J.L., Bennett, C.J., Ewing, C.M., Platz, E.A., Ferdinandusse, S., Wanders, R.J., Trent, J.M., Isaacs, W.B., and De Marzo, A.M., Alpha-methylacyl-CoA racemase: a new molecular marker for prostate cancer, *Cancer Res.*, 62, 2220–2226, 2002.

49. Loi, C., Chardigny, J.M., Berdeaux, O., Vatele, J.M., Poullain, D., Noel, J.P., and Sebedio, J.L., Effects of three trans isomers of eicosapentaenoic acid on rat platelet aggregation and arachidonic acid metabolism, *Thromb. Haemost.*, 80, 656–661, 1998.

50. Svahn, J.C., Feldl, F., Raiha, N.C., Koletzko, B., and Axelsson, I.E., Different quantities and quality of fat in milk products given to young children: effects on long chain polyunsaturated fatty acids and trans fatty acids in plasma, *Acta Paediatr.*, 91, 20–29, 2002.

51. Weiland, S.K., von Mutius, E., Husing, A., and Asher, M.I., Intake of trans fatty acids and prevalence of childhood asthma and allergies in Europe. ISAAC Steering Committee, *Lancet*, 353, 2040–2041, 1999.

52. Olivera, T., Carpentier, Y.A., Hansen, I. et al., Triacyglyceride hydrolysis of soy vs. fish oil LCT emulsions, *Clin. Nutr.*, 11 (Suppl.), 44–50, 1992.

53. Fresenius-Kabi. Omegaven Product Information, http://www.fresenius-kabi.com/internet/kabi/corp/fkintpub.nsf/AttachmentsByTitle/SPC_OMEGAVEN.pdf/$FILE/SPC_OMEGAVEN.pdf (2004).

54. Delmas, M. and Walsh, G., Process for Refining Animal and Vegetable Oil, International Patent (PCT), WO 01/49814, 2001.

55. Shahidi, F. and Wanasundara, U.N., Seal blubber oil and its nutraceutical products, *ACS Symp. Ser.*, 788, 142–150, 2001.

56. Department of Fisheries and Oceans, The 2001 Atlantic Seal Hunt 2001 Management Plan, http://www.dfo-mpo.gc.ca/seal-phoque/reports/Mgtplan2001/sealplan2001_e.htm

57. Zheng, S., Zheng, Y., Beissinger, R.L., Wasan, D.T., and McCormick, D.L., Hemoglobin multiple emulsion as an oxygen delivery system, *Biochim. Biophys. Acta*, 1158, 65–74, 1993.
58. Gould, M.J., Performing the LAL gel clot testing in facilities, *Nephrol. News Issues*, Nov., 26–29, 1988.
59. United States Pharmacopeia 27 National Formulary 22, United States Pharmacopeial Convention, Washington, DC, 2004, p. 1718.
60. Collins-Gold, L., Feichtinger, N., and Wärnheim, T., Are lipid emulsions the drug delivery solution? *Mdern Drug Discovery*, 3, 44–46, 48, 2000.
61. Perkins, W.R., Minchey, S.R., Boni, L.T., Swenson, C.E., Popescu, M.C., Pasternack, R.F., and Janoff, A.S., Amphotericin B-phospholipid interactions responsible for reduced mammalian cell toxicity, *Biochim. Biophys. Acta*, 1107, 271–282, 1992.
62. Lance, M.R., Washington, C., and Davis, S.S., Structure and toxicity of amphotericin B/triglyceride emulsion formulations, *J. Antimicrobiol. Chemother.*, 36, 119–128, 1995.
63. Prankerd, R.J., Frank, S.G., and Stella, V.J., Preliminary development and evaluation of a parenteral emulsion formulation of penclomedine: a novel, practically water insoluble cytotoxic agent, *J. Parenteral Sci. Technol.*, 42, 76–81, 1988.
64. Mackenzie, N. and Grant, I.S., Comparison of the new emulsion formulation of propofol with metho-hexitone and thiopentone for induction of anaesthesia in day cases, *Br. J. Anaesth.*, 57, 725–731, 1985.
65. Dundee, J.W. and Clark, R.S., Propofol, *Eur. J. Anaesth.*, 6, 5–22, 1989.
66. Mizushimna, Y., Yanagawa, A., and Hoshi, K., Prostaglandin E1 is more effective when incorporated in lipid microspheres for treatment of peripheral vascular diseases in man, *J. Pharm. Pharmacol.*, 35, 666–667, 1983.
67. Nanda, N.C., Kitzman, D.W., Dittrich, H.C., Hall, G., Imagent Clinical Investigators Group, Imagent improves endocardial border delineation, inter-reader agreement, and the accuracy of segmental wall motion assessment, *Echocardiography*, 20, 151–161, 2003.
68. Ivancev, K., Lunderquist, A., McCuskey, R., McCuskey, P., and Wretland, A., Effect of intravenously injected iodinated lipid emulsion on the liver, *Acta Radiol.*, 30, 291–298, 1989.
69. Bakan, D.A., Weichert, J.P., Longino, M.A., and Counsell, R.E., Polyiodinated triglyceride lipid emulsions for use as hepatoselective contrast agents in CT: effects of physicochemical properties on biodistribution and imaging profiles, *Invest. Radiol.*, 35, 158–169, 2000.
70. Kang, Z., Scott, T.M., Wesolowski, C., Feng, L., Wang, J., Wang, L., and Liu, H., *Ex vivo* evaluation of a novel polyiodinated compound for early detection of atherosclerosis, *Radiat. Res.*, 160, 460–466, 2003.
71. Miller, C.B., Waller, E.K., Klingemann, H.G., Dignani, M.C., Anaissie, E.J., Cagnoni, P.J., McSweeney, P., Fleck, P.R., Fruchtman, S.M., McGuirk, J., and Chao, N.J., Lipid formulations of amphotericin B preserve and stabilize renal function in HSCT recipients, *Bone Marrow Transpl.*, 33, 543–548, 2004.

19 Modified Oils

Frank D. Gunstone
Scottish Crop Research Institute, Invergowrie, Dundee, Scotland, U.K.

CONTENTS

19.1 INTRODUCTION

Statistics of oil and fat production are based mainly on vegetable oils and sometimes include animal fats. For example, Oil World data cover 13 vegetable oils and fats from four animal sources while reports from the U.S. Department of Agriculture are confined to nine vegetable oils. A small part of total production is used for animal feed (estimated at 6%, but not including feeding of whole oilseeds or seed meals containing residual oil) and for industrial purposes (14%), but the majority is used for food consumption (80%).

The 17 commercial sources of oils and fats are dominated by four vegetable oils followed by three animal fats. Production figures for 2004/05 are detailed in Table 19.1 where the 17 sources are listed in order of production and range from soybean oil (32.6 million metric tons, MMT) to sesame, linseed, and sesame oils each below 1 MMT. Dietary intake is not the same as these figures because of the consumption of fat-containing food that is not included in these statistics. For example, most nuts are rich in oil. Nor do the statistics include cocoa butter used to make chocolate, dairy products other than butter such as cheese, and the fat consumed when eating red meat, poultry, or fish.

Each of these 17 sources has its own fatty acid composition (Table 19.1). The oils show small variations from season to season and according to the locality where they are grown. It is clear that three acids predominate in the vegetable oils: palmitic (16:0), oleic (18:1), and linoleic (18:2). But these materials do not meet the nutritional recommendations nor the technical requirements to make good spreads or good lubricants. This chapter is devoted to the methods that lipid scientists and technologists have used in the past and are studying today to widen the range of fatty acid and triacylglycerol composition so that oils and fats will better meet the nutritional and technical requirements of the 21st century.

TABLE 19.1
Production (MMT) of Oils and Fats in the Harvest Year 2004/05 and Typical Fatty Acid Composition

Commodity	Production	16:0	18:0	18:1	18:2	18:3	
World total	136.40						
Soybean	32.57	11	4	23	53	8	
Palm	32.50	44	4	39	11		
Rapeseed/canola	16.11	4	2	62	22	10	
Sunflower	9.08	6	5	20	69		
Tallow	8.19	27	7	48	2		a
Lard	7.43	26	11	44	11		b
Butter	6.47	26	11	28	2		c
Groundnut/peanut	4.50	13	3	38	41		
Cottonseed	5.00	23	2	17	56		
Palm kernel	3.80	9	3	15	2		d
Coconut	3.01	9	2	7	2		e
Olive	2.73	10	2	78	7	1	
Corn	2.05	13	3	31	52	1	
Fish	1.07						f
Sesame	0.77	9	6	41	43		
Linseed	0.60	6	3	17	14	60	
Castor	0.52						g

Note: Data cover 13 vegetable oils and 4 animal fats and are listed in production order.
[a] Also 16:1 (11%).
[b] Also 16:1 (5%).
[c] Also C_4–C_{14} saturated acids (22%).
[d] Also 8:0 (3%), 10:0 (4%), 12:0 (45%), and 14:0 (18%).
[e] Also 8:0 (8%), 10:0 (7%), 12:0 (48%), and 14:0 (16%).
[f] These vary in composition but contain saturated, monounsaturated (16:1–22:1), and polyunsaturated C_{20} and C_{22} acids.
[g] Contains ~90% ricinoleic acid.
Source: *Oil World Annual 2005*, ISTA Mielke GmbH, Hamburg, Germany, 2003.

It is convenient to divide these techniques into two groups: technological and biological. In the first we accept what nature provides and seek to change fatty acid and/or triacylglycerol composition thereby modifying nutritional, chemical (mainly oxidative stability), and physical (mainly melting behavior) properties to make them more suitable for their end-use. In the biological procedures we interfere at an earlier stage and either seek new and better sources or we take plants which already produce large quantities of oil efficiently and try to improve the oils by conventional methods of seed breeding or by exploiting newer methods based on increasing genetic understanding.

None of the biological methods is to be undertaken frivolously. They are time-consuming and expensive and before undertaking large-scale studies it is necessary to be sure that the selected objectives are correct. These are based on perceived demand factors. For both food and nonfood purposes there is a demand for oils high in oleic acid with a balancing reduction in both saturated and/or polyunsaturated acids. Saturated acids may be undesirable on nutritional grounds and also because they promote crystallization which should be avoided in salad oils and in lubricants. Polyunsaturated acids oxidize more readily than their monounsaturated analogs and there are food and nonfood situations where this is undesirable.

Linolenic acid with three double bonds raises some interesting problems. It oxidizes more quickly than linoleic acid with two double bonds and the volatile products resulting from oxidation of n-3 acids produce undesirable flavors which are stronger than those of n-6 acids. This leads to a reduced shelf life for food products containing linolenic acid and this acid is therefore not favored by the food industry. Linolenic acid is present in two important vegetable oils: soybean with 8% linolenic acid and rapeseed/canola with 10% linolenic acid. The level of linolenic acid in such oils can be reduced by short partial hydrogenation often referred to as brush hydrogenation. In the longer term, the level of linolenic acid could be reduced by seed breeding, but this may require genetic engineering which still causes concern in some parts of the world. There is, however, another matter of greater concern. It is accepted increasingly that the dietary ratio of n-6 to n-3 acids is too high, particularly in those countries with a high consumption of (partially hydrogenated) soybean oil or of sunflower oil. There is still some uncertainty how far the dietary deficiency of n-3 acids can be met by metabolism of linolenic acid and how far the longer-chain polyunsaturated fatty acids (PUFAs) (eicosapentaenoic acid, EPA (n-3 20:5) and docosahexaenoic acid, DHA (n-3 22:6)) should themselves be constituents in the diet. It would be wiser to settle this question before putting a lot of resources into production of seeds with lower levels of linolenic acid in seed oils. Serious attempts are also being made develop plant sources of arachidonic acid (AA; n-6 20:4), EPA, and DHA, but progress is slow and commercialization is still many years away[1,2].

Many aspects of the topics to be developed in the following sections are covered in references[3-9].

19.2 BLENDING

Blending is the simplest of the modification procedures involving merely the mixing of two or more oils to combine the properties of the original oils. The blended fats may already be fractionated or hydrogenated and blending may be followed by chemical or enzymatic interesterification. When oils are mixed the composition (fatty acids, triacylglycerols, minor components) of the mix is a weighted mean of the original components but subsequent interesterification will result in a change in triacylglycerol composition (Section 19.5).

Blending may be undertaken for a number of reasons including the following:

- To produce oil with greater oxidative stability as in the addition of low levels of sesame oil or rice bran oil. Both of these oils are characterized by minor components which lead to high oxidative stability.
- To produce an oil with higher nutritive value. Such blends are usually low in saturated acids, rich in n-3 acids to give a more favorable n-6/n-3 ratio, do not contain acids with *trans* unsaturation, and are enriched in antioxidant to ensure good shelf life.
- To produce oils with appropriate melting behavior for margarines and shortenings. For some purposes it is necessary to promote crystallization and for others to inhibit crystallization.
- To produce spreads and other products with low or zero levels of *trans* acids.
- To produce the desired physical and nutritional properties at minimum cost[10].

For example, it has recently been reported that a range of oils consisting of blended base oil stocks, naturally stable oils, tropical oils, and interesterified shortenings and margarines can be tailor made for virtually any food use[11-14].

19.3 FRACTIONATION

Fractionation is the name given to a range of procedures whereby oils and fats are divided into two or more fractions with differing properties. The most common form in which this procedure is now applied is dry fractionation. This term implies an absence of solvent and crystallization of solid components from the wholly liquid oil, described as the melt. More efficient separations are obtained with a solvent such as acetone or hexane but this is only used for products of sufficient value to justify the higher costs[15–17].

A recent dossier devoted to fractionation of fats describes both existing procedures and new procedures not yet applied on a commercial scale. The latter include fractionation with supercritical carbon dioxide and fractionation by membranes. Industrial chromatography is used to produce products with very low color and odor for use in cosmetics requiring addition of expensive pigment or perfume. Only dry fractionation is discussed here[18].

Fractionation is a cheaper process than either hydrogenation or interesterification because there is no loss of oil and no posttreatment is required. One problem with fractionation is that there are two products and it may be difficult to find an economic use for both. Refractionation gives yet other materials but is only commercially practicable when high-value products are obtained, such as cocoa butter replacers[19].

Dry fractionation starts with molten oil free of any crystalline components and involves two stages each of which must be carried out as efficiently (economically) as possible. The first stage is crystallization in which equilibrium is established between solid and liquid phases and the second stage aims at the complete separation of the solid and liquid phases. Partial failure at the second stage negates, in some part, the success of the first stage.

Fractionation produces two (or more) fractions differing in fatty acid and triacylglycerol composition. These changes affect physical properties and so increase the oil's range of usefulness. The crystals (stearin) will be less soluble, higher melting, and more saturated while the liquid portion (olein) is more soluble, lower melting, and more unsaturated. In simple terms triacylglycerols of type SSS and SUS will concentrate in the stearin and those of type SUU and UUU will concentrate in the olein (S = saturated acyl chains, U = unsaturated acyl chains). However, this simple view will be modified by the chain length of the saturated acids and by the degree of unsaturation of the unsaturated acids.

There are three stages in the crystallization step: supercooling of the melt, nucleation, and crystal growth. Crystallization is an exothermic process and since oil is a poor heat conductor gentle agitation is essential to dissipate the heat without fragmenting any of the crystals. The rate of crystallization depends on the nature of the fatty oil (partially hydrogenated fats crystallize more readily than palm oil) and on crystallizer design, particularly insofar as this affects the efficiency of the removal of the heat of crystallization. Other components in the oil (triacylglycerol mixture) may affect crystallization. Diacylglycerols, such as those present in palm oil, retard nucleation. Waxes accelerate nucleation but result in unsatisfactory crystal morphologies, leading to more difficult filtration.

Successful and efficient fractionation requires that the production of good-quality crystals (easily separable from the liquid fraction) in the correct proportion be followed by efficient separation of the solid and liquid phases. The latter requires that the minimum amount of olein remains in the stearin fraction. Both stages are now more fully understood and improved equipment has been devised. Good filtration — aiming at complete separation of solid and liquid — is important and may be carried out under reduced pressure using a Florentine filter or under pressure with a membrane filter. Membrane filter presses are usually operated at 4 to 8 bar. Higher pressures (up to 30 bar) may be needed for more critical separations such as the production of cocoa butter replacers.

Fractionation is applied mainly to palm oil but also to lauric oils (coconut and palmkernel), butter oil, beef tallow, other animal fats, hardened soybean, and cottonseed oil. Palm oil is fractionated more than any other oil and plants handling up to 3000 tons/day are in operation. The olein is the more valuable product and Malaysia exports most of its palm oil in the form of palm olein for use as frying or salad oils. Palm stearin serves as hardstock in the production of spreads. Palm mid-fraction is used to produce cocoa butter equivalent fats. A single fractionation converts palm oil (iodine value (IV) 51–53) to palm olein (IV 56–59) and to hard stearin (IV 32–36). Each of these can be fractionated a second or a third time to give a range of products including top-olein (IV 70–72), superolein

(IV 64–66), soft palm mid-fraction (IV 42–48), soft stearin (IV 40–42), hard palm mid-fraction (IV 32–36), and superstearin (IV 17–21). These materials have a wide range of food and nonfood uses and extend considerably the use of palm oil. IV is a measure of average unsaturation.

Palm kernel oil of IV 18 is fractionated to give a stearin of IV ~7 which can be used as a cocoa butter substitute and an olein of IV ~25. These fractions are also useful after complete hydrogenation.

Anhydrous milk fat (AMF) contains many more triacylglycerol species than palm oil so its fractions are less distinct and useful fractionation is more difficult. Nevertheless, the fractions, either alone or after addition back to AMF, find specific uses including production of spreading butter (mixing hard stearin with top olein) and of baking products such as puff pastry (AMF and stearin).

Partially hydrogenated soybean oil (IV 77, melting point (MP) 35°C) can be crystallized to produce a stearin with IV reduced to 61 and MP raised to 45°C and an olein with an IV raised to 81 and MP 31°C. Both of these fractions can be crystallized a second time to give a mid-stearin (IV 73, MP 35°C) and a mid-olein (IV 88, MP 18°C).

A recent directive allows European manufactures to replace part of the cocoa butter in chocolate with other oils with appropriate melting behavior. These alternatives are restricted to six tropical vegetable oils and it is of interest in this connection that three of them (mango, sal, and shea) are used as stearins[20].

Fractionation of animal fats (mainly tallow, but also lard and chicken fat) gives higher melting stearins and lower melting oleins. For the tallows both fractions maintain their full beefy flavor that is considered an advantage for some purposes. The oleins are used for frying and the stearins as biscuit fats[21].

19.4 HYDROGENATION

Hydrogenation (more correctly, partial hydrogenation) differs from fractionation in that unsaturated acids are changed in a number of ways so that hydrogenated oils, in contrast to fractionated oils, have a changed fatty acid composition. Catalytic hydrogenation was the major method of modifying oils throughout the 20th century but recently it has come under suspicion because of the presence of acids with *trans* unsaturation which are considered to have undesirable nutritional properties.

When an oil is submitted to partial hydrogenation it becomes more solid because of an increase in the content of saturated acids and through the presence of *trans* monoene acids with higher melting points than their *cis* isomers. This increase in solid components makes the liquid oil more suitable for use in spreads requiring certain levels of solid fat at refrigerator temperature, at ambient temperature, and at mouth temperature. Only if these are correct will the product spread when taken from the refrigerator, remain as an apparent solid at ambient temperature, and melt completely in the mouth. Changes in melting behavior resulting from the increased level of solids affect spreadability, oral response, and baking performance. As a consequence of its changed composition the hydrogenated oil (and products containing it) is more resistant to oxidation and thus has a longer shelf life. This is linked mainly to a reduction in the content of PUFAs. But these two potential advantages are accompanied by the disadvantage that the product has reduced nutritional value because of increased levels of acids that are saturated or have *trans* unsaturation and reduced levels of PUFAs.

At the molecular level one or more of the following changes may occur:

- Hydrogenation (saturation) of unsaturated centers
- Stereomutation of natural *cis* olefins to their *trans* isomers
- Double bond migration
- Conversion of PUFAs to monounsaturated and saturated acids

These are the consequences of reaction between a liquid (fatty oil) and a gas (hydrogen) occurring at a solid surface (the catalyst).

Dikstra[16] has proposed a modification of the Horiuti–Polanyi mechanism to explain the changes that occur during partial hydrogenation of fatty acids and their esters. In the following sequence the horizontal line shows the conversion of diene (D) to monoene (M) and of monoene to saturated acid/ester (S) via the half hydrogenated states DH and MH. The steps shown vertically are the reverse processes whereby DH goes back to D and MH goes back to M. It is during these reverse steps that *trans* and positional isomers are formed. There are six stages altogether and it is important to understand the relative rates of these:

- In the two-step conversion of D to M, the first step is rate-determining and the second step is fast. The conversion of DH back to D is slow and is only important in the unusual situation that hydrogen is present in very low concentration.
- In the two-step conversion of M to S, the final stage is slow and rate-determining thus making it more likely that there will be considerable recycling of M and MH leading to formation of stereochemical and positional isomers.

$$D \rightarrow DH \rightarrow M \rightarrow MH \rightarrow S$$
$$\downarrow \qquad\qquad \downarrow$$
$$D \qquad\qquad M$$

The catalyst used on a commercial scale is nickel on an inert support at a 17 to 25% level encased in hardened fat. This preserves the activity of the nickel in a form which is easily and safely handled. The reaction is generally conducted at 180 to 200°C and 3 bar pressure in vessels containing up to 30 tons of oil. To minimize the use of catalyst it is desirable to use refined oil and the highest quality of hydrogen. Through improvements in the quality of catalyst and in equipment the requirement for catalyst has been gradually reduced. In 1960, 0.2% nickel was required but by the end of the 20th century this was reduced to between 0.025 and 0.05% (i.e., a 4- to 8-fold reduction). Reaction may proceed in a batchwise manner with up to 8 to 10 batches in a 24-hour day or in a semicontinuous fashion.

The variables that have to be considered are:

- The nature of the oil being treated.
- The extent of hydrogenation which is desired.
- The selectivity to be achieved in terms of the PUFA–MUFA–saturated ratios and the ratio of *cis* to *trans* isomers.
- The quality and quantity of catalyst in terms of pore length, pore diameter, activity, and amount used.
- The reaction conditions of temperature, pressure, and degree of agitation.

The competition between hydrogenation (a change incorporating hydrogen) and isomerization (a change not involving additional hydrogen in the product) is an important factor. The ratio between these depends on the availability of hydrogen at the catalyst surface in relation to the demand. A plentiful supply of hydrogen will promote hydrogenation, and an inadequate supply of hydrogen will allow isomerization to become more significant. Availability of hydrogen is enhanced by increased pressure and increased agitation and demand for hydrogen is increased with higher temperatures, higher catalyst quantity, higher catalyst activity, and a more highly unsaturated oil. The reaction is exothermic and appropriate cooling as well as stirring is required to distribute the heat.

Most of the above comments refer to the normal procedure of partial hydrogenation whereby oils are converted to the plastic fats required to make spreads. An alternative ("brush hydrogenation") is designed to enhance shelf life of oils containing linolenic acid (soybean and rapeseed/canola oil) by reducing slightly the level of this triene acid which is easily oxidized to compounds with very undesirable odors and flavors. Brush hydrogenation requires only a low level of hydrogenation.

Attempts are being made to improve the partial hydrogenation process and, in particular, to reduce the levels of acids with *trans* unsaturation. These include further improvement of catalyst, replacement of nickel by other metals, replacement of hydrogen by compounds that act as hydrogen donors, and reactions in solvents such as propane or carbon dioxide[22–25].

19.5   INTERESTERIFICATION

Interesterification is a catalytic procedure during which acyl groups are redistributed among glycerol hydroxyl groups. With one oil there is no change of fatty acid composition but the rearrangement of acyl groups results in a change of triacylglycerol composition. With a chemical catalyst the change is from the nonrandom distribution of fatty acids in the natural oil or fat towards a completely randomized distribution. With a lipase as catalyst the distribution is controlled according to the specificity of the enzyme. With blends of two or more oils and fats the fatty acid composition will be a weighted mean of the original components and the triacylglycerol composition will be fully or partially randomized. Changes in triacylglycerol composition affect the physical (especially melting behavior), chemical (mainly oxidative stability), and nutritional properties (depending, in particular, on the distribution of fatty acids between the *sn*-1/3 and *sn*-2 positions). Any improvement or deterioration in oxidative stability is partly a consequence of altered fatty acid distribution but perhaps, more importantly, in the changed ratios of pro- and antioxidants following the reaction process.

On an industrial scale, the catalyst is most commonly a base such as sodium methoxide but different products can be obtained with enzymatic catalysts (lipases). A wide range of lipases is now commercially available usually in immobilized form. The specificity of lipases varies in many ways. Some are regiospecific in that they distinguish between the *sn*-1/3 and *sn*-2 positions of glycerol and glycerol esters, some distinguish between mono-, di-, and triacylglycerol esters, some are fatty acid specific depending on chain length or on double bond position and/or configuration, and some combine more than one of these specificities.

19.5.1   INTERESTERIFICATION WITH CHEMICAL CATALYSTS

The production of fat spreads as an alternative to butter led to an increased demand for solid fats. This demand was met first by animal fats and then by the use of partially hydrogenated vegetable oils but concern about the health effects of *trans* unsaturated acids has raised interest in an alternative way of producing fats with the required melting behavior. This can be achieved by interesterification of blends of natural or fractionated fats. Products obtained in this way will probably contain slightly more saturated acids than their partially hydrogenated equivalents but they will have no *trans* acids.

Chemical interesterification is generally effected in 10 to 15 ton batches at 80 to 90°C over 30 to 60 min at a cost not very different from that for partial hydrogenation. It does not require expensive equipment nor the use of explosive gases but the catalyst has to be washed out and there is some loss of product leading to increased cost. To get a product with the desired properties a soft oil is generally interesterified with a hardstock such as a fractionated stearin, a lauric oil, or a fully hydrogenated seed oil. The last is a scientifically acceptable choice but has the disadvantage that the word "hydrogenated" will have to appear on the label. The average customer does not appreciate the difference between partially hydrogenated (with *trans* acids) and fully hydrogenated (without any unsaturated acids)[19].

The catalyst most commonly employed is an alkali metal at a level of 0.1 to 0.2% or a sodium alkoxide such as sodium methoxide at a 0.2 to 0.3% level. The true catalyst may be a diacyl glyceroxide anion produced in the following equilibrium process:

$$Gl(Oacyl)_3 + MeO^- \rightarrow GlO^-(Oacyl)_2 + MeOacyl$$

$(Gl(OH)_3$ represents glycerol)

The catalyst is easily destroyed by acid or water or peroxides so it is desirable that the oil to be interesterified are free of these impurities. Reaction occurs between 25 and 125°C but is usually carried out between 80 and 90°C for 30 min in a batch operation.

Directed interesterification is a modification of the normal process in which the reaction is conducted at a lower temperature (25 to 35°C). Under these conditions the less soluble triacylglycerols crystallize from the solution. This disturbs the equilibrium in the liquid phase that has to be reestablished. The consequence is to increase the formation of SSS and of UUU triacylglycerols.

The level of acids with *trans* unsaturation in margarines and spreads has been reduced considerably in recent years, especially in Europe. Denmark has introduced an upper limit of 2% and has expressed the hope that other European countries will follow this lead. This change has been achieved by avoiding partial hydrogenation and by developing the desired melting behavior and crystalline formulation through interesterification of liquid oils with appropriate hardstock (hydrogenated oil, palm oil, or palm stearin). A U.S. company has launched zero and low *trans* oil blends suitable for virtually any food application[11,26–28].

The following are typical applications of interesterification:

- Lard, with its unusually high level of palmitic acid in the β-position, crystallizes naturally in the β form. When randomized, the content of 2-palmitoglycerol esters is reduced from around 64% to 24% and the interesterified product crystallizes in the β' form with consequent improvement in shortening properties.
- The crystal structures of margarines based on sunflower or canola oil (rapeseed) along with hydrogenated oil are stabilized in the β' form by interesterification leading to randomization of the glycerol esters.
- Solid fats with about 60% of essential fatty acids can be obtained by directed interesterification of sunflower oil and about 5% of hard fat such as palm stearin. The product contains no hydrogenated fat and therefore no *trans* acids.

19.5.2 INTERESTERIFICATION WITH ENZYMIC CATALYSTS

Interesterification can also be catalyzed by lipases, many of which show useful specificities. The 1,3-specific lipases, such as those derived from *Aspergillus niger*, *Mucor javanicus*, *M. miehei*, *Rhizopus arrhizus*, *R. delemar*, and *R. niveus*, are particularly useful for interesterification. They are used to effect acyl exchange at the sn-1 and 3 positions whilst leaving acyl groups at the sn-2 position unchanged. Many interesting changes of this type have been effected on a bench scale but as yet only a few have been commercialized and then only for products of high value[29].

Unilever developed a method for upgrading palm mid-fraction (PMF) as a cocoa butter equivalent. PMF is rich in palmitic acid and has too little stearic acid for this purpose but this deficiency can be repaired by enzyme-catalyzed acidolysis with stearic acid. Reaction is confined to the exchange of palmitic acid by stearic acid at the sn-1 and 3 positions with no movement of oleic acid from the sn-2 position. A similar product is produced enzymatically from high-oleic sunflower oil (rich in triolein) and stearic acid.

$$POP + St \rightarrow POSt + StOSt$$
$$OOO + St \rightarrow StOO + StOSt$$

This result would not be attained by chemical interesterification which would lead to randomization of all the acyl chains and to products having different melting behavior from that required by a cocoa butter equivalent.

Another product named Betapol consisting mainly of triacylglycerols of the type UPU (U = unsaturated acyl chains, P = palmitic acyl group) is used as a constituent of infant formulas. Human milk fat is unusual in that it contains a significant proportion of its palmitic acid in the sn-2 position. This is true also for lard (pig fat) but is not a feature of vegetable oils. Betapol is made from tripalmitin and oleic acid using the lipase from *Mucor miehei* to promote 1,3-acyl exchange. These starting materials are expensive and in practice fractionated palm oil, rich in tripalmitin, is reacted with "oleic acid" from high-oleic sunflower or safflower or from olive oil.

Bohenin (BOB) is the name given to glycerol 1,3-behenate 2-oleate which inhibits fat bloom when added to chocolate. It is produced in Japan by enzymic interesterification of triolein and behenic (22:0) acid or ester in the presence of a 1,3-stereospecific lipase.

Diacylglycerols are being produced and used in growing quantities because of their beneficial effects in the management of obesity. Cooking oil with at least 80% of diacylglycerols (mainly the 1,3-isomer) has been marketed in Japan since 1999 and thereafter in other countries. Fatty acids from natural edible oils are reacted with glycerol or 1-monoacylglycerol in the presence of a 1,3-regiospecific lipase to produce a product that is 80% diacylglycerol with 70% of this being the 1,3-isomer[30].

There are many reports showing how, with an appropriate enzyme (those of *Mucor miehei* and *Candida antarctica* are frequently used), long-chain PUFAs such as EPA and/or DHA can be introduced into vegetable oils or synthetic glycerides to give products with enhanced nutritional value. In a similar way C_8 and C_{10} acyl chains can be introduced into vegetable oils or fish oils with a consequent change in nutritional properties and energy values. The products are triacylglycerols with either one long and two short chains (LS_2) or two long and one short chain (L_2S). This reaction can be used to produce triacylglycerols with easily metabolizable short- and medium-chain acids at the sn-1 and 3 positions and an essential fatty acid at the crucial sn-2 position.

For example, there are several reports of the synthesis of triacylglycerols that can be represented as Cap-PUFA-Cap where Cap is the caprylic (C_8) acyl chain and PUFA represents a PUFA chain including conjugated linoleic acid (CLA), γ-linolenic acid (GLA; n-6 18:3), EPA, or DHA. These are made by enzymatically controlled processes. Most of them are laboratory procedures but there is a desire to develop commercial products.

In one example[31] 1,3-dicapryloyl-2-docosahexaenoyl glycerol (CDC) was made from tricaprylin (99% pure), ethyl docosahexenoate (92% pure), and ethyl caprylate (8:0) in two enzymic steps. Molecular distillation is used to remove unwanted volatile compounds (tricaprylin and ethyl esters). In the first step tricaprylin reacts with ethyl docosahexaenoate in the presence of a nonregiospecific lipase (*Alcaligenes* spp.). The nonvolatile product (a mixture of DCC, CDC, DDC, and DCD glycerol esters) is then reacted with ethyl caprylate and a 1,3-regiospecific lipase (*Candida antarctica*) and after distillation the residue is a concentrate of CDC (63% pure with some CCD and some CD_2 isomers). The volatile components can be recycled. See also references[32–35].

The fatty acid specificity shown by some lipases can be exploited to separate or concentrate particular fatty acids and these can be subsequently incorporated into glycerol esters. Examples include the separation of the 9c11t and 10t12c 18:2 isomers present in the conjugated linoleic acid made by alkali isomerization of oils rich in linoleic acid and the isolation of individual PUFAs from fish oils or other appropriate sources. Many such procedures have been described and the following is but a typical example[36]. Tuna oil, containing EPA (5.2%) and DHA (24.5%), was used to prepare fatty acid fractions with these two acids at concentrations of 6.0 and 71.1% (recovery 28.6 and 89.8%) or of 2.9 and 77.5% (recovery 9.1 and 78.7%). When fish oil fatty acids are esterified with glycerol in the presence of *Rhizomucor miehei*) EPA with its Δ5 double bond and, more particularly, DHA with its Δ4 double bond, do not react easily and remain in the unreacted fatty acids at levels indicated above. Other examples are given in references[37–42].

19.6   DOMESTICATION OF WILD CROPS

The oils and fats required as food, feed, and for industrial purposes are produced almost entirely from a limited number of sources (around 20) even though many other potential sources exist (see Section 19.1 and Chapter 5). Most developments have involved the modification and improvement of existing crops because the agronomical issues for such plants are well understood. However, significant changes in lipid composition raise problems of identity preservation (IP). Plants, seeds, and oils have to be kept separate from their close relatives. One successful example of this is the production of high-erucic rapeseed oil even in areas where low-erucic rapeseed oil (canola) is a major crop.

The domestication of wild crops that contain interesting lipids has proved to take longer than was originally expected. Estimates of 10 years have been changed to 20 years or more. Domestication involves breeding out traits that allow the plant and its seeds to survive in the wild but are not suitable for plants that are to be sown, grown, and harvested under commercial conditions. This requires the development of seeds that will grow in economic yields, produce lipids of uniform composition year after year, and for which agronomic factors relating to yield and composition are understood[43].

There have been many attempts to domesticate wild plants that produce interesting lipids, particularly in Europe and North America. Examples include oils from cuphea, calendula, crambe, meadowfoam, and jojoba to name only a few. Most of these are sources of a less common fatty acid rather than yet another source of the common palmitic–oleic–linoleic complex. These and others have been described in Chapter 5.

19.7   OILS MODIFIED BY CONVENTIONAL SEED BREEDING OR BY GENETIC ENGINEERING

Over many years efforts have been devoted to the improvement of plants already providing oilseeds on a commercial scale. To increase yields, attempts have been made to develop strains more resistant to pests, diseases, weeds, and adverse weather conditions. Strains suitable for autumn sowing and over-wintering are important in areas where winters are not too severe. This leads to higher yields and spreads the harvest period through the summer months.

Changes in fatty acid composition may also be important and the best-known example of this is the development of double zero brassica species, low in both erucic acid and glucosinolates. Double zero varieties have become the standard with high-erucic rapeseed now being a minor product used for industrial purposes. More recently, Canadian investigators have modified *Brassica juncea* (oriental mustard) from an Australian line low in erucic acid and in glucosinolate to give it a fatty acid composition (palmitic 3%, stearic 2%, oleic 64%, linoleic 17%, and linolenic acid 10%) similar to that of canola oil from *B. napus* and *B. rapa*. This makes it possible to extend the canola-growing area of western Canada.

There are many reports of major seed oils with modified fatty acid composition but it is not always clear how many of these are commercially available and which have been obtained by genetic modification. This last is of major concern to some groups. In deciding between modifying an existing crop and domesticating a wild plant a factor to be considered is that the latter, but not the former, promotes crop diversity[44,45].

Moellers[46] has reported several rapeseed oils with modified fatty acid composition including those with changed levels of palmitic, stearic, medium-chain saturated, oleic, α-linolenic, γ-linolenic, erucic, and long-chain polyunsaturated fatty acids. The level of these acids may be raised or lowered compared to the standard oil. All these targets have been achieved but not all of them have been commercially exploited. Important changes in minor components may also occur during seed breeding. These changes may have significant effects on oxidative stability, but the control and manipulation of the minor components is not well understood.

Sunflower seed oils with three different fatty acid compositions have been produced by conventional seed breeding. Regular sunflower oil is rich in linoleic acid but new varieties (high-oleic and mid-oleic, with the latter called NuSun) with higher levels of oleic acid have the composition shown in Table 19.2. The mid-oleic oil was developed in the United States and is now the commodity sunflower oil of that country. The high oleic oil is a valuable commodity and is frequently used as source of "pure" oleic acid in lipid modification and synthesis.

Safflower exists in a wide range of varieties. The conventional oil is linoleic rich (75%) but an oleic-rich variety (74%), traded under the name of saffola, is now available.

Linseed oil is a highly unsaturated oil, rich in linolenic acid (~50%). Its use in paints and linoleum manufacture is based on its oxidative polymerization. In a carefully handled form and renamed flaxseed oil it is now offered as a healthy oil rich in n-3 acids, probably as a pressed rather than solvent extracted oil. Plant breeders in both Australia and Canada have used chemical mutation to develop varieties of linseed which yield an oil rich in linoleic rather than linolenic acid. The Australian seeds yield an oil designated linola while the Canadian oil is called solin.

TABLE 19.2
Fatty Acid Composition (%) of Selected New Varieties Compared with the Relevant Commodity Oil

	16:0 + 18:0	18:1	18:2	18:3	22:1
Soybean					
Regular	15	22	53	8	
High-oleic	12	79	3	6	
Sunflower					
Regular	11–13	20–30	60–70	—	
High-oleic	9–10	80–90	5-9	—	
Mid-oleic	<10	55–75	15–35	—	
Safflower					
Regular	10	14	75	—	
High-oleic	8	74	16		
Rapeseed/canola					
Regular	4[a]	60	20	10	<2
High-erucic	3[a]	15	13	9	58[b]
Low-linolenic	4[a]	60	30	2	<2
High-oleic	3[a]	84	5	4	<2
High-oleic/low linolenic	3[a]	84	7	2	<2
Linseed					
Regular	11	21	17	51	
Linola	11	16	70	2	
Groundnut (peanut)					
Regular[c]	10–18	36–67	15–43		
High oleic[d]	7–9	78–82	3–6		

Note: The names monola (rapeseed), soyola (soybean), saffola (safflower), and sunola (sunflower) have been given to the high-oleic varieties.
[a] Palmitic acid only.
[b] Also 8% of 20:1.
[c] Also 2–3% of 20:0 and 20:1.
[d] Also 3–4% of 20:0 and 20:1.

Palm oil is second only to soybean oil in its level of production. It differs from the common range of vegetable oils in that it contains almost equal amounts of saturated (mainly palmitic) and unsaturated acids (mainly oleic). As a tree crop, plant breeding is slower; nevertheless steady work has been and is undertaken to improve its agronomy, its economic return, and its fatty acid composition. According to Jalani et al. the following objectives are being pursued: (1) shorter plants that are easier to harvest and do not need to be replanted as frequently, (2) oils with less palmitic acid and more oleic acid having a higher iodine value and yielding more olein and less stearin when fractionated, and (3) plants with larger kernels (raised from 6% to around 12%) since palmkernel oil commands a higher price than palm oil[47].

19.8 ANIMAL FATS MODIFIED THROUGH NUTRITIONAL CHANGES

Animal fats are perceived to suffer from several disadvantages. They are not generally acceptable to vegetarians and products from particular animals are not acceptable to some ethnic groups. Also, they contain high levels of saturated acids (including myristic, the saturated acid with greatest enhancing effect on blood cholesterol levels), of monounsaturated acids with *trans* configuration, and of cholesterol, and have only low levels of essential fatty acids. Nevertheless, because meat and meat products are significant sources of fat intake and contain low levels of PUFAs they supply an important share of the dietary intake of these acids. A recent survey, published in 2002, relating to U.K. adults of 19 to 64 years old, in 2000–2001, suggests that meat and meat products provide an average of 18% of the British intake of n-6 acids and 17% of n-3 acids[48].

In the case of the ruminant animals (cattle, sheep, goats) the highly unsaturated dietary fatty acids in grass or animal feed based on oilseed meals are subject to biohydrogenation. Linoleic and linolenic acids are thereby converted into saturated acids and monounsaturated acids in both *cis* and *trans* forms. There are several ways in which the polyunsaturated acids can be protected from this process thereby leading to more unsaturated milk fat and body fat. These include coating the oil or feeding either calcium salts or oils hardened to the extent that they are solid in the rumen but melt in the abomasum. These changes have negative as well as positive consequences. The cooked meat has a different flavor from normal and this is not generally desirable.

In seeking to introduce acids such as CLA and DHA into human diets attempts have been made to incorporate these acids into milk, eggs, or broiler meat by adjustment of the animal diet. Eggs are now sold with enhanced levels of DHA. The recommended intake of DHA is important but small and therefore the level of DHA in eggs does not need to be high. In the study of Smuts et al.[49] pregnant women consumed regular or OmegaTech eggs with DHA levels of 18 and 135 mg/egg. This led to daily DHA intakes of 35.1 ± 13.2 mg and 183.9 ± 71.4 mg, respectively, which was sufficient to increase length of gestation and birth weight. In another study[50] addition of fish oil to the rations of laying hens raised the DHA level from 2 to 6mg/g of yolk; see also references[51–53].

19.9 SINGLE-CELL OILS

This topic has been reviewed by Ratledge[54,55]. The demand for long-chain PUFAs such as EPA and DHA is met mainly from fish oils such as cod liver oil and tuna oil but there is some concern about the presence of undesirable environmental contaminants (particularly organomercury compounds, dioxins, and polychlorinated biphenyls (PCBs)) and about the adequacy of fish supplies to meet a growing demand. Fish oils are not generally a convenient source of EPA or DHA alone and free of the other acid, nor are fish oils good sources of AA. For infant formula it is desirable to include AA and DHA but to exclude EPA. Many of these problems can be overcome using selected microorganisms which provide these acids in useful concentrations (~40%). The products are available commercially but only for high-priced products such as infant formula. Products obtained in this

TABLE 19.3
Fatty Acid Composition of Single-Cell Oils as Commercial Sources of Arachidonic and Docosahexaenoic Acids

Producing organism	14:0	16:0	16:1	18:0	18:1	18:2 n-6	20:4 n-6	22:5 n-3	22:6 n-3
Mortierella alpina[a]	0.4	8	0	11	14	7	49	0	0
Crypthecodinium cohnii[b]	20	18	2	0.4	15	0.6	0	0	39
Schizochrytrium sp.[c]	13	29	12	1	1	2	0	12	25

[a] ARASCO also 18:3 n-6 4%, 20:3 n-6 4%, and 24:0 1%.
[b] DHASCO-C also 12:0 4%.
[c] DHASCO-S also18:3 n-3 3%.
Adapted from Ratledge, C. and Wynn, J.P., *Lipid Technol.*, 16, 2004, in press.

way are free of cholesterol, heavy metals, and pesticides. Single-cell oils have been developed to meet the need for GLA, AA, and DHA.

GLA was the first PUFA to be produced on a commercial scale from *Mucor javanicus* (now called *M. circinelloides*) in the U.K. and from *M. isabellina* in Japan at concentrations of 18 to 20% and 8% in the recovered oil. The U.K. production has now become uneconomic through cheaper supplies of evening primrose and borage seeds (see Chapter 5) and has been discontinued.

AA is produced by strains of *Mortierella alpina* and commercial production is carried out in Japan and China and on a much larger scale in Italy (around 95% of total supplies). This is marketed as ARASCO™, a product with 40% AA. It is also available, with added DHA, as Formulaid™ for incorporation into infant formula. Mutants of *M. alpina* produce other PUFAs such as stearidonic acid (n-3 18:4), Mead's acid (n-9 20:3), and dihomo-γ-linolenic acid (n-6 20:4) and could probably be produced on a commercial scale if there was a demand for these acids.

Several sources of DHA have been identified and three are used to produce oils with useful levels of this acid. *Crypthecodinium cohnii* and a *Schizochytrium* sp. yield products which are marketed as DHASCO™ and as DHASCO-S or DHAGold. A third product from *Ulkenia* is scheduled to appear as DHActive during 2004. Production of all single-cell oils was considered to have exceeded 500 tons in 2003 and this was expected to rise considerably in 2004 and 2005.

REFERENCES

1. Drexler, H., Spiekermann, P., Meyer, A., Domergue, F., Kank, T., Sperling, P., Abbadi, A., and Heintz. E., Metabolic engineering of fatty acids for breeding of new oilseed crops: strategies, problems and first results, *J. Plant Physiol.*, 160, 779–802, 2003.
2. Sayanova, O.V. and Napier, J., Eicosapentaenoic acid: biosynthetic routes and the potential for synthesis in transgenic plants, *Phytochemistry,* 65, 147–158, 2004.
3. Gunstone, F.D. and Padley, F.B., Eds., *Lipid Technologies and Applications*, Marcel Dekker, New York, 1997.
4. Hamm, W. and Hamilton, R.J., Eds., *Edible Oil Processing*, Sheffield Academic Press, Sheffield, U.K., 2000.
5. Gunstone, F.D., Ed., *Structured and Modified Lipids*, Marcel Dekker, New York, 2001.
6. Rossell, B., Ed., *Oils and Fats Volume 2 Animal Carcass Fats*, Food RA Leatherhead Publishing, Leatherhead, U.K., 2001.
7. Rajah, K.K., Ed., *Fats in Food Technology*, Sheffield Academic Press, Sheffield, U.K., 2002.
8. Akoh, C.C. and Min, D.B., Eds., *Food Lipids: Chemistry, Nutrition, and Biotechnology*, 2nd ed., Marcel Dekker, New York, 2002.

9. O'Brien, R.D., *Fats and Oils: Formulating for Processing and Applications*, 2nd ed., CRC Press, Boca Raton, FL, 2004.

10. Block, J.M., Barrera-Arellano, D., Figueiredo, M.F., and Gomide, F.A.C., Blending process optimisation into special fat formulation by neural networks, *J. Am. Oil Chem. Soc.*, 74, 1537–1541, 1997.

11. Anon., ADM launches zero/low *trans*-fat oils, *Inform* 14, 550, 2003.

12. Shukla, V.K.S. and Bhattacharya, K., Nutridan bakes a recipe for health, *Oils Fats Int.*, 19, 26–27, 2003; see *Inform*, 14, 340–341, 2003.

13. Stewart, I. and Kristott, J., European Union Chocolate Directive defines vegetable fats for chocolate, *Lipid Technol.*, 16, 11–14, 2004.

14. Danthine, S. and Deroanne, C., Blending of hydrogenated low-erucic acid rapeseed oil, low-erucic acid rapeseed oil, and hydrogenated palm oil or palm oil in the preparation of shortenings, *J. Am. Oil Chem. Soc.*, 80, 1069–1075, 2003.

15. Hamm, W., Fractionation techniques, guest contribution, <britanniafoodingredients.com> (2000).

16. Dikstra, A.J., Hydrogenation and fractionation, in *Fats in Food Technology,*Rajah, K.K., Ed., Sheffield Academic Press, Sheffield, U.K., 2002, pp. 123–158.

17. Kellens, M., Oil modification processes, in *Edible Oil Processing*, Hamm, W. and Hamilton, R.J., Eds., Sheffield Academic Press, Sheffield, U.K., 2000, pp.129–173.

18. Parmentier, M., Fractionation of fats: a dossier covering dry fractionation, industrial chromatographic separation, fractionation with supercritical carbon dioxide, and fractionation of fats by membranes, *Eur. J. Lipid Sci. Technol.*, 102, 233–248, 2000.

19. Lim, S., Interesterification's future examined, *Oils Fats Int.*, 17, 24–25, 2001.

20. Stewart, I. and Kristott, J., *Lipid Technol.*, 16, 11–14, 2004.

21. Deffense, E., Fractionation of animal carcass fats into hard and soft fats, in *Oils and Fats Volume 2: Animal Carcass Fats*, Rossell, B., Ed., Food RA Leatherhead Publishing, Leatherhead, U.K., 2001, pp. 175–196.

22. King, J.W., Hollida, R.L., List, G.R., and Snyder, J.M., Hydrogenation of vegetable oils with supercritical carbon dioxide and hydrogen, *J. Am. Oil Chem. Soc.*, 78, 107–113, 2001.

23. List. G.R., Neff, W.E., Holliday, R.L., King, J.W., and Holser, R., Hydrogenation of soybean oil triglycerides: effect of pressure on selectivity, *J. Am. Oil Chem. Soc.*, 77, 311–314, 2000.

24. Mondal, K. and Lalvani, S.B., Electrochemical hydrogenation of canola oil using a hydrogen transfer agent, *J. Am. Oil Chem. Soc.*, 80, 1135–1141, 2003.

25. Prabhavathi Devi, B.L.A., Karuna, M.S.L., Narasimha Rao, K., Saiprasad, P.S., and Prasad, R.B.N., Microwave-assisted catalytic transfer hydrogenation of safflower oil, *J. Am Oil Chem. Soc.*, 80, 1003–1005, 2003.

26. Leth, T., Bysted, A., Hansen, K., and Ovesen, L., *Trans* FA content in Danish margarines and shortenings, *J. Am. Oil Chem. Soc.*, 80, 475–478, 2003.

27. Alonso, L., Fraga, M.J., and Juarez, M., Determination of *trans* fatty acid profiles in margarines marketed in Spain, *J. Am. Oil Chem. Soc.*, 77, 131–136, 2000.

28. Kellens, M., Production of low/zero *trans* fats by combined use of hydrogenation-interesterification and fractionation, in *Edible Oil Processing*, Hamm, W. and Hamilton, R.J., Eds., Sheffield Academic Press, Sheffield, U.K., 2000, pp. 169–171.

29. Xu, X., Production of specific-structured triacylglycerols by lipase catalysed reactions: a review, *Eur. J. Lipid Sci. Technol.*, 102, 287–303, 2000.

30. Watanabe, H. and Matsuo, M., Diacylglycerols, in *Lipids for Functional Foods and Nutraceuticals*, Gunstone, F.D., Ed., The Oily Press, Bridgewater, U.K., 2003, pp. 113–148.

31. Negishi, S., Arai, Y., Arimoto, S., Tsuchiye, K., and Takahashi, I., Synthesis of 1,3-dicaprryloyl-2-docosahexaenoyl glycerol by a combination of nonselective and *sn*-1,3-selective lipase reactions, *J. Am. Oil Chem. Soc.*, 80, 971–974, 2003.

32. Kim, I.-H., Yoon, C.-S., Cho, S.-H., Lee, K.-W., Chung, S.-H., and Tae, B.-S., Lipase-catalyzed incorporation of conjugated linoleic acid into tricaprylin, *J. Am. Oil Chem. Soc.*, 78, 547–551, 2001.

33. Kawashima, A., Shimada, Y., Nagao, T., Ohara, A., Matsuhisa, T., Sugihara, A., and Tominaga, Y., Production of structured TAG rich in 1,3-dicaprryloyl-2-γ-linolenoyl glycerol from borage oil, *J. Am. Oil Chem. Soc.*, 79, 871–877, 2002.

34. Irimescu, R., Furihata, K., Hata, K., Iwasaki, Y., and Yamane, T., Utilization of reaction medium-dependent regiospecificity of *Candida antarctica* lipase (Novozym 435) for the synthesis of 1,3-dicaprryloyl-2-docosahexaenoyl (or eicosapentaenoyl) glycerol, *J. Am. Oil Chem. Soc.*, 78, 285–289, 2001.

35. Bornscheuer, U.T., Adamczak, M., and Soumanou, M.M., Lipase-catalysed synthesis of modified lipids, in *Lipids for Functional Foods and Nutraceuticals*, Gunstone, F.D., Ed., The Oily Press, Bridgewater, U.K., 2003, pp. 149–182.

36. Halldorsson, A., Kristinsson, B., Glynn, C., and Haraldsson, G.G., Separation of DHA and EPA in fish oil by lipase-catalysed esterification with glycerol, *J. Am. Oil Chem. Soc.*, 80, 915–921, 2003.

37. McNeill, G.P., Rawlins, C., and Peilow, A.C., Enzymic enrichment of CLA isomers and incorporation into triacylglyceides, *J. Am. Oil Chem. Soc.*, 76, 1265–1268, 1999.

38. Ju Y.-H. and Chen, T.-C., High purity γ-linolenic acid from borage oil fatty acids, *J. Am. Oil Chem. Soc.*, 79, 29–32, 2002.

39. Nagao, T., Shimada, Y., Yamauchi-Sato, Y., Yamamoto, T., Kasai, M., Tsutsumi, K., Sugihara, A., and Tominaga, Y., Fractionation and enrichment of CLA isomers by selective esterification with *Candida rugosa* lipase, *J. Am. Oil Chem. Soc.*, 79, 303–308, 2002.

40. Yamauchi-Sato, Y., Nagao, T., Yamamoto, T., Terai, T., Sugihara, A., and Shimada, Y., Fractionation of CLA isomers by selective hydrolysis with *Candida rugosa* lipase, *J. Oleo Sci.*, 52, 367–374, 2003.

41. Shimada, Y., Use of lipase catalysed reactions to purify fatty acids for nutritional and pharmaceutical applications, *Lipid Technol. Newsletter*, 9, 101–104, 2003.

42. Shimada, Y., Application of lipase reactions to separation and purification of useful materials, *Inform*, 12, 1168–1174, 2001.

43. Gunstone, F.D., Oils modified by conventional seed breeding, in *Structured and Modified Lipids*, Gunstone, F.D., Ed., Marcel Dekker, New York, 2001, pp. 25–28.

44. Gunstone, F.D., Oilseed crops with modified fatty acid composition, *J. Oleo Sci.*, 50, 269–279, 2001.

45. Gunstone, F.D. and Pollard, M.R. Vegetable oils with fatty acid composition changed by plant breeding or by genetic modification, in *Structured and Modified Lipids*, Gunstone, F.D., Ed., Marcel Dekker, New York, 2001, pp. 155–184.

46. Moellers, C., Potential and future prospects for rapeseed oil, in *Rapeseed and Canola Oil: Production, Processing, Properties and Uses*, Gunstone, F.D., Ed., Blackwell, Oxford, 2004, pp. 186–217.

47. Jalani. B.S., Cheah, S.C., Rajanaidu, N., and Darus, A., Improvement of palm oil through breeding and technology, *J. Am. Oil Chem. Soc.*, 74, 1451–1455, 1997.

48. Survey of fat consumption in the UK, www.nutrition.org.uk/ndnsadults and <www.food.gov.uk/science/101717/ndnsdocuments/ndnsv2>

49. Smuts, C.M., Borod, E., Peeples, J.M., and Carlson, S.E., High-DHA eggs; feasibility as a means to enhance circulating DHA in mother and infant, *Lipids*, 38, 407–414, 2003.

50. Grune.T., Kramer, K., Hoppe, P.P., and Siems, W., Enrichment of eggs with n-3-polyunsaturated fatty acids: effects of vitamin E supplementation, *Lipids*, 36, 833–838, 2001.

51. Fernie, C.E., Conjugated linoleic acid, in *Lipids for Functional Foods and Nutraceuticals*, Gunstone, F.D., Ed., The Oily Press, Bridgewater, U.K., 2003, pp. 291–317.

52. Howe, P.R.C., Downing, J.A., Grenyer, B.F.S., Grigonis-Deane, E.M., and Bryden, W.L., Tuna fishmeal as a source of DHA for n-3 PUFA enrichment of pork, chicken and eggs, *Lipids*, 37, 1067–1076, 2002.

53. Shimasaki. H., Yamashita, E., Shimizu, Y., and Ise, S., Dietary effect of DHA-rich eggs on serum lipids in healthy human subjects: a randomised double-blind controlled study, *J. Oleo Sci.*, 52, 109–119, 2003.

54. Ratledge, C., Microorganisms as sources of polyunsaturated fatty acids, in *Structured and Modified Lipids*, Gunstone, F.D., Ed., Marcel Dekker, New York, 2001, pp. 351–399.

55. Ratledge, C., Single cell oils: a coming of age, *Lipid Technol.*, 16, 34–39, 2004.

20 Fat Replacers: Mimetics and Substitutes

Barry G. Swanson
Food Science and Human Nutrition, Washington State University,
Pullman, Washington

CONTENTS

20.1 INTRODUCTION

Fats and oils contribute flavor, palatability, a creamy mouthfeel, and lubricating action to foods[1,2]. Frying in fats or oils transmits heat rapidly and uniformly, evaporates moisture, and provides high temperatures promoting browning and crispness. The desirable functional properties that fats and oils provide in foods are important to consumer expectations, desires, and acceptance[2]. Consumer demand for reduced-fat food products fluctuates in response to public health and media information that relates calorie and fat restrictions to a healthy lifestyle and predictable longevity. Even though the consumer demand, industry marketing, and retail sales of reduced-fat foods are described as inconsistent and fluctuating, consumers still expect reduced-fat foods to exhibit the appearance, texture, and flavor of their full-fat counterparts[3].

20.2 FAT MIMETICS

Consumer expectations for reduced-fat foods resulted in the development of technologies and chemistry that led to the incorporation of fat mimetics that imitate organoleptic or physical properties of triacylglycerols and are labeled as stabilizer or humectant ingredients. Fat mimetics generally absorb a substantial amount of water and provide a subset of the properties and functionalities of fat with reduced caloric contribution to the diet[4,5]. The development of fat mimetics from natural or modified polysaccharide sources such as gums, pectins, starches, or cellulose introduces stability through water binding in foods. The development of fat mimetics from physical modification or microparticulation of proteins provides partial replacement of the creamy mouthfeel of fats and oils in foods. However, fat mimetics derived from polysaccharides or proteins cannot deliver thermal stability or heat transfer during high-temperature cooking or frying because they caramelize or denature at high temperatures. Many fat mimetics, however, are suitable for baking or retorting[4]. Fat mimetics are generally less flavorful than the fats or oils they are intended to replace because fat mimetics carry water-soluble flavors but not the more flavorful and more easily diffusible lipid-soluble flavor compounds. Successful incorporation of lipophilic flavors into reduced-fat foods may therefore require emulsifiers[4]. Fat mimetics are generally required in larger or smaller concentrations than the fats and oils they are replacing, and cannot replace fats on a one-to-one, gram-for-gram basis. Fat mimetic caloric contribution is related to concentration, digestion, and absorption, but is in the range of zero to 4 kcal/g expected from carbohydrates and proteins[3,6].

20.2.1 Fat Mimetics May Be Classified By Their Macronutrient Base[5]

The largest number are carbohydrate-based plant polysaccharides and include cellulose, gums, dextrins, dietary fibers, maltodextrins, and starches. Fat mimetics are added to foods to thicken and add density, bulk, and opacity thereby producing a mouthfeel and glossy appearance similar to the smoothness, creaminess, and appearance of fat-containing foods[7]. Polyols such as sorbitol and maltitol, as well as fructooligosaccharides such as inulin, may also be used to control water activity[4].

20.2.1.1 Polydextrose is a Poorly Digested, Randomly Bonded Polymer of Glucose, Sorbitol, and Citric or Phosphoric Acid

Polydextrose acts as a bulking agent, humectant, and texturizer in many food formulations[4]. Fat mimetics also contribute desirable emulsification and structural properties, as well as undesirable laxative properties derived from large concentrations in formulated foods[4,5].

TABLE 20.1
Selected Applications and Functions of Fat Replacers

Specific application	Fat replacer	General functions[a]
Baked goods	Lipid based	Emulsify, provide cohesiveness, tenderize, carry flavor, replace shortening, prevent staling, prevent starch retrogadation, condition dough
	Carbohydrate based	Retain moisture, retard staling
	Protein based	Texturize
Frying	Lipid based	Texturize, provide flavor and crispiness, conduct heat
Salad dressing	Lipid based	Emulsify, provide mouthfeel, hold flavorants
	Carbohydrate based	Increase viscosity, provide mouthfeel, texturize
	Protein based	Texturize, provide mouthfeel
Frozen desserts	Lipid based	Emulsify, texturize
	Carbohydrate based	Increase viscosity, texturize, thicken
	Protein based	Texturize, stabilize
Margarine, shortening, spreads, butter	Lipid based	Provide spreadability, emulsify, provide flavor and plasticity
	Carbohydrate based	Provide mouthfeel
	Protein based	Texturize
Confectionery	Lipid based	Emulsify, texturize
	Carbohydrate based	Provide mouthfeel, texturize
	Protein based	Provide mouthfeel, texturize
Processed meat products	Lipid based	Emulsify, texturize, provide mouthfeel
	Carbohydrate based	Increase water-holding capacity, texturize, provide mouthfeel
	Protein based	Texturize, provide mouthfeel, water holding
Dairy products	Lipid based	Provide flavor, body, mouthfeel, and texture; stabilize, increase overrun
	Carbohydrate based	Increase viscosity, thicken, aid gelling, stabilize
	Protein based	Stabilize, emulsify
Soups, sauces, gravies	Lipid based	Provide mouthfeel and lubricity
	Carbohydrate based	Thicken, provide mouthfeel, texturize
	Protein based	Texturize
Snack products	Lipid based	Emulsify, provide flavor
	Carbohydrate based	Texturize, aid formulation
	Protein based	Texturize

[a] Functions are in addition to fat replacement.
From Akoh, C.C., *Food Technol.*, 52, 48, 1998.

20.2.1.2 Protein Fat Mimetics Blended with Selected Gums From Gels That Provide Structure and Functionality Similar to the Structure and Mouthfeel of Fats[5]

Low molecular weight proteins may act like fats to alter the smooth texture of cheeses that are normally composed of large molecular weight proteins[5]. Whey and egg protein solutions homogenized at relatively high temperatures microparticulate the protein mixture into large concentrations of 1.0 to 1.5 μm spheres that function as a handful of spherical tapioca beads providing fat-like lubricity, creaminess, and mouthfeel[5,8]. Most protein-based fat replacers cannot be used at high temperatures because the protein denatures, coagulates, and loses its structural functionality. Protein fat mimetics may also carry the antigenic and potential sensitivity associated with particular protein sources.

20.2.1.3 Fat "Barriers" Constitute a Third Type of Fat Replacement System[5]

Inhibition of fat absorption into foods by altering the surface tension, structure, or area can be a singular-constituent or structural approach, or involve combined technologies to reduce the amount of oil absorbed during processing or preservation.

20.3 FAT SUBSTITUTES

Fat substitutes are macromolecules that physically and chemically resemble triacylglycerols, the primary component of natural fats and oils[4]. Fat substitutes theoretically replace the fats or oils in foods on a gram-for-gram, one-to-one basis. The critical factors that influence the ultimate energy value of fats and oils in foods ingested are (1) the heat of combustion, (2) the coefficient of digestibility, and (3) the degree of utilization by the body[9]. The heat of combustion of fats is dependent on the ratio of hydrogen to oxygen in the molecule: the smaller this ratio the smaller the heat

TABLE 20.2
Synthetic Fat Substitutes

Fat substitute	Chemical composition	FDA approval	Potential applications	Ref.
Polydextrose	Modified glucose polymers	GRAS	Bulking agent, humectant, texturizer	4
Caprenin®	Structured medium-chain triacylglycerols	GRAS	Cocoa butter equivalent	19
Salatrim®: Benefat™	Two medium-chain fatty acids and one long-chain fatty acid in a triacylglycerol	GRAS	Cocoa butter equivalent	20
Olestra: Olean®	Sucrose fatty acid polyester	Approved 01/1996 07/2004	Savory snacks Prepackaged ready-to-eat popcorn	26
Sucrose fatty acid esters (SFE)		Approved	Emulsifiers	22
Sorbestrin®	Cyclic sorbitol fatty acid polyesters	No approval	Frying or baking oil, salad dressings	22
Polyglycerol	Polyglycerol fatty acid esters	No approval	Shortenings, margarines, confectionaries	31
Retrofat	Trialkoxytricarballyate (TATC)	No approval	Shortenings, mayonnaise	36
Trialkoxycitrate (TAC)	Ester bonds reversed from triacylglycerol	No approval	Similar to corn oil	10
Trialkoxyglycerol ethers (TGE)	Dialkyl glycerol ethers, glycerol monoester diethers	No approval	Vegetable oils	10
Dialkyl dihexadecylmalonate (DDM)	Fatty alcohol dicarboxylic acid esters of malonic acid and alkylmalonic acid	No approval	Mayonnaise, margarine, frying oil	2
Polysiloxane (PS), phenylmethyl polysiloxane, and phenyldimethyl polysiloxane (PDMS)	Organic derivatives of silica	No approval	Functionally similar to soybean oil	42
Esterified propoxylated glycerol (EPG)	Polyether polyol fatty acids	No approval	Vegetable oils	22
Jojoba oil	Linear esters of C20:1 to C22:1	Toxic?	Edible oils	45
Membrane lipids	Extracts of *Halobacterium halobium*	Expensive?	Fat-like materials	41

of combustion. Short-chain fatty acids exhibit a smaller energy contribution than long-chain fatty acids, e.g., butyric acid, 5.95 cal/g and stearic acid, 9.54 cal/g[9]. The coefficient of digestibility assesses the difference between the quantity of fat ingested and the quantity of fat excreted in the feces. The digestibility of fat is dependent on the fatty acid chain length and saturation, but the relationship is varied relative to emulsification, partial hydrolysis, absorption, and transport of the fatty acids of triacylglycerols to the liver[9]. Naturally occurring fats and oils generally exhibit a coefficient of digestibility of 95% or greater[9]. Thus, there are two principal routes to low-calorie fats: (1) dilution of an existing fat with a noncaloric functionally compatible entity which results in a "fat-like" material having a smaller available energy value per gram than the fat it contains. Unfortunately, dilution frequently leads to a reduction in the perceived quality of the food because the texture and flavor characteristics of the food are changed significantly by the dilution[10]. (2) The synthetic chemical approach creating fat functional and fat-like materials which exhibit a smaller caloric value because of their structural design which reduces or prevents digestion or absorption of the fat substitutes and restricts the contribution of fat-derived calories[9,10]. By development of a compound that is not susceptible to digestion, it may be possible to formulate foods in which the high calorie conventional fat is replaced with a less absorbable "low-calorie" fat-like product[10].

Mieth[11] described efforts to reduce the high-energy properties of fats in foods: (1) replace fats with combinations of water and surface-active lipids or nonlipid additives with smaller energy contributions, including selected proteins or polysaccharides exhibiting emulsifying or gelation properties; (2) utilize compounds such as acetoglycerols and medium-chain triacylglycerols that contribute less energy per gram than many natural fats and oils; (3) replace fats with acaloric compounds with structures that differ significantly from the structures of triacylglycerols such as silicones and paraffins, yet contribute similar functionalities in foods; and (4) replace fats with acaloric compounds having fat-like functionality and structures similar to common lipids with modified ester linkages including such compounds as glycerol ethers, pseudofats, and carbohydrate fatty acid esters[12].

20.3.1 Structured Lipids

Structured lipids are triacylglycerols obtained by the hydrolysis and catalyzed transesterification or interesterification of hydrogenated vegetable oils with triacylglycerols of medium- or short-chain fatty acids[2,13–16].

The medium-chain fatty acids are defined as fatty acids made up of 12 carbon atoms or less, while long-chain fatty acids contain greater than 12 carbons in the hydrocarbon chain[14]. The resulting triacylglycerols contain fatty acid distributions representative of the initial triacylglycerols with the medium-, short-, and long-chain fatty acids randomly distributed on the glycerol backbone[15,17]. Structured triacylglycerols provide the physical and functional properties of medium- and short-chain saturated fatty acids and are efficiently digested and absorbed, contributing approximately one half the calories of edible oils and fats containing long-chain fatty acids[15,17].

Medium-chain triacylglycerols contain predominantly saturated fatty acids of chain length C8:0 (caprylic) to C10:0 (capric), with traces of C6:0 (caproic) and C12:0 (lauric) fatty acids. Triacylglycerols containing medium-chain fatty acids are produced from vegetable oils such as coconut and palm kernel oils through hydrolysis and fractionation to concentrate C8 and C10 fatty acids prior to resterification with glycerol[18–20]. Caprocaprylobehenic triacylglycerol, commonly known as Caprenin®, is manufactured by Procter & Gamble from glycerol by esterification with caprylic, capric, and behenic (C22:0) fatty acids. Caprenin® provides approximately 5 kcal/g because capric and caprylic fatty acids are more readily metabolized than longer chain fatty acids and behenic acid is only partially absorbed. Caprenin® exhibits functional properties similar to cocoa butter and is used as a cocoa butter equivalent (CBE) in soft candies and confectionary coatings. Salatrim®, **S**hort **a**nd **L**ong **A**cyl **T**riacylglycerol **M**olecule, is the generic name for a family of saturated triacylglycerols developed by Nabisco Foods Group (Parsippany, NJ) comprised of a mixture containing at least one short-chain (C2:0, C3:0, or C4:0) fatty acid and at least one long-chain (predominantly C18:0, stearic acid) randomly esterified to glycerol[17]. Salatrim® was given the trade name Benefat™ upon licensure to Cultor Food Science for manufacture and marketing purposes. Although Salatrim® is not suitable for frying, compositions with differing quantities of short- and long-chain fatty acids provide select functional and physical properties with a variety of applications[21]. The objective in developing Salatrim® was to replace cocoa butter in confectionary applications.

20.3.2 BASIC STRATEGY FOR DEVELOPING LOW-CALORIE SUBSTITUTES FOR FATS AND OILS

The basic strategy for developing low-calorie substitutes for fats and oils in foods is to reengineer, redesign, chemically alter, or synthesize conventional fats and oils to retain the conventional functional and physical properties of fats and oils in foods, but contribute few or no calories because of reduction or removal of susceptibility towards digestive hydrolysis and subsequent absorption[10,22].

Possible strategies, rationales, and examples include:

1. Replace the glycerol moiety of conventional triacylglycerols with alternative alcohols such as carbohydrates, sugar alcohols, polyols, or neopentyl alcohol[10,22]. Replacement of the glycerol moiety promotes steric hindrance or protection for the ester bonds from water-soluble enzyme hydrolysis. The presence of numerous branches of fatty acids, and the formation of a hydrophobic cloud or affinity will prevent enzyme encroachment and interfere with hydrolysis by pancreatic lipase. Fat substitutes of this type include sucrose fatty acid esters, sucrose polyesters, other carbohydrates, sugar alcohol and polyol fatty acid polyesters, alkyl glycoside polyesters, and polyglycerol esters[10,22]. Examples of this type of fat substitute utilized in published research include trehalose, raffinose, and stachyose fatty acid polyesters, alkyl glycoside (glucose) fatty acid polyesters, sorbitol fatty acid polyesters (Sorbestrin®)[22], and polyglycerol fatty acid polyesters[31].

2. Replace the long-chain fatty acids with alternative acids to confer steric protection to the ester bonds. Examples are branched carboxylic acid esters of glycerol and structured lipids containing short-, medium-, and select long-chain fatty acids esterified to glycerol[10,22].

3. Reverse the ester linkage in triacyglycerols by replacing the glycerol moiety with a polycarboxylic acid, amino acid, or other polyfunctional acid and esterify with a long-chain alcohol. Examples include trialkoxytricaballylate (TATCA) and trialkoxycitrate (TAC)[10,22].

4. Reduce the ester linkage of the glycerol moiety to an ether linkage. The ether linkage is not a good substrate for lipases which do not hydrolyze ether bonds as rapidly as ester bonds. Examples include diether monoesters of glycerol, triglyceryl ethers, and trialkoxyglyceryl ether (TGE)[10,22].

Another approach for the development of low-calorie fat substitutes may be to explore means to produce, synthesize, or discover a compound with the physical and functional properties of a fat or oil from chemistry unrelated to the triacylglycerol structure[10]. Examples may be:

1. Evaluate existing or develop new nonabsorbable polymeric materials which exhibit physical characteristics similar to edible oils such as phenyl methyl siloxane, silicon oil, or paraffins[10].

2. Develop microcapsules that will replace the discontinuous lipid droplets in emulsified foods[10].

3. Introduce oxypropylene groups between the glycerol and fatty acids to form propoxy-lated compounds such as esterified propoxylated glycerol (EPG) to restrict lipase hydrolysis[22].

4. Assess the potential of naturally occurring fats or oils that exhibit constrained hydroly-sis, absorption, or caloric contribution because of their unusual fatty acid components or structure[10,22]. Jojoba oil is an excellent example.

20.3.2.1 Sucrose Polyesters

Sucrose polyesters (SPE) are synthesized by transesterification or interesterification of sucrose with six to eight selected fatty acids, producing a lipid with physical and organoleptic properties equivalent to the physical and organoleptic properties of triacylglycerols[4,23].

The idea behind this synthesis was that since sucrose is hydrophilic, a lipid tail on sucrose would result in a molecule that is amphiphilic and able to serve as a surfactant in detergents. Adding this idea to the concept of finding a means of reducing fat-derived calories without resorting to dilution with water, air, carbohydrates, or proteins, and the objective was to somehow synthesize a fat-like molecule that would significantly reduce fat calories by preventing fat-like hydrolysis and absorption in the intestines. The initial recognition and synthesis of a molecule that fitted this nondigestible and nonabsorbable model was by Mattson and Volpenheim[24] while researching the absorption of fat-like compounds by infants[22]. The model compounds were a mixture of hexa-, hepta-, and octaesters of sucrose with long-chain fatty acids isolated from edible fats and oils[25,26], now recognized as olestra or Olean® as described in the patent granted to Proctor and Gamble in 1971.

The synthesis of SPE involves hydrolyzing and methylating fatty acids to form fatty acid methyl esters. The fatty acid methyl esters are added to sucrose for transesterification or to sucrose octaac-etate for interesterification using alkali metal or alkali soap catalysis under anhydrous conditions and a high vacuum[4,22]. The resulting sucrose fatty acid polyesters are purified by washing, bleaching, and deodorizing to remove free fatty acids and volatile organic constituents, followed by distillation to remove unreacted fatty acid methyl esters and sucrose fatty acid esters with small numbers of esterified fatty acids. Olestra is defined by specifications that include the fatty acid composition and degree of esterification[4,27]. The functionality, physical properties, and potential applications of the synthesized sucrose fatty acid polyesters are dependent on the chain lengths and saturation of selected fatty acids used in the syntheses[4,22,24].

Olestra is the only sucrose fatty acid polyester approved for replacing up to 100% of the fats in savory snacks and for frying of savory snacks[4,27,28]. Olestra is not hydrolyzed or absorbed in the intestines because of steric hindrance of digestive lipases by the impeding concentration of hydrophobic fatty acid constituents that make up olestra[4]. Because olestra exhibits the potential to produce undesirable physiological and nutritional responses following ingestion of large quantities of the indigestible fat substitute, many studies were conducted prior to U.S. Food and Drug Administration (FDA) approval. The conclusion were that (1) olestra is not toxic, carcinogenic, genotoxic, or teratogenic; (2) all safety issues were addressed; and (3) there is reasonable certainty that no harm will result from the use of olestra in savory snacks[27].

20.3.2.2 Other Carbohydrate Fatty Acid Polyesters and Polyol Fatty Acid Polyesters also Exhibit Potential as Fat Substitutes

The development and syntheses of carbohydrate- and alkyl glycoside-based fatty acid polyesters as fat substitutes are described in detail by Akoh and Swanson[29]. Digestion and absorption of the carbohydrate and polyol fatty acid polyesters are reduced by saturating a selected carbohydrate, sugar alcohol, or alkyl glycoside with selected fatty acids esterified to available hydroxyl groups to provide desirable physical and functional properties. Carbohydrate or polyol fatty acid polyesters are generally synthesized by one of four methods: (1) transesterification of the saccharides or sugar alcohols with methyl, ethyl, or glycerol fatty acid esters; (2) acylation with fatty acid anhydrides; (3) acylation with fatty acid chlorides; or (4) acylation with fatty acids *per se*[2,30]. Direct esterification of reducing sugars such as glucose or galactose results in extensive sugar degradation and charring. Therefore, glycosylation or alkylation is necessary to convert reducing sugars to nonreducing less reactive anomeric sugars[29]. Synthesis and potential utilization of sorbitol, glucose, trehalose, raffinose, and stachyose fatty acid polyesters as fat substitutes are described by Akoh[22].

Increasing the molecular weight and polymer length of polyglycerol increases the hydrophilicity of the polymer. However, increasing the chain lengths of the fatty acids esterified to the polyglycerol decreases hydrophilicity, digestion, and absorption of the polyglycerol esters. Babayan[31] described the synthesis and purification of polyglycerol and polyglycerol esters for use in shortenings, margarines, bakery products, frozen desserts, and confectionaries.

20.3.2.3   Sucrose Fatty Acid Esters

Sucrose fatty acid esters (SFE) with a degree of substitution of 1 to 3 fatty acid esters per molecule are hydrophilic, digestible, and absorbable. Because of the amphiphilic character, SFE provide solubilization, wetting, dispersion, emulsifying, and stabilization functionality in foods[22], and are suggested as potential antimicrobials[32,33] or protective coatings for fruit[34].

20.3.3   PARTIALLY AND POORLY DIGESTED ORGANIC COMPOUNDS AS POTENTIAL FAT SUBSTITUTES

Many partially and poorly digested organic compounds were investigated as potential fat substitutes[2,4,22,35]; however none are approved for use in foods to date.

Esterification of pentaerythritol and other polyhydric alcohols with selected fatty acids produces noncaloric, nondigestible, heat-resistant organic compounds that retain the functional attributes of fats or oils. Alcohols with the neopentyl nucleus ($-(CH_2)_4C$), pentaerythritol, trimethylolethane, trimethylolpropane, trimethylolbutane, or neopentyl alcohol can be esterified with select fatty acids to produce potentially acceptable fat substitutes[2,35].

20.3.3.1   Trialkoxytricarballyate

Trialkoxytricarballyate (TATCA), described by Dziezak[36] as retrofat, is similar in structure to triacylglycerols with polycarboxylic acids with two to four carboxylic acid groups such as tricarballic acid replacing glycerol and saturated or unsaturated alcohols replacing the fatty acids[2,6,37].

Hamm[10] describes the synthesis of TATCA as well as discussing several potential food uses. TATCA is not digestible and is therefore noncaloric, but when fed at dietary concentrations exceeding 9% resulted in anal leakage, depression, weakness, and fatalities. The fatalities were attributed to starvation or the laxative effects rather than to toxicity[10,12].

20.3.3.2   Trialkoxycitrate

Trialkoxycitrate (TAC) is not susceptible to lipase hydrolysis and is noncaloric because the ester bonds are reversed from the corresponding esters in triacylglycerols[10,38].

TAC exhibits polymorphic behavior during melting. TAC exhibits viscosity and surface tension similar to corn oil, but thermal decomposition may prevent TAC from being used as a frying oil[2,6,10].

20.3.3.3 Dialkyl dihexadecylmalonate

Dialkyl dihexadecylmalonate (DDM) is a mixture of hexadecyl dioleylmalonate and dihexadecyl dioleylmalonate fatty acid esters of malonic and alkylmalonic acids which are nondigestible and noncaloric[2,36,39,40].

Small molecular weight DDM is synthesized from malonyl dihydride and fatty alcohols, with larger molecular weight DDM requiring the addition of an alkyl halide in an alkaline solvent[35]. DDM exhibits thermal stability, is noncaloric, and its absorption is negligible[2,39].

20.3.3.4 Trialkoxyglycerol Ethers, Dialkyl Glycerol Ethers, and Glycerol Monoester Diethers

Trialkoxyglycerol ethers (TGE), dialkyl glycerol ethers, and glycerol monoester diethers[2,41] exhibit viscosities and surface tensions similar to vegetable oils at room temperature. However, commercial synthesis and production of TGE and other ethers is difficult and time consuming. Although synthesized ethers exhibit functional properties similar to vegetable oils, the difficult and time-consuming syntheses will make their commercial production difficult[10,12,35].

20.3.3.5 Polysiloxane, Phenylmethylpolysiloxane, and Phenyldimethylpolysiloxane

Polysiloxane (PS), phenylmethylpolysiloxane (a substituted polysiloxane), and phenyl-dimethylpolysiloxane (PDMS) are organic derivatives of silica (SiO_2) with a linear polymer structure consisting of the generic formula $[-R_2SiO]$, where R is an organic radical such as a methyl, phenyl, or other aliphatic or aromatic hydrocarbon[2,22,42]. Polysiloxanes are chemically inert, nonabsorbable, nontoxic, and oil-soluble, exhibiting lipid-like character in organic solvents and functionality similar to soybean oil. The polysiloxanes possess physical, functional, and organoleptic properties of fats and oils that are inherent to unique silicon chemistry and totally unrelated to the triacylglycerol structure[2]. PDMS exhibits oxidative and thermal stability, minimal changes in viscosity over a broad temperature range, water-repellent ability, and biological inertness[2]. Polysiloxane is a safe and effective calorie diluent in foods fed to experimental rats leading to satiety and weight reduction. The rats did not compensate for the caloric dilution by increasing their food intake[42].

20.3.3.6 Esterified Propoxylated Glycerols

Esterified propoxylated glycerols (EPG) are synthesized from glycerine and propylene oxide to form a polyether polyol subsequently esterified with fatty acids to yield an oil-like product[2,22,37,40,43].

The structure of EPG is similar to triacylglycerols except that an oxypropylene is inserted between the glycerol and the fatty acids. Although many polyols are acceptable, glycerol as a triol is preferred. The physical properties of EPG are dependent on the chain length and saturation of the fatty acids esterified onto the polyether polyol. Fatty acids with a chain length of C14 to C18 are preferred to produce EPG that resemble soybean, olive, cottonseed, and corn oils[43]. EPG are partially digestible[43], low to noncaloric, and heat stable[2,36].

20.3.3.7 Jojoba Oil

Jojoba oil is a naturally extracted plant oil composed of a mixture of poorly digestible linear esters of C20:1 to C22:1 monounsaturated fatty acids and fatty alcohols contributing less than 4 kcal/g[2,10,22].

The viscosity and interfacial tension of jojoba oil are similar to other edible oils, but the poor digestibility and absorption leads to potential problems with anal leakage, diarrhea, reduced growth, nutrient absorption, and death with consumption exceeding 16% of the diet[44,45]. Jojoba oil and its derivatives are used in a variety of nonfood cosmetic and pharmaceutical applications[2,10]. The potential for jojoba oil may be limited due to its expense and unavailability[35,45].

20.3.3.8   Membrane Lipids from Microorganisms

Membrane lipids from microorganisms such as *Halobacterium halobium* may also provide potential nondigestible fat-like materials that contribute few if any calories. The chemical composition of
these natural membrane lipids is complex and extraction, fractionation, purification, and yield may
constrain utilization in foods[41].

20.4   OTHER FAT REPLACERS

There are many more common traditional methods for restricting and replacing fats in foods
such as (1) substituting water or air for fat, (2) selecting lean meats or skim milk, and (3) broiling,
baking, or grilling foods instead of frying[4,46]. Fat may also be replaced by reformulating foods
with lipid, protein, or carbohydrate ingredients individually or in combination. Monoacylglycerol
emulsifiers and diacylglycerols stabilize emulsions and foams, provide mouthfeel and lubricity, as
well as carrying lipid-soluble flavors while replacing fat in emulsions and aerated foods. Recently
introduced 1,3-diacylglycerols (Enova®) are metabolized more readily than triacylglycerols, while
providing volume and functional properties equivalent to triacylglycerols in foods. Other emulsifiers
such as sodium stearol 2-lactylate, phosphospholipids (e.g., lecithin), polysorbates, and caseinates
stabilize fat and oil during baking of pastries, bread, and biscuits, and also provide stability to full- and
reduced-fat dairy products and frozen desserts. Fat replacers represent a variety of chemical structures
and constituents with diverse physical, functional, and sensory properties and physiological consequences[4]. The Institute of Food Technologists Scientific Status Summary[4] provides a convenient and
readable guide to the development, composition, source, application, and physiological consequences
of fat replacers[47]. It is clear the "ideal fat replacer" will recreate all the attributes of fat, while also
significantly reducing fat and calorie content. Unfortunately, the "ideal fat replacer" does not exist.
Consumers, however, can benefit from a variety of ingredients used as fat replacers to capitalize on
the unique qualities of each in the most appropriate food formulations[47,48].

   Future developments in fat replacement will rely on successful combined systems approaches
to achieve acceptable functionality and sensory characteristics in reduced-fat foods. Fat replacers,
dietary fibers, proteins, monoacylglycerol, diacylglycerol, and phospholipid emulsifiers, diacylglycerols, structured medium- and short-chain fatty acid triacyglycerols, and indigestible fat substitutes such as the chemically and functionally unique and FDA-approved fat substitute olestra will
provide the functionality, variety, and improved caloric balance in reduced-fat foods so desirable to
nutritional well being[47].

   "Nutritionists recognize that focusing on fat alone won't achieve better overall nutrition, or even
improved fat intake"[49]. The ultimate goal of nutritionists is to encourage and educate consumers to:
(1) find more convenient and easy ways to achieve a balanced diet and more variety in their food
choices; (2) select and consume additional servings of fruits and vegetables (five to seven or
even nine servings) that will improve variety and may replace consumption of dietary fats; and
(3) participate in regular physical activity[47]. It is recognized that the girth of Americans is growing
even though consumption surveys suggest that fat and energy ingestion in the United States is
decreasing, and that we sometimes prefer to ignore the evidence that physical activity and energy
expenditure are declining even more remarkably. Regular exercise is a high priority to the maintenance of a healthy lifestyle[47].

   Limiting fats does not excuse the need for the basic nutrients: water, calories, proteins, vitamins,
and minerals. Fat replacement will not replace the need for dietary moderation and balanced nutrition, or to improve the amount and regularity of physical activity among Americans. Fat replacers,
however, are a logical technological step in the direction of improving nutritional health and well-
being[50]. Fat replacer technology and utilization are improving, providing palatable alternative low-
and reduced-fat, as well as fat-free, foods making compliance with reduced-calorie diets easier and
more enjoyable[47].

REFERENCES

1. Swanson, B.G. and Akoh, C.C., *Carbohydrate Polyesters as Fat Substitutes*, Marcel Dekker, New York, 1994.
2. Swanson, B.G., Low calorie fats and synthetic fat substitutes, in *Handbook of Fat Replacers*, Roller, S. and Jones, S.A., Eds., CRC Press, Boca Raton, FL, 1996, p. 265.
3. Drake, M.A., Identification and Characterization Methods to Improve the Sensory Quality of Reduced Fat Cheeses, Ph.D. dissertation, Washington State University, 1996.
4. Akoh, C.C., Fat replacers, *Food Technol.*, 52, 47, 1998.
5. Mattes, R.D., Fat replacers: ADA position, *J. Am. Diet. Assoc.*, 98, 463, 1998.
6. Roller, S. and Jones, S.A., *Handbook of Fat Replacers*, CRC Press, New York, 1996.
7. Giese, J., Fats, oils, and fat replacers, *Food Technol.*, 50, 78, 1996.
8. Thayer, A.M., Food additives, *Chem. Eng. News*, 70, 26, 1992.
9. Merten, H.J., Low calorie lipids, *J. Agric. Food Chem.*, 18, 1002, 1970.
10. Hamm, D.J., Preparation and evaluation of trialoxytricarballyate, trialkoxycitrate, trialkoxyglycerlether, jojoba oil and sucrose polyester as low calorie replacements of edible fats and oils, *J. Food Sci.*, 49, 419, 1984.
11. Mieth, G., Alkalorische verbindugen mit fettahnlichen funktionelbn eigenschaften (pseudofette), *Die Nahrung*, 27, 853, 1983.
12. LaBarge, R.G., The search for a low-caloric oil, *Food Technol.*, 42, 84, 1988.
13. Mascioli, E.A., Babayan, B.K., Bistrian, B.R., and Blackburn, G.L., Novel triglycerides for special medical purposes, *J. Parenter. Enteral Nutr.*, 12, 1285, 1988.
14. Matthews, D.M. and Kennedy, J.P., Structured lipids, *Food Technol.*, 44, 127, 1990.
15. Klemann, L.P., Aji, K., Chrysam, M.M., D'Amelia, M.D., Henderson, J.M., Huang, A.S., Otterburn, M.S., and Yarger, R.G., Random nature of triacylglycerols produced by the catalyzed interesterification of short and long chain fatty acid triglycerides, *J. Agric. Food Chem.*, 42, 442, 1994.
16. Akoh, C.C., Structured lipids: enzymatic approach, *Inform*, 6, 1055, 1995.
17. Smith, R.E., Finley, J.W., and Leveille, G.A., Overview of Salatrim, a family of low calorie fats, *J. Agric. Food Chem.*, 42, 461, 1994.
18. Babayan, V.K., Specialty lipids and their biofunctionality, *Lipids*, 22, 417, 1987.
19. Bach, A.C., Ingenbleek, Y., and Frey, A., The usefulness of dietary medium chain triglycerides in body weight control: fact or fancy?, *J. Lipid Res.*, 37, 708, 1996.
20. Megremis, C.J., Medium chain triacylglycerols: a nonconventional fat, *Food Technol.*, 45, 108, 1991.
21. Kosmark, R., Salatrim: properties and applications, *Food Technol.*, 50, 98, 1996.
22. Akoh, C.C., Lipid-based synthetic fat substitutes, in *Food Lipids*, Akoh, C.C. and Min, D.B., Eds., Marcel Dekker, New York, 1998, p. 559.
23. Jandacek, R.J., Studies with sucrose polyester, *Int. J. Obesity*, 8 (Suppl. 1), 13, 1984.
24. Mattson, F.H. and Volpenheim, R.A., Rate and extent of absorption of the fatty acids of fully esterified glycerol, erythritol, xylitol and sucrose as measured in the thoracic duct of cannulated rats, *J. Nutr.*, 102, 1177, 1972.
25. Rizzi, G.P. and Taylor, H.M., A solvent free synthesis of sucrose polyester, *J. Am. Oil Chem. Soc.*, 55, 398, 1978.
26. Peters, J.C., Lawson, K.D., Middleton, S.J., and Triebwasser, K.C., Assessment of the nutritional effects of olestra, a nonabsorbed fat replacement: introduction and overview, *J. Nutr.*, 127, 1539S, 1997.
27. FDA, Food additives permitted for direct addition to food for human consumption: Olestra. Final Rule. Food and Drug Administration, U.S. Dept. Health and Human Services, *Fed. Reg.*, 61, 3118, 1996.
28. Lawson, K.D., Middleton, S.J., and Hassall, C.D., Olestra, a nonabsorbed, noncaloric replacement for dietary fat: a review, *Drug Metab. Rev.*, 29, 651, 1997.
29. Akoh, C.C. and Swanson, B.G., *Carbohydrate Polyesters as Fat Substitutes*, Marcel Dekker, New York, 1994.
30. McCoy, S.A., Madison, B.L., Self, P.M., and Weisgerber, D.J., Sucrose Polyesters Which Behave Like Cocoa Butters, U.S. Patent 4, 822, 875, 1989.
31. Babayan, V.K., Polyglycerol esters: unique additives for the bakery industry, *Cereal Foods World*, 27, 510, 1982.
32. Marshall, D.L. and Bullerman, L.B., Antimicrobial properties of sucrose fatty acid esters, in *Carbohydrate Polyesters as Fat Substitutes,* Akoh, C.C. and Swanson, B.G., Eds., Marcel Dekker, New York, 1994, p. 149.

33. Yang, C.-M., Luedecke, L.O., Swanson, B.G., and Davidson, P.M., Inhibition of microorganisms in salad dressing by sucrose and methylglucose fatty acid monoesters, *J. Food Proc. Pres.*, 27, 285, 2003.

34. Drake, S.R., Fellman, J.K., and Nelson, J.W., Postharvest use of sucrose polyesters for extending the shelf life of stored "Golden Delicious" apples, *J. Food Sci.*, 52, 1283, 1987.

35. Artz, W.E. and Hansen, S.L., Other fat substitutes, in *Carbohydrate Polyesters as Fat Substitutes*, Akoh, C.C. and Swanson, B.G., Eds., Marcel Dekker, New York, 1994, p. 197.

36. Dziezak, J.D., Fats, oils and fat substitutes, *Food Technol.*, 43, 66, 1989.

37. Schlicker, S.A. and Regan, C., Innovations in reduced calorie foods: a review of fat and sugar replacement technologies, *Topics Clin. Nutr.*, 6, 50, 1990.

38. Hamm, D.J., Low Calorie Edible Oil Substitutes, U.S. Patent 4, 508, 746, 1985.

39. Fulcher, J., Synthetic Cooking Oils Containing Dicarboxylic Acid Esters, U.S. Patent 4, 582, 927, 1986.

40. Gillis, A., Fat substitutes create new ideas, *J. Am. Oil Chem. Soc.*, 65, 1708, 1988.

41. Mela, D.J., Nutritional implications of fat substitutes, *J. Am. Diet. Assoc.*, 92, 472, 1992.

42. Bracco, E.F., Baba, N., and Hashim, S.A., Polysiloxane: potential non-caloric fat substitute; effects on body composition of obese Zucker rats, *Am. J. Clin. Nutr.*, 46, 784, 1987.

43. White, J.F. and Pollard, M.R., Non-digestible Fat Substitutes of Low Caloric Value, U.S. Patent 4, 861, 613, European Patent 325,010, 1989.

44. Decombaz, J., Ananthraman, K., and Hesie, C., Nutritional investigations on jojoba oil, *J. Am. Oil Chem. Soc.*, 61, 702, 1984.

45. Ranhotra, G.S. and Gelroth, J.A., Nutritional considerations of jojoba oil, *Cereal Foods World*, 34, 876, 1989.

46. Calorie Control Council, *Fat Replacers: Food Ingredients for Healthy Eating*, Calorie Control Council, Atlanta, GA, 1992.

47. Swanson, B.G., Fat replacers: part of a bigger picture, *Food Technol.*, 52, 16, 1998.

48. Nabors, L.O., Fat replacers: opinions for controlling fat and calories in the diet, *Food Nutr. News*, 64, 5, 1992.

49. Fat facts and fiction, *Food Insight*, Sept.–Oct., 2–3, 1997; International Food Information Council, Washington, DC.

50. Hassel, C.A., Nutritional implications of fat substitutes, *Cereal Foods World*, 38, 142, 1993.

21 Application of Functional Lipids in Foods

Charlotte Jacobsen, Maike Timm-Heinrich,
and Nina Skall Nielsen
Department of Seafood Research, Danish Institute for Fisheries Research,
Lyngby, Denmark

CONTENTS

21.1   INTRODUCTION

Fats and oils are major nutritional compounds and the main source of energy for humans. Dietary lipids may be divided into three major groups: saturated, monounsaturated, and polyunsaturated fatty acids. The polyunsaturated fatty acids can be subdivided into n-6 (omega-6) and n-3 (omega-3) fatty acids, which have different functions in the human body. Vegetable oils such as soybean, corn, sunflower, and safflower oils are rich in n-6 fatty acids, whereas fatty fish and algal oil as well as soybean oil and canola oil are good sources of n-3 fatty acids. The marine n-3 long-chain polyunsaturated fatty acids (n-3 LC PUFA) for the last three decades have received ever increasing attention because of their association with a lower risk of atherosclerosis[1–3] and for a number of other beneficial health effects. The beneficial effects of n-3 LC PUFA have been ascribed to eicosapentaenoic acid (EPA, C20:5 n-3) and docosahexaenoic acid (DHA, C22:6 n-3) (Figure 21.1).

Recent research has indicated that not only the fatty acid composition, but also the molecular structure of fats influences their function in the human body. Therefore, there has also been an increased interest in using so-called specific structured lipids (SSL) for nutritional applications. SSL can be defined as lipids restructured or modified to change the fatty acid composition and/or their positional distribution in the glycerol molecule. As an example specific fatty acids such as LC PUFA or conjugated linoleic acid (CLA) may be incorporated into the lipids. SSL for human consumption are mostly triacylglycerols (TAG) or diacylglycerols (DAG). SSL may for example be produced to have medium-chain fatty acids (M) in the *sn*-1,3 positions and long-chain fatty acids (L) in the *sn*-2 position, that is an MLM structure (Figure 21.2). Structured lipids may also be produced in a nonspecific manner, where the fatty acids are randomly distributed on the glycerol backbone. These types of lipids are termed randomized lipids (RL).

This chapter summarizes the health benefits associated with both n-3 LC PUFA and SSL, termed "functional lipids," due to their specific functions in the human body. Moreover, possibilities of using functional lipids in foods and problems associated therewith, with particular emphasis on aspects related to lipid oxidation are discussed.

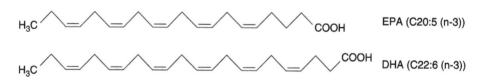

FIGURE 21.1 Molecular structure of EPA and DHA.

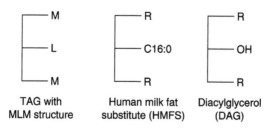

FIGURE 21.2 Examples of specific structured lipids: TAG with MLM structure, human milk fat substitute and DAG. M, medium-chain fatty acid; L, long-chain fatty acid; R, any fatty acid.

21.2 HEALTH ASPECTS OF n-3 LC PUFA AND STRUCTURED LIPIDS

21.2.1 n-3 LC PUFA

The first evidence regarding the beneficial effects of n-3 LC PUFA came in the 1970s when Bang et al.[1] discovered that Greenland Inuits had a much lower incidence of cardiovascular diseases than the Danes, who consumed much lower amounts of fish and marine animals than the Inuits. These findings led to investigations on the role of n-3 LC PUFA in human nutrition.

21.2.1.1 n-3 LC PUFA and Cardiovascular Health

Intervention trials directly support the improvement of cardiovascular health associated with intake of n-3 LC PUFA. Several of these trials have demonstrated a beneficial effect of n-3 LC PUFA intake on fatal and nonfatal myocardial infarction[4,5].

Results from epidemiological trials have demonstrated that intake of fish resulted in a lower mortality rate from coronary heart disease probably due to the high content of EPA and DHA in fish[6,7]. A large study involving 36 countries showed an association between fish consumption and reduced risk from ischemic heart disease and stroke[8]. Other studies have found a reduced risk of sudden cardiac death in men[9], a reduced risk of coronary heart disease in women[10], a significant reduction in the incidence of certain types of stroke caused by blood clots[11], and a reduced risk of primary cardiac arrest[12] following consumption of fatty fish rich in n-3 LC PUFA. However, some epidemiological studies have not shown any beneficial effect of fish intake. These conflicting results may, for example, be due to different populations examined (high risk versus low risk populations) or to the type of fish consumed (lean versus fatty fish).

Results from clinical studies have indicated that n-3 LC PUFA positively affect coronary disease risk factors by decreasing plasma TAG levels[13], improving platelet function[14,15], lowering blood viscosity[16,17], and modulating inflammatory processes[18], as reviewed by Simopoulos[19]. Moreover, n-3 LC PUFA reduce cardiovascular disease risk by decreasing the risk of arrhythmia and the growth rate of atherosclerotic plaque and by lowering blood pressure. Some of the different effects of the n-3 PUFA on cardiovascular disease risk factors are explained below.

Ambiguous effects of n-3 LC PUFA intake on plaque formation due to changes in the oxidative resistance of lipid-rich particles[20,21] and on cell adhesion molecules[22,23] have been observed. However, positive effects of n-3 LC PUFA on inhibition of platelet aggregation have been found[24,25]. This inhibition is due to the production of biologically less active components from EPA compared with the components produced from n-6 LC PUFA[26,27], while DHA reduces the binding of ligands to the thromboxane/prostaglandin receptor[28].

Prevention of arrhythmia is important because it may result in sudden cardiac death. Beneficial effects of n-3 LC PUFA on arrhythmia have been established in humans[29,30], experimental animals[31,32], and cultured cells[33]. The cardioprotective effect may be ascribed to direct effect of n-3 LC PUFA on myocytes cell membrane structure and calcium channels. The n-3 LC PUFA may also prevent arrythmia via their role in eicosanoid metabolism and in cell signaling as well as via their effect on various enzymes and receptors.

21.2.1.2 Other Effects of n-3 LC PUFA Consumption

Besides the effects of n-3 LC PUFA in cardiovascular diseases a positive role on immune functions has been demonstrated. For example, patients suffering from rheumatoid arthritis were observed to have a reduced incidence of joint tenderness and morning stiffness during supplementation with fish oil[34]. Intake of n-3 LC PUFA may also confer some protection against immune-compromised conditions including asthma[35,36], cystic fibrosis[37], Crohn's disease[38], and cancer[39].

DHA is known to promote the function of the retina. Thus, consumption of fish is associated with a decreased risk of development of age-related macular degeneration in adults and improved cognitive function and enhanced information processing in infants[40,41].

21.2.2   Specific Structured Lipids (SSL)

In the intestine, TAG are hydrolyzed in the *sn*-1 and *sn*-3 positions to free fatty acids and 2-monoacylglycerols (MAG) by pancreatic lipase. Therefore, the position of the fatty acids in the TAG has a great impact on how the fatty acids are metabolized. Moreover, medium-chain fatty acids are hydrolyzed faster than long-chain fatty acids[42]. Due to their small size, released medium-chain free fatty acids are oxidized directly to provide energy. The TAG is resynthesized from MAG and free fatty acids and transferred to the peripheral tissue. Thus, the fatty acid in the 2-position of the absorbed fat is mainly stored in the body. SSL with medium-chain fatty acids in the *sn*-1 and *sn*-3 positions and long-chain fatty acids in the *sn*-2 position may thus give the benefits of providing fast energy by the medium-chain fatty acid and the specific nutritional effects of the long-chain fatty acid, for example in patients suffering from malabsorption[42,43]. SSL containing medium-chain fatty acids and PUFA provide significant metabolic benefits in malabsorbing animals, such as a protein-sparing effect and improved nitrogen balance[44].

21.2.2.1   Human Milk Fat Substitute

The structure of human milk TAG is unique as 60 to 70% of palmitic acid is located at the *sn*-2 position and stearic, oleic, and linoleic acids are preferentially esterified in the *sn*-1 and *sn*-3 positions[45,46]. Due to the above-mentioned specificity of pancreatic lipase, palmitic acid is mainly present as 2-monopalmitin together with mainly C18 free fatty acids. 2-Monopalmitin is efficiently absorbed while free palmitic acid is poorly absorbed probably due to the formation of poorly absorbed calcium soaps in the intestine[47,48]. This formation of calcium soaps results in reduced absorption of both calcium and fat. Thus, the position of palmitic acid in TAG is of considerable significance for the absorption of fat and minerals in infants[48,49]. Therefore, the fatty acid composition and distribution in TAG in infant formulas are most important. Human milk fat substitutes have been developed by enzyme technology to mimic human milk fat composition and structure[50,51].

21.2.2.2   Diacylglycerols (DAG)

DAG are minor natural components of edible oils. DAG can be produced technologically and have been used as emulsifiers and recently as cooking oil. The interest in DAG as cooking oil is due to recent findings, which indicate that DAG of the *sn*-1,3 isomer may reduce body fat accumulation. Most characteristics of DAG oil are similar to those of TAG oil including the taste and viscosity, and the energy value of DAG is only approximately 2% lower than that of TAG of the same fatty acid composition[52]. Several studies on the effect on lipid metabolism and body weight by substituting DAG for TAG have been performed. In most studies the DAG oil used was the *sn*-1,3 isomer and it had the same fatty acid profile as the reference TAG oil.

   The effect of DAG on body fat accumulation is not clear. Thus, some animal and human studies revealed a reduction in body fat accumulation[53] or weight loss[54,55] whereas others did not[56,57]. The effect of DAG compared to TAG on weight is not related to differences in energy value or absorption[52], but rather to the fact that DAG is digested in a different manner. Thus, *sn*-1,3 DAG is hydrolyzed to *sn*-1 or *sn*-3 MAG[58]. Resynthesis of TAG from *sn*-1 or *sn*-3 MAG is slower than from *sn*-2 MAG[59,60]. Thus, the reduced accumulation of fat is a result of postabsorptional phenomena. The possible reduction in body fat accumulation after DAG intake may also be due to the ability of DAG to regulate the synthesis of enzymes involved in fatty acid synthesis[56,58,61] and fatty acid oxidation[62], probably at the gene expression level[56,62]. In addition, some studies indicate inhibited fatty acid synthesis or enhanced fatty acid oxidation that may enhance appetite suppression[63,64].

   Other nutritional effects of DAG intake seem to be a diminished increase in postprandial TAG and remnant particles (another potent risk factor for cardiovascular diseases)[65–67], possibly due to the above-mentioned slower resynthesis of TAG from *sn*-1 and *sn*-3 MAG[66]. The long-term effect of DAG consumption on reduction of fasting TAG levels was observed in diabetics[55,68] but not in healthy subjects[54].

Negative nutritional effects of DAG have also been observed. Thus, a lower activity of the enzyme responsible for reassembly of TAG from sn-1 or sn-3[56,58] may lead to an elevated level of free fatty acids and thereby plasma glucose concentrations. This condition may subsequently lead to glucose intolerance[56,69,70].

21.3 OXIDATIVE STABILITY OF n-3 LC PUFA AND STRUCTURED LIPIDS

21.3.1 LIPID OXIDATION

Lipid oxidation is one of the primary mechanisms of sensory deterioration in lipid-bearing foods. Moreover, reactions of generated radicals with proteins, pigments, and lipids in the food system contribute to decreased nutritional quality and the presence of potentially toxic lipid oxidation products[71,72].

The basic substrates for lipid oxidation reactions are unsaturated fatty acids. Different oxidation processes include autoxidation, photooxidation, and enzymatic oxidation. Autoxidation is a spontaneous free radical reaction with oxygen and includes three main stages of initiation, propagation, and termination. Meanwhile, photooxidation results from exposure of substrate to light in the presence of photosensitizers, while enzymatic oxidation is important in both plant and animal systems.

In case of autoxidation, the presence of initiators (e.g., metal ions, heat, protein radicals) causes unsaturated fatty acids (LH) to form carbon-centered alkyl radicals (L·, Equation (21.1)). These radicals propagate in the presence of oxygen by a free radical chain mechanism to form peroxyl radicals (LOO·, Equation (21.2)) and later hydroperoxides (LOOH, Equation (21.3)) as the primary products of autoxidation. The free radical chain reaction propagates until two free radicals combine and form a nonradical product to terminate the chain[73,74].

$$LH \rightarrow L\cdot + H\cdot \tag{21.1}$$

$$L\cdot + O_2 \rightarrow LOO\cdot \tag{21.2}$$

$$LOO\cdot + LH \rightarrow LOOH + L\cdot \tag{21.3}$$

Thermal dissociation of hydroperoxides as well as decomposition of hydroperoxides in the presence of traces of transition metals form alkoxy and peroxyl radical intermediates (LO· and LOO·, Equations (21.4)–(21.6)). These radicals propagate the free radical chain reaction[73]. Propagation reactions may though be more complicated than shown in Equations (21.2) and (21.3), as fragmentation, cyclization, and rearrangement have also been observed[75].

$$LOOH \rightarrow LO\cdot + \cdot OH \tag{21.4}$$

$$LOOH + M^n \rightarrow LO\cdot + OH^- + M^{n+1} \tag{21.5}$$

$$LOOH + M^{n+1} \rightarrow LOO\cdot + H^+ + M^n \tag{21.6}$$

In the case of photooxidation, the oxidation of unsaturated fatty acids is accelerated by exposure to light in the presence of photosensitizers. The latter will be activated by absorbing visible or near-UV light. Type I sensitizers then react with the substrate, generating substrate radicals, which can react with oxygen. Type II sensitizers react directly with triplet oxygen, transforming it into the short-lived, but highly reactive, high-energy form of singlet oxygen $^1O_2$, which reacts with the double bond of unsaturated fatty acids to form hydroperoxides (LOOH)[76]. This is not a free radical process. In food systems, chlorophyll, riboflavin, or hemoproteins serve as photosensitizers[73,77]. Hydroperoxides formed during photooxidation can also serve as initiators of free radical formation by reactions (21.4)–(21.6) and thereby propagate the autoxidation of lipids[78].

Hydroperoxides are essentially tasteless and odorless. However, at high temperatures and/or in the presence of traces of transition metals or light, they are readily decomposed to form a variety of nonvolatile and volatile secondary oxidation products[73]. The latter are termed "volatiles"; these are very important, as the wide range of carbonyl compounds (aldehydes, ketones, and alcohols), hydrocarbons, and furans contribute to flavor deterioration[71,72,77,79].

21.3.2 ANTIOXIDANTS

Many studies have indicated that lipid oxidation can be effectively controlled or minimized by the use of antioxidants[71]. Antioxidants are usually classified as either primary or secondary. The primary antioxidants (AH) include hindered phenols such as the synthetic antioxidants butylated hydroxyanisole (BHA), butylatedhydroxytoluene (BHT), propyl gallate (PG), *tert*-butylhydroquinone (TBHQ) or as the naturally occurring compounds such as tocopherols and plant polyphenols like carnosic acid. They are also referred to as free radical scavengers as they act as chain-breaking antioxidants by competing with the unsaturated lipid substrates (Equation (21.3)) and acting as electron/hydrogen donors to terminate the free radical chain (Equations (21.7) and (21.8)). The secondary antioxidants act by a number of different mechanisms, including metal chelation, oxygen scavenging and replenishing hydrogen to primary antioxidants. Therefore, the secondary antioxidants often exert synergistic effects together with primary antioxidants.

$$LOOC + AH \rightarrow LOOH + A \cdot \tag{21.7}$$

$$LOC + AH \rightarrow LOH + A \cdot \tag{21.8}$$

As consumers have become more health conscious, there has been a growing interest in natural food components such as ascorbic acid and tocopherols as well as naturally occurring antioxidants, such as extracts from vegetables, fruits, grains, spices, and herbs[80], mainly containing phenolic compounds[81]. The mechanism of action for different antioxidants is summarized in Table 21.1.

21.3.3 n-3 FATTY ACIDS

EPA and DHA are highly susceptible to oxidation reactions due to the presence of five and six double bonds in their molecular structure, respectively. The oxidation occurs readily at room temperature. Due to the high number of double bonds, oxidation processes may lead to a complex mixture of hydroperoxides and a myriad of volatile, nonvolatile, and polymeric secondary oxidation products. The structures of some cleavage products are known (Figure 21.3), but the exact

---

TABLE 21.1
Antioxidant Mechanisms of Different Antioxidants

Antioxidant mechanism	Examples of antioxidants
Metal chelation	Citric acid, EDTA[a], lactoferrin, phytic acid, (phosphates)
Oxygen scavenging	Ascorbic acid, glucose oxidase-catalase
Singlet oxygen quenching	Carotenoids, tocopherols
Active oxygen scavenging	Superoxid dismutase, catalase, mannitol
Primary radical chain breaking	Tocopherols, ascorbic acid and derivatives, gallic acid andgallates[a], BHA[a], BHT[a], several natural polyphenols, rosemary and sage antioxidants
Alkoxyl radical interruption	Tocopherols, (some rosemary compounds)
Secondary chain breaking	Glutathione peroxidase and glutathione-S-transferase, thiopropionic acid and its derivatives[a]

[a] These are all "synthetic" antioxidants. Thiopropionic acid and its derivatives are permitted as food additives in the U.S., but not in the E.U.

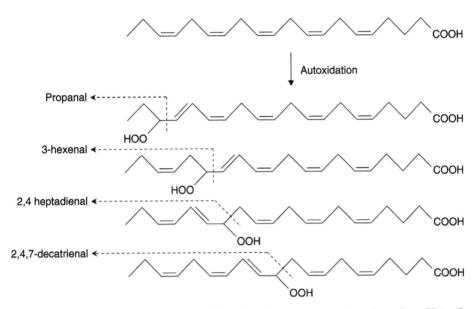

FIGURE 21.3 Autoxidation sites associated with major aldehydes expected to form from EPA. (Source: Kulås, E., in *Lipid Oxidation Pathways*, Kamal-Eldin, A., Ed., AOCS Press, Champaign, IL, 2003, pp. 37–69.)

mechanisms for the formation of others are not yet completely understood. Most n-3 LC PUFA though follow the main cleavage mechanisms recognized for linolenic acid.

Some of the oxidation products expected from autoxidation of n-3 LC PUFA are propanal, 2-pentenal, 3-hexenal, 4-heptenal, 2,4-heptadienal, 2,6-nonadienal, 2,4,7-decatrienal, as well as 1-penten-3-one and 1,5-octadien-3-one[78,82]. Aldehydes have very low flavor threshold values and are therefore the source of the characteristic odor of rancid fat, which in general is called "off-flavor"[77]. Several of the above-mentioned volatile components have been characterized in fish oil[83–85] and in fish itself[86,87]. The compound 1-penten-3-one as well as (Z)-4-heptenal, (E,Z)-2,6-nonadienal and 2,4,7-decatrienals have been associated with sharp-burnt-fishy off-flavors in oxidized fish oil[83,88]. Due to different flavor threshold values, concentration of oxidation products is not the determining factor for measurement of off-flavors. The most potent odorants can only be identified by gas chromatography-olfactometry. Moreover, Venkateshwarlu et al.[89] found that fish oil-enriched milk contained 1-penten-3-one, (Z)-4-heptenal, (E,E)-2,4-heptadienal, and (E,Z)-2,6-nonadienal as potent odorants, but despite their potency, none of the separated individual volatiles produced a fishy or metallic odor. In contrast, the combination of (E,Z)-2,6-nonadienal and 1-penten-3-one could be a useful marker for fishy and metallic off-flavors in fish oil and fish oil-enriched foods[90]. The only other model available in the literature describing a quantitative relationship of volatiles and fishy taste in fish oil was developed by MacFarlane et al.[91] and was based on three volatiles, namely 2,6-nonadienal, 4-heptenal, and 3,6-nonadienal.

The oxidative stability of fish and algae oils varies widely according to their fatty acid composition, the contents of tocopherols and other antioxidants, and the presence and activity of transition metals[92]. If fish oils are intended for human consumption, especially when the sensory quality is of major importance, efforts must be taken to minimize oxidation of LC PUFA, by stabilization with antioxidants among others[82]. Ethylenediaminetetraacetic acid (EDTA), lactoferrin, and citric acid are examples of metal chelators that have been shown to reduce lipid oxidation in bulk and emulsions of fish oil[93–95]. Tocopherols are able to retard the formation of both hydroperoxides and volatile secondary oxidation products when added to fish oil. However, the antioxidant activity and optimum concentration are very dependent on the isomer (blend) employed. It is recommended to use tocopherols in combination with other antioxidants, e.g., ascorbyl plamitate[82]. Besides the application of antioxidants, fish oils have to be handled and stored at low temperatures in the absence of light and oxygen. The shelf-life of fish oil is believed to be about two months when stored at 10°C.

21.3.4 STRUCTURED LIPIDS

Apart from the degree of unsaturation of fatty acids in a TAG molecule, several studies have also indicated that their position in the TAG may influence their oxidative stability[73,96,97]. Producers of structured lipids therefore have to be aware that they may change the oxidative stability of their products upon randomization or restructuring of the fatty acids. The results from different publications, though, are not always in agreement because they refer to different types of oils, TAG, or fatty acids used for production. Different methods for production of TAG (e.g., physical blends, randomized or regiospecific interesterification) may be employed; differences in purification (e.g., solid phase extraction, short-path distillation) and purity may also affect oil stability. In addition, different analytical methods are often employed for measurement of oxidative stability (e.g., peroxide value, conjugated dienes, volatiles) and this may make comparison of results difficult.

SSL based on caprylic acid and either sunflower oil or fish oil were found to be significantly less stable than their RL counterparts or the original oils[98,99], as shown in Figure 21.4. This is in agreement with findings of Akoh and Moussata[100] who found that SSL produced from caprylic acid and either canola oil or fish oil were less stable than the unmodified canola or fish oil, although caprylic acid was mainly replacing PUFA at *sn*-1,3 positions. On the other hand, when comparing the oxidative stability of products from soybean oil and methyl stearate, Konishi et al.[101] observed that the SSL had the best stability in comparison with the RL and soybean oil. This observation was ascribed to incorporation of stearic acid in *sn*-1,3 positions, which stabilized linoleic acid. This effect should be independent on the chain length of the saturated fatty acid[96]. The lower oxidative stability of RL compared with SSL may additionally be due to easier access of oxygen to the higher levels of PUFA located in the *sn*-1,3 positions in RL[102,103]. Accordingly, Senanayake and Shahidi[104] found that RL were less stable than their unmodified counterparts when incorporating EPA or DHA into borage and evening primrose oils, which was explained by the higher content of highly unsaturated fatty acids as well as the removal of endogenous antioxidants. A similar finding was reported by Senanayake and Shahidi[105] when capric acid was incorporated into three different algal oils.

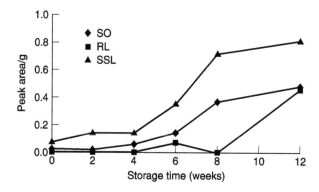

FIGURE 21.4 Development of 2,4-decadienal in lipids during storage at 50°C for up to 12 weeks. SSL, specific structured lipid based on sunflower oil and caprylic acid; RL, randomized lipid based on sunflower oil and caprylic acid; SO, sunflower oil. (Source: Timm-Heinrich, M., Xu, X., Nielsen, N.S., and Jacobsen, C., *Eur. J. Lipid Sci. Technol.*, 105, 436–448, 2003.)

The stability of structured lipids cannot be solely explained by their fatty acid composition and the location of the fatty acids on the glycerol backbone. Other factors, such as the production process itself and possible removal of natural antioxidants and phospholipids during the purification process may contribute to the decreased stability of structured lipids compared with the original lipids[98–100,106]. The structured lipids undergo further heat and mechanical treatment and may be exposed to oxygen, light, and trace metals during production and purification. This leads to higher initial concentrations of primary and secondary oxidation products, which are the basis for lower oxidative stability during storage, if not properly removed. Deodorization is the most effective purification process, as lipid hydroperoxides are decomposed and free fatty acids and volatiles are removed. However, deodorization might cause acyl migration and could thereby compromise the specific structure of SSL. The specific structured lipids therefore have to be purified under very mild conditions, e.g., by short-path distillation. Short-path distillation conserves the desired structure, but volatiles are not removed as effectively. Short-path distillation in comparison with batch deodorization gave a product of inferior sensory quality, as indicated for randomized lipid from fish oil and tricaprin[107]. Moreover, tocopherol levels did drop to approximately one tenth of the original value[98–100], which resulted in loss of protection of oil against oxidation.

21.4   APPLICATION OF n-3 LC PUFA AND STRUCTURED LIPIDS IN FOODS AND THEIR OXIDATIVE STABILITY

Due to the health benefits of functional lipids, the food industry is interested in incorporating these lipids into food products. Foods are often complex systems comprised of different phases such as air, water, lipid, and solid particles. Oil-in-water emulsions such as milk, mayonnaise, and salad dressing constitute one important group of complex food systems. In these types of emulsions, oil droplets are dispersed in the continuous aqueous phase. In water-in-oil emulsions, such as margarine and butter, the opposite is the case, i.e., water droplets are dispersed in the continuous oil phase. The two phases are separated by an interface comprised of amphiphilic compounds (emulsifiers) (Figure 21.5). The above mentioned food emulsions are important in relation to application of functional lipids, because they are consumed in large quantities and/or they have a high content of lipids, meaning that relatively large amounts of functional lipids may be incorporated into some of these foods. Infant formula constitutes another important complex food system in relation to functional lipids as already discussed in an earlier section. Due to the complexity of emulsions and infant formula, their oxidation mechanisms may be very different from those in bulk oils. Hence, to apply successfully functional lipids in foods it is important to understand the oxidation mechanisms in the food system in question. The major factors influencing lipid oxidation in food

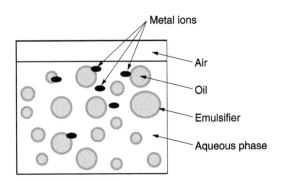

FIGURE 21.5 Schematic of the structure of an oil-in-water emulsion.

emulsions and infant formula are summarized; examples of applications of functional lipids and their protection via antioxidants are also provided below. For emulsions, most focus has been placed on oil-in-water emulsions.

21.4.1   PHYSICOCHEMICAL PROPERTIES

21.4.1.1   Metal Catalysis

Even after refining and deodorization most oils contain trace levels of lipid hydroperoxides. Furthermore, several food ingredients contain trace levels of metal ions. Therefore, metal-catalyzed decomposition of lipid hydroperoxides is probably the most important initiator of lipid oxidation in many foods. Reactions (21.5) and (21.6) not only generate free radicals, which may initiate further oxidation reactions, but may also lead to the formation of secondary volatile oxidation compounds, as described earlier.

In fish oil-enriched mayonnaise, iron present in the egg yolk, which is used as an emulsifier, seems to be the most important oxidation catalyst[108–110]. This hypothesis was supported by the fact that EDTA, as a metal chelator, efficiently inhibited lipid oxidation in mayonnaise enriched with fish oil or with structured lipid based on sunflower oil[111–113]. Very low concentrations of EDTA (6 mg/kg) were enough to inhibit efficiently the formation of lipid hydroperoxides and volatiles[113]. In fish oil-enriched milk, EDTA was able to reduce lipid oxidation, when fish oil with a peroxide value (PV) of 1.5 meq/kg was used. However, when fish oil with a PV of 0.1 meq/kg was used, EDTA did not significantly reduce lipid oxidation[114]. This finding was probably due to the fact that milk produced from the latter fish oil had a high oxidative stability in itself and therefore it was difficult to observe any effect of EDTA. Therefore, these data indicate that trace metals are important catalysts of lipid oxidation in fish oil-enriched milk.

21.4.1.2   Oxygen

As oxygen is required for oxidation to occur, lipid oxidation may be reduced by flushing the food product with nitrogen as observed in fish oil-enriched mayonnaise[115], by packaging in an air-tight container, or by packaging under modified atmosphere. The total amount of oxygen in the headspace above the product and the oxygen dissolved in the product are factors of great importance for lipid oxidation and should therefore be minimized. Off-flavor products may form even after all oxygen is consumed[116] and decomposition of the volatiles to smaller compounds may occur.

21.4.1.3   pH

The pH plays an extremely important role in lipid oxidation as it affects reactivity, solubility, partitioning, and interactions of several of the reactive compounds that participate in oxidation and antioxidant reactions. Metal ions are generally more soluble at low pH than at high pH[78]. This may explain why lipid oxidation is generally slowest at high pH values. Furthermore, pH influences the emulsifier charge and this may significantly affect oxidation as will be discussed later. In fish oil-enriched mayonnaise, lipid oxidation increased with decreasing pH from 6.0 to 3.8[108,109]. In contrast, the opposite effect of pH was observed in salmon oil-in-water model emulsions (5% oil). Thus, lipid oxidation was greater and more rapid at pH 7.0 than at pH 3.0[117]. These contradicting results demonstrate that in complex multiphase systems, pH may affect lipid oxidation in different directions through a number of different mechanisms.

21.4.1.4   Emulsifiers (Proteins and Surfactants)

Two types of emulsifiers are used for emulsion formation and stabilization. The first type are macromolecules such as proteins, while the second type are small amphiphilic molecules such as phospholipids, free fatty acids, monoacylglycerols, and synthetic surfactants. Among the first type of emulsifiers, proteins are used both to facilitate the formation and enhance the stability of food emulsions. During homogenization, proteins are absorbed to the oil droplet surface. Here, they will lower the surface tension and prevent coalescence of droplets by forming protective membranes around the droplets. Proteins may also provide the emulsion droplets with a positive or negative electrical charge at pH values below or above their pI, respectively. Since all droplets will either have a negative or a positive charge, the droplets repel each other whereby coalescence is prevented and the physical stability of the emulsion increases. The electrical charge of the interfacial layer around the oil droplet may also significantly influence oxidation in emulsions in the presence of metal ions[93,118]. If the charge of the interface is positive, metal ions will be repelled from the interface and thereby their ability to promote lipid oxidation will be low. In contrast, if the charge of the interface is negative, metal ions will be attracted to the interface and promote oxidation. The electrical charge of the interface depends on the type of emulsifier used and on the pH in the emulsion. However, factors other than the charge of the emulsion droplets seem to influence the effect of proteins on the oxidative stability[119]. Other factors could include differences in how the proteins influence the thickness or packing of the emulsion droplet interface[119–121]; the amino acid composition of proteins may affect their antioxidative activity[122].

Similar to proteins, surfactants may also affect the charge of the interface. However, surfactants are known to influence the location of the metal ions by forming micelles. Hence, under normal conditions, surfactants are present in excess in emulsions, and surfactants not associated with the emulsion droplets may form micelles in the continuous phase. Lipid hydroperoxides and/or metal ions could become associated with or solubilized in the micelles and this may in turn reduce lipid oxidation by preventing these components from reacting with lipid components in the oil phase[123,124].

21.4.1.5 Carbohydrates/Viscosity

Some carbohydrates show antioxidative activity in high concentrations due to their ability to scavenge free radicals[125]. Certain carbohydrates may decrease oxidation by decreasing the concentration of oxygen in the aqueous phase as suggested for sucrose. Furthermore, sucrose has an increasing effect on the viscosity of the emulsion and this may decrease the diffusion coefficient of oxygen[126], metals, other reaction products, and reactants, which may in turn slow down oxidation rates. However, the effect of viscosity on oxidation rates is not fully understood and requires further research.

21.4.1.6 Moisture

Lipid oxidation rates in foods are significantly influenced by the water activity (a_w), particularly in powdered products. The reason is that water interacts with a wide range of components. Under conditions of low a_w, i.e., <0.2 to 0.3, corresponding to <4 to 5% moisture, oxidation has generally been shown to be high, but decreases with increasing a_w (up to 0.2–0.4). When a_w increases from 0.4 to 0.7, oxidation will increase, probably due to increased mobility of metal catalysts[127]. In powders, a decrease in water activity leads to a decrease in the so-called glass transition temperature, which is the temperature where the food matrix changes from the glassy state to the rubbery state. Lipids have been suggested to be more susceptible to oxidation in the rubbery state. Hence, oxidation may be reduced by bringing the powder into the glassy state by optimizing the recipe to increase the glass transition temperature or by decreasing the storage temperature.

21.4.2 Physical Structure

In oil-in-water emulsions, the oil droplets are surrounded by an interface and are not in direct contact with air. Oxygen, antioxidants, and prooxidants must therefore diffuse through the aqueous phase before reaching the oil droplets or the oil–water interface. Lipid oxidation has frequently been described as an interfacial phenomenon, because prooxidant concentrations are often highest in the aqueous phase, and therefore oxidation may be initiated here or at the interface and not in the oil phase. Oxidation generally occurs much faster in emulsions than in bulk oils[92] and this may be attributed to the formation of a large interfacial area in emulsions compared with the relatively small interface between air and oil in bulk oils. Moreover, the emulsification process itself may lead to oxidation. Hence, during mechanical homogenization, the temperature may increase locally and oxygen could be incorporated into the emulsion, which leads to increased oxidation. Therefore, when homogenizing polyunsaturated functional lipids it is important to have a precise temperature control and to minimize incorporation of oxygen into the product, by purging the oil/emulsion with nitrogen during homogenization.

As mentioned earlier, a large interfacial area is created during emulsification. The total interfacial area depends on the size distribution of the oil droplets and can be calculated by $A_s = 6/D_{32}$, where D_{32} is the mean surface diameter. D_{32} is calculated using $\sum n_i d_i^3 / \sum n_i d_i^2$, where d_i is the diameter of the droplet and n is the number of droplets with that diameter. The increased interfacial area increases the potential contact area between the oil droplet and trace metals in the continuous, aqueous phase. Contradicting reports are available on the effect of the droplet size on oxidation in different food systems[128,129], indicating that the impact of the droplet size on oxidation in food emulsions is dependent on the composition of the food system.

21.4.3 ANTIOXIDANTS IN FOOD EMULSIONS

The antioxidant efficacy in multiphase systems depends on many factors. The partitioning of the antioxidant into different phases of the systems appears to be one of the most important factors. This is due to the so-called polar paradox theory described first by Porter[130] and later used by Frankel et al.[131] and Huang et al.[132]. According to this paradox, polar antioxidants such as ascorbic acid and Trolox will be more active in nonpolar media like bulk oil than their more nonpolar counterparts ascorbyl palmitate and tocopherol, respectively. This is probably due to the fact that polar antioxidants are located at the air–oil interface where oxidation is taking place. On the other hand, tocopherol and ascorbyl palmitate are more active in more polar systems like emulsions, because they are located in the oil phase where oxidation propagates. The partitioning of antioxidants has mainly been studied in model emulsions. However, the partitioning of Trolox, tocopherol, propyl gallate (PG), gallic acid, ferulic acid, caffeic acid, and catechin in fish oil-enriched mayonnaise has also been reported[133]. These antioxidants partitioned into different phases of mayonnaise according to their chemical structure. For example, a polar antioxidant such as gallic acid was mainly found in the aqueous phase (80%), whereas only 24% of its less polar counterpart PG was found in this phase. Several antioxidants, including tocopherol could also partly be found in the fraction that was suggested to represent the oil–water interface consisting of egg yolk components. This observation shows that the antioxidants interact with the interfacial layer in mayonnaise and that the egg yolk constituents participate in these interactions. Such interactions could affect the activity of the antioxidants.

21.4.4 OPTIMIZING THE OXIDATIVE STABILITY OF FOODS ENRICHED WITH FUNCTIONAL LIPIDS

Some examples of applications of functional lipids have been mentioned above. In this section, more examples are given with the main emphasis on attempting to reduce lipid oxidation by addition of antioxidants.

21.4.4.1 Mayonnaise with Functional Lipids

Lipid oxidation in fish oil-enriched mayonnaise was prevented by TBHQ (0.02%) and EDTA (0.075%) in such a manner that a sensory panel could not distinguish it from a soybean oil mayonnaise after storage at 2°C for 14 weeks[115]. Since TBHQ is not permitted in Europe, we evaluated the antioxidative effect of EDTA, PG, gallic acid, tocopherol, ascorbic acid, or a mixture of ascorbic acid (8.6% w/w), lecithin (86.2% w/w), and tocopherol (5.2% w/w) (the so-called A/L/T system) in fish oil-enriched mayonnaise by sensory profiling, measurements of lipid hydroperoxides and volatiles, and in some cases by measuring free radical formation).[108,109,134–136] Weak prooxidative effects of PG and gallic acid were observed [111]. Tocopherol was inactive as an antioxidant and it even seemed to have prooxidative effects at higher concentrations (>140 mg/kg)[135,136]. Ascorbic acid (40 to 800 mg/kg) and the A/L/T system (200 mg/kg total concentration) showed a strong prooxidative effect[108,109,135]. The prooxidative effect of these antioxidant systems was suggested to be due to the ability of ascorbic acid to promote the release of iron from egg yolk located at the oil–water interface. The released iron would then be able to decompose preexisting lipid hydroperoxides located near the oil–water interface or in the aqueous phase. In contrast, EDTA (75 mg/kg) was observed to be a strong antioxidant that totally inhibited oxidative flavor deterioration during storage at 20°C[111]. The strong antioxidative effect of EDTA was proposed to be due to its metal chelating properties. Subsequently, it was shown that lower concentrations (i.e., 6 mg/kg) of EDTA are sufficient to retard lipid oxidation in mayonnaise[113]. In the same study, it was shown that other metal chelators such as lactoferrin (700 to 2800 mg/kg) and phytic acid (15 to 116 mg/kg) are far less effective in preventing lipid oxidation in mayonnaise[113] (Figure 21.6).

The oxidative stability of mayonnaise enriched with SSL based on caprylic acid and either fish oil or sunflower oil has been compared with that of mayonnaise based on the original oils or a RL from the same oils[138]. The specific structured lipid was produced to have MLM structure; caprylic acid in sn-1 and sn-3 positions and long-chain fatty acids in the sn-2 position (see also Section 21.1). In both studies, mayonnaise based on the SSL had a reduced oxidative stability compared with mayonnaise produced from the original oil or RL. This difference in oxidative stability could be due to differences in the molecular structure of lipids (as described in an earlier section) or due to the different processing history that affected the initial PV and tocopherol levels in the mayonnaise. The antioxidative effect of PG (200 mg/kg), lactoferrin (800 mg/kg), and EDTA (75 mg/kg) was also investigated in mayonnaise based on specific structured sunflower oil. Only EDTA efficiently prevented lipid oxidation, whereas neither PG nor lactoferrin had any clear effect on lipid oxidation at the concentrations employed.

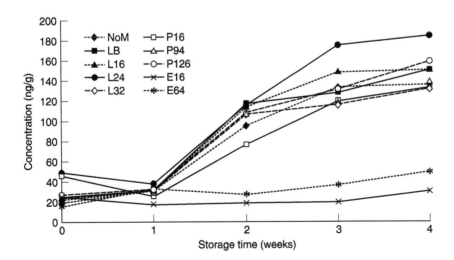

FIGURE 21.6 Development of 2-E-pentenal in mayonnaise during storage at 20°C for up to four weeks. NoM, no metal chelator added; L, lactoferrin addition; P, phytic acid addition; E, EDTA addition. Numbers after the letter refer to approximate concentrations in μM, e.g., L8 refers to addition of 8 μM lactoferrin. (Source: Nielsen, N.S., Petersen, A., Meyer, A.S., Timm-Heinrich, M., and Jacobsen, C., *J. Agric. Food Chem.*, 52, 7690–7699, 2004.)

21.4.4.2   Margarine/Spreads with Functional Lipids

Kolanowsky et al.[139] have shown that spreads could only be enriched with up to 1.5% fish oil (i.e., 0.5% EPA + DHA) without deteriorating the sensory quality of the product. In another study, it was concluded that low-calorie spreadable fats with 55% fat (soft margarine and a mix of butter and vegetable oil) could be enriched with up to 1% EPA + DHA without significantly affecting their sensory quality[140]. The margarine spread may be stored up to 6 weeks and the spread based on butter and vegetable oil up to 3 weeks without any significant decrease of quality. No antioxidants were added to any of these spreads.

Another margarine study dealt with optimization of antioxidant addition to fish oil-enriched low-calorie (40% fat) spreads in which 20% of the fat phase consisted of fish oil[141]. By adding 150 mg/kg EDTA, 300 mg/kg Grindsted 117 (ascorbyl palmitate (7.5%), PG (17.5%), citric acid (10.0%), and carrier (65%)), and 1000 mg/kg Toco 50 (50% tocopherol mixture in vegetable oil) it was possible to produce margarines of commercially acceptable quality and stability. To the best of our knowledge there are no reports available in the literature for structured lipids in margarine or spreads.

21.4.4.3   Milk Products with Functional Lipids

The oxidative deterioration of fish oil-enriched milk depends strongly on the quality of the fish oil. When cod liver oil with a relatively low PV (1.5 meq/kg) was added to milk at a concentration of 1.5%, strong fishy off-flavors were noted after one day of storage[114]. In contrast, milk containing a more unsaturated tuna oil (20.1% EPA + DHA in cod liver oil versus 30.2% in tuna oil) with a lower PV (0.1 meq/kg) did not give rise to off-flavor formation. Lipid oxidation was also slower when evaluated by PV and formation of volatiles[114]. Hence, the presence of both lipid hydroperoxides and trace metals in relatively low levels unavoidably leads to oxidative flavor deterioration in fish oil-enriched milk due to the mechanism described in Equations (21.5) or (21.6). However, the results indicated that it was possible to enrich milk with fish oil (0.5%) and obtain a product of satisfactory sensory quality provided that the fish oil used was of high quality[114].

It has been shown that mixing rapeseed oil with fish oil before adding it to the milk significantly retarded lipid oxidation both during homogenization and storage[142,143]. A triangle test showed that an untrained "consumer panel" could not distinguish a traditional milk from a milk enriched with the rapeseed oil–fish oil mixture (0.5% rapeseed oil–fish oil mixture (1:1) in milk with 1.0% fat)[143]. Rapeseed oil is thus able to protect the fish oil against oxidation. Research is underway to investigate how and why rapeseed oil is able to protect fish oil against oxidation when used in combination with milk. Interestingly, it was observed that even in the milk produced with the fish oil–rapeseed oil mixture the quality of the oil had a significant impact on sensory perception of milk. Milk containing the oil mixture with a PV between 0.5 and 2.0 meq/kg was perceived by the sensory panel to have a significantly higher fishy off-flavor intensity after one day of storage[143] (Figure 21.7). In contrast, the panel could not discriminate milk containing an oil mixture with a PV of 0.1 meq/kg from milk devoid of fish oil.

A strawberry flavored milk drink with 5% lipid has been used for evaluating the oxidative stability of specific structured lipids based on caprylic acid and either sunflower oil or fish oil compared with the original oils or with a randomized lipid produced from the same oil sources[138,144]. In the milk drinks with fish oil, only 10% of the lipid was from fish oil or structured lipid produced from it; the remaining lipid was rapeseed oil. Similar to the mayonnaise experiments, it was observed that the milk drink with SSL had a lower oxidative stability than milk drinks with the original or randomized lipids. Again it was suggested that different processing history and possibly the different structure of SSL could explain the lower oxidative stability of the milk drink with this lipid.

The antioxidative effect of EDTA (240 mg/kg) and gallic acid (200 mg/kg) was evaluated in milk drink with SSL based on sunflower oil. EDTA exhibited a strong antioxidant effect that efficiently inhibited the formation of peroxides, volatiles, and off-flavors to such an extent that the

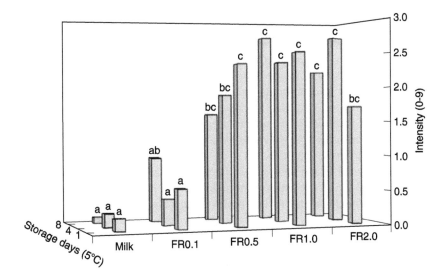

FIGURE 21.7 Development of fishy off-flavor in fish oil-enriched milk produced from fish oil:rapeseed oil mixtures (1:1) with different PV. The emulsions followed by the same letter were not significantly different in the Bonferroni multiple comparison test using 0.05 level of significance. FR, fish oil and rapeseed oil mixture. The number after FR refers to the PV in meq/kg. (Source: Let, M.B., Jacobsen, C., and Meyer A.S., *Int. Dairy J.*, 15, 173–182, 2005.)

oxidative stability was equal to that of the control containing the original sunflower oil[144]. In contrast, gallic acid did not reduce the formation of volatiles and off-flavors. In the experiment with milk drink containing SSL based on fish oil, EDTA (240 mg/kg) also significantly reduced the formation of off-flavor and of most of the volatiles determined[138]. In this experiment, it was also found that lactoferrin (1000 mg/kg) did not reduce lipid oxidation. The antioxidative effect of different concentrations of lactoferrin (500 to 2000 mg/kg) was subsequently evaluated in milk drink with structured lipid based on sunflower oil[113]. In this system, the addition of 1000 mg/kg lactoferrin slightly reduced lipid oxidation, whereas higher lactoferrin concentrations slightly promoted oxidation. Thus, these data indicate that EDTA is a better antioxidant in milk drink with structured lipids than lactoferrin or gallic acid, possibly due to the fact that oxidation in milk drink is metal catalyzed and that the ferric–EDTA complex has a higher binding constant (1.3×10^{25}) than that of the ferric–lactoferrin complex (10^{20})[145]. Moreover, EDTA has a better heat stability than lactoferrin and this could also have affected the results, as the milk drink was heated during production.

The oxidative stability of a milk drink containing SSL with MLM structure containing CLA was compared with that of four milk drinks with structured lipids based on other C18 fatty acids (stearic acid, oleic acid, linoleic acid, or linolenic acid)[146]. The lipid sources for different SSL were tri-CLA, tristearin, triolein, safflower oil, or linseed oil. Interestingly, the oxidative stability did not decrease with increasing number of double bonds in the fatty acids of the SSL, as the order of stability was linoleic acid < CLA < oleic acid ~ linolenic acid < stearic acid. The extent of lipid oxidation was measured by PV, volatiles, and sensory profiling. The lipids used to produce different SSL were of different oxidative status, because they were manufactured by different procedures, and this affected the oxidative status of the SSL used for milk drink production. Therefore, these data indicate that the quality of lipids has a greater influence on the oxidative stability of milk drink than the fatty acid composition/degree of unsaturation of lipids, in agreement with the findings of Let et al.[114]

21.4.5　Encapsulated Fat Powders and Infant Formula

When n-3 LC PUFA are used in infant formula, they are often added in the form of microencapsulated fat powders. Therefore, factors affecting the oxidative stability of microencapsulated fats are also described. Microencapsulated fats are essentially dried, homogenized emulsions of an oil or fat where modified starches or hydrocolloids and proteins are used as emulsifying materials. A non-emulsifying water-soluble material such as sugar or hydrolyzed starch is used as filler[147]. During processing and storage, controlling of lipid oxidation is dependent on the ability to control molecular diffusion through the protective wall matrix and maintain structural integrity that keeps emulsified lipids within each powder particle.

The effect on the shelf life and sensory quality of emulsifier type (three different casein types), free fat, surface fat, and the air content of spray dried fish oil powder has been reported by Keogh et al.[147]. These authors concluded that fish oil powder with a low level of off-flavor can be produced with a shelf life of 31 weeks at 4°C using dairy ingredients alone as encapsulating ingredients. They also found that the shelf life increased when the free nonencapsulated fat and vacuole volume of the powder decreased. They did not find any effect of the surface fat. Recently, Kagami et al.[148] observed that sodium caseinate in combination with highly branched cyclic dextrin produced from waxy corn starch resulted in more stable encapsulates compared with encapsulates made with maltodextrin, sodium caseinate, or combinations of highly branched cyclic dextrin and whey protein.

The effect of the relative humidity on the oxidative stability and physical stability in encapsulated fish oil has not been reported. However, studies on encapsulated milk fat have been carried out by Hardas et al.[149], who found that the effect of a_w on lipid oxidation was different in the surface fat and in the encapsulated fractions. The encapsulated fat exhibited a lower oxidation rate at lower a_w and the highest oxidation rate at the highest a_w (0.52), whereas the surface fraction oxidized at the slowest rate at $a_w = 0.52$. It was suggested that the wall matrix protected the encapsulated lipids from moisture and oxygen in the air. Therefore, the controlling factor is molecular diffusion through the wall matrix. When moisture levels increased, the wall compounds were hydrated and this led to plasticization (glassy–rubbery transition, as described in an earlier section). Thereby, the local viscosity is reduced and smaller molecules including oxygen are allowed to diffuse through the wall, leading to increased oxidation. In the case of the surface lipids, which are directly exposed to air and humidity, oxidation is not limited by molecular diffusion through the wall matrix. Oxidation of the surface fat is promoted by drier surface air and this could be related to less interference from water molecules, which otherwise would hydrate the oxidation catalyst in the local environment of the surface fat. The highest PV was determined in the surface fat.

Interestingly, Velasco et al.[150] found that oxidation was slower in the free oil fraction compared to the encapsulated fraction when A/L/T was used as an antioxidant and when the powder was stored in open Petri dishes. When A/L/T was used as an antioxidant system for protection of microencapsulated fish oil in vacuum bags the samples did not oxidize upon storage. In a study by Heinzelmann et al.[151], it was shown that the best shelf life was obtained when the encapsulated fish oil contained a combination of ascorbic acid, lecithin, and tocopherol (i.e., the A/L/T system) and when the freezing rate was slow. No sensory evaluation was performed on these powders.

Hogan et al.[152s] investigated the antioxidative effects of α-tocopherol (100 and 500 mg/kg) and its hydrophilic analog Trolox C (100 and 500 mg/kg) in fish oil encapsulates prepared from herring oil (37.5% oil) and sodium caseinate/maltodextrin (1:19). They observed that all antioxidants had reduced oxidation in the powders after 14 days of storage at 4°C. Tocopherol at 100 mg/kg was the most efficient antioxidant. It should be noticed that PV was very high in this study ranging from 10 meq/kg in the freshly produced powders to 60 meq/kg after 4 weeks. It is possible that the effect of tocopherol would be less clear in powders with lower starting PV.

In a study by Satué-Gracia et al.[94] on traditional infant formula, lactoferrin was able to reduce lipid oxidation, possibly due to its ability to chelate metal ions. These findings indicate that lactoferrin could be an important antioxidant in infant formula containing functional lipids.

21.4.6 EXAMPLES OF OTHER APPLICATIONS OF FUNCTIONAL LIPIDS

The sensory stability of bread baked with either fish oil made from sand eel, a specialty tuna oil, or an algal oil was evaluated by Becker and Kyle[153]. Sensory off-flavors seemed to be less pronounced in the DHA bread made with algal oils compared to those made with fish oils. Therefore, it was suggested that algal oil had a better stability than fish oils. This hypothesis was later reevaluated by Frankel et al.[92], who found that the high oxidative stability of the commercial DHA-rich algal oil was lost when the triacylglycerols were purified to remove tocopherols and other antioxidants. Moreover, an oil-in-water emulsion with the same algal oil had a lower oxidative stability than the corresponding fish oil emulsion.

21.5 CONCLUSIONS

The functional lipids described in this chapter have a number of beneficial effects that should be further exploited by incorporating them into foods. However, negative health aspects of markedly increasing PUFA in the diet, in particular n-3 LC PUFA, are also possible if high-quality oils are not used. Potential human health risks are associated with increased consumption of n-3 LC PUFA due to possible increased consumption of lipid oxidation products, increased *in vivo* production of lipid oxidation products, and depletion of tissue levels of tocopherols. Therefore, the storage stability, shelf life, and safety of foods enriched with functional lipids should be optimized. It was demonstrated that it is possible to incorporate highly unsaturated functional lipids into foods and obtain food products of good quality with acceptable storage stability provided that the lipids used are of high initial quality and stabilized with appropriate antioxidant(s). Keeping this and the consumers increased interest in functional lipids in mind, it is expected that several new food products enriched with functional lipids will appear in the market place in the coming years.

REFERENCES

1. Bang, H.O., Dyerberg, J., and Hjørne, N., The composition of food consumed by Greenland Eskimos, *Acta Med. Scand.*, 200, 69–73, 1976.
2. Kromann, N. and Green, A., Epidemiological studies in the Upernavik district, Greenland, *Acta. Med. Scand.*, 208, 401–406, 1980.
3. Bjerregaard, P. and Dyerberg, J., Mortality from ischemic heart disease and cerebrovascular disease in Greenland, *Int. J. Epidemiol.*, 17, 514–519, 1988.
4. Singh, R.B., Niaz, M.A., Sharma, J.P., Kumar, R., Rastogi, V., and Moshiri, M., Randomized, double-blind, placebo-controlled trial of fish oil and mustard oil in patients with suspected acute myocardial infarction: the Indian experiment of infarct survival 4, *Cardiovasc. Drug Ther.*, 11, 485–491, 1997.
5. Dietary supplementation with n-3 polyunsaturated fatty acids and vitamin E after myocardial infarction: results of the GISSI-Prevenzione trial, *Lancet*, 354, 447–455, 1999.
6. Kromhout, D., Bosschieter, E.B., and Coulander, C.D., The inverse relation between fish consumption and 20-year mortality from coronary heart-disease, *N. Engl. J. Med.*, 312, 1205–1209, 1985.
7. Dolecek, T.A. and Granditis, G., Dietary polyunsaturated fatty acids and mortality in the multiple risk factor intervention trial (MRFIT), *World Rev. Nutr. Diet*, 66, 205–216, 1991.
8. Zhang, J.J., Sasaki, S., Amano, K., and Kesteloot, H., Fish consumption and mortality from all causes, ischemic heart disease, and stroke: an ecological study, *Prev. Med.*, 28, 520–529, 1999.
9. Albert, C.M., Hennekens, C.H., O'Donnell, C.J., Ajani, U.A., Carey, V.J., Willett, W.C., Ruskin, J.N., and Manson, J.E., Fish consumption and risk of sudden cardiac death, *J. Am. Med. Assoc.*, 279, 23–28, 1998.
10. Hu, F.B., Bronner, L., Willett, W.C., Stampfer, M.J., Rexrode, K.M., Albert, C.M., Hunter, D., and Manson, J.E., Fish and omega-3 fatty acid intake and risk of coronary heart disease in women, *J. Am. Med. Assoc.*, 287, 1815–1821, 2002.

11. Iso, H., Rexrode, K.M., Stampfer, M.J., Manson, J.E., Colditz, G.A., Speizer, F.E., Hennekens, C.H., and Willett, W.C., Intake of fish and omega-3 fatty acids and risk of stroke in women, *J. Am. Med. Assoc.*, 285, 304–312, 2001.

12. Sisovick, D.S., Raghunathan, T.E., King, I., Weinmann, S., Wicklund, K.G., Albright, J., Bovbjerg, V., Arbogast, P., Smith, H., Kushi, L.H., Cobb, L.A., Copass, M.K., Psaty, B.M., Lemaitre, R., Retzlaff, B., Childs, M., and Knopp, R.H., Dietary-intake and cell-membrane levels of long-chain n-3 polyunsaturated fatty-acids and the risk of primary cardiac-arrest, *J. Am. Med. Assoc.*, 274, 1363–1367, 1995.

13. Harris, W.S., Fish oils and plasma-lipid and lipoprotein metabolism in humans: a critical review, *J. Lipid Res.*, 30, 785–807, 1989.

14. Prisco, D., Filipini, M., Paniccia, R., Gensini, G.F., and Serneri, G.G.N., Effect of n-3 fatty-acid ethyl-ester supplementation on fatty-acid composition of the single platelet phospholipids and on platelet functions, *Metabolism*, 44, 562–569, 1995.

15. Shahar, E., Folsom, A.R., Wu, K.K., Dennis, B.H., Shimakawa, T., Conlan, M.G., Davis, C.E., and Williams, O.D., Associations of fish intake and dietary n-3 polyunsaturated fatty-acids with a hypocoagulable profile: the atherosclerosis risk in communities (Aric) study, *Arterioscler. Thromb.*, 13, 1205–1212, 1993.

16. Green, P., Fuchs, J., Schoenfeld, N., Leibovici, L., Lurie, Y., Beigel, Y., Rotenberg, Z., Mamet, R., and Budowski, P., Effects of fish-oil ingestion on cardiovascular risk-factors in hyperlipidemic subjects in Israel: a randomized, double-blind crossover study, *Am. J. Clin. Nutr.*, 52, 1118–1124, 1990.

17. Pauletto, P., Puato, M., Caroli, M.G., Casiglia, E., Munhambo, A.E., Cazzolato, G., Bon, G.B., Angeli, M.T., Galli, C., and Pessina, A.C., Blood pressure and atherogenic lipoprotein profiles of fish-diet and vegetarian villagers in Tanzania: the Lugalawa study, *Lancet*, 348, 784–788, 1996.

18. Schmidt, E.B., Varming, K., Pedersen, J.O., Lervang, H.H., Grunnet, N., Jersild, C., and Dyerberg, J., Long-term supplementation with n-3 fatty-acids: 2. Effect on neutrophil and monocyte chemotaxis, *Scand. J. Clin. Lab. Inv.*, 52, 229–236, 1992.

19. Simopoulos, A.P., Omega-3 fatty acids in health and disease and in growth and development, *Am. J. Clin. Nutr.*, 54, 438–463, 1991.

20. Bonanome, A., Pagnan, A., Biffanti, S., Opportuno, A., Sorgato, F., Dorella, M., Maiorino, M., and Ursini, F., Effect of dietary monounsaturated and polyunsaturated fatty-acids on the susceptibility of plasma low-density lipoproteins to oxidative modification, *Arterioscler. Thromb. Vasc. Biol.*, 12, 529–533, 1992.

21. Sørensen, N.S., Marckmann, P., Høy, C.-E., van Duyvenvoorde, W., and Princen, H.M.G., Effect of fish-oil-enriched margarine on plasma lipids, low-density-lipoprotein particle composition, size, and susceptibility to oxidation, *Am. J. Clin. Nutr.*, 68, 235–241, 1998.

22. De Caterina, R., Liao, J.K., and Libby, P., Fatty acid modulation of endothelial activation, *Am. J. Clin. Nutr.*, 71, 213S–223S, 2000.

23. Johansen, O., Seljeflot, I., Hostmark, A.T., and Arensen, H., The effect of supplementation with omega-3 fatty acids on soluble markers of endothelial function in patients with coronary heart disease, *Arterioscl. Thromb. Vasc. Biol.*, 19, 1681–1686, 1999.

24. Kristensen, S.D., Schmidt, E.B., and Dyerberg, J., Dietary supplementation with n-3 polyunsaturated fatty acids and human platelet function: a review with particular emphasis on implications for cardio-vascular disease, *J. Intern. Med.*, 225, 141–150, 1989.

25. Hirai, A., Hamazaki, T., Terano, T., Nishikawa, T., Tamura, Y., Kumagi, A., and Sajiki, J., Eicosapentaenoic acid and platelet function in Japanese, *Lancet*, 2, 1132–1133, 1980.

26. Ahmed, A.A. and Holub, B.J., Alteration and recovery of bleeding times, platelet aggregation and fatty acid composition of individual phospholipids in platelets of human subjects receiving a supplement of cod-liver oil, *Lipids*, 19, 617–624, 1984.

27. Uauy, R., Mena, P. and Valenzuela, A., Essential fatty acids as determinants of lipid requirements in infants, children and adults, *Eur. J. Clin. Nutr.*, 53, S66–S77, 1999.

28. Bayon, Y., Croset, M., Daveloose, D., Guerbette, F., Chirouze, V., Viret, J., Kader, J.C., and Lagarde, M., Effect of specific phospholipid molecular-species incorporated in human platelet membranes on thromboxane A(2) prostaglandin H-2 receptors, *J. Lipid Res.*, 36, 47–56, 1995.

29. Christensen, J.H., Korup, E., Aaroe, J., Toft, E., Moller, J., Rasmussen, K., Dyerberg, J., and Schmidt, E.B., Fish consumption, n-3 fatty acids in cell membranes, and heart rate variability in survivors of myocardial infarction with left ventricular dysfunction, *Am. J. Cardiol.*, 79, 1670–1673, 1997.

30. Christensen, J.H., Gustenhoff, P., Korup, E., Aaroe, J., Toft, E., Moller, J., Rasmussen, K., Dyerberg, J., and Schmidt, E.B., Effect of fish oil on heart rate variability in survivors of myocardial infarction: a double blind randomised controlled trial, *Br. Med. J.*, 312, 677–678, 1996.

31. Pepe, S. and McLennan, P.L., Dietary fish oil confers direct antiarrhythmic properties the myocardium of rats, *J. Nutr.*, 126, 34–42, 1996.

32. Billman, G.E., Kang, J.X., and Leaf, A., Prevention of ischemia-induced cardiac sudden death by n-3 polyunsaturated fatty acids in dogs, *Lipids*, 32, 1161–1168, 1997.

33. Kang, J.X. and Leaf, A., Antiarrhythmic effects of polyunsaturated fatty acids: recent studies, *Circulation*, 94, 1774–1780, 1996.

34. Kremer, J.M., Lawrence, D.A., and Petrillow, G.F., Effects of high-dose fish oil on rheumatoid arthritis after stopping nonsteroidal antiinflammatory drugs, *Arthritis Rheum.*, 38, 1107–1114, 1995.

35. Broughton, K.S., Johnson, C.S., Pace, B.K., Liebman, M., and Kleppinger, K.M., Reduced asthma symptoms with n-3 fatty acid ingestion are related to 5-series leukotriene production, *Am. J. Clin. Nutr.*, 65, 1011–1017, 1997.

36. Shahar, E., Folsom, A.R., Melnik, S.L., Tockman, M.S., Comstock, G.W., Gennars, V., Higgins, M.W., Sorlie, P.D., Ko, W.-J., and Szklo, M., Dietary n-3 polyunsaturated fatty acids and smoking-related chronic obstructive pulmonary disease, *N. Engl. J. Med.*, 331, 228–233, 1994.

37. Lawrence, R. and Sorrell, T., Eicosapentaenoic acid in cystic fibrosis: evidence of a pathogenetic role for leukotriene B4, *Lancet*, 342, 465–469, 1993.

38. Billuzzi, A., Brignola, C., Campieri, M., Pera, A., Boschi, S., and Miglioli, M., Effect of an enteric-coated fish-oil preparation on relapses in Crohn's disease, *N. Engl. J. Med.*, 334, 1557–1560, 1996.

39. Gogos, C., Ginopoulos, P., Salsa, B., Apostolidou, E., Zoumbos, N.C., and Kalfarentzos, F., Dietary omega-3 polyunsaturated fatty acids plus vitamin E restore immunodeficiency and prolong survival for severely ill patients with generalized malignancy, *Cancer*, 82, 395–402, 1998.

40. Cho, E., Hung, S., Willett, W.C., Spiegelman, D., Rimm, E.B., Seddon, J.M., Colditz, G.A., and Hankinson, S.E., Prospective study of dietary fat and the risk of age-related macular degeneration, *Am. J. Clin. Nutr.*, 73, 209–218, 2001.

41. Werkman, S.H. and Carlson, S.E., A randomized trial of visual attention of preterm infants fed docosa-hexaenoic acid until nine months, *Lipids*, 31, 91–97, 1996.

42. Jandacek, R.J., Whiteside, J.A., Holcombe, B.N., Volpenhein, R.A., and Taulbee, J.D., The rapid hydrolysis and efficient absorption of triglycerides with octanoic acid in the 1 and 3 positions and long-chain fatty acid in the 2 position, *Am. J. Clin. Nutr.*, 45, 940–945, 1987.

43. Christensen, M.S., Mullertz, A., and Høy, C.E., Absorption of triglycerides with defined or random structure by rats with biliary and pancreatic diversion, *Lipids*, 30, 521–526, 1995.

44. Straarup, E.M. and Høy, C.-E., Structured lipids improve fat absorption in normal and malabsorbing rats, *J. Nutr.*, 130, 2802–2808, 2000.

45. Innis, S.M., Dyer, R., Quinlan, P., and Diersenschade, D., Palmitic acid is absorbed as sn-2 monopalmitin from milk and formula with rearranged triacylglycerols and results in increased plasma triglyceride sn-2 and cholesteryl ester palmitate in piglets, *J. Nutr.*, 125, 73–81, 1995.

46. Filer, L.J., Mattson, F.H., and Fomon, S.J., Triglyceride configuration and fat absorption by the human infant, *J. Nutr.*, 99, 293–298, 1969.

47. Tomarell, R.M., Meyer, B.J., Weaber, J.R., and Bernhart, F.W., Effect of positional distribution on the absorption of the fatty acids of human milk and infant formulas, *J. Nutr.*, 95, 583–590, 1968.

48. Lucas, A., Quinlan, P., Abrams, S., Ryan, S., Meah, S., and Lucas, P.J., Randomised controlled trial of a synthetic triglyceride milk formula for preterm infants, *Arch. Dis. Child*, 77, F178–F184, 1997.

49. Lien, E.L., The role of fatty acid composition and positional distribution in fat absorption in infants, *J. Pediatr.*, 125, 562–568, 1994.

50. Christensen, T.C. and Hølmer, G., Lipase catalyzed acyl-exchange reactions of butter oil, *Milchwissenschaft/Milk Sci. Int.*, 48, 543–547, 1993.

51. Yang, T., Xu, X., He, C., and Li, L., Lipase-catalysed modification of lard to produce human milk fat substitutes, *Food Chem.*, 80, 473–481, 2003.

52. Taguchi, H., Nagao, T., Watanabe, H., Onizawa, K., Matsuo, N., Tokimitsu, I., and Itakura, H., Energy value and digestibility of dietary oil containing mainly 1,3-diacylglycerol are similar to those of triacylglycerol, *Lipids*, 36, 379–382, 2001.

53. Murase, T., Mizuno, T., Omachi, T., Onizawa, K., Komine, Y., Kondo, H., Hase, T., and Tokimitsu, I., Dietary diacylglycerol suppresses high fat and high sucrose diet-induced body fat accumulation in C57BL/6J mice, *J. Lipid Res.*, 42, 372–378, 2001.

54. Maki, K.C., Davidson, M.H., Tsushima, R., Matsuo, N., Tokimitsu, I., Umporowicz, D.M., Dicklin, M.R., Foster, G.S., Ingram, K.A., Anderson, B.D., Frost, S.D., and Bell, M., Consumption of diacylglycerol oil as part of a reduced-energy diet enhances loss of body weight and fat in comparison with consumption of a triacylglycerol control oil, *Am. J. Clin. Nutr.*, 76, 1230–1236, 2002.

55. Nagao, T., Watanabe, H., Goto, N., Onizawa, K., Taguchi, H., Matsuo, N., Yasukawa, T., Tsushima, R., Shimasaki, H., and Itakura, H., Dietary diacylglycerol suppresses accumulation of body fat compared to triacylglycerol in men in a double-blind controlled trial, *J. Nutr.*, 130, 792–797, 2000.

56. Sugimoto, T., Kimura, T., Fukuda, H., and Iritani, N., Comparisons of glucose and lipid metabolism in rats fed diacylglycerol and triacylglycerol oils, *J. Nutr. Sci. Vitaminol.*, 49, 47–55, 2003.

57. Sugimoto, T., Fukuda, H., Kimura, T., and Iritani, N., Dietary diacylglycerol-rich oil stimulation of glucose intolerance in genetically obese rats, *J. Nutr. Sci. Vitaminol.*, 49, 139–144, 2003.

58. Watanabe, H., Onizawa, K., Taguchi, H., Kobori, M., Chiba, H., Naito, S., Matsuo, N., Yasukawa, T., Hattori, M., and Shimasaki, H., Nutritional characterization of diacylglycerol in rats, *J. Jpn. Oil Chem. Soc.*, 46, 301–308, 1997.

59. Bierbach, H., Triacylglycerol biosynthesis in human small intestinal-mucosa – Acyl-CoA – monoglyceride acyltransferase, *Digestion*, 28, 138–147, 1983.

60. Friedman, H.I. and Nylund, B., Intestinal fat digestion, absorption, and transport: a review, *Am. J. Clin. Nutr.*, 33, 1108–1139, 1980.

61. Tada, N., Watanabe, H., Matsuo, N., Tokimitsu, I., and Okazaki, M., Dynamics of postprandial remnant-like lipoprotein particles in serum after loading of diacylglycerols, *Clin. Chim. Acta*, 311, 109–117, 2001.

62. Murata, M., Ide, T., and Hara, K., Reciprocal responses to dietary diacylglycerol of hepatic enzymes of fatty acid synthesis and oxidation in the rat, *Br. J. Nutr.*, 77, 107–121, 1997.

63. Van Wymelbeke, V., Louis-Sylvestre, J., and Fantino, M., Substrate oxidation and control of food intake in men after a fat-substitute meal compared with meals supplemented with an isoenergetic load of carbohydrate, long-chain triacylglycerols, or medium-chain triacylglycerols, *Am. J. Clin Nutr.*, 74, 620–630, 2001.

64. Strauss, R.S., Fatty acid synthase inhibitors reduce food intake and body weight, *Pediatr. Res.*, 48, 422, 2000.

65. Hara, K., Onizawa, K., Honda, H., Otsuji, K., Ide, T., and Murata, M., Dietary diacylglycerol-dependent reduction in serum triacylglycerol concentration in rats, *Ann. Nutr. Metab.*, 37, 185–191, 1993.

66. Murata, M., Hara, K., and Ide, T., Alteration by diacylglycerols of the transport and fatty-acid composition of lymph chylomicrons in rats, *Biosci. Biotechnol. Biochem.*, 58, 1416–1419, 1994.

67. Taguchi, H., Watanabe, H., Onizawa, K., Nagao, T., Gotoh, N., Yasukawa, T., Tsushima, R., Shimasaki, H., and Itakura, H., Double-blind controlled study on the effects of dietary diacylglycerol on postprandial serum and chylomicron triacylglycerol responses in healthy humans. *J. Am. Coll. Nutr.*, 19, 789–796, 2000.

68. Yamamoto, K., Asakawa, H., Tokunaga, K., Watanabe, H., Matsuo, N., Tokimitsu, I., and Yagi, N., Long-term ingestion of dietary diacylglycerol lowers serum triacylglycerol in Type II diabetic patients with hypertriglyceridemia, *J. Nutr.*, 131, 3204–3207, 2001.

69. Frayn, K.N., Insulin resistance and lipid metabolism, *Curr. Opin. Lipidol.*, 4, 197–204, 1993.

70. Hubert, P., Bruneauwack, C., Cremel, G., Lemarchandbrustel, Y., and Staedel, C., Lipid-induced insulin resistance in cultured hepatoma-cells is associated with a decreased insulin-receptor tyrosine kinase-activity, *Cell Regul.*, 2, 65–72, 1991.

71. Gray, J.I., Gomaa, E.A., and Buckley, D.J., Oxidative quality and shelf life of meats, *Meat Sci.*, 43, S111–S123, 1996.

72. Kitts, D.D., An evaluation of the multiple effects of the antioxidant vitamins, *Trends Food Sci. Technol.*, 8, 198–203, 1997.

73. Frankel, E.N., Recent advances in lipid oxidation. Review, *J. Sci. Food Agric.*, 54, 495–511, 1991.

74. Gutteridge, J.M.C. and Halliwell, B., The measurement and mechanism of lipid peroxidation in biological systems, *Trends Biochem. Sci.*, 15, 129–135, 1990.

75. Porter, N.A., Caldwell, S.E., and Mills, K.E., Mechanisms of free radical oxidation of unsaturated lipids, *Lipids*, 30, 277–289, 1995.

76. Belitz, H.-D. and Grosch, W., *Food Chemistry*, 2nd ed., Springer-Verlag, Berlin/Heidelberg, Germany, 1999.
77. Ahmad, J.I., Free radicals and health: is vitamin E the answer?, *Food Sci. Technol. Today*, 10, 147–152, 1996.
78. Frankel, E.N., *Lipid Oxidation*, The Oily Press, Dundee, U.K., 1998.
79. Kanner, J., Oxidative processes in meat and meat products: quality implications, *Meat Sci.*, 36, 169–189, 1994.
80. Buckley, D.J. and Morrissey, P.A., Vitamin E and Meat Quality. Animal Production Highlights, Hoffmann-La Roche Ltd, 1992.
81. Halliwell, B., Aeschbach, R., Löliger, J., and Aruoma, O.I., The characterization of antioxidants, *Food Chem. Toxic.*, 33, 601–617, 1995.
82. Kulås, E., Oxidation of fish lipids and its inhibition with tocopherols, in *Lipid Oxidation Pathways*, Kamal-Eldin, A., Ed., AOCS Press, Champaign, IL, 2003, pp. 37–69.
83. Karahadian, C. and Lindsay, R.C., Evaluation of compounds contributing characterizing fishy flavors in fish oils, *J. Am. Oil Chem. Soc.*, 66, 953–960, 1989.
84. Hsieh, T.C.Y., Williams, S.S., Vejaphan, W., and Meyers, S.P., Characterization of volatile components of menhaden fish (Brevoortia tyrannus) oil, *J. Am. Oil Chem. Soc.*, 66, 114–117, 1989.
85. Aidos, I., Jacobsen, C., Jensen, B., Luten, J.B., van der Padt, A., and Boom, R.M., Volatile oxidation products formed in crude herring oil under accelerated oxidative conditions, *Eur. J. Lipid Sci. Technol.*, 104, 808–818, 2002.
86. Milo, C. and Grosch, W., Detection of odor defects in boiled cod and trout by gas chromatography-olfactometry of headspace samples, *J. Agric. Food Chem.*, 43, 459–462, 1995.
87. Milo, C. and Grosch, W., Changes in the odorants of boiled salmon and cod as affected by the storage of the raw material, *J. Agric. Food Chem.*, 44, 2366–2371, 1996.
88. Grosch, W., Low-MW products of hydroperoxide reactions, in *Autoxidation of Unsaturated Lipids*, Chan, H.W.S., Ed., Academic Press, London, 1987, pp. 95–139.
89. Venkateshwarlu, G., Let, M.B., Meyer, A.S., and Jacobsen, C., Chemical and olfactometric characterization of volatile flavour compounds in a fish oil enriched milk emulsion, *J. Agric. Food Chem.*, 52, 311–317, 2004.
90. Venkateshwarlu, G., Let, M.B., Meyer, A.S., and Jacobsen, C., Modeling the sensory impact of defined combinations of volatile lipid oxidation products on fishy and metallic off-flavors, *J. Agric. Food Chem.*, 52, 1635–1641, 2004.
91. Macfarlane, N., Salt, J., and Birkin, R., The FAST index. A fishy scale. *Int. News Fats, Oils Relat. Mater.*, 21, 244–249, 2001.
92. Frankel, E.N., Satué-Gracia, T., Meyer, A.S., and German, J.B., Oxidative stability of fish and algae oils containing long chain polyunsaturated fatty acids in bulk and in oil-in-water emulsions, *J. Agric. Food Chem.*, 50, 2094–2099, 2002.
93. Mei, L., Decker, E.A., and McClements, D.J., Evidence of iron association with emulsion droplets and its impact on lipid oxidation, *J. Agric. Food Chem.*, 46, 5072–5077, 1998.
94. Satué-Gracia, M.T., Frankel, E.N., Rangavajhyala, N., and German, J.B., Lactoferrin in infant formulas: effect on oxidation, *J. Agric. Food Chem.*, 48, 4984–4990, 2000.
95. Irwandi, J., Man, Y.B.C., Kitts, D.D., Bakar, J., and Jinap, S., Synergies between plant antioxidant blends in preventing peroxidation reactions in model and food oil systems, *J. Am. Oil Chem. Soc.*, 77, 945–950, 2000.
96. Wada, S. and Koizumi, C., Influence of the position of unsaturated fatty-acid esterified glycerol on the oxidation rate of triglyceride, *J. Am. Oil Chem. Soc.*, 60, 1105–1109, 1983.
97. Neff, W.E. and El-Agaimy, M., Effect of linoleic acid position in triacylglycerols on their oxidative stability, *Food Sci. Technol.*, 29, 772–775, 1996.
98. Timm-Heinrich, M., Xu, X., Nielsen, N.S., and Jacobsen, C., Oxidative stability of structured lipids produced from sunflower oil and caprylic acid, *Eur. J. Lipid Sci. Technol.*, 105, 436–448, 2003.
99. Nielsen, N.S., Xu, X., Timm-Heinrich, M., and Jacobsen, C., Oxidative stability during storage of structured lipids produced from fish oil and caprylic acid, *J. Am. Oil Chem. Soc.*, 81, 375–384, 2004.
100. Akoh, C.C. and Moussata, C.O., Characterization and oxidative stability of enzymatically produced fish and canola oil-based structured lipids, *J. Am. Oil Chem. Soc.*, 78, 25–30, 2001.
101. Konishi, H., Neff, W.E., and Mounts, T.L., Oxidative stability of soybean products obtained by regio-selective chemical interesterification, *J. Am. Oil Chem. Soc.*, 72, 1393–1398, 1995.

102. Ledóchowska, E. and Wilczynska, E., Comparison of the oxidative stability of chemically and enzymatically interesterified fats, *Fett/Lipid*, 100, 343–348, 1998.

103. Moussata, C.O. and Akoh, C.C., Influence of lipase-catalyzed interesterification on the oxidative stability of melon seed oil triacylglycerols, *J. Am. Oil Chem. Soc.*, 75, 1155–1159, 1998.

104. Senanayake, S.P.J.N. and Shahidi, F., Chemical and stability characteristics of structured lipids from borage (Borago officinalis L.) and evening primrose (Oenothera biennis L.) oils, *J. Food Sci.*, 67, 2038–2045, 2002.

105. Senanayake, S.P.J.N. and Shahidi, F., *J. Food Sci.*

106. Yankah, V.V. and Akoh, C.C., Batch enzymatic synthesis, characterization and oxidative stability of DHA-containing structured lipids, *J. Food Lipids*, 7, 47–261, 2000.

107. Xu, X., Nielsen, N.S., Timm Heinrich, M., Jacobsen, C., and Zhou, D., Purification and deodorization of structured lipids by short path distillation, *Eur. J. Lipid Sci. Technol.*, 104, 745–755, 2002.

108. Jacobsen, C., Adler-Nissen, J., and Meyer, A.S., The effect of ascorbic acid on iron release from the emulsifier interface and on the oxidative flavor deterioration in fish oil enriched mayonnaise, *J. Agric. Food Chem.*, 47, 4917–4926, 1999.

109. Jacobsen, C., Timm, M., and Meyer, A.S., Oxidation in fish oil enriched mayonnaise: ascorbic acid and low pH increase oxidative deterioration, *J. Agric. Food Chem.*, 49, 3947–3956, 2001.

110. Thomsen, M.K., Jacobsen, C., and Skibsted, L.H., Initiation mechanisms of oxidation in fish oil enriched mayonnaise, *Eur. Food Res. Technol.*, 211, 381–86, 2000.

111. Jacobsen, C., Hartvigsen, K., Thomsen, M.K., Hansen, L.F., Lund, P., Skibsted, L.H., Hølmer, G., Adler-Nissen, J., and Meyer, A.S., Lipid oxidation in fish oil enriched mayonnaise: calcium disodium ethylenediaminetetraacetate, but not gallic acid, strongly inhibited oxidative deterioration, *J. Agric. Food Chem.*, 49, 1009–1019, 2001.

112. Jacobsen, C., Xu, X., Nielsen, N.S., and Timm-Heinrich, M., Oxidative stability of mayonnaise containing structured lipids produced from sunflower oil and caprylic acid, *Eur. J. Lipid Sci. Technol.*, 105, 449–458, 2003.

113. Nielsen, N.S., Petersen, A., Meyer, A.S., Timm-Heinrich, M., and Jacobsen, C., The effects of lactoferrin, phytic acid and EDTA on oxidation in two food emulsions enriched with long chain polyunsaturated fatty acids, *J. Agric. Food Chem.*, 52, 7690–7699, 2004.

114. Let, M.B., Jacobsen, C., Frankel, E.N., and Meyer, A.S., Oxidative flavour deterioration of fish oil enriched milk: Effects of oil type and Ethylenediaminetetraacetate (EDTA), *Eur. J. Lipid Sci. Technol.*, 105, 518–528, 2003.

115. Li Hsieh, Y.T. and Regenstein, J.M., Factors affecting quality of fish oil mayonnaise, *J. Food Sci.*, 56, 1298–1307, 1991.

116. Min, D.B. and Wen, J., Effects of dissolved free oxygen on the volatile compounds of oil, *J. Food Sci.*, 48, 1429–1430, 1983.

117. Mancuso, J.R., McClements, D.J., and Decker, E.A., The effects of surfactant type, pH, and chelators on the oxidation of salmon oil-in-water emulsions, *J. Agric. Food Chem.*, 47, 4112–4116, 1999.

118. Mei, L., McClements, D.J., Wu, J., and Decker, E.A., Iron-catalyzed lipid oxidation in emulsion as affected by surfactant, pH and NaCl, *Food Chem.*, 61, 307–312, 1998.

119. Hu, M., McClements, D.J., and Decker, E.A., Impact of whey protein emulsifiers on the oxidative stability of salmon oil-in-water emulsions, *J. Agric. Food Chem.*, 51, 1435–1439, 2003.

120. Hu, M., McClements, D.J., and Decker, E.A., Lipid oxidation in corn oil-in-water emulsions stabilized by casein, whey protein isolate, and soy protein isolate, *J. Agric. Food Chem.*, 51, 1696–1700, 2003.

121. Silvestre, M.C.P., Chaiyasit, W., Brannan, R.G., McClements, D.J., and Decker, E.A., Ability of surfactant headgroup size to alter lipid and antioxidant oxidation in oil-in-water emulsions, *J. Agric. Food Chem.*, 48, 2057–2061, 2000.

122. Tong, L.M., Sasaki, S., McClements, D.J., and Decker, E.A., Mechanisms of the antioxidant activity of a high molecular weight fraction of whey, *J. Agric. Food Chem.*, 48, 1473–1478, 2000.

123. Cho, Y.J., McClements, D.J., and Decker, E.A., Ability of surfactant micelles to alter the physical location and reactivity of iron in oil-in-water emulsion, *J. Agric. Food Chem.*, 50, 5704–5710, 2002.

124. Nuchi, C.D., Hernandez, P., McClements, D.J., and Decker, E.A., Ability of lipid hydroperoxides to partition into surfactant micelles and alter lipid oxidation rates in emulsions, *J. Agric. Food Chem.*, 50, 5445–5449, 2002.

125. Coupland, J.N. and McClements, D.J., Lipid oxidation in food emulsions, *Trends Food Sci. Technol.*, 7, 83–91, 1996.

126. Sims, R.J., Fioriti, J.A., and Trumbetas, J., Effect of sugar and sugar alcohols on autoxidation of safflower oil in emulsions, *J. Am. Oil Chem. Soc.*, 56, 742–745, 1979.

127. Nelson, K.A. and Labuza, T.P., Relationship between water and lipid oxidation rate: water activity and glass transition theory, in *Lipid Oxidation in Food*, St. Angelo, A.J., Ed., ACS Symposium Series 500, American Chemical Society, Washington, DC, 1992, pp. 93–103.

128. Jacobsen, C., Hartvigsen, K., Lund, P., Thomsen, M.K., Skibsted, L.H., Adler-Nissen, J., Hølmer, G. and Meyer, A.S., Oxidation in fish oil-enriched mayonnaise: 3. Assessment of the influence of the emulsion structure on oxidation by discriminant partial least squares regression analysis, *Eur. Food Res. Technol.*, 211, 86–98, 2000.

129. Allen, J.C. and Wrieden, W.L., Influence of milk proteins on lipid oxidation in aqueous emulsion: I. Casein, whey protein and α-lactalbumin, *J. Dairy Res.*, 79, 239–248, 1982.

130. Porter, W.L., Paradoxical behavior of antioxidants in food and biological systems, *Toxicol. Ind. Health*, 9, 93–122, 1993.

131. Frankel, E.N., Huang, S.-W., Kanner, J., and German, J.B., Interfacial phenomena in the evaluation of antioxidants: bulk oils vs. emulsions, *J. Agric. Food Chem.*, 42, 1054–1059, 1994.

132. Huang, S.-W., Hopia, A., Frankel, E.N., and German, J.B., Antioxidant activity of α-tocopherol and Trolox in different lipid substrates: bulk oils vs. oil-in-water emulsions, *J. Agric. Food Chem.*, 44, 444–452, 1996.

133. Jacobsen, C., Schwarz, K., Stoeckmann, H., Meyer, A.S., and Adler-Nissen, J., Partitioning of selected antioxidants in mayonnaise, *J. Agric. Food Chem.*, 47, 3601–3610, 1999.

134. Jacobsen, C., Hartvigsen, K., Lund, P., Meyer, A.S., Adler-Nissen, J., Holstborg, J., and Hølmer, G., Oxidation in fish oil enriched mayonnaise: 1. Assessment of propyl gallate as antioxidant by discriminant partial least squares regression analysis, *Z. Lebensm. Unters. Forsch.*, 210, 13–30, 1999.

135. Jacobsen, C., Hartvigsen, K., Lund, P., Adler-Nissen, J., Hølmer, G., and Meyer, A.S., Oxidation in fish oil enriched mayonnaise: 2. Assessment of the efficacy of different tocopherol antioxidant systems by discriminant partial least squares regression analysis, *Eur. Food Res. Technol.*, 210, 242–257, 2000.

136. Jacobsen, C., Hartvigsen, K., Lund, P., Thomsen, M.K., Skibsted, L.H., Hølmer, G., Adler-Nissen, J., and Meyer, A.S., Oxidation in fish oil enriched mayonnaise: 4. Effect of tocopherol concentration on oxidative deterioration, *Eur. Food Res. Technol.*, 212, 308–318, 2001.

137. Reference deleted.

138. Timm-Heinrich, M., Xu, X., Nielsen, N.S., and Jacobsen, C., Oxidative stability of mayonnaise and milk drink produced with structured lipids based on fish oil and caprylic acid, *Eur. Food Res. Technol.*, 219, 32–41, 2004.

139. Kolanowski, W., Swiderski, F., and Berger, S., Possibilities of fish oil application for foodproducts enrichment with omega-3 PUFA, *Int.. J. Food Sci. Nutr.*, 50, 39–49, 1999.

140. Kolanowski, W., Swiderski, F., Lis, E., and Berger, S., Enrichment of spreadable fats with polyunsaturated fatty acids omega-3 using fish oil, *Int. J. Food Sci. Nutr.*, 52, 469–476, 2001.

141. Yong, F.V.K., Using unhydrogenated fish oil in margarine, *INFORM*, 1, 731–741, 1990.

142. Let, M.B., Jacobsen, C., and Meyer A.S., Effects of fish oil type, lipid antioxidants and presence of rapeseed oil on oxidative flavour deterioration of fish oil enriched milk, *Eur. J. Lipid Sci. Technol.*, 106, 170–182, 2004.

143. Let, M.B., Jacobsen, C., and Meyer A.S., Sensory stability and oxidation of fish oil enriched milk is affected by milk storage temperatures and oil quality, *Int. Dairy J.*, 15, 173–182, 2005.

144. Timm-Heinrich, M., Xu, X., Nielsen, N.S., and Jacobsen, C., Oxidative stability of milk drink containing structured lipids produced from sunflower oil and caprylic acid, *Eur. J. Lipid Sci. Technol.*, 105, 459–470, 2003.

145. Huang, S.-W., Satué-Gracia, M.T., Frankel, E.N., and German, J.B., Effect of lactoferrin on oxidative stability of corn oil emulsions and liposomes, *J. Agric. Food Chem.*, 47, 1356–1361, 1999.

146. Timm-Heinrich, M., Nielsen, N.S., Xu, X., and Jacobsen, C., Structured lipids containing C18:0, C18:1, C18:2, C18:3 or CLA in *sn*2-position: oxidative stability as bulk lipids and in milk drinks, *Innov. Food Sci. Emerging Technol.*, 5, 249–261, 2004.

147. Keogh, M.K., O'Kennedy, B.T., Kelly, J., Auty, M.A., Kelly, P.M., Fureby, A., and Haahr, A.M., Stability to oxidation of spray-dried fish oil powder microencapsulated using milk ingredients, *J. Food Sci.*, 66, 217–224, 2001.

148. Kagami, Y., Sugimury, S., Fujishima, N., Matsuda, K., Kometani, T., and Matsumura, Y., Oxidative stability, structure and physical characteristics of microcapsules formed by spray drying of fish oil with protein and dextrin wall materials, *J. Food Sci.*, 68, 2248–2255, 2003.

149. Hardas, N., Danviriyakul, S., Foley, J.L., Nawar, W.W., and Chinachoti, P., Effect of the relative humidity on the oxidative stability and physical stability of encapsulated milk fat, *J. Am. Oil Chem. Soc.*, 79, 151–158, 2002.

150. Velasco, J., Dobarganes, M.C., and Márquez-Ruiz, G., Oxidation of free and encapsulated oil fractions in dried microencapsulated fish oils, *Grasas y Aceites*, 51, 439–446, 2000.

151. Heinzelmann, K., Franke, K., Jensen, B., and Haahr, A.M., Protection of fish oil from oxidation by microencapsulation using freeze-drying techniques, *Eur. J. Lipid Sci. Technol.*, 102, 114–121, 2000.

152. Hogan, S.A., O'Riordan, E.D., and O'Sullivan, Microencapsulation and oxidative stability of spray-dried fish oil emulsions, *J. Microencapsulation*, 20, 675–688, 2003.

153. Becker, C.C. and Kyle, D.J., Developing functional foods containing algal docosahexaenoic acid, *Food Technol.*, 52, 68–71, 1998.

22 Application of Multistep Reactions with Lipases to the Oil and Fat Industry

Yuji Shimada, Toshihiro Nagao, and Yomi Watanabe
Osaka Municipal Technical Research Institute, Osaka, Japan

CONTENTS

22.1 INTRODUCTION

The first generation of food lipids was natural oils and fats, but their structures, physical properties and/or fatty acid compositions, were not most suitable for humans. Hence, improvement of oils and fats has been much desired, and a second generation of lipids has emerged. The second generation of lipids includes single-cell oils produced from microorganisms, and oils and fats modified by lipases.

Industrial production of oils and fats modified by lipases historically started from cocoa fat substitute at the beginning of the 1980s: 1,3-stearoyl-2-oleoylglycerol, which has a sharp melting point around body temperature, was produced by exchange of fatty acids at the 1,3-positions of 2-oleoyl triacylglycerols (TAGs) with stearic acid[1]. The new process with a fixed-bed bioreactor packed with an immobilized 1,3-position-specific lipase attracted attention in those days, and significantly affected subsequent oil processing with lipases. In the 1990s an oil containing a high concentration of docosahexaenoic acid (DHA) was produced by selective hydrolysis of tuna oil with a lipase that acted weakly on DHA[2], and human milk fat substitute, 1,3-oleoyl-2-palmitoylglycerol, was also developed by exchange of fatty acids at the 1,3-positions of tripalmitin with oleic acid[3]. In addition, diacylglycerols (DAGs)[4] and TAGs containing medium- and long-chain fatty acids[5] were recently produced through lipase-catalyzed esterification and interesterification, respectively.

When lipases are used for production of functional lipids and purification of oil- and fat-related compounds, it is desirable to construct a reaction system by which the aim can be achieved in one step. One-step reactions may, however, suffer from the following problems: (1) a desired compound may not be synthesized, (2) the purity may not attain a target value, and (3) the yield maybe low. Some of these problems can be eliminated by adopting a multistep reaction. This chapter therefore deals with multistep reactions with lipases which are applicable for industrial oil and fat processing.

22.2   PURPOSE OF MULTISTEP REACTIONS

The purposes of multistep reactions are summarized in Table 22.1, in which examples described in this chapter are also listed. First of all, a multistep process is useful when a desired compound cannot be synthesized by a single reaction (Table 22.1, A). Repeated reaction is also effective when the purity of a desired compound does not reach a target value by a single reaction (Table 22.1, B).

One of the features of enzymic reaction is substrate specificity. This property is effective upon using a starting material including contaminants. Contaminants are removed from the material, and the resulting preparation is then converted to a desired form (Table 22.1, C).

Equilibrium exists in enzyme reactions as well as in chemical reactions. The reaction yield increases significantly by addition of a process removing a product out of the reaction system into the main reaction. For example, when a product is water or a short-chain alcohol, the equilibrium shifts to the right by removing it under reduced pressure (Table 22.1, D.1). When the melting point of a product is the highest among the components in a reaction medium, the reaction at low temperature allows it to solidify. Because enzymes recognize poorly solid-state substrates, the product is eliminated out of the reaction system, resulting in a high reaction yield (Table 22.1, D.2). In addition, the yield will increase if a product is changed to a different molecular form, which the enzyme does not recognize, by a side reaction (Table 22.1, D.3; Figure 22.1).

When a substrate inhibits or inactivates the enzyme, stepwise addition of the substrate may increase the yield. Also, in a two-step *in situ* reaction, when a substrate in the first step inhibits the second step, the reaction proceeds efficiently by addition of the substrate to the reaction mixture after the first-step reaction reaches equilibrium (Table 22.1, E). Production of functional lipids and purification of useful materials using these multistep reactions are described in the following sections.

22.3   PRODUCTION OF OILS CONTAINING A HIGH
CONCENTRATION OF POLYUNSATURATED FATTY ACIDS

Polyunsaturated fatty acids (PUFAs) have various physiological functions. Because oils containing a high concentration of PUFAs can be expected to display greater physiological effects even at a small

TABLE 22.1
Multistep Reaction Applicable for Oil and Fat Processing

A. Multistep process for synthesis of a desired compound
Section 22.4.2: Production of structured lipids rich in MLM[a]
Section 22.4.4: Synthesis of high purity of MLM

B. Repeated reaction for increase in the purity of a product
Section 22.3: Production of PUFA-rich oils
Section 22.4.1: Production of structured lipids rich in MLM
Section 22.6: Purification of functional fatty acids

C. Removal of contaminants and conversion of a desired compound
Section 22.4.3: Production of structured lipids rich in MLM containing a desired fatty acid at the 2-position

D. Process comprising main reaction and *in situ* product removal
1. Removal of a product (water or short-chain alcohols) under reduced pressure
Section 22.5.3: Production of MAG by esterification at low temperature followed by dehydration
2. Removal of a product at low temperatures
Section 22.5.1: Production of MAG by stepwise decrease of temperature
Section 22.5.2: Production of MAG by two-step *in situ* reaction: esterification and glycerolysis
3. Removal of a product by its changing to different molecular form which the enzyme recognizes weakly
Section 22.9: Conversion of fatty acid ester to free form

E. Elimination of inhibition or inactivation of the enzyme by stepwise addition of a substrate
Section 22.7: Alcoholysis of TAG with short-chain alcohol
Section 22.8: Purification of tocopherols and sterols

Note: Each example is described in the section shown in the table.
[a] Triacylglycerol with medium-chain fatty acid at the 1,3-positions and long-chain fatty acid at the 2-position.

$$A + B \rightleftharpoons C + D \quad (1)$$

$$D + E \longrightarrow F \quad (2)$$

FIGURE 22.1 Increase in reaction yield by *in situ* product removal. Main reaction (1) reaches equilibrium state. When a product (D) is changed to a different molecular form (F) which the lipase catalyzing reaction (1) recognizes weakly, the equilibrium of reaction (1) shifts to the right. Consequently, the yield of the product (C) increases significantly.

amount of intake, they have widely been used as nutraceuticals. An oil rich in PUFAs is industrially produced by selective hydrolysis of a natural oil with a lipase[2].

PUFAs can be enriched in the acylglycerol fraction by hydrolysis of PUFA-containing oil with a lipase which acts weakly on the PUFA[6]. For example, when tuna oil containing DHA was hydrolyzed with *Candida rugosa* lipase, the content of DHA in acylglycerols increased with increasing the degree of hydrolysis (Figure 22.2a). The degree of hydrolysis was increased by addition of larger amounts of the enzyme or by extension of the reaction period, but the content of DHA in acylglycerols depended only on the degree of hydrolysis[2]. In addition, the content of DHA did not exceed 54% at above 80% hydrolysis. Similarly, hydrolysis of borage oil with *C. rugosa* lipase enriched γ-linolenic acid (GLA) in acylglycerols (Figure 22.2b)[7,8]. The content of GLA increased

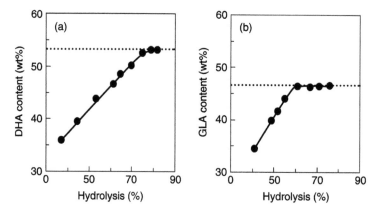

FIGURE 22.2 Selective hydrolysis of tuna and borage oils with *Candida rugosa* lipase. (A) Correlation of the degree of hydrolysis of tuna oil and the content of DHA in undigested acylglycerols. Tuna oil was hydrolyzed at 35°C for 24 h in a mixture containing 50% water and 20 to 2000 U/g mixture of *C. rugosa* lipase. (B) Correlation of the degree of hydrolysis of borage oil and the content of GLA in undigested acylglycerols. Borage oil was hydrolyzed at 35°C for 0.25 to 24 h in a mixture containing 50% water and 100 U/g mixture of *C. rugosa* lipase.

with increasing the degree of hydrolysis but was not raised over 46% at above 60% hydrolysis (Figure 22.2b)[2,8].

Repetition of the reaction is effective for increasing the content of PUFAs. After a single hydrolysis, acylglycerols were recovered and were hydrolyzed again under similar conditions. When selective hydrolyses of tuna and borage oils were repeated three to four times, the contents of DHA and GLA increased to 70 and 60%, respectively[2,8].

22.4   PRODUCTION OF STRUCTURED TAGS

In general, natural oils and fats contain saturated or monoenoic fatty acids at the 1,3-positions and highly unsaturated fatty acids at the 2-position. However, the distribution of fatty acids along the glycerol backbone is not specified. Meanwhile, a TAG having particular fatty acids at specific positions of glycerol is referred to as a structured TAG. TAGs with medium-chain fatty acid at the 1,3-positions and long-chain fatty acid at the 2-position (MLM-type) are hydrolyzed to 2-monoacylglycerols (MAGs) and fatty acids faster than TAGs with long-chain fatty acids (LLL-type), resulting in efficient absorption into intestinal mucosa[9,10]. Because PUFAs play a role in the prevention of a number of human diseases, MLM-containing PUFAs are expected as nutrition for patients with maldigestion and malabsorption of lipids, and as high-value-added nutraceuticals for the elderly. Studies on production of structured TAGs have been conducted at many laboratories since 1995, and many processes proposed, several of which are discussed from the view point of multistep process.

22.4.1   REPEATED ACIDOLYSIS

MLM-type structured TAGs are produced by acidolysis of natural oils with medium-chain fatty acids or by their interesterification with medium-chain fatty acid ethyl esters using immobilized 1,3-position-specific lipases (e.g., lipases from *Rhizopus oryzae*, *Rhizomucor miehei*, and *Thermomyces lanuginosa*) (Figure 22.3)[2,11–13]. To produce MLM-type structured TAGs containing

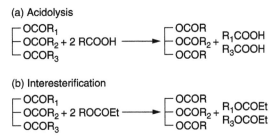

(a) Acidolysis

(b) Interesterification

FIGURE 22.3 Production of MLM-type structured TAGs. (A) Acidolysis of TAG with medium-chain fatty acid using an immobilized 1,3-position-specific lipase. (B) Interesterification of TAG with medium-chain fatty acid ethyl ester using an immobilized 1,3-position-specific lipase.

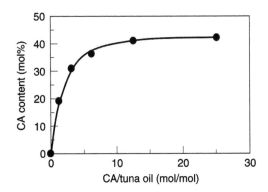

FIGURE 22.4 Effect of caprylic acid (CA) content in acidolysis of tuna oil with immobilized *Rhizopus oryzae* lipase. The reaction was conducted at 30°C for 48 h with shaking in a mixture containing different ratios of CA/tuna oil mixture (15 g) and immobilized lipase (0.3 g). nd not deleted

DHA, tuna oil underwent acidolysis with different amounts of caprylic acid (CA) using immobilized *R. oryzae* lipase (Figure 22.4). Incorporation of CA into acylglycerols increased with increasing amount of CA, and reached a constant value (42 mol%) at 13 mol of CA per mol of TAG[14]. If all fatty acids at the 1,3-positions are exchanged with CA, the content of CA will reach 66.7 mol%. Hence, the degree of acidolysis by this single reaction was only 63% based on the fatty acids at the 1,3-positions.

Repeated reaction may be effective for complete exchange of fatty acids at the 1,3-positions of natural oils with CA. To evaluate the efficiency of repeated reaction, tuna oil is not suitable as a substrate because it contains DHA at the 1,3-position, on which the lipase acts very weakly. Therefore, safflower and linseed oils were selected as substrates, and underwent acidolysis with CA using immobilized *R. oryzae* lipase (Table 22.2). Single reactions of safflower and linseed oils achieved 49 and 47 mol% incorporation of CA into acylglycerols, respectively. After the reaction, their acylglycerol fractions were recovered and then allowed to react again, resulting in 60 mol% of CA content. Threefold repetitions of acidolysis of the two oils reached 67 mol% of CA content, showing that all fatty acids at the 1,3-positions were exchanged with CA. Actually, HPLC analysis indicated that all TAGs were of the MLM type[15,16]. From these results, repeated reaction is evaluated to be effective for increasing the content of MLM-type TAGs.

TABLE 22.2
Increase in the Degree of Acidolysis by Repeated Reaction

		Fatty acid composition (mol%)						
Oil	Treatment	8:0	16:0	18:0	18:1	18:2	18:3	Acidolysis (%)
Safflower	None	nd	7.7	2.5	13.5	74.3	nd	—
	First	48.9	1.4	0.5	7.2	41.9	nd	73.3
	Second	59.8	0.5	nd	5.7	41.9	nd	89.7
	Third	67.4	nd	nd	5.2	27.4	nd	101.2
Linseed	None	nd	6.0	2.9	16.7	15.4	57.6	—
	First	46.6	1.3	0.6	10.0	10.5	30.9	69.9
	Second	60.8	0.4	nd	7.6	8.4	22.7	91.2
	Third	66.7	nd	nd	6.3	7.3	19.7	100.0

Note: Safflower and linseed oils underwent acidolysis at 30°C for 48 h with two weight parts of CA using 4% immobilized *Rhizopus oryzae* lipase. After the single acidolysis, acylglycerols were recovered and then allowed to react again under similar conditions. The acidolysis was repeated three times in total. nd, not deleted.

22.4.2 Two-Step Process Comprising Ethanolysis and Acylation

Natural oils containing PUFAs do not carry all PUFAs at the 2-position, and some of PUFAs are located at the 1,3-positions. When natural oils containing PUFAs are converted to MLM-type structured TAGs using a 1,3-position-specific lipase, it is difficult to convert all of the oils to MLM. Indeed, DHA at the 1,3-positions of tuna oil was converted with CA at only 30%, even though acidolysis of tuna oil with CA was repeated three times[16].

This problem was solved by a two-step process which was developed by Irimescu et al.[17]. The process comprised two steps: conversion of TAG to 2-MAG and acylation of the resulting 2-MAG (Figure 22.5). Bonito oil underwent ethanolysis with excess amounts of ethanol (EtOH) (>40 mol/mol of TAGs) using immobilized *Candida antarctica* lipase. The lipase acted strongly on C_8 to C_{24} of saturated and unsaturated fatty acids, and also acted on only fatty acids at the 1,3-positions in the presence of excess amounts of EtOH. Hence, 2-MAGs were efficiently prepared (yield, 93%). Acylation of the 2-MAGs was then performed with CA ethyl ester using immobilized *R. miehei* lipase. Because this reaction produced EtOH as a byproduct, removing EtOH under reduced pressure shifted the reaction to the right. Synthesis of MLM-type TAGs at 71% yield through this two-step process was achieved.

22.4.3 Two-Step Process Comprising Selective Hydrolysis and Repeated Acidolysis

Several kinds of fatty acids are located at the 2-position in natural oils. Even though all TAGs in a natural oil are converted to MLM-type TAGs through repeated reaction (Section 22.4.1) or the two-step process (Section 22.4.2), the content of MLM containing a desired fatty acid is determined by the content of the fatty acid esterified at the 2-position of the original oil. Including a pretreatment for increasing the content of a desired fatty acid at the 2-position, MLM-type TAGs with the fatty acid are produced in a high yield. Production of structured TAGs rich in 1,3-capryloyl-2γ-linolenoyl glycerol (CGC) by a process comprising selective hydrolysis of borage oil and repeated acidolysis of the resulting acylglycerols with CA is presented here.

The content of GLA at the 2-position of borage oil was 49 mol% (Table 22.3). Hence, even though all TAGs are converted to MLM, the content of CGC should not exceed 49 mol%. The

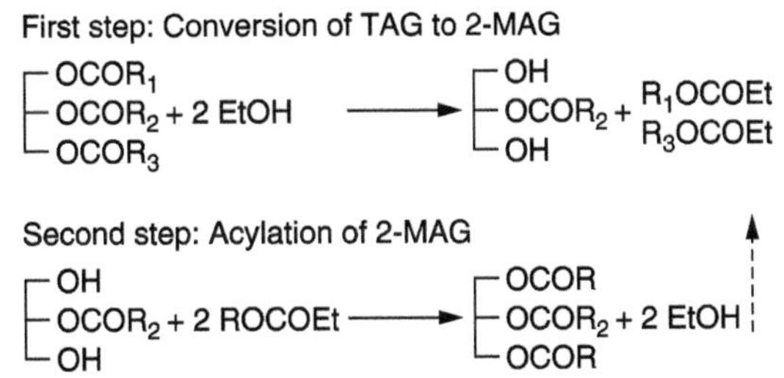

First step: Conversion of TAG to 2-MAG

$$
\begin{array}{l}
\text{OCOR}_1 \\
\text{OCOR}_2 + 2\,\text{EtOH} \\
\text{OCOR}_3
\end{array}
\longrightarrow
\begin{array}{l}
\text{OH} \\
\text{OCOR}_2 + \begin{array}{l}\text{R}_1\text{OCOEt}\\ \text{R}_3\text{OCOEt}\end{array} \\
\text{OH}
\end{array}
$$

Second step: Acylation of 2-MAG

$$
\begin{array}{l}
\text{OH} \\
\text{OCOR}_2 + 2\,\text{ROCOEt} \\
\text{OH}
\end{array}
\longrightarrow
\begin{array}{l}
\text{OCOR} \\
\text{OCOR}_2 + 2\,\text{EtOH} \\
\text{OCOR}
\end{array}
$$

FIGURE 22.5 Production of MLM-type structured TAGs from a natural oil by a two-step process comprising conversion of TAG to 2-MAG and acylation of the 2-MAG. The equilibrium in the second-step acylation shifted efficiently to the right by evaporating a product, ethanol (EtOH), using a vacuum pump.

TABLE 22.3
Fatty Acid Composition at the 1,3- and 2-Positions of Borage Oil and GLA45 TAGs

Oil	Position	Fatty acid composition (mol%)							
		16:0	18:0	18:1	18:2	18:3	20:1	22:1	24:1
Borage	1,3	16.1	5.8	17.8	36.6	9.3	5.5	3.3	2.1
	2	0.6	nd	12.3	34.8	48.9	nd	0.3	nd
GLA45 TAGs	1,3	9.1	5.4	15.1	28.5	25.3	7.7	4.5	2.4
	2	nd	0.3	2.4	10.8	84.8	0.5	0.3	0.2

Note: nd, not detected.

content of GLA at the 2-position can be increased by previous removal of TAGs with fatty acids except for GLA at the 2-position. Selective hydrolysis described in Section 22.3 is suitable for this purpose. An oil containing 45% GLA (GLA45 oil) was produced by selective hydrolysis of borage oil with *C. rugosa* lipase, and the content of GLA at the 2-position of GLA45 TAGs increased to 85 mol% (Table 22.3). Acidolysis of GLA45 TAGs with CA using immobilized *R. oryzae* lipase produced structured TAGs containing 45 mol% CGC, and the content of CGC increased to 61 mol% by repeating the acidolysis three times[18]. Because threefold repetition of acidolysis of borage oil produced structured TAGs containing only 35 mol% CGC, the process including selective hydrolysis was shown to be effective for increasing the content of MLM containing a desired fatty acid.

This two-step process was adopted for production of structured TAGs rich in 1,3-capryloyl-2-arachidonoyl glycerol (CAC) from a single-cell oil containing 40% arachidonic acid (AA). While acidolysis of the single-cell oil was repeated three times, the content of CAC increased to only 36 mol%. The process including selective hydrolysis of the oil, however, increased the content of CAC to 51 mol%[19], confirming that this two-step process is valuable for increasing the content of a desired MLM.

22.4.4 TWO-STEP PROCESS COMPRISING PRODUCTION OF TAG AND ITS ACIDOLYSIS

MLM with only one fatty acid at the 2-position cannot be synthesized as far as natural oils are used as starting materials, even though the reaction yield is raised. A high-purity MLM can be synthesized by preparation of a simple TAG, followed by exchange of fatty acids at the 1,3-positions with a medium-chain fatty acid (Figure 22.6).

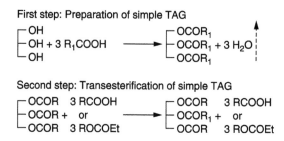

FIGURE 22.6 Synthesis of high-purity MLM by a process comprising preparation of a simple TAG and transesterification of the TAG. The equilibrium in the first-step esterification shifted efficiently to the right by evaporating a product, water, using a vacuum pump.

Chemical reactions can be employed in the synthesis of a simple TAG with saturated or monoenoic acids, but this is not suitable for synthesis of TAG with unstable PUFA. For this purpose, an enzymatic reaction is desirable. The TAG is synthesized by esterification of 1 mol of glycerol with 3 mol of free fatty acids (FFAs) using immobilized lipases. The yield may be significantly improved by the removal of generated water under reduced pressure. Because high reaction temperatures (>50°C) induce spontaneous acyl migration, even 1,3-position-specific lipases are available for the synthesis of TAG[20]. Immobilized *C. antarctica* lipase efficiently catalyzes the synthesis of TAG from glycerol and FFAs even at 40°C. In particular, this lipase is suitable for the synthesis of TAG with PUFA (TAG-PUFA) because of its strong activity on PUFAs[20,21]. High-purity MLM with PUFA at the 2-position can be synthesized by acidolysis of TAG-PUFA with CA or by interesterification of the TAG with CA ethyl ester (CAEE) using immobilized 1,3-position-specific lipases.

According to this strategy, 1,3-capryloyl-2-eicosapentaenoylglycerol (CEC) and CAC were synthesized. Trieicosapentaenoin was first synthesized from 1 mol of glycerol and 3 mol of eicosapentaenoic acid (EPA) using immobilized *C. antarctica* lipase, and then underwent interesterification with 100 mol CAEE using immobilized *R. miehei* lipase (yield of CEC, 88%)[22]. Acidolysis of triarachidonin and triesicosapentaenoin with 13 mol CA using immobilized *R. oryzae* lipase was repeated three times, and CEC and CAC were synthesized with yields of 87 and 86%, respectively[20].

As described in Section 22.4.2, transesterification with 1,3-position-specific lipases includes a drawback that the enzymes act weakly on PUFA. Tridocosahexaenoin actually underwent acidolysis with CA using immobilized *R. oryzae* lipase. While the reaction was repeated three times, the yield of 1,3-capryloyl-2-docosahexaenoylglycerol (CDC) was only 19%[20]. High-yield synthesis of CDC was achieved through a two-step process comprising ethanolysis of tridocosahexaenoin using immobilized *C. antarctica* lipase for preparing 2-monodocosahexaenoin and acylation of the 2-MAG with CAEE using immobilized *R. miehei* lipase (Figure 22.5). The yield of CDC increased to 85%[23].

22.5   PRODUCTION OF MAGs

MAGs are very good emulsifiers and are widely used as food additives. MAGs with saturated and monoenoic fatty acids are produced industrially by chemical alcoholysis of oils and fats with 2 mol of glycerol at high temperatures of 210 to 240°C[24,25]. However, the process cannot be used to synthesize MAGs with unstable fatty acids. Many research groups have therefore directed their attention to enzymatic reactions which catalyze efficiently the reaction under mild conditions, and have engaged in the synthesis of MAGs through hydrolysis of TAGs, esterification of FFAs with glycerol, glycerolysis of TAGs, and ethanolysis of TAGs in organic solvent systems[26]. Here, multistep reactions for production of MAGs that do not include any organic solvents and are applicable to the industrial production are presented.

The melting point of MAGs is the highest among FFAs, MAGs, DAGs, and TAGs, if the constituent fatty acids are the same. In addition, enzymes catalyze reactions below room temperature, and are active even at temperatures below the freezing point if the reaction efficiency is ignored. MAGs are produced efficiently by taking advantage of these two features. Considering the highest melting point of MAGs among those of the components in the reaction mixture, the following processes were successfully conducted.

22.5.1 Stepwise Decrease of Reaction Temperature: Glycerolysis

A mixture of TAG/glycerol (1:2, mol/mol) and a lipase was allowed to react at around the melting point of the TAG. The melting point of the MAG synthesized was higher than the reaction temperature; thus, the MAG was solidified. Because lipase acts poorly on solid-state substrates, the synthesized MAG was eliminated from the reaction mixture, resulting in a high reaction yield. According to this principle, beef tallow (reaction temperature, 42°C), palm oil (40°C), palm stearin (40°C), and olive oil (10°C) underwent glycerolysis. These reactions reached equilibrium after 4 to 5 days, and the yields of MAG from their oils and fats were 76, 67, 86, and 90%, respectively[27].

The yields were increased further by decreasing the temperature during the reaction. The glycerolysis of beef tallow, palm oil, and palm stearin began at 42°C, and the temperature was decreased to 5°C after 8 h. Continuation of these reactions for a further 4 days afforded good yields of 90, 91, and 94%, respectively[28]. This result shows that temperature programming for easy solidification of MAGs is very effective for an increase in yield.

22.5.2 Two-Step *In Situ* Reaction: Esterification and Glycerolysis

Conjugated linoleic acids (CLA) are a group of C_{18} fatty acids containing a pair of conjugated double bonds in either *cis* or *trans* configuration. A typical commercial product (referred to as FFA-CLA) contains almost equal amounts of 9*cis*, 11*trans* (9*c*,11*t*)-CLA and 10*t*,12*c*-CLA. FFA-CLA has been reported to have various physiological activities, such as reduction of the incidence of cancer[29,30], decrease in body fat content[31,32], beneficial effects on atherosclerosis[33,34], and improvement of immune function[35]. The activities have attracted a great deal of attention, and synthesis of MAG from FFA-CLA and glycerol was attempted because the first product in the industrial process is the FFA mixture containing CLA.

Monoacylglycerol lipase synthesizes MAG from FFAs and glycerol, and may not catalyze conversion of MAG to DAG. However, since there is no monoacylglycerol lipase available as an industrial enzyme, *Penicillium camembertii* mono- and diacylglycerol lipase (referred to as lipase) was used as a catalyst for production of MAG of CLA by a two-step *in situ* reaction system (Figure 22.7)[36].

The first step is esterification of FFA-CLA with 5 mol of glycerol. This reaction required 2% water by weight of the reaction mixture for maximal expression of activity of *P. camembertii* lipase. However, this water and that generated from esterification prevents a high degree of esterification, thus the reaction reached a steady state at 80% esterification. Hence, after the reaction at 30°C reached the steady state (10 h), the water was removed continuously by evaporation at 5 mmHg pressure using a vacuum pump. The degree of esterification reached was 95% after 24 h (34 h in total), and the contents of MAG and DAG were 49 and 46%, respectively.

The second step is glycerolysis of DAG. The reaction mixture in the first-step esterification was solidified by vigorous agitation on ice. When the solidified mixture was allowed to stand at 5°C for 2 weeks, glycerolysis of DAG proceeded successfully, and the content of MAG in the reaction mixture increased to 89%. Hydrolysis did not occur during the glycerolysis, and the content of FFAs decreased slightly from 5 to 4%.

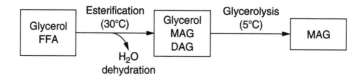

FIGURE 22.7 Production of MAG by a two-step *in situ* reaction with *Penicillium camembertii* lipase. The equilibrium in the first-step esterification proceeded efficiently by dehydration after the reaction reached the steady state.

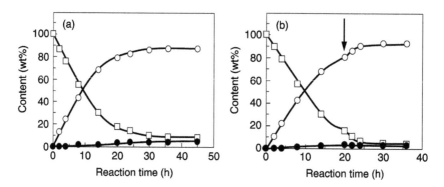

FIGURE 22.8 Production of MAG-CLA by esterification of FFA-CLA with glycerol followed by dehydration. A mixture of 294 g FFA-CLA/glycerol (1:5, mol/mol) and 6 mL (60,000 U) *P. camembertii* lipase solution was agitated at 5°C and 200 rpm. (A) Reaction was conducted without dehydration. (B) The reaction was conducted with dehydration: the reactor was connected to a vacuum pump at 20 h (indicated with arrow), and the reaction was continued with dehydration at 5 mmHg. □, FFAs; ○, MAGs; ●, DAGs.

22.5.3 ESTERIFICATION AT LOW TEMPERATURE AND DEHYDRATION

MAG-CLA was efficiently produced by the two-step *in situ* reaction described in Section 22.5.2. However, this process required more than 2 weeks for its completion. If *P. camembertii* lipase does not recognize MAG synthesized by esterification of FFA-CLA with glycerol, the yield of MAG would increase because the MAG does not convert to DAG. Thus, esterification of FFA-CLA with glycerol at low temperatures was attempted.

When FFA-CLA was esterified at 5°C with 5 mol of glycerol using *P. camembertii* lipase in the presence of 2% water, the degree of esterification reached 90% after 45 h and the contents of MAGs and DAGs were 87 and 5%, respectively (Figure 22.8a)[37]. TAGs were not synthesized in this *P. camembertii* lipase-catalyzed esterification. After the esterification was conducted for 20 h (the degree of esterification was 81%), dehydration was started by evaporation at 5 mmHg using a vacuum pump. The degree of esterification increased concomitantly with dehydration and reached 95% after 16 h (36 h in total). The contents of MAGs and DAGs were 93 and 3%, respectively (Figure 22.8b)[37].

Other triacylglycerol lipases (from *R. oryzae* and *C. rugosa*) were also effective for production of MAG-CLA by esterification at 5°C, followed by dehydration. When *C. rugosa* lipase was used instead of *P. camembertii* lipase and the reaction was conducted under the same conditions, the degree of esterification reached 95% and the content of MAGs was 93%[38]. These results show that removal of products from the reaction mixture (MAGs, at low temperature; water, under reduced pressure) is effective for an increase in the reaction yield because the equilibrium shifts to the right.

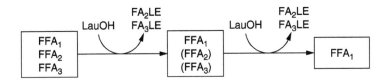

FIGURE 22.9 Purification of PUFA by repetition of selective esterification of a FFA mixture containing a desired PUFA with LauOH. FFA$_1$, desired PUFA; FFA$_2$ and FFA$_3$, contaminating FFAs; LauOH, lauryl alcohol; FA$_x$LE, fatty acid lauryl esters.

22.6   PURIFICATION OF FUNCTIONAL FATTY ACIDS

A mixture of FFAs containing a desired fatty acid is esterified with an alcohol by taking advantage of fatty acid specificity of lipase. When only contaminating fatty acids are esterified, the desired fatty acid is enriched in the FFA fraction. The purity can be raised by repeated esterification of FFAs recovered from the reaction mixture (Figure 22.9). The reaction mixture in the esterification consisted of alcohol, FFAs, and fatty acid esters. When these compounds were separated by short-path distillation, lauryl alcohol (LauOH) was reported to be the most suitable substrate[39,40]. In this section, repeated selective esterification is shown to be effective for purification of PUFAs and CLA isomers.

22.6.1   PURIFICATION OF PUFAS BY HYDROLYSIS AND REPEATED ESTERIFICATION

GLA was purified from borage oil by a process comprising repeated selective esterification and distillation (Figure 22.10)[41,42]. A 10 kg oil containing 45% GLA (GLA45 oil) was used as a starting material, which was prepared by selective hydrolysis of borage oil with *C. rugosa* lipase[8]. GLA45 oil was first hydrolyzed with *Burkholderia cepacia* lipase which acted strongly on GLA. The reaction mixture was subjected to short-path distillation, and FFAs were recovered in the distillate fraction. The FFA mixture was then esterified with LauOH using *R. oryzae* lipase which acted weakly on GLA: 96% of the contaminating FFAs were esterified, and GLA was enriched to 90% in the FFA fraction. To further increase the purity of GLA, FFAs recovered by distillation were reesterified with LauOH, resulting in an increase in the purity to 98%. In addition, fatty acid lauryl esters (FALEs) contaminated the FFA fraction at 14%, which were completely removed by urea adduct formation. The FFAs with 98% GLA were prepared with a recovery of 49% of the initial content of GLA45 oil through a series of purification procedures[42]. It was confirmed from this result that repeated selective reaction is useful for increasing the purity of a desired compound to a high level.

The process described above is effective for purification of other PUFAs. When tuna oil was used as a starting material, DHA was purified to 91% with 60% recovery by nonselectively hydrolyzing tuna oil with *Pseudomonas* lipase, followed by repeated selective esterification of the resulting FFAs with *R. oryzae* lipase twice[43]. In addition, n-6 PUFAs were purified from a single-cell oil containing 40% arachidonic acid (AA). The FFAs were prepared by nonselective hydrolysis of the oil with *B. cepacia* lipase, and underwent selective esterification with *C. rugosa* lipase twice. Because *C. rugosa* lipase acted weakly not only on AA, but also on GLA and dihomo-GLA[44], the total content of n-6 PUFAs increased to 96% with a 52% recovery (AA purity, 81%; AA recovery, 53%)[45].

22.6.2   PURIFICATION OF CLA ISOMERS BY REPEATED ESTERIFICATION

A typical commercial product of CLA contains almost equal amounts of 9*c*,11*t*-CLA and 10*t*,12*c*-CLA. It was recently reported that 9*c*,11*t*-CLA has anticancer activity[46], and that 10*t*,12*c*-CLA decreases

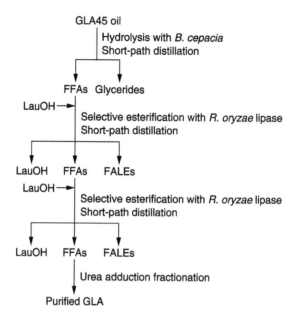

FIGURE 22.10 Purification process of GLA from GLA45 oil.

body fat content[47–49] and suppresses the development of hypertension[50]. These studies created a great deal of interest in the fractionation of CLA isomers.

A mixture of CLA isomers was first prepared by alkali conjugation of linoleic acid; referred to as CLAmix. The FFA mixture contained 45% 9*c*,11*t*-CLA, 47% 10*t*,12*c*-CLA, and 5% other CLA isomers. The two CLA isomers can be purified by the process comprising repeated esterification with LauOH and short-path distillation (Figure 22.11)[51,52].

CLAmix was esterified with LauOH using *C. rugosa* lipase which acts strongly on 9*c*,11*t*-CLA and weakly on 10*t*,12*c*-CLA, and the FFA fraction containing 78% 10*t*,12*c*-CLA and the FALE fraction containing 85% 9*c*,11*t*-CLA were recovered by short-path distillation. The FFA and FALE fractions were used for further purification of 10*t*,12*c*-CLA and 9*c*,11*t*-CLA, respectively.

To purify 10*t*,12*c*-CLA, the FFA fraction was esterified again with LauOH. The FFA fraction was recovered by distillation and subsequently subjected to urea adduct fractionation to remove the contaminating CLA isomers. Consequently, the purity of 10*t*,12*c*-CLA reached 95% (the content of 10*t*,12*c*-CLA based on the total content of 9*c*,11*t*- and 10*t*,12*c*-isomers, 97%). The recovery of 10*t*,12*c*-CLA by a series of purification procedures was 31% of the initial amount[52].

Another isomer, 9*c*,11*t*-CLA, enriched in the FALE fraction was also purified. The FALEs were hydrolyzed under alkaline conditions, and FFAs were recovered. The FFAs were reesterified with LauOH. After the reaction, FALEs were recovered by distillation. The FALEs were hydrolyzed and LauOH and FFAs separated by distillation. These procedures increased the purity of 9*c*,11*t*-CLA to 93% (the content of 9*c*,11*t*-CLA based on the total content of 9*c*,11*t*- and 10*t*,12*c*-isomers, 96%). The recovery of 9*c*,11*t*-CLA was 34% of the initial amount[52].

Success in purification of CLA isomers also indicated that repeated esterification with LauOH was effective for achieving a high-purity product.

22.7 ALCOHOLYSIS OF TAGS WITH SHORT-CHAIN ALCOHOLS

Fatty acid methyl esters (FAMEs) produced from waste edible oil and surplus oil are used as biodiesel fuel. Presently, the industrial production of biodiesel fuel is conducted by methanolysis of

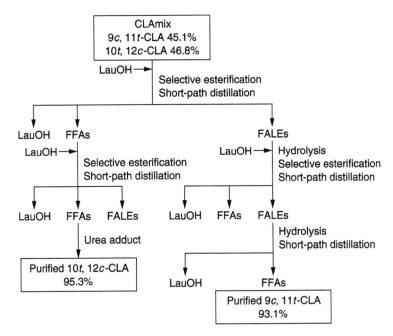

FIGURE 22.11 Purification process of CLA isomers.

waste oil using alkaline catalysts. A byproduct, glycerol, thus contains the alkali, and has to be treated as a waste material. In addition, because waste oils contain a small amount of water and FFAs, the reaction generates fatty acid alkaline salts (soaps). The soaps are removed by washing with water, but the resulting water contains glycerol, methanol (MeOH), and alkaline catalyst. Hence, disposal of the alkaline water creates other environmental concerns. On the other hand, since enzymatic methanolysis of waste oil does not generate any waste material, production of biodiesel fuel with a lipase is strongly desired. An enzymatic system with stepwise addition of MeOH is shown here as being effective for production of biodiesel fuel from waste oil.

22.7.1 INACTIVATION OF LIPASE BY INSOLUBLE MeOH

In general, lipases efficiently catalyze the reactions when substrates are miscible. The solubility of MeOH in a vegetable oil is therefore of interest. It was shown that the solubility of MeOH was 0.5 mol against total fatty acids in the oil. Disregarding the low solubility of methanol in the oil, all methanolysis reactions of TAGs so far reported were conducted with more than the required stoichiometric amount of MeOH. It is known that proteins are generally unstable in short-chain alcohols, such as MeOH and EtOH. Therefore, it was assumed that low methanolysis may arise from inactivation of lipase by contact with insoluble MeOH which exists as micelles in the oil. Actually, MeOH was completely consumed in methanolysis of vegetable oil with <1/3 mol equivalent of MeOH for the stoichiometric amount using immobilized *C. antarctica* lipase, but the methanolysis was decreased significantly by adding >1/2 mol equivalent of MeOH (Figure 22.12)[53]. In addition, the decreased activity was not restored in subsequent reactions with 1/3 mol equivalent of MeOH, showing that the immobilized lipase was irreversibly inactivated by contact with insoluble MeOH in the oil[53].

22.7.2 STEPWISE METHANOLYSIS OF VEGETABLE OIL

Inactivation of lipase by excess amounts of a substrate should be eliminated by stepwise addition of the substrate. Thus the methanolysis of vegetable oil by three successive additions of 1/3 mol

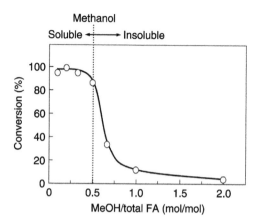

FIGURE 22.12 Methanolysis of vegetable oil with different amounts of methanol (MeOH) using immobilized *Candida antarctica* lipase. A mixture of 10 g vegetable oil/MeOH and 0.4 g immobilized lipase was shaken at 30°C for 24 h. The conversion is expressed as the amount of MeOH consumed for methanolysis of the oil (when the molar ratio of MeOH/oil is less than 1.0) and as the ratio of FAMEs to oil (more than 1.0).

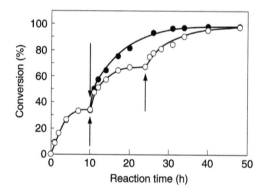

FIGURE 22.13 Stepwise methanolysis of vegetable oil with immobilized *C. antarctica* lipase. Three-step reaction (○): a mixture of 28.95 g vegetable oil, 1.05 g MeOH (1/3 molar equivalent for the stoichiometric amount), and 4% immobilized *C. antarctica* lipase was incubated at 30°C with shaking at 130 oscillations/min, and 1.05 g MeOH was added at 10 and 24 h. Two-step reaction (●): the reaction was started under the same conditions as those of three-step reaction. After 10 h, 2.10 g MeOH (2/3 molar equivalent) was added in the reaction mixture. Upward and downward arrows indicate the addition of 1/3 and 2/3 molar equivalent of MeOH, respectively.

equivalent of MeOH was attempted (Figure 22.13)[53]. The first step of the methanolysis was conducted in a mixture of the oil, 1/3 mol equivalent of MeOH, and immobilized *C. antarctica* lipase. After 7 h, conversion of the oil reached 33%. The addition of a second 1/3 mol equivalent of MeOH at 10 h converted 66% of the oil to its corresponding FAMEs after 10 h (total 20 h). A third 1/3 mol equivalent of MeOH was added again after a total of 24 h, and the reaction was continued. After 24 h (48 h in total), the reaction reached 97% conversion, showing that three-step reaction was effective for nearly complete conversion of the oil.

Solubility of MeOH in TAG is low, but high in FAME. The reaction mixture after the first-step methanolysis is composed of 67% acylglycerols and 33% FAMEs. In this reaction mixture, 2/3 mol equivalent of MeOH was completely dissolved, and *C. antarctica* lipase was found not to inactivate even in the mixture of acylglycerols/FAMEs and the 2/3 mol equivalent of MeOH. This finding led to success in a two-step methanolysis of the oil (Figure 22.13)[54]. The first-step methanolysis was started in a mixture of the oil, 1/3 mol equivalent of MeOH, and immobilized *C. antarctica* lipase. The content of FAMEs reached 33% at 7 h. The addition of a second 2/3 mol equivalent of MeOH at 10 h converted 97% of the oil to its corresponding FAMEs after 24 h (34 h in total). Achievement of this two-step reaction shortened the reaction time by approximately 30%.

The three- and two-step reactions were repeated using the same lipase to study its stability, indicating that immobilized *C. antarctica* lipase can be used for >100 days without significant loss of activity[53–55]. These results show that a reaction system with stepwise addition of the substrate, which inactivates the enzyme, is useful for achieving a good reaction yield.

The stepwise methanolysis was applied to conversion of waste edible oil to FAMEs. The conversion of the waste oil reached only 91%, although the lipase was stable for >100 days[56]. The low conversion may be attributed to the oxidized fatty acids in waste oil, such as epoxides, aldehydes, and polymers, among others. Also, a stepwise reaction was used for ethanolysis of tuna oil. The degree of ethanolysis reached >95% by adding 1/3 mol of EtOH three times; the enzyme was useable for >100 days in this stepwise ethanolysis process[57].

22.8 PURIFICATION OF TOCOPHEROLS AND STEROLS

The final process in vegetable oil refining is deodorization by steam distillation. The resulting distillate (vegetable oil deodorizer distillate, VODD) contains tocopherols, sterols, and steryl esters which are utilized as important starting materials for purification of tocopherols. In a purification process, however, sterols cause inefficient yield of tocopherols. Sterols have been purified from the residue obtained after purification of tocopherols, but the yield is generally low. However, tocopherols and sterols are shown to be highly purified from VODD by a two-step *in situ* reaction in which a substrate is added after the first-step reaction reaching the equilibrium state.

A process for purification of tocopherols and sterols from VODD is shown in Figure 22.14. VODD was first fractionated into a low boiling point fraction (not including steryl esters; VODD tocopherols and sterols concentrate, VODDTSC) and a high boiling point fraction (not including tocopherols and sterols); tocopherols and sterols were purified from VODDTSC.

VODDTSC contained FFAs, tocopherols, sterols, partial acylglycerols, and unknown hydrocarbons. Among them, tocopherols, sterols, and partial acylglycerols are not efficiently fractionated by short-path distillation. Removal of FFAs in the tocopherol fraction is also difficult to achieve. If hydrolysis of partial acylglycerols, conversion of sterols to steryl esters, and methyl esterification of FFAs are achieved, the resulting mixture will be composed of tocopherols, steryl esters, and FAMEs. The boiling point of FAMEs is lower than that of FFAs, and the molecular weight of steryl ester becomes higher than that of tocopherols. Hence, the three components (FAMEs, tocopherols, and steryl esters) are presumably efficiently fractionated by short-path distillation.

To construct a system in which the three reactions (hydrolysis of acylglycerols, conversion of sterols to steryl esters, and methyl esterification of FFAs) proceed in one batch, *C. rugosa* lipase was selected because this enzyme efficiently catalyzes esterification of sterols with fatty acids and methyl esterification of fatty acids, even in the presence of 50% water. VODDTSC was first treated with *C. rugosa* lipase in a mixture containing 20% water and 2 mol of MeOH per total fatty acids. Consequently, hydrolysis of acylglycerols and methanolysis of fatty acids proceeded efficiently, but conversion of sterols to steryl esters proceeded poorly (Figure 22.15a)[58]. This observation was presumably due to the preference for methanolysis of fatty acids rather than conversion of sterols to steryl esters. To solve this problem, a two-step *in situ* reaction was attempted (Figure 22.15b). The

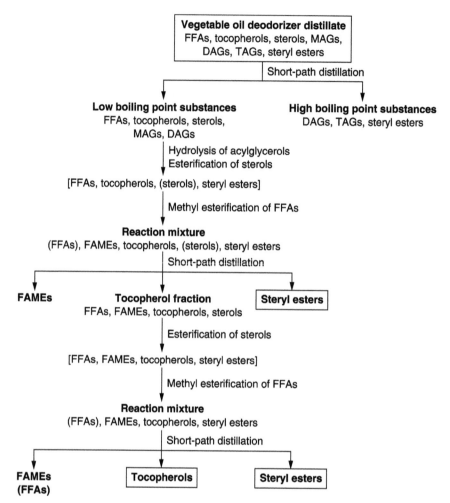

FIGURE 22.14 Purification of tocopherols and sterols by a process comprising two-step *in situ* reaction and short-path distillation.

first step was hydrolysis of acylglycerols and conversion of sterols to steryl esters; VODDTSC was treated with the lipase in the presence of 20% water. After the reaction reached equilibrium, MeOH was added to the mixture and the reaction was continued. This two-step *in situ* reaction successfully proceeded, and achieved complete hydrolysis of acylglycerols; 80% conversion of sterols to steryl esters, and 78% methyl esterification of fatty acids[58].

The two-step *in situ* reaction with *C. rugosa* lipase was adopted for purification of tocopherols and sterols from VODDTSC (Figure 22.14). A single *in situ* reaction attained only 80% conversion of sterols to steryl esters. To increase the conversion, after the single reaction, FAMEs and steryl esters were removed by short-path distillation, and the fraction containing tocopherols was subjected again to the two-step *in situ* reaction. Consequently, conversion of sterols to steryl esters reached 60% (total conversion, 92%). The resulting mixture was finally subjected to short-path distillation and fractionated into FAMEs, tocopherols, and steryl esters. Through a series of purification procedures, tocopherols were purified to 76% with 90% recovery, and sterols were purified as steryl esters to 97% with 86% recovery[58].

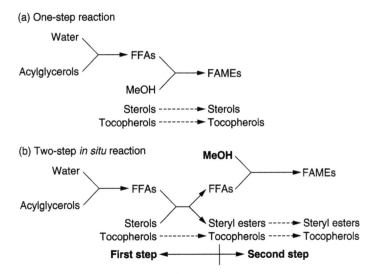

FIGURE 22.15 Two-step *in situ* reaction applied to purification of tocopherols and sterols from VODDTSC. (A) One-step reaction. VODDTSC was treated with *C. rugosa* lipase in a mixture containing 20% water and 2 mol of MeOH per the total fatty acids. (B) Two-step *in situ* reaction. VODDTSC was treated with *C. rugosa* lipase in a mixture containing 20% water. After esterification of sterols with fatty acids reached a steady state, 2 mol of MeOH per FFAs was added to the mixture, and the reaction was continued.

22.9 CONVERSION OF FATTY ACID ESTERS TO FREE-FORM PRODUCTS

22.9.1 CONVERSION OF STERYL ESTERS TO FREE STEROLS

Tocopherols and steryl esters in VODDTSC may be highly purified with high recovery through the process shown in Figure 22.14. However, sterols were purified as steryl esters. Since free sterols are also useful food additives, conversion of steryl esters to free sterols was next attempted.

Various industrial lipases were screened to find a suitable lipase for hydrolysis of steryl esters at a high yield. *C. rugosa*, *Geotrichum candidum*, *Pseudomonas aeruginosa*, *Pseudomonas stutzeri*, *B. cepacia*, and *Burkholderia glumae* lipases hydrolyzed steryl esters in the presence of 50% water, but the hydrolysis reached the equilibrium state around 50%. Products in hydrolysis of steryl esters are sterols and FFAs. The equilibrium in the reaction should shift in the direction of hydrolysis by the removal of FFAs. Lipase acts strongly on FFAs, but weakly on FAMEs. Therefore, conversion of FFAs to FAMEs along with hydrolysis of steryl esters was attempted (Figure 22.16; *in situ* product removal). When a mixture of steryl esters/MeOH (1:2, mol/mol) was treated with *P. aeruginosa* lipase in the presence of 50% water, 98% steryl esters were converted to free sterols. Sterols are not soluble in *n*-hexane, but FAMEs, FFAs, and steryl esters are soluble. Hence, *n*-hexane fractionation of the reaction mixture purified steryl esters to 99% with 92% recovery[59].

22.9.2 CONVERSION OF ASTAXANTHIN ESTERS TO FREE ASTAXANTHIN

Astaxanthin (3,3′-dihydroxy-β,β-carotene-4,4′-diene) is widely distributed in shellfish exoskeleton and has various physiological functions, such as serving as a precursor of vitamin A[60], quenching of free radicals and active oxygen species[61], rendering anticancer activity[62], and enhancing immune response[63]. These activities have attracted a great deal of attention, and astaxanthin has been used as a nutraceutical and cosmetic ingredient.

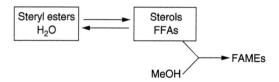

FIGURE 22.16 Conversion of steryl esters to free sterols by a reaction system with *in situ* product removal.

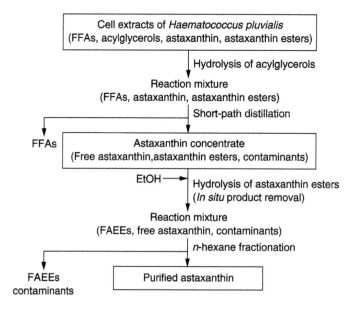

FIGURE 22.17 Purification process of astaxanthin from *Haematococcus pluvialis* cell extracts.

An industrially available preparation of astaxanthin from *Haematococcus pluvialis* cells contained 42% acylglycerols, 25% FFAs, and 15% astaxanthin in the free and esterified forms (free astaxanthin/astaxanthin monoesters/astaxanthin diesters = 5:80:15 (mol)). Astaxanthin was purified from this preparation by a process including a reaction (*in situ* product removal) in which astaxanthin esters were converted to the free form (Figure 22.17).

Contaminating acylglycerols were first hydrolyzed with *C. rugosa* lipase in the presence of 50% water, and FFAs were removed by short-path distillation. This procedure increased the purity of astaxanthin from 15 to 41%. Insolubility of free astaxanthin in *n*-hexane was noted, and astaxanthin esters in the concentrate were attempted for conversion to the free form. When a mixture of astaxanthin concentrate, 5 mol of EtOH against fatty acids, and 50% water was agitated with *P. aeruginosa* lipase, 90% of astaxanthin esters were converted to the free form. The free form was efficiently recovered by precipitation with *n*-hexane. The purity of astaxanthin was thereby increased to 70% with 64% overall recovery from the initial content in the *H. pluvialis* cell extracts[64].

The two reactions (hydrolysis of fatty acid esters and methyl (ethyl) esterification of FFAs) seemed to proceed simultaneously in conversion of steryl esters or astaxanthin esters to the free form. Multistep reactions generally proceed successively, but may also proceed at the same time, as discussed in this section.

22.10 CONCLUSIONS

Lipases have been used as important tools for modification of lipids in the oil and fat industry. Also, the lipase-catalyzed reactions have recently been reported to be effective for purification of oil- and fat-related compounds because lipases have substrate specificity. To apply the reactions industrially, high yields have to be achieved, and multistep processes have been proposed. Naturally, multistep processes require removal of contaminants or recovery of products. The most effective system for achieving a high reaction yield was *in situ* reaction in which more than one reaction proceeds successively or at the same time. Such processes may become useful for the development of reaction systems with high yields.

REFERENCES

1. Yokozeki, K., Yamanaka, S., Takinami, K., Hirose, Y., Tanaka, A., Sonomoto, K., and Fukui, S., Application of immobilized lipase to regio-specific interesterification of triglyceride in organic solvent, *Eur. J. Appl. Microbiol. Biotechnol.*, 14, 1–5, 1982.
2. Shimada, Y., Sugihara, A., and Tominaga, Y., Production of functional lipids containing polyunsaturated fatty acids with lipase, in *Enzymes in Lipid Modification*, Bornscheuer, U.T., Ed., Wiley-VCH, Weinheim, 2000, pp. 128–147.
3. Akoh, C.C. and Xu, X., Enzymatic production of Betapol and other specialty fats, in *Lipid Biotechnology*, Kuo, T.M. and Gardner, H.W., Eds., Marcel Dekker, New York, 2002, pp. 461–478.
4. Matsuo, N. and Tokimitsu, I., Metabolic characteristics of diacylglycerol, *INFORM*, 12, 1098–1102, 2001.
5. Negishi, S., Shirasawa, S., Arai, Y., Suzuki, J., and Mukataka, S., Activation of powdered lipase by cluster water and the use of lipase powders for commercial esterification of food oils, *Enzym. Microb. Technol.*, 32, 66–70, 2003.
6. Hoshino, T., Yamane, T., and Shimizu, S., Selective hydrolysis of fish oil by lipase to concentrate n-3 polyunsaturated fatty acids, *Agric. Biol. Chem.*, 54, 1459–1467, 1990.
7. Rahmatsullah, M.S.K.S., Shukla, V.S.K., and Mukherjee, K.D., Enrichment of γ-linolenic acid from evening primrose oil and borage oil *via* lipase-catalyzed hydrolysis, *J. Am. Oil. Chem. Soc.*, 71, 569–573, 1994.
8. Shimada, Y., Fukushima, N., Fujita, H., Honda, Y., Sugihara, A., and Tominaga, Y., Selective hydrolysis of borage oil with *Candida rugosa* lipase: two factors affecting the reaction, *J. Am. Oil. Chem. Soc.*, 75, 1581–1586, 1998.
9. Ikeda, I., Tomari, Y., Sugano, M., Watanabe, S., and Nagata, J., Lymphatic absorption of structured glycerolipids containing medium-chain fatty acids and linoleic acid, and their effect on cholesterol absorption in rats, *Lipids*, 26, 369–373, 1991.
10. Christensen, M.S., Hoy, C.E., Becker, C.C., and Redgrave, T.G., Intestinal absorption and lymphatic transport of eicosapentaenoic (EPA), docosahexaenoic (DHA), and decanoic acids: dependence on intramolecular triacylglycerols structure, *Am. J. Clin. Nutr.*, 61, 56–61, 1995.
11. Xu, X., Enzymatic production of structured lipids: process reactions and acyl migration, *INFORM*, 11, 1121–1131, 2000.
12. Iwasaki, Y. and Yamane, T., Lipase-catalyzed synthesis of structured triacylglycerols containing polyunsaturated fatty acids: monitoring the reaction and increasing the yield, in *Enzymes in Lipid Modification*, Bornscheuer, U.T., Ed., Wiley-VCH, Weinheim, 2000, pp. 148–169.
13. Akoh, C.C., Sellappan, S., Fomuso, L.B., and Yankah, V.V., Enzymatic synthesis of structured lipids, in *Lipid Biotechnology*, Kuo, T.M. and Gardner, H.W., Eds., Marcel Dekker, New York, 2002, pp. 433–460.
14. Shimada, Y., Sugihara, A., Maruyama, K., Nagao, T., Nakayama, S., Nakano, H., and Tominaga, T., Production of structured lipid containing docosahexaenoic and caprylic acids using immobilized *Rhizopus delemar* lipase, *J. Ferment. Bioeng.*, 81, 299–303, 1996.
15. Shimada, Y., Sugihara, A., Nakano, H., Yokota, T., Nagao, T., Komemushi, S., and Tominaga, Y., Production of structured lipids containing essential fatty acids by immobilized *Rhizopus delemar* lipase, *J. Am. Oil. Chem. Soc.*, 73, 1415–1420, 1996.

16. Shimada, Y., Sugihara, A., Nakano, H., Nagao, T., Suenaga, M., Nakai, S., and Tominaga, T., Fatty acid specificity of *Rhizopus delemar* lipase in acidolysis, *J. Ferment. Bioeng.*, 83, 321–327, 1997.

17. Irimescu, R., Furihata, K., Hata, K., Iwasaki, Y., and Yamane, T., Two-step enzymatic synthesis of docosahexaenoic acid-rich symmetrically structured triacylglycerols *via* 2-monoacylglycerols, *J. Am. Oil Chem. Soc.*, 78, 743–748, 2001.

18. Kawashima, A., Shimada, Y., Nagao, T., Ohara, A., Matsuhisa, T., Sugihara, A., and Tominaga, Y., Production of structured TAG rich in 1,3-dicapryloyl-2γ-linolenoyl glycerol from borage oil, *J. Am. Oil Chem. Soc.*, 79, 871–877, 2002.

19. Nagao, T., Kawashima, A., Sumida, M., Watanabe, Y., Akimoto, K., Fukami, H., Sugihara, A., and Shimada, Y., Production of structured TAG rich in 1,3-capryloyl-2-arachidonoyl glycerol from *Mortierella* single-cell oil, *J. Am. Oil Chem. Soc.*, 80, 867–872, 2003.

20. Kawashima, A., Shimada, Y., Yamamoto, M. Sugihara, A., Nagao, T., Komemushi, S., and Tominaga, Y., Enzymatic synthesis of high-purity structured lipids with caprylic acid at 1,3-positions and polyunsaturated fatty acid at 2-position, *J. Am. Oil Chem. Soc.*, 78, 611–616, 2001.

21. Haraldsson, G.G., Gudmundsson, B.O., and Almarsson, O., The preparation of homogeneous triglycerides of eicosapentaenoic acid and docosahexaenoic acid by lipase, *Tetrahedron Lett.*, 34, 5791–5794, 1993.

22. Irimescu, R., Yasui, M., Iwasaki, Y., Shimizu, N., and Yamane, T., Enzymatic synthesis of 1,3-dicapryloyl-2-eicosapentaenoyl glycerol, *J. Am. Oil Chem. Soc.*, 77, 501–506, 2000.

23. Irimescu, R., Furihata, K., Hata, K., Iwasaki, Y., and Yamane, T., Utilization of reaction medium-dependent regiospecificity of *Candida antarctica* lipase (Novozym 435) for the synthesis of 1,3-dicapryloyl-2-docosahexaenoyl (or eicosapentaenoyl) glycerol, *J. Am. Oil Chem. Soc.*, 78, 285–289, 2001.

24. Sonntag, N.O.V., New developments in the fatty acid industry in America, *J. Am. Oil Chem. Soc.*, 61, 229–232, 1984.

25. Lauridsen, J.B., Food emulsifiers: surface activity, edibility, manufacture, composition, and application, *J. Am. Oil Chem. Soc.*, 53, 400–407, 1976.

26. Aha, B., Berger, M., Jakob, B., Machmüller, G., Waldinger, C., and Schneider, M.P., Lipase-catalyzed synthesis of regioisomerically pure mono- and diglycerides, in *Enzymes in Lipid Modification*, Bornscheuer, U.T., Ed., Wiley-VCH, Weinheim, 2000, pp. 100–115.

27. McNeill, G.P., Shimizu, S., and Yamane, T., High-yield enzymatic glycerolysis of fats and oils, *J. Am. Oil Chem. Soc.*, 68, 1–5, 1991.

28. McNeill, G.P. and Yamane, T., Further improvements in the yield of monoglycerides during enzymatic glycerolysis of fats and oils, *J. Am. Oil Chem. Soc.*, 68, 6–10, 1991.

29. Pariza, M.W., CLA, a new cancer inhibitor in dairy products, *Bull. Int. Dairy Fed.*, 257, 29–30, 1991.

30. Ip, C., Chin, S.F., Scimeca, J.A., and Pariza, M.W., Mammary cancer prevention by conjugated dienoic derivatives of linoleic acid, *Cancer Res.*, 51, 6118–6124, 1991.

31. Park, Y., Albright, K.J., Liu, W., Storkson, J.M., Cook, M.E., and Pariza, M.W., Effect of conjugated linoleic acid on body composition in mice, *Lipids*, 32, 853–858, 1997.

32. Ostrowska, E., Muralitharan, M., Cross, R.F., Bauman, D.E., and Ddunshea, F.R., Dietary conjugated linoleic acids increase lean tissue and decrease fat deposition in growing pigs, *J. Nutr.*, 129, 2037–2042, 1999.

33. Lee, K.N., Kritchevsky, D., and Pariza, M.W., Conjugated linoleic acid and atherosclerosis in rabbits, *Atherosclerosis*, 108, 19–25, 1994.

34. Nicolosi, R.J., Rogers, E.J., Kritchevsky, D., Scimeca, J.A., and Huth, P.J., Dietary conjugated linoleic acid reduces plasma lipoproteins and early aortic atherosclerosis in hypercholesterolemic hamsters, *Artery*, 22, 266–277, 1997.

35. Sugano, M., Tsujita, A., Yamasaki, M., Noguchi, M., and Yamada, K., Conjugated linoleic acid modulates tissue levels of chemical mediators and immunoglobulins in rats, *Lipids*, 33, 521–527, 1998.

36. Watanabe, Y., Shimada, Y., Yamauchi-Sato, Y., Kasai, M., Yamamoto, T., Tsutsumi, K., Tominaga, Y., and Sugihara, A., Synthesis of MAG of CLA with *Penicillium camembertii* lipase, *J. Am. Oil Chem. Soc.*, 79, 891–896, 2002.

37. Watanabe, Y., Yamauchi-Sato, Y., Nagao, T., Yamamoto, T., Ogita, K., and Shimada, Y., Production of monoacylglycerol of conjugated linoleic acid by esterification followed by dehydration at low temperature using *Penicillium camembertii* lipase, *J. Mol. Catal. B: Enzym.*, 27, 249–254, 2004.

38. Watanabe, Y., Yamauchi-Sato, Y., Nagao, T., Yamamoto, T., Tsutsumi, K., Sugihara, A., and Shimada, Y., Production of MAG of CLA in a solvent-free system at low temperature with *Candida rugosa* lipase, *J. Am. Oil Chem. Soc.*, 80, 909–914, 2003.

39. Shimada, Y., Sugihara, A., Nakano, H., Kuramoto, T., Nagao, T., Gemba, M., and Tominaga, Y., Purification of docosahexaenoic acid by selective esterification of fatty acids from tuna oil with *Rhizopus delemar* lipase, *J. Am. Oil Chem. Soc.*, 74, 97–101, 1997.

40. Shimada, Y., Sugihara, A., and Tominaga, Y., Enzymatic enrichment of polyunsaturated fatty acids, in *Lipid Biotechnology*, Kuo, M. and Gardner, H.W., Eds., Marcel Dekker, New York, 2002, pp. 433–460.

41. Shimada, Y., Sugihara, A., Shibahiraki, M., Fujita, H., Nakano, H., Nagao, T., Terai, T., and Tominaga, T., Purification of γ-linolenic acid from borage oil by a two-step enzymatic method, *J. Am. Oil Chem. Soc.*, 74, 1465–1470, 1997.

42. Shimada, Y., Sakai, N., Sugihara, A., Fujita, H., Honda, Y., and Tominaga, Y., Large-scale purification of γ-linolenic acid by selective esterification using *Rhizopus delemar* lipase, *J. Am. Oil Chem. Soc.*, 75, 1539–1543, 1998.

43. Shimada, Y., Maruyama, K., Sugihara, A., Moriyama, S., and Tominaga, Y., Purification of docosahexaenoic acid from tuna oil by a two-step enzymatic method: hydrolysis and selective esterification, *J. Am. Oil Chem. Soc.*, 74, 1441–1446, 1997.

44. Shimada, Y., Sugihara, A., Minamigawa, Y., Higashiyama, K., Akimoto, K., Fujikawa, S., Komemushi, S., and Tominaga, Y., Enzymatic enrichment of arachidonic acid from *Mortierella* single-cell oil, *J. Am. Oil Chem. Soc.*, 75, 1213–1217, 1998.

45. Shimada, Y., Nagao, T., Kawashima, A., Sugihara, A., Komemushi, S., and Tominaga, Y., Enzymatic purification of n-6 polyunsaturated fatty acids, *Kagaku to Kogyo*, 73, 125–130, 1999 (in Japanese).

46. Ha, Y.L., Storkson, J.M., and Pariza, M.W., Inhibition of benzo(α)pyrene-induced mouse forestomach neoplasia by conjugated dienoic derivatives of linoleic acid, *Cancer Res.*, 50, 1097–1101, 1990.

47. Park, Y., Albright, K.J., Storkson, J.M., Liu, W., Cook, M.E., and Pariza, M.W., Change in body composition in mice during feeding and withdrawal of conjugated linoleic acid, *Lipids*, 34, 243–248, 1999.

48. Park, Y., Albright, K.J., Storkson, J.M., Liu, W., and Pariza, M.W., Evidence that the *trans*-10, *cis*-12 isomer of conjugated linoleic acid induces body composition changes in mice, *Lipids*, 34, 235–241, 1997.

49. DeDeckere, E.A., Van Amelsvoort, J.M., McNeill, G.P., and Jones, P., Effect of conjugated linoleic acid (CLA) isomers on lipid levels and peroxisome proliferation in the hamster, *Br. J. Nutr.*, 82, 309–317, 1999.

50. Nagao, K., Inoue, N., Wang, Y-M., Hirata, J., Shimada, Y., Nagao, T., Matsui, T., and Yanagita, T., The 10*trans*, 12*cis* isomer of conjugated linoleic acid suppresses the development of hypertension in Otsuka Long-Evans Tokushima fatty rats, *Biochem. Biophys. Res. Commun.*, 306,134–138, 2003.

51. Nagao, T., Shimada, Y., Yamauchi-Sato, Y., Kasai, M., Tsutsumi, K., Sugihara, A., and Tominaga, Y., Fractionation and enrichment of conjugated linoleic acid isomers by selective esterification with *Candida rugosa* lipase, *J. Am. Oil Chem. Soc.*, 79, 303–308, 2002.

52. Nagao, T., Yamauchi-Sato, Y., Sugihara, A., Iwata, T., Nagao, K., Yanagita, T., Adachi, S., and Shimada, Y., Purification of conjugated linoleic acid isomers through a process including lipase-catalyzed selective esterification, *Biosci. Biotechnol. Biochem.*, 67, 1429–1433, 2003.

53. Shimada, Y., Watanabe, Y., Samukawa, T., Sugihara, A., Noda, H., Fukuda, H., and Tominaga, Y., Conversion of vegetable oil to biodiesel using immobilized *Candida antarctica* lipase, *J. Am. Oil Chem. Soc.*, 76, 789–793, 1999.

54. Watanabe, Y., Shimada, Y., Sugihara, A., Noda, H., Fukuda, H., and Tominaga, Y., Continuous production of biodiesel fuel from vegetable oil using immobilized *Candida antarctica* lipase, *J. Am. Oil Chem. Soc.*, 77, 355–360, 2000.

55. Shimada, Y., Watanabe, Y., Sugihara, A., and Tominaga, Y., Enzymatic alcoholysis for biodiesel fuel production and application of the reaction to oil processing, *J. Mol. Catal. B: Enzym.*, 17, 133–142, 2002.

56. Watanabe, Y., Shimada, Y., Sugihara, A., and Tominaga, Y., Enzymatic conversion of waste edible oil to biodiesel fuel in a fixed-bed bioreactor, *J. Am. Oil Chem. Soc.*, 78, 703–707, 2001.

57. Watanabe, Y., Shimada, Y., Sugihara, A., and Tominaga, Y., Stepwise ethanolysis of tuna oil using immobilized *Candida antarctica* lipase, *J. Biosci. Bioeng.*, 88, 622–626, 1999.

58. Watanabe, Y., Nagao, T., Hirota, Y., Kitano, M., and Shimada, Y., Purification of tocopherols and phytosterols by a two-step *in situ* enzymatic reaction, *J. Am. Oil Chem. Soc.*, 81, 339–345, 2004.

59. Shimada, Y., Nagao, T., Watanabe, Y., Takagi, Y., and Sugihara, A., Enzymatic conversion of steryl esters to free sterols. *J. Am. Oil Chem. Soc.*, 80, 243–247, 2003.

60. Matsuno, T., Xanthophylls as precursors of retinoids, *Pure Appl. Chem.*, 63, 81–88, 1991.

61. Naguib, Y.M.A., Antioxidant activities of astaxanthin and related carotenoids, *J. Agric. Food Chem.*, 48, 1150–1154, 2000.

62. Jyonouchi, H., Sun, S., Iijima, K., and Gross, M.D., Antitumor activity of astaxanthin and its mode of action, *Nutr. Cancer*, 36, 59–65, 2000.

63. Jyonouchi, H., Zhang, L., and Tomita, Y., Studies of immunomodulating actions of carotenoids: II. Astaxanthin enhances *in vivo* antibody production to T-dependent antigens without facilitating polyclonal B-cell activation, *Nutr. Cancer*, 19, 269–280, 1993.

64. Nagao, T., Fukami, T., Horita, Y., Komemushi, S., Sugihara, A., and Shimada, Y., Enzymatic enrichment of astaxanthin from *Haematococcus pluvialis* cell extracts, *J. Am. Oil Chem. Soc.*, 80, 975–981, 2003.

23 Structure-Related Effects on Absorption and Metabolism of Nutraceutical and Specialty Lipids

Armand B. Christophe
Department of Internal Medicine, Division of Nutrition,
Ghent University Hospital, Ghent, Belgium

CONTENTS

23.1 INTRODUCTION

Dietary lipid classes usually have an alcohol and fatty acids as building blocks but other constituents can also be present. The alcohol moiety of the usual dietary fats is mainly glycerol but fats with other alcohols such as phytosterols, phytostanols, and sucrose have been synthesized and incorporated into foods with special health effects. Health-related nutritional effects of food lipids depend on their digestibility, fatty acid composition, and identity of their alcohol moiety. In the case where the alcohol moiety is a polyalcohol, the number and position of the fatty acids on the alcohol backbone is also important. Other lipid components such as choline, which is present in phospholipids, can also have distinct nutritional effects. The specificity of the lipases that are present in the intestinal tract is a major determinant of the digestibility of fat and of the digestion products that will be formed. Indigestible lipids will not be absorbed and may affect the extent of absorption of lipid-soluble components which are present in the dietary fat. The chemical nature of fat digestion products determines their reactivity in the intestinal lumen. This in turn affects their uptake and that of other food constituents with which they react. Likewise the lipid digestion products that are absorbed also affect intestinal transformations and metabolism. This in turn can have an influence on the physicochemical nature of the fats that leave the intestinal cells, their chyloportal repartition, and further metabolism. Specificity of lipases in combination with different handling of different fat digestion products is at the basis of structure-related effects of food lipids. The nature of the alcohol backbone of food fats is also extremely important for their health related nutritional effects. Phytosterol/-stanol esters for instance have cholesterol-lowering effects and sucrose polyesters cannot be absorbed from the intestinal tract.

23.2   LONG-CHAIN TRIACYLGLYCEROLS

Interesterification of food fats is a procedure often used in industry to modify physicochemical properties of a fat mixture. This can be done chemically or enzymatically. For triacylglycerols (TAGs) this results in a redistribution of different fatty acids over different stereochemical positions of glycerol. As common food fats contain components other than TAGs, other reactions can occur depending on the method used. For instance, base-catalyzed interesterification converts cholesterol into cholesterol esters whereas this is not the case with enzymatic interesterification. Of practical importance is the question whether natural fats or interesterified fats have the same nutritional properties. More general is the question whether the stereochemical position of fatty acids in TAGs, or the way they are combined in individual TAG molecules have an effect on their nutritional properties.

The position of fatty acids in acylglycerols can be unequivocally described by the stereospecific numbering convention (prefix *sn*-) of Hirschmann[1]. A number of other prefixes are commonly used to designate the position of substituents in acylglycerols. Alpha (α) refers to the two primary hydroxyl groups and β to the secondary hydroxyl groups, *rac* precedes the names of acylglycerols that are equal mixtures of enantiomers, and X is used when the positioning of the substituents is either unknown or unspecified.

The position or combination of fatty acids in TAGs can have an effect on digestibility, chyloportal repartition, conversion to higher unsaturated fatty acids (from parent essential fatty acids), chylomicronemia, cholesterolemia, atherogenicity, etc., depending on the identity of the fatty acids. Long-chain saturated fatty acids for instance will be more completely absorbed when they are esterified at the inner than at the outer positions of TAGs. The well-known cholesterolemic and atherogenic effects of different fatty acids are more pronounced when these fatty acids are esterified at the inner position of triacylglycerols.

Digestion of common food fats and subsequent absorption of their digestion products have recently been reviewed[2]. The *in vitro* digestion profiles of long- and medium-chain acylglycerols and the phase behavior of their lipolytic products have also been reported[3]. In this chapter, the stereospecific aspects are highlighted.

23.2.1   THE MAIN LIPASES THAT ACT ON TAGS ARE GASTRIC LIPASE AND PANCREATIC LIPASE

Gastric lipase starts the hydrolysis of dietary lipids. It has positional, substrate, but probably no fatty acid specificity[4]. Gastric lipase only splits primary ester bonds. It catalyzes the hydrolysis of the fatty acid at the *sn*-3 position in preference to that at the *sn*-1 position and prefers TAGs with short- and medium-chain lengths[5]. Thus TAGs with these fatty acids in the *sn*-3 position as are present in bovine milk and its products are preferential substrates. The main digestion products of gastric lipase are free fatty acids (FFAs) and *sn*-1,2 diacylglycerols (DAGs)[6]. It has been demonstrated that free medium-chain fatty acids can be absorbed directly from the stomach into the portal blood and thus can affect hepatic metabolism. About 10% of the dietary TAG fatty acids are liberated by gastric lipase from a meal containing long-chain fatty acids but considerably higher amounts from a test meal containing short- or medium-chain containing fatty acids[7]. Even for long-chain fatty acid containing TAGs, this corresponds to the breakdown of about one third of the TAGs into DAGs and FFAs in the stomach. There is evidence that gastric lipase is further active in the upper small intestine[8]. The importance of gastric digestion is that the hydrolysis products it generates induce the secretion of hormones that stimulate biliary and pancreatic flow, help to stabilize the surface of the TAG emulsion, promote the binding of pancreatic colipase later in the small intestine, and thus make the emulsion particles a better substrate for pancreatic lipase[9].

In healthy individuals, the rate of gastric emptying can be rate limiting for fat absorption[10]. The lipids that are emptied from the stomach into the small intestine are diluted by lipids from endogenous origin (~10 to 25 g/day when fasting). The two major sources of endogenous lipids originate from slaughtered intestinal cells and from biliary secretions. The amount of phopholipids from biliary origin (7 to 22 g/day; mainly phosphatidylcholine, PC) is considerably larger than from dietary origin (3 to 8 g/day). The same is also often true for cholesterol (~1 g/day of endogenous origin). In bile, bile salt–phospholipid mixed micelles exist in cylindrical arrangements. In the intestinal lumen bile salts and PC molecules are likely to partly adsorb to the lipid emulsion droplets[11]. Such bile salt-stabilized lipid particles are a substrate for pancreatic lipase which reacts at the oil/water interface provided that colipase is present. Colipase is a protein that is secreted together with lipase by the pancreas except in specific diseases.

Human pancreatic lipase catalyzes the equilibrium reaction between ester formation and ester hydrolysis. It is specific for primary alcohols and esters of such alcohols. Thus under certain conditions pancreatic and other lipases can be used for synthesis of TAGs with well-defined stereospecific structure[12]. During fat digestion *in vivo*, the digestion products are continuously removed from the interface favoring lipolysis. Due to its positional specificity, pancreatic lipase sets free fatty acids from the *sn*-1 and *sn*-3 positions of TAGs. It is more active against *sn*-1,2 and *sn*-2,3 DAGs, which are formed at the site where the enzyme acts, than against TAGs[13]. Thus there is no major accumulation of these intermediary digestion products, which are further hydrolyzed into *sn*-2 monoacylglycerols (MAGs). These are no longer substrates for pancreatic lipase.

Digestion, chemical isomerizations, and absorption are processes that are integrated and occur simultaneously in the intestine. If there were neither isomerization nor absorption of TAGs and DAGs the ultimate digestion products of TG would be a 2/1 mixture of FFAs and *sn*-2 MAGs. However, there is evidence that small amounts of TAGs, especially of medium-chain TAGs, and of DAGs can be absorbed without further digestion and that partial acylglycerols undergo isomerization.

"Linear" partial acylglycerols are thermodynamically favored over "branched" partial acylglycerols. MAGs with equilibrium isomeric composition consist of about 45% *sn*-1, 45% *sn*-3, and 10% *sn*-2 MAGs[14]. For DAGs, the equilibrium isomer ratio is about 70/30[15] for the "linear" (*sn*-1,3) and branched (*sn*-1,2 plus 2,3) isomers, respectively. Isomerization of partial acylglycerols yields products with primary ester bonds, which are substrates for pancreatic lipase and thus results in more complete digestion. The extent of isomerization of the DAGs formed by long-chain TAG digestion is probably small as the rate by which they are further digested is fast. As MAGs are readily absorbed, the rate of isomerization relative to that of uptake is important as it determines their further intraluminal digestion. Isomerization rate of partial acylglycerols increases with fatty acid unsaturation and decreases with fatty acid chain length[16]. Thus, spontaneous isomerization of long-chain fatty acids containing *sn*-2 MAGs in model systems is low compared to their intestinal uptake[17]. After TAG feeding about 15 to 20% of long-chain *sn*-2-MAGs generated by lipase action would isomerize in the intestinal lumen[18].

The chemical composition of the acylglycerols which are present in the intestinal lumen after TAG administration depends on several factors such as their fatty acid composition and time of sampling. For example, 10 min after intraduodenal infusion of triolein in the rat, 76% of the TAGs were digested and the major digestion products in the intestine were FFAs (~40%), MAGs (~30%), and DAGs (~10%). For long-chain TAGs it is fair to state that almost all of the fatty acids esterified in the outer positions come free in the intestinal lumen whereas most of the fatty acids from β-positions remain esterified and are absorbed as sn-2 MAGs. The amount of fatty acids from β-positions in TAGs that come free as FFAs is larger when the chain length is lower or the unsaturation higher.

23.2.2   IN THE INTESTINAL LUMEN OF THE SMALL INTESTINE FREE AND ESTERIFIED FATTY ACIDS HAVE DIFFERENT REACTIVITY

Saturated fatty acids can form poorly absorbable hydrated acid soaps and complex with calcium and other divalent ions[19]. The effect is more pronounced when the amount of calcium in the diet is high[20]. In the presence of insoluble soaps, cholesterol absorption may be reduced[21]. Thus the stereospecific position of fatty acids in TAGs can have an effect on their own absorption and on the absorption of minerals and other lipid constituents. This is not the case when they remain bound to glycerol. In general, the digestibility of a saturated fatty acid is higher when it is esterified at the internal than at the external positions of glycerol. Intact absorption of TAGs, which contain only medium-chain fatty acids, has been demonstrated (see below). Such a phenomenon is probably very limited for TAGs containing only long-chain fatty acids if it occurs at all. There is some evidence that intact DAGs with long-chain fatty acids can be absorbed as such in limited amounts (see below). Absorption of intact DAGs could possibly explain the finding that the combination of fatty acids in the dietary TAG molecules affects some of their physiological effects[6,22].

In the cells of the small intestine different enzymes can act on free and on esterified fatty acids. In contrast to MAGs, FFAs are diluted with fatty acids originating from the plasma FFA pool[23]. After activation they can be oxidized, elongated, chain desaturated, and converted into complex lipids[24,25]. The relative rates depend on the nature of the fatty acids and on the presence of other components in the intestinal cells[26]. Conversion of saturated fatty acids to monounsaturated ones when they are absorbed as FFAs, i.e., when they were present in the outer position of the dietary TAGs, could contribute to their lower hypercholesterolemic effect than when originally present at the inner position. FFAs also affect intestinal gene expression[27] and the production of apolipoproteins. These effects are different for different fatty acids[28].

Intestinal cells excrete the TAGs formed in the intestine as chylomicrons. The composition of the fatty acids in the sn-2 position of chylomicron TAGs is similar to that of the sn-2 position of the long-chain fatty acids of the ingested fat. *Chylomicron clearing* is slowed down by saturated fatty acids in the sn-2 position compared to unsaturated fatty acids[29]. More generally, plasma clearing of chylomicrons depends on the specific arrangement of acyl chains of the constituting TAGs and not necessarily on their overall saturation[30]. The rate of chylomicron clearing relative to uptake of chylomicron remnants is of importance as it affects remnant concentration in plasma. Chylomicron remnant removal is strongly influenced by the type of dietary fat with slower clearance with saturated fats with high melting points[31]. High concentrations of chylomicron remnants have been correlated with accelerated atherosclerosis[32].

Relative to fatty acids esterified in the 1(3) position of chylomicron TAGs, fatty acids in the sn-2 position are delivered to a greater extent to the liver whereas the fatty acids originally present at the outer positions are taken up to a greater extent by peripheral tissues[33]. As the biological effects of fatty acids are dependent on their identity, their tissue targeting could be of importance.

The effect different fatty acids have on cholesterolemia is well known. Whatever the kind of effect (hyper- or hypocholesterolemic) it seems to be more pronounced when the fatty acids are esterified at the inner rather than at the outer positions of TAGs. The lower hypercholesterolemic effect of saturated fatty acids at the outer positions[34] can be the result of a combination of different factors such as reduced absorption by unabsorbable soap formation which in turn interferes with cholesterol absorption in the intestinal lumen, partial desaturation and oxidation in the small intestinal cells, and reduced targeting into the liver[33]. The stronger hypocholesterolemic effect of polyunsaturated fatty acids at the *sn*-2 position[35] could be the result of increased influx in the liver[36].

The atherogenicity of fats containing palmitic acid is higher when this fatty acid is esterified at the *sn*-2 position than at the outer positions, even when cholesterolemia is not affected[37].

Higher conversion of "parent" essential fatty acids into their longer and more unsaturated metabolites has been described when the parent essential fatty acids were esterified into the *sn*-2 position[36]. This may be explained by higher influx in the liver[36].

Significantly lower values of high-density lipoprotein (HDL) cholesterol and of apolipoprotein A and significantly higher levels of apoB were found in infants when the same amount of palmitic acid was fed in TAGs in the *sn*-2 position than in the *sn*-1(3) position[38]. This illustrates that lipoprotein metabolism can be affected by fatty acid distribution in TAGs.

Lipids with palmitic acid mainly in the 1(3) position caused a larger increase in triacylglyceridemia and lower increase in insulinemia in the beginning of the postprandial period in postmenopausal women[39].

Symmetrical stearic acid-containing TAGs with oleic acid in the *sn*-2 position (cocoa butter) are absorbed faster than asymmetrical TAGs with saturated fatty acids in the *sn*-2 position, leading to higher postprandial lipemia and plasma levels of activated factor VIIa[40].

Clear differences between feeding seal oil, with eicosapentaenoic and docosahexaenoic acids mainly at the outer positions of TAGs, or fish or squid oil with these fatty acids mainly at the inner position, were obtained for eicosanoid production[41], reduction of triacylglyceridemia[40], and serum phospholipid arachidonic acid content[42]. In general, the positive effects of these fatty acids seem to be more pronounced when they are esterified at the outer position of dietary TAGs. Moreover fatty acid effects are not only dependent on the identity of the fatty acid itself and its positional incorporation in acylglycerols but also on coingestion with other fats[43] and other dietary macronutrients[44].

23.3   MEDIUM-CHAIN TRIACYLGLYCEROLS

Some natural fats such as coconut oil contain fatty acids of medium chain length. These fatty acids can be isolated and esterified to glycerol to produce medium-chain TAGs (MCTs). It has to be realized that when these fats or other specialty fats are incorporated in a diet, usually considerable amounts of long-chain TAGs are present, even in low-fat diets. MCTs are more easily digested than long-chain TAGs which confers advantages to them for oral nutrition in patients in whom digestive capacity is reduced. Medium-chain fatty acids which are their major digestion products interfere somewhat with the digestion of long-chain TAGs. Especially in patients with compromised digestion for whom MCT administration forms an indication, incorporation of MCT in the diet may have a negative impact on the essential fatty acid status. Administration of MAGs rich in essential fatty acids proved successful to improve essential fatty acid deficiency in patients with cystic fibrosis[45].

After absorption, medium-chain fatty acids leave the intestinal cells mainly as FFAs which are transported by the portal vein, making them useful in patients with disturbed lymphatic transport and with impaired chylomicron clearing. They are easily oxidized and increase energy expenditure showing potential for the prevention of obesity[46]. Their use for enteral and parenteral nutrition has been reviewed[47,48]. Their safety for human consumption up to 1 g/kg has been shown in several trials[49].

In a patient with obstruction of the thoracic duct and accumulation of ascites fluid, incorporating MCTs in an otherwise lipid-poor diet resulted in the appearance of medium-chain fatty acids in ascites fluid TAGs[50]. TAG species analysis demonstrated that these fatty acids were present as trimedium-chain TAGs (MCTs), dimediummonolong-chain TAGs, and monomediumdilong-chain TAGs. This suggests that small amounts of intact MCTs and medium chain-containing partial acylglycerols can be absorbed intact and converted with long-chain fatty acids into "mixed-chain"

TAGs. "Mixed chain" TAGs were more abundant than MCTs suggesting that these are more physiological than MCTs for parenteral nutrition. In agreement with our findings in humans, increased levels of dimedium-chain monolong-chain TAGs were demonstrated in lymph of rats when MCTs were fed together with long-chain TAGs[51].

23.4 STRUCTURED LIPIDS AND SALATRIMS

TAGs containing in the same molecule medium- and long-chain fatty acids are called structured lipids (SL) whereas those containing short- and long-chain fatty acids are called salatrims. Modern technology makes it possible to prepare SL with medium (M)-chain fatty acids and long (L)-chain fatty acids at well defined stereochemical positions. For instance, TAGs with the following combination of fatty acids can be prepared: MLM, LMM, MML. In interesterified mixtures of MCTs and long-chain TAGs, all the possible combinations are present. The applications of SL have been reviewed[52]. The safety of salatrims as a food additive has been evaluated[53]. They are low-energy fats (~5 kcal/g).

A major reason why physical mixtures of MCTs and long-chain TAGs have different nutritional properties from SL is the fast isomerization of medium-chain fatty acids in partial glycerides and the fast hydrolysis of medium-chain fatty acids under the catalytic action of pancreatic lipase. For instance, LMM-TAGs are expected to be digested first to LM-DAGs, which are isomerized to sn-1 L, sn-3 M-DAGs which are further hydrolyzed to L-MAGs. Long-chain MAGs have been found to have special nutritional properties (see below). Thus incorporating LMM- or MML-TAGs in the diet may be a practical way to generate sn-1(3) MAGs in the intestinal lumen. An application of LML with stable isotope-labeled octanoic acid has been found in clinical diagnosis of maldigestion. Octanoic acid will only come free after digestive removal of the long-chain fatty acid followed by isomerization of the medium-chain fatty acid from the sn-2 position to the sn-1(3) position followed by fast removal. Free octanoic acid is absorbed and oxidized very fast leading to enrichment of the isotope of carbon dioxide in the exhaled air[54].

23.5 MONOACYLGLYCEROLS

MAGs are used as emulsifiers in some foods. They are present in small amounts in the diet. MAG mixtures with thermodynamic equilibrium composition can easily be prepared industrially, for instance by glycerolysis and molecular distillation. Some dietary MAGs will be digested and some will be taken up as such.

Sn-2 MAGs are absorbed in large quantities after consumption of the usual food fats together with an excess of FFAs. Their further metabolism in the intestinal cells under these conditions is well documented. The long-chain fatty acids are converted into acyl-coenzyme A derivatives and become "activated." Most of the sn-2 MAGs react with activated fatty acids to be converted by a multienzyme complex into TAGs. The excess of FFAs reacts with sn-3 glycerophosphate, which is a glucose metabolite, to form lysophosphatidic acids. These are further converted into phosphatidic acids. Dephosphorylation results in sn-1,2 DAGs which are precursors of both TAGs and phospholipids.

What will happen after influx of sn-1(3) MAGs in the intestinal cell is less clear. A series of competing reactions must be considered such as complete digestion, acylation, transacylation, phoshorylation, intact excretion out of the intestinal cells, etc.

Lipases that hydrolyze MAGs are present in the intestinal cells. Some of them show considerable fatty acid specificity but their action seems to be reduced by other intestinal cell constituents[55]. Both sn-1 and sn-3 MAGs can be acylated to form sn-1,3 DAGs. This transformation is fatty acid and species dependent[52]. However, sn-1,3 DAGs are poor substrates for intestinal TAG synthesis[56,57]. Sn-1 but not sn-3 MAGs can be phosphorylated into lysophosphatidic acid[58]. This is a normal intermediate for intestinal TAG synthesis. If MAGs would escape intestinal metabolism and could leave the cell intact, they could be transported by portal blood bound to albumin[59].

23.5.1 HEALTH-RELATED DIFFERENCES IN POSTPRANDIAL EFFECTS AFTER TAG OR MAG FEEDING

Long-chain fatty acids leave the intestinal cells mainly as TAGs incorporated in intestinal lipoproteins with the size of chylomicrons after TAG feeding. These are transported via the lymphatic route. When they enter the bloodstream, there is an increase in chylomicronemia, triglyceridemia, and a change of the fatty acid composition of plasma TAGs towards that of the fat fed. Chylomicron TAGs are hydrolyzed by lipoprotein lipase resulting in the formation of remnants and liberation of free fatty acids. As a consequence, the concentration of the nonesterified fatty acids in plasma increases and their composition also changes towards that of the fat fed. Chylomicron remnants are taken up by the liver. Medium-chain fatty acids provided in the diet as MCTs or as structured lipids are transported mainly as FFAs by the portal vein.

After administration of a MAG mixture with thermodynamic equilibrium composition (about 90% *sn*-1(3) isomers) the well-known postprandial chylomicronemia, triglyceridemia, and shift of the fatty acid composition of serum triacylglycerols and nonesterified fatty acids towards that of the fat that occur after natural fat feeding are much less pronounced[60]. Postprandial effects of TAG and MAG loading are visualized in Figure 23.1.

This reduced response cannot be due to reduced absorption, as fecal fat was not increased. Neither is it probable that the attenuated response can be attributed to lymphatic transport of MAGs or DAGs, as these could not be demonstrated in the general circulation after MAG feeding. Another possibility would be portal absorption of MAGs or unidentified MAG derivatives formed in the intestinal cells. Circulating TAGs in the fasting state are mainly derived from the liver where they are formed by the *sn*-3 glycerophosphate metabolic route (see above). Fatty acid specificities of the enzymes involved in hepatic TAG synthesis generate TAG mixtures of which the major fraction contains palmitic acid at the *sn*-1 position and fatty acids with 18 carbon atoms at the other positions. The composition of TAGs according to the combination of fatty acids they contain ("carbon number analysis") can be determined by chromatographic techniques. The major TAG species in serum of subjects consuming usual food fats in the fasting state has a carbon number of 52. This

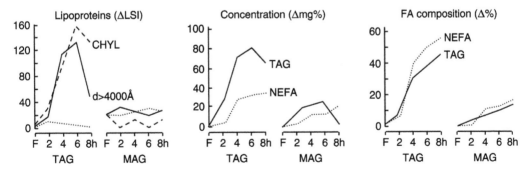

FIGURE 23.1 Differential postprandial effects of loading with usual food fats or monoacylglycerols (1 g/kg body weight; fatty acid composition of sunflower seed oil). LSI, light scattering intensity (arbitrary units); TAG, triacylglycerol; NEFA, nonesterified fatty acids; F, fasting; VLDL, very low-density lipoproteins; Δ%, percent change in fatty acid composition towards that of the fat fed (0% is no change from fasting composition; 100% change means that the fatty acid composition becomes identical to that of the fat fed).

corresponds to a combination of two fatty acids, one with 18 carbon atoms and one with 16. After long-term feeding of monolinolein the fraction of serum TAGs with carbon number 54 (= 18 + 18 + 18) increased and that with carbon number 52 decreased. This suggests that phosphorylated monolinolein was used in the liver with structure retention of the *sn*-1 ester bond as a substrate for TAG synthesis. This implies that *sn*-1 MAGs in free or phosphorylated form can be transported directly to the liver by the hepatic portal vein. In the intestinal cells, *sn*3-MAGs cannot be phosphorylated but can be acylated at the *sn*-3 position. The DAGs thus formed are poor substrates for intestinal TAG synthesis and no evidence for intact secretion could be found. There is evidence that they may be hydrolyzed in the intestinal cells.

If the MAGs contain at the *sn*-1 position a fatty acid such as linoleic acid, which normally is not incorporated into lysophosphatidic acid by *sn*-3 glycerophosphate acyltransferase, TAGs may be formed with unusual structures. Such plasma TAGs, which may also occur after DAG feeding in the case where these are taken up as MAGs after partial digestion, may be turned over at different rates and consequently affect their fasting concentration. Lower fasting concentrations of TAGs have been found when TAGs in the diet were partially replaced by DAGs.

The concept of portal transport of MAGs based on findings in humans contrasts with findings in rats. Docosahexaenoic acid administered intragastrically as an emulsion in lymph canulated rats resulted in higher lymphatic output, mainly as TAGs, when administered as MAGs than as TAGs[61]. It is not known whether this is due to the differences in experimental set up or to species differences.

MAGs have potential for clinical nutrition. However, feeding MAGs as food fat has practical drawbacks, as MAGs form sticky phases in aqueous environments[62] giving an unpleasant mouthfeel when eaten. Even tube feeding can cause problems because the tube can become blocked when the MAGs come in contact with water. They have been administered as such or together with MCTs[45]. Compared to TAGs with the fatty acid composition of sunflower seed oil, long-term MAG administration resulted in higher triglyceridemia and lower cholesterolemia[14]. Feeding structured lipids may be a practical way of generating long-chain MAGs in the intestinal lumen. Rectal absorption of MAGs has been demonstrated in the rat[63].

23.6 DIACYLGLYCEROLS

DAGs are natural components that occur in various edible oils up to about 10%. DAGs with thermodynamic equilibrium composition are easily prepared industrially in large quantities. Properties of commercial DAG oil have been reviewed recently[64]. They are on the market as edible oils in Japan where they have FOSHU status (foods of special health use) and in the U.S. where they have GRAS status (generally recognized as safe). In these oils 1,3 DAGs are the major components. When DAGs are present in the diet, it is expected on theoretical grounds that they will form emulsion droplets together with the TAGs that are invariably present in a human diet. *rac*-1,2 isomers are normal intermediates of TAG digestion. The *sn*-1,3 isomers would only be formed from TAGs in very small quantities by isomerization of their intermediary *sn*-1,2 and *sn*-2,3 DAG digestion products. If pancreatic lipase preferentially attacks the *sn*-1 position of 1,3 DAGs as it does in TAGs, the major first digestion product would be *sn*-3 MAGs, but *sn*-1 MAGs would also be formed. This would be of no consequence if both MAG stereoisomers, which are substrates for pancreatic lipase, were broken down completely to glycerol and fatty acids before absorption. However, if there is substantial absorption of the intact MAGs, their isomeric composition could be of importance because it has been shown that they are handled differently in the intestinal cells (see above).

23.6.1 HEALTH-RELATED DIFFERENCES BETWEEN NATURAL FAT AND DAG OIL FEEDING

DAG oil administration has postprandial effects distinct from feeding TAGs with the same fatty acid composition. For instance decreased activities of enzymes involved in fatty acid synthesis and increased activity in those involved in fatty acid oxidation were found after DAG compared to TAG feeding in rat liver. It is conceivable that when 1,3-*sn* DAGs are fed, they will be partially digested and taken up as 1(3)-MAG and FFAs. Indeed some of their postprandial effects mimic those of isomeric equilibrium MAG feeding (e.g., lower increase of chylomicron TAG[65]).

Although the energy value and digestibility of DAGs are similar to those of TAGs[66], replacing the latter by the former in a mildly reduced-energy diet enhances loss of body weight and fat[67]. This could be due to uncoupling of phosphorylation from oxidation. Increased levels of uncoupling proteins have been found in the liver after DAG feeding. Uncoupling phosphorylation from oxidation leads to increased thermogenesis and energy dissipation. Long-term DAG feeding results in lower triacylglyceridemia than TAG feeding[68] and enhances the cholesterol-lowering effect of phytosterols[69]. DAG feeding does not affect fat-soluble vitamin status[62]. In type 2 diabetics, replacing TAGs by DAGs in the diet resulted in reduced levels of glycosylated hemoglobin without affecting fasting blood glucose levels. They also lead to lower adiposity, insulinemia, and leptinemia[64].

23.7 PHYTOSTEROL/PHYTOSTANOL ESTERS

Plant sterols and stanols are mainly present in the diet in vegetables and vegetable oils. Their intake in a Western diet is estimated to be about 160 to 360 mg/day[70]. Stanols make up about 10% of this amount. In order to increase their lipid solubility for incorporation in dietary lipids they can be esterified. The effects of phytostanol esters and phytosterol esters are mediated by the free phytostanol/sterol that is formed in the small intestine by the action of cholesterol esterase which hydrolyzes sterol esters. Dietary cholesterol is absorbed as free sterol[71]. During intestinal transit, phytosterol esters are hydrolyzed to a large extent and reduce hydrolysis and subsequent absorption of the fraction of the dietary cholestrol that was esterified.

The effects of phytosterols/stanols in the intestinal lumen on cholesterol absorption can be best understood by their influence on the physicochemical phases that are present during fat digestion.

For a lipid to be a candidate for absorption it must be able to reach the absorptive surface of the small intestine. This is the surface of the microvilli. Microvilli are so close to each other that emulsion droplets are too large to gain access between them for geometrical reasons. Moreover, an "unstirred" water layer lines the microvilli. Passage through this layer is rate limiting for lipid absorption. In the presence of bile salts, MAGs and FFAs leave the emulsion/water interface where they were formed by the action of pancreatic lipase and form mixed micelles with conjugated bile acids. Only in this form can they easily access the absorptive surface of the small intestine and be absorbed. To the best of our knowledge, no data are available on the effect of isomer structure or fatty acid composition of the MAGs on the rate at which they are transferred from the emulsion phase to the micellar phase if any. Partitioning of fatty acids between the emulsion interface and the micellar phase depends on pH. When pH is low such as in the absence of pancreatic juice protonated fatty acids remain at the interface of the emulsion and inhibit further digestion. The interior of the MAG/FFA/conjugated bile salt mixed micelle is a lipophilic environment in which lipophilic substances such as cholesterol, carotenes, and fat-soluble vitamins also can dissolve. It is in this way that they can reach the absorptive surface and are candidates for uptake. The composition of mixed micelles that exist in the intestinal lumen after TAG consumption and several of their physicochemical properties have been described[72]. Phytosterols/-stanols can compete with cholesterol for incorporation in the micelles[73] and thus interfere with cholesterol absorption[74]. Plant sterols and stanols derived from wood pulp and vegetable oils lower total and LDL cholesterol by inhibiting cholesterol absorption from the intestine in humans. The effect is dose dependent and significant reductions of apoB are obtained with doses as low as 1 g/day[75]. By affecting the physicochemical composition of the micellar phase, they also reduce the absorption of carotenes and fat-soluble vitamins[76]. A healthy diet rich in carotenoids is effective, however, in maintaining normal blood carotenoid levels[77] and there are no changes in serum fat-soluble vitamins when a normal diet is consumed[78]. Except in some rare diseases, phytosterols/-stanols are poorly absorbed and are resecreted into the intestinal lumen through bile[79]. Sterol esters and stanol esters seem to have comparable cholesterol-lowering activity[80]. The serum cholesterol-lowering effect of phytosterols is higher when fed together with DAGs than with TAGs[69]. In addition to possessing cholesterol-lowering potency, phytosterols have been suggested to have anticancer properties[81]. Potential uses and benefits of phytosterols in diet have been reviewed[82].

23.8   PHOSPHOLIPIDS

The presence of intact phospholipids in the intestinal lumen retards TAG digestion[83], affects the partitioning of lipid digestion products between the oil and micellar phase[84], and slows down the uptake of lipid digestion products from the micellar phase[85]. Phosphatidylcholine (PC) is digested, however, under the action of phospholipase A_2 which hydrolyzes the fatty acid esterified at the *sn*-2 position to yield lyso-PC and free fatty acids[85]. In contrast to PC, the resulting lyso-PC favors micellar solubilization and uptake of fat digestion products[86] and other dietary lipophilic substances[87]. Lyso-PC itself is taken up by the cells of the small intestine, affects intestinal metabolism[88], and can be converted into PC again. There is evidence that PC can be taken up intact and excreted as part of intestinal lipoproteins in the lymph[89].

23.8.1   NUTRITIONAL DIFFERENCES BETWEEN DIETARY PHOSPHOLIPIDS AND TAGS

Differences in nutritional effects between phospholipids and TAGs can be due to several factors not related to their fatty acid composition such as the presence of a phosphate group and a nitrogen base (mainly choline) that may interact in several metabolic pathways[90]. Moreover, several glycerophospholipid preparations studied can contain other components such as cholesterol, cerebrosides, and sphingomyelins also depending on their source, method of isolation, and purification. These components may also affect the nutritional properties.

Enrichment of dietary fat with phospholipids or TAGs may or may not affect fecal excretion of fat and minerals and may increase or decrease saturated fat absorption depending on the phospholipid and TAG source[91]. Long-chain polyunsaturated fatty acids were better absorbed in preterm infants when fed as phospholipids than as TAGs[92]. Feeding long-chain polyunsaturates as phospholipids or TAGs influences the distribution of these fatty acids in plasma lipoprotein fractions, affects their content in different plasma lipid fractions, and affects the composition of HDL and LDL phospholipids[91]. Arachidonic acid was shown to be more effective for brain arachidonic acid accretion when fed as phospholipids compared with TAG[93].

23.9   SUCROSE POLYESTERS

Sucrose polyesters have no primary ester bonds and are not digested. As a result they remain in the oil phase, are not taken up and are excreted with stools. In the small intestine they have some effect on the partitioning of fat soluble components between the emulsion and micellar phase and as a consequence on their absorption. Reduced uptake of fat-soluble vitamins can be avoided by enriching the sucrose polyesters with these vitamins. Their main use is related to the fact that they can replace usual food fats in many prepared foods but they do not provide for calories.

23.10   CONCLUSIONS

Modern methodology allows the synthesis of special food fats, which are not digested, or with stereochemical structures that are formed only in very small quantities after feeding common food fats.

TAGs with different combination of fatty acids and/or with different stereochemical structure can also be produced. These fats can either have improved long-term health effects or may be useful for clinical nutrition.

REFERENCES

1. Hirschmann, H., The nature of substrate asymmetry in stereoselective reactions, *J. Biol. Chem.*, 235, 762–2767, 1960.
2. Phan, C.T. and Tso, P., Intestinal lipid absorption and transport, *Front. Biosci.*, 6, D299–D319, 2001.
3. Sek, L., Porter, C.J., Kaukonen, A.M., and Chapman, W.N., Evaluation of the in vitro digestion profiles of long and medium chain glycerides and the phase behavior of their lipolytic products, *J. Pharm. Pharmacol.*, 54, 29–41, 2002.
4. Jensen, R.G., De Jong, F.A., Lambert-Davis, L.G., and Hamosh, M., Fatty acid and positional selectivities of gastric lipase from premature human infants: in vitro studies, *Lipids*, 29, 433–435, 1994.
5. Levy, R., Goldstein, H. Stankievicz et al., Gastric handling of medium-chain triglycerides and subsequent metabolism in the suckling rat, *J. Pediatr. Gastroenterol. Nutr.*, 5, 784–789, 1984.
6. Hamosh, M., Klaeveman, H.L., Wolf, R.O., and Scow, R.O., Pharyngeal lipase and digestion of dietary triglyceride in man, *J. Clin. Invest.*, 55, 908–913, 1975.
7. Carriere, F., Barrowman, J.A., Verger, R., and Laugier, R., Secretion and contribution of lipolysis by gastric and pancreatic lipases during a test meal in humans, *Gastroenterology*, 105, 876–888, 1993.
8. Liao, T.H., Hamosh, M., and Hamosh, P., Fat digestion by lingual lipase: mechanism of lipolysis in the stomach and upper small intestine, *Pediatr. Res.*, 18, 402–4090, 1984.
9. Jian, R., Vigneron, N., Najean, Y., and Bernie, J.J., Gastric emptying and intagastric distribution of lipids in man. New scintographic method of study, *Dig. Dis. Sci.*, 27, 705–711, 1982.
10. Maes, B.D., Ghoos, Y.F., Geypens, B.J. et al., Gastric emptying of liquid, solid and oil phase of a meal in normal volunteers and patients with Bilroth II gastrojejustomy, *Gut*, 38, 23–27, 1996.
11. Wickham, M., Wilde, P., and Fillerely-Travis, A., A physiochemical investigation of two phosphatylcholine/bile salt interfaces: implications for lipase activation, *Biochim. Biophys. Acta*, 1580, 110–122, 2002.
12. Xu, X., Fomuso, L.B., and Akoh, C.C., Synthesis of structured triacylglycerols by lipase-catalyzed acydolysis in a packed bed bioreactor, *J. Agric. Food Chem.*, 48, 3–10, 2000.
13. Carriere, F., Rogalska, E., Cudry, C. et al., In vivo and in vitro studies on the stereoselective hydrolysis of tri- and diglycerides by gastric and pancreatic lipases, *Bioorg. Med. Chem.*, 5, 429–435, 1997.
14. Christophe, A. and Verdonk, G., Effects of substituting monoglycerides for natural fats in the diet on serum lipids in fasting man, *Biochem. Soc. Trans.*, 5, 1041–1042, 1977.
15. Murase, T., Mizuno, T., Omachi, T. et al., Dietary diacylglycerol suppresses high fat and high-sucrose diet-induced body fat accumulation in C57BL/6J mice, *J. Lipid Res.* 42, 372–378, 2001.
16. Boswinkel, J.P.T., Derksen, K., van't Riet, and Cuperus, F.P., Kinetics of acyl migration in monoglycerides and dependence on acyl chain length, *J. Am. Oil Chem. Soc.*, 73, 707–711, 1996.
17. Lyubachesskaya, G. and Boyle-Rode, E., Kinetics of 2-monoacylglycerol acyl migration in model chylomicra, *Lipids*, 35, 1353–1358, 2000.
18. Skipski, V.P., Morehous, M.G., and Duel, H.J., The absorption in the rat of a dioleoyl-2-deuteriostearoyl glyceride C^{14}, *Arch. Biochem. Biophys.*, 81, 93–104, 1959.
19. Small, D.M., The effects of glyceride structure on the absorption and metabolism, *Ann. Rev. Nutr.*, 11, 413–434, 1991.
20. Shahkhalili, Y., Murset, C., Meirim, I. et al., Calcium supplementation of chocolate: effect on cocoa butter digestibility and blood lipids in humans, *Am. J. Clin. Nutr.*, 73, 246–252, 2001.
21. Aoyama, T., Fukui, K., Tanagushi, K. et al., Absorption and metabolism of lipids in rat depend on fatty acid isometric position, *J. Nutr.*, 126, 225–231, 1996.
22. Redden, P.R., Lin, X., Fahey, J., and Horrobin, D.F., Stereospecific analysis of the major triacylglycerol species containing γ-linolenic acid in evening primrose oil and borage oil, *J. Chromatogr. A*, 704, 99–111, 1995.
23. Boucrot, B. and Clement, J., Participation des acides gras du sang et de la bile à la formation des lipides endogénes de la lymphe chez le rat qui ingère un repas contenant des graisses, *Biochim. Biophys. Acta*, 187, 59–72, 1969.

24. Nilsson, and Becker, B., Uptake and interconvertion of plasma unesterified n-3 polyunsaturated fatty acids by the GI tract of rats., *Am. J. Physiol.*, 268, G732–G738, 1995.

25. Huang, Y.S., Lin, J.W., Koba, K., and Anderson, S.N., N-3 and n-6 metabolism in undifferentiated and differentiated human intestinal cell line (Caco-2), *Mol. Cell. Biochem.*, 18, 121–130, 1995.

26. Chen, Q. and Nillson, A., Interconvertion of alpha-linolenic acid in rat intestinal mucosa: studies in vivo and in isolated villus and crypt cells, *J. Lipid Res.*, 35, 601–609, 1994.

27. Murase, T., Kondo, H., Hase, T., Tokimitsu, I., and Saito, M., Abundant expression of uncoupling protein-2 in the small intestine: upregulation by dietary fish oil and fibrates, *Biochim. Biophys. Acta*, 1530, 15–22, 2001.

28. Sanderson, I.R. and Naik, S., Dietary regulation of intestinal gene expression, *Ann. Rev. Nutr.*, 20, 311–338, 2000.

29. Mortimer, B.C., Kenrick, M.A., Holthouse, D.J., Stick, R.V., and Redgrave, T.G., Plasma clearance of model lipoproteins containing saturated and polyunsaturated monoacyl glycerols injected intravenously in the rat, *Biochim. Biophys. Acta*, 1127, 67–73, 1992.

30. Mortimer, B.C., Holthouse, D.J., Martins, I.J., Stick, R.V., and Redgrave, T.G., Effect of triacylglycerol-saturated acyl chains on the clearance of chylomicron-like emulsions from the plasma of the rat, *Biochim. Biophys. Acta*, 1211, 171–180, 1994.

31. Phan, C.T., Mortimer, B.C., Martins, B.J., and Redgrave, T.G., Plasma clearance of chylomicrons from butterfat is not dependent on saturation: studies with butterfat fractions and other fats containing triacylglycerols with low and high melting points, *Am. J. Clin. Nutr.*, 69, 1151–1161, 1999.

32. Yu, K.C. and Cooper, A.D., Postprandial lipoproteins and atherosclerosis, *Front. Biosci.*, 6, D332–D354, 2001.

33. Pufal, D.A., Quinlan, P.T., and Salter, A.M., Effect of dietary triacylglycerol structure on lipoprotein metabolism: a comparison of the effects of dioleylpalmitoylglycerol in which palmitate is esterified to the 2 or 1(3)-position of the glycerol, *Biochim. Biophys. Acta*, 1258, 41–48, 1995.

34. Yamamoto, I., Sugano, M., and Wada, M., Hypocholesterolaemic effect of animal and plant fats in rats, *Atheroscl.* 13, 171–184, 1971.

35. Elson, C.E., Dugan, L.R., Bratzler, L.J., and Parson, A.M., Effect of isoessential fatty acid lipids from animal and plant sources on cholesterol levels in mature male rats, *Lipids*, 1, 322–324, 1967.

36. Renaud, S.C., Ruf, J.C., and Pethitory, D., The positional distribution of fatty acids in palm oil and lard influences their biological effects in rats, *J. Nutr.*, 125, 229–237, 1995.

37. Kritchevski, D., in *Structural Modified Food Fats: Synthesis, Biochemistry and Use*, Christophe, A.B., Ed., AOCS Press, Champaign, IL, 1998, pp. 183–188.

38. Nelsson, C.M. and Innis, S.M., Plasma lipoprotein fatty acids are altered by the positional distribution of fatty acids in infant formula and human milk, *Am. J. Clin. Nutr.*, 70, 62–69, 1999.

39. Yli-Jokipii, K., Kalio, H., Schwab, U. et al., Effects of palm oil and transesterified palm oil on chylomicron and VLDL triacylglycerol structures and postprandial lipid response, *J. Lipid Res.*, 42, 1618–1628, 2001.

40. Sanders, T.A., Berry, S.E., and Miller, G.J., Influence of triacylglycerol structure on the postprandial response of factor VII to stearic acid-rich fats, *Am. J. Clin. Nutr.*, 74, 777–782, 2003.

41. Ikeda, I., Yoshida, H., Tomooka, M. et al., Effects of long-term feeding of marine oils with different positional distribution of eicosapentaenoic and docosahexaenoic acid on lipid metabolism, eicosanoid production and platelet aggregation in hypercholesterolemic rats, *Lipids*, 33, 897–904, 1998.

42. Yoshida, H., Ikeda, I., and Tomooka, M., Effect of dietary seal and fish oils on lipid metabolism in hamsters, *J. Nutr. Sci. Vitaminol.*, 47, 242–247, 2001.

43. Lawson, L.D. and Hughes, B.G., Absorption of eicosapentaenoic acid and docosahexaenoic acid from fish oil triacylglycerols or fish oil ethyl esters co-ingested with a high fat meal, *Biochim. Biophys. Res. Commun.*, 156, 960–963, 1988.

44. Kaku, S., Yunoki, S., Ohkura, K. et al., Interaction of dietary fats and proteins on fatty acid composition of immune cells and LTB4 production by peritoneal exudate of rats, *Biosci. Biotechnol. Biochem.*, 65, 315–351, 2001.

45. Christophe, A., Verdonk, G., Robberecht, E., and Mahathanakhun, R., Effect of supplementing medium chain triglycerides with linoleic acid rich monoglycerides on severely disturbed serum lipid fatty acid patterns in patients with cystic fibrosis, *Ann. Nutr. Metab.*, 29, 239–245, 1985.

46. St-Onge, M.P. and Jones, P.J., Physiological effects of medium-chain triglycerides: potential agents in the prevention of obesity, *J. Nutr.*, 132, 329–332, 2002.

47. Babayan, V.K., Medium chain triglycerides and structured lipids, *Lipids*, 22, 417–420, 1987.
48. Ulrich, H., Pastores, S.M., Katz, D.P., and Kvetan, V., Parenteral use of medium-chain triglycerides: a reappraisal, *Nutrition*, 12, 231–238, 1996.
49. Traul, K.A., Driedger, A., Ingle, D.L., and Nakhasi, D., Review of the toxicological properties of medium-chain triglycerides, *Food Chem. Technol.*, 38, 79–98, 2000.
50. Christophe, A., Verdonk, G., Mashali, M., and Sandra, P., Fatty acid chain length combinations in ascitic fluid triglycerides containing lymphatic absorbed medium chain fatty acids, *Lipids*, 17, 759–761, 1982.
51. Mu, H. and Hoy, C.E., Distribution of medium-chain FA in different lipid classes after administration of specific structured TAG in rat, *Lipids*, 37, 329–331, 2002.
52. Stein, J., Chemically defined structured lipids: current status and future directions in gastrointestinal diseases, *Int. J. Colorectal Dis.*, 14, 79–85, 1999.
53. Anonymous, World Health Organization Technical Report Series 913, 2002, pp. 1–153.
54. Ghoos, Y., Geypens, B., and Rutgeers, P. Stable isotopes and 13CO2 breath tests for investigating gastrointestinal functions. *Food Nutr. Bull.* 23, Suppl 3, 166–168, 2002.
55. De Jong, B.J. and Hulsmann, B.C., Monoacylglycerol hydrolase activity of isolated rat small intestinal cells, *Biochim. Biophys. Acta*, 27, 36–46, 1978.
56. Hanel, A.M. and Gelb, M.H., Multiple enzymatic activities of the human cytosolic 85-kDa phospholipase A2: hydrolysis reactions and acyl transfer to glycerol, *Biochemistry*, 20, 7807–7818, 1995.
57. Lehner, R., Kuksis, A., and Itabashi, Y., Stereospecificity of monoacylglycerol and diacylglycerol acyltransferases from rat intestine as determined by chiral phase high performance liquid chromatography, *Lipids*, 28, 29–34, 1993.
58. Paris, R. and Clement, G., Biosynthesis of lysophosphatidic acid from 1-monoolein by ATP by subcellular particles from intestinal mucosa, *Proc. Soc. Exp. Biol. Med.*, 131, 363–365, 1969.
59. Duff, S.M., Kalambur, S., and Boyle-Roden, E., Serum albumin binds beta- and alpha-monoolein in vitro, *J. Nutr.*, 131, 774–778, 2001.
60. Christophe, A. and Verdonk, G., Postprandial effects of feeding triglycerides or their digestion products on the quantity and fatty acid pattern of the majorserum lipid classes, *Arch. Intern. Physiol. Biochim.*, 80, 954–955, 1972.
61. Banno, F., Doisaki, S., Shimizu, N., and Fujimoto, K., Lymphatic absorption of docosahexaenoic acid given as monoglyceride, diglyceride, triglyceride and ethyl ester in rats, *J. Nutr. Sci. Vitaminol.*, 48, 30–35, 2002.
62. Wanatabe, K. Onizowa, S., Naito et al., Fat-soluble vitamin status is not affected by diacylglycerol consumption, *Ann. Nutr. Metab.*, 45, 259–264, 2001.
63. Christophe, A., Hill, E.G., and Holman, R.T., Efficacy of linoleic acid administered rectally as monoglyceride, *Lipids*, 22, 328–332, 1987.
64. Katsuragi, Y., Yasukawa, T., Matsuo, N. et al., Eds., *Diacylglycerol Oil*, AOCS Press, Champaign, IL, 2004.
65. Tagushi, H., Watanabe, H., Onizawa, K. et al., Double-blind controlled study on the effects of dietary diacylglycerol on postprandial serum and chylomicron triacylglycerol responses in healthy humans, *J. Am. Coll. Nutr.*, 19, 789–796, 2000.
66. Taguchi, H., Nagao, T., Watanabe, H. et al., Energy value and digestibility of dietary oil containing mainly 1,3-diacylglycerol are similar to those of triacylglycerol, *Lipids*, 36, 379–382, 2001.
67. Maki, K.C., Davidson, M.H., Tsushim, R. et al., Consumption of diacyl glycerol oil as part of a reduced-energy diet enhances loss of body weight and fat in comparison with consumption of a triacylglycerol control oil, *Am. J. Clin. Nutr.*, 76, 1230–1236, 2002.
68. Yamamoto, K., Asakawa, H., Tokunaga, K. et al., Long-term ingestion of dietary diacylglycerol lowers serum triacylglycerol in Type II diabetic patients with hypertriglyceridemia, *J. Nutr.*, 131, 3204–3207, 2001.
69. Meguro, S., Higashi, K., Hase, T. et al., Solubilization of phytosterols in diacylglycerol versus triacylglycerol improves the serum-cholesterol-lowering effect, *Eur. J. Clin. Nutr.*, 55, 513–517, 2001.
70. Björkhem, I., Boberg, K.M., and Leitersdorf, E., in *The Metabolic and Molecular Basis of Inherited Disease*, McGraw-Hill, New York, 2001, pp. 2961–2988.
71. Shamir, R., Johnson, W.J., Zolfaghari, R. et al., Role of bile salt-dependent cholesteryl ester hydrolase in the uptake of micellar cholesterol by intestinal cells. *Biochemistry*, 16, 6351–6358, 1995.
72. Mansbach, C.M., Cohen, R.S., and Leff, P.B., Isolation and properties of the mixed lipid micelles present in intestinal content during fat digestion in man, *J. Clin. Invest.*, 56, 781–791, 1975.

73. Nissinen, M., Gylling, H., Vuoristo, M., and Miettinen, T.A., Micellar distribution of cholesterol and phytosterols after duodenal plant stanol ester infusion, *Am. J. Physiol.*, 282, G1009–G1015, 2002.

74. Jones, P.J., Raeini-Sarjaz, M., Ntanios, F.Y. et al., Modulation of plasma lipid levels and cholesterol kinetics by phytosterol versus phytostanol esters, *J. Lipid Res.*, 41, 679–705, 2000.

75. Hallikainen, M.A., Sarkinen, E.S., and Uusitupa, M.I., Comparison of the effects of plant sterol ester and plant stanol ester-enriched margarines in lowering serum cholesterol concentrations in hypercholesterolaemic subjects on a low-fat diet, *J. Nutr.*, 130, 767–776, 2000.

76. Richelle, M., Enslen, M., Hager, C. et al., Both free and esterified plant sterols reduce cholesterol absorption and the bioavailability of beta-carotene and alpha-tocopherol in normocholesterolemic humans, *Am. J. Clin. Nutr.*, 80, 171–177, 2004.

77. Ntanios, F.Y., Duchateau, G.S., A healthy diet rich in carotenoids is effective in maintaining normal blood carotenoid levels during the daily use of plant sterol-enriched spreads, *Int. J. Vitam. Nutr. Res.*, 72, 32–39, 2002.

78. Raeini-Sarjaz, F.Y., Ntanios, C.A. Vanstone, and Jones, P.J., No changes in serum fat-soluble vitamin and carotenoid concentrations with the intake of plant sterol/stanol esters in the context of a controlled diet, *Metabolism*, 51, 652–656, 2002.

79. Miettinen, T.A., Vuoristo, M., Nissinen, M. et al., Serum, biliary, and fecal cholesterol and plant sterols in colectomized patients before and during consumption of stanol ester margarine, *Am. J. Clin. Nutr.*, 71, 1095–1102, 2000.

80. Hallikainnen, M.A., Sarkkinen, E.S., Gylling, H. et al., Comparison of the effects of plant sterol ester and plant stanol ester-enriched margarines in lowering serum cholesterol concentrations in hypercholesterolaemic subjects on a low-fat diet, *Eur. J. Clin. Nutr.*, 54, 715–725, 2000.

81. Awad, A.B. and Fink, C.S., Phytosterols as anticancer dietary components: evidence and mechanism of action, *J. Nutr.*, 130, 2127–2130, 2000.

82. Quilez, J., Garcia- Lorda, P., and Salas-Salvado, J., Potential uses and benefits of phytosterols in diet: present situation and future directions, *Clin. Nutr.*, 22, 343–351, 2003.

83. Saunders, D.R. and Sillery, J., Lecithin inhibits fatty acid and bile salt absorption from rat small intestine in vivo, *Lipids*, 11, 830–832, 1976.

84. Nalbone, G., Lairon, D., Charbonnier-Augeire, M. et al., Pancreatic phospholipase A2 hydrolysis of phosphatidylcholines in various physicochemical states, *Biochim. Biophys. Acta*, 620, 612–625, 1980.

85. Borgstrom, B., Importance of phospholipids, pancreatic phospholipase A2, and fatty acid for the digestion of dietary fat: in vitro experiments with the porcine enzymes, *Gastroenterology*, 78, 954–962, 1984.

86. Rampone, A.J. and Long, L.R., The effect of phosphatidylcholine and lysophosphatidylcholine on the absorption and mucosal metabolism of oleic acid and cholesterol in vitro, *Biochim. Biophys. Acta*, 486, 500–510, 1977.

87. Sugawara, T., Tushiro, M., Zhang, H. et al., Lysophosphatidylcholine enhanced carotenoid uptake from mixed micelles by Caco-2 human intestinal cells, *J. Nutr.*, 131, 2921–2927, 2001.

88. O'Doherty, J.A., Yousef, I.M., and Kuksis, A., Effect of phosphatidylcholine on triacylglycerol synthesis in rat intestinal mucosa, *Can. J. Biochem.*, 52, 726–733, 1974.

89. Zierenberg, O., Odenthal, J., and Betzing, H., Incorporation of PPC into serum lipoproteins after oral or i.v. administration, *Atherosclerosis*, 34, 259–276, 1979.

90. Zeisel, S.H. and Blusztajn, Choline and human nutrition, *Annu. Rev. Nutr.*, 14, 269–296, 1994.

91. Amate, A. Gil, and Ramirez, M., Dietary long-chain polyunsaturated fatty acids from different sources affect fat and fatty acid excretion in rats, *J. Nutr.*, 131, 3216–3221, 2001.

92. Amate, L. and Ramirez, M., Feeding infant piglets formula with long-chain polyunsaturated fatty acids as triacylglycerols or phospholipids influences the distribution of these fatty acids in plasma lipoprotein fractions, *J. Nutr.*, 131, 1250–1255, 2001.

93. Wijendran, V., Huang, M.C., Diau, G.Y. et al., Efficacy of dietary arachidonic acid provided as triglyceride or phospholipid as substrates for brain arachidonic acid accretion in baboon neonates, *Pediatr. Res.*, 51, 265–272, 2002.

24 Lipid Oxidation in Specialty Oils

Karen Schaich

Department of Food Science, Rutgers University,
New Brunswick, New Jersey

CONTENTS

24.1   INTRODUCTION

Lipid oxidation or rancidity is clearly the major challenge for stabilizing specialty oils, particularly since oils with special nutraceutical properties have predominantly polyunsaturated fatty acids. While there is relatively little data yet available regarding oxidation of specialty oils *per se*, all oils follow the same fundamental processes, modified by endogenous pro- and antioxidants and innate differences in fatty acid composition. What has been learned about lipid oxidation in conventional food oils can be applied to predict stability of specialty oils and to explain observed behaviors. This chapter thus begins with an overview of important initiation, propagation, and termination reactions that need to be followed to provide full information about oxidation of lipids, and then reviews specific behavior patterns that have been identified in specialty oils.

24.2   CLASSIC RADICAL CHAIN REACTION SCHEME

It has been recognized since the 1940s that lipids oxidize by classic free radical chain reactions[1–5], one version of which is shown in Figure 24.1. The classic chain reaction scheme with three phases has been repeated in many forms. Sometimes secondary abstraction reactions of lipid alkoxyl radicals (LO$^{\bullet}$) and peroxyl radicals (LOO$^{\bullet}$) are presented as initiation reactions because they form lipid radicals (L$^{\bullet}$). That is true when lipid oxyl radicals are from outside sources, e.g., lipoxygenase reactions followed by Fe^{2+} and Fe^{3+} reactions with LOOH. However, in the following discussion LO$^{\bullet}$ and LOO$^{\bullet}$ arising from the initial L$^{\bullet}$ or its subsequent reactions are considered to mediate propagation or chain branching (initiation of secondary chains) rather than *ab initio* initiation.

The driving force in the chain reaction is the repeated abstraction of hydrogen atoms by LOO$^{\bullet}$ to form hydroperoxides plus free radicals on a new fatty acid. The process continues indefinitely until no hydrogen source is available or the chain is intercepted, imparting several unique characteristics that present distinct challenges in measuring and controlling lipid oxidation and partially explain why lipid oxidation is a major problem *in vivo* and in storage stability of foods:

1. Lipid oxidation is autocatalytic – once started, the reaction is self-propagating and self-accelerating.
2. *G* (product yield) $\gg$ 1, i.e., many more than one LOOH is formed and more than one lipid molecule is oxidized per initiating event. Chain lengths as long as 200 to 300 lipid molecules have been measured[6–10] showing how effective a single initiating event can be. However, this also points out one reason why it has been so difficult to study initiation processes — initiators become the proverbial needle in a haystack once oxidation chains become established.
3. Very small amounts of pro- or antioxidants cause large rate changes.
4. The reaction produces multiple intermediates and products that change with reaction conditions and time.

Despite continued application of the classic chain reaction to lipid oxidation, there is now considerable evidence that only Reactions (1), (2), and (5) (and perhaps also (6)) of Figure 24.1 are always present and hydrogen abstraction alone does not fully account for oxidation kinetics or products. Research has clearly shown that although hydrogen abstraction ultimately occurs to

Classical free radical chain reaction mechanism of lipid oxidation

Initiation *(formation of ab initio lipid free radical)*

$$L_1H \xrightarrow{k_i} L_1^{\bullet} \tag{1}$$

Propagation

Free radical chain reaction established

$$L_1^{\bullet} + O_2 \underset{k_{\beta}}{\overset{k_o}{\rightleftharpoons}} L_1OO^{\bullet} \tag{2}$$

$$L_1OO^{\bullet} + L_2H \xrightarrow{k_{p1}} L_1OOH + L_2^{\bullet} \tag{3}$$

$$L_2OO^{\bullet} + L_3H \xrightarrow{k_{p1}} L_2OOH + L_3^{\bullet} \; etc. \dashrightarrow L_nOOH \tag{4}$$

Free radical chain branching (initiation of new chains)

$$L_nOOH \xrightarrow{k_{d1}} L_nO^{\bullet} + OH^{-} \; (reducing \; metals) \tag{5}$$

$$L_nOOH \xrightarrow{k_{d2}} L_nOO^{\bullet} + H^{+} \; (oxidizing \; metals) \tag{6}$$

$$L_nOOH \xrightarrow{k_{d3}} L_nO^{\bullet} + {}^{\bullet}OH \; (heat \; and \; uv) \tag{7}$$

$$\left.\begin{array}{l} L_nO^{\bullet} \\ L_nOO^{\bullet} \\ HO^{\bullet} \end{array}\right\} + L_4H \begin{array}{l} \xrightarrow{k_{p2}} \\ \xrightarrow{k_{p1}} \\ \xrightarrow{k_{p3}} \end{array} \left.\begin{array}{l} L_nOH \\ L_nOOH \\ HOH \end{array}\right\} + L_4^{\bullet} \qquad \begin{array}{l} (8a) \\ (8b) \\ (8c) \end{array}$$

$$L_1OO^{\bullet} + L_nOOH \xrightarrow{k_{p4}} L_1OOH + L_nOO^{\bullet} \tag{9}$$

$$L_1O^{\bullet} + L_nOOH \xrightarrow{k_{p5}} L_1OH + L_nOO^{\bullet} \tag{10}$$

Termination *(formation of non-radical products)*

$$\left.\begin{array}{l} L_n^{\bullet} \\ L_nO^{\bullet} \\ LnOO^{\bullet} \end{array}\right\} + \left.\begin{array}{l} L_n^{\bullet} \\ L_nO^{\bullet} \\ LnOO^{\bullet} \end{array}\right\} \begin{array}{l} \text{Radical recombinations} \\ \xrightarrow[k_{t2}]{k_{t1}} \text{Polymers, non-radical monomer products} \\ k_{t3} \quad \text{(ketones, ethers, alcohols, alkanes, etc.)} \end{array} \qquad \begin{array}{l} (11a) \\ (11b) \\ (11c) \end{array}$$

$$\left.\begin{array}{l} LOO^{\bullet} \\ LO^{\bullet} \end{array}\right\} \begin{array}{l} \text{Radical scissions} \\ \xrightarrow[k_{ts2}]{k_{ts1}} \text{Non-radical products} \\ \quad \text{(aldehydes, ketones, alcohols, alkanes, etc.)} \end{array} \qquad \begin{array}{l} (12a) \\ (12b) \end{array}$$

i-initiation; o-oxygenation; β-O_2 scission; p-propagation; d-dissociation;
t-termination; ts-termination/scission

FIGURE 24.1 Classic free radical chain reaction mechanism of lipid oxidation.

terminate a radical, it is not always the fate of the initial lipid peroxyl or alkoxyl radicals. Indeed, lipid alcohols from H abstraction are relatively minor products of lipid oxidation. Competing alternative reactions for LOO$^{\bullet}$ and LO$^{\bullet}$ also propagate the radical chain but lead to different kinetics and different products than expected from the classic reaction sequence[11–13].

In reality, the classic radical chain sequence portrays only a small part of the lipid oxidation process and products. Variations in initiation mechanisms as well as radical reactions other than hydrogen abstraction must be considered in interpreting oxidation kinetics and even deciding what products to analyze. Indeed, multiple pathways and reaction tracks need to be evaluated simultaneously to develop an accurate picture of lipid oxidation in model systems, foods, and biological tissues.

24.3 INITIATION (LH → L•)

24.3.1 REACTIONS

The initiation process that produces the *ab initio* lipid free radicals, L•, in lipid oxidation is not well understood, so it is usually denoted as an "X" or "?" over the reaction arrow. Although lipid oxidation is a very facile reaction that is nearly ubiquitous in foods and biological systems, it is *not a spontaneous reaction*! Thermodynamically, oxygen cannot react directly with double bonds because the spin states are different. Ground state oxygen is in a triplet state (two free electrons in separate orbitals have the same spin direction, net positive angular momentum) while the double bond is in a singlet state (no unpaired electrons, paired electrons are in the same orbital and have opposite spin, no net angular momentum). Quantum mechanics requires that spin angular momentum be conserved in reactions, so triplets cannot invert (flip spins) to singlet states. Reaction then demands that the double bond be excited into a triplet state, which requires prohibitive amounts of energy (E_a = 35 to 65 kcal/mol). Thus, no direct reaction occurs.

$$\overset{\cdot\uparrow}{O}\!\!-\!\!\overset{\cdot\uparrow}{O} \; + \; -C\overset{\uparrow\cdot\cdot\uparrow}{=\!=}C- \quad\xrightarrow{\quad\times\quad}\quad ROOH$$

Triplet   Singlet

$$(24.1)$$

To overcome this spin barrier, initiators or catalysts are *required* to start the lipid oxidation process by removing an electron from either the lipid or oxygen or by changing the electron spin of the oxygen. Only trace amounts of catalysts are needed, so many situations that appear to be spontaneous or uncatalyzed are actually driven by contaminants or conditions which have gone undetected or unconsidered. Indeed, in most foods and biological systems and laboratory experiments, it is fair to say that multiple catalysts and initiators are always operative. The most common initiators are described in Table 24.1.

24.3.2 SITES OF RADICAL INITIATION BY HYDROGEN ABSTRACTION AND CONFIGURATION OF PEROXYL RADICALS

Hydrogen abstraction by free radicals is generally quite specific, occurring preferentially at allylic hydrogens where the C–H bond energies are lowest[62]. The order of reactivity is: doubly allylic hydrogens between two double bonds > singly allylic hydrogens next to double bonds ≫ hydrogens α to the –COOH group > hydrogens on methylene groups farther down the acyl chains. The one exception to this "rule" is the hydroxyl radical, HO•, which is so electrophilic and reactive that it abstracts hydrogens indiscriminately from all positions along the acyl chain[49]. Radicals thus formed then either migrate to the acyl carbon with the weakest bonding, i.e., the allylic hydrogens, or abstract allylic hydrogens from a neighboring lipid molecule.

Following hydrogen abstraction from the allylic hydrogens, the free electron becomes distributed across a resonance stabilized double bond system. The highest electron density is in the center, and the outside positions are relatively electron deficient. Thus, oxygen preferentially adds at the outermost points. When oxygen pressures are greater than 100 mmHg, the addition occurs at diffusion-controlled rates ($k > 10^8$ L M^{-1} s^{-1})[33] so is essentially instantaneous as long as oxygen is available — one reason lipid radicals (L•) are so difficult to detect, even by electron paramagnetic resonance.

$$(24.2)$$

e⁻deficient points

TABLE 24.1
Mechanisms for initiation of lipid oxidation.

Initiator	Simplified reactions	Conditions	Ref.
Metals		Generally all but especially at $< 10^{-4}$ M; high conc. can be antioxidant	14–18
Direct	$RCH=CHR + M^{(n+1)+} \xrightarrow{LH} RCH-CHR + M^{n+} [\xrightarrow{LH} L^{\bullet} + RH]$		
	$M^{n+} + O_2 \longrightarrow [M^{(n+1)+}...O_2^{-\bullet}] \xrightarrow{LH} L^{\bullet} + M^{n+} + HO_2^{\bullet}$	Moderate to high pO_2	18, 19
	$\xrightarrow{L'H} L'^{\bullet} + H_2O_2$		
Indirect	(a) Autoxidation of reduced metals		18, 19
	$Fe^{2+} + O_2 \longrightarrow Fe^{3+} + O_2^{-\bullet} \xrightarrow[H^+]{} HOO^{\bullet} \xrightarrow{LH} L^{\bullet} + H_2O_2$		
	$O_2^{-\bullet} + HOO^{\bullet} \longrightarrow H_2O_2 \xrightarrow{Fe^{2+}} Fe^{3+} + HO^- + {}^{\bullet}OH \xrightarrow{LH} H_2O + L^{\bullet}$	Dominates with low metal, substrate, and oxygen conc.	20–26
	(b) Reduction or oxidation of hydroperoxides (ROOH, H_2O_2, LOOH)		
	$Fe^{2+} + \begin{cases} ROOH \\ LOOH \end{cases} \xrightarrow{fast} Fe^{3+} + \begin{cases} RO^{\bullet} + OH^- \\ LO^{\bullet} + OH^- \end{cases} \xrightarrow{LH} \begin{cases} ROH + L^{\bullet} \\ LOH + L^{\bullet} \end{cases}$		
	$Fe^{3+} + ROOH \xrightarrow{extremely\ slow} Fe^{2+} + ROO^{\bullet} + H^+ \xrightarrow{LH} ROOH + L^{\bullet}$		
Cycling	$LH + M^{(n+1)+} \longrightarrow L^{\bullet} + H^+ + M^{n+} \xrightarrow[LH]{O_2} LOOH + L^{\bullet}$	Presence of both valence states or $M^{(n+1)+}$ plus reducing agent AH continuously regenerates active catalysts, accelerates oxidation	16, 129
	$LOOH + M^{n+} \longrightarrow LO^{\bullet} + OH^- + M^{(n+1)}$		
	$\underline{\quad AH \quad}$		
Ferryl	See Figure 24.2	Requires electrophilic Lewis base ligand, low water; Fe:ROOH ratio determines rx path	16, 27–30, 40–42
Lipoxygenase	$E(Fe^{2+}) \xrightarrow[LH\ H^+]{} E(Fe^{3+}) ... L^{\bullet} \xrightarrow{O_2} E(Fe^{2+}) ... LOO^{\bullet} \longrightarrow E(Fe^{3+}) ... LOO^- \xrightarrow{E(H^+)} E(Fe^{3+}) + LOOH$	LOOH formed directly in enzyme complex, no free radicals released	31–33
Free radicals	$LH + R^{\bullet}, RO^{\bullet}, ROO^{\bullet} \longrightarrow L^{\bullet} + RH, ROH, ROOH$	See Table 24.2	14

TABLE 24.1 (Continued)
Mechanisms for initiation of lipid oxidation.

Initiator	Simplified reactions	Conditions	Ref.
Light	$R_1CHR_2 \xrightarrow{h\nu} R_1\dot{C}R_2 + H^\bullet$ or $R_1\dot{C}H + {}^\bullet R_2$	Little C-C scission at >200 nm	14, 34
UV	$LOOH \xrightarrow{h\nu} LO^\bullet + {}^\bullet OH \xrightarrow{L'H} L'^\bullet + LOH$ or HOH	Major effect under normal conditions	
Visible Light, Photosensitized	${}^1S \xrightarrow{h\nu} {}^3S^*$ *Type I – Redox/Free radical* $L \xrightarrow[-H^+]{LH} {}^\bullet SH + {}^\bullet L \xrightarrow{O_2} {}^0S + LOO^\bullet \xrightarrow{H^+} LOOH$ $L \xrightarrow{-e^-} (S^{-\bullet}+L^{+\bullet})$ or $(S^{+\bullet}+L^{-\bullet}) \xrightarrow[H^+]{O_2} {}^0S + LOOH$ *Type II – Oxygenation* A. Direct ${}^3O$ ${}^1S + {}^1O_2 \xrightarrow{LH} {}^0S + LOOH$ B. Indirect ${}^3O_2$ $S\!-\!O_2^* \xrightarrow{LH} {}^1S + {}^1O_2 \rightarrow LOOH$ ${}^0S + AOO^\bullet \xrightarrow{H^+} LOOH$	Chlorophyll is major sensitizer	14, 222, 225
		Most pigments are sensitizers, carotenoids are ${}^1O_2$ quenchers	14, 35, 36, 223–236
Ozone	$RCH{=}CHR \xrightarrow{O_3} RCH{-}CHR$ (O–O–O ring) $\rightarrow RCH{-}CHR$ $\rightarrow$ epoxides, cyclic peroxides, free radical chain rxs $LOOH + O_3 \rightarrow LOO^\bullet + HO^\bullet + O_2$	Slow rx ($k \sim 10^5$), important mainly as source of radicals in induction, does not change product mix, rx increases with number of double bonds, An and higher PUFAs particularly susceptible	37–39
Hemes	see Figure 24.3	Dominant effect at >~40°C	40–42
Heat	$LOOH \rightarrow LO^\bullet + {}^\bullet OH$ $LOOH..HOOL \rightarrow LOO^\bullet + H_2O + {}^\bullet OL$ $RCH_2{-}CH_2R \rightarrow RCH_2^\bullet + {}^\bullet CH_2R$	C chain scissions dominate at high temps.	14, 318 43, 44

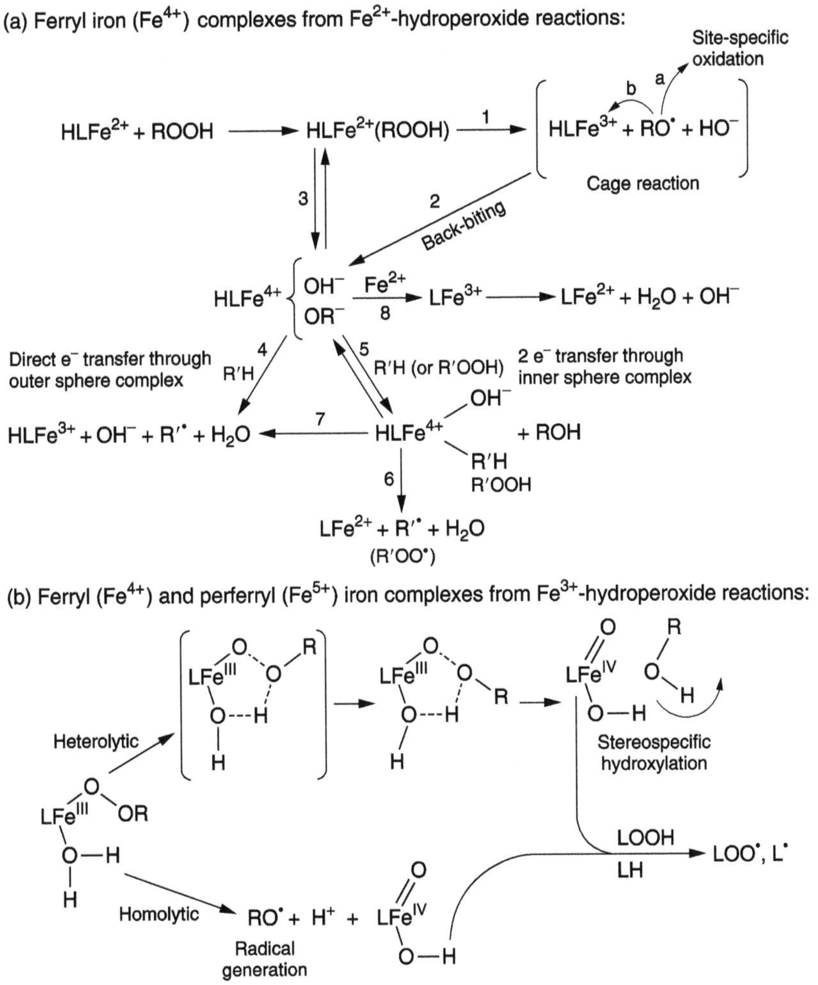

(a) Ferryl iron (Fe^{4+}) complexes from Fe^{2+}-hydroperoxide reactions:

$$HLFe^{2+} + ROOH \longrightarrow HLFe^{2+}(ROOH) \xrightarrow{1} \left[HLFe^{3+} + RO^{\bullet} + HO^{-} \right]$$

Site-specific oxidation

b a

Cage reaction

3 ↕ 2 Back-biting

$$HLFe^{4+} \begin{Bmatrix} OH^- \\ OR^- \end{Bmatrix} \xrightarrow[8]{Fe^{2+}} LFe^{3+} \longrightarrow LFe^{2+} + H_2O + OH^-$$

Direct e^- transfer through outer sphere complex 4 R'H 5 R'H (or R'OOH) 2 e^- transfer through inner sphere complex

$$HLFe^{3+} + OH^- + R'^{\bullet} + H_2O \longleftarrow 7 \; HLFe^{4+} \begin{array}{l} OH^- \\ \\ R'H \\ R'OOH \end{array} + ROH$$

6 ↓

$$LFe^{2+} + R'^{\bullet} + H_2O$$
$$(R'OO^{\bullet})$$

(b) Ferryl (Fe^{4+}) and perferryl (Fe^{5+}) iron complexes from Fe^{3+}-hydroperoxide reactions:

Heterolytic

$$LFe^{III} \begin{array}{c} O\cdots O\diagdown R \\ O---H \\ H \end{array} \longrightarrow LFe^{III} \begin{array}{c} O\cdots O \\ \diagdown R \\ O---H \\ H \end{array} \longrightarrow LFe^{IV} \begin{array}{c} O \\ \diagdown R \\ O-H \end{array}$$

Stereospecific hydroxylation

$$LFe^{III} \diagdown OR$$
$$O-H$$
$$H$$

Homolytic $\longrightarrow RO^{\bullet} + H^+ + LFe^{IV}$

Radical generation

$$LFe^{IV}$$
$$O-H$$

LOOH / LH $\longrightarrow LOO^{\bullet}, L^{\bullet}$

FIGURE 24.2 Formation of ferryl iron in initiation and catalysis of lipid oxidation. Reaction schemes for formation of hypervalent iron states by Fe^{2+} and Fe^{3+} complexes and subsequent reactions leading to radicals that can initiate lipid oxidation. L, metal ligand; R, alkyl or acyl group. (Fe^{2+} sequence adapted from Rush, J.D. and Koppenol, W.H., *J. Am. Chem. Soc.*, 110, 4957–4963, 1988; Tung, H.-C., Kang, C., and Sawyer, D.T., *J. Am. Chem. Soc.*, 114, 2445–3555, 1992. Fe^{3+} sequence adapted from Bassan, A., Blomberg, M.R.A., Siegbahn, P.E.M., and Que, L., Jr., *J. Am. Chem. Soc.*, 124, 11056–11063, 2002.)

Hydroperoxide positional distributions in unsaturated fatty acids undergoing autoxidation and photosensitized oxidation are presented in Table 24.3. Isolated double bonds behave as if there were two separate resonant systems of equal probability, so oleic acid yields (C9 + C11) and (C8 + C10) hydroperoxides from the two resonance systems, respectively, in approximately equivalent amounts. In 1,4-diene systems, H abstraction occurs preferentially at the doubly allylic hydrogen between the two double bonds, and the resonance system with the unpaired electron extends across both double bonds with electron density focused at the central carbon C11 and electron deficient positions at external carbons 9 and 13 (18:2). In higher polyunsaturated fatty acids with multiple 1,4-diene structures, the resonant systems from multiple doubly allylic radicals overlap. In theory, then, hydroperoxides should form at internal positions in equal proportion to the external positions.

TABLE 24.2

Radical	Half-life with typical substrate, 10^{-3} M, 37°C			Ave. rx rate, k (L mol^{-1} s^{-1})		Ref.
HO$^\bullet$	10^{-9} s			10^9–10^{10}		45
RO$^\bullet$	10^{-6} s			10^6–10^8		45
ROO$^\bullet$	10 s			10^1–10^3		45
L$^\bullet$	10^{-8} s			10^4–10^8		45
AnOO$^\bullet$	10^{-5} s					46[a]
O$_2^-$$^\bullet$				~1		47[b]
HOO$^\bullet$				10^0–10^3		48[b]

	18:1	18:2	18:3	20:4	
HO$^\bullet$	~10^9	9.0×10^9	7.3×10^9	~10^{10}	49,50
Monomer		8×10^9	8×10^9		48
Micellar		1.3×10^9	2.5×10^9		50
Nonallylic H	4×10^2	3.4×10^3	7×10^3	1×10^4	48
RO$^\bullet$	3.3×10^6	8.8×10^6	1.3×10^7	2.0×10^7	49
t-BuO$^\bullet$	3.8×10^6	9.1×10^6	1.3×10^7	2.1×10^7	51
	(trans) 3.3×10^6	(trans) 8.8×10^6			51
	aqueous 6.8×10^7	1.3×10^8	1.6×10^8	1.8×10^8	52
ROO$^\bullet$	1.1	6×10^1	1.2×10^2	1.8×10^2	53–55
O$_2^-$$^\bullet$	no rx	no rx	< 1	< 1	48, 54
	(MLOOH) 7.4×10^3				56
HOO$^\bullet$	no rx.	1.1×10^3	1.7×10^3	3.1×10^3	47
		< 3×10^2			54
O$_3$ CCl$_4$	6.4×10^5	6.9×10^5			57
O$_3$ aq SDS	9.5×10^5	1.1×10^6			57
SO$_3^-$$^\bullet$		1.8×10^6	2.8×10^6	3.9×10^6	58
GS$^\bullet$ < 2×10^6	8×10^6	1.9×10^7	3.1×10^7		59
^{1}O$_2$	0.74×10^5	1.3×10^5	1.9×10^5	2.4×10^5	60
O$^-$$^\bullet$	7.5×10^2	9.7×10^3	1.2×10^4	1.9×10^4	48
NO$_2$$^\bullet$	1.2×10^6	6.2×10^6	6.6×10^6		61

[a] Aqueous solution.
[b] H abstraction from unsaturated alkenes.

Nevertheless, only minor amounts of internal hydroperoxides are observed, and then only with three or more double bonds, because they undergo rapid β-elimination of the oxygen to regenerate the original 1,4-diene radical[63] and also have a very strong tendency towards cyclization[64–67]. Consequently, the dominant hydroperoxides of autoxidizing fatty acids are always found at the external positions, regardless of the number of double bonds, except under two circumstances: (1) in autoxidation, equal distribution of LOOH at all positions without any cyclic products is found only in media of high H donating power — e.g., when 3 to 5% tocopherol is added[12], and (2) internal hydroperoxides are characteristic of singlet oxygen-photosensitized oxidation.

Hydroperoxides have geometric as well as positional isomers on lipid chains. When the hydrogen is abstracted at an allylic carbon, the double bond shifts one carbon to a position β to the abstraction site, and it reforms in the *trans* rather than *cis* configuration. The *trans,cis*-conjugated diene structure of 18:2 and higher unsaturated fatty acids is retained whether oxygen adds or not, and provides the first detectable intermediate in lipids during autoxidation[75].

(24.3)

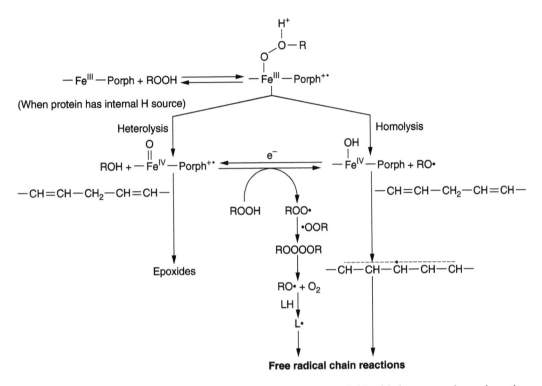

FIGURE 24.3 Heme-catalyzed formation of species that can initiate lipid oxidation: generation and reaction of ferryl iron complexes [$Fe^{IV} = O$, $Fe^{IV}(OH)$]. (Adapted from Nam, W., Han, H.J., Oh, S.-Y., Lee, Y.J., Choi, M.-H., Han, S.-Y., Kim, C., Woo, S.K., and Shin, W., *J. Am. Chem. Soc.*, 122, 8677–8684, 2000; Nam, W., Lim, M.H., Lee, H.J., and Kim, C., *J. Am. Chem. Soc.*, 122, 6641–6647, 2000.)

Porter and colleagues[13,63,76–90] have shown that both positional and geometric isomerism proceed through the delocalized allyl radical for oleate or the dienyl radical for linoleate and higher polyunsaturated fatty acids (PUFAs) (Figure 24.4) via alternating removal of the peroxyl oxygen by β-scission, migration of the free radical, and readdition of the oxygen at a new carbon position or orientation. β-scission of oxygen is very fast (27 to 430 s^{-1})[89] and competes effectively with H abstraction. There can be interconversion of peroxyl position and orientation indefinitely as long as the radical is in the manifold (Figure 24.4). Once the peroxyl radical is protonated via H abstraction, it becomes fixed as the hydroperoxide, but can return to the manifold if the LOOH hydrogen is abstracted.

Both thermodynamic and kinetic processes influence proportions of *trans,cis* and *trans,trans* isomers in linoleic acid[13,63,84]. Kinetically, hydroperoxides form whenever an abstractable hydrogen atom is available, but thermodynamically the system equilibrium moves toward *trans,trans* isomers in the absence of good H donors, as in organic solvents[91]. The observed isomer mix thus reflects the balance and competition between these two processes in a given system. In the presence of good H donors — a protic solvent, an antioxidant, a cosubstrate, or the allylic hydrogens of the fatty acid chains themselves — the *trans,cis* isomers kinetically form first. For oleic and linoleic acids with only slightly bent chains, *trans,cis* formation is favored in oriented systems or at high concentrations that increase interchain contact. *Trans,trans* isomers are favored in dilute solutions, aprotic solvents, and at elevated temperatures where there is less interchain contact and decreased H availability. With linolenic, arachidonic, and higher polyunsaturated fatty acids, the fatty acid chains bend back on each other, bringing double bonds and allylic hydrogens from opposite ends of the chain into proximity with the peroxyl radicals. When oxidized neat, higher PUFAs thus have an immediate internal H source and characteristically yield high proportions of *trans,cis* peroxides

TABLE 24.3

	5-OOH	6-OOH	8-OOH	9-OOH	10-OOH	11-OOH	12-OOH	13-OOH	14-OOH	15-OOH	16-OOH	Ref.
18:1Δ9												
Autoxidation			26.4	24.2	22.8	26.6						68
Photo-ox $^1O_2$				47.5	52.3							69
Photo-ox Chl*				49.1	50.8							70
Thermal oxidation			25.1	25.1	24.9	24.9						68
18:2Δ9,12												
Autoxidation				51	tr		tr	49				71
Photo-ox $^1O_2$				31.9	16.7		17	34.5				69
Photo-ox Chl*				30.2	19.8		19.8	30.1				70
18:3Δ9,12,15												
Autoxidation				33.4			10.1	12.5			43.9	72
Photo-ox $^1O_2$				22.7	12.7		12.0	14.0		13.4	25.3	69
Photo-ox Chl*				21.6	14.3		15.3	15.7		12.0	21.1	70
20:4Δ5,8,11,14												
Autoxidation	27		7	9		11	6			40		49
Photo-ox $^1O_2$	14.4	4.8	12.9	13.2		14.4	13.3		6.9	20.3		73
22:6Δ4,7,10,13,16,19												
Autoxidation												74

% of –OOH at C20 (27.1), C17 (7.9), C16 (9.2), C14 (10.8), C13 (8.9), C11 (7.3), C10 (7.3), C8(7.9), C7(7.0), C4(6.5)

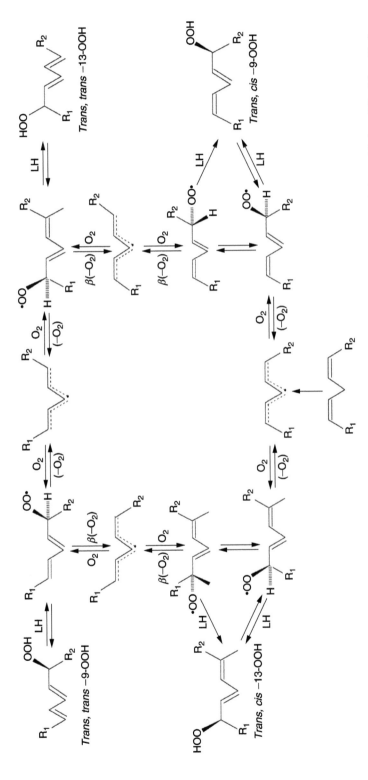

FIGURE 24.4 Reaction scheme for positional isomerization of double bonds and formation of *trans,trans*-hydroperoxides during oxidation of linoleic acid via reversible β-scission of oxygen. (Adapted from Porter, N.A., Weber, B.A., Weenan, H., and Kahn, J. A., *J. Am. Chem. Soc.*, 102, 5597–5601, 1980; Porter, N.A. and Wujek, D.G., *J. Am. Chem. Soc.*, 106, 2626–2629, 1984; Weenan, H. and Porter, N.A., *J. Am. Chem. Soc.*, 104, 5216–5221, 1982.)

(kinetic products). However, when an H donor is lacking (e.g., low concentrations, aprotic solvent, elevated temperature), *trans,trans* cyclic hydroperoxides become dominant[82].

The *cis* to *trans* ratio changes with reaction system and with temperature. *Cis* isomers are enhanced by the presence of antioxidants such as tocopherol and by high concentrations of lipids, while *trans* isomers are enhanced by even mild heating which reduces contact between lipid and potential H donors. Contrary to earlier reports, the *cis* to *trans* ratio does not vary with the extent of oxidation unless reaction conditions are changing or H abstraction from LOOH is occurring, allowing LOO[•] to undergo β-scission.

24.4 PROPAGATION

The classic free radical chain depicts propagation as proceeding directly and entirely by hydrogen abstraction. In reality, however, H abstraction by LOO[•] is very slow ($k = 36$ to 62 L mol^{-1} s^{-1})[53,92] and selective, abstracting only hydrogens with low bond energy (e.g., doubly allylic –CH$_2$–, thiols, phenols)[93]. Consequently, there is plenty of time for alternative reaction pathways to compete and change the direction of oxidation[94], yielding distinctly different products at different rates and having significant consequences to the ultimate mixture of products. Addition, cyclization, and scission reactions compete with H abstraction to reroute LO[•] and generate products and additional radical species. Ultimately, radicals are always transferred between molecules by hydrogen abstraction, but the original LOO[•] may not be the propagating radical, and the product mix is much more complicated than implied by the simple free radical chain. Consideration of the multiple competing pathways can explain complicated oxidation kinetics, account for complex product mixes, enable more accurate evaluation of the extent of oxidation, and facilitate design of more effective antioxidant strategies.

24.4.1 Chain Propagation by Peroxyl Radicals, LOO[•]

Lipid peroxyl radicals (LOO[•]) are the chain carriers in early stages of lipid oxidation. Four reactions of LOO[•] — hydrogen abstraction, rearrangement/cyclization, addition, and disproportionation — all contribute to chain propagation, though under different conditions. Simplified reactions involved in these propagations are shown in Table 24.4. The mechanism(s) occurring in any given system are determined by the ease of H abstraction and double bond structure in the target molecule, solvent, and reaction conditions, particularly temperature. Contrary to earlier expectations, data are now available demonstrating that LOOH is not the only product in early stages of lipid oxidation. In some cases, hydroperoxides may ultimately form only after addition, cyclization, or other rearrangement; in some systems conventional monohydroperoxides may not form at all.

24.4.2 Propagation by Alkoxyl Radicals, LO[•]

Alkoxyl radicals (LO[•]) are responsible for propagation of the radical chain during the very rapid oxidation that ensues after the induction period ends. Because LO[•] react faster than LOO[•] by several orders of magnitude, LO[•] become dominant almost as soon as LOOH breaks down and their reactions become important as secondary events in oxidation. Alkoxyl radicals propagate radical chain reactions by hydrogen abstraction, rearrangement/cyclization, addition to double bonds, and β-scission as outlined in Table 24.5. The mechanism dominating in a given system is determined largely by double bond structure, solvent conditions, and steric factors[11].

24.4.3 Propagation Reactions of LOOH: Mono- versus Bimolecular LOOH Decomposition and Chain Branching

While production of lipid hydroperoxides via the reactions given in Table 24.4 and Table 24.5 is the major process driving initial lipid oxidation, decomposition of hydroperoxides provides the alkoxyl radicals responsible for chain branching that greatly amplifies and broadcasts effects of initiation.

Lipid oxidation gathers steam, increasing in rate and extent as LO$^\bullet$ becomes the dominant chain carrier and secondary chains are initiated. Indeed, a large proportion of the LOO$^\bullet$ and all of the LO$^\bullet$ involved in propagation are not *ab initio* radicals but arise from some form of LOOH decomposition[152]. The net effect of LOOH decomposition is, therefore, a transition in mechanism and kinetics.

In early stages of oxidation, LOOH decomposition to propagating radicals is monomolecular. Heterolytic decomposition of LOOH by metals to LO$^\bullet$ and LOO$^\bullet$ is shown in Table 24.1. Homolytic decomposition of LOOH by heat is much more catastrophic because two propagating radicals are released per hydroperoxide: LO$^\bullet$ is more reactive and more selective than LOO$^\bullet$; HO$^\bullet$ is extremely reactive and rather unselective, abstracting hydrogen atoms all along the acyl chain, and it can also readily add to double bonds (still generating a radical). The O–O bond in organic hydroperoxides begins decomposing at about 50°C and is completely decomposed at 160°C[153]. A less recognized third mechanism, H abstraction from LOOH by LOO$^\bullet$ and LO$^\bullet$ (H abstraction reaction (b) in Table 24.4 and Table 24.5, respectively), actually occurs more rapidly than the traditional propagation via allylic hydrogens. It recycles peroxyl radicals and propagates the radical chain without producing secondary products.

Propagation becomes more complicated as LOOH accumulates. At [LOOH] greater than 1% oxidation[154], hydroperoxides interact and bimolecular decomposition leads to greatly accelerated oxidation[112,155]. The traditional Russell explanation for this acceleration involves decomposition of hydrogen bonded LOOH transition dimers to generate equal amounts of alkoxyl and peroxyl radicals[156]:

$$2 \text{ LOOH} \longrightarrow \text{LOOH...HOOL} \longrightarrow \text{LO}^\bullet + \text{H}_2\text{O} + {}^\bullet\text{OOL} \tag{24.4}$$

The dramatic increase in oxygen consumption rates and production of LOOH during the bimolecular rate period was attributed to the faster reactions of LO$^\bullet$, although LOO$^\bullet$ also produces secondary chains. Recently, however, thermodynamic considerations and EPR evidence for free radical intermediates have indicated that the dramatic increase in oxidation more likely results from LOO$^\bullet$ radical-induced decomposition of LOOH where one slowly reacting radical reacts with a nonpropagating hydroperoxide to generate a powerful cascade of reactive radicals: *three* very reactive radicals, LO$^\bullet$, epoxy-LO$^\bullet$, and $^\bullet$OH[115]. The heavy arrows in the following scheme indicate the favored pathway:

$$\tag{24.5}$$

24.4.4   SUMMARY OF PROPAGATION

Substantial experimental evidence now makes it clear that there is no fixed sequence of reaction pathways for lipid oxidation. Rather, the pathways that are most active change with reaction system, determined by the type and concentration of lipid, the solvent, phase distributions of catalysts, surface and interfaces, and numerous other factors. Consequently, we are usually looking at totally different processes when lipid systems are oxidized under different conditions, and interpretations of product data and designs of antioxidant strategies must recognize and account for alternative oxidation pathways.

TABLE 24.4
Major Competing Propagation Mechanisms of Lipid Peroxyl Radicals, LOO•

Mechanism and simplified reaction	Controlling conditions	Ref.
1. H abstraction → hydroperoxides (a) LOO• + LH ⟶ LOOH + L• $k = 62$ L M⁻¹s⁻¹ (b) LOO• + L'OOH ⟶ LOOH + L'OO• $k = 620$ L M⁻¹s⁻¹	Relative availability of H sources in lipids and solvent; (a) rx slow, highly selective for doubly allylic hydrogens; (b) 10 times faster; facilitated by neat lipids, aprotic solvents, high [lipid], elevated temp., low-viscosity media	12, 14, 49, 54, 58, 92, 95–104
2. Rearrangement/cyclization → epidioxides, hydroperoxy epidioxides	Requires *cis* db homo-allylic to –OOH, as in Ln and higher PUFAs (top) and ¹O₂-oxidized L (middle); favored by low pO₂, aprotic solvents, dilute lipids at room temp. Competitive, $(k \sim 10^3$ sec⁻¹) is faster than β-scission and H abstraction; ↓ number of molecules oxidized per chain, ↓ overall propagation rate	14, 67, 79, 83, 85, 89, 101, 105–107, 170
	Endo cyclization in fatty acids with three or more dbs requires enzyme catalysis	14, 67, 89, 103, 108, 109
3. Addition to double bonds → dimers, polymers, hydroperoxy epoxides (a) L,OO• + R₁CH₂CH = CHR₂ ⟶ R₁CH₂ĊHCHR₂ ⟶ polymers QOL₁ (b) L₁O• + R₁ĊHCH–CHR₂ —O₂→ L₂(epoxy)OO•	Favored by conjugated dbs and heat, competes with H abstraction when abstractable H's limited (aprotic solvents, neat lipids, low temperature); steric restrictions slow reactions with lipids ($k = 20$–1130 M⁻¹ s⁻¹); conjugated dbs yield mostly polymers via (a), isolated and nonconjugated dbs yield alkoxyl radicals and epoxides (b); most important in late stages of oxidation	14, 110–123

4. Disproportionation → peroxides, alkoxyl radicals
           (a)
R₁OO• + R₂OO• ⟶ [R₁OOOOR₂] ⟶ R₁O• + •OO)OR₂ ⟶ R₁O• + O₂ + •OR₂
                   (b)
              R₁OOR₂ + O₂
	(a) Production of LO• favored in neat lipids, aprotic solvents, $2k = 9.5 \times 10^3$ L M⁻¹ s⁻¹, contributes to rapid ↑ oxidation in bimolecular rate period;
	(b) LOOL dominates in aq. soln, $2k = 2 \times 10^7$ L M⁻¹ s⁻¹, and in viscous media
		14, 124–126, 318–321

TABLE 24.5
Propagation Mechanisms of Lipid Alkoxyl Radicals, LO•

Mechanism and simplified reaction	Conditions	Ref.
1. H abstraction (a) $R_1-\overset{\overset{\displaystyle O\bullet}{\|}}{C}H-R_2 + LH \longrightarrow R_1-\overset{\overset{\displaystyle OH}{\|}}{C}H-R_2 + L\bullet$	Very fast ($k \sim 10^7-10^8$ L M^{-1} s^{-1}) but less selective than LOO• — both allylic and bis-allylic H abstracted; most effective in neat lipids, aprotic solvents, high [lipid], elevated temp.; rate $\propto$ no. of allylic H's (dbs); (b) $\gg$ (a), dominant chain carrier	14, 65, 67, 103, 107, 111, 113, 127–134
(b) $LO\bullet + LOOH \longrightarrow LOH + LOO\bullet$	(a) Fast but not effective chain propagator in aq. soln — ↓ selectivity, ↑ abstraction from nonlipid H sources, ↑ competition from β-scission and internal H-abstraction (cyclization)	
(c) $RCH_2CH{=}CH{-}CH{=}CHR_1 \longrightarrow$ [cyclic internal abstraction scheme] $\longrightarrow \dot{R}CHCH{=}CH{-}CH{=}CHR_1$ and $R\dot{C}HCH{=}CH{-}CH{=}CH{-}CHR_1$ (OH)	(c) Internal 1,5 and 1,6 abstraction mimics HO• rx	14, 135, 136
2. Cyclization $R_1HCH{=}CH{-}CH{-}CH{-}CHR_2 \longrightarrow R_1HCH{=}CH{-}CH{-}CH{-}CHR_2$ (epoxide O) $\xrightarrow{O_2}$ $R_1HCH{=}CH{-}CH{-}CH{-}CHR_2$ (•OO, epoxide O) $HOO \; \overset{L_2^\bullet}{\underset{L_2H}{\rightleftarrows}} \; R_1HCH{=}CH{-}CH{-}CH{-}CHR_2$	Very fast, dominant rx in neat lipids and aprotic solvents, low [lipid], high surface dispersion, room temp, low pO$_2$; strongly favored with internal LO• Rx accelerates with ↑ polarity of aprotic solvent. Not important in polar and aq. solvents (LO• stability ↓↓, H abstraction very fast)	14, 107, 113, 137–144
3. Addition to double bonds $LO\bullet +$ [diene] $\longrightarrow$ [LO–adduct radical]	Facilitated by absence of allylic hydrogens and by conjugation, neat lipids, organic solvents, heat ∴ most active in catalyzing chain branching in secondary stages of oxidation; rx with cis $\gg$ trans dbs	11, 14, 122, 139, 145, 146
4. β-Scission $R_1 \overset{\beta}{\underset{\underset{\displaystyle O\bullet}{\|}}{C}} H \overset{\alpha}{-} \overset{\beta}{C} H - R_2 \longrightarrow \dot{R}_1 + \overset{\overset{\displaystyle O}{\|}}{C}H-R_2 \;\; \mathbf{OR} \;\; R_1\overset{\overset{\displaystyle O}{\|}}{C}H + \overset{\alpha}{\bullet}R_2$	Requires H$^+$ for product stabilization, very fast rx in aq/polar solvents ($k = 10^6-10^7$ s^{-1}); greatest contribution at elevated temps, minor process in neat lipids or aprotic solvents at room temp.	11, 14, 147–151

The distinction between propagation mechanisms is important because shifts in propagation pathways critically affect the kinetics of oxidation, whether determined by oxygen consumption or appearance of specific products, and can induce large differences in the ultimate mix of products, particularly volatiles. This has several important implications. Analytically, if the dominant pathway is not being monitored, an inaccurate picture of the rate, extent, and character of lipid oxidation is generated and reactivity is misinterpreted. For example, when peroxide values alone are used to follow oxidation under conditions favoring cyclization or scission, much of the lipid change may be missed altogether. Second, changes in the product distributions critically alter flavors and odors from lipid oxidation, and also the potential for secondary effects such as nonenzymatic browning and reactions with proteins. Finally, without information about dominant and active propagation pathways, the most effective strategies for inhibition of the oxidation may not be applied. For example, using only phenolic antioxidants in systems where scission is dominant will probably not be sufficient to stop production of off-flavors and odors. To achieve long-term stability, antioxidant approaches must be tailored specifically to control all active propagation pathways.

24.5   TERMINATION

Termination is a nebulous term that implies a process is coming to a close. In lipid oxidation, "termination" is an even fuzzier concept in that, from a practical standpoint, the lipid oxidation chains probably never fully stop. In addition, a *specific radical* may be terminated and form some product, but if this occurs by H abstraction or rearrangement, another radical is left behind *so the chain reaction continues*. Net oxidation slows down when H abstractions or other radical quenching processes exceed the rate of new chain production, but it would be difficult indeed to totally stop the entire radical chain reaction. Thus, in the discussion below, "termination" refers to an individual lipid radical, not the overall reaction.

Lipid free radicals terminate to form nonradical products by four major mechanisms: radical recombinations, scission reactions when proton sources (e.g., water) are present to stabilize products, group eliminations, and cooxidation of nonlipid molecules such as proteins. These reactions and their controlling conditions are outlined in Table 24.6. The mechanisms dominating in a given system are influenced by the nature and concentration of the radicals, the temperature and oxygen pressure, and the solvent.

24.6   MAJOR FACTORS AFFECTING LIPID OXIDATION

The course and extent of lipid oxidation change with conditions and the differences can often be dramatic. Thus, effects of conditions on reaction pathways must be considered when determining mechanisms, deciding how to analyze for lipid oxidation, or developing stabilization strategies. Factors affecting lipid oxidation can be divided into the four categories shown in Figure 24.6: nature of the lipids, surfaces, presence of other pro- and antioxidant compounds, and the environment and solvent system. Only major effects will be discussed here as a basis for understanding oxidation in specialty oils. For additional details, the reader is referred to other reviews[190–192].

24.6.1   NATURE OR FORM OF LIPID

24.6.1.1   Degree of Unsaturation

Oxidizability of fatty acids in oils increases with the number of double bonds (Table 24.7), due to greater availability of highly abstractable doubly allylic hydrogens and the associated increase in rates of propagation and chain branching. Thus, the presence of fatty acids with three or more double bonds in particular — linolenic acid in vegetable oils, arachidonic acid in animal fats, and docosahexaenoic acid (DHA) and eicosapentaenoic acid (EPA) in fish oils — markedly sensitizes oils to rapid oxidation. The degree of increase varies with the system and product(s) measured.

TABLE 24.6
Typical Termination Processes in Lipid Oxidation

Termination mechanism and reaction	Conditions	Ref.
1a. Peroxyl radical recombinations Concerted addition: "Russell" tetroxide	Rapid ($2k = 10^8$–10^9 M^{-1} s^{-1}) but not favored due to competition from other rxs. Recombination patterns nonrandom — determined by temp., pO$_2$ and radical reactivity. L$^•$ rxs dominate at pO$_2$ < 100 mmHg, high temp., low oxidation. LOO$^•$ self-reaction facilitated in neat oils, aprotic solvents, high pO$_2$, high [LOO$^•$]. Products include alcohols and ketones, alkanes, acyl peroxides, dialkyl peroxides	8, 9, 14, 46, 101, 104, 129, 159–163, 318
$2 ROO^• \longrightarrow [RO^•O_2^•OR] \longrightarrow ROOR + O_2$	Debatable rx, not important at room temp.	9, 14, 84, 101, 166–169, 318, 319
Stepwise addition:	Dominant rx at elevated temps.	9, 14, 112, 318
1b. Alkoxyl and alkyl radical recombinations $R_1O^• + R_2O^• \longrightarrow R_1OOR_2$ peroxides $R_1O^• + R_2^• \longrightarrow R_1OR_2$ ethers $R_1{-}CH{-}R_2 + R^• \longrightarrow R_1{-}C{-}R_2 + RH$ ketones, alkanes $R_1{-}CH{-}R_2 + RO^• \longrightarrow R_1{-}C{-}R_2 + ROH$ ketones, alcohols $R_1^• + R_2^• \longrightarrow R_1{-}R_2$ alkane polymers	LO$^•$ rxs fast ($k = 10^9$ M^{-1} s^{-1}), dominate when LOOH or LOO$^•$ decompositions are faster than their formation, i.e., in secondary stages of oxidation, at moderate temp. and pO$_2$. Recombinations of the fragment radicals from α and β scissions of LO$^•$ generate low levels of volatile compounds and flavor components; favored in neat lipids and high [lipid] in aprotic solvents, unimportant in dilute soln. and polar/aq. solvents	14, 52, 111, 150, 158
2. Scission reactions See Figure 24.5 for general scission patterns of linoleic acid	Requires protons, rate increases with water and heat, unsaturated radical fragments continue to oxidize and undergo secondary scissions	14, 97, 136, 157, 171
	MDA requires 3 or more dbs, internal hydroperoxides, heat, acid; facilitated by aprotic solvents, low lipid concentrations, limited oxygen pressures; usually < 0.1% except in photosensitization	14, 81, 172, 173

TABLE 24.6 (Continued)
Typical Termination Processes in Lipid Oxidation

Termination mechanism and reaction	Conditions	Ref.
3. Eliminations $R_1CH=CH-CH=CH-\overset{\overset{OOH}{\mid}}{CH}-CH_2-R_2$ $\xrightarrow{-OH} R_1CH=CH-CH=CH-\overset{\overset{O}{\parallel}}{C}-CH_2-R_2$ $\xrightarrow{-OOH} R_1CH=CH-CH=CH-CH=CH-R_2$	Minor reaction, forms internal carbonyl (ketone) (left) and desaturated product with additional double bond (right)	14, 174, 175
4. Co-oxidations with nonlipid molecules $\left.\begin{array}{l}LOO^{\bullet}\\ LO^{\bullet}\end{array}\right\} + RSH \longrightarrow RS^{\bullet} + \left\{\begin{array}{l}LOOH\\ LOH\end{array}\right. \longrightarrow$ RS-OOL, RS-OL, RS-epoxy-L adducts $\left.\begin{array}{l}LOO^{\bullet}\\ LO^{\bullet}\end{array}\right\} + \left\{\begin{array}{l}-NH_2\\ >NH\end{array}\right. \longrightarrow \left\{\begin{array}{l}-\overset{\bullet}{NH}\\ >N^{\bullet}\end{array}\right. + \left\{\begin{array}{l}LOOH\\ LOH\end{array}\right. \longrightarrow$ -N-OOL, -N-OL adducts	Any molecule with abstractable H or db is potential target; radicals formed in nonlipid molecules combine with lipid radicals to generate co-oxidation products that are foot prints of LOOH reactions; co-oxidation products limit extractability of lipids for analysis and remove lipids from product streams normally analyzed	14, 15, 176–189

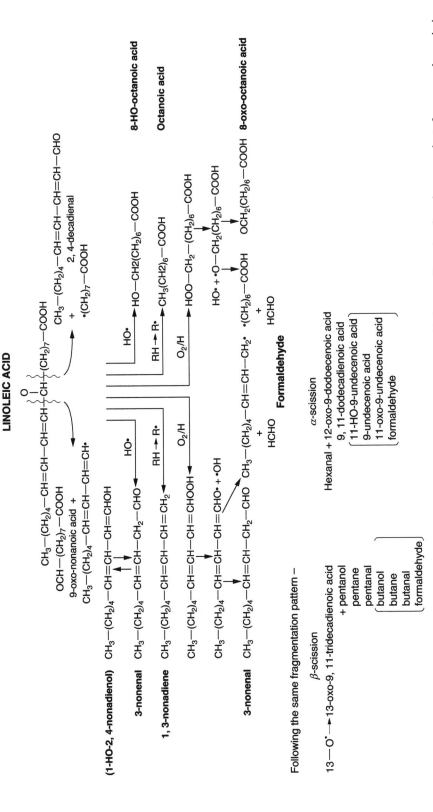

FIGURE 24.5 Typical initial scission patterns of oxidizing linoleic acid. Parentheses indicate unstable intermediates; brackets denote products from secondary scissions. (Data from Neff, W.E., Frankel, E.N., and Fujimoto, K., *J. Am. Oil Chem. Soc.*, 65, 616–623, 1988; Frankel, E.N., *Prog. Lipid Res.*, 22, 1–33, 1982; Frankel, E.N., Neff, W.E., and Selke, E., *Lipids*, 19, 790–800, 1984; Frankel, E.N., *Chem. Phys. Lipids*, 44, 73–85, 1987; Grosch, W., in *Autoxidation of Unsaturated Lipids*, Chan, H.W.-S., Ed., Academic Press, London, 1987, pp. 95–139.)

Factors affecting lipid oxidation

Nature of lipids	Degree of unsaturation	
	trans versus *cis*	
	Conjugation	
	Phospholipids	
	Free fatty acids versus esters versus TAGs	
Surfaces	Bulk oil – exposed surface	
	Emulsions	
	Dispersal on solid surface	
Presence of other components	Pro-oxidants – Preformed hydroperoxides	
		Metals
		Porphyrins, e.g. chlorophyll
		Heme compounds
		Lipoxygenase
		Cyclooxygenase
		Amino acids
		Ascorbic acid (low concentrations)
	Antioxidants (endogenous and added)	
		Polyphenols
		Amino acids
		Metal chelators and complexers
		Synergists
		Glutathione peroxidase
	Interceptors – proteins	
		DNA
		Vitamins
		Pigments
Environment and solvent system	Temperature	
	Light	
	Oxygen pressures	Processing
	Solvent	and storage
	Water	conditions
	pH	
	Packaging	

FIGURE 24.6 Factors affecting lipid oxidation.

TABLE 24.7
Effects of Degree of Unsaturation on Relative Oxidizability of Fatty Acids

	Relative rate of oxidation[a]		Oxidizability (10^2 M$^{-1/2}$ sec$^{-1/2}$)				
Fatty acid	Fatty acid	Ester	Ref.[7b]	Ref.[60c]	Ref.[92]	Ref.[194]	Ref.[195]
18:1	1	1		0.3	0.9	1	1
18:2	28	41	2.03	6	21	12	10.4
18:3	77	98	4.07	14	39	25	22
20:4			5.75				
22:6			10.15				

[a] Ref. 193.
[b] In chlorobenzene, initiated by AIBN [2,2'-azobis(2-methylpropionitrile)] or AMVN [2,2'-azobis(2,4-dimethylvaleronitrile)], oxidizability $= k_p/(2k_t)^{1/2}$.
[c] NO$_2$ dilute, neat fatty acids.

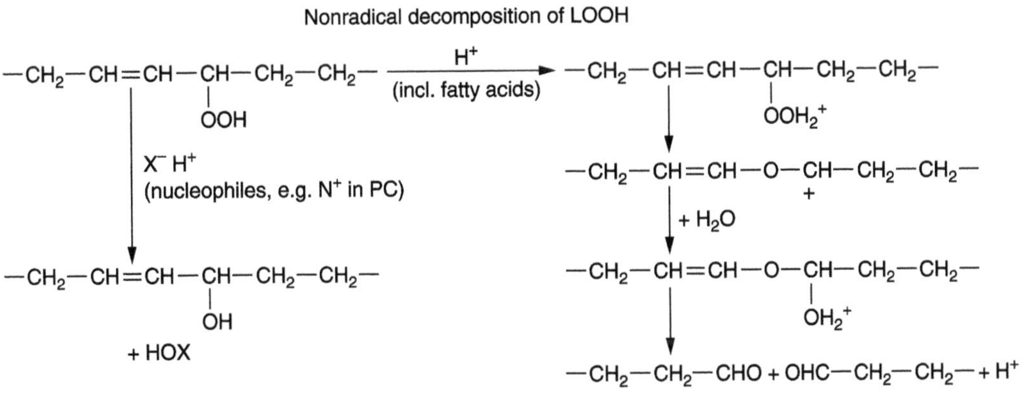

Nonradical decomposition of LOOH

FIGURE 24.7 Nonradical decomposition of LOOH by acids and nucleophiles. (Adapted from O'Brien, P.J., *Can. J. Biochem.*, 47, 485–493, 1969; Corliss, G.A. and Dugan, L.R., Jr., *Lipids*, 5, 846–853, 1971.)

24.6.1.2 *Trans* versus *Cis* Isomers

Altered accessibility of H atoms in geometric isomers affects rates of lipid oxidation. *Cis* fatty acids oxidize more readily than their *trans* counterparts due to increased exposure of allylic hydrogens[196], and this has distinct consequences to secondary oxidations and to *in vivo* action of dietary *trans* fatty acids[197].

24.6.1.3 Presence of Phospholipids

Phospholipids follow the pattern of triacylglycerols (TAGs) and fatty acids/esters, i.e., increasing oxidation with increasing unsaturation of component fatty acids. However, the oxidative behavior of phospholipids in oils as contaminants or as emulsifiers is complicated. Phospholipids can be both pro- and antioxidant, depending on the system and concentration[198]. The underlying causes are not completely understood, but current evidence suggests that phospholipids exert their major influence in the induction period and early stages of oxidation[199], while the final maximum oxidation rate is governed by fatty acid composition and degree of unsaturation[199].

Prooxidant effects: Fatty acids in phospholipids oxidize rapidly because they are highly unsaturated and orientation of fatty acid acyl chains at interfaces facilitates electron transfer via the aligned double bond systems[91]. Thus, in mixed systems, phospholipids are preferentially oxidized and can seed oxidation of TAGs[198]. Phospholipids also mobilize catalysts in their hydration water, and they complex and activate metals[198].

Antioxidant effects: In mixed systems, phospholipids are exposed to catalysts at interfaces[200] and preferentially oxidized in order of unsaturation[201], thus exerting a sparing effect on TAG oxidation in oils. The association of phospholipids into bilayers or mesophases isolates lipid radicals and limits interaction with the oil phase. Bilayers have high viscosity, which reduces migration of catalysts, slows radical transfers and chain propagation, and accelerates terminations[202]. Phospholipids also enhance activity of any tocopherol present by increasing accessibility of tocopherol to chain-initiating radicals in aqueous microenvironments[199]. However, the major antioxidant effect is from nonradical decomposition of LOOH by the nucleophilic $-N^+(CH_3)_3$ of phosphatidylcholine (PC), thus limiting both O_2 uptake and LOOH accumulation[203,204] (Figure 24.7).

Mixed effects: Phospholipids bind both metals and water, and thus can be pro- or antioxidant depending on the system. Binding of metals is antioxidant if the metal ligand orbitals are filled, the redox potential is decreased out of the range of any reducing agents present, or if the phospholipids move the metal away from reactive sites[205,206]. However, metal binding can be prooxidant if the redox potential is lowered in a system with LOOH and the metal and LOOH are both concentrated at interfaces. The phosphate group of phospholipids binds water. It is not known yet whether this hydration water is protective by stabilizing LOOH and metals or is detrimental by activating metals or facilitating contact between metals and LOOH. Clearly, much more research is needed to fully understand effects of phospholipids on lipid oxidation.

24.6.2 FREE FATTY ACIDS VERSUS FREE ESTERS VERSUS TAGS

In pure systems, fatty acids oxidize more slowly than free esters[18,193] (Table 24.7), primarily because the acid groups facilitate nonradical decomposition of LOOH[197]. The decomposition can be acid (H^+) induced with ionized fatty acids as shown in Figure 24.6[18,193] or involve hydrogen bonding between undissociated –COOH and –OOH[155]. However, fatty acids also complex metals, carry them into bulk oil phases, and *increase* the initiation rate of lipid oxidation[207], so when even traces of metals are present, fatty acids may oxidize faster than their esters because –COOH (or –COO⁻) can also catalyze oxidation of TAGs in oils or emulsions. Decomposition of lipid hydroperoxides to radicals by fatty acid–metal complexes is the most likely explanation for observations of Miyashita and Takagi[208] that free fatty acids oxidized more rapidly than their esters and oxidation of linoleic acid was accelerated by stearic acid. Metal complexation can also account for the prooxidant action of free fatty acids in unfiltered olive oil[209].

In the absence of a free carboxyl group to induce LOOH decomposition, oxidation increases with the degree of esterification in acylglycerols[7]:

Lipid	L oxidizability ($\times 10^2 M^{-1/2} sec^{-1/2}$)
FFA	2.03
MAG	2.83
DAG	5.89
TAG	7.98

Oxidation in TAGs is usually faster than in free fatty acids but slower than in free fatty esters due to reduced access to double bonds[210]. The alignment of acyl chains is more irregular in TAG structures, which reduces the efficiency of radical transfer between acyl chains and limits access of initiators to double bonds. In sunflower oil, initiation of oxidation at 100°C is six times higher in methyl esters than in TAGs[211]. Arrangement of fatty acids in TAG crystals also influences oxidation through chain organization. Concentration of unsaturated fatty acids at *sn*-2 stabilizes TAGs, while *sn*-1 and 3 unsaturation enhances oxidation; randomization of TAGs decreases oxidation[212]. Symmetrical SUS (saturated–unsaturated–saturated) or USU (unsaturated–saturated–unsaturated) structure favors β crystal structures which exist in ortho arrays with close association in the liquid state, thus facilitating radical transfer between chains. However, randomization shifts chains to α-structures with less organization, and this interferes with radical transfer.

24.6.2.1 Surfaces

Surfaces, both solid and liquid, govern lipid exposure and molecular organization as well as contact with pro- and antioxidants and thus have among the strongest influences on lipid oxidation rates and product distributions. Lipid oxidation rates increase with surface area, whether liquid or solid[200]. For lipids on a solid surface (such as a dehydrated food matrix), maximum oxidation occurs with dispersal at just about a molecular monolayer over the solid matrix[213] (Figure 24.8). With lower coverage, lipid molecules are separated and radical transfers do not easily occur. With more lipid, multiple layers of lipid molecules compete for radical transfers and side reactions reduce the efficiency of the primary radical chain. The system begins to behave more like bulk lipid, and oxygen diffusion and interfacial character become the controlling factors. Monolayer coverage is the point of maximum surface area and oxygen exposure for a lipid film; adding more lipid layers does not

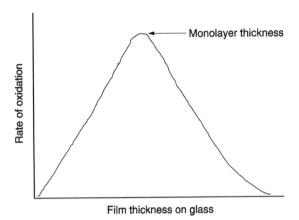

FIGURE 24.8 Effects of film thickness on oxidation of lipids spread on a surface. Maximum oxidation occurs at a film thickness approximating a molecular monolayer of lipids. (Redrawn from Koch, R.B., in *Symposium on Foods: Lipids and Their Oxidation*, Schultz, H., Day, E.A., and Sinnhuber, R.O., Eds., AVI Publishing, Westport, CT, 1962, pp. 230–249.)

increase surface exposure but does increase the thickness through which oxygen must diffuse, and the declining oxidation rate reflects this[214].

Spreading lipid in films has two main consequences. The first is the alteration of product mix. Especially when oxygen is limited, large surface areas favor rearrangements of LOO$^\bullet$ and LOOH to hydroxyepoxides, endoperoxides, and similar products rather than decomposition to LO$^\bullet$[215]:

The second consequence is that migrational mobility of hydrogen donors or acceptors is strongly restricted on silica so quantum yields are reduced by radical recombination and fewer product species are found[143,216,217].

Lipid oxidation is generally considered to be fastest in emulsions where there is greatest surface area of lipids, greatest diffusion of oxygen, and greatest interface with the aqueous phase which is a source of protons and catalysts. Oxidation rates in emulsions increase with proportion of water and concentration of emulsifying agent[218] due to increased effectiveness of initiation at the oil droplet surface (Table 24.8)[154]; propagation and termination occur within droplet at normal rates. Nevertheless, oxidation rates in emulsions can be less than on spread surfaces that increase access to decomposers and water.

Lipid oxidation is slowest in bulk (neat) oils. In oils, oxygen solubility is high but diffusion-limited, and surface to volume ratio is usually low. As noted above, lipid acyl chains are only loosely associated in TAGs and are not oriented for efficient radical transfer. In addition, oxidation products are more hydrophilic than the parent TAG, and they migrate to water and less hydrophobic interfaces. Thus, propagating radicals from hydroperoxide decomposition may not be formed in close proximity to abstractable hydrogens. However, oxidation in neat lipids also favors extensive polymerization[219].

24.6.3 PRESENCE OF PRO- AND ANTIOXIDANTS

From a practical standpoint, metals are ubiquitous and the major prooxidants present in most foods, including oils. Metals can be particular problems in oils that are pressed and not refined. It was

TABLE 24.8

Effects of Surface on Overall Rate Constants for Lipid Oxidation during the Monomolecular (K_m) and Bimolecular (K_B) Rate Periods

Methyl linoleate system	K_M ($\times 10^{-3}$)	K_B ($\times 10^{-2}$)
Bulk	13	6.4
Filter paper, wet	20	6.8
Filter paper, dry	7	4.6
Emulsion	32	8.5
TAG dispersed on cellulose	1.7	0.8

Adapted from Labuza, T.P., *CRC Crit. Rev. Food Sci. Nutr.*, 2, 355–405, 1971.

previously thought that most LOOH decomposition (and hence reaction propagation) occurs at interfaces due to contact with metals and other reactants in the aqueous phase, and polar products are then released into the aqueous phase[15]. However, initiation is most effective in the oil phase[220]. Lipid-soluble metal complexers such as ADP (adenosine diphosphate) and citric acid can carry metals into oil phases and both initiate radical production and decompose hydroperoxides there[221].

Pressed oils also contain a number of other compounds that alter stability relative to extracted, refined oils. Refining removes most contaminants, but oils that are pressed or minimally processed can retain the potent prooxidant chlorophyll at considerable levels, and lipoxygenase may also be present. Cholorophyll sensitizes light initiation of lipid oxidation by both free radical (Type 1) and singlet oxygen (Type 2) mechanisms as shown in Table 24.1, and it is a very effective catalyst even in trace concentrations[222]. Type 1 reactions are oxidations, while Type 2 reactions are oxygenations (oxygen insertions) that are 1500 times faster than with normal triplet oxygen[223]. Other photosensitizers potentially present at trace levels in pressed oils include flavins (especially riboflavin), porphyrins, aromatic amino acids, and any molecules with carbonyls and/or an extended conjugated double bond system[224]. Whether the dominant reactions are free radical or singlet oxygen, or both, depend on substrate and reaction conditions[222,225,226] and the specificity of photosensitizers present[227]. However, effects in pressed oils can be complex because carotenoids (singlet oxygen quenchers), polyphenols (free radical scavengers), and other sensitizing pigments may all be present. Antioxidants may counterbalance the effects of chlorophyll and other prooxidants, or in some cases may actually enhance oxidation. The net effects depends on the absolute and relative concentrations of the various agents, and the competition must be considered in all oils, both pressed and extracted/refined.

24.6.4 ENVIRONMENT AND SOLVENT SYSTEM (PROCESSING AND STORAGE CONDITIONS)

24.6.4.1 Temperature

The activation energy E_a for lipid oxidation is relatively high (16.2 kcal mol^{-1}) so oxidation increases with temperature[154]. The main effect of heat, except at frying temperatures, is to increase the rate of LOOH decomposition and initiation of secondary chains, and oxidation kinetics is where heat effects are most often observed. Equally important to food oil stability, however, may be the heat-induced shift in dominant reaction pathways:

- H abstraction (LOH and LOOH) dominant – produces 10% small volatile products (from α and β scission) and 90% polymers[228].
- Altered bond scission patterns – hexanal is a major product of lipids under mild oxidation conditions, but 2,4-decadienal becomes dominant when systems are heated[136].

TABLE 24.9

Comparative Characteristics of Free Radical Autoxidation and Type 1 Photosensitization versus Singlet Oxygen-Photosensitized Oxidation

	Free radical: Type 1 autoxidation	$^1O_2$: Type 2 photosensitization
Mechanism	Conventional free radical chain rx.	"Ene" rx, concerted addition of O_2, no free radicals involved. Radicals generated by LOOH decomposition
Kinetics	Induction period present	No induction period
	Dependent on pO_2	Independent of O_2 (when O_2 not limiting)
		Dependent on sensitizer concentration
	Rate not ∝ number of double bonds	Rate directly ∝ number of double bonds
Relative reactivity	18:1: ~1	18:1: ~1
	18:2: 17	18:2: 2
	18:3: 25	18:3: 3
Products	Conjugated LOOHs	Both nonconjugated and conjugated LOOHs
	External 9, 13, 16 positions dominant	LOOHs at all positions
		High proportion of cyclics, endoperoxides, epoxides, di- and trihydroperoxides at internal positions
Sensitizers	Chlorophyll, pheophytin	Chlorophyll
	Hemes: protoporphyrin, myoglobin, hemoglobin	Hemes
		Erythrosine
		Rose bengal
	Flavins (especially riboflavin)	Flavins (low substrate concentration)
	Xanthenes	Methylene blue
	Anthracenes	Proflavine
	Anthroquinones	Eosin
	Crystal violet	Food dyes
	Food dyes	
Quenchers	Phenols, polyphenols	β-Carotene
	Tocopherol (carotenes)	Tocopherols

Data from Refs.[35,69,97,144,158,226,227,230–236].

- Dimer and polymer formation in fatty acids and oils increase with heating temperature and time (150 to 240°C)[228].
- Shifts in type of dimer formed – β scission of peroxyl oxygen increases, leading to more C–O–C and C–C dimers and fewer C–O–O–C crosslinks[99,171].
- Increased proportion of *trans* isomers[52], which might become more important with the advent of new labeling requirements for *trans* fats.

24.6.4.2 Light

The role of ultraviolet (UV) light as a direct initiator and visible light in photosensitization is outlined in Table 24.1, but light also changes some pathways and products of lipid oxidation, so further discussion is warranted here.

Visible light starts lipid oxidation by both free radical and $^1O_2$ mechanisms, and essentially at the same time UV light propagates oxidation by decomposing hydroperoxides. Thus, initiating events rapidly become obscured as radical chains and branches become established, and it can then be difficult to distinguish autoxidation from photosensitized sequences. While it is impossible to distinguish autoxidation from Type 1 (free radical) photosensitized oxidation, several markers have proven to be quite useful in detecting Type 2 (singlet oxygen) light catalysis, particularly when trouble-shooting sources of lipid oxidation in oils or foods. Table 24.9 compares characteristics of free radical- and singlet oxygen-initiated lipid oxidation. The most notable product difference — increased nonconjugated and internal 10- and 12-hydroperoxides and cyclic products — can be used diagnostically to differentiate photosensitization from autoxidation[35,69,225,229]. Cyclic products, in particular, are thought to be important precursors of light-induced off-flavors[230].

Free radical autoxidation or photosensitization may start with a low percentage of non-conjugated products from initial H abstractions, but conjugated forms increase with oxidation as secondary H abstractions quite selectively occur at doubly allylic sites. In contrast, $^1O_2$-sensitized oxidations begin with higher levels of nonconjugated products and these increase with extent of oxidation, providing a clear indicator of Type 2 photosensitization. When both free radical and $^1O_2$ sensitization occur, as with chlorophyll, conjugated and nonconjugated products are initially about equal, but the conjugated forms become dominant with increasing reaction as hydroperoxides break down and initiate radical branching reactions.

However, a cautionary note must be offered about using product mixes to determine causes of oxidation or mechanisms of a given agent. Many photosensitizers change mechanisms with solvent and concentration. For example, riboflavin normally photosensitizes by a Type I mechanism (H abstraction), but at sufficient substrate dilution it converts to the $^1O_2$ mechanism[231]. Thus, sensitivity to light often changes with extraction conditions that alter the mix and concentrations of photosensitizers and quenchers. This needs to be kept in mind when tracking and comparing causes of oxidation in different oils.

24.6.4.3 Oxygen Pressure

It is a common (mis)perception that lipid oxidation increases with oxygen pressures. In general oxygen *does* increase with oxygen pressure, but not without limits and qualifications. The following complexities in oxygen effects need to be considered when evaluating oxidation of specialty oils:

- Oxygen by itself does not accelerate oxidation. Oxygen effects *require* an initiation to create the *ab initio* radicals as sites for addition. Formation of LOO$^\bullet$ depends on O_2 only up to 10 mmHg pO_2 and further increases in oxygen have no incremental effects[130]. Particularly in the low pO_2 range (1 to 80–100 mmHg), lipid oxidation depends more on the nature of initiators and their reactions than absolute oxygen concentrations[237]. For example, Fe^{2+} requires oxygen for autoxidation to Fe^{3+} and production of $O_2^{\bullet-}$ and downstream radicals, so the presence or absence of a metal chelator can have a huge effect on apparent oxygen dependence of lipid oxidation without fatty acids being directly involved.
- Oxygen has different effects on initiation and propagation, so the oxygen dependence observed often varies with the products measured. In particular, oxygen alters the balance between alternative reaction paths, e.g., internal rearrangement versus abstraction, and directs the proportion of oxygenated recombination products formed.
- Oxygen catalysis of lipid oxidation is proportional to exposed surface area in both liquid and solid foods.
- The effects of oxygen on oxidation decrease at elevated temperatures where increased initiation drives oxidation reactions and increased β-elimination shifts products to less oxygenated versions.

24.7 LIPID OXIDATION IN SPECIALTY OILS

Oxidation of any oil is determined primarily by its fatty acid composition, natural antioxidant content, presence of prooxidant substances from pressing or extraction, and presence of any other system components that can compete for H abstractions in the radical chain[238]. It is also affected by processing and storage conditions. What was discussed about initiation, propagation, and

TABLE 24.10
Typical Fatty Acid Compositions of Specialty Oils in Comparison to Conventional Food Oils

Oil	Unsat./sat. ratio	16:0	18:0	18:1	18:2	18:3	γ-Ln	Ref.[a]
Specialty								
Borage							22	240
Rapeseed (low 22:1)[b]	15.7	3	1	58	24	10		241
Canola	15.7	4	2	62	22	10		
Hazelnut	11.5	6	2	75	17	tr		242
Evening primrose	11.5	6	2	11	81	—	9	240
Safflower	10.1	7	2	13	78	—		
Walnut	9.9	7	2	17	60	12		242
Flaxseed oil	9.0	3	7	21	16	53		
Grapeseed	7.3	8	4	15	73	—		
Meadowfoam[c]		2		[14		]		243
Sunflower	7.3	7	5	19	68	—		
Sesame oil	6.6	9	4	41	45	—		
Olive	5.1	13	3	71	10	1		
Rice bran	4.9	—	17	47	35	1		240
Palm	1.0	45	4	40	10	—		
Conventional								
Corn	6.7	11	2	28	58	1		
Soy	5.7	11	4	24	54	7		
Peanut	4.0	11	2	48	32	—		
Cottonseed	2.8	22	3	19	54	1		

[a] Ref.[244], unless otherwise noted.
[b] 0–5% 20:1, 0–5% 22:1, plus 1% each 20:0 and 22:0.
[c] 69% 20:4, 14% 22:1Δ5, 1.4% 24:1Δ5

termination reactions in model systems and soybean oils should apply fully to all food oils, and thus we can use the reactions and factors outlined above to predict oxidation and explain behavior of any specialty oil once its composition is known. It must be noted that while soybean oil has been studied intensively for decades, specialty oils have received attention only recently since their potential unique health effects were recognized. Consequently, research on oxidation of specialty oils is still in the "descriptive" stage and essentially no specific mechanistic information is yet available.

Table 24.10 lists typical fatty acid compositions of specialty oils and conventional food oils. Assignment of oils to the specialty category is somewhat arbitrary. Olive, safflower, sunflower, and sesame oils have been marketed as food oils and used in cooking for a very long time, but are included with specialty oils because of their newly recognized relationships to health and disease prevention. Canola is listed with specialty oils because it was genetically engineered to modify fatty acid compositions. Reviewing the data in this table, given that lipid oxidation increases with number of double bonds and particularly with linolenic acid in oils[239], three classes of relative oxidation rates can be predicted: high (flaxseed, borage, walnut, canola, rapeseed, soy); moderate (evening primrose, safflower, grapeseed, sunflower, corn, sesame seed, cottonseed, peanut); low (hazelnut, olive, rice bran, palm).

Numerous studies generally validate this expectation, showing a correlation between degree of unsaturation and rate of oxidation. For example, the Rancimat time for walnut oil is shorter than for hazelnut oil in line with its higher 18:2 content[245]. The relative stability of rapeseed > sunflower > safflower > walnut oils as determined by Rancimat, conductivity, and chemiluminescence corresponds inversely to iodine value[246]. Breeding peanuts to increase oleic acid contents markedly increases induction periods for thermally induced oxidation of peanut oil[247]. As will be shown repeatedly in the discussion below, fatty acid composition is probably the most important factor influencing oxidative stability of specialty oils.

However, research also has revealed complex variations due to differences in endogenous prooxidants (pigments and metals) and antioxidants (phenolics, carotenoids) as well as extraction and processing methods. Fatty acid composition accounts for only about 50% of the oxidative stability of canola oil[248]. Differences in oxidizability of conventional, specialty, and frying oils as determined by thermogravimetric analysis could not be accounted for by fatty acid compositions alone and must reflect influences of other oil components[249]. The presence of trace metals has a pronounced effect on oil oxidation[250] and is particularly problematic with pressed oils that are not refined. Levels and distributions of phenols and other antioxidants in oils alter radical-trapping capabilities of oil independently of fatty acid composition[251]. Non-TAG contaminants are largely removed in refining, but because specialty oils are used more as pharmaceuticals or nutraceuticals and less as food ingredients or cooking media they are generally minimally processed to retain natural antioxidants and other minor components. The studies reviewed below thus also provide some interesting perspectives on effects of plant growth conditions, handling, and processing methods on lipid oxidation through their modification of pro- and antioxidants in specialty oils.

24.7.1 Walnut Oil and Pecan Oil

High unsaturation (> 70% PUFA) imparts significant nutritional value to walnut oil but is also responsible for exceptionally high susceptibility to oxidation and production of off-flavors during even short periods of storage[252]. Oxidizability varies among cultivars due to differences in proportions of unsaturated fatty acids and antioxidant composition. Six heterogeneous cultivars of Portuguese walnuts all had low stability when heated, showing only 3 h induction period due to high PUFA content (18:2 and others)[253]. The cultivars had only minor differences in fatty acid contents but showed large variations in oxidation patterns and level due to differences in endogenous pro and antioxidants. In oils extracted from 13 homogeneous cultivars of New Zealand walnuts and five U.S. or European cultivars, there was surprisingly weak correlation between peroxide values and fatty acid unsaturation or tocopherol contents, particularly γ-tocopherol[254,255]. Rather, oxidation was driven by minor components such as lower contamination with chlorophyll and differences in trace metals (not measured) when hexane:isopropanol was used as the extraction solvent, thus reducing the radical load and need for tocopherol[256].

Extraction and refining methods can have marked effects on oxidation of walnut oil due to variation in nonlipid components (both pro- and antioxidant). In comparisons of walnut oil degumming by salting out, phosphoric acid, and potassium aluminum sulfate, phosphoric acid produced the most stable oils[257]. Pressing of walnuts gives higher tocopherol levels[258] but <50% yield of solvents (CH_2Cl_2, $CHCl_3$–MeOH, hexane, supercritical CO_2)[259] and lower stability due to increased chlorophyll and metal contamination[258]. Although fatty acid composition was the same in all extraction methods, chloroform–methanol most effectively removed all lipid classes, oxidation products, and total volatiles; hexane and CO_2 lost significant amounts of volatiles[259]. This study found no difference in oxidation between grades of CO_2, but Calvo[260] observed that using ultrapure CO_2 decreased oxidation in sunflower oil.

Because of its strong hydrophobicity, supercritical CO_2 does not extract pigments, antioxidants, and metals along with oils. While this property is an advantage in generating stable high-volume commercial food oils such as soybean, it can be a detriment in specialty oils where endogenous antioxidants are critical for stabilization. In walnut oils oxidized at 60°C in the dark, peroxide values of oils extracted by supercritical extractions (SCE) were higher than in pressed oils due to reduced tocopherol and phospholipid contents. However, oxidation at 35°C under light was reduced in the SCE oil because significantly less chlorophyll was extracted by CO_2[261]. For the same reasons, walnut oil was less stable when extracted by CO_2 than by hexane[262]. It should be noted that supercritical extractions were performed at 353 K/68.9 MPa[261]. Under milder extraction conditions (308 to 321 K, 18 to 23.4 MPa), CO_2 extracted 30% more tocopherol from walnuts than did hexane without changing TAG and fatty acid composition[262], so there may be potential for modifying supercritical extraction conditions to both maintain high yields of oil and improve stabilization by endogenous antioxidants.

As would be expected, oxidation of walnut oil during storage is affected by temperature, time, and water activity (a_w)[263]. Walnut oil was coated onto cellulose and incubated at various a_w levels. At relatively low oxidation levels (PV < 100), a_w altered the distribution of products between classes but did not affect total carbonyl content of walnut oil. However, with more extensive oxidation (PV > 100), increasing a_w preferentially enhanced scission reactions; total carbonyl levels rose steadily and there was an irregular shift in product distribution: 2-alkanones were dramatically reduced, hexanal increased at the expense of other alkanals, and short chain alkenals increased[264]. Moisture contents of nuts before pressing or extraction, as well as moisture and a_w of foods in which walnut and other specialty oils are used, are important factors affecting oil stability.

In contrast to the highly unsaturated walnut oil, pecan and almond oils are both predominantly monounsaturated oils (pecan: 60 to 70% 18:1, 19 to 30% 18:2; almond: 60 to 80% 18:1, 8 to 28% 18:2). This plus their natural tocopherol levels generally make them much more stable, particularly at elevated temperatures[265,266]. The interaction between fatty acid unsaturation and types and levels of endogenous tocopherols is a major factor controlling stability in specialty oils, albeit a somewhat controversial one because observations have been inconsistent. Some studies have suggested that the type of tocopherol and tocopherol/unsaturated fatty acid ratio are even more important than absolute tocopherol levels or the degree of fatty acid unsaturation in walnut and other seed oils, accounting for the oxidation order walnut > pecan > sunflower oil[258]. Most pecans and sunflower oils contain predominantly γ-tocopherol, while walnuts have both lower total levels[266] and what is present is divided between α-, γ-, and δ-tocopherols[258]. γ-Tocopherol is a better antioxidant than α- and δ-tocopherols. At least part of the stability of almond oil to thermal oxidation is due to its favorable tocopherol/PUFA ratio[266].

24.7.2 BORAGE AND EVENING PRIMROSE OILS

Borage and evening primrose (EP) oils are receiving considerable attention and increased nutraceutical use due to their high contents of ω-6 fatty acids. With, respectively, 22% and 9% γ-linolenic acid (GLA)[267], these two oils have the potential to oxidize rapidly, but substantial levels of endogenous antioxidants provide stabilization. Crude borage extracts contain 413 mg total phenolics/g oil as sinapic acid equivalents, 89% hydrophilic and 11% hydrophobic phenolics (w/w), while evening primrose contains 304 mg/g phenolics as catechin equivalents, 3:2 (w/w) hydrophilic:hydrophobic ratio[268]. Six phenolic fractions with varying radical scavenging and electron donating capacities were separated but individual components were not identified. Most fractions also had strong metal complexing capability. EP phenols have greater Fe(II) chelating capacity than borage phenols[269].

These endogenous antioxidants play important roles in stabilizing borage and EP oils. Although borage oil has a much higher 18:3 n-6 content, it also is protected with more than double the amounts of tocopherols of EP oil. Consequently, EP is less stable when oxidized in the dark at 60°C under forced air[267]. However, when tocopherols were equalized (200 and 500 ppm) in stripped oils, effects of fatty acid composition dominated. EP oil had higher conjugated dienes due to more 18:2, but borage had higher thiobarbituric acid reactive substances (TBARS) due to its high 18:3 n-6 content. Tocopherols were most effective at 500 ppm. All synthetic antioxidants (200 ppm) were more effective in reducing oxidation than tocopherols, perhaps because their concentrations were 2 to 2.5 times higher than tocopherols on a molar basis. *tert*-Butylhydroquinone (TBHQ) reduced oxidation most in both oils, due either to greater antioxidant activity of its two *para* hydroxyl groups[270] or to its heat stability and lower volatility.

Minor components other than antioxidants also influence oxidation of borage and EP oils[271]. When cold-pressed borage and EP oils stripped of minor components were emulsified with Tween 40 by sonication, relative rates of oxidation were emulsions > neat oils, EP oil > borage oil, and non-stripped > stripped for both heat-accelerated and photosensitized oxidation. EP oil was destabilized by higher total PUFA (> 70% 18:2 versus 36% in borage) and higher endogenous levels of carotenoids and chlorophyll, not all of which were removed by stripping. Stripped EP oil also retained lipid oxidation products plus some γ-tocopherol, but the tocopherol levels present were insufficient to counteract the increased surface area in emulsions and dominant prooxidant effects of the chlorophyll photosensitizer. In contrast, borage oil had more tocopherol to start with and had no trace components or lipid oxidation products after stripping, so despite its high GLA, it was more stable than EP oil. The water used in emulsions was distilled, not deionized, so metal contamination of water may contribute to increased oxidation in the emulsions, in addition to increased surface exposure.

Extraction methods used to prepare borage oils influence oxidative stability by variations in pro- and antioxidants coextracted and by differences in processing stresses. Borage oil extracted with supercritical CO_2 gave lower iodine values and saponification numbers as well as higher PV (33.5 versus 7.8) than with conventional hexane extraction[272]. Hexane extracted more nonlipid compounds, including antioxidants, and was conducted at lower temperatures than the 40 to 60°C required for supercritical extractions.

Oxidation of borage and EP oils can be modified by interesterification with other fatty acids. Sensitivity to oxidation increases if the new fatty acids introduced are more unsaturated. Structured lipids synthesized by inserting DHA into TAGs to increase ω-3 fatty acid content of borage and EP oils showed decreased oxidative stability by all measures (PV, conjugated dienes, hexanal, volatiles, double bond and methylene bridge indices, etc.)[273], as would be expected from introduction of six double bonds. Supplementation with antioxidants thus was recommended to protect modified oils during processing and storage.

24.7.3 Sesame Oil

Production of sesame oil involves optional dehulling, roasting, grinding, cooking, and pressing or solvent extraction[274]. Of these steps, dehulling, roasting and cooking temperatures and moisture, and extraction methods critically affect the oxidative stability of sesame oil[275]. Dehulling is necessary to increase oil yields. Oils from whole seeds are more stable than those from dehulled seeds regardless of heating or extraction method[275-278]. Levels of peroxides, free fatty acids, carotenoids, and δ- and γ-tocopherols were not changed[278], so the presence of some polar antioxidant factors or synergists such as the lignans sesamol, sesamin, and sesamolin in sesame hulls was proposed[275].

Sesame seeds are heat-treated by roasting, steaming, or microwaving before extraction to improve flavor and color and to inactivate enzymes, and oxidative stability depends on both roasting temperature and moisture. Higher roasting temperatures give stronger characteristic flavors but also greatly increase oxidation and deterioration of oils, particularly above 220°C where hydrolysis of TAGs and degradation of phospholipids increases dramatically. Destruction of antioxidants and chlorophyll also increases with temperature[279]. In a comparison of dry oven roasting, steam roasting, and microwaving, oil from raw and roasted-steamed seeds had greater stability than that from other methods, and microwave heating was not recommended[276]. PV and destruction of tocopherols increased with time of microwave heating[280]. High localized heat in microwaved whole seed oils increases hydrolysis and release of free fatty acids[280] as well as destruction of phospholipids[280]; it also breaks down lipid hydroperoxides faster, yielding lower PVs and higher aldehydes[280]. An optimum microwave heating treatment of 16 to 20 min at 200°C was recommended to develop flavor fully, retain highest levels of sesamol and γ-tocopherol, and minimize degradation[280]. A temperature of 180°C was recommended for oven roasting[281].

For oil extraction, fine crushing facilitates release of oils as well as phenols and tocopherols, but heat from pressing and Soxhlet increases extraction of prooxidants, particularly copper and iron

TABLE 24.11
Effects of Reducing or Eliminating Higher PUFA on Thermooxidation of Sunflower Oil for 10 h at 100°C

Oil	Fatty acid composition (mol%)						Total polar products (wt%)
	16:0	16:1	18:0	18:1[a]	18:2	18:3	
Sunflower oil	6.3	—	4.9	35.8	51.0	—	24.6
High-oleic sunflower oil	3.7	—	3.8	87.4	2.5	0.3	21.3
CAS-8	6.9	—	9.8	39.2	41.3	—	22.3
CAS-12	30.2	6.4	2.0	56.8	1.9	—	12.9

[a] 18:1(n-9) plus 18:1(n-7).
Data from Marquez-Ruiz, G., Garces, R., Leon-Camacho, M., and Mancha, M., *J. Am. Oil Chem. Soc.*, 76, 1169–1174, 1999.

ions. Solvent extraction with heptane–isopropanol (3:1, v/v) yields more stable oils than hexane due to reduced extraction of porphyrins and increased extraction of phospholipids, other antioxidant phosphatides, and polar phenolics[275]. Refining removes browning pigments formed during roasting but decreases oil stability[282]. Stabilization of hulled sesame oils is counteracted during refining by removal of the lignans and other antioxidants on filtering adsorbents, particularly Trisyl 300[278]. In oils treated by filtration or centrifugation to remove cloudiness, highest γ-tocopherol was retained in oils receiving both treatments[283]. Filtered oils were lightest in color but also had highest peroxide values. Centrifuged oil had best color and oil stability at 70°C. Blending sesame oil with cottonseed, safflower, or peanut oil improved its stability at 180°C[284].

24.7.4 SUNFLOWER OIL

Sunflower oil has been marketed as a commercial cooking oil valued for its mild flavor for decades, but it has recently been rediscovered with the development of high oleic acid varieties. With 68% 18:2 (Table 24.10) and being largely stripped of natural antioxidants during refining, traditional sunflower oil had some stability problems. The high-oleic sunflower oils (HOSO) are more stable and mimic the physiological effects of olive oil.[285]

In studies of thermooxidative stability of genetically modified sunflower oils, resistance to oxidation was increased by replacing 18:1 and/or 18:2 with varying levels of 16:0 and 18:0, but not as much as would be expected (Table 24.11). Nearly all the 18:2 had to be replaced with 18:1 to counterbalance rapid oxidation of trace amounts of 18:3 that were still present[286]. Mutant CAS-8 with only 20% of the 18:2 removed but totally lacking 18:3 showed stabilities similar to HOSO. However, when nearly all the 18:2 was shifted to 18:1, 16:1, and 16:0, oil stability increased by more than 50%. It was thus proposed that this approach of shifting fatty acid compositions could be quite useful for genetically engineering oils specifically designed for cooking or frying applications.

Since antioxidants are largely removed during refining of sunflower oil, various combinations of stabilizers are usually added to ensure shelf stability. Hras et al.[287] tested antioxidant effectiveness of rosemary (0.02%), α-tocopherol (0.01%), ascorbyl palmitate (0.01%), and citric acid (0.01%) in sunflower oil in accelerated oxidation at 60°C for 11 days. Monitoring peroxide value, anisidine value, and induction period (time to PV = 20), the greatest protection was observed with rosemary. No reduction in oxidation was apparent with citric acid or tocopherol, although synergism was found between rosemary and ascorbyl palmitate or citric acid. Rosemary is an active phenolic radical scavenger, but in this case its marked effectiveness was more likely due to metal chelation via its vicinal diphenol structure, or to the higher concentrations used. It is difficult to

TABLE 24.12

Effects of Packaging and Temperature on Stability of Commercial Sunflower and Olive Oils

	Shelf life (months)			
	Sunflower oil		Olive oil	
Package type	10°C	20°C	10°C	20°C
PET	10.4	4.8	17.5	8.0
Clear glass	11.2	5.1	17.8	8.4
Colored glass	11.8	6.5	18.3	8.7

	Induction period of fresh oils (h)					
	Sunflower oil			Olive oil		
Temp. °C	100	110	120	110	120	130
	9.0	4.4	2.2	12.4	6.2	2.8

Adapted from Kaya, A., Tekin, A.R., and Oner, M.D., *Lebensm.-Wiss. u.-Technol.*, 26, 464–468, 1993.

judge apparent ineffectiveness of tocopherol and other compounds because lower amounts were added on a molar basis.

As has been discussed above, extraction and processing can have marked influences on stability of specialty oils. In an excellent example of how theory developed from model systems can apply directly to food oils, Crapiste et al.[288] evaluated stability of unrefined sunflower oil obtained by pressing versus solvent extraction (hexane) plus water degumming. Solvent-extracted oils had slightly higher initial levels of oxidation and free fatty acids, possibly due to the high 18:2 content, but they showed lower oxidation during storage. Although no trace components were analyzed, it is likely that more prooxidants such as metals or porphyrins were released from the seed meats during pressing, and these became active during storage. In both oils, formation of polar and polymer oxidation products and free fatty acids from hydrolysis increased with storage temperature (30 < 47 < 67°C) and exposure to oxygen (surface area and pO_2 in headspace), as would be expected. No oxidation or hydrolysis was observed with storage at 30°C or under nitrogen.

If addition of antioxidants to replace those removed during refining provides a first line of defense in protecting processed oils, packaging provides the second. The effects of packaging on shelf life of sunflower and olive oils were evaluated by comparing accelerated testing at elevated temperature (> 100°C) with long-term storage at 10 or 20°C under fluorescent light[289]. Sunflower oil deteriorated due to oxygen permeability and transparency of packaging, high 18:2 content (major factor), and storage temperature. Shelf life of olive oil was determined by metal content (metals released during pressing), chlorophyll content, oxygen dissolved in the oil during packaging, and temperature. As might be expected, oils that were older and of lower quality (higher PV) at the start of testing were less stable during storage; shelf life and induction periods both decreased with initial PV of tested oils, showing the importance of proper handling during processing. Oxidation was greatest in PET (polyethyleneterephthalate) packaging due to its oxygen permeability (Table 24.12). Similar effects of oxygen permeability were observed with PVC (polyvinyl chloride), PP (polypropylene), and PS (polystyrene), and leaching of BHA (butylated hydroxyanisole) and BHT (butylated hydroxytoluene) from the plastics did not reduce oxidation in oils[290]. Glass packaging alone improved oil stability by eliminating oxygen, but greatest protection was offered by colored glass that also screened out light. Stability of both oils decreased as temperature increased. Packaging exerted stronger effects on the sunflower oil because it was inherently more sensitive, while temperature was a greater catalyst for olive oil which is inherently more stable. Effects of both protective and destabilizing factors increased with unsaturation and can become quite dramatic in high-PUFA oils.

24.7.5 GRAPESEED OIL

Very high 18:2 content (73%) makes grapeseed oil prized as a source of essential fatty acids but also creates challenges for stability. The hard coat on the seeds limits accessibility to the oil; to obtain reasonable oil yields it is thus necessary to pretreat the seeds, e.g., with air drying, microwaving, or tannin removal in ethanol before solvent extraction[291.] Microwave pretreatment for short times (< 10 min) improves oil yields, increases tocotrienol content, and significantly reduces chlorophyll, carotenoids, and other pigments, and resulting oils are generally light in color[251]. However, any treatment that involves more extensive heating gives darkened oils with high viscosity and increased lipid oxidation. Cold-pressed oils have lower yields (e.g., 36% oil) but produce higher quality oils that are of lighter color and more stable[251].

24.7.6 RAPESEED OIL (*BRASSICA* SPP.)

Rapeseed oil is predominantly monounsaturated, but its high 18:3 n-3 content (10%) makes it highly susceptible to oxidation[292]. Oil from traditional rapeseed varieties contains erucic acid that is toxic to humans, so its uses are limited to animal feeds and industrial applications such as paints in the West. Low erucic acid varieties, including canola, are now in common use and have increased the interest in rapeseed oil as a nutritional source of 18:2 and 18:3 n-3[241].

As has been noted throughout this review, fatty acid composition is the major driver of oxidation. Oxidative stability of sunflower and rapeseed oils was evaluated by electron paramagnetic resonance (EPR) spin trapping (60°C) and compared with differential scanning calorimetry (DSC) and Rancimat (100°C)[293]. By all tests, sunflower oil with >60% linoleic acid and only traces of linolenic acid oxidized faster than rapeseed oil that contained 60% oleic acid and 10% linolenic acid. In a further test of unsaturation effects, stability of low Ln (linolenic acid), low Ln–high O (oleic acid), high O, and regular canola with soy, sunflower, and low-Ln flax oils was compared during accelerated storage at 65°C without light and at 35°C under fluorescent light. Highest oxidative stability to both heat and light was found in low Ln–high O samples[294]. That stability improved in all samples with reduced Ln showed the value of genetic engineering for manipulating oxidizability, although the change was not proportional to the Ln content due to variations in tocopherol levels.

Rapeseed itself is inherently protected by more phenolic substances than any other oilseed, but few of these antioxidants are transferred to the oil during pressing or solvent extraction[292]. Screw-pressed rapeseed oil is most stable postexpelling, before refining. At this stage the main phenolics are sinapic acid derivatives; the most active compound was identified as vinylsyringol formed via decarboxylation of sinapic acid during heating. Tocopherols are partially removed in refining, but phenols are completely removed, so addition of antioxidants is required to stabilize processed oils. However, this is not always a straightforward solution. When α-, γ-, and δ-tocopherols were added at 0.01 to 0.025% to purified rapeseed oil oxidized at 40°C in the dark with low oxygen, α- and γ-tocopherols acted as prooxidants and δ-tocopherol offered no protection except at the highest concentration[295]. Stability could be improved by careful removal of prooxidant compounds (mainly metals and chlorophyll) through refining, but these procedures also increased removal of antioxidant phenols. Addition of propyl gallate substantially increases resistance of rapeseed oil to thermooxidation[296].

24.7.7 HAZELNUT OIL

Hazelnut oil was little used until recently when the striking similarity of its fatty acid composition to olive oil and its cardiovascular health properties were recognized. In the past, low-priced hazelnut oil was frequently mixed with olive oil to economic advantage. Indeed, adulteration of olive oil with hazelnut oil is almost impossible to detect and has become quite a regulatory problem[297,298]. However, this practice may change now that hazelnut oil itself commands a premium price in retail food markets.

Hazelnut oil has two characteristics that markedly enhance its oxidative stability: its predominant fatty acid is oleic (83%), and it has the highest α-tocopherol level among nut oils, 38 to 43 mg/100 g[253,299,300]. Hazelnut oil also has a high content of nontocopherol phenolics including many phenolic aglycones (vanillic, gallic, caffeic, ferulic, coumaric, and sinapic acids), plus catechin, epicatechin, quercetin, and rutin[301]. Nevertheless, considerable differences in susceptibility to oxidation arises from variations of linoleic and linolenic acids contents with cultivar, growing conditions and climate, and storage conditions of the nuts before oil extraction[252]. Irrigation decreases total oil content and increases linoleic acid content and oxidation rates in Catalonian hazelnuts[252]. While only present in small amounts, PUFA contents drive oxidation of hazelnut oil so completely that unsaturated:saturated ratios have been used as a tool to predict shelf life of hazelnut oils from different sources[302]. Polyunsaturates to saturates (P:S) ratios ranging from 7.6 to 9.7 were found in comparisons of traditional hazelnut varieties and five new hybrids with up to 4% increase in 18:1 and about 1% increase in 18:2[303]. Oxidation rates generally followed P:S ratios, but were also modified by endogenous antioxidant levels and mineral content (Fe, Mn, Cu), and a second indicator — the stability index (α-tocopherol × S:P ratio) — more accurately predicted stability. With low unsaturation and high tocopherol content, oils from commercial varieties (particularly Tombul) showed greater stability than the hybrids.

Since hazelnut oil is increasingly being used like olive oil in salad dressings, its stability in emulsions must also be considered. Alasalvar et al.[300] stripped hazelnut oil of tocopherols and polars, then oxidized the oil in emulsions at room temperature and by the oven test (60°C, dark). Crude oil had an induction period of 20 days while stripped oils showed a slow steady oxygen uptake over that period, suggesting that endogenous antioxidants were protecting the oil. However, stripped oil had no measurable peroxide values during that time and TBARS production was negligible in both oils, so some nonlipid oxidation must have been active. Oxidation was considerably faster in emulsions than in bulk oils by all measures. Most notably, secondary aldehydes were produced in emulsions where water was present to facilitate scission reactions, but not in bulk oils. During the 21-day oxidation, tocopherols decreased by about 50% in bulk oils but nearly 90% in emulsions. The rapid onset of oxidation observed at 18 to 21 days appears to correspond to the loss of tocopherol: i.e., tocopherol protects oils until it is consumed, then oil oxidation takes off. These results reiterate the importance of endogenous antioxidants and demonstrate a critical need for antioxidant supplementation in refined oils and especially in emulsions.

24.7.8   OLIVE OIL

Perhaps the best known and most used of the "specialty" oils having unique health properties is olive oil. In addition to its unique characteristic flavor, olive oil offers a food use advantage in being more stable than other vegetable oils commonly used in cooking[239,285]. Olive oils are inherently stabilized by high oleic acid and tocopherol and other phenolics contents, but their susceptibility to oxidation varies tremendously with variety[304], growth conditions and season[305], maturity[306], plant health[307], and processing[308] due to changes in both fatty acid composition and phenol content.

Salvador et al.[305] found that for the Cornicabra variety of Castillian virgin olive oil stability varied considerably with seasons. The oil contained high O (>80%) and low L (linoleic acid; 4.5%), high sterol and phenol levels, also high chlorophyll and carotenoid levels. The PV and Rancimat induction period correlated with total phenolic content. Oxidative stability also changes with ripening of olives[306]. Oleic acid increases and palmitic acid decreases with ripening, and phenolics, particularly oleuropein and hydroxytyrosol, also decrease with ripening. Thus, oil from riper olives generally has lower resistance to oxidation. However, there are varietal exceptions to this rule, at least in part, because the interactions between fatty acid composition, antioxidant types and levels, and processing effects are so complex. For example, in a study covering a wide range of oxidation conditions and storage times, levels and loss of α-tocopherol were not related to stability[306]. Of late-ripening Italian varieties, Coratina oils retained high phenol contents and were most resistant to oxidation while Leccino oils changed phenol and fatty acid composition rapidly during ripening and deteriorated rapidly under oxidizing conditions. Stability of Greek olive oils showed a linear correlation with total phenol content and the hydroxytryosol:tyrosol ratio but not with tyrosol, the major phenolic component.

The variable presence of prooxidants such as chlorophyll and metals (not measured in these studies) or antioxidant phenolics other than the standard three noted above may account for some of these discrepancies in olive oil oxidation. For example, two hydroxytyrosol derivatives and an isomer of oleoeuropeine aglycone identified by HPLC chemometric analyses accounted for most antioxidant activity in the polar phenol fraction of ripening and diseased olives[307]. Future research must therefore examine stability contributions of minor phenols as well as major fractions, and both polar and nonpolar phenols must be considered because they stabilize oil-based systems differently.

Processing methods affected olive oil stability by differential removal of pro- and antioxidants and in some cases by heat stress. Continuous processing with a stone mill–integral decanter produces oils with higher tocopherol (mostly α), carotenoids, xanthophylls, and chlorophylls but lower total phenolics and oxidative stability (probably due to high chlorophyll) with difficult-to-process olives. With the same process, easy-to-process olives yielded oils with higher flavor volatiles and tocopherols, and comparable level of phenolics. Stability was due mostly to o-diphenols, not tocopherol[308]. Chlorophyll contents of pressed oils tend to be high, and this sensitizes extra virgin olive oil to rapid photooxidation[309]. Stripping the oil during processing can eliminate the chlorophyll but also removes important phenols, tocopherols, and carotenoids. Commercial olive oil therefore requires protection from light to prevent deterioration[309]. Tocopherols lost in the refining process can be replaced, but since tocopherols can become a prooxidant at high levels amounts added should be titrated to the degree of oxidative stress in the particular olive oil and kept as low as possible. When olive oil was stored at 40°C in the dark, more than 100 ppm α-tocopherol accelerated oxidation[310]. However, with more established oxidation or increased free radical loads due to catalysts — conditions that consume antioxidants more rapidly — higher levels of tocopherol may be needed for effective inhibition of lipid oxidation.

In early stages of oxidation, o-diphenols are more effective than tocopherols, probably due to metal binding and inhibition of initiation[310]. This observation has been used to advantage by adding garlic, rosemary, and oregano to improve keeping quality and add flavor to olive oils[311]. Rosemary and oregano both contain o-diphenols; when added at 1 and 2% they significantly lower oxidation of olive oil during storage while providing interesting flavors for consumers. Garlic, with primarily sulfur-based antioxidants, has no effect on olive oil stability.

24.7.9 MEADOWFOAM

Meadowfoam oil is unusual in both composition and stability. Over 98% fatty acids have 20 or more carbons and these are mostly monounsaturated (C20:1, 63%; C22:1, 16%; C22:2, 17%)[312]. As a result, crude meadowfoam oil is one of the most stable lipids known, resisting oxidation nearly 18 times longer than soybean oil[313]. Oxidative stability index comparisons at 100°C showed stability decreased in the order crude and refined meadowfoam > jojoba > castor > high-oleic sunflower > soybean > triolein > high-erucic rapeseed[314]. With this high oxidative stability, adding even 5 to 10% of crude meadowfoam oil improves stability of other oils[313].

The source of meadowfoam's exceptional stability is still being debated. The unusual fatty acid composition, the double bond positions, and the TAG structures appear to be key contributors to the exceptional stability of this oil. Most critical for stability are the Δ5-monounsaturated fatty acids. The main fatty acid from meadowfoam oil, 5-eicosenoic acid, is nearly 5 times more stable than oleic acid and 16 times more stable than other monounsaturated fatty acids. However, when meadowfoam fatty acids were tested as a mixture, they gave the same stability as other fatty acids, even some with higher unsaturation[313,314]. Thus, specific fatty acid organization in TAGs must also be important.

Lipid composition alone accounts for only part of the stability of meadowfoam oil because much of the stability is lost after refining. Seed conditioning at elevated temperature and moisture prior to pressing are necessary to inactivate enzymes that degrade glucosinolates to nitrile and isocyanate compounds responsible for off-flavors and odors[315,316]. Conditioning at 4, 6, or 8% moisture at temperatures of 70, 90, or 110°C increased free fatty acid levels but stabilized the oil to oxidation without altering oil yield or fatty acid composition[243]. However, this stabilization was lost during degumming and refining[317]. Meadowfoam does have a moderate tocopherol content but not enough to account for the huge difference in stability of crude meadowfoam oil compared to other oils[314], and the tocopherols are largely removed during refining[317]. Other unidentified antioxidants in meadowfoam may contribute to stabilization, or more likely, there is a combination of factors that cooperatively inhibit oxidation of meadowfoam oil.

24.8   SUMMARY AND CONCLUSIONS

Specialty oils clearly follow the general patterns expected for lipid oxidation, i.e., stability decreases with degree of unsaturation in component fatty acids, contamination with chlorophyll and metals, heat, and exposure to light. Conversely, stability generally increases with content of endogenous antioxidants (tocopherols, anthocyanins, carotenoids, other phenols, etc.) and with protection in the dark, in UV-impermeable containers, at low temperatures, and under inert atmospheres. Nevertheless, specialty oils present considerable challenge in elucidating the complex interactions between pro- and antioxidant factors that dictate their final susceptibility to oxidation.

Deciding what extraction and processing methods are best for various oils would be easy if each oil had a single consistent factor that controlled its stability, but that is not the case. At the first line of delineation, fatty acid compositions vary between cultivars, and single cultivars vary with growth environment and processing methods. Beyond this, a broad range of trace components, both pro- and antioxidant, also vary with oilseed species, varieties, growth conditions, and processing treatments. Trying to sort out which interactions or interferences counterbalance each other productively to stabilize oils and which ones aggravate oxidation and decrease shelf life often seems like aiming at a constantly shifting target. Largely empirical research testing a range of conditions or compositions for stability has revealed some common behaviors and, as importantly, identified critical areas of discrepancies and uncertainty. Future research now needs to consider in depth the competition and balance between key individual components of the specialty oils now receiving intense nutraceutical and pharmacological attention. Indeed, these applications entirely depend on the ability of the food oil industry to provide highly stable oils that retain their nutritional and nutraceutical value and do not contribute oxidation products with potentially toxic physiological effects. More focus must thus be given to elucidating mechanisms of specialty oil oxidation under various conditions in order to control processing and to produce oils of consistently high quality and long-term stability.

REFERENCES

1. Farmer, E.H., Peroxidation in relation to olefinic structure, *Trans. Faraday Soc.,* 42, 228–236, 1946.
2. Bolland, J.L., Kinetics of olefin oxidation, *Quart. Rev.,* 3, 1–21, 1949.
3. Bolland, J.L., The course of autoxidation reactions in polyisoprenes and allied compounds: IX. The primary thermal oxidation products of ethyl linoleate, *J. Chem. Soc.,* 445–447, 1945.
4. Swern, D., Primary products of olefinic autoxidation, in *Autoxidation and Antioxidants,* Vol. 1, Lundberg, W.O., Ed., Interscience, New York, 1961, pp. 1–54.

5. Farmer, E.H., Koch, H.P., and Sutton, D.A., The course of autoxidation reactions in polyisoprenes and allied compounds: VII. Rearrangement of double bonds during autoxidation, *J. Chem. Soc.*, 541–547, 1943.

6. Hyde, S.M. and Verdin, D., Oxidation of methyl oleate induced by cobalt-60 γ-irradiation: I. Pure methyl oleate, *Trans. Faraday Soc.*, 64, 144–154, 1968.

7. Cosgrove, J.P., Church, D.F., and Pryor, W.A., The kinetics of the autoxidation of polyunsaturated fatty acids, *Lipids*, 22, 299–304, 1987.

8. Howard, J.A. and Ingold, K.U., Absolute rate constants for hydrocarbon oxidation: XII. Rate constants for secondary peroxy radicals, *Can. J. Chem.*, 46, 2661–2665, 1968.

9. Howard, J.A. and Ingold, K.U., The self-reaction of *sec*-butylperoxy radicals: confirmation of the Russell mechanism, *J. Am. Chem. Soc.*, 90, 1056–1058, 1968.

10. Howard, J.A. and Ingold, K.U. Rate constants for the self-reaction of *n*- and *sec*-butylperoxy radicals and cyclohexylperoxy radicals. The deuterium isotope effect in the termination of secondary peroxy radicals, *J. Am. Chem. Soc.*, 90, 1058–1059, 1968.

11. Kochi, J. K., Oxygen radicals, in *Free Radicals*, Vol. 2, Kochi, J.K., Ed., John Wiley, New York, 1973, pp. 665–710.

12. Gardner, H.W., Oxygen radical chemistry of polyunsaturated fatty acids, *Free Radic. Biol. Med.*, 7, 65–86, 1989.

13. Porter, N.A. and Wujek, D.G., The autoxidation of polyunsaturated fatty acids, in *Reactive Oxygen Species in Chemistry, Biology, and Medicine*, Quintanilla, A., Ed., Plenum, New York, 1987, pp. 55–79.

14. Schaich, K. M. Lipid oxidation in Fats and Oils: An integrated view, in *Bailey's Industrial Fats and Oils*; Shahidi, F., Ed.; John Wiley, New York, 2005, pp. 2681-2767.

15. Schaich, K.M., Free radical initiation in proteins and amino acids by ionizing and ultraviolet radiations and lipid oxidation: III. Free radical transfer from oxidizing lipids, *CRC Crit. Rev. Food Sci. Nutr.*,13, 189–244, 1980.

16. Schaich, K.M., Metals and lipid oxidation: contemporary issues, *Lipids*, 27, 209–218, 1992.

17. Waters, W.A., Roles of cobaltic salts as catalysts of oxidation, *Discuss. Faraday Soc.*, 46, 158–163, 1968.

18. Smith, P. and Waters, W.A., Oxidation of organic compounds by cobaltic salts: XII. Oxidations of unsaturated acids, *J. Chem. Soc. B., Phys Org. Chem.* 462–467, 1969.

19. Uri, N., Physico-chemical aspects of autoxidation, in *Autoxidation and Antioxidants*, Vol. 1, Lundberg, W.O., Ed., Interscience, New York, 1961, pp. 55–106.

20. Heaton, F.W. and Uri, N., The aerobic oxidation of unsaturated fatty acids and their esters: cobalt stearate-catalyzed oxidation of linoleic acid, *J. Lipid Res.*, 2, 152–160, 1961.

21. Bawn, C.E.H., Free radical reactions in solution initiated by heavy metal ions, *Discuss. Faraday Soc.*,14, 181–190, 1953.

22. Banks, A., Eddie, E., and Smith, J.G.M., Reactions of cytochrome c with methyl linoleate hydroperoxide, *Nature* (London), 190, 908–909, 1961.

23. Ochiai, E., Mechanisms of catalysis by metal complexes in autoxidation of an olefin, *Tetrahedron*, 20, 1819–1829, 1964.

24. Tkac, A., Vesely, K., Omelka, L. and Prikryl, R., Complex-bonded and continuously generated peroxy and alkoxy radicals, *Coll. Czech. Chem. Commun.*, 40, 117–128, 1975.

25. Prikryl, R., Tkac, A., Omelka, L., and Vesely, K., Decomposition mechanism of peroxides in presence of Co(II) acetylacetonate, *Coll. Czech. Chem. Commun.*, 40, 104–116, 1975.

26. Kochi, J.K., The mechanism of the copper salt catalysed reactions of peroxides, *Tetrahedron*, 18, 483–497, 1962.

27. Bray, W.C. and Gorin, M.H., Ferryl ion, a compound of tetravalent iron, *J. Am. Chem. Soc.*, 54, 2124–2125, 1932.

28. Rush, J.D. and Koppenol, W.H., Reactions of Fe^{II}nta and Fe^{II}edda with hydrogen peroxide, *J. Am. Chem. Soc.*, 110, 4957–4963, 1988.

29. Tung, H.-C., Kang, C., and Sawyer, D.T., Nature of the reactive intermediates from the iron-induced activation of hydrogen peroxide: agents for the ketonization of methylenic carbons, the monooxygenation of hydrocarbons, and the dioxygenation of arylolefins, *J. Am. Chem. Soc.*, 114, 2445–3555, 1992.

30. Bassan, A., Blomberg, M.R.A., Siegbahn, P.E.M., and Que, L., Jr., A density functional study of the O–O bond cleavage for a biomimetic non-heme iron complex demonstrating an Fe^V-intermediate, *J. Am. Chem. Soc.*, 124, 11056–11063, 2002.

31. Egmond, M.R., Vliegenthart, J.F.G., and Boldingh, J., Stereospecificity of the hydrogen abstraction in carbon atom n-8 in the oxygenation of linoleic acid by lipoxygenases from corn germs and soya beans, *Biochim. Biophys. Res. Commun.*, 48, 1055–1060, 1972.

32. deGroot, J.J.M.C., Garssen, G.J., Vliegenthart, J.F.G., and Boldingh, J., Demonstration by EPR spectroscopy of the functional role of iron in soybean 1-lipoxygenase, *Biochim. Biophys. Acta*, 377, 71–79, 1975.

33. deGroot, J.J.M.C., Garssen, G.J., Vliegenthart, J.F.G., and Boldingh, J., On the interaction of soybean lipoxygenase-1 and 13-L-hydroperoxylinoleic acid, involving yellow and purple coloured enzyme species, *FEBS Lett.*, 56, 50–54, 1975.

34. Chiba, T., K. Fujimoto, and T. Kaneda, Radicals generated in autoxidized methyl linoleate by light irradiation, *J Am. Oil Chem. Soc.*, 58, 587-590, 1981.

35. Khan, A.U., Activated oxygen: Singlet molecular oxygen and superoxide anion, *Photochem. Photobiol.*, 28, 615-627, 1978.

36. Geoffroy, M., Lambelet,P., and Richert, P., Role of hydroxyl radicals and singlet oxygen in the formation of primary radicals in unsaturated lipids: a solid state electron paramagnetic resonance study, *J. Agric. Food Chem.*, 48, 974-978, 2000.

37. Balchum, O. J. and O'Brien, J. S., Ozone and unsaturated fatty acids, *Arch. Environ. Health*, 22, 32-34, 1971.

38. Nickell, E. C., Albi, M., and Privett, O. S., Ozonization products of unsaturated fatty acid methyl esters, *Chem. Phys. Lipids*, 17, 378-388, 1976.

39. Pryor, W. A., Stanley, J. P., Blair, E., and Cullen, G. B., Autoxidation of polyunsaturated fatty acids. Part I. Effect of ozone on the autoxidation of neat methyl linoleate and methyl linolenate, *Arch. Environ. Health*, 31, 201-210, 1976.

40. Nam, W., Lim, M.H., Lee, H.J., and Kim, C., Evidence for the participation of two distinct reactive intermediates in iron(III) porphyrin complex-catalyzed epoxidation reactions, *J. Am. Chem. Soc.*, 122, 6641–6647, 2000.

41. Nam, W., Han, H.J., Oh, S.-Y., Lee, Y.J., Choi, M.-H., Han, S.-Y., Kim, C., Woo, S.K., and Shin, W., New insights into the mechanism of O-O bond cleavage of hydrogen peroxide and *tert*-alkyl hydroperoxides by iron(III) porphyrin complexes, *J. Am. Chem. Soc.*, 122, 8677–8684, 2000.

42. Rao, S.I., Wilks, A., Hamberg, M., and Ortiz de Montellano, P.R., The lipoxygenase activity of myoglobin. Oxidation of linoleic acid by the ferryl oxygen rather than the protein radical, *J. Biol. Chem.*, 269, 7210–7216, 1994.

43. Nawar, W.W., Thermal degradation of lipids, *J. Agric. Food Chem.*, 17, 18–21, 1969.

44. Nawar, W.W., Lipids, in *Food Chemistry*, Fennema, O.R., Ed., Academic Press, New York, 1966, pp. 225–320

45. Pryor, W.A., Oxy-radicals and related species: their formation, lifetimes and reactions, *Annu. Rev. Physiol.*, 48, 657–667, 1986.

46. Rao, P.S., Ayres, S.M., and Mueller, H.S., Identity of peroxy radicals produced from arachidonic acid in oxygenated solutions as studied by pulse radiolysis technique, *Biochem. Biophys. Res. Commun.*, 104, 1532–1536, 1982.

47. Bielski, B.H.J., Cabelli, D.E., Arudi, R.L., and Ross, A.B., Reactivity of HO_2/O_2^- radicals in aqueous solution, *J. Phys. Chem. Ref. Data*, 14, 1041–1100, 1985.

48. Hasegawa, K. and Patterson, L.K., Pulse radiolysis studies in model lipid systems: formation and behavior of peroxy radicals in fatty acids, *Photochem. Photobiol.*, 28, 817–823, 1978.

49. Simic, M.G., Jovanovic, S.V., and Niki, E., Mechanisms of lipid oxidative processes and their inhibition, in *Lipid Oxidation in Food*; St. Angelo, A.J., Ed., American Chemical Society, Washington, DC, 1992, pp. 14–32.

50. Patterson, L.K. and Hasegawa, K., Pulse radiolysis studies in model lipid systems. The influence of aggregation on kinetic behavior of OH induced radicals in aqueous sodium linoleate, *Ber. Bunsenges. Phys. Chem.*, 82, 951–956, 1978.

51. Small, R.D., Jr., Scaiano, J.C., and Patterson, L.K., Radical processes in lipids. A laser photolysis study of t-butoxy radical reactivity toward fatty acids, *Photochem. Photobiol.*, 29, 49–51, 1979.

52. Erben-Russ, M., Michael, C., Bors, W., and Saran, M., Absolute rate constants of alkoxyl radical reactions in aqueous solution, *J. Phys. Chem.*, 91, 2362–2365, 1987.

53. Piretti, M.V., Cavani, C., and Zeli, F., Mechanism of the formation of hydroperoxides from methyl oleate, *Revue francaise des Corps Gras*, 25, 73–78, 1978.

54. Gebicki, J.M. and Bielski, B.H.J., Comparison of the capacities of the perhydroxyl and the superoxide radicals to initiate chain oxidation of linoleic acid, *J. Am. Chem. Soc.*, 103, 7020–7022, 1981.

55. Hamilton, R.J., Kalu, C., Prisk, E., Padley, F.B., and Pierce, H., Chemistry of free radicals in lipids, *Food Chem.*, 60, 193–199, 1997.

56. Thomas, M.J., Mehl, K.S. and Pryor, W.A., The role of superoxide in xanthine oxidase-induced autoxidation of linoleic acid, *J. Biol. Chem.*, 257, 8343–8347, 1982.

57. Giamalva, D.H., Church, D.F., and Pryor, W.A., Kinetics of ozonation: 4. Reactions of ozone with *a*-tocopherol and oleate and linoleate esters in carbon tetrachloride and in aqueous micellar solvents, *J. Am. Chem. Soc.*, 108, 6646–6651, 1986.

58. Erben-Russ, M., Michael, C., Bors, W., and Saran, M., Determination of sulfite radical (SO_3^-) reaction rate constants by means of competition kinetics, *Radiat. Environ. Biophys.*, 26, 289–294, 1987.

59. Schöneich, C. and Asmus, K.-D., Reaction of thiyl radicals with alcohols, ethers, and polyunsaturated fatty acids: a possible role of thiyl free radicals in thiol mutagenesis, *Radiat. Environ. Biophys.*, 29, 263–271, 1990.

60. Doleiden, F.H., Fahrenholtz, S.R., Lamola, A.A., and Trozzolo, A.M., Reactivity of cholesterol and some fatty acids toward singlet oxygen, *Photochem. Photobiol.*, 20, 519–521, 1974.

61. Pryor, W.A., Lightsey, J.W., and Church, D.F., Reaction of nitrogen dioxide with alkenes and polyunsaturated fatty acids: addition and hydrogen abstraction mechanisms, *J. Am. Chem. Soc.*, 104, 6685–6692, 1982.

62. Kerr, J.A., Bond dissociation energies by kinetic methods, *Chem. Rev.*, 66, 465–500, 1966.

63. Tallman, K.A., Pratt, D.A., and Porter, N.A., Kinetic products of linoleate peroxidation: rapid *β*-fragmentation of nonconjugated peroxyls, *J. Am. Chem. Soc.*, 123, 11827–11828, 2001.

64. Coxon, D.T., Price, K.R., and Chan, H.W.-S., Formation, isolation and structure determination of methyl linolenate diperoxides, *Chem. Phys. Lipids*, 28, 365–378, 1981.

65. Neff, W.E., Frankel, E.N., and Weisleder, D., High-pressure liquid chromatography of autoxidized lipids: II. Hydroperoxy-cyclic peroxides and other secondary products from methyl linolenate, *Lipids*, 16, 439–448, 1981.

66. Pryor, W.A. and Stanley, J.P., A suggested mechanism for the production of malonaldehyde during the autoxidation of polyunsaturated fatty acids. Nonenzymatic production of prostaglandin endoperoxides during autoxidation, *J. Org. Chem.*, 40, 3615–3617, 1975.

67. Porter, N.A., Funk, M.O., Gilmore, D., Isaac, R., and Nixon, J., The formation of cyclic peroxides from unsaturated hydroperoxides: models for prostaglandin biosynthesis, *J. Am. Chem. Soc.*, 98, 6000–6005, 1976.

68. Garwood, R.F., Khambay, B.P.S., Weedon, B.C.L., and Frankel, E.N., Allylic hydroperoxides from the autoxidation of methyl oleate, *J. Chem. Soc. Chem. Commun.*, 364–365, 1977.

69. Frankel, E.N., Neff, W.E., and Bessler, T.R., Analysis of autoxidized fats by gas chromatography-mass spectrometry: V. Photosensitized oxidation, *Lipids*, 14, 961–967, 1979.

70. Terao, J. and Matsushita, S., The isomeric compositions of monohydroperoxides produced by oxidation of unsaturated fatty acids esters with singlet oxygen, *J. Food Process. Preserv.*, 3, 329–337, 1980.

71. Haslbeck, F. and Grosch, W., Autoxidation of phenyl linoleate and phenyl oleate: HPLC analysis of the major and minor monohydroperoxides as phenyl hydroxystearates, *Lipids*, 18, 706–713, 1983.

72. Chan, H.W.-S. and Levett, G., Autoxidation of methyl linolenate: analysis of methyl hydroxylinolenate isomers by high performance liquid chromatography, *Lipids*, 12, 837–840, 1977.

73. Terao, J. and Matsushita, S., The isomeric composition of hydroperoxides produced by oxidation of arachidonic acid with singlet oxygen, *Agric. Biol. Chem.*, 45, 587–593, 1981.

74. VanRollins, M. and Murphy, R.C., Autooxidation of docosahexaenoic acid: analysis of ten isomers of hydroxydocosahexaenoate, *J. Lipid Res.*, 25, 507–517, 1984.

75. Parr, L.J. and Swoboda, P.A.T., The assay of conjugable oxidation products applied to lipid deterioration in stored foods, *J. Food Technol.*, 11, 1–12, 1976.

76. Porter, N.A. and Nixon, J.R., Stereochemistry of free-radical substitution on the peroxide bond, *J. Am. Chem. Soc.*, 100, 7116–7117, 1978.

77. Porter, N.A., Roberts, D.H., and Ziegler, C.B., Jr., A new route to lipid hydroperoxides: orbital symmetry controlled ring opening of vinylcyclopropyl bromides, *J. Am. Chem. Soc.*, 102, 5912–5913, 1980.

78. Porter, N.A., Weber, B.A., Weenan, H., and Kahn, J.A., Autoxidation of polyunsaturated lipids. Factors controlling the stereochemistry of product hydroperoxides, *J. Am. Chem. Soc.*, 102, 5597–5601, 1980.

79. Porter, N.A., Lehman, L.S., Weber, B.A., and Smith, K.J., Unified mechanism for polyunsaturated fatty acid autoxidation. Competition of peroxy radical hydrogen atom abstraction, *β*-scission, and cyclization, *J. Am. Chem. Soc.*, 103, 6447–6455, 1981.

80. Roe, A.N., McPhail, A.T., and Porter, N.A., Serial cyclization: studies in the mechanism and stereochemistry of peroxy radical cyclization, *J. Am. Chem. Soc.*, 105, 1199–1203, 1983.

81. Porter, N.A. and Wujek, D.G., Autoxidation of polyunsaturated fatty acids, an expanded mechanistic study, *J. Am. Chem. Soc.,* 106, 2626–2629, 1984.

82. Porter, N.A., Lehman, L.S., and Wujek, D.G., Oxidation mechanisms of poly-unsaturated fatty acids, in *Oxygen Radicals in Chemistry and Biology,* Bors, W., Saran, M., and Tait, D., Eds., Walter de Gruyter, Berlin, 1984, pp. 235–237.

83. Porter, N.A. Mechanisms for the autoxidation of polyunsaturated lipids, *Acc. Chem. Res.,* 19, 262–268, 1986.

84. Porter, N.A. and Wujek, J.S., Allylic hydroperoxide rearrangement: β-scission or concerted pathway?, *J. Org. Chem.,* 52, 5085–5089, 1987.

85. Porter, N.A., Autoxidation of polyunsaturated fatty acids: initiation, propagation, and product distribution (basic chemistry), in *Membrane Lipid Oxidation,* Vol. I, Vigo-Pelfrey, C., Ed., CRC Press, Boca Raton, FL, 1990, pp. 33–62.

86. Mills, K.A., Caldwell, S.E., Dubay, G.R., and Porter, N.A., An allyl radical-dioxygen caged pair mechanism for *cis*-allylperoxyl rearrangements, *J. Am. Chem. Soc.,* 114, 9689–9691, 1992.

87. Porter, N.A., Mills, K.A., Caldwell, S.E., and Dubay, G.R., The mechanisms of the [3,2]-allylperoxyl rearrangement. A radical-dioxygen pair reaction that proceeds with stereochemical memory, *J. Am. Chem. Soc.,* 116, 6697–6705, 1994.

88. Porter, N.A., Mills, K.A., and Carter, R.L., A mechanistic study of oleate oxidation: competing peroxyl H-atom abstraction and rearrangement, *J. Am. Chem. Soc.,* 116, 6690–6696, 1994.

89. Porter, N.A., Caldwell, S.E., and Mills, K.A., Mechanisms of free radical oxidation of unsaturated lipids, *Lipids,* 30, 277–290, 1995.

90. Pratt, D.A., Mills, J.H., and Porter, N.A., Theoretical calculations of carbon-oxygen bond dissociation enthalpies of peroxyl radicals formed in the autoxidation of lipids, *J. Am. Chem. Soc.,* 125, 5801–5810, 2003.

91. Weenan, H. and Porter, N.A., Autoxidation of model membrane systems: cooxidation of polyunsaturated lecithins with steroids, fatty acids, and α-tocopherol, *J. Am. Chem. Soc.,* 104, 5216–5221, 1982.

92. Howard, J.A. and Ingold, K.U., Absolute rate constants for hydrocarbon autoxidation. VI. Alkyl aromatic and olefinic hydrocarbons, *Can. J. Chem.,* 45, 793-802, 1967.

93. Sheldon, R.A. and Kochi, J.K., *Metal-Catalyzed Oxidations of Organic Compounds,* Academic Press, New York, 1981.

94. Simic, M., in *Fast Processes in Radiation Chemistry and Biology,* Fielden, E.M. and Michael, B.D., Eds., Wiley, New York, 1975, pp. 162–179.

95. Simic, M. G. and Karel, M., eds., *Autoxidation in Food and Biological Systems,* Plenum, New York, 1980.

96. Simic, M.G., Free radical mechanisms in autoxidation processes, *J. Chem. Ed.,* 58, 125–131, 1981.

97. Frankel, E.N., Chemistry of free radical and singlet oxidation of lipids, *Prog. Lipid Res.,* 23, 197–221, 1985.

98. Chan, H. W.-S., ed.,*Autoxidation of Unsaturated Lipids,* Academic Press, London, 1987.

99. Frankel, E.N., Recent advances in lipid oxidation, *J. Sci. Food Agric.,* 54, 495–511, 1991.

100. St. Angelo, A. J., ed., *Lipid Oxidation in Food,* Amer. Chem. Soc., Washington, DC, 1992.

101. Ingold, K.U., Peroxy radicals, *Acc. Chem. Res.,* 2, 1–9, 1969.

102. Bockman, T.M., Hubig, S.M., and Kochi, J.K., Kinetic isotope effects for electron-transfer pathways in the oxidative C–H activation of hydrocarbons, *J. Am. Chem. Soc.,* 120, 2826–2830, 1998.

103. Bors, W., Erben-Russ, M., and Saran, M., Fatty acid peroxyl radicals: their generation and reactivities, *J. Electroanal. Chem.,* 232, 37–49, 1987.

104. Factor, A., Russell, C.A., and Traylor, T.G., Bimolecular combination reactions of oxy radicals, *J. Am. Chem. Soc.,* 87, 3692–3696, 1965.

105. Howard, J.A., Homogeneous liquid-phase autoxidations, in *Free Radicals,* Vol. II, Kochi, J.K., Ed., Wiley-Interscience, New York, 1973, pp. 3–62.

106. Porter, N.A., Zuraw, P.J., and Sullivan, J.A., Sterochemistry of hydroperoxide cyclization reactions, *J. Org. Chem.,* 49, 1345-1348, 1984.

107. Van Sickle, D.E., Mayo, F.R., Gould, E.S., and Arluck, R.M., Effects of experimental variables in oxidations of alkenes, *J. Am. Chem. Soc.,* 89, 977–984, 1967.

108. Hamberg, M. and Samuelsson, B., Novel biological transformations of 8,11,14-eicosatrienoic acid, *J. Am. Chem. Soc.,* 88, 2349–2350, 1966.

109. Porter, N.A. and Funk, M.O., Peroxy radical cyclization as a model for prostaglandin synthesis, *J. Org. Chem.,* 40, 3614–3615, 1975.

110. Haynes, V., *J. Chem. Soc. Chem. Commun.*, Iron(III) and copper(II) catalyzed transformation of fatty acid hydroperoxides: efficient generation of peroxy radicals with Cu(II) trifluoromethane sulfonate, 1102–1104, 1990.

111. Neff, W.E., Frankel, E.N., and Fujimoto, K., Autoxidative dimerization of methyl linolenate and its monohydroperoxides, hydroperoxy epidioxides and dihydroperoxides, *J. Am. Oil Chem. Soc.*, 65, 616–623, 1988.

112. Hiatt, R. and McCarrick, T., On "bimolecular initiation" by hydroperoxides, *J. Am. Chem. Soc.*, 97, 5234–5237, 1975.

113. Mayo, F.R., The oxidation of unsaturated compounds: IX. The effects of structure on the rates and products of oxidation of unsaturated compounds, *J. Am. Chem. Soc.*, 80, 2500–2507, 1958.

114. Mayo, F.R., Free-radical autoxidation of hydrocarbons, *Acc. Chem. Res.*, 1, 193–201, 1968.

115. Elson, I.H., Mao, S.W., and Kochi, J.K., Electron spin resonance study of addition of alkoxy radicals to olefins, *J. Am. Chem. Soc.*, 97, 335–341, 1975.

116. Lewis, S.E. and Mayo, F.R., Copolymerization: IX. A comparison of some *cis* and *trans* isomers, *J. Am. Chem. Soc.*, 70, 1533–1536, 1958.

117. Kochi, J. K., Addition of peroxides to conjugated olefins catalyzed by copper salts, *J. Am. Chem. Soc.*, 84, 2785–2793, 1962.

118. Hendry, D.G. and Schuetzle, D., Reactions of hydroperoxy radicals. Comparison of reactivity with organic peroxides, *J. Org. Chem.*, 41, 3179–3182, 1976.

119. Witting, I.A., Chang, S.S., and Kummerow, F.A., The isolation and characterization of the polymers formed during the autoxidation of ethyl linoleate, *J. Am. Oil Chem. Soc.*, 34, 470–473, 1957.

120. Sims, R.P. and Hoffman, W.H., Oxidative polymerization, in *Autoxidation and Antioxidants*, Vol. II, Lundberg, W.O., Ed., Interscience, London, 1962, pp. 629–694.

121. Privett, O.S., Autoxidation and autoxidative polymerization, *J. Am. Oil Chem. Soc.*, 36, 507–512, 1959.

122. Mounts, T.L., McWeeny, D.J., Evans, C.D., and Dutton, H.J., Decomposition of linoleate hydroperoxides: precursors of oxidative dimers, *Chem. Phys. Lipids*, 4, 197–202, 1970.

123. Frankel, E.N., Neff, W.E., Selke, E., and Brooks, D.D., Analysis of autoxidized fats by gas chromatography-mass spectrometry: X. Volatile thermal decomposition products of methyl linolenate dimers, *Lipids*, 23, 295–298, 1988.

124. Thomas, J.R., The self-reaction of *t*-butylperoxy radicals, *J. Am. Chem. Soc.*, 87, 3935–3940, 1965.

125. Adamic, K., Howard, J.A., and Ingold, K.U., Absolute rate constants for hydrocarbon autoxidation: XVI. Reactions of peroxy radicals at low temperatures, *Can. J. Chem.*, 1969, 3803–3808, 1969.

126. Hiatt, R. and Traylor, T.G., Cage recombination of *t*-butoxy radicals, *J. Am. Chem. Soc.*, 87, 3766–3768, 1965.

127. Neff, W.E., Frankel, E.N., Schofield, C.R., and Weisleder, D., High-pressure liquid chromatography of autoxidized lipids: I. Methyl oleate and linoleate, *Lipids*, 13, 415–421, 1978.

128. Selke, E., Frankel, E.N., and Neff, W.E., Thermal decomposition of methyl oleate hydroperoxides and identification of volatile components by gas chromatography-mass spectrometry, *Lipids*, 13, 511–513, 1978.

129. Hiatt, R., Mill, T., and Mayo, F.R., Homolytic decompositions of hydroperoxides: I. Summary and implications for autoxidation, *J. Org. Chem.*, 33, 1416–1420, 1968.

130. Mayo, F.R., The oxidation of unsaturated compounds: VIII. The oxidation of aliphatic unsaturated compounds, *J. Am. Chem. Soc.*, 80, 2497–2500, 1958.

131. Kim, S.S., Kim, S.Y., Ryou, S.S., Lee, C.S., and Yoo, K.H., Solvent effects in the hydrogen abstractions by *tert*-butoxy radical: veracity of the reactivity/selectivity principle, *J. Org. Chem.*, 58, 192–196, 1993.

132. Hiatt, R. and Zigmund, L., Interaction of s-butyl peroxy and alkoxy radicals, *Can. J. Chem.*, 48, 3967, 1970.

133. Hiatt, R. and Szilagyi, S., *Can. J. Chem.*, 48, 616, 1970.

134. Gilbert, B.C., Holmes, R.G.G., Norman, R.O.C., *J. Chem. Res.*, 8, 1, 1977.

135. Walling, C. and Padwa, A., Positive halogen compound: VII. Intramolecular chlorination with long chain hypochlorites, *J. Am. Chem. Soc.*, 85, 1597–1601, 1963.

136. Frankel, E.N., Volatile lipid oxidation products, *Prog. Lipid Res.*, 22, 1–33, 1982.

137. Acott, B. and Beckwith, A.L.J., Reactions of alkoxy radicals: IV. Intramolecular hydrogen-atom transfers in the presence of cupric ion: a novel directive effect, *Austral. J. Chem.*, 17, 1342–1353, 1964.

138. Dix, T.A. and Marnett, L.J., Free radical epoxidation of 7,8-dihydroxy-7,8-dihydrobenzo[a]pyrene by hematin and polyunsaturated fatty acid hydroperoxides, *J. Am. Chem. Soc.*, 103, 6744–6746, 1981.

139. Bors, W., Tait, D., Michel, C., Saran, M., and Erben-Russ, M., Reactions of alkoxyl radicals in aqueous solutions, *Isr. J. Chem.*, 24, 17–24, 1984.

140. Kochi, J.K., Chemistry of alkoxyl radicals, *J. Am. Chem. Soc.*, 84, 1193–1197, 1962.

141. Walling, C. and Wagner, P.J., Positive halogen compounds: X. Solvent effects in the reactions of *t*-butoxy radicals, *J. Am. Chem. Soc.*, 86, 3368–3375, 1964.

142. Wu, G.-S., Stein, R.A., and Mead, J.F., Autoxidation of fatty acid monolayers adsorbed on silica gel: II. Rates and products, *Lipids*, 12, 971–978, 1977.

143. Wu, G.-S., Stein, R.A., Mead, J.F., Autoxidation of fatty acid monolayers adsorbed on silica gel: III. Effects of saturated fatty acids and cholesterol, *Lipids*, 13, 517–524, 1978.

144. Neff, W.E., Quantitative analyses of hydroxystearate isomers from hydroperoxides by high pressure liquid chromatography of autoxidized and photosensitized-oxidized fatty esters, *Lipids*, 15, 587–590, 1980.

145. Chan, H. W. S.; Prescott, F. A. A.; Swoboda, P. A. T. Thermal decomposition of individual positional isomers of methyl linoleate hydroperoxide: Evidence of carbon-oxygen bond scission. *J. Am. Oil Chem. Soc.*, 53, 572-576, 1976.

146. Walling, C. and Thaler, W., Positive halogen compounds: III. Allyl chlorination with *t*-butyl hypochlorite. The stereochemistry af allylic radicals, *J. Am. Chem. Soc.*, 83, 3877–3884, 1961.

146. Schauenstein, E., Autoxidation of polyunsaturates esters in water: chemical structure and biological activity of the products, *J. Lipid Res.*, 8, 417–428, 1967.

147. Walling, C., Wagner, P.J., Effects of solvents on transition states in the reactions of *t*-butoxy radicals, *J. Am. Chem. Soc.*, 85, 2333–2334, 1963.

148. Walling, C. and Padwa, A., Positive halogen compounds: VI. Effects of structure and medium on the β-scission of alkoxyl radicals, *J. Am. Chem. Soc.* 85, 1593–1597, 1963.

149. Avila, D.V., Brown, C.E., Ingold, K.U., and Lusztyk, J., Solvent effects on the competitive β-scission and hydrogen atom abstraction reactions of the cumyloxyl radical. Resolution of a long-standing problem, *J. Am. Chem. Soc.*, 115, 466–470, 1993.

150. Tsentalovich, Y.P., Kulik, L.V., Gritsan, N.P., and Yurkovskaya, A.V., Solvent effect on the rate of β-scission of the *tert*-butoxyl radical, *J. Phys. Chem.*, 102, 7975–7980, 1998.

151. Bateman, L., Hughes, H., and Morris, A.L., Hydroperoxide decomposition in relation to the initiation of radical chain reactions, *Discuss. Faraday Soc.*, 14, 190–199, 1953.

152. Chan, H.W.S., Prescott, F.A.A., and Swoboda, P.A.T., Thermal decomposition of individual positional isomers of methyl linoleate hydroperoxide: evidence of carbon–oxygen bond scission, *J. Am. Oil Chem. Soc.*, 53, 572–576, 1976.

153. Labuza, T.P., Kinetics of lipid oxidation in foods, *CRC Crit. Rev. Food Sci. Nutr.*, 2, 355–405, 1971.

154. Sliwiok, J., Kowalska, T., Kowalski, W., and Biernat, A., The influence of hydrogen-bond association on the destruction of hydroperoxides in the autoxidation process of oleyl alcohol, oleic acid, and methyl oleate, *Microchem. J.*, 19, 362–372, 1974.

155. Russell, G.A., Fundamental processes of autoxidation, *J. Chem. Ed.*, 36, 111–118, 1959.

156. Frankel, E.N., Neff, W.E., and Selke, E., Analysis of autoxidized fats by gas chromatography-mass spectrometry: IX. Homolytic vs heterolytic cleavage of primary and secondary oxidation products, *Lipids*, 19, 790–800, 1984.

157. Frankel, E.N., Secondary products of lipid oxidation, *Chem. Phys. Lipids*, 44, 73–85, 1987.

158. Grosch, W., Reactions of hydroperoxides: products of low molecular weight, in *Autoxidation of Unsaturated Lipids*, Chan, H.W.-S., Ed., Academic Press, London, 1987, pp. 95–139.

159. Scheiberle, P., Tsoukalas, B., and Grosch, W., Decomposition of linoleic acid hydroperoxides by radicals, *Z. Lebensm. Unters. Forsch.*, 168, 448–456, 1979.

160. Scheiberle, P., Grosch, W., Kexel, H., and Schmidt, H.-L., *Biochim. Biophys. Acta*, 666, 322–326, 1981.

161. Howard, J.A. and Ingold, K.U., Absolute rate constants for hydrocarbon oxidation XI. The reactions of tertiary peroxy radicals, *Can. J. Chem.*, 46, 2655–2660, 1968.

162. Howard, J.A., Self-reactions of alkylperoxy radicals in solution, in *Organic Free Radicals*, Gould, R.F., Ed., American Chemical Society, Washington, DC, 1978, pp. 413–432.

163. Howard, J.A., Schwalm, W.J., and Ingold, K.U., Absolute rate constants for hydrocarbon autoxidation: VII. Reactivities of peroxy radicals toward hydrocarbons and hydroperoxides., *Adv. Chem. Series*, 75, 6–23, 1968.

164. Kamiya, Y., Beaton, S., Lafortune, A., and Ingold, K.U., The metal-catalyzed autoxidation of tetralin: I. Introduction. The cobalt-catalyzed autoxidation in acetic acid, *Can. J. Chem.*, 41, 2020–2032, 1963.

165. Lundberg, W.O. and Chipault, J.R., The oxidation of methyl linoleate at various temperatures, *J. Am. Chem. Soc.,* 69, 833–836, 1947.

166. Schieberle, P. and Grosch, W., Decomposition of linoleic acid hydroperoxides. II. Breakdown of methyl 13-hydroperoxy-cis-9-trans- 11-octadecadeinoate by radicals or copper II ions, *Lebensm. Unters Forsch,* 173, 192–198, 1981.

168. Schieberle, P. and Grosch, W., Detection of monohydroperoxides with unconjugated diene systems as minor products of the autoxidation of methyl linoleate, *Z. Lebensm. Unters Forsch,* 173, 199–203, 1981.

169. Chan, H.W.-S., Levett, G., and Matthew, J.A., Thermal isomerization of methyl linoleate hydroperoxides. Evidence of molecular oxygen as a leaving group in a radical rearrangement, *J. Chem Soc. Chem. Commun.,* 756–758, 1978.

170. Baldwin, J.E., Rules for ring closure, *J. Chem Soc. Chem. Commun.,* 734–736, 1975.

171. Frankel, E.N., Lipid oxidation: mechanisms, products, and biological significance, *J Am. Oil Chem. Soc.,* 61, 1908–1917, 1984.

172. Shamberger, R.J., Shamberger, B.A., and Willis, C.E., Malonaldehyde content of food, *J. Nutr.,* 107, 1404–1409,1977.

173. Frankel, E.N. and Neff, W.E., Formation of malonaldehyde from lipid oxidation products, *Biochim. Biophys. Acta,* 754, 264–270, 1983.

174. Terao, J., Ogawa, T., and Matsushita, S., Degradation process of autoxidized methyl linoleate, *Agric. Biol. Chem.,* 39, 397–402, 1975.

175. Bothe, E., Schuchmann, M.N., Schulte-Frohlinde, D., and von Sonntag, C., HO$_2$$^\bullet$ elimination from α-hydroxyalkylperoxyl radicals in aqueous solution, *Photochem. Photobiol.,* 28, 639–644, 1978.

176. Gardner, H.W. and Weisleder, D., Addition of N-acetylcysteine to linoleic acid hydroperoxide, *Lipids,* 11, 127–134, 1976.

177. Gardner, H.W., Kleiman, R., Weisleder, D., and Inglett, G.E., Cysteine adds to liquid hydroperoxide, *Lipids,* 12, 655–660, 1977.

178. El-Magoli, S.B., Karel, M., and Yong, S., Acceleration of lipid oxidation by volatile products of hydroperoxide decomposition, *J. Food Biochem.,* 3, 111–123, 1980.

179. Yong, S.H. and Karel, M., Reaction of histidine with methyl linoleate: characterization of the histidine degradation products, *J. Am. Oil Chem. Soc.,* 55, 352–357, 1978.

180. Yong, S.H. and Karel, M., Cleavage of the imidazole ring in histidyl residue analogs reacted with peroxidizing lipids, *J. Food Sci.,* 22, 568–574, 1979.

181. Schaich, K.M. and Karel, M., Free radicals in lysozyme reacted with peroxidizing methyl linoleate, *J. Food Sci.,* 40, 456–459, 1975.

182. Schaich, K.M. and Karel, M., Free radical reactions of peroxidizing lipids with amino acids and proteins: an ESR study, *Lipids,* 11, 392–400, 1976.

183. Yang, M.-H. and Schaich, K.M., Factors affecting DNA damage by lipid hydroperoxides and aldehydes, *J. Free Radic. Biol. Med.,* 20, 225–236, 1996.

184. Schaich, K.M. and Borg, D.C., Radiomimetic effects of peroxidizing lipids on nucleic acids and their bases, in *Oxygen Radicals in Chemistry and Biology,* Bors, W., Saran, M., and Tait, D., Eds., Walter de Gruyter, Berlin, 1984, pp. 603–606.

185. Gardner, H.W., Eskins, K., Grams, G.W., and Inglett, G.R., Radical addition of linoleic hydroperoxides to a-tocopherol or the analogous hydroxychroman, *Lipids,* 7, 324–334, 1972.

186. Gardner, H.W., Effects of hydroperoxides on food components, in *Xenobiotics in Foods and Feeds,* American Chemical Society, Washington, DC, 1983, pp. 63–84.

187. Pryor, W.A., On the detection of lipid hydroperoxides in biological samples, *Free Radic. Biol. Med.,* 7, 177–178, 1989.

188. Borg, D.C. and Schaich, K.M., Cytotoxicity from coupled redox cycling of autoxidizing xenobiotics and metals, *Isr. J. Chem.,* 24, 38–53, 1984.

189. Karel, M., Schaich, K.M., and Roy, R.B., Interaction of peroxidizing methyl linoleate with some proteins and amino acids, *Agric. Food Chem.,* 23, 159–164, 1975.

190. Pokorny, J., Major factors affecting the autoxidation of lipids, in *Autoxidation of Unsaturated Lipids,* Chan, H.W.-S., Ed., Academic Press, London, 1987, pp. 141–206.

191. Eriksson, C.E., Lipid oxidation catalysts and inhibitors in raw materials and processed foods, *Food Chem.,* 9, 3–19, 1982.

192. Eriksson, C.E., Oxidation of lipids in food systems, in *Autoxidation of Unsaturated Lipids,* Chan, H. W. S., Ed., Academic Press, London, 1987, pp. 207–232.

193. Scott, G., *Atmospheric Oxidation and Antioxidants,* Elsevier, London, 1965.
194. Holman, R.T. and Elmer, O.C., The rates of oxidation of unsaturated fatty acids and esters, *J. Am. Oil Chem. Soc.,* 24, 27, 1947.
195. Fatemi, S.H. and Hammond, E.G., Analysis of oleate, linoleate and linolenate hydroperoxides in oxidized ester mixtures, *Lipids,* 15, 379–385, 1980.
196. Sargis, R.M. and Subbaiah, P.V., *Trans* unsaturated fatty acids are less oxidizable than *cis* unsaturated fatty acids and protect endogenous lipids from oxidation in lipoproteins and lipid bilayers, *Biochemistry,* 42, 11533–11543, 2003.
197. Pokorny, J., Rzepa, J., and Janicek, G., Lipid oxidation: 1. Effect of free carboxyl group on the decomposition of lipid hydroperoxide, *Die Nahrung,* 20, 1–6, 1976.
198. Nwosu, C.V., Boyd, L.C., and Sheldon, B., Effect of fatty acid composition of phospholipids on their antioxidant properties and activity index, *J. Am. Oil Chem. Soc.,* 74, 293–297, 1997.
199. Koga, T., Terao, J., Phospholipids increase radical-scavenging activity of vitamin E in a bulk oil model system, *J. Agric. Food Chem.,* 43, 1450–1454, 1995.
200. Bishov, S.J., Henick, A.S., and Koch, R.B., Oxidation of fat in model systems related to dehydrated foods, *Food Res.,* 25, 174–181, 1960.
201. Sugino, H., Ishikawa, M., Nitoda, T., Koketsu, M., Juneja, L.R., Kim, M., and Yamamoto, T., Antioxidative activity of egg yolk phospholipids, *J. Agric. Food Chem.,* 45, 551–554, 1997.
202. Barclay, L.R.C. and Ingold, K.U., Autoxidation of biological membranes: 2. The autoxidation of a model membrane. A comparison of the autoxidation of egg lecithin phosphatidylcholine in water and chlorobenzene, *J. Am. Chem. Soc.,* 103, 6478–6485, 1981.
203. O'Brien, P.J., Intracellular mechanisms for the decomposition of a lipid peroxide: I. Decomposition of a lipid peroxide by metal ions, heme compounds, and nucleophiles, *Can. J. Biochem.,* 47, 485–493, 1969.
204. Corliss, G.A. and Dugan, L.R., Jr., Phospholipid oxidation in emulsions, *Lipids,* 5, 846–853, 1971.
205. Yoshida, K., Terao, J., Suzuki, T., and Takama, K., Inhibitory effect of phosphatidylserine on iron-dependent lipid peroxidation, *Biochem. Biophys. Res. Commun.,* 179, 1077–1081, 1991.
206. Chen, Z.Y. and Nawar, W.W., The role of amino acids in the autoxidation of milk fat, *J. Am. Oil Chem. Soc.,* 68, 47–50, 1991.
207. Chalk, A.J. and Smith, J.F., Catalysis of cyclohexene autoxidation by trace metals in non-polar media: 2. Metal salts in the presence of chelating agents, *Trans. Faraday Soc.,* 53, 1235–1245, 1957.
208. Miyashita, K. and Takagi, T., Study on the oxidative rate and prooxidant activity of free fatty acids, *J. Am. Oil Chem. Soc.,* 63, 1380–1384, 1986.
209. Frega, N.G., Mozzon, M., and Lercker, G., Effects of free fatty acids on oxidative stability of vegetable oil, *J. Am. Oil Chem. Soc.,* 76, 325–329, 1999.
210. Carless, J.E. and Nixon, J.R., The oxidation of solubilized and emulsified oils: III. The oxidation of methyl linoleate in potassium laurate and cetomacrogol dispersions, *J. Pharm. Pharmacol.,* 12, 348–359, 1960.
211. Yanishlieva, N.V. and Marinova, E.M., Effects of antioxidants on the stability of triacylglycerols and methyl esters of fatty acids of sunflower oil, *Food Chem.,* 54, 377–382, 1995.
212. Raghuveer, K.G. and Hammond, E.G., The influence of glyceride structure on the rate of autoxidation, *J. Am. Oil Chem. Soc.,* 44, 239–243, 1967.
213. Koch, R.B., Dehydrated foods and model systems, in *Symposium on Foods: Lipids and Their Oxidation,* Schultz, H., Day, E.A., and Sinnhuber, R.O., Eds., AVI Publishing, Westport, CT, 1962, pp. 230–249.
214. Mikula, M. and Khayat, A., Reaction conditions for measuring oxidative stability of oils by thermo-gravimetric analysis, *J. Am. Oil Chem. Soc.,* 62, 1694–1698, 1985.
215. Slawson, V., Adamson, A.W., and Mead, J.F., Autoxidation of polyunsaturated fatty esters on silica, *Lipids,* 8, 129–134, 1973.
216. Leermakers, P.A., Thomas, H.T., Weis, L.D., and James, P.C., Spectra and photochemistry of molecules adsorbed on silica gel. IV, *J. Am. Chem. Soc.,* 88, 5075–5083, 1966.
217. Porter, W.L., Levasseur, L.A., Jeffers, J.I., and Henick, A.S., UV spectrophotometry of autoxidized lipid monolayers while on silica gel, *Lipids,* 6, 16–25, 1971.
218. Hyde, S.M., Oxidation of methyl oleate induced by cobalt-60 γ-radiation: II. Emulsions of methyl oleate in water, *Trans. Faraday Soc.,* 64, 155–162, 1968.
219. Treibs, W., Autoxidation of oxygen-active acids, XI. The O_4 stage of the autoxidation of linoleic and linolenic acid esters, *Chem Ber.,* 81, 472–477, 1948.

220. Orlien, V., Andersen, A.B., Sinkho, T., and Skibsted, L.H., Hydroperoxide formation in rapeseed oil encapsulated in a glassy food model as influenced by hydrophilic and lipophilic radicals, *Food Chem.*, 68, 191–199, 2000.

221. Schaich, K.M., Fenton reactions in lipid phases, *Lipids*, 23, 570–578, 1988.

222. Sastry, Y.S.R. and Lakshminarayana, G., Chlorophyll-sensitized peroxidation of saturated fatty acid esters, *J. Am. Oil Chem. Soc.*, 48, 452–454, 1971.

223. Rawls, H.R., Van Santen, P.J., A possible role for singlet oxygen in the initiation of fatty acid autoxidation, *J. Am. Oil Chem. Soc.*, 47, 121–125, 1970.

224. Schaich, K.M., Free radical initiation in proteins and amino acids by ionizing and ultraviolet radiations and lipid oxidation: II. Ultraviolet radiation and photolysis, *CRC Crit. Rev. Food Sci. Nutr.*, 13, 131–159, 1980.

225. Terao, J. and Matsushita, S., Structures of monohydroperoxides produced from chlorophyll sensitized photooxidation of methyl linoleate, *Agric. Biol. Chem.*, 41, 2467–2468, 1977.

226. Foote, C.S., Photosensitized oxidation and singlet oxygen: consequences in biological systems, in *Free Radicals in Biology*, Pryor, W.A., Ed., Academic Press, New York, 1976, pp. 85–134.

227. Murray, R.W., Chemical sources of $^1O_2$, in *Singlet Oxygen*, Wasserman, H. and Murray, R.W., Eds., Academic Press, New York, 1979, chap. 3.

228. Pokorny, J., Kundu, M.K., Pokorny, S., Bleha, M., and Coupek, J., Lipid oxidation: 4. Products of thermooxidative polymerization of vegetable oils, *Die Nahrung*, 20, 157–163, 1976.

229. Terao, J. and Matsushita, S., Geometrical isomers of monohydroperoxides formed by autoxidation of methyl linoleate, *Agric. Biol. Chem.*, 41, 2401–2405, 1977.

230. Neff, W.E., Frankel, E.N., and Weisleder, D., Photosensitized oxidation of methyl linolenate. Secondary products, *Lipids*, 17, 780–790, 1982.

231. Kearns, D.R., Hollins, R.A., Khan, A.U., and Radlick, P., Evidence for the participation of $^1\mathrm{Sigma_g}^+$ and $^1\mathrm{delta_g}$ oxygen in dye-sensitized photooxygenation reactions: II, *J. Am. Chem. Soc.*, 89, 5456–5457, 1967.

232. Foote, C.S., Photosensitized oxygenation and the role of singlet oxygen, *Acc. Chem. Res.*, 1, 104–110, 1968.

233. Foote, C.S., Denny, R.W., Weaver, L., Chang, Y., and Peters, J., Quenching of singlet oxygen, *Ann. N.Y. Acad. Sci.*, 171, 139–148, 1970.

234. Neff, W.E. and Frankel, E.N., Photosensitized oxidation of methyl linolenate monohydroperoxides: hydroperoxy cyclic peroxides, dihydroperoxides and hydroperoxy bis-cyclic peroxides, *Lipids*, 19, 952–957, 1984.

235. Frankel, E.N., Neff, W.E., Selke, E., and Weisleder, D., Photosensitized oxidation of methyl linoleate: secondary and volatile thermal decomposition products, *Lipids*, 17, 11–18, 1982.

236. Umehara, T., Terao, J., and Matsushita, S., Photosensitized oxidation of oils with food colors, *J. Agric. Chem. Soc. Jpn.*, 53, 51–56, 1979.

237. De Groot, H. and Noll, T., The role of physiological oxygen partial pressures in lipid peroxidation. Theoretical considerations and experimental evidence, *Chem. Phys. Lipids*, 44, 209–226, 1987.

238. Shahidi, F., Oxidative stability of edible oils as affected by their fatty acid composition and minor constituents, in *Freshness and Shelf Life of Foods*, American Chemical Society, Washington, DC, 2003, pp. 201–211.

239. Naz, S., Sheikh, H., Siddiqi, R., and Sayeed, S.A., Oxidative stability of olive, corn and soybean oil under different conditions, *Food Chem.*, 88, 253–259, 2004.

240. Inchbald, G., Fat content and fatty acid composition of seeds and seed oils, http://www.queenhill.demon.co.uk/seedoils/oilcomp.htm, 2000.

241. Typical composition of fats and oils, http://www.uniqema.com/chem/lit/pi9.pdf, 2001.

242. Fatty acid composition of some plant oils, http:/www.cyberlipid.org/glycer/glyc0064.htm#top, 2004.

243. Holser, R.A., Seed conditioning and meadowfoam press oil quality, *Indust. Crops Prod.*, 17, 23–26, 2003.

244. Fatty acid composition of some common edible fats and oils, http://www.scientificpsychic.com/fitness/fattyacids.html, 2004.

245. Savage, G.P., McNeil, D.L., and Dutta, P.C., Lipid composition and oxidative stability of oils in hazelnuts *(Corylus avellana* L.) grown in New Zealand, *J. Am. Oil Chem. Soc.*, 74, 755–759, 1997.

246. Matthäus, B.W., Determination of the oxidative stability of vegetable oils by rancimat and conductivity and chemiluminescence measurements, *J. Am. Oil Chem. Soc.*, 73, 1039–1043, 1996.

247. O'Keefe, S.F., Wiley, V.A., and Knauft, D.A., Comparison of oxidative stability of high- and normal oleic peanut oils, *J. Am. Oil Chem. Soc.,* 70, 489–492, 1993.

248. Zambiazi, R.C., The role of endogenous lipid components on vegetable oil stability, PhD Thesis, 1997, Univ. of Manitoba, Winnepeg, Manitoba, Canada.

249. Coni, E., Podesta, E., and Catone, T., Oxidizability of different vegetable oils evaluated by thermogravimetric analysis, *Thermochim. Acta,* 418, 11–15, 2004.

250. Knothe, G. and Dunn, R.O., Dependence of oil stability index of fatty compounds on their structure and concentration and presence of metals, *J. Am. Oil Chem. Soc.,* 80, 1021–1026, 2003.

251. Cabrini, L., Barzanti, V., Cipollone, M., Fiorentini, D., Grossi, G., Tolomelli, B., Zambonin, L., and Landi, L., Antioxidants and total peroxyl radical-trapping ability of olive and seed oils, *J. Agric. Food Chem.,* 49, 6026–6032, 2001.

252. Bonvehi, J.S., Coll, F.V., Oil content, stability, and fatty acid composition of the main varieties of Catalonian hazelnuts *(Orylus avellana L.),* *Food Chem.,* 48, 237–241, 1993.

253. Amaral, J.S., Casal, S., Pereira, J.A., Seabra, R.M., and Oliviera, B.P.P., Determination of sterol and fatty acid compositions, oxidative stability, and nutritional value of six walnut (*Juglans regia* L.) cultivars grown in Portugal, *J. Agric. Food Chem.,* 51, 7698–7702, 2003.

254. Savage, G.P., McNeil, D.L., and Dutta, P.C., Vitamin E content and oxidative stability of fatty acids in walnut oils, *Proc. Nutrit. Soc. New Zealand,* 23, 64–69, 1998.

255. Savage, G.P., Dutta, P.C., and McNeil, D.L., Fatty acid and tocopherol contents and oxidative stability of walnut oils, *J. Am. Oil Chem. Soc.,* 76, 1059–1063, 1999.

256. Savage, G.P. and McNeil, D.L., Oxidative stability of walnuts during long term in shelf storage, *Acta Hortic.,* 544, 591–597, 2001.

257. Fan, J., Chen, C., and Huang, Y., Effect of degumming treatment on oxidative stability of aqueous extracted walnut oil, *Zhongguo Youzhi,* 25, 49–51, 2000.

258. Demir, C. and Cetin, M., Determination of tocopherols, fatty acids, and oxidative stability of pecan, walnut, and sunflower oils, *Deutsche Lebensmittel-Rundschau,* 95, 278–282, 1999.

259. Crowe, T.D., Crowe, T.W., Johnson, L.A., and White, P.J., Impact of extraction method on yield of lipid oxidation products from oxidized and unoxidized walnuts, *J. Am. Oil Chem. Soc.* 79, 453–456, 2002.

260. Calvo, L., Cocero, M.J., and Diez, J.M., Oxidative stability of sunflower oil extracted with supercritical carbon dioxide, *J. Am. Oil Chem. Soc.,* 71, 1251–1254, 1994.

261. Crowe, T.D. and White, P.J., Oxidative stability of walnut oils extracted with supercritical carbon dioxide, *J. Am. Oil Chem. Soc.,* 80, 575–578, 2003.

262. Oliveira, R., Rodrigues, M.F., and Bernardo-Gil, M.G., Characterization and supercritical carbon dioxide extraction of walnut oil, *J. Am. Oil Chem. Soc.,* 79, 225–230, 2002.

263. Zhao, S., Li, T., Cai, S., Wei, D., Yu, X., and Chen, Z., Study on autoxidation and antioxidation of walnut oil, *Shipin Gongye Keji,* 22, 27–29, 2001.

264. Prabhakar, J.V. and Amla, B.L., Influence of water activity on the formation of monocarbonyl compounds in oxidizing walnut oil, *J. Food Sci.,* 43, 1839–1843, 1978.

265. Toro-Vasquez, J.F., Charó-Alonso, M.A., and Pérez-Briceño, F., Fatty acid composition and its relationship with physicochemical properties of pecan *(Carya illinoensis)* oil, *J. Am. Oil Chem. Soc.,* 76, 957–965, 1999.

266. Rikhter, A.A., Study of oxidation of almond and walnut oils, *Prikladnaya Biokhimiya i Mikrobiologiya,* 16, 603–608, 1980.

267. Khan, M.A. and Shahidi, F., Effects of natural and synthetic antioxidants on the oxidative stability of borage and evening primrose triacylglycerols, *Food Chem.,* 75, 431–437, 2001.

268. Wettasinghe, M., and Shahidi, F., Scavenging of reactive-oxygen species and DPPH free radicals by extracts of borage and evening primrose meals, *Food Chem.,* 70, 17–26, 2000.

269. Wettasinghe, M., and Shahidi, F., Iron(II) chelation activity of extracts of borage and evening primrose meals, *Food Res. Int.,* 35, 65–71, 2002.

270. Madhavi, D.L., Singhal, R.S., and Kulkarani, P.R., Technological aspects of food antioxidants, in *Food Antioxidants: Technological, Toxicological, and Health Perspectives,* Madhavi, D.L., Dephapande, S.S., and Salunke, D.K.M., Eds., Marcel Dekker, New York, 1995, pp. 159–265.

271. Khan, M.A. and Shahidi, F., Photooxidative stability of stripped and non-stripped borage and evening primrose oils and their emulsions in water, *Food Chem.,* 79, 47–53, 2002.

272. Gomez, A.M. and de la Ossa, E. M., Quality of borage seed oil extracted by liquid and supercritical carbon dioxide, *Chem. Eng. J.,* 88, 103–109, 2002.

273. Senanayake, S.P.J.N. and Shahidi, F., Oxidative stability of structured lipids produced from Borage (Borago officinalis L.) and evening primrose (Oenothera biennis L.) oils with docosahexaenoic acid, *J. Am. Oil Chem. Soc.,* 79, 1003–1013, 2002.

274. Fukuda, Y. and Namiki, M., Recent studies on sesame seed oil, *J. Jpn. Soc. Food Sci. Technol.,* 35, 552–562, 1988.

275. Kamal-Eldin, A. and Appelqvist, L.-A., The effects of extraction methods on sesame oil stability, *J. Am. Oil Chem. Soc.,* 72, 967–969, 1995.

276. Abou-Gharbia, H.A., Shahidi, F., Shehata, A.A.Y., and Youssef, M.M., Oxidative stability of extracted sesame oil from raw and processed seeds, *J. Food Lipids,* 3, 59–72, 1996.

277. Abou-Gharbia, H.A., Shahidi, F., Shehata, A.A.Y., and Youssef, M.M., Effects of processing on oxidative stability of sesame oil extracted from intact and dehulled seed, *J. Am. Oil Chem. Soc.,* 74, 215–221, 1997.

278. Rocha Uribe, A. and Almendarez Camarillo, A., Effect of the level of dehulling and adsorption on the oxidative stability of sesame oil, *Avances en Ingenieria Quimica,* 7, 40–43, 1997.

279. Yen, G.C., Influence of seed roasting process on the changes in composition and quality of sesame *(Sesame indicum)* oil, *J. Sci. Food Agric.,* 50, 563–570, 1990.

280. Yoshida, H. and Kajimoto, G., Microwave heating affects composition and oxidative stability of sesame (Sesamum indicum) oil, *J. Food Sci.,* 59, 613–616, 625, 1994.

281. Yoshida, H. Composition and quality characteristics of sesame seed *(Sesamum indicum)* oil roasted at different temperatures in an electric oven, *J. Sci. Food Agric.,* 65, 331–336, 1994.

282. Han, J. and Ahn, S.Y., Effect of oil refining process on oil characteristics and oxidation stability of sesame oil, *Han'guk Nonghwa Hakhoechi,* 36, 284–289, 1993.

283. Choe, E. and Moon, S., Effects of filtration or centrifugation on the oxidative stabilities of sesame oil, *Han'guk Nonghwa Hakhoechi,* 37, 168–174, 1994.

284. Millwalla, R.H.H. and Subrahmanyam, V.V.R., Thermal stability of blended oils having sesame oil, *J. Oil Technol. Assoc. India,* 18, 87–89, 1986.

285. Guillén, M.D. and Cabo, N., Fourier transform infrared spectra data versus peroxide and anisidine values to determine oxidative stability of edible oils, *Food Chem.,* 77, 503–510, 2002.

286. Marquez-Ruiz, G., Garces, R., Leon-Camacho, M. and Mancha, M., Thermoxidative stability of triacylglycerols from mutant sunflower seeds, *J. Am. Oil Chem. Soc.,* 76, 1169–1174, 1999.

287. Hras, A.R., Hadolin, M., Knez, Z., and Bauman, D., Comparison of antioxidative and synergistic effects of rosemary extract with *a*-tocopherol, ascorbyl palmitate and citric acid in sunflower oil, *Food Chem.,* 71, 229–233, 2000.

288. Crapiste, G.H., Brevedan, M.I.V., and Carelli, A.A., Oxidation of sunflower oil during storage, *J. Am. Oil Chem. Soc.,* 76, 1437–1443, 1999.

289. Kaya, A., Tekin, A.R., and Oner, M.D., Oxidative stability of sunflower and olive oils: comparison between a modified active oxygen method and long term storage, *Lebensm.-Wiss. u.-Technol.,* 26, 464–468, 1993.

290. Tawfik, M.S. and Huyghebaert, A., Interaction of packaging materials and vegetable oils: oil stability, *Food Chem.,* 64, 451–459, 1999.

291. Oomah, B.D., Liang, J., Godfrey, D., and Mazza, G., Microwave heating of grapeseed: effect on oil quality, *J. Agric. Food Chem.,* 46, 4017–4021, 1998.

292. Koski, A., Pekkarunin, S., Hopia, A., Wahala, K., and Heinonen, M., Processing of rapeseed oil: effects on sinapic acid derivative content and oxidative stability, *Eur. Food Res. Technol.* 217, 110–114, 2003.

293. Velasco, J., Andersen, M.L., and Skibsted, L.H., Evaluation of oxidative stability of vegetable oils by monitoring the tendency to radical formation. A comparison of electron spin resonance spectroscopy with the Rancimat method and differential scanning calorimetry, *Food Chem.,* 85, 623–632, 2004.

294. Przybylski, R. and Zambiazi, R., Storage stability of genetically modified canola oils, Conference Proceedings: New horizons for an old crop, Canberra Australia, The Regional Institute Ltd., http://www.regional.org.au/au/gcirc/1/347.htm, 1999.

295. Isnardy, B., Wagner, K.-H., and Elmadea, I., Effects of *a, γ,* and *d*-tocopherols on the autoxidation of purified rapeseed oil triacylglycerols in a system containing low oxygen, *J. Agric. Food Chem.,* 51, 7775–7780, 2003.

296. Kowalski, B., Thermal-oxidative decomposition of edible oils and fats. DSC studies, *Thermochim. Acta,* 184, 49–57, 1991.

297. Cercaci, L., Rodriguez-Estrada, M.T., and Lercker, G., Solid-phase extraction-thin-layer chromatography-gas chromatography method for the detection of hazelnut oil in olive oils by determination of esterified sterols, *J. Chromatogr. A.,* 985, 211–220, 2003.

298. Zabaras, D. and Gordon, M.H., Detection of pressed hazelnut oil in virgin olive oils by analysis of polar components: improvement and validation of the method, *Food Chem.,* 84, 475–483, 2004.

299. Alasalvar, C., Shahidi, F., Liyanapathirana, C.M., and Ohshima, T., Turkish Tombul hazelnut *(Corylus avellana L.):* 1. Compositional characteristics, *J. Agric. Food Chem.,* 51, 3790–3796, 2003.

300. Alasalvar, C., Shahidi, F., Ohshima, T., Wanasundara, U., Yurttas, H.C., Liyanapathirana, C.M., and Rodrigues, F.B., Turkish Tombul hazelnut *(Corylus avellana L.):* 2. Lipid characteristics and oxidative stability, *J. Agric. Food Chem.,* 51, 3797–3805, 2003.

301. Yurttas, H.C., Schafer, H.W., and Wartheson, J.J., Antioxidant activity of nontocopherol hazelnut *(Corylus spp.)* phenolics, *J. Food Sci.,* 65, 276–280, 2000.

302. Pershern, A.S., Breene, W.M., and Lulai, E.C., Analysis of factors influencing lipid oxidation in hazelnuts *(Corylus sp.), J. Food Process. Preserv.,* 19, 9–25, 1995.

303. Özdemir, M., Açkurt, F., Kaplan, M.L., Yildiz, M., Löker, M., Gürcan, T., Biringen, G., Okay, A., and Seyhan, F.G., Evaluation of new Turkish hybrid hazelnut *(Corylus avellana L.)* varieties: fatty acid composition, *a*-tocopherol content, mineral composition and stability, *Food Chem.,* 73, 411–415, 2001.

304. Tsimidou, M., Papadopooulos, G., and Boskou, D., Phenolic compounds and stability of virgin olive oils: I, *Food Chem.,* 45, 141–144, 1992.

305. Salvador, M.D., Aranda, F., Gomez-Alonso, S., and Fregapane, G., Cornicabra virgin olive oil: a study of five crop seasons. Composition, quality and oxidative stability, *Food Chem.,* 74, 267–274, 2001.

306. Conquanta, I., Esti, M., and Di Matteo, M., Oxidative stability of virgin olive oils, *J. Am. Oil Chem. Soc.,* 78, 1197–1202, 2001.

307. Evangelisti, F., Zunin, P., Tiscornia, E., Petacchi, R., Drava, G., and Lanteri, S., Stability to oxidation of virgin olive oils as related to olive conditions: study of polar compounds by chemometric methods, *J. Am. Oil Chem. Soc.,* 74, 1017–1023, 1997.

308. Ranalli, A., Cabras, P., Iannucci, E., and Contento, S., Lipochromes, vitamins, aromas and other components of virgin olive oil are affected by processing technology, *Food Chem.,* 73, 445–441, 2001.

309. Khan, M.A. and Shahidi, F., Rapid oxidation of commercial extra virgin olive oil stored under fluorescent light, *J. Food Lipids,* 6, 331–339, 1999.

310. Blekas, G., Tsimidou, M., and Boskou, D., Contribution of *a*-tocopherol to olive oil stability, *Food Chem.,* 52, 289–294, 1995.

311. Antoun, N. and Tsimidou, M., Gourmet olive oil: stability and consumer acceptability studies, *Food Res. Int.,* 30, 131–136, 1997.

312. Meadowfoam seed oil, http://www.meadowfoam.com/technical3.htm, 2004.

313. Isbell, T.A., Oxidative stability index of vegetable oils in binary mixtures with meadowfoam oil, http://www.nal.usda.gov/ttic/tektran/data/000008/42/0000084244.html, 2004.

314. Isbell, T.A., Abbott, T.P., and Carlson, K.D., Oxidative stability index of vegetable oils in binary mixtures with meadowfoam oil, *Ind. Crops Prod.,* 9, 115–123, 1999.

315. Vaughn, S.F., Boydston, R.A., and Mallory-Smith, C.A., Isolation and identification of (3-methoxyphenyl) acetonitrile as a phytotoxin from meadowfoam *(Limnanthes alba)* seedmeal, *J. Chem. Ecol.,* 22, 1939–1949, 1996.

316. Shahidi, F., Daun, J.K., and De Clerq, D.R., Glucosinolates in *Brassica* oilseed: processing effects and extraction, in *Antinutrients and Phytochemicals in Food,* American Chemical Society, Washington, DC, 1997, pp. 152–170.

317. Holser, R.A. and Isbell, T.A., Effect of processing conditions on the oxidative stability of meadowfoam press oil, *J. Am. Oil Chem. Soc.,* 79, 1051–1052, 2002.

318. Lindsay, D. A., Howard, J. A., Horswill, E. C., Iton, L., Ingold, K. U., Cobbley, T., and Li, A., The bimolecular self-reaction of secondary peroxy radicals: product studies, *Can. J. Chem.,* 51, 870-880, 1973.

319. Traylor, T. G. and Russell, C. A., Mechanism of autoxidations: Terminating radicals in cumene autoxidations, *J. Am. Chem. Soc,.* 87, 3698-3706, 1965.

320. Walling, C., Waits, H. P., Milanovic, J., and Pappiaonnou, C. G., Polar and radical paths in the decomposition of diacyl peroxides, *J. Am. Chem. Soc.,* 92, 4927-4932, 1970.

321. Heijman, M. G. J., Nauta, H., and Levine, Y. K., A pulse radiolysis study of the dienyl radical in oxygen-free linoleate solutions: time and linoleate concentration dependence, *Radiat. Phys. Chem.,* 26, 73-82, 1985.

25 *Trans* Fatty Acids in Specialty Lipids

G.R. List, R.O. Adlof, and J.W. King
Food and Industrial Oil Research, National Center for Agricultural
Utilization Research, ARS, USDA, Peoria, Illinois

CONTENTS

25.1   INTRODUCTION

An edible oil quality triangle has been described as having oxidative stability, functionality, and nutrition as the three critical parameters[1]. The ideal fat or oil should have excellent oxidative stability at both high and ambient temperatures, and have enough solid fat for margarines, shortenings,

---

Names are necessary to report factually available data: the USDA neither guarantees nor warrants the standard of the product, and the use of the name USDA implies no approval of the product to the exclusion of others that may also be suitable.

and frying fats. It should also be both low in saturated fats and high in polyunsaturated acids, thus satisfying all of the requirements of the edible oil triangle. However, no single fat or oil fully satisfies all of these requirements.

Worldwide consumption of edible fats and oils amounted to just over 92 million metric tons for the year 2001–2002 of which 59% can be accounted for by soybean (29 million tons) and palm (25.4 million metric tons) oils[2]. While soybean oil is relatively low in saturated acids (15%), it represents a major source of *trans* fatty acids (TFA) in our food supply. Soybean oil must be hydrogenated in order to achieve functional properties and oxidative stability for use in salad/cooking, shortenings, and margarine/spreads. Palm oil, while containing no *trans* acids, contains about 50% saturated acids making it also attractive in the aforementioned products particularly when modified by interesterification and fractionation. Thus, just two fats and oils dominate and dictate oil processing worldwide and any discussion of strategies to reduce *trans* and saturated acids in the food supply must focus on soybean and palm oils.

In order to improve the quality triangle for functionality and oxidative stability, various fat modification techniques, including hydrogenation, interesterification, fractionation and combinations thereof, are employed by the industry. A decade ago it was estimated that about 33% of the edible oils produced worldwide were hydrogenated, while 10% were processed by interesterification/fractionation[3]. While more recent data are lacking, it is anticipated that the latter techniques will see increased usage in the future.

Over the past several decades, a number of oilseeds have been introduced with modified fatty acid composition. These include canola and soybean oils with low linolenic acid content; corn, soybean, and sunflower oils with high oleic acid content; and soybean oils with low and high saturated fatty acid content[1,4–6]. Many of the above oils show promise in reducing both *trans* and/or saturated acids in food oils. Improvements in oil processing technology also provide options for *trans*/saturated acid reduction. Traditional random interesterification, in which fatty acids are redistributed across the glycerol backbone, is usually accomplished with a chemical catalyst, and is thus nonspecific. Numerous advances in interesterification have been made by use of enzymes that are stereospecific, allowing production of tailor-made fats[7]. Indeed, shortening and confectionery fats are now produced commercially by this technology. The time-honored hydrogenation process offers potential for reducing *trans* acids in food oils[8–10]. This chapter reviews nutrition and aspects of *trans*/saturated fats, briefly overviews source oils, covers processing reformulation strategies to reduce TFA, and surveys *trans* acid content of oils worldwide.

25.2   NUTRITION LABELING/HEALTH

In 1990 the U.S. Congress passed the Nutrition Labeling and Education Act (NELA) and by 1994 the guidelines had been published. Of particular interest to the fats and oils industry are statements concerning fat. Total fat must be in bold print and listed in grams. Saturated fat must be listed in grams, but is not required if the food contains less than 0.5 grams of total fat per serving and if no claims are made about fat or cholesterol content. If not required and not declared, the statement, "Not a significant source of saturated fat" must be included at the bottom of the nutritional label. "Polyunsaturated fat" or "Polyunsaturated" in grams is voluntary, unless "Monounsaturated fat" or "Monounsaturated" is declared, or a cholesterol or saturated fat claim is made and the total fat is declared is greater than zero. "Cholesterol" must be in bold print and listed in milligrams but is not required if the product contains less than 2 milligrams of cholesterol per serving and makes no claim about fat, saturated fat, or cholesterol, and if not declared, the statement, "Not a significant source of cholesterol" must be included at the bottom of the nutrient label.

Shortly thereafter, the Center for Science in the Public Interest (CSPI) petitioned the Food and Drug Administration (FDA) to include TFA as part of the nutrition labeling for saturated acid content. The Palm Oil Institute of Malaysia followed suit. During the period 1994–1999, 14 trade associations and 4 food science nutrition groups opposed such labeling, citing little evidence to suggest that current levels of *trans* acids consumption are harmful. The coalition included the American Dietetic Association, the Society for Clinical Nutrition, and the Institute of Food Technologists[11]. Indeed, such opposition was based upon sound science reached by a distinguished panel of scientists. Published in the *Journal of Clinical Nutrition*[12], the panel concluded, in part, "Data supporting a link between *trans* fatty acid intake and coronary heart disease (CHD) are equivocal compared with extensive data from epidemiologic observation and animal and human studies that support a direct effect of saturated fat intake on CHD risk." In addition, they concluded that, "Additional research is needed to adequately determine whether *trans* fatty acids independently affect plasma lipoprotein cholesterol concentrations. Studies are needed to explore mechanisms and dose-response issues in experimental animals that have lipoprotein responses similar to those in humans, such as hamsters, non-human primates and pigs. Human metabolic studies are needed to better assess effects of *trans* fatty acids on serum lipid concentrations and CHD risk."

Another task force concluded, "Compared with saturated acids, the issue of TFA is less significant because U.S. diets provide a smaller proportion of TFA and the data on their biological effects are limited. The debate about TFA should not detract from the body of scientific evidence linking the intake of saturated acids to cardiovascular risk, thereby providing the basis for dietary recommendations limiting the intake of saturated fat. It is perhaps premature to make new dietary recommendations for the population at large or to change nutrition policy (such as nutrition labeling) to mandate that TFA be listed separately or be included with saturated fatty acids on nutrition labels, especially in view of the inadequate data base for making or implementing such a change"[13]. However, certain countries, such as Canada, have made it mandatory to declare *trans* fat content on the labels.

Epidemiological data were published in 1997[14] in which over 80,000 women aged 34 to 59 with no known coronary disease, stroke, cancer, high cholesterol levels, or diabetes were monitored for 14 years beginning in 1980. It was observed that 939 cases of nonfatal myocardial infarction or death resulted from coronary heart disease. Multivariate analyses included age, smoking status, total energy intake, dietary cholesterol intake, percentage of energy obtained from protein and specific types of fat, and other risk factors. Total fat intake was not significantly related to the risk of coronary disease, but it was estimated that replacement of 5% of energy from saturated fats with unsaturated fats would reduce the risk by 42% and replacement of 2% of energy from *trans* fat with energy from unhydrogenated, unsaturated fats would reduce risk by 53%. They concluded, "Our findings suggest that replacing saturated and *trans* fats with unhydrogenated, monounsaturated and polyunsaturated fats is more effective in preventing coronary heart disease in women than reducing overall fat intake." Others, however, have called into question the results of the study[15,16]. Inconsistencies in the data suggest the method of assessing either diet or exercise as flawed as well as drawing conclusions from a study where there was much lower risk of heart disease (27 deaths per 100,000 per year) in the group studied, who were all nurses, than the overall risk in the population of women in the United States with rates of death from heart disease among white women 45 to 64 years of age. In addition, it was pointed out that the range of total fat intake recorded in the study is limited and has little relevance to the protective effect of really low-fat diets, but supported the conclusion that composition of dietary fat, rather than level, is of primary importance[16].

In 1999, the FDA issued proposed *trans* acid labeling regulations which were opened for public comment several times. Under this proposal *trans* acids would be included on labels with grams/serving listed as a footnote and a statement to the effect that *trans* acid consumption should be as low as possible[17]. On July 11, 2003 the FDA issued final *trans* labeling requirements. By January 1, 2006, *trans* acids must be listed as a separate line on food labels; however, the statement regarding *trans* acid consumption is not required[18]. Canada has followed suit.

25.2.1 SOURCE OILS: *TRANS* AND SATURATED ACIDS

Nearly 18 billion pounds of soybean oil were consumed domestically broken down as follows: shortening, 8.57 billion pounds; margarine/spreads, 1.24 billion pounds; salad/cooking oil, 7.9 billion

TABLE 25.1
***Trans* and Saturated Acid Consumption in the United States**

Oil/fat	Total consumption (billion lb)[a]	Billion lb *Trans* acids	Saturated acids
Soybean spreads	1.24	0.186	0.38
Soybean shortening	8.57	1.714	1.97
Soybean salad/cooking oil[b]	7.9	0.79	1.185
Corn oil	1.345	—	0.18
Cottonseed oil	0.767	—	0.203
Canola oil	1.793	—	0.129
Coconut oil	1.1	—	1
Palm oil	0.455	—	0.205
Palm kernel oil	0.471	—	0.409
Lard	0.989[c]	0.01	0.392
Tallow	1.47	0.07	0.401

[a] 2001–2002.
[b] May contain *trans* acids; 10%.
[c] 1 to 2% *trans*.
[d] 4 to 5% *trans*.

pounds; other edible, 125 million pounds[19]. The shortening and margarine categories contain hydrogenated components. Based on the assumption in that soft and stick margarine/spreads contain an average of 15% *trans* acids and shortenings contain an average of 20% *trans* acids, about 1.9 billion pounds of *trans* acids are produced by hydrogenation.

During 2000–2001, corn oil consumption amounted to 1.41 billion pounds, canola oil to 1.75 billion pounds, and cottonseed oil to 767 million pounds. No data are available on the amounts of these oils going into shortenings, margarines, and salad cooking oils. Significant amounts of edible food and oils high in saturated acids are consumed in the United States including coconut oil at 1.1 billion pounds, lard at 989 million pounds, palm and palm kernel at 826 million pounds, and edible tallow at 1.47 billion pounds (2.48 billion pounds sat.). Based upon their usage and fatty acid composition about 2.4 billion pounds of saturated acids are also consumed from coconut, palm and palm kernel, lard, and edible tallow. Thus, the U.S. consumptions of *trans* and saturated acids are about equal: 1.9 billion vs. 2.4 billion pounds, respectively. The value for *trans* consumption is low because of lack of data for the amounts of corn, cottonseed, and canola processed by hydrogenation (Table 25.1).

25.2.2 *Trans* Acids Surveys

Reviews of margarine/spread formulations in North America have been authored by Mag[20], Chrysam[21], and Pelloso[22]. A survey of the *trans* content of soft tub margarine/spreads taken over a seven-year period in the United States is shown in Table 25.2[23]. The samples reported are premium products taken from grocery store shelves, and, according to their labels, were formulated from hydrogenated and liquid soybean oil. Over the seven-year period, *trans* reduction ranged from about 25 to 82% showing an average reduction of nearly 56%. The 1992 data indicate an average of nearly 20% *trans*, whereas the 1999 data show less than 9% *trans* in soft tub products. Stick/spreadable products have shown decreased *trans* acid contents over the past decade. In 1989[24] *trans* acids averaged 26.8%, but by 1999 the average had dropped to 16.9% (or a 37% overall reduction)[23].

TABLE 25.2
***Trans* Contents (%) of Soft Margarines/Spreads by Year**

| | | Year | | |
Brand	1992	1995	1999	Overall reduction
1	19.4	15.9	14.5	25.3
2	10.3	18.2	7.9	23.4
3	12.2	7.8	2.6	79.7
4	31.4	16.2	14.6	53.6
5	11.6	19.5	6.1	47.5
6	24.6	14.7	10.5	57.4
7	29.5	20.2	5.3	82.1
Average	19.9	16.1	8.8	55.8

The domestic edible oil industry has thus made a concerted effort to reduce *trans* acids in edible products. This has been accomplished through reformulation methods in which a multiple component basestock system, employing three hydrogenated oils, has been replaced with a single component. Typically, soybean oil hydrogenated to an iodine value of about 65 (40% *trans*) is blended with liquid soybean oil (25 to 50%) to yield components suitable for a wide variety of spreads, including soft tub, spreadable stick, or stick products.

The *trans* acid content of margarines and spreads taken from the literature over a seven-year period (1995–2002) are shown in Table 25.3[23,25–30]. Low-*trans* products are arbitrarily defined as those having 5% or less while zero-*trans* products may have smaller amounts of *trans* acids (1 to 2%). Of the 228 samples reported, 87, or 38.2%, may be considered zero or low *trans*. The largest number of samples reported include those from Canada and the United States, where 10 of the 126 are zero-/low-*trans* type. These results indicate that margarine/spreads formulated in North America are formulated primarily from hydrogenated components rather than by interesterification. The products produced in Denmark are formulated from interesterified oil components and represent the only country surveyed where hydrogenation has been replaced entirely. However, if the 59 samples from Denmark are excluded, 28 of 169, or 16.6%, are of the low-/zero-*trans* composition. Thus, it would appear that hydrogenation remains the technology of choice to formulate margarine/spread products throughout most of the world.

In a recent report[31], margarines marketed in 11 countries including Europe, Scandinavia, and the United States were analyzed for *trans* fatty acids and other compositional data. Their data confirm that any decrease in *trans* fatty acids is achieved at the expense of increased saturated acids content. Print margarines produced in the United States had the lowest *trans*/saturated acid content (41.4%) and the highest ratio of polyunsaturated fatty acids (PUFA)/TFA + SAT compared to the rest of the world. The same trend was observed for soft tub margarines.

25.2.3 Formulation of Food Oils by Hydrogenation (Soybean Based)

A discussion of the fundamentals of hydrogenation is beyond the scope of this chapter. For further information the reader should consult reviews listed in the references[32–34].

Temperature, pressure, agitation, and catalyst concentration are the most important factors governing the course and speed of fat/oil hydrogenation. However, with other factors being equal, temperature has the largest effect on *trans* acid formation. Industrially, hydrogenation is carried out under selective conditions favoring reduction of polyunsaturated groups over that of monoenic acids. Typically, selective conditions involve high temperature (160 to 220°C) and low hydrogen

TABLE 25.3
Zero-/Low-*Trans* Margarine Formulation Trends Across the World

Year	Country	No. samples reported	Zero/low trans[a] No.	Zero/low trans[a] %
1995	France	12	5	41.7
1998	Denmark[b]	59	59	100
1998	Bulgaria	5	3	60
1998	Canada	109	8	73
2000	United States	17	2	11.8
2000	Spain	12	6	50
2002	Turkey	14	4	28.6
	Total	228	87	38.2

[a] 5% or less.
[b] Excluding Denmark, only 16.6% are low/zero *trans*.

pressure (10 to 40 psi) in the presence of 0.02 to 0.04% nickel metal (Table 25.4). Typically the catalysts are supplied as 25% nickel on a support. Although selective conditions promote *trans* fatty acid formation, it is imperative that the amount of stearic acid formed at lower iodine values be kept at a minimum since any tristearin formed in the reaction will unduly raise the melting point of the finished oil such that the sharply melting "*trans*" functionality is not compromised.

25.2.4 PARTIALLY HYDROGENATED WINTERIZED SOYBEAN SALAD OIL (PHWSBO)

Partially hydrogenated winterized soybean salad/cooking oil (PHWSBO) has been produced commercially in the U.S. since the early 1960s. Typically, PHWSBO is produced by hydrogenation under selective conditions (160 to 170°C, low hydrogen pressure, and 0.02% nickel) where the iodine value (IV) is reduced from approximately 130–132 to 110–115. This feedstock is then chilled slowly to allow the higher melting triacylglycerols to crystallize. The liquid oil is then recovered by filtration. The higher melting fraction, or stearine, may be incorporated into margarine/spread or shortening products. The composition and properties of PHWSBO are given in Table 25.4, along with hydrogenation conditions used in its manufacture. PHWSBO (IV 110–112) will remain clear at refrigerator temperatures and is superior to unhydrogenated oil for light frying in the home[35].

25.2.5 MARGARINE BASESTOCK AND FORMULATION OF SPREADS

Excellent reviews of basestock formulations have been published by O'Brien[36] Latondress[37], and Erickson and Erickson[34]. Margarine oil basestocks require "selective conditions" where stearic acid formation needs to be controlled very carefully. Typically, low pressures and high temperatures are used. The preparation of multipurpose margarine basestocks generally involves reducing the IV value of soybean oil from about 130 to 65–70 in the presence of a nickel catalyst at temperatures of about 220°C and low hydrogen pressures. The solid fat content of such a product is shown in Figure 25.1.

Margarine oils typically show the following solid fat index (SFI) contents: tub products 10°C 8–14, 21.1°C 4–8, and 33.3°C 1–2; stick products 10°C 20–25, 21.1°C 10–12, and 33.3°C 2–4. The hydrogenated soybean oil, when blended with 25 to 50% liquid oil, results in solid fat contents suitable for tub and stick products, respectively[38]. The melting points of the blended oils will usually be from 32 to 35°C. The *trans* isomer content of the basestock is about 40% which, after blending, yields final *trans* values for tub and stick products of 10 and 20%, respectively. Figure 25.1 demonstrates that *trans*-containing triacylglycerols are very sharply melting materials at higher temperatures. They

TABLE 25.4
Properties of Partially Hydrogenated Soybean Oils

Oil	Fatty acid composition (%)					Iodine value	% trans	Melting point (°C)	Solid fat content (%)				
	C16	C18	C18:1	C18:2	C18:3				10°C	21.1°C	26.7°C	33.3°C	40°C
Original soybean oil	11.2	3.7	22.1	55.0	6.8	132.0	0.0	−14.3	0.0	0.0	0.0	0.0	0.0
Hydrogenated soybean oil[a]	11.4	4.7	40.3	40.5	3.0	112.6	8.7	16.7	0.0	0.3	0.0	0.2	0.2
Stearine	12.8	6.5	48.4	30.3	1.9	97.2	13.3	27.6	14.2	5.5	1.7	0.0	0.0
HW SBO	10.5	4.4	42.0	4.0	2.9	109.5	9.1	−7.5	0.9	0.0	0.1	0.0	0.0
Margarine oil[b]	11.3	13.6	75.2	0.0	0.0	64.7	39.7	41.3	73.7	54.1	44.7	22.3	3.3
Shortening oil[c]	11.3	5.1	72.6	11.0	0.0	81.5	31.8	29.3	36.3	13.7	4.6	0.0	0.0

[a] 160°C, 40 psi H_2 pressure, 0.02% nickel.
[b] 216°C, 20 psi H_2 pressure, 0.03% nickel.
[c] 221°C, 20 psi H_2 pressure, 0.03% nickel.

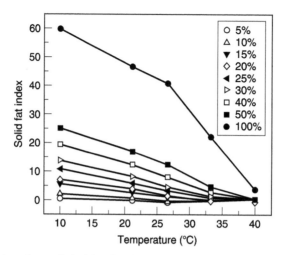

FIGURE 25.1 Formulation of spreads by blending hydrogenated and liquid soybean oils.

provide functionality for spreadability and resistance to oil at room temperature, yet they melt very quickly at body temperature to yield a pleasant, cooling sensation in the mouth.

The baking industry employs a number of shortenings[39]; their *trans* acid contents are shown in Table 25.5. The *trans* acid contents vary from about 12 to nearly 25%. Reduction in the TFA contents of shortenings is possible by alternative processing techniques such as interesterification. However, zero-/low-*trans* shortenings, while equivalent in performance compared to hydrogenated products, may be more expensive and will have elevated saturated acid contents.

25.2.6 SHORTENING BASESTOCK AND FORMULATION OF SHORTENING

A shortening basestock is illustrated in Figure 25.2. Typically shortening basestocks are prepared by hydrogenation of soybean oil to an IV of about 80. Lower temperatures and higher pressures are used compared to margarine basestocks. To formulate shortenings, liquid oil is blended to decrease the 10°C solid fat content and addition of completely hydrogenated hardstock (10 to 12%) raises the solid fat content values for 33.3 and 40°C.

25.2.7 HIGH-STABILITY OILS (HSO)

High-stability oils (HSO) were developed over 30 years ago[40–42]. Compared to commodity oils they are expensive, extremely stable, yet fill definite needs for the food industry. They are liquids at ambient temperature and perform well as spray oils and in applications in products with large surface areas and/or where long shelf life is required. HSO are at least four times more resistant to oxidation and hydrolysis than commodity salad oils which translates into slower development of off-flavors, and the color stability shows marked improvements. Typical applications include roasting of nuts, carriers of flavors, moisture barriers, viscosity modifiers, glass enhancers, lubricating/releasing agents, antidusting agents, and frying operations. The market for HSO use was estimated at 45,000 to 57,000 metric tons in 2000 with a breakdown as 70% low-end HSO, 10% mid-range, and 20% high-end. These designations refer to their relative stability under active oxygen method conditions (100°C) or hours to reach a peroxide value of 100. Low, mid-range, and high-end oils have values of 50 to 100, 100 to 300, and >300, respectively[43].

HSO can be processed from both commodity and genetically/plant breeding modified oils. High-oleic corn, soybean, and sunflower oils meet requirements for low-end HSO without processing

TABLE 25.5
Composition and *Trans* Acid Content of Commercial Baking Shortenings

Sample	Solid fat content (%)								Drop melting point (°C)	% *trans* acids	Fatty acids by FAME-GC (%)							
	10°C	21.1°C	26.7°C	33.3°C	40°C	45°C	50°C	55°C			14:1	16:0	18:0	18:1	18:1 *trans*	18:2	18:2 *trans*	18:3
Pie shortening	27.2	20.3	17.2	10.3	3.2	0.2	0.4	0.0	41.3	12.8	0.18	13.93	11.50	30.14	11.87	27.99	1.45	2.94
Cake	22.9	15.9	12.9	8.9	3.0	0.6	0.8	0.2	41.1	11.8	0.00	11.56	18.16	36.98	9.15	18.34	4.34	1.47
Cake and icing	37.7	21.9	19.5	14.4	8.1	3.7	0.8	0.0	45.8	24.4	0.22	15.14	14.06	36.36	19.23	7.29	7.51	0.19
Puff pastry	40.1	29.5	27.7	21.5	11.4	5.0	1.1	0.0	47.7	18.0	0.20	13.59	17.23	32.51	16.19	16.75	2.83	0.70
Veg./butter, all purpose	32.7	18.1	15.2	9.5	3.8	0.8	0.0	0.3	42.4	11.6	4.77	21.75	14.85	29.39	11.17	15.32	0.81	1.95
All purpose, butter flavor	30.0	17.8	16.5	14.4	9.4	6.0	2.0	0.5	48.4	20.4	0.19	12.87	13.31	44.32	17.49	7.06	4.43	0.34
Icing	38.6	23.1	20.7	15.9	8.9	3.2	0.6	0.3	45.6	22.7	0.20	14.77	13.39	40.58	20.50	7.00	3.27	0.29
Cake and icing	37.1	18.9	16.2	10.7	5.8	3.6	0.9	0.5	45.0	24.7	0.00	12.07	11.94	43.61	22.26	6.62	3.20	0.31
All purpose	37.0	18.8	15.5	10.4	6.0	3.8	1.6	0.0	45.6	24.9	0.09	11.64	11.24	44.57	22.78	6.52	2.83	0.34

Note: 18:1 *cis* had high amounts of C11 and C13 double bonds, up to 10% of total fatty acid.

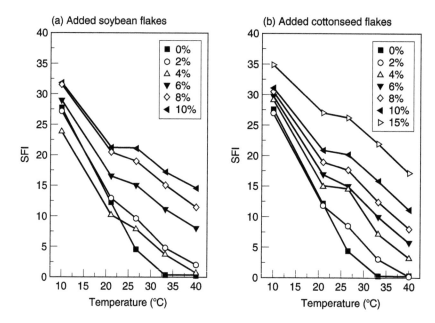

FIGURE 25.2 Formulation of shortenings by blending hydrogenated soybean oil with soybean/cottonseed flakes.

beyond the customary refining, bleaching, and deodorizing, whereas soybean oil requires partial hydrogenation and dry fractionation as well. Both medium and high-end HSO can be prepared from both commodity and genetically/structurally modified oils. However, light, partial, or heavy hydrogenation and dry fractionation must be employed with the usual refining, bleaching, and deodorization steps.

Low-end HSO prepared from hydrogenated, dry-fractionated soybean oil contains 15% saturates and 32% *trans*, whereas high-oleic sunflower contains 9.5% saturates and 1% *trans*. Canola and high-oleic-canola based HSO offer opportunity for *trans* reduction in mid-range applications compared to soybean oil. Canola HSO shows *trans* values of 18 to 29% compared to 51% for soy based oil. Similarly, high-oleic-based, high-end oils have 33% *trans* compared to 48% for cottonseed/soybean-based oils[43].

HSO are liquids at ambient temperature, highly functional, and convenient to use. They are used at low levels often between 0.2 and 1% and are most commonly sprayed onto the surface of the food or ingredient. Although more expensive than commodity oils, processing costs are reduced making the final pricing competitive. It is expected that increased use of HSO will occur in the future to achieve fat reduction in foods, to improve product shelf life, and to improve health and nutrition.

25.2.8 CANOLA OIL

Hydrogenation of canola oil has been studied extensively in the laboratory[44–47] and has been reviewed by Koseoglu and Lusas[48]. Although canola oil is widely used in margarines/spreads and shortenings, the *trans* isomer content of these products tends to be somewhat higher than that of soybean-based products[24].

Compared to soybean oil, hydrogenation of canola oil results in more TFA at a given IV level. For example, canola oil (IV 113), when hydrogenated to an iodine value of 80 for shortening stock, contains about 45% TFA compared to 31 to 32% for soybean at the same IV. Canola margarine stock

(IV 75) contains 50% *trans* compared to 40% for soybean oil at an IV of 65. Lightly hydrogenated canola (IV 90) used for frying fats contains 20% *trans*, whereas soybean oil at IV 110 to 116 contains about 10 to 12% trans[35,49]. These properties can be accounted for by the differences in fatty acid composition of the two oils. Canola oil contains 60% oleic acid and about 32% polyunsaturated acids (linoleic acid 22%, linolenic acid 10%), whereas soybean oil contains about 22% oleic acid and 63% polyunsaturated acids. Thus, the high content of oleic acid renders canola oil more amenable to isomerization/TFA formation during hydrogenation coupled with the fact that sulfur-containing glucosinolate breakdown products found in canola oil tend to poison the catalyst, thus promoting TFA formation. Canola oil tends to crystallize in the undesirable beta form owing to the low levels of palmitic acid found in the oil. To improve the crystal habit, the addition of palm oil prior to hydrogenation reportedly stabilizes the oil in the beta prime form often desired for margarines and shortenings[49].

25.2.9 COTTONSEED OIL

An excellent review of food uses and processing of cottonseed oil has been published by O'Brien and Wan[50] and Jones and King[51]. Cottonseed oil has an IV of about 110 and contains about 25% saturated acids. A basestock system for formulation of shortenings, margarines, and spreads consists of five hydrogenated oils along with a hard stock or completely hydrogenated component (IV > 5). Oils hydrogenated to an IV of 75 to 80 under nonselective conditions (350°F, 30 to 45 psi, 0.02% nickel catalyst) possess flat SFI curves suitable for shortenings, whereas the 58, 65 and 70 IV oils hydrogenated under selective conditions (440°F, 11 to 15 psi, 0.04 to 0.08% nickel catalyst) possess steep SFI curves needed for margarine/spreads. The IV 58 oils have the solid fat and melting properties very similar to IV 65 soybean oil and are similar in *trans* acid content. In 1997 only 8 million pounds of cottonseed oil was used in margarines/spreads in the U.S.

25.2.10 CORN OIL

During the period 1955–1985, corn oil usage in margarine/spreads increased from less than 500,000 pounds to 210 million pounds, but by 1997 this had dropped to about 61 million pounds. Corn oil contains about 13% saturated fatty acids and has an IV of approximately 127. Soft and stick margarines formulated with hydrogenated soybean oil and liquid corn oil are available commercially. Soybean oil basestocks are produced by hydrogenation of the oil to IV of 65 to 67 under selective conditions and contain about 48% *trans* acids. The 65 IV oil, when blended with 25 to 30% liquid corn oil, is suitable for soft products while the 67 IV oil is blended with 45 to 47% liquid corn oil is suitable for stick products. The *trans* acid content of soft and stick products ranges are 12 to 14% and 20 to 24%, respectively.

25.2.11 SUNFLOWER OIL

Sunflower oil ranks fourth in global consumption behind soybean, palm, and canola. Three types include those from conventional varieties high in linoleic acid (~70%), high in oleic (~84%), and mid-oleic oils (50 to 65%). The latter is known as Nu-Sun. Conventional sunflower oil requires hydrogenation for oxidative stability in frying applications and for functionality, i.e., solid fat in spreads and shortenings. However, high-linoleic oil is used as the liquid component in soft margarines. High-oleic sunflower qualifies as a high-stability oil in applications where extended shelf life is desired. Nu-Sun does not require hydrogenation, therefore offering a *trans*-free oil for deep fat frying. A major U. S. snack food manufacturer has switched its premium brand potato chips from cottonseed oil to Nu-Sun with very satisfactory results[52].

## 25.3	ALTERNATIVE STRATEGIES TO REDUCE *TRANS* ACIDS

### 25.3.1	INTERESTERIFICATION

A thorough discussion of interesterification is beyond the scope of this chapter and the reader is referred to Rozendall and MaCrae[53]. The patent literature is replete with information on interesterification as a route to low-*trans* fats and oils. Gillies[54] has published an excellent resource covering the patent literature from 1960 to 1974. A search of the patent literature using margarine/shortening and interesterification as key words yields over 46,000 references since 1976 alone! Thus a comprehensive review would be a review in itself. Here, several illustrative examples will be given from the author's laboratory. Random interesterification of 80% liquid soybean oil with 20% completely hydrogenated soybean oil provides a route to soft margarine oil having suitable solid fat and crystal structure[55]. Further studies, in which the interesterified oils were formulated into soft margarines in pilot plant runs, demonstrated that the 80:20 blend produced a margarine that was harder and more difficult to spread than hydrogenated controls requiring addition of an additional 20% liquid oil to the formulation. This product had suitable spreadability, sensory properties, and resistance to oil/water loss[56].

Random interesterification of other liquid oils including corn, peanut, cottonseed, canola, and palm oils with completely hydrogenated soybean or cottonseed flakes yields basestocks suitable for formulation of zero-*trans* margarines and shortenings[57]. Shortening oils require higher, flatter solid fat curves and melting points than margarine oils. In order to obtain these properties, it is necessary to incorporate more stearine with the interesterified blend. The desired solid fat and melting points are achieved by blending the interesterified basestock with additional liquid oil. The effect is shown in Figure 25.3 where the solid fat indices of a 50:50 mixture of interesterified soybean oil and soybean stearine (IV 0) are shown along with mixtures (5 to 50%) of the basestock and liquid oil. Typically, all-purpose shortenings prepared from hydrogenated components show SFI values at 50, 70, 92, and 104°F of 18 to 23, 14 to 19, 13 to 14, and 7 to 11, respectively. The 50:50 blend of interesterified and liquid soybean oil closely match these values and the drop melting point of 42.2°C is close to the 45 to 47°C value observed for commercial products. Fluid shortenings can be prepared by blending 35% of the basestock with 65% soybean, corn, peanut, cottonseed, or canola oils.

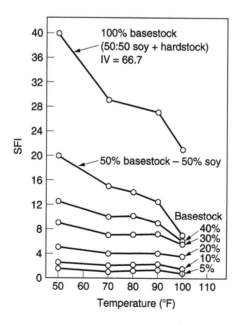

FIGURE 25.3 Formulation of shortening oils by blending interesterified basestock with liquid soybean oil.

25.3.2 FRACTIONATION

Rapid growth of the palm industry, beginning in the mid-1970s, prompted development of improved fractionation technology. Historically, solvent and detergent processes had been used, but today physical or dry fractionation is the industry standard. Palm oil (IV 51 to 53) is fractionated into olein (IV 56 to 59) and stearine fractions (IV 32 to 36). The olein fraction is further fractionated into midfractions and super oleins and top oleins. The palm midfractions (IV 42 to 48) are further processed into harder fractions (IV 32 to 36). Fractionation of the stearine (IV 32 to 36) yields soft and super stearines having IV 40 to 42 and 17 to 21, respectively[58].

A discussion of the fractionation process will not be attempted here. The reader is referred to a number of reviews[59–62] covering both the theoretical and practical aspects of fractionation technology.

25.3.3 PALM OIL

Excellent reviews of palm oil in food oils have been published[63–68]. Zero-*trans* margarine fats can be prepared via random interesterification of soybean oil with palm stearine or fully hydrogenated soybean oil in ratios of 80:20. It was concluded that increasing the amount of fully hydrogenated soy to 30 parts produced zero-*trans* oils is suitable for shortening. Forty parts palm stearine randomized with 60 parts soybean oil also produced shortening oil formulations very similar to commercial products[69].

Ozay et al.[70] described the formulation of *trans*-free margarines prepared from sunflower and cottonseed oils interesterified with palm oil, palm kernel oil, palm stearine, and palm kernel olein and compared them to products made by simply blending the components in various ratios. Palm oil crystallizes slowly compared to other fats and oils leading to a phenomenon known as posthardening in which products become harder upon storage. The authors observed minimal posthardening in their study and reported that the use of skim in preparing the emulsions prior to crystallization was effective in retarding posthardening in blends high in palm and palm kernel oil.

Yusoff et al.[70] reported formulation of *trans*-free tub, pastry, and bakery margarines from palm oil, palm olein, palm stearine, palm kernel olein interesterified with soybean and rapeseed oil. The normal level of palm oil in commercial soft tub products was reported to be 25%, but palm olein can be used up to a level of 40%.

Studies on the use of palm olein in milk fat blends as baking shortenings have been reported[72]. Blends consisting of 80% palm olein, 60% palm olein, and 40% palm olein with hard milk fat fraction have been processed in pilot-plant studies. Baking and sensory tests were compared against a commercial shortening. Noraini et al.[73] studied shortenings formulated from hydrogenated palm oil and palm sterine blended with liquid oils, blends of palm stearine with liquid oils and interesterified (100%) palm oil. They concluded that high palmitic acid oils are best for aeration of fat sugar mixtures but there is no direct relationship between creaming and baking performance. Palm stearine blended with cottonseed oil in a 3:2 ratio produced shortening best suited for aerated fillings. Palm stearine (IV 44) and low erucic acid rapeseed (11:1) are best in cake baking. For applications in both cream fillings and baking, interesterified palm is the most suitable moiety.

Berger[66] reported studies on palm oil usage in food products including margarine, vanispati (a food oil product used extensively in India and Pakistan)[74,75], baking shortenings, frying fats, and dairy products, i.e., whipped toppings, ice cream, and cheese. Vanispati, a butter-like product (melting point 37 to 38°C) is manufactured with a granular structure and a minimum of free oil at room temperatures. In the early 1980s the product (1 million tons/year), produced in Pakistan from hydrogenated components, contained about 30% *trans* acids. After reformulation with 80% palm oil and 20% liquid vegetable oil, *trans* acids were reduced to less than 4%. However, this saturated acid content is of the order of 40%.

Analysis of frying oils prepared from rapeseed, sunflower, soybean, peanut, and olive blended with 30% palm olein (IV minimum 56) show improvement over the 100% liquid oils. Analytical tests (including rancimat at 100°C, cloud point, free fatty acids, viscosity, polymers, oxidized acids, color, and smoke point) were made after 3 days of use. Since extended shelf-life is an important consideration in packaged snack foods, the improved oxidative stability of palm oil containing blends has prompted their use in Europe and Asia for preparing of instant and fried noodles[66].

25.3.4 Oils with Modified Composition

Over the past several decades a number of oilseeds with modified fatty acid compositions have been developed and commercialized. Most have resulted from traditional plant breeding techniques[1,4,76,77]. These include high- and mid-level oleic sunflower oil[5,52,78,79], high-oleic corn, soybean, and safflower oils, low-linolenic canola and soybeans oils, and high-oleic/low-linolenic canola oil. A number of laboratory frying studies have demonstrated superiority ever commodity oils[80–87]. Although low-linolenic acid canola and soybeans oils have reached commercialization, costs and production problems have impeded their success in the marketplace[88]. However, the new labeling regulations may provide new markets and demands for modified composition oils.

The A90, A6, and HS-1 developed by the Pioneer–Iowa State University–Hartz Seed Company consortium represent high-saturate soybean oils in which the normal 15% saturated acids have been elevated to as high as 40%. In their natural states these oils lack sufficient solid content at temperatures required of a margarine/spread oil. Addition of harder components, such as palm oil, interesterified palm oil, or cottonseed/soybean stearines, shows potential in soft margarine applications[89,90]. Owing to the symmetrical nature of their triacylglycerol structures, the high-saturate lines are low, sharply melting materials that, upon random interesterification, melt over a wide range, thereby rendering them more amenable to utilization in soft margarine formulation[91–93]. Typically, soft margarines formulated from hydrogenated and liquid soybean oils contain about 10% *trans* fatty acids and about 20% saturated acids. Studies have shown that approximately 25 to 30% saturated acids are required to formulate zero-*trans* soft margarine oils from soybean oil-based components[91–93]. Thus, any reduction in the *trans* fatty acids will be achieved at the expense of increased saturated acid content.

25.3.5 Fluid Shortenings

Fluid shortenings are stable suspensions of 2 to 20% hard fat in liquid vegetable oil which may or may not be hydrogenated[94–98]. Fluid shortenings have been used in the baking industry for many years where high solids are not required such as fillings, cakes, and breads. They serve the same function as solid shortenings by imparting tenderness and lubricity as well as serving as carriers for emulsifiers needed for aerating cake batter or giving crumb strength to bread. Other advantages include the fact that, since they are liquid and pumpable, they can be easily metered into batch and continuous processes. Some reduction of *trans* acids might be achieved by substituting liquid or unhydrogenated oil in formulations where hydrogenated oils have been traditionally used.

25.3.6 Cholesterol-Lowering/Zero-*Trans* Spreads

The introduction of low-/zero-*trans*, cholesterol-lowering spreads began in the mid- to late 1990s, both in Europe and the United States. These products fall into two categories with most fortified with phytosterols to lower low-density lipoprotein (LDL) cholesterol, while others employ a natural blend of palm, canola, menhaden, and olive oils to control high-density lipoprotein (HDL)/LDL ratios. Benecol (55% fat) is formulated from liquid canola oil, partially hydrogenated soybean oil, and plant sterol esters and, when consumed according to the label, lowers LDL up to 14% (U.S. Patent 5,502,045). Take Control (35% fat) is formulated from liquid canola and soybean oils, partially hydrogenated soybean oil, and is fortified with plant sterol esters (U.S. Patent 6,156,370). Smart Balance has introduced two products (67% fat) into the market, Omega Plus and Original.

The former is formulated from palm, canola, menhaden, and olive oils and natural plant sterols. Original is formulated from palm, soybean, canola, and olive oils and contains no plant sterols (U.S. Patents 5,843,497 and 5,578,334).

The Promise spreads Light (35% fat) and Promise (60% fat) are zero grams *trans* fat and are formulated from liquid soybean, canola, sunflower, palm, and palm kernel oils and partially hydrogenated soybean oils (U.S. Patent 6,156,370). Interesterification technology is employed where interesterified palm and palm kernel stearines are used as the hardstock and blended with liquid sunflower or other vegetable oils to formulate the oil blend for Promise Light. The 60% spread is formulated from hydrogenated and liquid oil components.

Land o' Lakes has entered the zero-*trans* market with soft baking and spreadable butter formulated from canola oil and cream.

Con-Agri Foods (Fleischmann's label) has introduced several no-*trans* spreads including an olive oil-based (60% fat) and a light spread (31% fat). The former employs a blend of soybean and olive oils as the liquid component and partially hydrogenated soybean oil for the harder component. The light spread is formulated from liquid canola and partially hydrogenated corn oil.

25.3.7 LOW-*TRANS* SHORTENINGS

Low-/zero-*trans* shortenings (where low means < 5% TFA) were introduced in 1998 by Cargill under the trade name of Transend. Formulated from interesterified canola/soybean oils, these products were shown to perform equivalently to conventional shortenings in baking operations. However, increased cost was a major deterrent to their success in the marketplace. However, in view of the TFA labeling regulations, the need for such products is apparent. In 2002 Bunge Foods introduced low-*trans* baking margarine/shortenings into the marketplace. These products are formulated from partially hydrogenated soybean and cottonseed oils and typically contain 4% TFA and 23% saturated acids.

In 2003 Archer Daniels Midland (ADM) introduced the Nova Lipid line of low-*trans* shortenings based on enzymatic interesterification of soybean and hydrogenated soybean and cottonseed oils. Products include an all-purpose baking shortening, cake shortening, and a baking margarine. TFA contents of 9% permit a 1 g TFA/serving statement on nutritional labels. Other products in the ADM line include all-purpose/pie shortenings, cake and icing shortenings, and all-purpose vegetable shortenings formulated from interesterified soybean and cottonseed oils with TFA contents of 4.5 to 5.5%.

REFERENCES

1. Lui, K., Soy oil modifications: products and applications, *Inform*, 10, 868–877, 1999.
2. Anon., *Oil Crops Situation and Outlook*, Yearbook, Economic Research Service, USDA, 2002, p. 53.
3. Haumann, B.F., Tools: hydrogenation, interesterification, *Inform*, 5, 668–678, 1994.
4. Gunstone, F.D., Oilseed crops with modified fatty acid composition, *J. Oleo. Sci.*, 50, 269–279, 2001.
5. Gupta, M., NuSun: the future generation of oils, *Inform*, 9, 1150–1154, 1998.
6. McDonald, B.E. and Fitzpatrick, K., *Designer Vegetable Oils in Functional Foods: Biochemical and Processing Aspects*, Mazza, G., Ed., Technomic, Lancaster, PA, 1999, pp. 265–291.
7. Macrae, A., Lipase catalyzed interesterification of oils and fats, *J. Am. Oil Chem. Soc.*, 60, 243A–246A, 1983.
8. Hasman, J., *Trans* suppression in hydrogenated oils, *Inform*, 6, 1206–1213, 1995.
9. King, J.W., Holliday, R.L., List, G.R., and Snyder, J.M., Hydrogenation of soybean oil in supercritical carbon dioxide and hydrogen, *J. Am. Oil Chem. Soc.*, 78, 107–113, 2001.
10. Warner, K., Neff, W.E., List, G.R., and Pintauro, P., Electrochemical hydrogenation of edible oils in a solid polymer electrolyte reactor. Sensory and compositional characteristics of low *trans* oils, *J. Am. Oil Chem. Soc.*, 77, 1113–1117, 2000.

11. Watkins, B., *Trans* fatty acids: aA health paradox, *Food Technol.*, 52, 120, 1998.

12. Kris-Etherton, P., *Trans* fatty acids and coronary heart disease risk, *Am. J. Clin. Nutr.*, 62, 655S–708S, 1995.

13. Anon., Special task force report: position paper of *trans* fatty acids, *Am. J. Clin. Nutr.*, 63, 663–670, 1996.

14. Hu, F.B., Stamper, M. S., Manson, J.E., Rimm, E., Colditz, G.A., Posner, B.A., Hennekeas, C.H., and Willet, W.C., Dietary fat intake and the risk of coronary heart disease in women, *N. Engl. J. Med.*, 337, 1491–1499, 1997.

15. Okene, I. and Nicolosi, R., Dietary fat intake and risk of coronary heart disease in women, *N. Engl. J. Med.*, 338, 917–919, 1998. (Also Hegsted, D.M.)

16. Hegsted, D.M., Dietary fat intake and the risk of coronary heart disease in women, *N. Engl. J. Med.*, 338, 917–918, 1998. (Also Johnson et al., ibid.)

17. Anon., Federal Register, Part II, Department of Health and Nutrition, Food and Drug Administration, 21CFR, Part 101, Food labeling: *trans* fatty acids in nutrition labeling, nutrient content claims and health claims: proposed rule, November 17, Federal Register, 1999.

18. Anon., Food labeling: *trans* fatty acids in nutrition labeling, *Fed. Reg.*, 68, 41433–41506, 2003.

19. Anon., *Soya and Oilseed Blue Book*, Soyatech, Bar Harbor, ME, 2003, pp. 313–358.

20. Mag, T.K., Margarine oils, blends in Canada, *Inform*, 5, 1350–1353, 1994.

21. Chrysam, M.M., Margarines and spreads, in *Bailey's Industrial Oil and Fat Products*, Hui, Y.H., Ed., John Wiley, New York, 1996, pp. 65–114.

22. Pelloso, T., Margarines and spreads, in *Proceedings of World Conference on Oilseed and Utilization*, AOCS Press, Champaign, IL, 2001, pp. 44–46.

23. List, G.R., Steidley, K.R., and Neff, W.E., Commercial spreads formulation, structures and properties, *Inform*, 11, 980–986, 2000.

24. Postmus, E., deMan, L., and deMan, J.M., Composition and physical properties of North American stick margarines, *Can. Inst. Food Sci. Technol. J.*, 22, 481–486, 1989.

25. Alonzo, L., Fraga, M.J., and Juarez, M., Determination of *trans* fatty acids in margarines marketed in Spain, *J. Am. Oil Chem. Soc.*, 77, 131–136, 2000.

26. Oveson, L., Leth, T., and Hansen, K., Fatty acid composition and contents of *trans* monounsaturated fatty acids in frying fats and margarines and shortenings marketed in Denmark, *J. Am. Oil Chem. Soc.*, 75, 1079–1083, 1998.

27. Ratnayake, W.M.N., Pelletier, G., Hollywood, R., Bacler, S., and Leyte, D., *Trans* fatty acids in Canadian margarines: recent trends, *J. Am. Oil Chem. Soc.*, 75, 1587–1594, 1998.

28. Tsanev, R., Russeva, A., Rizov, T., and Dontcheva, I.N., Content of *trans* fatty acids in edible margarines, *J. Am. Oil Chem. Soc.*, 75, 143–145, 1998.

29. Tekin, A., Cizmeci, M., Karabacak, H., and Kayaban, M., *Trans* fatty acids and solid contents of margarines marketed in Turkey, *J. Am. Oil Chem. Soc.*, 79, 443–445, 2002.

30. Bayard, C. and Wolff, R.L., *Trans* 18:1 acids in French tub margarines and shortenings, *J. Am. Oil Chem. Soc.*, 72, 1465–1489, 1995.

31. Matsuzaki, H., Okamoto, T., Oyama, M., Mavruyama, T., Niiya, I., Yanagita, T., and Sugano, M., Trans fatty acids marketed in eleven countries, *J. Oleo Sci.*, 51, 551–565, 2002.

32. Hastert, R.C., Hydrogenation, in *Bailey's Industrial Oil and Fat Products*, 5th ed., Hui, Y.H., Ed., John Wiley, New York, 1996, pp. 213–300.

33. Patterson, H.B.W., *Hydrogenation of Fats and Oils: Theory and Practice*, AOCS Press, Champaign, IL, 1994, pp. 1–267.

34. Erickson, D.R. and Erickson, M.D., Hydrogenation and base stock formulation, in *Practical Handbook of Soybean Processing and Utilization*, Erickson, D.R., Ed., AOCS Press, Champaign, IL, 1994, pp. 277–296.

35. List, G.R. and Mounts, T.L., Partially hydrogenated-winterized soybean oil, in *Handbook of Soybean Oil Processing and Utilization*, Erickson, D.R., Pryde, E.H., Mounts, T.L., Brekke, O.L., and Falb, R.A., Eds., AOCS Press, Champaign, IL, 1980, pp. 93–214.

36. O'Brien, R.D., *Fats and Oils, Formulating and Processing For Application*, Technomic, Lancaster, PA, 1998, pp. 437–458.

37. Latondress, E.G., Shortenings and margarines: basestock preparation and formulations, in *Handbook of Soy Oil Processing and Utilization*, Erickson, D.R., Ed., AOCS Press, Champaign, IL, 1980, pp. 145–154.

38. List, G.R., Steidley, K.R., Palmquist, D., and Adlof, R.O., Solid fat index (SFI) vs. solid fat content (SFC): a comparison of dilatometry and pulsed NMR for solids in hydrogenated soybean oil, in *Crystallization and Solidification of Lipids*, Widlak, N., Ed., AOCS Press, Champaign, IL, 2001.
39. Suhan, W.J., *Shortenings and Their Uses in Practical Baking*, AVI, Westport, CT, 1980, pp. 5–10.
40. Gooding, C. M., Production of High Stability Liquid Vegetable Oils, US Patent 3,674,821, 1972.
41. Carrick, V.A. and Yodice, R., Vegetable Oil Compositions, US Patent 5,260,077, 1993.
42. Miller, K.L., High stability oils, *Cereal Foods World*, 38, 478–482, 1993.
43. Lampert, D., High stability oils: what are they? How are they made? And why do we need them?, in *Physical Properties of Fats, Oils and Emulsifiers*, Widlak, N., Ed., AOCS Press, Champaign, IL, 2000, pp. 238–246.
44. El-Shattory, Y., Deman, L., and Deman, J.M., Hydrogenation of low erucic acid rapeseed oil (Zephyr) under selective conditions, *J. Food Sci.*, 18, 527–533, 1981.
45. Deman, J.M., El-Shattory, Y., and Deman, L., Hydrogenation of canola oil (Tower), *Chem. Mikrobiol. Technol. Lebensm.*, 7, 117–124, 1981.
46. Deman, J.M. and El-Shattory, Y., Formation of *trans* isomers during hydrogenation of canola oil (Tower), Chem. *Mikrobiol. Technol. Lebensm.*, 7, 33–336, 1981.
47. Bansal, J.D. and Deman, J.M., Effect of hydrogenation on the chemical composition of canola oil, *J. Food Sci.*, 47, 1338–1345, 1982.
48. Koseoglu, S. and Lusas, E., Hydrogenation of canola oil, in *Canola and Rapeseed*, Shahidi, F., Ed., AVI, New York, 1990, pp. 123–148.
49. Eskin, N.A.M., McDonald, B.E., Przybyski, R., Malcomson, L.J., Scarth, R., Mag, T., Ward, K., and Adloph, D., Canola oil, in *Bailey's Industrial Oil and Fat Products*, 5th ed., Hui, Y.H., Ed., John Wiley, New York, 1996, pp. 1–96.
50. O'Brien, R.D., and Wan. P., Cottonseed oil: processing and utilization, in *Proceedings of World Conference on Oilseed and Edible Oil Processing*, Wilson, R.F., Ed., AOCS Press, Champaign, IL, 2001, pp. 90–140.
51. Jones, L.A. and King, C.C., Cottonseed oil, in *Bailey's Industrial Oil and Fat Products*, 5th ed., Hui, Y.H., Ed., John Wiley, New York, 1996, pp. 159–240.
52. Gupta, M.K., NuSun: a healthy non-transgenic sunflower oil, in *Proceedings of World Conference on Oilseed and Edible Oil Processing*, Wilson, R.F., Ed., AOCS Press, Champaign, IL, 2001, pp. 80–83.
53. Rozendall, A. and Macrae, A.R., Interesterification of oils and fats, in *Lipid Technologies and Applications*, Gunstone, F., Ed., Marcel Dekker, New York, 1997, pp. 223–263.
54. Gillies, M.T., *Shortenings, Margarines and Food Oils*, Noyes Data Corporation, Park Ridge, NJ, 1974, pp. 142–226.
55. List, G.R., Emken, E.A., Kwolek, W.F., Simpson T.D., and Dutton, H.J., "Zero *trans*" margarines: preparation, structure, and properties of interesterified soybean oil-soy trisaturate blends, *J. Am. Oil Chem. Soc.*, 54, 408–413, 1977.
56. List, G.R., Pelloso, T., Orthoefer, F., Chrysam, M., and Mounts, T.L., Preparation and properties of zero *trans* soybean oil margarines, *J. Am. Oil Chem. Soc.*, 72, 383–384, 1995.
57. List, G.R., Mounts, T.L., Orthoefer, F., and Neff, W.E., Margarine and shortening oils by interesterification of liquid and trisaturated triglycerides, *J. Am. Oil Chem. Soc.*, 72, 379–382, 1995.
58. Tirtiaux, A., Dry fractionation: the beat goes on, in *Proceedings of World Conference on Oilseed Processing*, AOCS Press, Champaign, IL, 1998, pp. 92–98.
59. Krishnamurthy, R. and Kellens, M., Fractionation and winterization, in *Bailey's Industrial Oil and Fat Products*, 5th ed., Hui, Y.H., Ed., John Wiley, New York, 1996, pp 301–337.
60. Timms, R.E., Crystallization of fats, in *Developments in Oils and Fats*, Hamilton, R.J. Ed., Blackie Academic and Professional, London, 1995, pp. 204–223.
61. Timms, R.E., Fractionation, in *Lipid Technologies and Applications*, Gunstone, F. Ed., Marcel Dekker, New York, 1997, pp. 199–222.
62. Deffense, E., Fractionation of palm oils, *J. Am. Oil Chem. Soc.*, 62, 376–385, 1985.
63. Duns, M., Palm oil in margarines and shortenings, *J. Am. Oil Chem. Soc.*, 62, 408–410, 1985.
64. Berger, K., Recent development in palm oil, *Oleaeagineux*, 45, 437–447, 1990.
65. Berger, K., Food product formulation to minimize the content of hydrogenated fats, *Lipid Technol.*, 4, 37–40, 1993.

66. Berger, K., Recent results on palm oil uses in food products, in *Proceedings of World Conference on Oilseed and Edible Oil Processing*, Vol. 1, AOCS Press, Champaign, IL, 1998, pp. 151–155.

67. Traitler, H. and Dieffenbacher, A., Palm oil and palm kernel oil in food products, *J. Am. Oil Chem. Soc.*, 62, 417–421, 1985.

68. Ong, A., Choo, Y.M., and Ooi, C.K., Developments in palm oil, in *Developments in Oils and Fats*, Hamilton, R.J., Ed., Blackie Academic and Professional, London, 1995, pp. 153–191.

69. Petrauskite, V., DeGreyt, W., Kellens, M., and Huyghaebaert, A., Physical and chemical properties of *trans* free fats produced by chemical interesterification of vegetable oil blends, *J. Am. Oil Chem. Soc.*, 75, 489–493, 1998.

70. Ozay, G., Yildiz, M., Mahidin, M., Yusoff, M., Yurdagul, M., and Goken, N., in *Proceedings of World Conference on Oilseed and Edible Oil Processing*, Vol. 1, AOCS Press, Champaign, IL, 1998, pp. 143–146.

71. Yusoff, M., Kifli, H., Noorlida, H., and Rozig, M.P., Formulation of *trans*-free margarines, in *Proceedings of World Conference on Oilseed Processing*, Vol. 1, Koseglu, S., Rhee, K., and Wilste, R.F., Eds., AOCS Press, Champaign, IL, 1998, pp. 156–158.

72. Noraini, I., Chemaimon, C., and Hanriah, H., The use of palm olein: hard milk fat blends as baking shortenings, in *Proceedings of World Conference on Oilseed and Edible Oil Processing*, Vol. 1, AOCS Press, Champaign, IL, 1998, pp. 147–155.

73. Noraini, I., Berger, K., and Hong, S., Evaluation of shortenings based on various palm oil products, *J. Am. Oil Chem. Soc.*, 46, 481–493, 1989.

74. Majumdar, S. and Bhattacharyya, D.K., *Trans* free vanispatti from palm stearine and vegetable oils by interesterification process, *Oleaeagineux*, 41, 235–238, 1986.

75. Achaya, K.T., Vanispatti and special fats in India, in *Lipid Technologies and Applications*, Gunstone, F., Ed., Marcel Dekker, New York, 1997, pp. 369–390.

76. Loh, W., Biotechnology and vegetable oils: first generation products in the marketplace, in *Physical Properties of Fats, Oils and Emulsifiers*, Widlak, N., Ed., AOCS Press, Champaign, IL, 2000, pp. 247–253.

77. Wilson, R.F., Alternatives to genetically modified soybeans: the better bean initiative, *Lipid Technol.*, 11, 107–109, 1999.

78. Kleingartner, L. and Warner, K., A new look at sunflower oil, *Cereal Foods World*, 46, 399–404, 2001.

79. Kiatsrichart, S., Brewer, M., Cadwallader, K.R., and Artz, W.E., Pan frying stability of NuSun, a mid-oleic sunflower oil, *J. Am. Oil Chem. Soc.*, 80, 479–483, 2003.

80. Soheili, K.C., Artz, W.E., and Tippaywat, P., Pan heating of low linolenic and partially hydrogenated soybean oils using pan fried hash browns, *J. Am. Oil Chem. Soc.*, 79, 1197–1200, 2002.

81. Warner, K. and Gupta, M., Frying quality and stability of ultra low and low linolenic acid soybean oils, *J. Am. Oil Chem. Soc.*, 80, 275–280, 2003.

82. Warner, K. and Knowlton, S., Frying oils and oxidative stability of high-oleic corn oils, *J. Am. Oil Chem. Soc.*, 74, 1317–1322, 1997.

83. Warner, K., Orr, P., and Glynn, M., Effect of fatty acid composition of oils on flavor and stability of fried foods, *J. Am. Oil Chem. Soc.*, 74, 347–356, 1997.

84. Mounts, T.L., Warner, K., List, G.R., Neff, W.E., and Wilson, R.F., Low linolenic acid soybean oil: alternatives to frying oils, *J. Am. Oil Chem. Soc.*, 71, 495–499, 1994.

85. Su, C., Gupta, M., and White, P., Oxidative and flavor stabilities of soybean oils with low and ultra-low linolenic acid composition, *J. Am. Oil Chem. Soc.*, 80, 171–176, 2003.

86. Warner, K. and Mounts, T.L., Frying stability of soybean and canola oils with modified fatty acid compositions, *J. Am. Oil Chem. Soc.*, 70, 983–988, 1993.

87. Warner, K., Orr, P., Parrott, L., and Glynn, M., Effect of frying oil composition on potato chip stability, *J. Am. Oil Chem. Soc.*, 71, 1117–1121, 1994.

88. Krawczyk, T., Edible specialty oils: an unfulfilled promise, *Inform*, 10, 555–561, 1999.

89. List, G.R., Mounts, T.L., Orthoefer, F., and Neff, W.E., Potential margarine oils from genetically modified soybeans, *J. Am. Oil Chem. Soc.*, 73, 729–732, 1996.

90. List, G.R., Orthoefer, F., Pelloso, T., Warner, K., and Neff, W.E., Preparation and properties of low *trans* margarine and oils by interesterification, blending and genetic modification, in *Physical Properties of Fats, Oils and Emulsifiers with Application in Foods*, Widlak, N., Ed., AOCS Press, Champaign, IL, 2000, pp. 226–237.

91. Kok, L.L., Fehr, W.R., Hammond, E.G., and White, P.J., *Trans* margarine from highly saturated soybean oil, *J. Am. Oil. Chem. Soc.*, 76, 1175–1181, 1999.
92. List, G.R., Mounts, T.L., Orthoefer, F., and Neff, W.E., Effect of interesterification on the structure and physical properties of high-stearic acid soybean oils, *J. Am. Oil. Chem. Soc.*, 74, 327–329, 1997.
93. List, G.R., Pelloso, T., Orthoefer, F., Warner, K., and Neff, W.E., Soft margarines from high stearic acid soybean oils, *J. Am. Oil. Chem. Soc.*, 78, 103–104, 2001.
94. Widlak, N., Formulation and production of fluid shortenings, in *Proceedings of World Conference on Oilseed and Utilization*, AOCS Press, Champaign, IL, 2001, pp. 39–44.
95. Herzing, A.C., Fluid shortenings in bakery products, *Inform*, 7, 165–167, 1996.
96. Andre, J.R. and Going, L.H., Liquid Shortening, U.S. Patent 2,815,286, 1957.
97. Holman, G.W. and Quimby, O.T., Process of Preparing Suspensions of Solid Triglyceride and Liquid Oil, U.S. Patent 2,521,219, 1950.
98. Mitchell, P.J., Permanently Pumpable Oleaginous Suspensions, U.S. Patent 2,521,242, 1950.

26 Tocopherols and Tocotrienols as Byproducts of Edible Oil Processing

Vitamin E: A New Perspective

Andreas M. Papas
YASOO Health Inc., Johnson City, Tennessee

CONTENTS

26.1 INTRODUCTION

Eight natural compounds, four tocopherols, designated as α, β, γ, and δ, and four tocotrienols, also designated as α, β, γ, and δ, have vitamin E activity. α-Tocopherol has become synonymous with vitamin E because it is the predominant form in human and animal tissues. In addition, it is the most bioactive form based on the rat fetal resorption test, which is the classic assay for vitamin E activity. Recent research, however, shows that the other tocopherols and tocotrienols have important and unique antioxidant and other biological effects in nutrition and health and are now receiving increased attention[1].

This chapter provides an overview of the absorption, transport, metabolism, and biological function of tocopherols (with emphasis on non-α-tocopherols) and tocotrienols and their role in health and disease.

26.2   CHEMISTRY AND OCCURRENCE IN FOODS

Tocopherols consist of a chroman ring and a long, saturated phytyl chain. The four tocopherols differ only in the number and position of the methyl groups on the chroman ring. Tocotrienols have identical chroman rings to the corresponding tocopherols, but their side chain is unsaturated with double bonds in the 3', 7', and 11' positions (Figure 26.1). Tocopherols are found most abundantly in the oils extracted from seeds such as soy, corn, canola, cotton, and sunflower. γ-Tocopherol is the predominant tocopherol in soy and corn oils. Tocotrienols are found primarily in the oil fractions of cereal grains such as rice, barley, rye, and wheat, and the fruit of palm (Table 26.1). Commercial quantities of tocotrienols are extracted from palm oil and rice bran oil deodorizer distillates[2].

Two novel tocotrienols, namely desmethyl tocotrienol (3,4-dihydro-2-methyl-2-(4,8,12-trimethyltrideca-3'(E),7'(E), 11'-trienyl)-2H-1-benzopyran-6-ol) and didesmethyl tocotrienol (3,4-dihydro-2-(4,8,12-trimethyltrideca-3'(E),7'(E), 11'-trienyl)-2H-1-benzopyran-6-ol), were isolated from rice bran. Finally, an unusual vitamin E constituent (α-tocomonoenol) was isolated and reported to enhance antioxidant protection in marine organisms adapted to cold-water environments[3].

Only α-tocopherol is available commercially both as natural and synthetic. Tocopherols have three asymmetric carbons at the 2, 4', and 8' positions. Biosynthesis of tocopherols in nature yields only the RRR stereoisomer. For example, α-tocopherol derived from natural sources is 2R,4'R,8'R-α-tocopherol. In contrast, α-tocopherol produced by chemical synthesis, by condensing isophytol with tri-, di-, or monomethyl hydroquinone, is an equimolar racemic mixture (all-*rac*) of eight stereoisomers. In commercial product labeling natural RRR-α-tocopherol is designated as d-α-tocopherol and synthetic all-*rac*- is labeled as dl-α-tocopherol.

The National Research Council (NRC)[4] recommended that, for dietary purposes, vitamin E activity be expressed as d-α-tocopherol equivalents (α-TE; 1.0 mg RRR-α-tocopherol = 1.0 α-TE). For mixed tocopherols the NRC proposed the following biopotencies in α-TE: β-tocopherol 0.5, γ-tocopherol 0.1, α-tocotrienol 0.3. Commercial products are usually labeled in IU (1.0 α-TE = 1.49 IU). Vitamin E activity in commercial products is computed only from its α-tocopherol content; no IU or α-TE is included from other tocopherols or tocotrienols. The biopotency of synthetic dl-α-tocopherol is officially recognized as 0.74 α-TE or 1.1 IU versus 1.0 α-TE or 1.49 IU for the naturally occurring d-α-tocopherol. However, based on more recent research[5,6], the NRC recommended that the biopotency of the naturally occurring d-α-tocopherol is twice as that of the synthetic dl form[7].

26.3   ABSORPTION, TRANSPORT, AND BIOAVAILABILITY

Tocopherols are absorbed in the same path as other nonpolar lipids such as triacylglycerols and cholesterol (Figure 26.2). Bile, produced by the liver, emulsifies the tocopherols and incorporates them into micelles along with other fat-soluble compounds, thereby facilitating absorption. Tocopherols are absorbed from the small intestine and secreted into lymph in chylomicrons produced in the intestinal wall. Lipoprotein lipases catabolize chylomicrons rapidly and a small amount of tocopherol may be transferred from chylomicron remnants to other lipoproteins or tissues. During this process, apolipoprotein E binds to chylomicron remnants. Because the liver has specific apolipoprotein

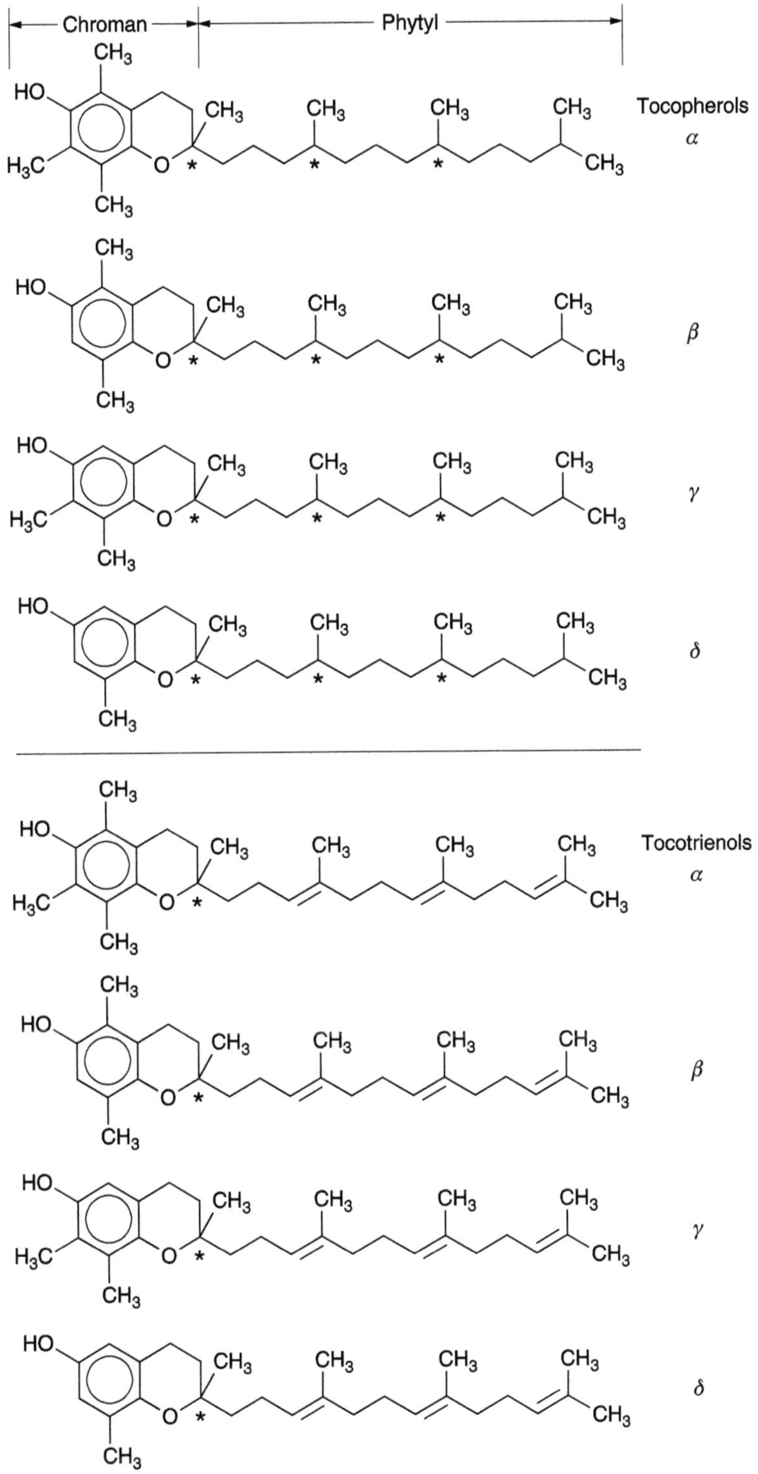

FIGURE 26.1 Structures of tocopherols and tocotrienols. Tocopherols and tocotrienols have the same chroman ring but the phytyl tail of tocotrienols contains three double bonds.

TABLE 26.1
Tocopherol and Tocotrienols in Vegetable Oils and Animal Fats

Oil	Tocopherols (mg/100 g)					Tocotrienols (mg/100 g)					Grand total (mg/100 g)
	α	β	γ	δ	Total	α	β	γ	δ	Total	
Soybean	10		59	26	96					0	96
Corn	11	5	60	2	78					0	78
Canola	17		35	1	53					0	53
Sunflower	49		5	1	55					0	55
Peanut	13		22	2	37					0	37
Cottonseed	39		39		78					0	78
Safflower	39		17	24	80					0	80
Palm	26		32	7	65	14	3	29	7	53	118
Coconut	0.5			0.6	1	0.5		2	0.6	3	4
Olive	20	1	1		22					0	22
Evening primrose	16		42	7	65					0	65
Wheat germ	121	65	24	25	235	2	17			19	254
Rice	12	4	5		21	18	2	57		77	98
Barley	35	5	5		45	67	12	12		91	136
Oats	18	2	5	5	30	18		3		21	51
Butter	2	0	0	0	2	0	0	0	0	0	2
Lard	1.2				1.2	0.7				0.7	1.9
Margarine	7	0	51	3	62	0	0	0	0	0	62

From Papas, A.M., *The Vitamin E Factor*, HarperCollins, New York, 1999; Papas, A.M., Ed., *Antioxidant Status, Diet, Nutrition and Health*, CRC Press, Boca Raton, FL, 1998.

E receptors, it retains and clears the majority of the chylomicron remnants. Tocopherols in the remnants are secreted into very low-density lipoprotein (VLDL) and circulated through the plasma. VLDL is hydrolyzed by lipoprotein lipase to low-density lipoprotein (LDL), which carries the largest part of plasma tocopherols and appears to exchange them readily with high-density lipoprotein (HDL). Tocopherols in HDL may be readily transferred back to chylomicron remnants as they pass through circulation returning plasma tocopherol to the liver[8].

Absorption of tocotrienols appears similar to that of tocopherols. Their transport and tissue uptake, however, appear to differ from those of α-tocopherol. Tocotrienols disappear from plasma with chylomicron clearance and are deposited, in conjunction with triacylglycerols, in the adipose tissue[9,10].

Our plasma and tissues contain at least two to three times more α- than γ-tocopherol even though the typical American diet contains more γ- than α-tocopherol[11,12]. Concentration of tocotrienols in plasma and tissues is low even when consumed in comparable amounts with α-tocopherol. All tocopherols and tocotrienols are apparently equally well absorbed. However, α-tocopherol is preferentially secreted into nascent VLDL. It was proposed that a tocopherol binding protein is responsible for incorporating preferentially α-tocopherol into nascent VLDL. This protein has been identified and characterized in the rat, rabbit, and humans. This protein has greater affinity for α-tocopherol than other tocopherols and tocotrienols[8].

Little information is available on the catabolism of different tocopherols and tocotrienols in tissues. Two metabolic products of α-tocopherol, called the Simon metabolites, have been isolated and characterized in the urine of rabbits, humans, and rats. These are 2-(3-hydroxy-3-methyl)-3,5,6-trimethyl-1,4-benzoquinone and its γ-lactone[13]. Two other tocopherol metabolites have been isolated from human urine. The first, 2,5,7,8-tetramethyl-2(2′-carboxyethyl)-6-hydroxychroman, a metabolite of α-tocopherol, has been suggested as an indicator of vitamin E supply[14]. The second,

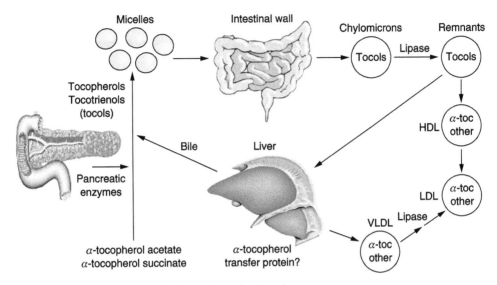

FIGURE 26.2 Absorption of tocopherols and tocotrienols in humans.

2,7,8-trimethyl-2-(β-carboxyethyl)-6-hydroxychroman, is believed to be a metabolite of γ-tocopherol. This metabolite, named LLU-α, has been suggested to be part of the natriuretic system, which controls extracellular fluids[15]. It was recently reported that γ-tocotrienol increases the production of LLU-α in rats[16]. It is believed that oxidized tocopherols (quinones) and other metabolic products are secreted in the bile but information on their structure and relative abundance is very limited[2].

26.3.1 BLOOD AND TISSUE LEVELS: INTERACTIONS OF TOCOPHEROLS AND TOCOTRIENOLS

Research from several laboratories, including ours, shows that high intake of α-tocopherol reduces the blood and tissue levels of γ-tocopherol[17–19]. Since there is evidence that all tocopherols are absorbed in a similar manner[8], it was suggested that this effect is due in part to the affinity of the tocopherol transfer protein for α-tocopherol. Ingestion of γ-tocopherol transiently increases plasma concentrations of γ-tocopherol as it undergoes sustained catabolism without markedly influencing the preexisting plasma pool of γ-tocopherol or the concentration and metabolism of α-tocopherol[20].

High levels of γ-tocopherol increase the blood levels of α-tocopherol which may be due to a sparing effect of γ-tocopherol for α-tocopherol[21]. A similar effect was observed by Stone et al.[19] (Figure 26.3). Cultured RAW 264.7 macrophages showed a greater uptake of γ-tocopherol compared to α-tocopherol. Surprisingly, γ-tocopherol promoted the cellular uptake of α-tocopherol[22].

The practical significance of the effect of high levels of α-tocopherol on the blood and tissue levels of γ-tocopherol (and possibly other tocopherols and tocotrienols) merits serious evaluation especially because α-tocopherol is the predominant form used in food fortification and in dietary supplements. This is particularly important in the light of the emerging understanding of the role of other tocopherols and tocotrienols.

26.4 TOCOPHEROLS AND TOCOTRIENOLS AS ANTIOXIDANTS

26.4.1 LIPID OXIDATION

A major antioxidant function of vitamin E in humans, studied mostly with α-tocopherol, is the inhibition of lipid peroxidation[23]. Vitamin E also plays a critical role in preventing lipid oxidation in lipoproteins. α-Tocopherol was thought to be primarily responsible for this effect in LDL because it is the most abundant and the best scavenger of peroxyl radicals. Chylomicrons, however, carry other tocopherols and tocotrienols in similar or even higher concentrations than α-tocopherol, depending on the diet, and they may also play an important role as lipid antioxidants. The other tocopherols and the tocotrienols may also function as important lipid antioxidants in adipose tissue and the liver. Some *in vitro* studies[2] suggested that tocotrienols are significantly more effective than tocopherols in inhibiting LDL oxidation; in contrast, studies with plasma obtained from rats fed tocotrienol-rich diets indicate approximately similar inhibition by α-tocotrienol and α-tocopherol; γ-tocopherol and γ-tocotrienol had similar effects but lower than the α forms. A recent study with rats indicated that γ-tocopherol and mixed tocopherols were more effective than α-tocopherol in inhibiting LDL oxidation[24-26]. Another recent study with humans evaluated the acetate esters of the tocotrienols. The results indicated that α-tocotrienol was the most potent tocotrienol for inhibiting LDL oxidation[27].

26.4.2   NITROGEN RADICALS

In addition to peroxy radicals, tocopherols and tocotrienols trap singlet oxygen and other reactive species and free radicals. The antioxidant effect of vitamin E on nitrogen reactive species has been receiving increasing attention. In biological systems, nitrogen dioxide (NO_2) is produced from the reaction of nitrogen oxide (NO) with oxygen. α-Tocopherol reacts with NO_2 to yield a nitrosating agent, but this does not occur with γ-tocopherol. In contrast, γ-tocopherol reduces NO_2 to NO or reacts with it without generating a nitrosating species[28,29]. This unique ability of γ-tocopherol, probably due to the number and positioning of the methyl groups on the chroman ring, may be particularly important in carcinogenesis, arthritis, and neurologic diseases because nitrosating agents can deaminate DNA bases causing mutations or interfere with important physiologic and immune functions[30,31].

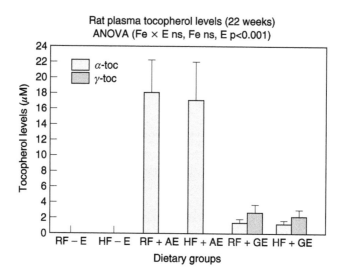

FIGURE 26.3 Plasma α- and γ-tocopherol levels after 22-week feeding. The trial included 6 dietary groups of 10 rats each. There were two levels of iron 35 mg/kg (low) and 280 mg/kg diet (high). Tocopherols were none, α-tocopherol (0.156 mmol/kg of feed), and γ-tocopherol (0.156 mmol/kg). Both tocopherols were in the naturally occurring RRR form. (Source: Stone, W.L., Papas, A.M., LeClair, I.O., Qui, M., and Ponder, T., *Cancer Detect. Prev.*, 26, 78–84, 2002.)

26.4.3 OTHER ANTIOXIDANT EFFECTS

Antioxidants function as components of a complex system with significant additive, synergistic, and antagonistic effects. Additive and synergistic effects of mixtures of tocopherols have been demonstrated in food systems and *in vitro*[2]. Vitamin C has been shown to regenerate α-tocopherol in such systems. These interactions are very important in evaluating the overall antioxidant properties. For example, it was reported that both α- and γ-tocopherol increased superoxide dismutase (SOD) activity in plasma and arterial tissues as well as Mn SOD and Cu/Zn SOD protein expression in arterial tissues. γ-Tocopherol and mixed tocopherols were more potent than α-tocopherol in these effects[25,26].

26.5 OTHER BIOLOGICAL EFFECTS OF TOCOPHEROLS AND TOCOTRIENOLS

Tocopherols and tocotrienols and their metabolites appear to have significant and sometimes quite different metabolic effects, which may be independent or only partially related to their function as antioxidants. The following are major examples of such effects with focus on γ-tocopherol and tocotrienols. The effects of α-tocopherol were reviewed previously.

26.5.1 ANTIINFLAMMATORY EFFECTS

Researchers at Berkeley evaluated the effects of α- and γ-tocopherol on γ-tocopherol reduced PGE2 synthesis in both lipopolysaccharide-stimulated macrophages and interleukin (IL)-1β-treated human epithelial cells. The major metabolite of dietary γ-tocopherol, 2,7,8-trimethyl-2-(β-carboxyethyl)-6-hydroxychroman (γ-CEHC or LLU-α), also exhibited an inhibitory effect in these cells. In contrast, α-tocopherol reduced slightly PGE2 formation in macrophages, but had no effect in epithelial cells. The inhibitory effects of γ-tocopherol and γ-CEHC stemmed from their inhibition of cyclooxygenase (COX)-2 activity, rather than affecting protein expression or substrate availability, and appeared to be independent of antioxidant activity[32]. In further work, γ-tocopherol, but not α-tocopherol, decreased proinflammatory eicosanoids and inflammation damage[33].

26.5.2 NATRIURETIC EFFECTS?

LLU-α, a metabolite of γ-tocopherol, has been suggested as an endogenous natriuretic factor[15] which may inhibit the 70 pS K channel in the apical membrane in the kidney. It was recently reported that γ-tocotrienol increased the concentration of this metabolite in the urine of rats[16]. If the proposed function of LLU-α is confirmed then γ-tocopherol and γ-tocotrienol may play a role in the control of extracellular fluids, which is important in hypertension and cardiovascular health.

26.5.3 SIGNAL TRANSDUCTION

Tocopherols and tocotrienols inhibit protein kinase C, an important isoenzyme in signal transduction. α-Tocopherol is a stronger inhibitor than β, γ, and δ-tocopherols and α-tocotrienol[34]. It was reported recently that α- and γ-tocopherol increased NO generation and nitric oxide synthase (cNOS) activity. However, only γ-tocopherol increased cNOS protein expression[35]. Nitric oxide plays a key role in signal transduction and arterial elasticity and dilation, and has antioxidant and other major functions in the body[36,37]. High levels of NO, however, can lead to production of nitrosating agent but this does not occur with γ-tocopherol[28,29]. Excess NO has been implicated in cerebral ischemic neurodegeneration and excitotoxicity[31]. If indeed γ-tocopherol stimulates the production of NO while preventing its conversion to NO_2, it could have a beneficial effect in cardiovascular health[36,37].

26.5.4 PLATELET ADHESION

This is the initial event in a cascade, which converts the soluble fibrinogen to insoluble fibrin and causes blood to clot. Platelet adhesion and aggregation are absolutely essential for prevention of hemorrhaging to death. Adhesion, however, of platelets to atherosclerotic lesions accelerates the formation of plaque and heart disease. α-Tocopherol and its quinone form reduce the rate of adhesion of platelets in a dose-dependent manner[38]. An antioxidant effect cannot explain the activity of the quinone. Vitamin E compounds appear to modulate the function of phospholipase A_2 and lipoxygenases and production of prostacyclin, a major metabolite of arachidonic acid[2]. It was reported recently that γ-tocopherol and mixed tocopherols were more effective than α-tocopherol in reducing platelet aggregation and thrombus formation in rats[25,26].

26.5.5 EFFECTS ON CHOLESTEROL SYNTHESIS AND REMOVAL

Tocotrienols, particularly γ-tocotrienol, appear to inhibit 3-hydroxy-3-methylglutaryl-coenzyme A (HMG-CoA) reductase in *in vitro* and animal models. This enzyme is important for the synthesis of cholesterol. Specifically it was suggested that tocotrienols inhibit the posttranscriptional suppression of HMG-CoA reductase[39]. Unlike tocopherols, tocotrienols, and particularly γ-tocotrienol, appear to reduce plasma apoB levels in hypercholesterolemic subjects. It has been suggested from preliminary data that γ-tocotrienol stimulates apoB degradation possibly as the result of decreased apoB translocation into the endoplasmic reticulum lumen[40].

26.5.6 EFFECTS ON CANCER CELLS

Research on the role of vitamin E in cancer has focused on α-tocopherol. Recent studies, primarily in cell cultures, suggest that other tocopherols and tocotrienols may affect the growth and/or proliferation of several types of human cancer cells. Tocopherol/tocotrienol blends reduced cell growth of breast cancer cells *in vitro*, while α-tocopherol alone had no effect[41]. Other researchers have reported that α, γ, and δ-tocotrienols and δ-tocopherol induce breast cancer cells to undergo apoptosis[42]. Another study suggested a role of γ-tocotrienol in inhibiting the growth of breast cancer, leukemia, and melanoma cells[43]. γ-Tocopherol appeared to be superior to α-tocopherol in inhibiting prostate cancer cells *in vitro*[44]. Our research indicated that γ-tocopherol, added to a sumipurified diet, was more effective than α-tocopherol in reducing *ras*-p21 oncogenes in the colonocyte of rats (Figure 26.4)[19]. Both α- and γ-tocopherols can upregulate the expression of peroxisome proliferator activator receptor gamma which is considered an important molecular target for colon cancer chemoprevention. The effect of γ-tocopherol, however, is several-fold more potent[45].

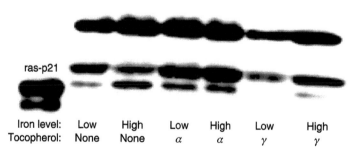

FIGURE 26.4 Effect of α- and γ-tocopherol on the expression of ras-p21 oncogenes in rat colonocytes. The trial included 6 dietary groups of 10 rats each as described in Figure 26.3. (Source: Stone, W.L., Papas, A.M., LeClair, I.O., Qui, M., and Ponder, T., *Cancer Detect. Prev.*, 26, 78–84, 2002.)

Although limited, the results of cancer cell studies conducted to date have been promising and suggest a need for more research to determine the role of all tocopherols and tocotrienols[46].

26.5.7 EFFECT ON C REACTIVE PROTEIN AND CELL APOPTOSIS

It was recently reported that high levels of α-tocopherol reduced C reactive protein in healthy and diabetic people[47]. The effect of other tocopherols and tocotrienols is not known. A recent study examined the role of γ-tocopherol in oxidized LDL-induced nuclear factor (NF)-kappaB activation and apoptosis in human coronary artery endothelial cells. Treatment of cells with γ-tocopherol attenuated degradation of IkappaB and activation of NF-kappaB and reduced apoptosis[24]. Another study reported that both α- and γ-tocopherols reduced oxidized LDL-induced activation and apoptosis in human coronary artery endothelial cells *in vitro* with α-tocopherol being more effective than γ-tocopherol[48]. As discussed above, tocotrienols and tocopherols may induce apoptosis of some cancer cells. This would suggest a different effect of tocopherols and tocotrienols on cancerous versus noncancerous cells. It also appears that the apoptotic effects of tocopherols and tocotrienols are not uniform.

26.6 TOCOPHEROLS AND TOCOTRIENOLS IN HEALTH AND DISEASE

26.6.1 EFFECT ON CHOLESTEROL

Because of their effect on HMG-CoA reductase, there has been significant interest on the role of tocotrienols in the lipoprotein profile. The tocotrienol-rich products available commercially are mixtures of tocopherols and tocotrienols and other phytochemicals such as sterols. For this reason, direct comparisons of results from human studies with effects observed *in vitro* are difficult.

In animal studies reviewed earlier, tocotrienols and tocotrienol-rich extracts were reported to decrease total and LDL cholesterol[49]. In a recent study with apolipoprotein ApoE +/– female mice, which develop atherosclerosis only when fed diets high in triacylglycerol and cholesterol, mice fed a palm oil tocotrienol-rich extract had 60% lower plasma cholesterol than groups fed the atherogenic diets. Mice fed the atherogenic diet had markedly higher VLDL, intermediate-density lipoprotein (IDL), and LDL cholesterol and markedly lower HDL cholesterol than the controls. Lipoprotein patterns in mice supplemented with α-tocopherol were similar to those of the mice fed an atherogenic diet alone, but the pattern in mice supplemented with the tocotrienol extract was similar to that of mice fed the control diet[50].

Human studies have produced conflicting results. In a recent study in humans, subjects were randomly assigned to receive placebo, α-, γ-, or δ-tocotrienyl acetate supplements (250 mg/d). Subjects followed a low-fat diet for 4 weeks, and then took supplements with dinner for the following 8 weeks while still continuing diet restrictions. α-Tocotrienyl acetate increased *in vitro* LDL oxidative resistance by 22%, and decreased its rate of oxidation. Neither serum or LDL cholesterol nor apolipoprotein B were significantly decreased by tocotrienyl acetate supplements[27].

In two other studies in humans with tocotrienol-rich palm oil there was no significant effect on cholesterol profile. In the first study conducted in The Netherlands, 20 men with slightly elevated lipid concentrations received daily for 6 weeks a palm oil-rich tocotrienol extract, which supplied 140 mg tocotrienols and 80 mg α-tocopherol; 20 other men received daily 80 mg α-tocopherol. The tocotrienol-rich extract had no marked favorable effects on the serum lipoprotein profile[51]. In the second study, a five-year clinical study described below, a palm oil-rich tocotrienol extract which supplied approximately 240 mg tocotrienols and 98 mg α-tocopherol, did not reduce cholesterol levels after 18 months' supplementation. There was, however, significant effect on LDL oxidation as indicated by thiobarbituric acid reactive substances (TBARS)[49,52]. In contrast, with the same experimental group, a rice bran oil tocotrienol-rich extract supplying 312 mg tocotrienols and 360 mg tocopherols reduced significantly total cholesterol, LDL, and triglycerides. HDL was higher in the treated group than in the control[53].

In another study supplementation with 200 mg tocotrienols per day from three commercially available sources has no beneficial effect on cholesterol parameters in adults with elevated blood lipid concentrations[54].

26.6.2   Cardiovascular Health

Epidemiological studies indicated a strong inverse association of blood vitamin E or vitamin E intake, measured as α-tocopherol, and heart disease[55,56]. Subsequent large intervention studies with high levels of α-tocopherol failed to show the beneficial effect especially in high-risk subjects[57,58].

It is important to consider whether these results are due, at least in part, to the use of α-tocopherol alone. This is an important consideration because, as discussed above, high levels of supplemental α-tocopherol reduce the blood and tissue levels of γ-tocopherol and probably of the other tocopherols and tocotrienols. Some epidemiological studies suggest an inverse association of blood γ-tocopherol and heart disease[59,60]. In this regard, the role of tocotrienols may also be very important.

A five-year study in humans evaluated 50 patients with previous cerebrovascular disease. These patients with confirmed carotid atherosclerosis were divided randomly into two groups. Twenty-five consumed a tocotrienol-rich extract which contained approximately 240 mg of tocotrienols (plus tocopherols)/day while 25 control patients received a placebo. After 18 months, patients receiving the tocotrienol supplement demonstrated a significant regression or slowed progression in the amount of blockage of their carotid artery as revealed by ultrasonography compared to the controls[52]. This trend continued through four years of supplementation. Those receiving a placebo had an overall worsening of the disease. These results are remarkable because over 40% of the patients showed regression of stenosis[49].

26.6.3   Cancer

A major intervention clinical study indicated that in elderly smokers receiving α-tocopherol as dl-α-tocopheryl acetate (50 mg/d) the incidence of prostate cancer was lower than in the controls[61]. There was also a trend for lower incidence of colon cancer. Researchers at Johns Hopkins University reported that there was a stronger inverse relationship between blood γ-tocopherol and prostate cancer than the association observed with α-tocopherol[62]. As discussed above, tocotrienols and δ-tocopherol inhibited the growth of several lines of cancer cells including those of breast, leukemia, and melanoma. In the light of these findings it is important to evaluate the role of tocopherols and tocotrienols in the prevention of some cancers, especially colon, breast, and prostate cancers.

The Selenium and Vitamin E Cancer Prevention Trial (SELECT) evaluates the role of supplemental selenium and α-tocopherol in 32,400 men in a 2×2 factorial design. The trial commenced in 2001 and was to be completed in 2003[63]. Results are expected to be communicated at a later date.

26.6.4   Other Chronic Diseases

The role of α-tocopherol in chronic disease has been evaluated for a variety of diseases including cataracts, Alzheimer's disease, and diabetes[1]. In the light of our emerging understanding of the role of the other tocopherols and tocotrienols it is important to determine whether individually or, more importantly, as mixtures they have a role in the prevention and/or treatment of these diseases. This is particularly important for γ-tocopherol which has been suggested to affect the production and metabolism of NO which can have either beneficial effects or harmful effects if it is metabolized to nitrogen reactive compounds.

26.7 CONCLUSIONS

Recent studies suggest that the non-α-tocopherols and the tocotrienols have antioxidant and other functions and may play an important role in nutrition and health. The understanding of their effects will help determining whether the current practice of using α-tocopherol in food fortification and in supplements is warranted.

REFERENCES

1. Papas, A.M., *The Vitamin E Factor*, HarperCollins, New York, 1999.
2. Papas, A.M., Ed., *Antioxidant Status, Diet, Nutrition and Health*, CRC Press, Boca Raton, FL, 1998.
3. Yamamoto, Y., Fujisawa, A., Hara, A., and Dunlap, W.C., An unusual vitamin E constituent (alpha-tocomonoenol) provides enhanced antioxidant protection in marine organisms adapted to cold-water environments, *Proc. Natl. Acad. Sci. USA*, 98, 13144–13148, 2001.
4. NRC, *Recommended Dietary Allowances*, 10th ed., National Academy Press, Washington, DC, 1989.
5. Burton, G.W., Traber, M.G., Acuff, R.V., Walters, D.N., Kayden, H., Hughes, L., and Ingold. K.U., Human plasma and tissue alpha-tocopherol concentrations in response to supplementation with deuterated natural and synthetic vitamin E, *Am. J. Clin. Nutr.*, 67, 669–684, 1998.
6. Acuff, R.V., Thedford S.S., Hidiroglou N.N., Papas, A.M., and Odom, T.A., Jr., Relative bioavailability of RRR- and all-rac-alpha-tocopheryl acetate in humans: studies using deuterated compounds, *Am. J. Clin. Nutr.*, 60, 397–402, 1994.
7. Food and Nutrition Board, National Academy of Sciences, *Dietary Reference Intakes for Vitamin C, Vitamin E, Selenium, and Carotenoids*, National Academy Press, Washington, DC, 2000.
8. Kayden, H.J. and Traber, M.G., Absorption, lipoprotein transport and regulation of plasma concentrations of vitamin E in humans, *J. Lipid Res.*, 34, 343–358, 1993.
9. Hayes, K.C., Pronczuk, A., and Liang, J.S., Differences in the plasma transport and tissue concentrations of tocopherols and tocotrienols: observations in humans and hamsters, *Proc. Soc. Exp. Biol. Med.*, 202, 353–359, 1993.
10. Ikeda, I., Imasato, Y., Sasaki, E., and Sugano, M., Lymphatic transport of alpha-, gamma- and delta-tocotrienols and alpha-tocopherol in rats, *Int. J. Vitam. Nutr. Res.*, 66, 217–221, 1996.
11. Bieri, J.G. and Evarts R.P., Tocopherols and fatty acids in American diets, *J. Am. Diet. Assoc.*, 62, 147–151, 1973.
12. Murphy, S.P., Subar, A.F., and Block, G., Vitamin E intakes and sources in the United States, *Am. J. Clin. Nutr.*, 52, 361–367, 1990.
13. Dutton, P.J., Hughes, L.A., Foster, D.O., Burton, G.W., and Ingold, K.U., Simon metabolites of alpha-tocopherol are not formed via a rate-controlling scission of the 3'C–H bond, *Free Radic. Biol. Med.*, 9, 435–439, 1990.
14. Schultz, M., Leist, M., Petrzika, M., Gassmann, B., and Brigelius-Flohé, R., A novel urinary metabolite of α-tocopherol, 2,5,7,8-tetramethyl-2(2'-carboxyethyl)-6-hydroxy chroman (α-CEHC) as an indicator of an adequate vitamin E supply?, *Am. J. Clin. Nutr.*, 62 (Suppl.), 1527S–1534S, 1995.
15. Wechter, W.J., Kantoci, D., Murray, E.D. Jr., D'Amico, D.C., Jung, M.E., and Wang, W.H., A new endogenous natriuretic factor: LLU-alpha, *Proc. Natl. Acad. Sci. USA*, 93, 6002–6007, 1996.
16. Hattori, A., Fukushima, T., Yoshimura, H., Abe, K., and Ima, K., Production of LLU-alpha following an oral administration of gamma-tocotrienol or gamma-tocopherol to rats, *Biol. Pharm. Bull.*, 23, 1395–1397, 2000.
17. Handelman, G.J., Epstein, W.L., Peerson, J., Spiegelman, D., and Machlin, L.J., Human adipose α-tocopherol and γ-tocopherol kinetics during and after 1 y of α-tocopherol supplementation, *Am. J. Clin. Nutr.*, 59, 1025–1032, 1994.
18. Traber, M.G. and Kayden, H.J., Preferential incorporation of alpha-tocopherol vs gamma-tocopherol in human lipoproteins, *Am. J. Clin. Nutr.*, 49, 517–526, 1989.
19. Stone, W.L., Papas, A.M., LeClair, I.O., Qui, M., and Ponder, T., The influence of dietary iron and tocopherols on oxidative stress and ras-p21 levels in the colon, *Cancer Detect. Prev.*, 26, 78–84, 2002.
20. Galli, F., Lee, R., Atkinson, J., Floridi, A., and Kelly, F.J., gamma-Tocopherol biokinetics and transformation in humans, *Free Radic. Res.* 37, 1225–1233, 2003.

21. Clement, M. and Bourre, J.M., Graded dietary levels of RRR-gamma-tocopherol induce a marked increase in the concentrations of alpha- and gamma-tocopherol in nervous tissues, heart, liver and muscle of vitamin-E-deficient rats, *Biochim. Biophys. Acta*, 1334, 173–181, 1997.

22. Gao, R., Stone, W.L., Huang, T., Papas, A.M., and Qui, M., The uptake of tocopherols by RAW 264.7 macrophages, *Nutr. J.*, 1, 2, 2002.

23. Burton, G.W., Joyce, A., and Ingold, K.U., First proof that vitamin E is major lipid-soluble, chain-breaking antioxidant in human blood plasma, *Lancet*, 8292ii, 327, 1982.

24. Li, D., Saldeen, T., and Mehta, J.L., gamma-Tocopherol decreases ox-LDL-mediated activation of nuclear factor-kappaB and apoptosis in human coronary artery endothelial cells. *Biochem. Biophys. Res. Commun.*, 259, 157–161, 1999.

25. Saldeen, T., Li, D., and Mehta, J.L., Differential effects of alpha- and gamma-tocopherol on low-density lipoprotein oxidation, superoxide activity, platelet aggregation and arterial thrombogenesis, *J. Am. Coll. Cardiol.*, 34, 1208–1215, 1999.

26. Liu, M., Wallin, R., Wallmon, A., and Saldeen, T., Mixed tocopherols have a stronger inhibitory effect on lipid peroxidation than alpha-tocopherol alone, *J. Cardiovasc. Pharmacol.*, 39, 714–21, 2002.

27. O'Byrne, D., Grundy, S., Packer, L., Devaraj, S., Baldenius, K., Hoppe, P.P., Kraemer, K., Jialal, I., and Traber M.G., Studies of LDL oxidation following alpha-, gamma-, or delta-tocotrienyl acetate supplementation of hypercholesterolemic humans, *Free Radic. Biol. Med.*, 29, 834–845, 2000.

28. Cooney, R.V., Franke, A.A., Harwood, P.J., Hatch-Pigott, V., Custer, L.J., and Mordan, L.J., Gamma-tocopherol detoxification of nitrogen dioxide: superiority to alpha-tocopherol, *Proc. Natl. Acad. Sci. USA*, 90, 1771–1775, 1993.

29. Christen, S., Woodall, A.A., Shigenaga, M.K., Southwell-Keely, P.T., Duncan, M.W., and Ames B.N., Gamma-tocopherol traps mutagenic electrophiles such as NO(X) and complements alpha-tocopherol: physiological implications, *Proc. Natl. Acad. Sci. USA*, 94, 3217–22, 1997.

30. Wink, D.A., Vodovotz, Y., Laval, J., Laval, F., Dewhirst, M.W., and Mitchell, J.B., The multifaceted roles of nitric oxide in cancer, *Carcinogenesis*, 19, 711–721, 1998.

31. Strijbos, P.J., Nitric oxide in cerebral ischemic neurodegeneration and excitotoxicity, *Crit. Rev. Neurobiol.*, 12, 223–243, 1998.

32. Jiang, Q., Elson-Schwab, I., Courtemanche, C., and Ames, B.N., gamma-Tocopherol and its major metabolite, in contrast to alpha-tocopherol, inhibit cyclooxygenase activity in macrophages and epithelial cells, *Proc. Natl. Acad. Sci. USA*, 97, 11494–11499, 2000.

33. Jiang, Q. and Ames, B.N., Gamma-tocopherol, but not alpha-tocopherol, decreases proinflammatory eicosanoids and inflammation damage in rats, *FASEB J.*, 17, 816–822, 2003.

34. Õzer, N.K. and Azzi, A., Beyond antioxidant function: other biochemical effects of antioxidants, in *Antioxidant Status, Diet, Nutrition and Health*, Papas, A.M., Ed., CRC Press, Boca Raton, FL, 1998, pp. 449–460.

35. Li, D., Saldeen, T., Romeo, F., and Mehta, J.L., Relative Effects of alpha- and gamma-tocopherol on low-density lipoprotein oxidation and superoxide dismutase and nitric oxide synthase activity and protein expression in rats, *J. Cardiovasc. Pharmacol. Ther.*, 4, 219–226, 1999.

36. Ignarro, L.J., Byrns R.E., Buga G.M., Wood, K.S., and Chaudhuri G., Endothelium-derived relaxing factor produced and released from artery and vein is nitric oxide, *Proc. Natl. Acad. Sci. USA*, 84, 9265–9269, 1987.

37. Dowd, P. and Zheng, Z.B., On the mechanism of the anticlotting action of vitamin E quinone, *Proc. Natl. Acad. Sci. USA*, 92, 8171–8175, 1995.

38. Murad, F., Nitric oxide signaling: would you believe that a simple free radical could be a second messenger, autacoid, paracrine substance, neurotransmitter, and hormone?, *Recent Prog. Horm. Res.*, 53, 43–59, 1998.

39. Parker, R.A., Pearce, B.C., Clark, R.W., Gordon, D.A., and Wright, J.J., Tocotrienols regulate cholesterol production in mammalian cells by post-transcriptional suppression of 3-hydroxy-3-methylglutaryl-coenzyme A reductase, *J. Biol. Chem.*, 268, 11230–11238, 1993.

40. Theriault, A., Wang, Q., Gapor, A., and Adeli, K., Effects of gamma-tocotrienol on ApoB synthesis, degradation, and secretion in HepG2 cells, *Arterioscler. Thromb. Vasc. Biol.*, 19, 704–712, 1999.

41. Nesaretnam, K., Stephen, R., Dils, R., and Darbre, P., Tocotrienols inhibit the growth of human breast cancer cells irrespective of estrogen receptor status, *Lipids*, 33, 461–469, 1998.

42. Yu, W., Simmons-Menchaca, M., Gapor, A., Sanders, B.G., and Kline, K., Induction of apoptosis in human breast cancer cells by tocopherols and tocotrienols, *Nutr. Cancer*, 33, 26–32, 1999.

43. Mo, H. and Elson, C.E., Apoptosis and cell-cycle arrest in human and murine tumor cells are initiated by isoprenoids, *J. Nutr.*, 129, 804–813, 1999.

44. Moyad, M.A., Brumfield, S.K., and Pienta, K.J., Vitamin E, alpha- and gamma-tocopherol, and prostate cancer, *Semin. Urol. Oncol.*, 17, 85–90, 1999.

45. Campbell, S.E., Stone, W.L., Whaley, S.G., Qui, M., and Krishnan, K., Gamma tocopherol upregulates peroxisome proliferator activated receptor (PPAR) gamma expression in SW 480 human colon cancer cell lines, *BMC Cancer*, 3, 25, 2003.

46. Stone, W.L. and Papas, A.M., Tocopherols and the etiology of colon cancer, *J. Natl. Cancer Inst.*, 89, 1006–1014, 1997.

47. Devaraj, S. and Jialal, I., Alpha tocopherol supplementation decreases serum C-reactive protein and monocyte interleukin-6 levels in normal volunteers and type 2 diabetic patients. *Free Radic. Biol. Med.*, 29, 790–792, 2000.

48. de Nigris, F., Franconi, F., Maida, I., Palumbo, G., Anania, V., and Napoli, C., Modulation by alpha- and gamma-tocopherol and oxidized low-density lipoprotein of apoptotic signaling in human coronary smooth muscle cells, *Biochem. Pharmacol.*, 59, 1477–1487, 2000.

49. Watkins, T.R., Bierenbaum, M.L., and Giampaolo, A., Tocotrienols: biological and health effects, in *Antioxidant Status, Diet, Nutrition and Health*, Papas A.M., Ed., CRC Press, Boca Raton, FL, 1998, pp. 479–496.

50. Black, T.M., Wang, P, Maeda, N., and Coleman, R.A., Palm tocotrienols protect ApoE +/– mice from diet-induced atheroma formation, *J. Nutr.*, 130, 2420–2426, 2000.

51. Mensink, R.P., van Houwelingen, A.C., Kromhout, D., and Hornstra, G., A vitamin E concentrate rich in tocotrienols had no effect on serum lipids, lipoproteins, or platelet function in men with mildly elevated serum lipid concentrations, *Am. J. Clin. Nutr.*, 69, 213–219, 1999.

52. Tomeo, A.C., Geller, M., Watkins, T.R., Gapor, A., and Bierenbaum, M.L., Antioxidant effects of tocotrienols in patients with hyperlipidemia and carotid stenosis, *Lipids*, 30, 1179–1183, 1995.

53. Watkins, T.R., Geller, M., Kooyenga, D.K., and Bierenbaum, M., Hypocholesterolemic and antioxidant effects if rice bran oil non-saponifiables in hypercholesterolemic subjects, *J. Environ. Nutr. Int.*, 3, 115–122, 1999.

54. Mustad, V.A., Smith, C.A., Ruey, P.P., Edens, N.K., and DeMichele, S.J., Supplementation with 3 compositionally different tocotrienol supplements does not improve cardiovascular disease risk factors in men and women with hypercholesterolemia, *Am. J. Clin. Nutr.*, 76, 1237–1243, 2002.

55. Rimm, E.B., Stampher, M.J., Ascherio, A., Giovannuci, E., Colditz, M.B., and Willet, W.C., Vitamin E consumption and the risk of coronary disease in men, *N. Engl. J. Med.*, 328, 1450–1456, 1993.

56. Stampfer, M.J., Hennekens, C.H., Manson, J.E., Colditz, G.A., Rosner, B., and Willett, W.C., Vitamin E consumption and the risk of coronary disease in women, *N. Engl. J. Med.*, 328, 1444–1449, 1993.

57. GISSI Group, Dietary supplementation with n-3 polyunsaturated fatty acids and vitamin E after myocardial infarction: results of the GISSI-Prevenzione trial, *Lancet*, 354, 447–455, 1999.

58. Yusuf, S., Dagenais, G., Pogue, J., Bosch, J., and Sleight, P., Vitamin E supplementation and cardiovascular events in high-risk patients. The Heart Outcomes Prevention Evaluation Study Investigators, *N. Engl. J. Med.*, 342, 154–160, 2000.

59. Kontush, A., Spranger, T., Reich, A., Baum, K., and Beisiegel U., Lipophilic antioxidants in blood plasma as markers of atherosclerosis: the role of alpha-carotene and gamma-tocopherol, *Atherosclerosis*, 144, 117–122, 1999.

60. Kristenson, M., Zieden, B., Kucinskiene, Z., Elinder, L.S., Bergdahl, B., Elwing, B., Abaravicius, A., Razinkoviene, L., Calkauskas, H., and Olsson, A.G., Antioxidant state and mortality from coronary heart disease in Lithuanian and Swedish men: concomitant cross sectional study of men aged 50, *BMJ*, 314, 629–633, 1997.

61. Heinonen, O.P., Albanes, D., and Virtamo, J., Prostate cancer and supplementation with alpha-tocopherol and beta-carotene: incidence and mortality in a controlled trial, *J. Natl. Cancer Inst.*, 90, 440–446, 1998.

62. Helzlsouer, K.J., Huang, H-Y, Alberg, A.J., Hoffman, S., Burke, A., Norkus E.P., Morris, J.S., and Comstock, G.W., Association between alpha-tocopherol, gamma-tocopherol, selenium, and subsequent prostate cancer, *J. Natl. Cancer Inst.*, 92, 2018–2023, 2000.

63. Klein, E.A., Thompson, I.M., Lippman, S.M., Goodman, P.J., Albanes, D., Taylor, P.R., and Coltman, C., SELECT: the Selenium and Vitamin E Cancer Prevention Trial: rationale and design, *Prostate Cancer Prostatic Dis.*, 3, 145–151, 2000.

27 Plant Sterols and Steryl Esters in Functional Foods and Nutraceuticals

Nikolaus Weber and Kumar D. Mukherjee
Institute for Lipid Research, Federal Research Centre for Nutrition and Food, Münster, Germany

CONTENST

27.1   INTRODUCTION

Phytosterols are analogs of cholesterol which occur in plant materials and, similar to sterols in mammalian cells, they are structural and functional constituents of cell membranes[1,2]. Sterols have the tetracyclic ring system specific to steroids, a hydroxyl group at C-3, and a side chain of different length at C-17. The IUPAC-IUB rules for numbering sterols and for ring lettering are based on the cholestane (cyclopenta[α]phenanthrene) skeleton shown for sitosterol in Figure 27.1.

Phytosterols can be divided, on both structural and biosynthetic bases, into 4-demethylsterols, 4-monomethylsterols, and 4,4-dimethylsterols based on the number of methyl groups at C-4 in the tetracyclic ring system (Figure 27.2). Plant sterols are not endogenously synthesized in the human body and are derived exclusively from dietary sources.c

4-Demethylsterols which do not carry methyl groups at C-4 are the major class of sterols in (desmethylsterols) plants. The 4-demethylsterols constitute >85% of total sterols in common crude vegetable oils[3]. Sitosterol (β-sitosterol), the major sterol in most plants, is usually accompanied by its 22-dehydro analog, stigmasterol. The molecular structures of the most important plant sterols and stanols are shown in Figure 27.3. Another abundant plant sterol is 24-methylcholesterol,

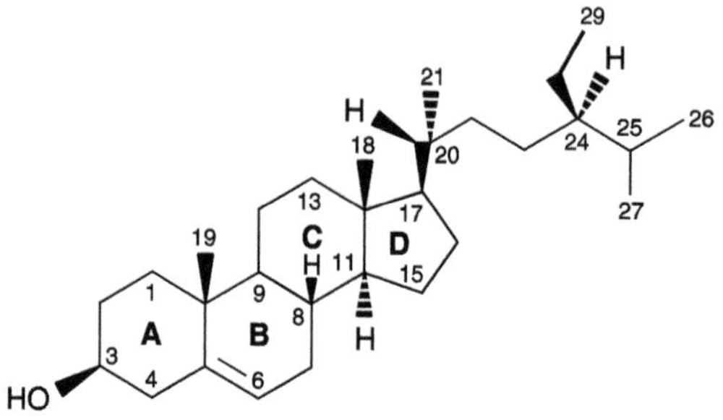

FIGURE 27.1 Chemical structure of sitosterol (β-sitosterol) and IUPAC-IUB rules for numbering sterols and for ring lettering based on the cholestane skeleton.

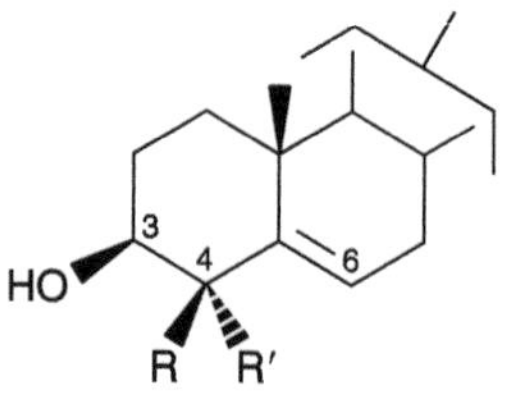

FIGURE 27.2 Chemical structure of 4,4-demethyl- (R, R′ = H), 4-methyl- (R, R′ = Me or H), and 4,4-dimethylsterols (R, R = Me).

frequently referred to as campesterol. 24-Methylcholesterol is usually a mixture of two epimers, i.e., 24α-methylcholesterol (campesterol) and 24β-methylcholesterol (22,23-dihydrobrassicasterol).

The similarity between the ring structure and the substitution at C-24 of cholesterol, also a 4-demethylsterol, to those of some plant sterols (Figure 27.3) seems to be the determining elements for the important effects of plant sterols on cholesterol absorption, e.g., inhibition of cholesterol absorption by plant sterols. It should be noted that only 4-demethylsterols — not 4-methyl or 4,4-dimethylsterols — exert this effect.

4-Methylsterols are minor constituents of plant sterols and they are precursors of the desmethylsterols. Thus the sterols from soybean are composed of 98.1% 4-demethylsterols, 1.3% 4-methylsterols, and 0.6% 4,4-dimethylsterols[4].

4,4-Dimethylsterols occur in trace amounts in many plants. Some plant materials such as rice bran oil contain substantial proportions of 4,4-dimethylsterols, e.g., cycloartenol and 24-methylcycloartenol. Crude rice bran oil contains 1.8, 0.4, and 1.2 g/100 g oil of 4-desmethyl, 4-methyl, and 4,4-dimethylsterols, respectively, which occur mainly as esters of ferulic and p-coumaric acids[5]. Phytostanols are saturated analogs of sterols lacking double bonds in the ring structure and the side chain (see Figure 27.3). The biosynthesis, biological function, and importance of plant sterols to human nutrition have been reviewed[3]. Sitostanol and campestanol, which are 5,6-saturated analogs of the major 4-demethylsterols (sitosterol and campesterol), occur in plants such as coniferous trees (pine, spruce) and certain grains (rye, wheat, and corn).

Phytosteryl and phytostanyl esters of fatty acids (Figure 27.4) are well established as useful supplements for functional foods. It is generally accepted that among other factors, increased concentrations in blood of both total cholesterol and low-density lipoprotein (LDL) cholesterol and reduced concentrations of high-density lipoprotein (HDL) cholesterol are associated with higher risk of cardiovascular diseases. The concentrations of total, LDL, and HDL cholesterol can be

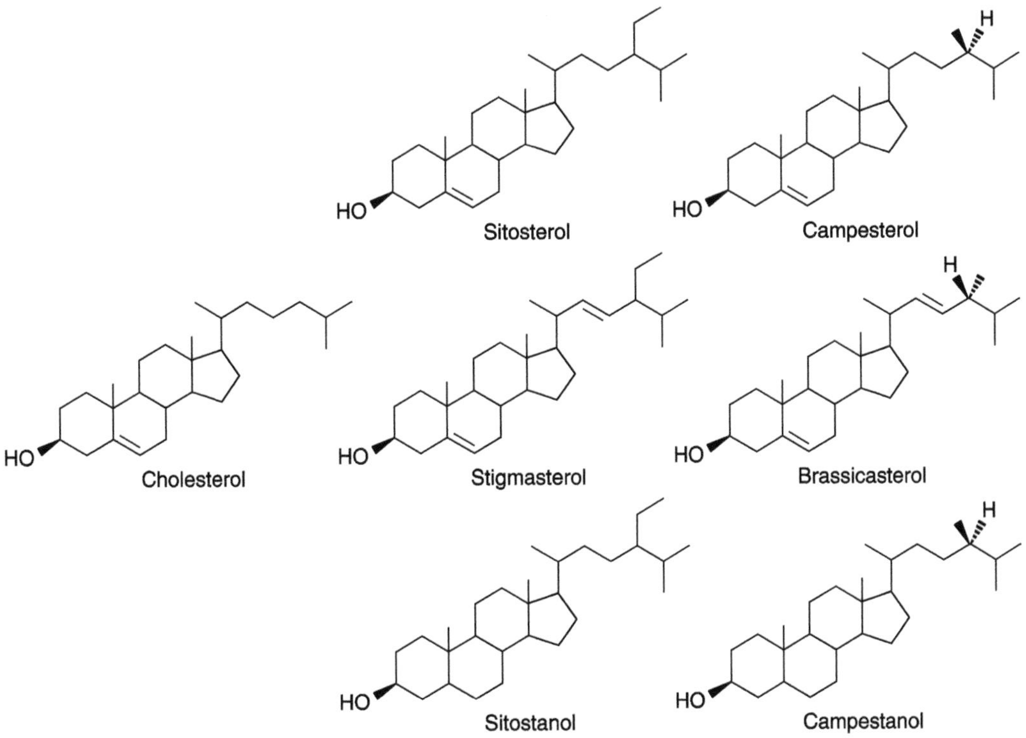

FIGURE 27.3 Chemical structure of various phytosterols and phytostanols as compared to cholesterol (stereochemical configuration as given in Figure 27.1 if not otherwise indicated).

affected beneficially by lifestyle and dietary changes as well as by medication[6,7]. It is well known that, in addition to the above measures, the ingestion of phytosterols, phytostanols, and their fatty acid esters can reduce total as well as LDL cholesterol concentrations of blood by inhibiting the intestinal absorption of cholesterol — both of dietary origin and that formed endogenously[8–11].

Many studies are known on the cholesterol-lowering effect of phytosterols and phytostanols added as their fatty acid esters to margarines and spreads (Table 27.1 and Table 27.2). Fatty acid esters of phytosterols and phytostanols, rather than the unesterified sterols, are commonly used as supplements in fat-containing functional foods, such as margarines, spreads, cooking and salad oils, mayonnaise, and salad dressings, due to a better solubility of the former in fat[12]. A meta-analysis of randomized, placebo-controlled, double-blind intervention studies has revealed that the daily intake of 2 g of phytosterols, phytostanols, or their esters reduces the LDL cholesterol concentration of blood by 9 to 14% without affecting the concentrations of HDL cholesterol and triacylglycerols[13].

Despite the close structural resemblance between cholesterol, phytosterols, and phytostanols (Figures 27.1–27.3), the intestinal absorption in humans of dietary phytosterols and phytostanols is far lower[41–43] than that of dietary cholesterol[23]. Moreover, phytosterols and phytostanols reduce the intestinal absorption of both dietary cholesterol and endogenously synthesized cholesterol which is excreted into the bile[44]. The above effects of dietary phytosterols and phytostanols lead to reduction of total cholesterol and LDL cholesterol of blood; however, the exact mechanism of this action is not known[2].

The intake of phytosterols and their esters in industrial countries amounts to 200 to 400 mg per day and up to 800 mg per day for vegetarians[45]. The intake of phytostanols and their esters is estimated to be about 20 to 50 mg per day[46]. Major dietary sources of phytosterols and their esters are plant oils.

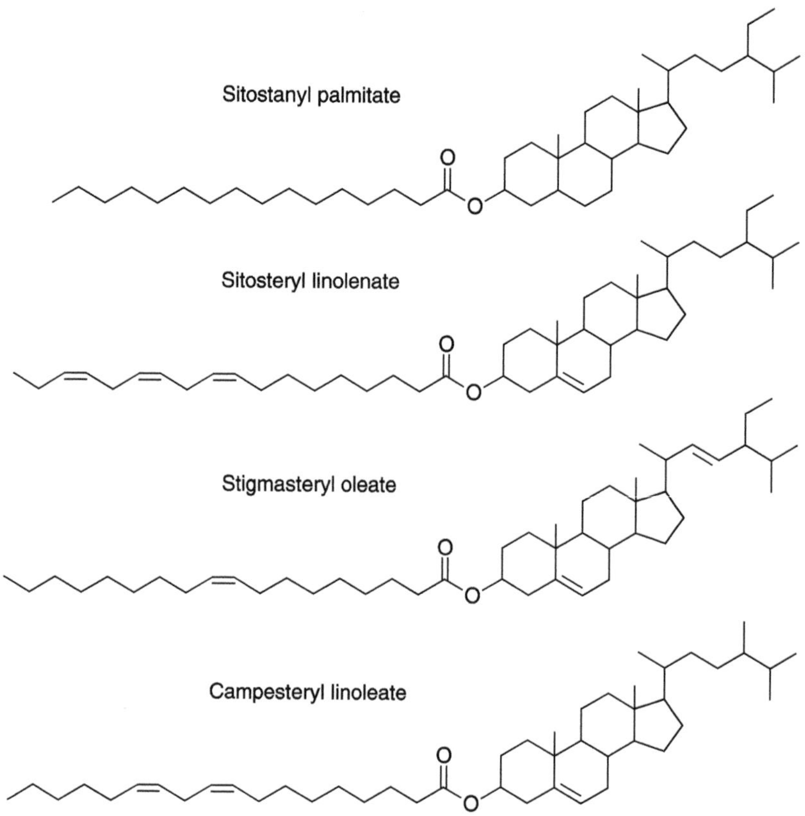

FIGURE 27.4 Chemical structure of various phytosteryl esters and phytostanyl esters.

27.2 OCCURRENCE OF PHYTOSTEROLS, PHYTOSTANOLS, AND THEIR ESTERS

Plant oils, such as corn oil and rapeseed oil, are major dietary sources of phytosterols, whereas phytostanols occur mainly in cereals, such as wheat and rye[47,48]. Phytosterols and phytostanols occur naturally both as unesterified sterols and steryl esters of carboxylic acids. Both phytosterols and phytostanols also occur in nature as glycosidic conjugates such as steryl glycosides and their fatty acid esters.

Table 27.3 summarizes the content of individual sterols of various fats and oils. It is obvious from the table that both the concentration and pattern of sterols vary widely in the fats and oils of different origin. The salad and cooking oils predominantly contain Δ^5- (and partly Δ^7-) sterols. The sterol composition of various Amazonian palm kernel oils has been reported[49]. Table 27.4 summarizes the content of phytosterols and phytostanols in selected foodstuffs.

Nutritional studies (Table 27.1 and Table 27.2) indicate that a new approach to increase the contents of phytosterols, phytostanols, and their fatty acid esters in plant oils is desirable[56]. Traditional breeding, however, has not been very effective to reach this goal in oil plants such as rapeseed or soybean[57]. Genetic engineering of oil plants might be able to increase the phytosterol content of oilseeds as was shown recently for transgenic tobacco seeds[58].

TABLE 27.1
Results of Selected Clinical Studies on Cholesterol-Reducing Action of Phytostanyl Esters

Number of patients	Duration of studies (weeks)	Dose of phytostanyl esters (g/day)	Decrease of LDL cholesterol (%)	Ref.
34	6	3.4	10	14
7	6	0.8	7	15
7	9	0.8	8	16
7	6	2	16	15
11	6	3	9	17
8	7	3.0	14	18
23	6	3.16	8	19
23	6	3.16	10	19
15	6	3.0	15	20
11	7	3.0	15	21
61	8	2	7	22
318	8	2.2	5	23
79	8	1.0	14	24
51	52	1.8	11	25
51	52	2.6	14	25
11	1	2	14	26
12	5	2.2	13	27
39	4	2.47	9	28
112	8	3.8	13	29
112	8	4	13	29
20	8	2.2	9	30
18	8	2.4	14	30
34	4	2.0	12	31
22	4	1.6	6	32
22	4	3.2	10	32
155	24	1.5	7.5–11.6	33

TABLE 27.2
Results of Selected Clinical Studies on Cholesterol-Reducing Action of Phytosteryl Esters

Number of patients	Duration of study (weeks)	Dose of phytosteryl esters (g/day)	Decrease of LDL cholesterol (%)	Ref.
100	3.5	3.2	13	34
80	3.5	3.0	7	35
90	5	1.1	8	36
21	3 or 4	8.6	23	37
15	3	1.8	13	38
34	4	2.0	10	31
30	4	2.0	15.6	39
62	8	2.5	10–15	40

TABLE 27.3
Sterol Composition (mg/kg) of Various Fats and Oils[50]

Fat or oil	Campesterol	Stigmasterol	Sitosterol	Other 4-demethyl sterols	Total 4-demethylsterols [range][51]	4-Methyl-sterols	Triterpene alcohols
Coconut	18	296	1322	478	[470–1100] 2000[52]	159	679
Cottonseed	170	42	3961	85	[2690–5915] 3500[52]	418	479
Maize (corn)	2691	702	7722	585	[7931–22137] 12400[53]	128	131
Olive	28	14	1310	87	[1820–1938][54] 1600[52]	152	1440
Palm	139[52]	73[52]	358[52]	30[52]	[389–481] 600[52]	358	313
Palm kernel[49]	59–130	35–174	746–2664	95–860	[1088–3719]	—	—
Peanut	360	2160	1536	288	[901–2854] 2520[53]	157	360
Rapeseed	1530	—	3549	1040	[4824–11276] 4500[53]	266	540
Rice bran	5056	2709	8849	1444	—	4200	11642
Shea butter	—	—	—	1357	—	1020	38251
Soybean	720	720	1908	252	[1837–4089] 2600[53], 3700[52]	658	845
Sunflower	313	313	2350	939	[2437–4545] 4170[53]	1119	700
Wheat germ	5702	—	17336	2850	—	1158	2106

TABLE 27.4
Phytosterol and Phytostanol Content of Selected Foodstuffs

Foodstuff	Total content of phytosterols and phytostanols (mg/100 g)
Potatoes	5
Tomatoes	7
Pears	8
Lettuce	10
Carrots	12
Apples	12
Onions	15
Bananas	16
Figs	31
Kidney beans	127
Soybeans	161
Almonds	143
Sesame seeds	714
Cashew nuts	158
Peanuts	220

Data from Moreau, R.A., Norton, R.A., and Hicks, K.B., *Int. News Fats Oils Relat. Mater.*, 10, 572–577, 1999.

27.3 ISOLATION OF PHYTOSTEROLS, PHYTOSTANOLS, AND THEIR ESTERS

Phytosterols and their fatty acid esters occur as byproducts in the steam distillate resulting from the deodorization or physical refining of plant oils. Phytostanols, phytosterols, and their esters are also present in tall oil which is a byproduct formed from wood during paper manufacture.

A considerable proportion of phytosterols and their esters present in plant oils is lost during oil refining[47,54,59]. Different losses of phytosterols and their esters occur in the individual processing steps during industrial refining of oils, such as corn, soybean, and rapeseed oils. A substantial proportion of phytosterols and their esters is lost by distillation during the deodorization process or by conversion to steradienes by dehydration[59,60]. The bleaching of oils using bleaching earths also results in the formation of steradienes[61] and disteryl ethers[62,63] from the phytosterols.

In the deodorization process plant oils are subjected to treatment with steam at 190 to 220°C under vacuum (5 to 20 hPa) for the removal of undesirable volatile constituents that affect the flavor and taste of oils. During deodorization, some phytosterols and their esters also enter the vapor phase together with the unesterified fatty acids and other volatile constituents. Subsequent cooling of the vapors yields a steam distillate (deodorizer distillate) containing the phytosterols and their fatty acid esters. Similarly, in the physical refining for the deacidification of plant oils they are treated with steam at 230 to 260°C under vacuum (<4 hPa) which yields a steam distillate containing the phytosterols and their esters in addition to unesterified fatty acids and volatile constituents of the oil.

The composition of sterols and steryl esters in steam distillate obtained by deodorization or physical refining of various fats is shown in Table 27.5. Such steam distillates contain, in addition to phytosterols and their fatty acid esters, unesterified fatty acids, tri-, di-, and monoacylglycerols, as well as tocopherols and tocotrienols[64,65]. The steam distillates from deodorization processes generally contain more phytosterols and their esters than those from physical refining. The steam distillates resulting from both deodorization and physical refining yield upon hydrolysis and/or

TABLE 27.5
Sterols and Steryl Esters (g/kg) of Deodorizer Distillates of Various Fats and Oils

	Soybean oil				Rapeseed oil		Sunflower oil		Corn oil
	Chemically refined (mean)[54]	Chemically refined[64]	Physically refined[54]	Refined by molecular distillation[73]	Chemically refined (mean)[54]	Physically refined[64]	Chemically refined[54]	Physically refined[54]	Physically refined (mean)[54]
Sterols (total)[65]	120–180	113.9	—	130	148 (101[74])	15.8 (25[74])	180 (97.2[75])	—	—
Campesterol	53.6	21.3	19.1	37	36.5	3.3	15.8	4.5	12.6
Stigmasterol	44.5	38.8	13.8	30	—	—	20.4	6.2	2.8
Sitosterol	81.2	53.8	30.3	64	51.5	7.8	86.0	26.0	25.3
Others		—			21.9	4.6	16.8	6.2	3.1
Steryl esters	24.6	137.3	44.5		33.4 (182[74])	9.1(26[74])	3.0	0.9	

interesterification followed by fractional distillation or molecular distillation and crystallization highly enriched or almost pure phytosterols[53,66–68]. Fractionation using acetone as a solvent followed by crystallization[69] or extraction with supercritical carbon dioxide[70–72] have also been employed for the enrichment of sterols and their esters from deodorizer distillates.

An important source of plant sterols is tall oil, which is a byproduct of the wood pulp industry formed when coniferous woods are digested under alkaline conditions. The fatty and resinous substances contained in the wood are thus converted to the sodium salts (soaps) of fatty acids and rosin acids, together with neutral substances such as plant sterols and long-chain fatty alcohols. After acidulation, the tall oil soap is processed by a sequence of distillations into different fatty acid and rosin acid products.

Sterols can be extracted from the soap with different solvent mixtures or they can be recovered from the tall oil pitch remaining after the first distillation. The sterol levels in tall oil soaps are 3 to 5% (w/w). The sterol content of the pitch fraction remaining after distillation can be as high as 10% (w/w)[76].

Phytosterols, phytostanols, and their esters can be recovered from tall oil soaps and tall oil residues by solvent extraction, followed by fractionation[77,78]. The content of sterols and steryl esters in lipophilic extracts from wood, tall oil, and tall oil fractions is shown Table 27.6. The sterols recovered from tall oil consist of phytosterols (sitosterol and campesterol) and phytostanols (sitostanol and campestanol) in a ratio of about 4:1[56]. Catalytic hydrogenation of phytosterols yields phytostanols[77–80]. Sterols can also be isolated from their mixtures with other lipids by mixing with methyl isobutyl ketone, followed by the addition of a solution of calcium bromide in methyl isobutyl ketone which results in the precipitation of the sterols as a complex that is recovered by filtration[81].

27.4 CHEMICAL AND ENZYMATIC METHODS FOR THE PREPARATION OF ESTERS OF PHYTOSTEROLS AND PHYTOSTANOLS

Fatty acid esters of phytosterols and phytostanols, rather than the unesterified sterols, are generally preferred as supplements in fat-containing functional foods, such as margarine, mayonnaise, and salad dressings, due to the higher solubility of the esters in fat[12,89]. Fatty acid esters of sterols and stanols can be prepared from the corresponding sterols by chemical esterification with fatty acids or interesterification with fatty acid methyl esters as well as by their reaction with fatty acid halides or anhydrides[90–92]. Fatty acid esters of phytosterols and phytostanols have been prepared efficiently by reacting them with N,N'-dicyclohexylcarbodiimide and a fatty acid in the presence of catalytic amounts of 4-N,N-dimethylaminopyridine[92]. This method has been applied for the preparation of *trans*-feruloyl-sitostanol[93].

Phytosterols and phytostanols are commonly converted to their fatty acid esters by chemical interesterification with fatty acid methyl esters[56,80,94] or triacylglycerols[95] using an alkaline catalyst, such as sodium methylate. Chemical esterification of phytosterols and phytostanols with fatty acids in the presence of calcium hydroxide, calcium salts of fatty acids, and calcium or magnesium oxides also yields their fatty acid esters[96].

Phytosterols and phytostanols have been converted to diesters and monoesters of lactic, tartaric, citric, and cinnamic acids by esterification with the corresponding acids using acidic catalysts, such as sulfuric or phosphoric acids[97]. Recently, solid-base catalysts such as magnesium and zinc oxides have been shown to be effective for transesterification of plant sterols[98].

Lipase-catalyzed hydrolysis of steryl esters in the presence of water or aqueous alcohol solutions yields free sterols efficiently[99–105]. Various commercially available microbial lipases have been tested for steryl ester hydrolase activity and highest activities were found for the enzymes from *C. rugosa*, *Pseudomonas aeruginosa*, and *Burkholderia cepacia*[106]. Little is known, however, on the enzymatic preparation of steryl esters, particularly fatty acid esters of phytosterols and phytostanols. It has been shown that phytosterols contained in plant oils are extensively converted to steryl esters of fatty acids during interesterification of the oils, catalyzed by triacylglycerol lipase

TABLE 27.6
Sterols and Steryl Esters Contained in Lipophilic Extracts from Wood, Pitch Deposits, Tall Oil, and Tall Oil Fractions

	Sterols and steryl esters in lipophilic extracts from[82]						Sterols and steryl esters in tall oil or tall oil fractions			
	Conifers		Hardwoods					Tall oil[86]	Tall pitch[87]	Tall pitch[88]
	Pinus silvestris	Picea abies	Betula verrucosa	Populus tremola	Eucalyptus globulus[83,84]	Eucalyptus globulus (pitch deposit)[85]	Raw	Unsaponifiable matter	Unsaponifiable matter	Acetone extract of unsaponifiable matter, crystallized from methanol and acetone
Sterols (total; g/kg)[b]	1.39	1.00	1.56	2.40	0.68	0.72	47	402	445	730
Campesterol (%)					1.9		7.4	12.3	12.4	13.0
Stigmastanol (%)					9.4	13.5	92.6	87.7	87.6	87.0
Sitosterol (%)					78.1	70.0				
Steryl esters and/or wax esters (g/kg)	0.89	0.87	1.96	3.07	0.57	0.31		86		

[a] Tall pitch = residue from tall oil distillation.

[b] A typical composition of (conifer) wood sterols is sitosterol 71%; campesterol 9%; sitostanol 18%; campestanol 2%; stigmasterol 0%[86].

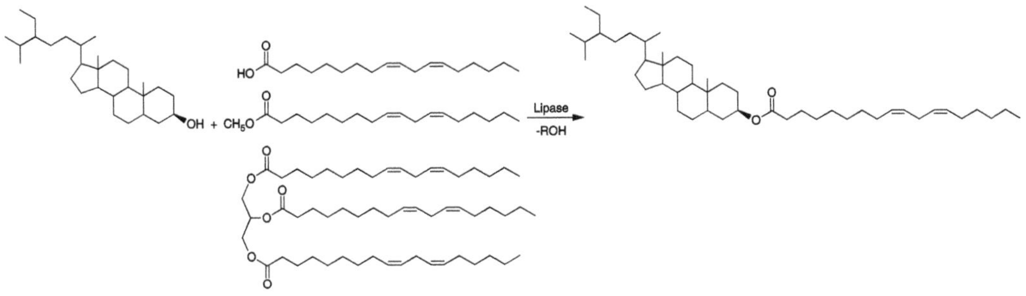

FIGURE 27.5 Lipase-catalyzed preparation of phytosteryl and phytostanyl esters, e.g., esterification of sitostanol with linoleic acid and its transesterification with methyl linoleate or trilinolein (R = H, methyl, or 1,2-dilinoleoylglyceryl).

from *Candida rugosa*[107]. It appears that various triacylglycerol lipases are able to catalyze the interesterification and esterification of sterols.

Lipase-catalyzed procedures for preparation of steryl esters of fatty acids in organic solvents in the presence of molecular sieves or other drying agents have been reported[108–117]. Moreover, preparation of steryl esters of saturated and polyunsaturated fatty acids by esterification in an aqueous system catalyzed by various lipases has been reported[118,119].

The reaction rates are rather low in most of the above methods, the extent of formation of steryl esters is moderate, and the use of organic solvents limits the application of such products as food supplements. It has recently been shown that phytosterols, such as sitosterol and stigmasterol, and phytostanols, such as sitostanol, are converted in high to near-quantitative yields to the corresponding long-chain acyl esters via esterification with fatty acids or transesterification with methyl esters of fatty acids or triacylglycerols using lipase from *C. rugosa* as biocatalyst *in vacuo* (20 to 40 hPa) at 40°C (Figure 27.5); neither an organic solvent nor water or a drying agent, such as molecular sieve, is added in these reactions[120,121].

The data on maximum conversion and enzyme activity in the esterification and transesterification of various phytosterols and phytostanols with different acyl donors catalyzed by *C. rugosa* lipase *in vacuo* are summarized in Table 27.7[121]. With sitostanol and sitosterol most of the acyl donors lead to near-quantitative conversion to the corresponding steryl and stanyl esters at high rates, as evident from the data on enzyme activity (Table 27.7). The esterification of the heat-sensitive diunsaturated sterols such as stigmasterol with oleic acid also approaches completion after 16 h of reaction at 40°C. Transesterifications of sitostanyl oleate with myristic acid yield only small proportions (2 to 4%) of sitostanyl myristate (Table 27.7).

Subsequent studies using immobilized lipases from *Rhizomucor miehei* (Lipozyme RM IM) and *Candida antarctica* (lipase B, Novozym 435) as biocatalysts have shown that sitostanol is converted in high to near-quantitative yields to the corresponding long-chain acyl esters via esterification with oleic acid or transesterification with methyl oleate or trioleoylglycerol *in vacuo* at 80°C[122]. Table 27.8 summarizes the data on maximum conversion and enzyme activity in esterification and transesterification reactions of sitostanol with different acyl donors, catalyzed by Novozym 435, papaya lipase, Lipozyme IM (*syn.* Lipozyme RM IM), and Lipozyme TL IM (data not shown) *in vacuo* at various temperatures. From these data it is obvious that sitostanol is converted in high to near-quantitative yields to the corresponding long-chain acyl esters via esterification with oleic acid or transesterification with methyl oleate or trioleoylglycerol using Novozym 435, Lipozyme IM, and Lipozyme TL IM (data not shown) as biocatalysts *in vacuo* (20 to 40 hPa) at 80°C, whereas the conversion is markedly lower at 60°C and very low at 40°C. Conversions observed with papaya lipase are generally lower at all temperatures than those with Novozym 435, Lipozyme IM, and Lipozyme TL IM. Highest conversion rates are observed in transesterification reactions of

TABLE 27.7

Esterification and Transesterification of Phytosterols and Phytostanols *in vacuo* Catalyzed by *Candida rugosa* Lipase[a]

Sterol or stanol	Fatty acid or fatty acid ester	Time (h)	Maximum conversion[b] (%)	Enzyme activity[c] (units)
Sitostanol	Oleic acid	2	96.8	25.3
Sitostanol	Methyl oleate	8	74.3	18.7
Sitostanol	Triolein	8	95.1	20.3
Sitostanol	Sunflower oil	4	97.8	8.2[d]
Sitostanyl oleate	Myristic acid	24	1.1	0.3
Sitosterol[e]	Oleic acid	48	73.0	15.7
Sitosterol[e]	Methyl oleate	48	52.9	12.3
Sitosterol[e]	Triolein	48	95.9	14.3
Stigmasterol	Oleic acid	16	98.8	4.0[f]

[a] Experimental conditions: 100 μmol sterol; molar ratio of sterol:fatty acid and sterol:fatty acid methyl ester 1:3, sterol: triacylglycerol including tributyrin 2:3; amount of *C. rugosa* lipase 50 mg; 20–40 mbar; 40°C.

[b] Determined by GC.

[c] Enzyme units were calculated as 1 μmol steryl ester formed × min^{-1} × g^{-1} *C. rugosa* lipase from the initial rates (1 h) of esterification or transesterification as described in[121].

[d] 4 h value was used for calculation of enzyme activity.

[e] Commercial sitosterol preparation contained around 55% sitosterol, 40% campesterol. and 5% stigmasterol. The conversions were calculated as total amount of steryl esters formed.

[f] 3 h value was used for calculation of enzyme activity.

Data from Weber, N., Weitkamp, P., and Mukherjee, K.D., *J. Agric. Food Chem.*, 49, 67–71, 2001.

phytostanol with methyl oleate or triolein at 80°C using Lipozyme IM and Lipozyme TL IM (data not shown) as catalysts. Recently, immobilized, heat-stable Lipase QLM from *Alcaligenes* spp. has been used for the preparation of plant steryl esters by transesterification of sunflower oil with a plant sterol mixture[123]. Interestingly, the enzyme activity of *C. rugosa* lipase in the esterification of sitostanol with oleic acid yielding sitostanyl oleate is by far higher (~25 units/g; Table 27.7) than the enzyme activity of Lipozyme IM or Novozym 435 in the transesterification reaction of sitostanol with methyl oleate (~2 to 3 units/g; Table 27.8).

Deodorizer distillate from soybean oil containing phytosterols, tocopherols, and fatty acids has been treated with lipase from *C. rugosa* which leads to selective esterification of sterols to steryl esters. However, the tocopherols are not esterified; the resulting steryl esters are separated from tocopherols by distillation of the latter[64].

Sterols present in steam distillates obtained from rapeseed oil by conventional deodorization and those from rapeseed oil or a mixture of soybean and rapeseed oils by physical refining have been converted to a high degree *in situ* to the corresponding long-chain acyl esters via esterification and/or transesterification with fatty acids and/or acylglycerols — also present in the steam distillates—using lipase from *C. rugosa* as biocatalyst *in vacuo* (20 to 40 hPa) at 40°C[74].

The relative proportions of sterols and steryl esters of deodorizer distillates after esterification/transesterification *in situ* of sterols catalyzed by *C. rugosa* lipase *in vacuo* at 40°C show that 87 to 97% of the sterols occur as steryl esters[75]. The constituent sterols of purified steryl esters contain 16.5% brassicasterol, 39.2% campesterol, and 44.3% sitosterol.

Sterols present in tall oil are also converted to a high degree to the corresponding long-chain acyl esters via *in situ* esterification and/or transesterification with fatty acids and/or triacylglycerols—also contained in the tall oil —using lipase from *C. rugosa* as biocatalyst *in vacuo* (20 to 40 hPa) at 40°C[125].

TABLE 27.8
Enzyme Activities of Various Lipases During Esterification and Transesterification of Sitotostanol *in vacuo* at Different Temperatures[a]

Enzyme	Fatty acid or fatty acid ester	Temperature (°C)	Time (h)	Maximum conversion[b] (%)	Enzyme activity[c] (units)
Novozym 435	Oleic acid	40	48	13.6	0.17
Novozym 435	Methyl oleate	40	48	11.2	0.08
Lipozyme IM	Oleic acid	40	48	8.6	0.08
Lipozyme IM	Methyl oleate	40	48	14.0	0.08[e]
Lipozyme IM	Triolein	40	48	13.9	0.08
Papaya[d]	Oleic acid	60	48	29.3	0.25
Papaya[d]	Methyl oleate	60	48	48.0	0.33
Novozym 435	Oleic acid	60	48	40.2	0.25
Novozym 435	Methyl oleate	60	48	69.0	0.63[f]
Lipozyme IM	Oleic acid	60	48	40.0	0.4
Lipozyme IM	Methyl oleate	60	48	78.3	1.0
Lipozyme IM	Triolein	60	48	64.8	0.5
Papaya[d]	Oleic acid	80	48	46.6	0.42
Papaya[d]	Methyl oleate	80	48	61.6	0.83
Novozym 435	Oleic acid	80	48	88.7	0.92
Novozym 435	Methyl oleate	80	48	99.2	1.8
Lipozyme IM	Oleic acid	80	48	63.8	1.3
Lipozyme IM	Methyl oleate	80	24	93.2	3.2
Lipozyme IM	Triolein	80	48	95.7	2.3

[a] Experimental conditions: 100 μmol sterol; molar ratio of sterol:fatty acid and sterol:fatty acid methyl ester 1:3, sterol: trioleoylglycerol 2:3; amount of various lipases 50 mg; 20–40 mbar.
[b] Determined by GC.
[c] Enzyme units were calculated as 1 μmol steryl ester formed $\times$ min^{-1} $\times$ g^{-1} lipase from the initial rates (4 h) of esterification or transesterification, as described in[122].
[d] A fine powder obtained from crude granular papaya latex preparation by grinding in a mortar was used.
[e] 8 h value was used for calculation of enzyme activity.
[f] 24 h value was used for calculation of enzyme activity.
Adapted and extended from Weber, N., Weitkamp, P., and Mukherjee, K.D., *J. Agric. Food Chem.*, 49, 5210–5216, 2001;

A process has been reported for the purification of steryl esters from a high-boiling-point fraction of soybean oil deodorizer distillate. This fraction containing steryl esters and various acylglycerols is reacted with water in the presence of *C. rugosa* lipase which leads to hydrolysis of acylglycerols, whereas steryl esters are transesterified in moderate yield and their content does not change. Steryl esters are subsequently purified by molecular distillation from the oil layer of the reaction mixture[126].

27.5   HEALTH EFFECTS OF PHYTOSTEROLS, PHYTOSTANOLS, AND THEIR ESTERS

27.5.1   Effect on the Absorption of Cholesterol

As mentioned earlier, despite the close resemblance in the chemical structure of phytosterols and phytostanols to that of cholesterol (Figures 27.1–27.3), their absorption in organisms is distinctly different. Humans can absorb up to 80% of the dietary cholesterol in the duodenum[23], but only about 5% of the ingested sitosterol, the most abundant phytosterol of dietary origin[41]. The absorption of sitostanol is even lower[42,43]. The absorption of phytosterols decreases with increasing length of the branched side chain, probably due to increasing hydrophobicity[127].

In the gastrointestinal tract both phytosterols and phytostanols impair the absorption of dietary cholesterol as well as cholesterol contained in the bile[44]. The exact mechanism of this phenomenon is not known[2]. It appears, however, that phytosterols and phytostanols, owing to their higher hydrophobicity as compared to cholesterol, have a stronger tendency to form lipid-containing micelles in the duodenum. This probably results in the displacement of cholesterol from the micelles of the bile salts and decrease in its absorption[128–131], whereas a mechanism involving cocrystallization is rather unlikely[132]. Phytostanyl esters are more effective in the reduction of cholesterol absorption than the unesterified phytostanols, possibly due to a better solubility of the former in fat as well as contents of the intestine[14,16]. However, it is difficult to interpret these results, because both phytostanyl esters and cholesteryl esters of dietary origin are hydrolyzed to free sterols prior to their absorption in the duodenum[133].

The absorption of phytosterols is reduced after the ingestion of phytostanols[42]. Intake of either phytosteryl esters or phytostanyl esters has been shown to result in a similar extent of reduction of cholesterol absorption[134]. Ingestion of unesterified phytostanols as lecithin micelles, as compared to phytostanols in powder form, has led to a larger reduction of cholesterol absorption, probably due to better mixing of the micellar phytostanols[135,136]. Similarly, intake of phytosterols dissolved in diacylglycerols has been found to be more effective in the reduction of blood cholesterol as compared to phytosterols dissolved in triacylglycerols, possibly due to better solubility of the phytosterols in diacylglycerols[137].

Comparison of the intestinal uptake of deuterated cholesterol, plant sterols, and stanols in mice supports the hypothesis that (1) discrimination probably takes place at the level of reverse transport back to the gut lumen, (2) plant stanols are taken up, but not absorbed to a measurable extent, and (3) the process of discrimination probably exists at the level of biliary excretion[138]. Current understanding of cholesterol absorption at the intestinal wall indicates that receptor and transporter proteins may be involved in brush border membrane cholesterol trafficking[139,140].

27.5.2 BLOOD CHOLESTEROL-REDUCING ACTION

The cholesterol-reducing (hypocholesterolemic) action of phytosterols, phytostanols, and their esters is very likely due to reduction of cholesterol absorption[141]. Moreover, a reduction of cholesterol absorption leads to an increase in the excretion of cholesterol and bile acids from the liver in the bile[142,143]. So far only one investigation has been reported in which no cholesterol-reducing effect has been observed after intake of low doses of unesterified sitostanol in the form of capsules[144].

In most of the studies on the hypocholesterolemic action after the intake of phytostanyl esters and phytosteryl esters in subjects, measured as a decrease in LDL cholesterol (Table 27.1 and Table 27.2), steryl esters were applied as spreads (margarine)[145], whereas studies on the inhibition of cholesterol absorption by vegetable foods containing phytosterols are rare[146]. Similar hypocholesterolemic action has been observed upon ingestion of margarines enriched with either phytosteryl esters or phytostanyl esters[31,34]. Very recently, sustained efficacy of cholesterol reduction and long-term compliance of spreads enriched with plant sterol esters have been demonstrated in a study with children and parents suffering from familial hypercholesterolemia[147]. It is interesting to note that not all phytosterols have hypocholesterolemic action. Thus, the intake of a margarine enriched with 4,4-dimethylsterols (α-amyrin and lupeol from shea butter) did not result in the reduction of LDL cholesterol, whereas unesterified 4-demethylsterols (e.g., sitosterol and campesterol) showed hypocholesterolemic action[4]. Despite different absorption of sitostanol and campestanol, similar LDL cholesterol-reducing action has been observed after daily intake of 2 to 4 g phytostanyl ester mixtures containing varying proportions of sitostanyl and campestanyl esters[19,29,30,148]. Phytosteryl esters[35] and unesterified phytosterols[4,149,150] showed after moderate intake (about 0.8 g per day) similar hypocholesterolemic action (about 6% reduction of LDL cholesterol).

Phytostanyl esters were shown to have a better hypocholesterolemic action than unesterified phytostanols[15].

In dose-dependent studies on cholesterol-reducing action of phytosteryl esters, intakes of 0.8, 1.6, and 3.2 g phytosteryl esters per day resulted in 6.7, 8.5, and 9.9% reduction in LDL cholesterol, respectively[35]. Similar reduction of plasma lipoprotein concentrations has been obtained with a phytosterol ester-enriched spread in children with familial hypercholesterolemia[151] and with a corresponding vegetable oil in healthy men[152]. Ingestion of 0.8, 1.6, 2.4, and 3.2 g phytostanyl esters resulted in 1.6, 6.1, 10.6, and 11.5% reduction in LDL cholesterol, respectively[32]. A meta-analysis revealed that the intake of phytosteryl or phytostanyl esters beyond 2 g per day did not result in a further reduction in LDL cholesterol[13].

Consumption of a margarine containing stanyl esters by subjects receiving either low-fat and low-cholesterol diets or diets containing more fat and cholesterol resulted in similar reduction of LDL cholesterol[30]. Similar results have been reported for subjects receiving an unesterified sterol mixture consisting of 80% sitosterol plus campesterol and 20% sitostanol[153]. In subjects receiving 1.5 to 3.0 g per day of a mixture of unesterified phytosterols a 7.5 to 11.6% decrease in total cholesterol and LDL cholesterol was observed[33]. Moreover, plant sterol-enriched spread enhances the cholesterol-lowering potential of fat-reduced diet in volunteers with moderately elevated plasma total cholesterol[154]. A decrease of serum total and LDL cholesterol in hypercholesterolemic adults was also described for food products containing both free tall oil-based phytosterols and oat β-glucan[155]. Similarly, soy protein has been found to enhance the cholesterol-lowering effect of plant steryl esters in cholesterol-fed hamsters[156].

The extent of reduction of LDL cholesterol has been found to be similar, irrespective of whether phytosteryl esters or phytostanyl esters are ingested only with lunch and dinner[34] or with every meal[13]. Moreover, it has been shown that the reduction of LDL cholesterol is similar whether a daily dose of 2.5 g phytosteryl esters or phytostanyl esters is taken at once with lunch or distributed among various meals (0.4 g with breakfast, 0.8 g with lunch, and 1.3 g with dinner)[28].

In a clinical study on slightly hypercholesterolemic subjects, ingestion of 1.5 g per day of Phytrol®, a phytosterol preparation obtained from tall oil, for 10 days has been found to reduce total cholesterol and LDL cholesterol to an extent of 6 and 10%, respectively[157]. Similar results were obtained by Hayes et al.[158] in a human study with nonesterified phytosterols which were applied to fried foods via the frying oil. CardioRex®, another phytosterol preparation obtained from tall oil, has been successfully tested in animal experiments for its cholesterol-reducing action[159]. Similarly, it has been shown in animal experiments that the beneficial nutritional properties of phytosterols and ω3 long-chain polyunsaturated fatty acids (LC-PUFA) are obtained when phytosteryl esters prepared from marine oils are fed[160].

The relation between intake of natural dietary plant sterols and serum lipid concentration, examined in a free-living population (EPIC-Norfolk study), has demonstrated that a high intake of plant sterols from foods is inversely associated with lower concentrations of total and LDL cholesterol[161].

27.5.3 ACTION ON THE ABSORPTION OF LIPID-SOLUBLE ANTIOXIDANTS AND VITAMINS

Intake of phytostanols has been found to reduce the α- and β-carotene content of blood[19]. Phytostanyl esters also reduce contents of α-carotene[162,163], β-carotene[28,30,162–164], and lycopene[28] of blood; however, the retinol (vitamin A) content of blood is not affected either by phytostanols[19] or by phytostanyl esters[22,28,30,32,163,164]. Intake of phytosteryl esters has also been shown to reduce the concentration of α-carotene, β-carotene, and lycopene in blood[34,35]. On the other hand, it has been demonstrated that bakery products enriched with phytosterol esters and β-carotene decrease plasma LDL cholesterol levels and maintain plasma β-carotene concentrations in normocholesterolemic volunteers[165].

The concentration of tocopherols in blood is not affected by the intake of phytostanols[19], phytostanyl esters[22,28,30,32,163,164], and phytosteryl esters[35]. The ingestion of phytostanols[19] and phytostanyl esters[22,30,32] does not alter the 25-hydroxy-vitamin D content of blood. Moreover, the intake of phytosteryl esters does not alter the relative proportions vitamin K_1 and 25-hydroxy-vitamin D of blood[35].

27.5.4 SAFETY ASPECTS CONCERNING THE ENRICHMENT OF FOODS WITH PHYTOSTEROLS, PHYTOSTANOLS, AND THEIR ESTERS

No detrimental effects of ingestion of recommended amounts of phytosterols, phytostanols, and their esters as food supplements in a hypocholesterolemic diet have become known so far[25,34,166–169]. Even at a high dose only small amounts of dietary phytosterols, phytostanols, and their esters are absorbed and taken up in circulating blood. Similar to cholesterol, these substances are completely excreted in the bile[34]. Toxicological examination of phytosteryl esters[37,170–173] and phytostanyl esters[174–177] has shown no detrimental action. Even in children no undesirable effects have been observed after the intake of margarines containing phytostanyl esters[164,178]. However, there are some indications that phytosterols might alter the metabolism of bile acids[179].

To determine which population groups are likely to be at risk of excessive intakes of foods liberally enriched with phytosterols and phytostanols, a cross-sectional simulation study among a representative sample ($n = 23,106$) of the Dutch population (MORGEN-project) has been performed. From this study it has been concluded that the daily intake of phytosterols and phytostanols might exceed the recommended maximum intake levels of human safety studies by far (> 8.6 g/d) if a liberal phytosterol/phytostanol fortification is allowed[180].

In a rare case of a congenital metabolic disease — phytosterolemia — foods rich in phytosterols must be avoided[181]. Subjects suffering from phytosterolemia, who are prone to atherosclerosis, can absorb large proportions of phytosterols and accumulate them in the blood. Since phytosterols, similar to cholesterol, are potentially atherogenic[182], phytosterolemia leads to higher risk of atherosclerosis. It might be noted that a mixture of phytosterols fed to mice slowed the growth and metastasis of mammary cancer cells[183].

Higher concentrations of phytosterol oxides may be expected in phytosterol-enriched food under extreme heating conditions. The corresponding cholesterol oxidation products (COP) have shown adverse biological effects. The results of genotoxicity and subchronic toxicity experiments have shown that a phytosterol oxide concentrate containing approximately 30% phytosterol oxides does not possess genotoxic potential and no obvious evidence of toxicity is observed when administered in the diet of rats for 90 days[184].

27.6 COMMERCIAL APPLICATIONS OF PHYTOSTEROLS, PHYTOSTANOLS, AND THEIR ESTERS

Meta-analysis of randomized, placebo-controlled, double-blind studies has revealed that a daily intake of 2 g phytosterols, phytostanols, or their fatty acid esters reduces the LDL cholesterol content of blood by 9 to 14% without affecting the HDL cholesterol and triacylglycerol concentration from which about 25% reduction of the risk of cardiovascular diseases has been derived[13].

Spreads were the first commercial application of foods enriched with plant stanyl esters, and they remain by far the most widely known applications of phytosterols, phytostanols, and their esters among consumers and food companies. Table 27.9 summarizes a selected list of commercial food products enriched with phytosterols, phytostanols, and their fatty acid esters. Margarines (Benecol™) enriched with phytostanyl esters have been commercially available in Finland since 1995, and so far there is no evidence of any health hazard from such a product[78]. A few years later, in 1999, Benecol™ products were launched in the U.S., the U.K., Belgium, The Netherlands, and Ireland. Since the recognition by the American Heart Association of the important contribution of phytosterols, phytostanols, and their esters to the reduction of risk of atherosclerosis[185], spreads

TABLE 27.9
Selected List of Commercial Food Products Enriched with Phytosterols, Phytostanols, and Their Fatty Acid Esters

Product	Constituent	Use	Manufacturer
Benecol™	Phytostanyl esters	Margarine (U.S.), spreads, milk, yogurt, mayonnaise, snack bars (Europe)	Raissio/McNeil
Take Control™	Phytosteryl esters	Margarine, salad dressings (U.S.)	Lipton/Unilever
Flora Pro-active™, Becel Pro-activ™	Phytosteryl esters	Margarine, salad dressings (Europe, Canada)	Unilever
Logicol™	Phytosterols	Spreads, milk (Australia)	Meadow Lea
CookSmart™	Phytosteryl esters	Cooking oil (U.S.)	Procter & Gamble
Danacol™	Phytosteryl esters	Yogurts	Danone

such as margarines (Take Control™) enriched with these substances have been permitted in the U.S. since 1999[186,187]. An edible oil (CookSmart™) containing such cholesterol-reducing supplements is commercially available[45]. The U.S. Food and Drug Administration (FDA) now permits health claims for such products that they reduce the risk of coronary heart diseases. According to the Novel Food Regulations of the European Union the use of spreads, salad dressings, milk, and cheese products as well as soy drinks enriched with phytosterols, phytostanols, and their esters is now permitted at a maximum dose of 3 g/day[188–191]. Since 2000 a low-fat diet margarine (Becel pro-activ™) has been commercially available in Germany.

Plant stanyl esters incorporated into mayonnaise are available in Finland; vegetable oils containing diacylglycerols and enriched with plant sterols are also commercially available in Japan as Econa™ brand[2]. Benecol™ yogurt enriched with stanyl esters is the first example of the application of plant steryl esters in a dairy product which appeared in the U.K. in 1999. Milk and drinks containing milk and juice, both enriched with plant steryl esters have been available in Argentina since 2000 under the brand name SereCol™. A drink enriched with unesterified plant sterols is available in South Korea under the brand name Ucole™. Snack bars, such as Benecol™ energy bars, enriched with plant sterols have been marketed in the U.S., the U.K., and Ireland. However, there are no food products containing plant sterol supplements in Canada as of yet.

The development of novel low-fat dairy products containing free plant sterols has been encouraged by a double-blind, randomized, placebo-controlled crossover study showing the reduction of LDL cholesterol levels in mildly hypercholesterolemic patients[192]. The results of a stable isotope double-blind crossover study show that nonesterified plant sterols solubilized in low-fat dairy products are able to inhibit cholesterol absorption[193]. However, it has been demonstrated that the cholesterol-lowering effects may differ according to the food matrix. Phytosteryl esters in low-fat milk are almost three times more effective than in bread and cereals as shown in a randomized study of three research centers[165], whereas free phytosterols in low- and nonfat beverages as part of a controlled diet have been shown to be unable to lower plasma lipid levels[194].

27.7 PERSPECTIVES

Although cholesterol-reducing medication using products such as statins is in common usage these days, ingestion of recommended amounts of food items enriched with phytosterols, phytostanols, and their esters should be generally acceptable for subjects with hypercholesterolemia[21]. Such products are, however, more expensive than the corresponding conventional spreads. Limiting factors in the use of phytosterols, phytostanols, and their esters in mass products might be their restricted availability from natural resources and the rather expensive processes involved in their isolation and purification.

REFERENCES

1. Akihisa, T., Kokke, W.C.M.C., and Tamura, T., Naturally occurring sterols and related compounds from plants, in *Physiology and Biochemistry of Sterols,* Patterson, G.W. and Nes, W.D., Eds., American Oil Chemists' Society, Champaign, IL, 1991, pp. 172–228.
2. Salo, P., Wester, I., and Hopia, H., Phytosterols, in *Lipids for Functional Foods and Nutraceuticals,* Gunstone, F.D., Ed., The Oily Press, Bridgwater, U.K., 2003, pp. 183–224.
3. Piironen, V., Lindsay, D.G., Miettinen, T.A., Toivo, J., and Lampi, A.M., Plant sterols: biosynthesis, biological function and their importance to human nutrition, *J. Sci. Food Agric.,* 80, 939–966, 2000.
4. Sierksma, A., Weststrate, J.A., and Maijer, G.W., Spreads enriched with plant sterols, either esterified 4,4-dimethylsterols or free 4-desmethylsterols and plasma total- and LDL-cholesterol concentrations, *Br. J. Nutr.,* 82, 273–282, 1999.
5. de Deckere, E.A.M. and Korver, O., Minor constituents of rice bran oil as functional foods, *Nutr. Rev.,* 54, S120–S126, 1996.
6. Hooper, L., Summerbell, C.D., Higgins, J.P., Thompson, R.L., Capps, N.E., Smith, G.D., Riemersma, R.A., and Ebrahim, S., Dietary fat intake and prevention of cardiovascular disease: systematic review, *Br. Med. J.,* 322, 757–763, 2001.
7. Hu, F.B., Manson, J.E., and Willett, W.C., Types of dietary fat and risk of coronary heart disease: a critical reviewm *J. Am. Coll. Nutr.,* 20, 5–19, 2001.
8. Pollak, O.J., Reduction of blood cholesterol in man, *Circulation,* 7, 702–706, 1953.
9. Moreau, R.A., Whitaker, B.D., and Hicks, K.B., Phytosterols, phytostanols, and their conjugates in foods: structural diversity, quantitative analysis, and health-promoting uses, *Prog. Lipid Res.,* 41, 457–500, 2002.
10. Ostlund, Jr., R.E., Phytosterols in human nutrition, *Annu. Rev. Nutr.,* 22, 533–549, 2002.
11. Ostlund, Jr., R.E., Phytosterols and cholesterol metabolism, *Curr. Opin. Lipidol.,* 15, 37–41, 2004.
12. Leesen, P. and Flöter, E., Solidification behaviour of binary sitosteryl esters mixtures, *Food Res. Int.,* 35, 983–991, 2002.
13. Law, M., Plant sterol and stanol margarines and health, *Br. Med. J.,* 320, 861–864, 2000.
14. Vanhanen, H.T., Blomqvist, S., Ehnholm, C., Hyvonen, M., Jauhiainen, M., Torstila, I., and Miettinen, T.A., Serum cholesterol, cholesterol precursors, and plant sterols in hypercholesterolemic subjects with different apoE phenotypes during dietary sitostanol ester treatment, *J. Lipid Res.,* 34, 1535–1544, 1993.
15. Vanhanen, H.T., Kajander, J., Lehtovirta, H., and Miettinen, T.A., Serum levels, absorption efficiency, faecal elimination and synthesis of cholesterol during increasing doses of dietary sitostanol esters in hypercholesterolaemic subjects, *Clin. Sci.* (Lond.), 87, 61–67, 1994.
16. Miettinen, T.A. and Vanhanen, H., Dietary sitostanol related to absorption, synthesis and serum level of cholesterol in different apolipoprotein E phenotype, *Atherosclerosis,* 105, 217–226, 1994.
17. Gylling, H. and Miettinen, T.A., Serum cholesterol and cholesterol and lipoprotein metabolism in hypercholesterolemic NIDDM patients before and during sitostanol ester-margarine treatment, *Diabetologia,* 37, 773–780, 1994.
18. Gylling, H. and Miettinen, T.A., Effects of inhibiting cholesterol absorption and synthesis on cholesterol and lipoprotein metabolism in hypercholesterolemic non-insulin-dependent diabetic men, *J. Lipid Res.,* 37, 1776–1785, 1996.
19. Gylling, H. and Miettinen, T.A., Cholesterol reduction by different stanol mixtures and with variable fat intake, *Metabolism,* 48, 575–580, 1999.
20. Gylling, H., Siimes, M.A., and Miettinen, T.A., Sitostanol ester margarine in dietary treatment of children with familial hypercholesterolemia, *J. Lipid Res.,* 36, 1807–1812, 1995.
21. Gylling, H., Radhakrishnan, R., and Miettinen, T.A., Reduction of serum cholesterol in postmenopausal women with previous myocardial infarction and cholesterol malabsorption induced by dietary sitostanol ester margarine: women and dietary sitostanol, *Circulation,* 96, 4226–4231, 1997.

22. Andersson, A., Karlström, B., Mohsen, R., and Vessby, B., Cholesterol-lowering effects of a stanol ester-containing low-fat margarine used in conjunction with a strict lipid-lowering diet, *Eur. Heart J.*, 1 (Suppl.), S80–S90, 1999.

23. Nguyen, T.T., The cholesterol-lowering action of plant stanol esters, *J. Nutr.*, 129, 2109–2112, 1999.

24. Nguyen, T.T., Recent clinical trial evidence for the cholesterol-lowering efficacy of a plant stanol ester spread in a USA population, *Eur. Heart J.*, 1 (Suppl.), S73–S79, 1999.

25. Miettinen, T.A., Puska, P., Gylling, H., Vanhanen, H., and Vartiainen, E., Reduction of serum cholesterol with sitostanol-ester margarine in a mildly hypercholesterolemic population, *N. Engl. J. Med.*, 333, 1308–1312, 1995.

26. Miettinen, T.A., Vuoristo, M., Nissinen, M., Jarvinen, H.J., and Gylling, H., Serum, biliary and faecal cholesterol and plant sterols in colectomized patients before and during consumption of stanol ester margarine, *Am. J. Clin. Nutr.*, 71, 1095–1102, 2000.

27. Niinikoski, H., Viikari, J., and Palmu, T., Cholesterol-lowering effect and sensory properties of sitostanol ester margarine in normocholesterolemic adults, *Scand. J. Nutr.*, 41, 9–12, 1997.

28. Plat, J., van Onselen, E.N.M., van Heugten, M.M., and Mensink, R.P., Effects on serum lipids, lipoproteins, and fat soluble antioxidant concentrations of consumption frequency of margarines and shortenings enriched with plant stanol esters, *Eur. J. Clin. Nutr.*, 54, 671–677, 2000.

29. Plat, J. and Mensink, R.P., Vegetable oil based versus wood based stanol ester mixtures: effects on serum lipids and hemostatic factors in non-hypercholesterolemic subjects, *Atherosclerosis*, 148, 101–112, 2000.

30. Hallikainen, M.A. and Uusitupa, M.I.J., Effects of 2 low-fat stanol ester-containing margarines on serum cholesterol concentrations as part of a low-fat diet in hypercholesterolemic subjects, *Am. J. Clin. Nutr.*, 69, 403–410, 1999.

31. Hallikainen, M.A., Sarkkinen, E.S., Gylling, H., Erkkilä, A.T., and Uusitupa, M.I.J., Comparison of the effects of plant sterol ester and plant stanol ester-enriched margarine in lowering serum cholesterol concentrations in hypercholesterolaemic subjects on a low-fat diet, *Eur. J. Clin. Nutr.*, 54, 715–725, 2000.

32. Hallikainen, M.A., Sarkkinen, E.S., and Uusitupa, M.I.J., Plant stanol esters affect serum cholesterol concentrations of hypercholesterolemic men and women in a dose-dependent manner, *J. Nutr.*, 130, 767–776, 2000.

33. Christiansen, L.I., Lähteenmäki, P.L.A., Mannelin, M.R., Seppänen-Laakso, T.E., Hiltunen, R.V.K., and Yliruusi, J.K., Cholesterol-lowering effect of spreads enriched with monocrystalline plant sterols in hypercholesterolemic subjects, *Eur. J. Nutr.*, 40, 66–73, 2001.

34. Weststrate, J.A. and Meijer, G.W., Plant sterol-enriched margarines and reduction of plasma total- and LDL-cholesterol concentrations in normocholesterolaemic and mildly hypercholesterolaemic subjects, *Eur. J. Clin. Nutr.*, 52, 334–343, 1998.

35. Hendriks, H.F.J., Weststrate, J.A., van Vilet, T., and Meijer, G.W., Spreads enriched with three different levels of vegetable oil sterols and the degree of cholesterol lowering in normocholesterolemic and mildly hypercholesterolemic subjects, *Eur. J. Clin. Nutr.*, 53, 319–327, 1999.

36. Maki, K., Davidson, M., Umporowicz, D., Schaefer, E., Dicklin, M., Ingram, K., Chen, S., Gebhart, B., and Franke, W., Lipid responses to plant sterol-enriched reduced-fat spreads incorporated into a step 1 diet, *Circulation*, 100 (Suppl. 1), 1115, 1999.

37. Ayesh, R., Weststrate, J.A., Drewitt, P.N., and Hepburn, P.A., Safety evaluation of phytosterol esters: 5. Faecal short-chain fatty acid and microflora content, faecal bacterial enzyme activity and serum female sex hormones in healthy normolipidaemic volunteers consuming a controlled diet either with or without a phytosterol ester-enriched margarine, *Food Chem. Toxicol.*, 37, 1127–1138, 1999.

38. Jones, P.J., Raeini-Sarjaz, M., Ntanios, F.Y., Vanstone, C.A., and Parsons, W.E., Modulation of plasma lipid levels and cholesterol kinetics by phytosterol versus phytostanol esters, *J. Lipid Res.*, 41, 697–705, 2000.

39. Volpe, R., Niittynen, L., Korpela, R., Sirtori, C., Bucci, A., Fraone, N., and Pazzucconi, F., Effect of yoghurt enriched with plant sterols on serum lipids in patients with moderate hypercholesterolaemia, *Br. J. Nutr.*, 86, 233–239, 2001.

40. Neil, H.A.W., Meijer, G.W., and Roe, L.S., Randomised controlled trial of use by hypercholesterolaemic patients of a vegetable oil sterol-enriched fat spread, *Atherosclerosis*, 156, 329–337, 2001.

41. Kritchevsky, D., Phytosterols, in *Dietary Fiber in Health and Disease*, Kritchevsky, D. and Bonfield, C., Eds., Plenum, New York, 1997, pp. 235–243.

42. Gylling, H., Puska, P., Vartiainen, E., and Miettinen, T.A., Serum sterols during stanol ester feeding in a mildly hypercholesterolemic population, *J. Lipid Res.*, 40, 593–600, 1999.

43. Subbiah, M.T.R., Dietary plant sterols: current status in human and animal sterol metabolism, *Am. J. Clin. Nutr.*, 26, 219–225, 1973.

44. Heinemann, T., Kullak-Ublick, G.A., Pietruck, B., and Bergmann, K., Mechanisms of action of plant sterols on inhibition of cholesterol absorption. Comparison of sitosterol and sitostanol, *Eur. J. Clin. Pharmacol.*, 40 (Suppl 1), S59–S63, 1991.

45. Hicks, K.B. and Moreau, R.A., Phytosterols and phytostanols: functional food cholesterol busters, *Food Technol.*, 55, 63–67, 2001.

46. Czubayko, F., Beumers, B., Lammsfuss, S., Lutjohann, D., and Bergmann, K., A simplified micro-method for quantification of fecal excretion of neutral and acidic sterols for outpatient studies in humans, *J. Lipid Res.*, 32, 1861–1867, 1991.

47. Weihrauch, J.L. and Gardner, J.M., Sterol content of foods of plant origin, *J. Am. Diet Assoc.*, 73, 39–47, 1978.

48. Wester, I., Cholesterol-lowering effect of plant sterols, *Eur. J. Lipid Sci. Technol.*, 102, 37–44, 2000.

49. Bereau, D., Benjelloun-Mlayah, B., Banoub, J., and Bravo, R., FA and unsaponifiable composition of five Amazonian palm kernel oils, *J. Am. Oil Chem. Soc.*, 80, 49–53, 2003.

50. Padley, F.B., Gunstone, F.D., and Harwood, J.L., Occurrence and characteristics of fats and oils, in *The Lipid Handbook*, Gunstone, F.D., Harwood, J.L., and Padley, F.B., Eds., Chapman and Hall, London, 1994, pp. 47–223.

51. Rossell, J.B., Vegetable oils and fats, in *Analysis of Oilseeds, Fats and Fatty Foods,* Rossell, J.B. and Pritchard, J.L.R., Eds., Elsevier Applied Science, London, 1991, pp. 261–327.

52. Homberg, E. and Bielefeld, B., Freie und gebundene Sterine in Pflanzenfetten, *Fette Seifen Anstrichm.*, 87, 61–64, 1985.

53. Daguet, D., Phytosterols: highly promising compounds, *Lipid Technol.*, 7, 77–80, 2000.

54. Verleyen, T., Sosinska, U., Ioannidou, S., Verhe, R., Dewettinck, K., Huygenbaert, A., and De Greyt, W., Influence of vegetable oil refining process on free and esterified sterols, *J. Am. Oil Chem. Soc.*, 79, 947–953, 2002.

55. Moreau, R.A., Norton, R.A., and Hicks, K.B., Phytosterols and phytostanols lower cholesterol, *Int. News Fats Oils Relat. Mater.*, 10, 572–577, 1999.

56. Miettinen, T.A., Phytosterols: what plant breeders should focus on, *J. Sci. Food Agric.*, 81, 895–903, 2001.

57. Vlahakis, C. and Hazebroek, J., Phytosterol accumulation in canola, sunflower, and soybean oils: effects of genetics, planting location, and temperature, *J. Am. Oil Chem. Soc.*, 77, 49–53, 2000.

58. Harker, M., Holmberg, N., Clayton, J.C., Gibbard, C.L., Wallace, A.D., Rawlins, S., Hellyer, S.A., Lanot, A., and Safford, R., Enhancement of seed phytosterol levels by expression of an N-terminal truncated *Hevea brasiliensis* (rubber tree) 3-hydroxy-3-methylglutaryl-CoA reductase, *Plant Biotechnol. J.*, 1, 113–121, 2003.

59. Ferrari, R.Ap., Schulte, E., Esteves, W., Brühl, L., and Mukherjee, K.D., Minor constituents of vegetable oils during industrial processing, *J. Am. Oil Chem. Soc.*, 73, 587–592, 1996.

60. Ferrari, R.Ap., Esteves, W., Mukherjee, K.D., and Schulte, E., Alteration of sterols and steryl esters in vegetable oils during industrial refining, *J. Agric. Food Chem.*, 45, 4753–4757, 1997.

61. Schulte, E., Determination of edible fat refining by HPLC of $\Delta^{3,5}$-steradienes, *Fat Sci. Technol.*, 96, 124–128, 1994.

62. Homberg, E., Veränderung der Sterine durch industrielle Verarbeitungsprozesse von Fetten und Ölen: II. Veränderungsprodukte bei der Behandlung von Cholesterin mit Bleicherden, *Fette Seifen Anstrichm.*, 77, 8–11, 1975.

63. Schulte, E. and Weber, N., Analysis of disteryl ethers, *Lipids*, 22, 1049–1052, 1987.

64. Ramamurthi, S. and McCurdy, A.R., Enzymatic pretreatment of deodorizer distillate for concentration of sterols and tocopherols, *J. Am. Oil Chem. Soc.*, 70, 287–295, 1993.

65. Ramamurthi, S., McCurdy, A.R., and Tyler, R.T., Deodorizer distillate: a valuable by-product, in *Proceedings of the World Conference on Oilseed and Edible Oils Processsing: Emerging Technologies, Current Practices, Quality Control, Technology Transfer, and Environmental Issues,* Vol. I, Koseoglu, S.S., Rhee, K.C., and Wilson, R.F. Eds., AOCS Press: Champaign, IL, 1998, pp. 130–134.

66. Daguet, D. and Coïc, J.-P., Phytosterol extraction: state of the art, *J. Franc. Oléagineux Corps Gras Lipides*, 6, 25–28, 1999.

67. Struve, A., Process for Isolating Sterols from Fat Processing Residues, U.S. Patent 4,148,810, April 10, 1979.

68. Albiez, W., Kozak, W.G., and Louwen, T., Verfahren zur Gewinnung von Sterinen und Tocopherolen, German Offenlegung. DE 100 38 457 A1, February 2002.

69. Laur, J., Castera, A., Mordret, F., Pages-Xatart-Pares, X., and Guichard, J.-M., Method of Preparing Fat Fractions of Vegetable Origin Enriched With Unsaponifiable Materials and Use of Said Fractions for Prepairing Cosmetic and/or Pharmaceutical Compositions in Particular Dermatological Compositions, U.S. Patent 5,679,393, October 21, 1997.

70. Dunford, N.T. and King, J.W., Phytosterol enrichment of rice bran oil by a supercritical carbon dioxide fractionation technique, *J. Food Sci.*, 65, 1395–1399, 2000.

71. Dunford, N.T., Teel, J.A., and King, J.W., A continuous countercurrent supercritical fluid deacidification process for phytosterol ester fortification in rice bran oil, *Food Res. Int.*, 36, 175–181, 2003.

72. Ibanez, E., Benavides, A.M.H., Senorans, F.J., and Reglero, G., Concentration of sterols and tocopherols from olive oil with supercritical carbon dioxide, *J. Am. Oil Chem. Soc.*, 79, 1255–1260, 2002.

73. Shimada, Y., Nakai, S., Suenaga, M., Sugihara, A., Kitano, M., and Tominaga, Y., Facile purification of tocopherols from soybean oil deodorizer distillate in high yield using lipase, *J. Am. Oil Chem. Soc.*, 77, 1009–1013, 2000.

74. Weber, N., Weitkamp, P., and Mukherjee, K.D., Cholesterol-lowering food additives: lipase-catalyzed preparation of phytosterol and phytostanol esters, *Food Res. Int.*, 35, 177–181, 2002.

75. Ghosh, S. and Bhattacharyya, D.K., Isolation of tocopherol and sterol concentrate from sunflower oil deodorizer distillate, *J. Am. Oil Chem. Soc.*, 73, 1271–1274, 1996.

76. Holmbom, B., and Erä, V., Composition of tall oil pitch, *J. Am. Oil Chem. Soc.*, 55, 342–344, 1978.

77. Miettinen, T., Vanhanen, H., and Wester, I., A Substance for Lowering High Cholesterol Level in Serum and a Method for Preparing the Same, WO Patent 92/19640, November 12, 1992.

78. von Hellens, S., Benecol margarine enriched with stanol esters, *Lipid Technol.*, 11, 29–31, 1999.

79. van Amerongen, M. and Lievense, L.C., Stanol Ester Composition and Production Thereof, U.S. Patent 6, 031, 118, February 29, 2000.

80. Amerongen, M. and van Lievense, L.C., Stanol Comprising Compositions, U.S. Patent 6,231,915 B1, May 5, 2001.

81. Crawford, R.R., Blum, W.P., and Naramore, D.C., Process for Separating 3-Hydroxy Steroids or Sterols from Mixtures Such as Lipids, U.S. Patent 4,425,275, October 10, 1984.

82. Gutierrez, A., del Rio, J.C., Martinez, J.M., and Martinez, A.T., The biotechnological control of pitch in paper pulp manufacturing, *Trends Biotechnol.*, 19, 340–348, 2001.

83. del Rio, J.C., Gutierrez, A., and Gonzalez-Vila, F.J., Analysis of impurities occurring in a totally chlorine-free bleached Kraft pulp, *J. Chromatogr.*, 830, 227–232, 1999.

84. del Rio, J.C., Romero, J., and Gutierrez, A., Analysis of pitch deposits produced in Kraft pulp mills using a totally chlorine-free bleaching sequence, *J. Chromatogr.*, 874, 235–245, 2000.

85. Gutierrez, A., Romero, J., and del Rio, J.C., Lipophilic extractives from *Eucalyptus globulus* pulp during Kraft cooking followed by TCF and ECF bleaching, *Holzforschung*, 55, 260–264, 2001.

86. Ivanov, S.A. and Biceva, P., Untersuchungen über die Zusammensetzung des Unverseifbaren verschiedener Tallprodukte (Tallöl, Tallseife und Tallpech) in technologischer Hinsicht, *Seifen Öle Fette Wachse*, 101, 475–478, 1975.

87. Zlatanov, M., Vasvazova, P., and Ivanov, S.A., Methode zur komplexen Verarbeitung des Unverseifbaren von Fettsäuredestillationspechen zur Gewinnung von Sterolen, höheren Fettalkoholen und Filmbildnern, *Seifen Öle Fette Wachse*, 105, 513–516, 1979.

88. Vasvazova-Biceva, P.I., Ivanov, S.A., Filipov, L.A., and Vytov, V.C., Neue Methode zur gleichzeitigen Gewinnung von Sterolen und pulverförmigem Leim aus Tallpech, *Seifen Öle Fette Wachse*, 102, 561–563, 1976.

89. Kochhar, S., Influence of processing of sterols of edible vegetable oils, *Prog. Lipid Res.*, 22, 161–188, 1983.

90. Mangold, H.K. and Muramatsu, T., Preparation of reference compounds, in *CRC Handbook of Chromatograph*, Vol. II, Mangold, H.K., Ed., CRC Press, Boca Raton, FL, 1984, pp. 319–329.

91. Spener, F., Preparation of common and unusual waxes, *Chem. Phys. Lipids*, 24, 431–448, 1979.

92. Serreqi, A.N., Leone, R., del Rio, L.F., Mei, S., Fernandez, M., and Breuil, C., Identification and quantification of important steryl esters in aspen wood, *J. Am. Oil Chem. Soc.* 77, 413–418, 2000.

93. Condo, A.M., Jr., Baker, D.C., Moreau, R.A., and Hicks, K.B., Improved method for the synthesis of *trans*-feruloyl-β-sitostanol, *J. Agric. Food Chem.*, 49, 4961–4964, 2001.

94. Baltes, J. and Merkle, R.m Verfahren zur Umwandlung von in pflanzlichen und tierischen Fetten enthaltenen Sterinen in ihre Fettsäureester, German Offenlegung, DE 2248921, April 11, 1974.

95. Miettinen, T., Vanhanen, H., and Wester, I., Use of Stanol Fatty Acid Ester for Reducing Serum Cholesterol Level, U.S. Patent 5,502,045, March 26, 1996.

96. Milstein, N.; Biermann, M., Leidl, P., and von Kries, R., Sterol Esters as Food Additives, U.S. Patent 6, 394, 230 B1, May 28, 2002.

97. Mikkonen, H., Heikkilä, E., Anttila, E., and Lindeman, A., Use of Organic Esters in Dietary Fat, U.S. Patent 6,441,206 B1, August 27, 2002.

98. Pouilloux, Y., Courtois, G., Boisseau, M., Piccirilli, A., and Barrault, J., Solid base catalysts for the synthesis of phytosterol esters, *Green Chem.*, 5, 89–91, 2003.

99. Baldessari, A., Maier, M.S., and Gros, E.G., Enzymatic deacetylation of steroids bearing labile functions, *Tetrahedron Lett.*, 36, 4349–4352, 1995.

100. Baldessari, A., Bruttomesso, A.C., and Gros, E.G., Lipase-catalysed regioselective deacetylation of androstane derivatives, *Helv. Chim. Acta*, 79, 999–1004, 1996.

101. Deykin, D. and Goodman, D.S., The hydrolysis of long-chain fatty acid esters of cholesterol with rat liver enzymes, *J. Biol. Chem.*, 237, 3649–3656, 1962.

102. Kamei, T., Suzuki, H., Asano, K., Matsuzaki, M., and Nakamura, S., Cholesterol esterase produced by *Streptomyces lavendulae*. Purification and properties as a lipolytic enzyme, *Chem. Pharm. Bull.*, 27, 1704–1707, 1979.

103. Taketani, S., Nishino, T., and Katsuki, H., Purification and properties of sterol-ester hydrolase from *Saccharomyces cerevisiae*, *J. Biochem.*, 89, 1667–1673, 1981.

104. Tenkanen, M., Kontkanen, H., Isoniemi, R., and Spetz, P., Hydrolysis of steryl esters by a lipase (Lip 3) from *Candida rugosa*, *Appl. Microbiol. Biotechnol.*, 60, 120–127, 2002.

105. Uwajima, T. and Terada. O., Purification and properties of cholesterol esterase from *Pseudomonas fluorescens*, *Agric. Biol. Chem.*, 40, 1957–1964, 1976.

106. Shimada, Y., Nagao, T., Watanabe, Y., Takagi, Y., and Sugihara, A., Enzymatic conversion of steryl esters to free sterols, *J. Am. Oil Chem. Soc.*, 80, 243–247, 2003.

107. Ferrari, R.Ap., Esteves, W., and Mukherjee, K.D., Alteration of steryl ester content and positional distribution of fatty acids in triacylglycerols by chemical and enzymatic interesterification of plant oils, *J. Am. Oil Chem. Soc.*, 74, 93–96, 1997.

108. Koshiro, A., Cholesterol ester and its manufacture using microbial lipase. Japan Kokai Tokkyo Koho 62, 296, 894, December 24, 1987 (*Chem. Abstr.*, 110, 210969p, 1989).

109. Faber, K. and Riva, S., Enzyme-catalyzed irreversible acyl transfer, *Synthesis*, 895–910, 1992.

110. Haraldsson, G.G., The application of lipases in organic synthesis, in *Supplement B: The Chemistry of Acid Derivatives*, Vol. 2, Patai, S., Ed., John Wiley, Chichester, 1992, pp. 1395–1473.

111. Hedström, G., Slote, J.P., Backlund, M., Molander, O., and Rosenholm, J.B., Lipase-catalyzed synthesis and hydrolysis of cholesterol oleate in aot/isooctane microemulsions, *Biocatalysis*, 6, 281–290, 1992.

112. Jonzo, M.D., Hiol, A., Druet, D., and Comeau, L.-C., Application of immobilized lipase from *Candida rugosa* to synthesis of cholesterol oleate, *J. Chem. Tech. Biotechnol.*, 69, 463–469, 1997.

113. Jonzo, M.D., Hiol, A., Zagol, I., Druet, D., and Comeau, L.-C., Concentrates of DHA from fish oil by selective esterification of cholesterol by immobilized isoforms of lipase from *Candida rugosa*, *Enzyme Microb. Technol.*, 27, 443–450, 2000.

114. Kosugi, Y., Tanaka, H., Tomizuka, N., Akeboshi, K., Matsufune, Y., and Yoshikawa, S., Enzymic manufacture of sterol fatty acid esters, Japan Kokai Tokkyo Koho 1,218,593, August 31, 1989 (*Chem. Abstr.*, 113, 4707k, 1990).

115. Myojo, K. and Matsufune, Y., Process for preparing sterol fatty acid esters with enzymes, *Yukagaku*, 44, 883–896, 1995.

116. Riva, S. and Klibanov, A.M., Enzymochemical regioselective oxidation of steroids without oxidoreductases, *J. Am. Chem. Soc.*, 110, 3291–3295, 1988.

117. Riva, S., Bovara, R., Ottolina, G., Secundo, F., and Carrea, G., Regioselective acylation of bile acid derivatives with *Candida cylindracea* lipase in anhydrous benzene, *J. Org. Chem.*, 54, 3161–3164, 1989.

118. Shimada, Y., Hirota, Y., Baba, T., Sugihara, A., Moriyama, S., Tominaga, Y., and Terai, T., Enzymatic synthesis of steryl esters of polyunsaturated fatty acids, *J. Am. Oil Chem. Soc.*, 76, 713–716, 1999.

119. Vu, P.-L., Shin, J.-A., Lim, C.-H., and Lee, K.-T., Lipase-catalyzed production of phytosteryl esters and their crystallization behavior in corn oil, *Food Res. Int.*, 37, 175–180, 2004.

120. Weber, N. and Mukherjee, K.D., Enzymatisches Verfahren zur Herstellung von Carbonsäure-sterylestern, German Offenlegung, DE 100,187,870, May 3, 2001.

121. Weber, N., Weitkamp, P., and Mukherjee, K.D., Fatty acid steryl, stanyl and steroid esters by esterification and transesterification in vacuo using *Candida rugosa* lipase as catalyst, *J. Agric. Food Chem.*, 49, 67–71, 2001.

122. Weber, N., Weitkamp, P., and Mukherjee, K.D., Steryl and stanyl esters of fatty acids by solvent-free esterification and transesterification in vacuo using lipases from *Rhizomucor miehei*, *Candida antarctica* and *Carica papaya*, *J. Agric. Food Chem.*, 49, 5210–5216, 2001.

123. Negishi, S., Hidaka, I., Takahashi, I., and Kunita, S., Transesterification of phytosterol and edible oil by lipase powder at high temperature, *J. Amer. Oil Chem. Soc.*, 80, 905–907, 2003.

124. Weber, N., Weitkamp, P., and Mukherjee, K.D., Steryl esters by transesterification reactions catalyzed by lipase from *Thermomyces lanuginosus*, *Eur. J. Lipid Sci. Technol.*, 105, 624–626, 2003.

125. Weber, N., Weitkamp, P., and Mukherjee, K.D., Steryl and stanyl esters by solvent-free lipase-catalyzed esterification and transesterification in vacuo, *Fresenius Environ. Bull.*, 12, 517–522, 2003.

126. Hirota, Y., Nagao, T., Watanabe, Y., Suenaga, M., Nakai, S., Kitano, M., Sugihara, A., and Shimada, Y., Purification of steryl esters from soybean oil deodorizer distillate, *J. Am. Oil Chem. Soc.*, 80. 341–346, 2003.

127. Heinemann, T., Axtmann, G., and Bergmann, K., Comparison of intestinal absorption of cholesterol with different plant sterols in man, *Eur. J. Clin. Investig.*, 23, 827–831, 1993.

128. Armstrong, M.J. and Carey, M.C., Thermodynamic and molecular determinants of sterol solubilities in bile salt micelles, *J. Lipid Res.*, 28, 1144–1155, 1987.

129. Ikeda, I. and Sugano, M., Inhibition of cholesterol absorption by plant sterols for mass intervention, *Curr. Opin. Lipidol.*, 9, 527–531, 1998.

130. Ling, W.H. and Jones, P.J.H., Dietary phytosterols: a review of metabolism, benefits and side effects, *Life Sci.*, 57, 195–206, 1995.

131. Mel'nikov, S.M., Seijen ten Hoorn, J.W.M., and Eijkelenboom, A.P.A.M., Effect of phytosterols and phytostanols on the solubilization of cholesterol by dietary mixed micelles: an *in vitro* study, *Chem. Phys. Lipids*, 127, 121–123, 2004.

132. Mel'nikov, S.M., Seijen ten Hoorn, J.W.M., and Bertrand, B., Can cholesterol absorption be reduced by phytosterols and phytostanols via a cocrystallization mechanism?, *Chem. Phys. Lipids*, 127, 15–33, 2004.

133. Wilson, M.D. and Rudel, L.L., Review of cholesterol absorption with emphasis on dietary and biliary cholesterol, *J. Lipid Res.*, 35, 943–955, 1994.

134. Normén, L., Dutta, P., Lia, A., and Andersson, H., Soy sterol esters and β-sitostanol ester as inhibitor of cholesterol absorption in human small bowel, *Am. J. Clin. Nutr.*, 71, 908–913, 2000.

135. Gremaud, G., Dalan, E., Piguet, C., Baumgartner, M., Ballabeni, P., Decarli, B., Leser, M.E., Berger, A., and Fay, L.B., Effects of non-esterified stanols in a liquid emulsion on cholesterol absorption and synthesis in hypercholesterolemic men, *Eur. J. Nutr.*, 41, 54–60, 2002.

136. Ostlund, Jr., R.E., Spilburg, C.A., and Stenson, W.F., Sitostanol administered in lecithin micelles potently reduces cholesterol absorption in humans, *Am. J. Clin. Nutr.*, 70, 826–831, 1999.

137. Meguro, S., Higashi, K., Hase, T., Honda, Y., Otsuka, A., Tokimitsu, I., and Itakura, H., Solubilization of phytosterols in diacylglycerol versus triacylglycerol improves the serum cholesterol-lowering effect, *Eur. J. Clin. Nutr.*, 55, 513–517, 2001.

138. Igel, M., Giesa, U., Lütjohann, D., and von Bergmann, K., Comparison of the intestinal uptakes of cholesterol, plant sterols and stanols in mice, *J. Lipid Res.*, 44, 533–538, 2003.

139. Trautwein, E.A., Duchateau, G.S.M.J.E., Lin, Y., Mel'nikov, S.M., Molhuizen, H.O.F., and Ntanios, F.Y., Proposed mechanisms of cholesterol-lowering action of plant sterols, *Eur. J. Lipid Sci. Technol.*, 105, 171–185, 2003.

140. Turley, S.D. and Dietschy, J.M., Sterol absorption by the small intestine, *Curr. Opin. Lipidol.*, 14, 233–240, 2003.

141. Plat, J., Kerckhoffs, D.A.J.M., and Mensink, R.P., Therapeutic potential of plant sterols and stanols, *Curr. Opin. Lipidol.*, 11, 571–57, 2000.

142. Grundy, S.M., Ahrens, Jr., E.H., and Salen, G., Dietary β-sitosterol as an internal standard to correct for cholesterol losses in sterol balance studies, *J. Lipid Res.*, 9, 374–387, 1968.

143. Grundy, S.M., Ahrens, Jr., E.H., and Davignon, J., The interaction of cholesterol absorption and cholesterol synthesis in man, *J. Lipid Res.*, 10, 304–315, 1969.

144. Denke, M.A., Lack of efficacy of low-dose sitostanol therapy as an adjunct to a cholesterol-lowering diet in men with moderate hypercholesterolemia, *Am. J. Clin. Nutr.*, 61, 392–396, 1995.

145. St-Onge, M.-P. and Jones, P.J.H., Phytosterols and human lipid metabolism: efficacy, safety and novel foods, *Lipids*, 38, 367–375, 2003.

146. Ostlund, Jr., R.E., Racette, S.B., and Stenson, W.F., Inhibition of cholesterol absorption by phytosterol-replete wheat germ compared with phytosterol-depleted wheat germ, *Am. J. Clin. Nutr.*, 77, 1385–1389, 2003.

147. Amundsen, A.L., Ntanios, F., van der Put, N., and Ose, E., Long-term compliance and changes in plasma lipids, plant sterols and carotenoids in children and parents with FH consuming plant sterol ester-enriched spread, *Eur. J. Clin. Nutr.*, 58, 1612–1620, 2004.

148. Miettinen, T.A. and Gylling, H., Regulation of cholesterol metabolism by dietary plant sterols, *Curr. Opin. Lipidol.*, 10, 9–14, 1999.

149. Vissers, M.N., Zock, P.L., Meijer, G.W., and Katan, M.B., Effect of plant sterols from rice bran oil and triterpene alcohols from sheanut oil on serum lipoprotein concentration in humans, *Am. J. Clin. Nutr.*, 72, 1510–151, 2000.

150. Lee, Y.-M., Haasterst, B., Scherbaum, W., and Hauner, H., A phytosterol-enriched spread improves the lipid profile of subjects with type 2 diabetes mellitus. A randomized controlled trial under free-living conditions, *Eur. J. Nutr.*, 42, 111–117, 2003.

151. Amundsen, A.L., Ose, L., Nenseter, M.S., and Ntanios, F.Y., Plant sterol ester-enriched spread lowers plasma total and LDL cholesterol in children with familial hypercholesterolemia, *Am. J. Clin. Nutr.*, 76, 338–344, 2002.

152. Seki, S., Hidaka, I., Kojima, K., Yoshino, H., Aoyama, T., Okazaki, M., and Kondo, K., Effects of phytosterol ester-enriched vegetable oil on plasma lipoproteins in healthy men, *Asia Pac. J. Clin. Nutr.*, 12, 282–91, 2003.

153. Jones, P.J., Ntanios, F.Y., Raeini-Saraz, M., and Vanstone, C.A., Cholesterol-lowering efficacy of a sitostanol-containing phytosterol mixture with a prudent diet in hyperlipidemic men, *Am. J. Clin. Nutr.*, 69, 1144–1150, 1999.

154. Cleghorn, C.L., Skeaff, C.M., Mann, J., and Chisholm, A., Plant sterol-enriched spread enhances the cholesterol-lowering potential of a fat-reduced diet, *Eur. J. Clin. Nutr.*, 57, 170–176, 2003.

155. Maki, K.C., Shinnick, F., Seeley, M.A., Veith, P.E., Quinn, L.C., Hallissey, P.J., Temer, A., and Davidson, M.H., Food products containing free tall oil-based phytosterols and oat β-glucan lower serum total and LDL cholesterol in hypercholesterolemic adults, *J. Nutr.*, 133, 808–813, 2003.

156. Lin, Y., Meijer, G.W., Vermeer, M.A., and Trautwein, E.A., Soy protein enhances the cholesterol-lowering effect of plant sterol esters in cholesterol-fed hamster, *J. Nutr.*, 134, 143–148, 2004.

157. Jones, P.J., Howell, T., MacDougall, D.E., Feng, J.Y., and Parsons, W., Short-term administration of tall oil phytosterols improves plasma lipid profiles in subjects with different cholesterol levels, *Metabolism*, 47, 751–756, 1998.

158. Hayes, K.C., Pronczuk, A., and Perlman, D., Nonesterified phytosterols dissolved and recrystallized in oil reduce plasma cholesterol in gerbils and humans, *J. Nutr.*, 134, 1395–1399, 2004.

159. Moghadasian, M.H., McManus, B.M., Godin, D.V., Rodrigues, B., and Frohlich, J.J., Proatherogenic and antiatherogenic effects of probucol and phytosterols in apolipoprotein E-deficient mice: Possible mechanism of action, *Circulation*, 99, 1733–1739, 1999.

160. Ewart, H.S., Cole, L.K., Kralovec, J., Layton, H., Curtis, J.M., Wright, J.L.C., and Murphy, M.G., Fish oil containing phytosterol esters alters blood lipid profiles and left ventricle generation of thromboxane A2 in adult guinea pigs, *J. Nutr.*, 132, 1149–1152, 2002.

161. Andersson, S.W., Skinner, J., Ellgard, L., Welch, A.A., Bingham, S., Mulligan, A., Andersson, H., and Khaw, K.-T., Intake of dietary plant sterols is inversely related to serum cholesterol concentration in men and women in the EPIC norfolk population: a cross-sectional study, *Eur. J. Clin. Nutr.*, 58, 1378–1385, 2004.

162. Judd, J.J., Baer, D.J., Chen, S.C., Clevidence, B.A., Muesing, R.A., Kramer, M., and Meijer, G.W., Plant sterol esters lower plasma lipids and most carotenoids in mildly hypercholesterolemic adults, *Lipids*, 37, 33–42, 2002.

163. Gylling, H., Puska, P., Vartiainen, E., and Miettinen, T.A., Retinol, vitamin D, carotenes and alpha-tocopherol in serum of a moderately hypercholesterolemic population consuming sitostanol ester margarine, *Atherosclerosis*, 145, 279–285, 1999.

164. Tammi, A., Rönnemaa, T., Gylling, H., Rask-Nissilä, L., Vükari, J., Tuominen, J., Pulkki, K., and Simell, O., Plant stanol ester margarine lowers serum total and low-density lipoprotein cholesterol concentrations of healthy children: the STRIP project, *J. Pediatr.*, 136, 503–51, 2000.

165. Quilez, J., Rafecas, M., Brufau, G., Garcia-Lorda, P., Megias, I., Bullo, M., Ruiz, J.A., and Salas-Salvado, J., Bakery products enriched with phytosterol esters, α-tocopherol and β-carotene decrease plasma LDL-cholesterol and maintain plasma β-carotene concentrations in normocholesterolemic men and women, *J. Nutr.*, 133, 3103–3109, 2003.

166. Becker, M., Staab, D., and von Bergmann, K., Treatment of severe familial hypercholesterolemia in childhood with sitosterol and sitostanol, *J. Pediatr*, 122, 292–296, 1993.

167. Jones, P.J.H., MacDougall, D.E., Ntanios, F., and Vanstone, C.A., Dietary phytosterols as cholesterol-lowering agents in humans, *Can. J. Physiol. Pharmacol.*, 75, 217–227, 1997.

168. Katan, M.B., Grundy, S.M., Jones, P., Law, M., Miettinen, T., and Paoletti, R., Efficacy and safety of plant stanols and sterols in the management of blood cholesterol levels, *Mayo Clin. Proc.* 78, 965–978, 2003.

169. Hendriks, H.F., Brink, E.J., Meijer, G.W., Princen, H.M.G., and Ntanios, F.Y., Safety of long-term consumption of plant sterol esters-enriched spread, *Eur. J. Clin. Nutr.*, 57, 681–692, 2003.

170. Baker, V.A., Hepburn, P.A., Kennedy, S.J., Jones, P.A., Lea, L.J., Sumpter, J.P., and Ashby, J., Safety evaluation of phytosterol esters: 1. Assessment of oestrogenicity using a combination of *in vivo* and *in vitro* assays, *Food Chem. Toxicol.*, 37, 13–22, 1999.

171. Hepburn, P.A., Horner, S.A., and Smith, M., Safety evaluation of phytosterol esters: 2. Subchronic 90-day oral toxicity study on phytosterol esters: a novel functional food, *Food Chem. Toxicol.*, 37, 521–532, 1999.

172. Sanders, D.J., Minter, H.J., Howes, D., and Hepburn, P.A., The safety evaluation of phytosterol esters: 6. The comparative absorption and tissue distribution of phytosterols in the rat, *Food Chem. Toxicol.*, 38, 485–491, 2000.

173. Waalkens-Berendsen, D.H., Wolterbeek, A.P.M., Wijnands, M.V.W., Richold, M., and Hepburn, P.A., Safety evaluation of phytosterol esters: 3. Two generation reproduction study in rats with phytosterol esters: a novel functional food, *Food Chem. Toxicol.*, 37, 683–696, 1999.

174. Slesinski, R.S., Turnbull, D., Frankos, V.H., Wolterbeek, A.P.M., and Waalkens-Berendsen, D.H., Developmental toxicity study of vegetable oil-derived stanol fatty acid esters, *Regul. Toxicol. Pharmacol.*, 29, 227–233, 1999.

175. Turnbull, D., Frankos, V.H., van Delft, J.H.M., and deVogel, N., Genotoxicity evaluation of wood-derived and vegetable oil-derived stanol esters, *Regul. Toxicol. Pharmacol.*, 29, 205–210, 1999.

176. Turnbull, D., Whittaker, M.H., Frankos, V.H., and Jonker, D., 13-Week oral toxicity study with stanol esters in rats, *Regul. Toxicol. Pharmacol.*, 29, 216–226, 1999.

177. Whittaker, M.H., Frankos, V.H., Wolterbeek, A.P., and Waalkens-Berendsen, D.H., Two-generation reproductive toxicity study of plant stanol esters in rats, *Regul. Toxicol. Pharmacol.*, 29, 196–204, 1999.

178. Vuorio, A.F., Gylling, H., Turtola, H., Kontula, K., Ketonen, P., and Miettinen, T.A., Stanol ester margarine alone and with simvastatin lowers serum cholesterol in families with familial hypercholesterolemia caused by the FH-North Karelia mutation, *Arterioscler. Thromb. Vasc. Biol.*, 20, 500–506, 2000.

179. Carr, T.P., Cornelison, R.M., Illston, B.J., Stuefer–Powell, C.L., and Gallaher, D.D., Plant sterols alter bile acid metabolism and reduce cholesterol absorption in hamsters fed a beef-based diet, *Nutr. Res.*, 22, 745–754, 2002.

180. de Jong, N., Pijpers, L., Bleeker, J.K., and Ocke, M.C., Potential intake of phytosterols/-stanols: results of a stimulation study, *Eur. J. Clin. Nutr.*, 58, 907–919, 2004.

181. Lütjohann, D. and Bergmann, K., Phytosterolaemia: diagnosis characterization and therapeutical approaches, *Ann. Med.*, 29, 181–184, 1997.

182. Glueck, C.J., Speirs, J., Tracy, T., Streicher, P., Illig, E., and Vandergrift, J., Relationships of serum plant sterols (phytosterols) and cholesterol in 595 hypercholesterolemic subjects, and familial aggregation of phytosterols, cholesterol, and premature coronary heart disease in hyperphytosterolemic probands and their first-degree relatives, *Metabolism,* 40, 842–848, 1991.

183. Awad, A.B., Downie, A., Fink, C.S., and Kim, U., Dietary phytosterol inhibits the growth and metastasis of MDA-MB-231 human breast cancer cells grown in SCID mice, *Anticancer Res.*, 20, 821–824, 2000.

184. Lea, L.J., Hepburn, P.A., Wolfreys, A.M., and Baldrick, P., Safety evaluation of phytosterol esters: 8. Lack of genotoxicity and subchronic toxicity with phytosterol oxides, *Food Chem. Toxicol.*, 42, 771–783, 2004.

185. Howard, B.V. and Kritchevsky, D., Phytochemicals and cardiovascular disease: a statement for healthcare professionals from the American Heart Association, *Circulation*, 95, 2591–2593, 1997.

186. Hollingsworth, P., Margarine: the over-the-top functional food, *Food Technol.*, 55, 59–62, 2001.

187. Yankah, V.V. and Jones, P.J.H., Phytosterols and health implications: commercial products and their regulation, *Int. News Fats Oils Relat. Mater.*, 12, 1011–1016, 2001.

188. 2004/333/EC: Commission decision of 31 March 2004 authorizing the placing on the market of yellow fat spreads, salad dressings, milk type products, fermented milk type products, soya drinks and cheese-type products with added phytosterols/phytostanols as novel foods or novel food ingredients under Regulation (EC) No 258/97 of the European Parliament and of the Council (notified under document number C(2004) 1243). Official Journal 2004, L 105, 14/04/2004, 0040–0042.

189. 2004/334/EC: Commission decision of 31 March 2004 authorizing the placing on the market of yellow fat spreads, milk-type products, yoghurt-type products, and spicy sauces with added phytosterols/phytostanols as novel foods or novel food ingredients under Regulation (EC) No 258/97 of the European Parliament and of the Council (notified under document number C(2004) 1244). Official Journal 2004, L 105, 14/04/2004, 0043–0045.

190. 2004/335/EC: Commission decision of 31 March 2004 authorizing the placing on the market of milk-type products and yoghurt-type products with added phytosterol esters as novel food ingredients under Regulation (EC) No 258/97 of the European Parliament and of the Council (notified under document number C(2004) 1245). Official Journal 2004, L 105, 14/04/2004, 0046–0048.

191. 2004/336/EC: Commission decision of 31 March 2004 authorizing the placing on the market of yellow fat spreads, milk-based fruit drinks, yoghurt-type products and cheese-type products with added phytosterols/phytostanols as novel foods or novel food ingredients under Regulation (EC) No 258/97 of the European Parliament and of the Council (notified under document number C(2004) 1246). Official Journal 2004, L 105, 14/04/2004, 0049–0051.

192. Thomsen, A.B., Hansen, H.B., Christiansen, C., Green, H., and Berger, A., Effect of free plant sterols in low-fat milk on serum lipid profile in hypercholesterolemic subjects, *Eur. J. Clin. Nutr.*, 58, 860–870, 2004.

193. Pouteau, E.B., Monnard, I.E., Piguet-Welsch, C., Groux, M.J.A., Sagalowicz, L., and Berger, A., Non-esterified plant sterols solubilized in low fat milks inhibit cholesterol absorption. A stable isotope double-blind crossover study. *Eur. J. Nutr.*, 42, 154–164, 2003.

194. Jones, P.J.H., Vanstone, C.A., Raeini-Sarjaz, M., and St.-Onge, M.-P., Phytosterols in low- and nonfat beverages as part of a controlled diet fail to lower plasma lipid levels, *J. Lipid Res.*, 44, 1713–1719, 2003.

28 Phospholipids/Lecithin: A Class of Neutraceutical Lipids

Frank T. Orthoefer
Food Science and Technology Consultants, Germantown, Tennessee

G.R. List
Food and Industrial Oil Research, National Center for Agricultural
Utilization Research, ARS, USDA, Peoria, Illinois

CONTENTS

28.1 INTRODUCTION

Phospholipids have a 50-year history of use as a nutrient, supplement, and food ingredient[1]. The major commercial source of phospholipids is called *lecithin*, a natural mixture of phospholipids[2]. Lecithin is a coproduct of soybean oil processing and consists of complex fat-like compounds obtained by water washing of crude vegetable oil. The hydrated phospholipids, called gums, are recovered as the dense phase, which is then dried to produce a shelf-stable product[3]. Phospholipids refer to the active components of the gums. About 75% of the gums are composed of phosphorus-containing lipids. Other components are glycolipids, tocopherols, sterols, and sugars. This widely used source of phospholipids is also utilized in many formulated foods acting as an emulsifier. It is also used in cosmetics as a processing aid and stabilizing agent.

As a supplement containing phospholipids, lecithin is a bioavailable source of fatty acids and active subunits such as choline and inositol. Some of the biological/physical functions of phosphatides are described as a cell signaling agent and structural component of cell membranes[4]. People using phospholipids as a supplement promote its role in preventing cardiovascular disease, improving vascular health and liver function, enhancing memory and learning, and improving reaction time, physical endurance, and reproductive health[1].

Phospholipids are important constituents of all cells and play a role in the pathophysiology of human disease. Phospholipids are critical mediators of normal cell function. Metabolites of phospholipids, fatty acids and diacylglycerols, influence transmembrane signaling and, therefore, affect a wide variety of hormones, growth factors, and neurotransmitters. Dietary phospholipids are a source of choline that is believed important for the prevention of cancer and other human diseases.

Phospholipids or phosphatides which include the compound phosphatidylcholine, the chemical term for lecithin, are lipids containing a phosphoric acid group. In humans, the phosphatides are concentrated in the vital organs such as brain, liver, and kidneys. In plants, they are present in the highest amounts in the seeds, nuts, and grains. They are essential to life.

Commercial lecithin consists of a variety of lipid components and surface-active compounds. Commercially, this mixture consists of several other chemical entities in addition to phosphatidylcholine[5].

Lecithin is an important coproduct of edible oil processing. In the 1930s, solvent extracted crude oil was produced using hexane as the solvent. Small amounts of water in the crude oil resulted in precipitation of gums. Water degumming or removal of the phospholipids from the oil was necessary to stabilize the oil. In the following years various applications of lecithin were investigated and developed. Viscosity control in chocolate and emulsification of margarines were two of the earliest uses and continue to be major markets for lecithin. Lecithin is now recognized as a functional and nutritional coproduct of edible oil refining. In this chapter, structure, function, composition nutrition, composition of commercial lecithin, and metabolism is reviewed.

28.2 DEFINITION OF PHOSPHOLIPIDS

Three polymeric alcohols provide the basic structure for the various phospholipids (Figure 28.1). These are glycerol, sphingosine, and inositol. The glycerol phosphatides include phosphatidylcholine (PC), phosphatidylethanolamine (PE), and phosphatidylserine (PS), and the acetal phosphatides or plasmalogens, lysophosphatides and phosphatidic acids. Sphingosine includes sphingomyelin and other glycolipids (sphingolipids). Inositol includes phosphatidyl inositol[2].

Phospholipids being zwitterionic will form complexes with proteins, carbohydrates, glycosides, alkaloids, minerals, enzymes, cholesterol, and other substances. Derivatives such as lysophosphatides represent a special class derived from chemical or enzymatic hydrolysis of phospholipids. The nomenclature is discussed by Sipos and Szuhai[2] and Scholfield[5].

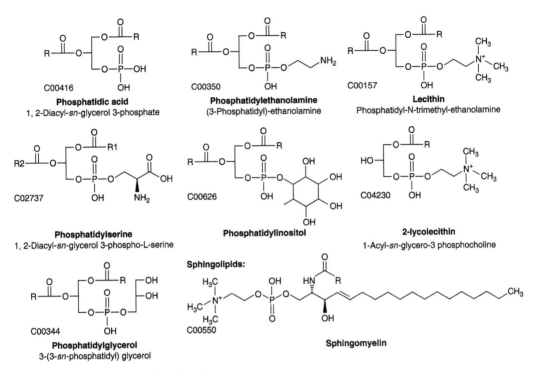

FIGURE 28.1 Structures of phospholipids.

28.3   COMMERCIAL SOURCES OF PHOSPHOLIPIDS

Both plants and animals serve as source materials for commercial phospholipid production. The most common plant source is soybeans supplying more than 90% of commercial lecithin, but corn, cottonseeds, and sunflower have also been used[6]. The various animal sources include eggs, milk, and various tissues.

28.4   PLANT SOURCES

Soybeans contain 0.3 to 0.6% phospholipids. The phospholipids are recovered along with the crude oil during solvent extraction or pressing of the oil from the seed (Figure 28.2). Recovery of the phospholipids from the crude oil occurs during water washing of the oil. Further processing includes vacuum drying and addition of components for standardizing. The lecithin obtained is a complex mixture of phosphatides, triglycerides, and other substances such as phytoglycolipids, phytosterols, tocopherols, and fatty acids (Table 28.1).

The commercial products from isolation of the lecithin are categorized into:

1. Natural lecithin
   - Plastic
   - Fluid
2. Refined lecithin
   - Blended products
   - Oil free
   - Fractionated (alcohol soluble, alcohol insoluble, chomatographically pure)
3. Chemically modified lecithin

TABLE 28.1
Approximate Composition of Commercial Soybean Lecithin[2]

Component	Content (%)
Soybean oil	35
Phosphatidylcholine	16
Phosphatidylethanolamine	14
Phosphatidylinositol	10
Phytoglycolipids	17
Carbohydrates	7
Moisture	1

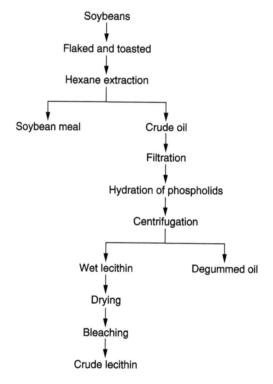

FIGURE 28.2 Processing of soybeans.

The range of composition of soybean lecithin is shown in Table 28.2. The fatty acid composition of soybean lecithin is shown in Table 28.3. During maturation of the seed, the major phospholipids (PC, PE, PI) increase while others decrease or remain constant.

Soybean lecithin is produced as a liquid, plastic, or free-flowing solid. The color, solubility, surfactancy, and chemical reactivity all can be modified. These alter the functional properties of the lecithin permitting its use in many applications[7].

The composition of the lecithin from other plant sources is similar to soybean lecithin[2]. Corn lecithin has a higher glycolipid content than soy and is lower in linolenic acid. Cottonseed phospholipids also have lower linolenic acid and higher saturated fatty acid content. Gossypol from cottonseed binds to lecithin during oil extraction limiting its utility as a commercial lecithin source.

TABLE 28.2
Composition of Soybean Lecithin[8]

		Range (%)		
	Component	Low	Intermediate	High
PC	Phosphatidylcholine	12.0–21.0	29.0–39.0	41.0–46.0
PE	Phosphatidylethanolamine	8.0–9.5	20.0–26.3	31.0–34.0
PI	Phosphatidylinositol	1.7–7.0	13.0–17.5	19.0–21
PA	Phosphatidic acid	0.2–1.5	5.0–9.0	14.0
PS	Phosphatidyl serine	0.2	5.9–6.3	—
LPC	Lysophosphatidylcholine	1.5	8.5	—
LPI	Lysophosphatidylinositol	0.4–1.8	—	—
LPS	Lysophospatidylserine	1.0	—	—
LPA	Lysophosphatidic acid	1.0	—	—
	Phytoglycolipids	—	14.3–15.4	29.6

TABLE 28.3
Fatty Acid Composition of Soybean Lecithin[8]

	Range (%)		
Fatty acid	Low	Intermediate	High
Myristic (C14:0)	0.3–1.9	—	—
Palmitic (C16:0)	11.7–18.9	2.5–26.7	42.7
Palmitoleic (C16:1)	7.0–8.6	—	—
Stearic (C18:0)	3.7–4.3	9.3–11.7	—
Oleic (C18:1)	6.8–9.8	17.0–25.1	39.4
Linoleic (C18:2)	17.1–20.0	37.0–40.0	55.0–60.8
Linolenic (C18:3)	1.6	4.0–6.2	9.2
Arachidic (C20:0)	1.4–2.3	—	—

Sunflower lecithin has a higher PC content (42.2 to 64.2%) and readily undergoes oxidative deterioration[8].

28.5 ANIMAL SOURCES OF LECITHIN

Lecithins from animal sources consist of sphingomyelins, PC, PE, PS, phosphatidylinositol, and glycerol phosphatides with complex fatty acid compositions[9]. These occur naturally in all membranes and lipoproteins serving both functional and structural purposes. The composition is dependent on the stage of growth and differentiation of the animal tissues as well as on dietary sources. The phospholipid content may also be modified by disease and lipid peroxidation. Production of animal-sourced phospholipids is highly variable depending on tissue source and metabolic state of the animal.

The exact composition of human and animal phospholipids depends on the source and method of extraction. The central nervous system has a very high phospholipid content, the liver is the site for biosynthesis, and the mitochondria consist of up to 90% phospholipids.

Only egg yolk, milk, and brain tissue have been utilized for commercial production of animal-sourced lecithin. Some isolated and purified lecithin as well as synthetic phospholipids have been developed for clinical use.

26.6   EGG PHOSPHOLIPIDS

Egg yolk contains a relatively high content of phospholipids[2]. The phospholipid composition of egg yolk compared to soy lecithin is shown in Table 28.4[10] and its fatty acid composition is shown in Table 28.5. Soybean lecithin containing 6 to 9% linolenic acid is less saturated than egg lecithin and contains much less PC (20 to 22% versus 68 to 72%).

28.7   MILK

The phospholipid content of milk is about 0.035%. Milk phospholipids are part of the milk fat globule membrane, a thin membrane that surrounds the milk fat globule. Five different phospholipids are included in milk fat: PE (33%), PC (34%), phosphatidylinositol (5%), PS (3%), and sphingomyelin (25%)[11].

28.8   BRAIN

The brain along with the spinal cord is the richest organ source of animal phospholipids. There are many different types of phospholipids in the central nervous system[12]. The major classes of brain phospholipids are shown in Table 28.6.

28.9   LIVER, KIDNEY, MUSCLE, AND OTHER TISSUES

Organ meats are a major source of dietary phospholipids[9]. Sphingomyelin is present in varying amounts in all animal organs. In blood, PC is quantitatively the most important phospholipid present. Total blood contains 0.2 to 0.3% phospholipids. In plasma and serum, PC predominates. In corpuscles, PE and sphingomyelin are dominant. Some 60 to 65% of the total lipids in red blood cells are phospholipids.

28.10   CHEMICAL AND PHYSICAL PROPERTIES

Phospholipid molecules in a mixture of water and oil will arrange at the interface into a layer with the fatty acids facing the oil surface and the phosphoric acid portion facing the water surface. This lowers the interfacial tension resulting in rapid wetting, lowering of the viscosity, and improving emulsion stability.

Soybean lecithin is soluble in aliphatic and aromatic solvents, partially soluble in ethanol and, insoluble in acetone and in water (Table 28.7)[2]. Lecithin is soluble in mineral oil and fatty acids. Soybean lecithin hydrates when mixed with water, forming a thick emulsion.

28.11   PREPARATION OF SOYBEAN LECITHIN

Almost all soybeans are processed into a high-protein meal and crude vegetable oil (Figure 28.2)[8]. The process consists of toasting and flaking of the beans followed by continuous solvent extraction using hexane. After distillation of the extraction solvent, the crude oil contains approximately 750 ppm

TABLE 28.4
Composition of Soy and Egg Lecithin[10]

Polar lipid	Soy	Egg
Phosphatidylcholine	20–22	68–72
Phosphatidylethanolamine	21–23	12–16
Phosphatidylinositol	18–20	0–2
Phosphatidic acid	4–8	—
Sphingomyelin	—	2–4
Other phospholipids	15	10
Glycolipids	9–12	—

TABLE 28.5
Fatty Acid Composition of Soy and Egg Lecithin[2]

Fatty acid	Soy	Egg
Palmitic (C16:0)	15–18	27–29
Stearic (C18:0)	3–6	14–17
Oleic (C18:1)	9–11	35–38
Linolenic (C18:2)	56–60	15–18
Linolenic (C18:3)	6–9	0–1
Arachidonic (C20:4)	0	3–5

TABLE 28.6
Major Classes of Brain Phospholipids (Myelin)[12]

Polar lipid	Amount (%)
Phosphatidylcholine	21.8
Phosphatidylethanolamine	35.4
Phosphatidylserine	18.8
Phosphatidylinositol	1.8
Phosphatidic acid	1.1
Lysophosphatidylcholine	2.0
Sphingomyelin	16.3

TABLE 28.7
General Characteristics of Soybean Lecithin[2]

Insoluble in water and acetone
Soluble in abs. alcohol, chloroform, ether, petroleum ether, mineral oil
Edible, digestible surfactant
Waxy mass of 20 acid value
Thick, pourable liquid at 30 acid value
Color yellow to brown
IV 95
Saponification value 196

TABLE 28.8
Soybean Lecithin Specifications[13]

	Fluid natural lecithin	Fluid bleached lecithin	Fluid double-bleached lecithin
Fluidized lecithin			
Acetone-insolubles, min. (%)	62	62	62
Moisture, max. (%)	1	1	1
Hexane-insolubles, max. (%)	0.3	0.3	0.3
Acid value, max.	32	32	32
Color, Gardner, max.	10	7	4
Viscosity at 77°F, max. (P)	150	150	150
Plastic lecithin			
Acetone-insolubles, min. (%)	65	65	65
Moisture, max. (%)	1	1	1
Hexane-insolubles, max. (%)	0.3	0.3	0.3
Acid value, max.	30	30	30
Color, Gardner, max.	10	7	4
Penetration, max. (mm)	22	22	22

phosphorus present as phospholipids. Water is added to isolate the phospholipids as lecithin from the crude oil. The phospholipids form gels that swell, become insoluble, and precipitate from the oil. The hydrated lecithin, called gums, can be separated by centrifugation. A degummed soybean oil contains 2 to 5 ppm phosphorus.

The wet lecithin gums are dried using thin-film evaporators at temperatures of 85 to 105°C under 25 to 30 mmHg vacuum. The residence time in the dryer is 1 to 2 minutes. The product is cooled after drying to 55 to 60°C if it is to be further processed or 35 to 50°C before packaging.

Lecithin is standardized to National Oilseed Processors Association (NOPA) standards (Table 28.8)[13]. Typically, fluid lecithin contains 62 to 64% acetone insolubles and an acid value of 30. The acetone insolubles (AI) value is representative of the phospholipid content (AOCS official method Ja 4-46) and the acid value (AV) is the titrable acidity (AOCS official method Ja 6-55). Fatty acids and soybean oil are added to the dried lecithin to meet the NOPA specifications and for standardization[14].

Dried lecithin is a dark to medium brown color. Carotenoids are the major pigments. Minor amounts of brown pigments and porphyrins are also present. The dark color of soybean lecithin is lightened by hydrogen peroxide addition or benzoyl peroxide bleaching. The hydrogen peroxide may be added to the oil before degumming or to the gums prior to drying, or to the dried gums. Benzoyl peroxide is added to the dried lecithin only.

The typical chemical characteristics of soybean lecithin are shown in Table 28.9. Lecithin is defined under 21CFR 184.1400 (Table 28.10)[15] and in the Food Chemical Codex, 3rd edition (Table 28.11 and Table 28.12)[16]. Most parameters are controlled during preparation of the lecithin by manipulation of bleaching, filtration, and standardization using fatty acids and oil addition.

Lecithin blends are prepared to meet specific functional properties such as water dispersibility or plasticity. Only minor quantities are marketed as blended products. Deoiled lecithin having in AI of about 98 is produced by acetone precipitation as shown in Figure 28.3. Specifications on lecithin to be deoiled are shown in Table 28.13[3,14]. Multiple acetone extractions are used in batch production. Continuous extraction units are also used. The acetone wet, deoiled lecithin is recovered by vacuum filtration, granulated, and desolventized in moving bed, forced air dryers. The acetone content of the deoiled lecithin is less than 25 ppm.

The phospholipid composition of deoiled lecithin is shown in Table 28.14. Alcohol may be used to produce lecithin fractions with different hydrophilic and lipophilic properties (Figure 28.4)[1]. The

TABLE 28.9
Chemical Parameters for Lecithin

Acetone insolubles	Phospholipid content AOCS method Ja 4-46
Acid value	Acidity of phospholipids plus acidity of the carrier (oil and fatty acids) = mg of KOH required to neutralize 1 g of lecithin *AOCS method Ja 6-55*
Moisture	By Karl–Fischer titration *AOCS method Jb2b-87*
Hexane insolubles	Insoluble impurities such as meal *AOCS method Ja 3-87*
Color	Gardener color
Viscosity at 25°C	Rotating viscometer
Total phosphorus	*AOCS method Ja 5-55*

TABLE 28.10
Code of Federal Regulations 184.1400[15]

Mixture of phosphatides of choline, ethanolamine, and inositol and other lipids
Isolated as gums from soybean, corn, or sunflower oils
May be bleached with H_2O_2 or benzoyl peroxide
Meets Food Chemical Codex, 3rd edition
Used in foods with no limitation

TABLE 28.11
Definition of Lecithin (Food Chemicals Codex, 3rd edition)[4]

Mixture of acetone-insoluble phosphatides combined with triglycerides, fatty acids, and carbohydrates
Refined grades vary in proportions and combinations
Oil-free form is 90% or more of phosphatides representing all or certain fractions
Consistency varies from fluid to plastic
Color is from yellow to brown
Partially soluble in water and alcohol, and insoluble in acetone
Will hydrate to form emulsions

TABLE 28.12
Lecithin Requirement Based on Food Chemicals Codex, 3rd edition[16]

Acetone insolubles (AI)	Not less than 50.0%
Acid value (AV)	Not more than 36
Arsenic (As)	Not more than 3 ppm
Heavy metals	Not more than 0.004%
Hexane insolubles	Not more than 0.3%
Lead	Not more than 10 ppm
Peroxide value (PV)	Not more than 100
Water	Not more than 1.5%

TABLE 28.13
Specifications for Crude Lecithin for Deoiling[14]

Acetone insolubles (%)	65	Min.
Acid value (mg KOH/g)	25	Max.
Moisture (%)	0.7	Max.
Hexane insolubles (%)	0.5	Max.
Color (Gardner undiluted)	12–16	Max.
Peroxide value (mg/kg)	2	Max.

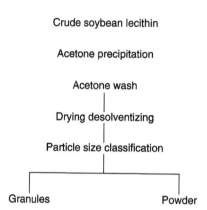

FIGURE 28.3 Production of deoiled lecithin[3,14].

compositions of the alcohol fractions are shown in Table 28.15[14]. Nonfractionated oil-free lecithin contains nearly equal portions of PC, PE, and phosphatidylinositol. The alcohol-soluble fraction is enriched in PC. The high-PC fraction favors oil-in-water emulsions whereas the fraction rich in PE and phosphatidylinositol favors water-in-oil emulsions. Further fractionation of either the alcohol-soluble fraction or crude lecithin using alumina chromatography yields fractions nearly pure in PC (Figure 28.5)[17].

Lecithin is storage stable when kept in closed containers. Lecithin may be kept for years at 20 to 25°C, although bleached lecithin may undergo color reversion. Deoiled granular lecithin can be stored for up to two years at less than 25°C.

28.12   MODIFICATION

Lecithin contains several functional groups that may be modified. The modification reactions are:

1. Hydroxylation
2. Acetylation
3. Hydrolysis
4. Hydrogenation
5. Halogenation
6. Phosphorylation
7. Sulfonation

Only acetylation, hydroxylation, and hydrolysis are used on a commercial scale[18].

TABLE 28.14
Phospholipid Composition of Deoiled Soybean Lecithin[1]

Phospholipid	Lecithin (concentrated) (%)	Deoiled lecithin (%)
Phospholipids (AI)	65	98
Phosphatidylcholine	16	23
Phosphatidylethanolamine	14	21
Phosphatidylinositol	13	19
Other phospholipids	10	15
Phosphatidic acid	4	6
Glycolipids	9	14
Other components	35	2

TABLE 28.15
Approximate Composition (%) of Alcohol Fractionated Lecithin[14]

Component	Oil-free lecithin	Alcohol-soluble	Alcohol-insoluble
Chemical lecithin	29	60	4
Chemical cephalin	29	30	29
Inositol and other phosphatides (including glycolipids)	32	1	55
Soybean oil	3	4	4
Miscellaneous	7	4	8
Emulsion type favored	Either oil-in-water or water-in-oil	Oil-in-water	Water-in-oil

Acetylation with acetic anhydride results in acetylation of the primary amine of PE and the amino group of PC. Acetylated lecithin has improved fluidity, water dispersibility, and enhanced oil-in-water emulsification properties[18]. Blocking the primary amine of PE with acetylation also improves the heat tolerance of lecithin making it ideal for kitchen pan sprays used in greaseless cooking.

Hydroxylated lecithin is desirable for applications requiring rapid water dispersibility, enhanced oil-in-water emulsifying properties, or a particularly "light" color. Hydroxylation is carried out by reacting crude lecithin with hydrogen peroxide and lactic or acetic acids. The active site for hydroxylation is likely the double bonds of the fatty acids as measured by the drop in iodine value (IV) and isolation of dihydroxystearic acid from the reaction mixture. The hydroxylation is carried out until a 10% drop in IV occurs. The ethanolamine group is also modified during hydroxylation.

Hydrolysis of the fatty acids of lecithin occurs upon exposure to strong acid, base, or enzymes. Functional products intended particularly for baked goods result when just one of the fatty acids is cleaved from the phospholipid molecule; however, acid or base hydrolysis is difficult to control. Enzymatic hydrolysis is the preferred method. Phospholipase A_2 cleaves the beta fatty acid from the phospholipid molecule. Partially hydrolyzed lecithins exhibit improve oil-in-water emulsification properties. While also used in foods, enzyme-modified lecithins have been used in formulation of calf milk replacers to improve emulsification and digestibility of feed fats.

An oxidatively stable, hydrogenated lecithin with its IV reduced to 10–30 may be produced. The process requires a nickel or palladium catalyst, a solvent such as ethyl acetate, temperatures of 70 to 80°C, and high hydrogen pressures (70 atm). The hydrogenated lecithin is a hard, wax-like

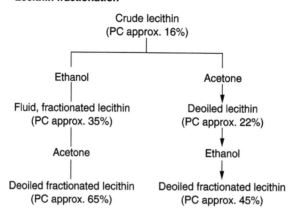

FIGURE 28.4 Alcoholic fractionation of lecithin. (Source: Orthoefer, F.T., *Lecithin and Health*, Vital Health Publishing, Bloomingdale, IL, 1998.)

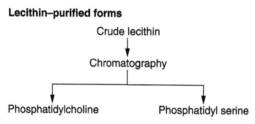

FIGURE 28.5 Lecithin purified forms.

material that does not oxidize and is less hydroscopic than regular lecithin. This is used to produce emulsions for intravenous injections, chocolate products, and special lubricants.

28.13   TRADITIONAL APPLICATIONS FOR LECITHIN

The applications for lecithin depend upon the multifunctional properties of the phospholipids. Some of the properties promoted are:

1. Emulsification
2. Film forming
3. Antioxidant
4. Viscosity reduction
5. Surfactancy[18]

The uses range from ingredients for foods to medicinal/nutraceutical products to nonfood, industrial applications. Most of the lecithin is used in foods/feeds, cosmetics, and pharmaceutical products.

The earliest uses for lecithin were for incorporation into chocolate and margarines. Without lecithin addition to chocolate, 25 to 40% more cocoa butter is required for acceptable viscosities of chocolate coatings. Margarines without lecithin will "weep" during storage and spatter when used for frying. Lecithin is used in bakery products as a low-cost emulsifier and also to prevent staling.

Today more than 150 million pounds of lecithin are produced in the U.S., almost all of which is derived from soybeans. Animal feeds, particularly that for infant animals, take advantage of lecithin for emulsification and enhanced nutrition. Even farm-raised shrimp, trout, and salmon have lecithin added to their diets for improved survival and growth.

Nonfood industrial applications of lecithin include agricultural pesticide spray formulations, vinyl paints, and nonstick coatings for conveyors.

Lecithin is nature's most widely distributed emulsifier. In heterogeneous systems, it has the ability to spread producing a monomolecular layer on the surface of water[19]. Specially formulated lecithins may be used to encapsulate oil in moisturizing creams[2]. Giant unilamellar liposomes have been formulated with lecithin to be used as delivery and sustained release vehicles for drugs, cosmetics, hormones, and vaccines.

In pharmaceuticals and injectibles, lecithin is added as an emulsifying agent to produce an oil-in-water emulsion. The lecithin may be isolated from plant or animal sources. Lecithin seems to protect against mucosal lesions, perforations, and bleeding found for some pharmaceuticals.

28.14   LECITHIN AS A SUPPLEMENT

"The role of lecithin in nutrition is complex."

Lecithin is contained in all cell membranes naturally. We therefore consume phospholipids as part of our everyday diet. Foods such as eggs, liver, peanuts, and soybeans are particularly rich in lecithin. Many consumers of supplemental dietary lecithin claim it plays a role in[1]:

1. Cardiovascular health
2. Liver and cell function
3. Fat transport and fat metabolism
4. Reproduction and child development
5. Physical performance and muscle endurance
6. Cell communication
7. Improvement of memory, learning, and reaction time
8. Relief of arthritis
9. Healthy hair and skin
10. Treatment for gallstones

The use of lecithin as a treatment for certain diseases has increased public interest in it as a dietary supplement. One turning point came with the use of lecithin for humans with Tardive dyskinesia, a neurological disorder, and the belief that normal memory may be influenced by the choline content of the diet.

Healthy humans normally consume about 6 g of lecithin per day but this figure is decreasing because of lower fat foods and reduced red meat consumption[20]. The lecithin content of several foods is shown in Table 28.16[21]. Choline is an important constituent of lecithin. Total dietary choline intake is estimated at 600 to 1000 mg/day. Lecithin is also available as a retail dietary supplement. In most supplements, the PC content is less than 35%.

28.15   LECITHIN AS A CHOLINE SOURCE

Mammals do not have to consume lecithin to survive because they can synthesize phospholipids. PC is not an essential component of their metabolism. PC is a major bioavailable delivery form of choline, an essential nutrient. Several biochemical pathways lead to lecithin synthesis in mammals, but only one generates choline. The methylation pathway catalyzed by the enzyme phosphatidylethanolamine-*N*-methyl tranferase (PEMT) makes new choline by sequentially methylating

TABLE 28.16
Choline and Lecithin Content of Selected Food[21]

Food	Lecithin (mg)	Choline (mg)
Calf liver	850	—
Lamb chop	753	—
Beef	453	—
Ham	800	—
Trout	580	—
Cheese	100	—
Egg	394	0.4
Oatmeal	650	131
Soybean	1480	237
Wheat germ	2820	—
Peanuts	1113	—
Cauliflower	2	78
Kale	2	89
Potato	1	40

PE using *S*-adensylmethionine (SAM) as the methyl donor[21]. This enzyme is most active in the liver and also found in the kidney, testis, heart, adrenal gland, lung, erythrocytes, brain, and spleen. *In vitro* data estimates that 15 to 40% of the lecithin in the liver is synthesized via the PEMT pathway. The methylation pathway makes new free choline as the lecithins are hydrolyzed.

Choline, however, is essential. Choline is a precursor in the synthesis of the neurotransmitter acetylcholine, the methyl donor betaine, and phospholipids, including PC and sphingomyelin, among others. Choline prevents fat accumulation in the liver. A choline deficiency may disturb lecithin synthesis which is needed to export triglycerides as part of lipoproteins. PC is involved in the hepatic export of very low-density lipoprotein (VLDL). Renal function is compromised with choline deficiency with abnormal concentration ability, free water reabsorption, sodium excretion, glomerular filtration rate, and renal plasma flow, and gross renal hemorrhaging. Infertility, growth impairment, bony abnormalities, decreased hematopoiesis, and hypertension are associated with diets low in choline. In mice, choline deficiency is associated with impaired memory[22]. The complex relationships between choline, SAM, and lecithin make it difficult to determine how much choline should be included in the diet. The expression of choline deficiency varies with dietary calories and amino acid content and the rate of growth of animals. Methyl donors spare the choline requirement since choline can be formed by the methylation of PE within the liver and other organs. It is apparent that during choline deficiency, SAM in the liver is depleted limiting the rate of choline neogenesis.

Malnourished humans have very low plasma choline[21]. Humans fed with amino acid solutions during total parental nutrition (TPN) that take in no choline do not rehabilitate their low choline stores. In animals, choline added to TPN solutions reverses fat accumulation in the liver.

Supplemental choline can also influence organ function. Choline acetyltranferase, which catalyzes the acetylation of choline, is normally not saturated with either substrate.

28.16   CHOLINE AND CANCER

Eating a choline-devoid diet is a promoter of carcinogenesis. Choline is a major source of methyl groups that are involved in formation of methionine from homocysteine. Animals on a choline-deficient diet for prolonged periods develop hepatocarcinoma. Earlier changes include accumulation of lipid in the liver followed by single-cell necrosis and liver fibrosis progressing to cirrhosis and carcinoma. Liver cancer occurs without the presence of any known carcinogen[23]. Choline addition seems to prevent the development of cancer.

The mechanisms suggested for cancer promotion by choline-deficient diets are:

1. Increased cell death with cell regeneration and cell proliferation
2. Decreased DNA methylation
3. Increased lipid oxidation
4. Excessive diacylglycerols resulting in protein kinase C (PKC) activation

Agents that result in stimulating PKC are mitogens and tumor promoters. Several oncogenes have been linked to activation of PKC.

Fatty liver from choline deficiency occurs because lecithin availability at the site of fat export from the liver is limited. Lecithin is an essential component of VLDL, the blood lipid transport molecule.

28.17 DIETARY ABSORPTION OF LECITHIN

Lecithin upon ingestion is hydrolyzed during absorption. Both pancreatic secretions and intestinal mucosal cells have enzymes that hydrolyze lecithin. Phospholipase A_2 cleaves the beta fatty acid. Within the gut mucosal cell, phospholipase A_1 cleaves the alpha fatty acid. Phospholipase B cleaves both fatty acids. These cleave less lecithin than pancreatic lipase. Most dietary lecithin is absorbed as lysolecithin (deacylated in the beta position). Reacylation of lysolecithin takes place in the intestinal mucosal cells, reforming PC which is transported by the lymphatics in the form of chylomicrons to the blood. It is then distributed to various tissues. Some of the PC is incorporated into the cell membranes. Within the erythrocyte, lysolecithin can be deacylated to form glycerol phosphoryl choline (GPC) or reacylated to reform lecithin.

Free choline is absorbed from the upper small intestine, chiefly in the duodenum and jejunum. Transport is mediated by a carrier system located in the brush border of the jejunal mucosal cells. Injected lecithin does not liberate free choline within the gut lumen. Within the gut mucosal cell, the conversions of GPC to glycerol phosphate and free choline occur. The free choline then enters the portal circulation of the liver.

28.18 TRANSPORT OF LECITHIN

Lecithin synthesized within the gut mucosal cell enters the lymphatic circulation. Lecithin synthesized in the liver enters the blood. Transport liproproteins (VLDL, high- (HDL) and low-density lipoprotein (LDL), chylomicrons) are formed both in the intestine and the liver.

Chylomicrons are rich in triacylglycerols, VLDL contains more phospholipids, and LDL contains more cholesterol and phospholipid than VLDL. HDL contains esterified cholesterol and little triacylglycerols. Lecithin metabolism results in lysolecithins being formed and transferred to albumen for transport within the blood.

Many tissues possess enzymes capable of degrading lecithin and lysolecithin. In addition to the phospholipases A_1, A_2, and B, phospholipases C and D form phosphoryl choline and free choline.

28.19 CELL SIGNALING

Various lipid and lipid-derived bioregulatory molecules have been recognized. Phospholipids, lysophospholipids, and derived compounds are within this group. The messenger molecules are produced by phospholipases and lipid kinases whose activity is regulated by extracellular signals acting via cell surface receptors.

Phosphatidylinositol is also involved in the transmission of messages from the cell surface to the cell interior. Simplistically, phosphatidylinositol breakdown is triggered after a membrane receptor is excited. Breakdown products act to increase intracellular calcium concentrations and to activate the enzyme PKC, which modifies other proteins (channels, pumps, ion exchange proteins, regulatory proteins). These can then turn on or turn off important cell functions. PKC is ubiquitous in tissues of all mammals. Messengers that induce inositol phospholipid turnover are acetylcholine, norepinephrine, epinephrine, dopamine, histamine, serotonin, vasopressin, and many more. The phosphorylated derivatives and degradation products are also active in signaling.

Choline phospholipids also are involved in signal transduction pathways, especially those related to cell death. PC is a substrate for the formation of biologically active molecules that can amplify external signals or can terminate the signaling process. Cell death or apoptosis is a regulated cell suicide through endonuclease activity. Choline deficiency induces apoptosis. Reactive oxygen species are intermediates in apoptosis.

The phospholipid bilayer is asymmetric to the phospholipids present. Choline phospholipids are extracytosolic and the amino phospholipids are cytosolic. Loss of the asymmetry results in appearance of PS. The pathway and signal recognition is complex. The reader is referred to Zeisel and Szuhaj[24].

Nerve cell functions have also been linked to PS that includes conduction of the nerve impulse, accumulation, storage, and release of the nerve transmitter substances, and nerve transmitter action by receptors on the target cell surface[25]. Nerve cell membranes are particularly high in PS. This outermost membrane is a master switch for controlling the cell for a number of functions. These include:

1. Entry of nutrients into the cell and exit of waste products
2. Movement of ions into and out of the cell
3. Passage of molecular messages from outside the cell to its interior
4. Cell movement, shape changes
5. Cell-to-cell communication

PS carries a negatively charged amino group which tends to associate with ATPases, kinases, receptors, and other key membrane proteins.

28.20 LECITHIN IN TREATMENT OF DISEASE

Lecithin has been used as a therapy in the treatment of diseases of the nervous, cardiovascular, and immune systems (Figure 28.6).

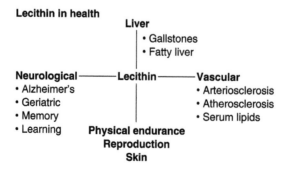

FIGURE 28.6 Lecithin in health. (Source: Orthoefer, F.T., *Lecithin and Health*, Vital Health Publishing, Bloomingdale, IL, 1998.)

28.21 NERVOUS SYSTEM

Lecithin therapy benefits patients with Tardive dyskinesis, an involuntary choreic movement of the tongue, lips, and jaw[25]. It is perhaps a side effect of antipsychotic drug use. Positive results with lecithin ingestion show that it may play a role in prevention of Parkinson's and Alzheimer's diseases and Tourette's and Freidreich's ataxia syndromes[26]. Lecithin may improve memory[24]. More recent information shows lecithin supplementation delays age-related mental aging. Choline administration results in improved memory. Lecithin as a source of choline acts as a "warehouse" that can deliver choline upon demand. Dietary lecithin supplementation sustains plasma choline at higher levels for a longer period of time than choline chloride when taken orally (Figure 28.7)[27,28]. This may also be important for physical endurance.

Evidence from animal studies shows lecithin and choline improve memory and learning[22,29]. Rats born to mothers supplemented with lecithin demonstrate improved learning. Human studies also suggest lecithin can improve learning and memory[30].

A chronic, inadequate intake of choline may be involved in the development of Alzheimer's disease[31]. Alzheimer's disease mostly affects cholinergic neurons. These neurons use choline to synthesize acetylcholine and membrane lecithin. The choline is from the bloodstream or metabolism of PC in their membranes. Little is obtained by methylation of PE. Consumption of choline-rich foods can increase plasma choline levels resulting in increases in brain choline and acetylcholine levels. It is believed that prolonged inadequate supply of choline could lead to membrane disruption and cell death. Adverse beta amyloid precursor protein is lodged in the cell membrane and found in plaque deposits characteristic of Alzheimer's patients. While some studies suggest benefits from dietary PC supplements to patients with Alzheimer's disease, others have led to disappointing results[31,34]. Current investigations do not support the use of choline in Alzheimer's disease patients.

PS is most concentrated in the brain. Clinical studies have suggested that PS can support brain functions that decline with age. Clinical studies on mental function are difficult to assess, but supplements showed trends favoring PS in:

1. Learning names and faces
2. Recalling names and faces
3. Facial recognition[32,33]

PS had rolled back the clock by about 12 years. Heiss[34] demonstrated that PS-treated subjects exhibited a greater activation upon taking a test than subjects who were not given PS and that PS

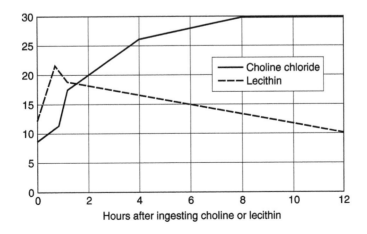

FIGURE 28.7 Effects of lecithin and choline on serum choline in normal subjects.

stabilized cognitive decline during the six months of the study. PS may also benefit young, healthy subjects in lowering stress hormone production.

PS has good bioavailability when taken orally. After consumption it appears in rat blood in 30 minutes; uptake by the liver and increased levels in the brain appear shortly after. PS also can be converted to PE.

28.22   CARDIOVASCULAR SYSTEM

Lecithin and choline perform various functions affecting the cardiovascular system. Choline is important in metabolism of homocysteine, a sulfur-containing amino acid implicated in cardiovascular disease risk. Lecithin may lower serum cholesterol levels and is a component of lipoproteins that transport fat and cholesterol.

High homocysteine levels increase the risk of cardiovascular disease[35]. The actual defect in homocysteine metabolism may occur by several pathways but therapy that includes choline can reduce homocysteine levels. Dietary choline deficiency, in addition, causes arteriosclerosis in rats suggesting lecithin may protect against high homocysteine levels.

Other efforts show that lecithin may be a source of cholesterol-lowering polyunsaturated fats, inhibiting intestinal absorption of cholesterol and increasing the excretion of cholesterol. Mostly, lecithin is a source of linoleic and linolenic acids. Lecithin is a component of lipoproteins and is needed for the HDL enzyme lecithin cholesterol acyltransferase. Soy lecithin increased HDL cholesterol more than egg lecithin or a comparable quantity of fatty acids[36]. Reduction in plasma total cholesterol and non-HDL cholesterol was observed for monkeys on the American Heart Association (AHA) step 1 diet[37]. In hamsters, the diet containing soy lecithin produced a greater hypocholesterolemic effect than diets containing equivalent amounts of linoleate and choline suggesting other components in soy lecithin are hypocholesterolemic. Further research is needed to determine the mechanism of the effect of lecithin on lipoproteins.

28.23   REPRODUCTION

Lecithin and choline are important in brain and mental development in both the fetus and infant[31,38,45]. Choline in mothers' milk is at a level of over 100 times that in the maternal bloodstream. Platelet activating factor, a choline phospholipid, is involved in implanting the egg in the uterine wall, fetal maturation, and inducing labor. Choline may also play a role in preventing neural tube defects.

Lecithin also plays a role in male fertility. Lecithin *in vitro* restores normal structure and movement to abnormal sperm cells and about doubles the acrosomal response[31].

28.24   LIVER FUNCTION AND HEALTH

Choline-deficient diets promote liver carcinogenesis. During lecithin or choline deficiency, the accumulated triglycerides in the liver may be metabolized to form diacylglyceroles (DAG). Without choline, DAG accumulate. Choline deficiency disturbs the PKC transmembrane signaling in the liver and contributes to development of hepatic cancer. Supplemental choline, as lecithin, has been shown to reverse fatty liver in TPN patients[24].

Lecithin may also protect the liver independently of providing choline[31]. PC has hepatoprotective effects independent of its choline content. PC is a better delivery form and is also more tolerated than choline. Diets high in alcohol fed to baboons either supplemented with or without lecithin showed nonsupplemented diets to cause liver cirrhosis or fibrosis[39]. No animals fed lecithin developed fibrosis or cirrhosis. Choline substitutions for the PC had no protective effects on the liver. PC also protects the liver from various toxic substances and benefits viral hepatitis patients with quicker recoveries, fewer relapses, and faster restoration of liver function.

28.25 PHYSICAL ENDURANCE

Intense exercise of long duration lowers plasma choline levels. Lecithin supplements prevent this decline and enhances physical performance. For example, Boston marathon runners had an about 40% decline in plasma choline levels after the race[40]. Significant decreases in plasma choline have also been shown in swimmers and triathletes. Supplementation improves race time in runners and swimmers and decreases the feeling of fatigue. Lecithin supplementation seems to prevent the decrease of plasma choline and improves performance[41].

28.26 DRUG INTERACTIONS

Phospholipids have a range of applications in drug delivery based on their encapsulation behavior and surfactancy[42]. Phospholipids have had an important role in the formulation of parenteral products. These products may be used for drug administration, delivery of vaccines, and for nutritional purposes. Specific drugs and target destination are foremost in formulation.

Various drug interactions with lecithin supplements occur. Nonsteroidal antiinflammatory drugs (NSAIDs) cause gastrointestinal mucosal lesions, perforation, and bleeding that may result in significant morbidity[43]. Enteric coatings have had limited success.

NSAIDs associate with phospholipids on the extracellular surface of mucosal cells. The mucosal surface becomes wettable and open to stomach action and irritation. Addition of phospholipids prevented gastrointestinal irritation and enhanced therapeutic activity. Soy PC relieves the pain from NSAIDs.

Aspirin with lecithin has been shown to exert a greater antipyretic, antiinflammatory, and analgesic efficacy[44]. The improved efficacy may be due to increased uptake by target cells, increased binding to cyclooxygenase, or prolonged half-life.

28.27 SAFETY AND TOXICOLOGICAL LIMITATIONS

As a food ingredient, lecithin from plant sources is "generally recognized as safe" (GRAS) by the U.S. Food and Drug Administration provided that good manufacturing practice is followed in its production. The minimum content of acetone-insoluble matter is 50% in the U.S. and 60% in European markets. Strict "limits" are placed on heavy metals and acid used in the purification. International specifications also regulate the residual peroxides in commercial lecithin products.

There are no known major side effects from using PC. Extensive toxicological testing has shown no adverse side effects even at doses up to 10 g/kg body weight[46]. No other side effects or reports of over dosage have been reported. No contraindications for PC have been found, nor known interactions. Lecithin is not toxic at typical levels of intake. Some people have experienced mild adverse gastrointestinal effects from taking more than 40 g of PC per day[31]. Many clinical studies with PS supplements show minimal side effects even at high dosages[47,48].

28.28 THERAPEUTIC SUPPLEMENT ROLE

The many clinical trials studying lecithin over the past few years have looked at its efficacy in treatment of neurological disorders, manic symptoms, cognitive disorders, and Alzheimer's disease. Only a few studies have suggested a small benefit in memory restoration. A modest improvement in lowering serum cholesterol may occur with PC intake. However, most trials have not shown significant cholesterol reduction.

PS has been marketed as a memory improver. Many double-blind trials support the use of this lecithin form[25,32,34,47,50]. PS is now a supplement accepted by the public.

Several PC supplements are promoted. Normal commercial lecithin contains 20 to 30% PC. Soft gel capsules are available that contain 20 to 90% PC. Liquid supplements have 3 g PC per teaspoon. Granular lecithin is a convenient form to be mixed with other foods. Two tablespoonfuls of lecithin contain about 3450 mg of PC. This quantity also contains about 500 mg of choline.

PS is also available as a mixture with other phospholipids or enhanced products. Most trials use 100 to 300 mg of PS supplement per day.

28.29 SUMMARY

Soybeans and soybean products have been popular over the past few years as a healthful addition to the diet. A coproduct of soybean oil production is lecithin, a complex mixture of phospholipids with minor amounts of tocopherols, sterols, and fatty acids.

Soybean lecithin has been used primarily for the production of formulated foods capturing the emulsifying characteristics in products such as chocolate, margarines, and baked goods. More recently, lecithin has been used as a supplement with attributes for improving cardiovascular health, liver health, physical endurance, and even enhanced memory. As a source of choline, lecithin is a bioavailable source providing timed release.

Lecithin is sold in fluid capsule, powder, and granular forms in drug and healthfood stores. The "activity" varies with the form chosen, but the powder or granular forms are most concentrated in phospholipids. Fractionated lecithin with enhanced PC content is also available.

New lecithin combination products have recently been formulated in solid-like applications using deoiled lecithin with soluble components[49]. Products with this technology include lecithin/phytosterols, lecithin/tocopherols, lecithin/cosmetic ingredients, and lecithin/pharmaceutical products. Some of these combinations are more active than their individual components when in a separate delivery system. Research on lecithin as a supplement continues. Equivocal results still require validation.

REFERENCES

1. Orthoefer, F.T., *Lecithin and Health*, Vital Health Publishing, Bloomingdale, IL, 1998.
2. Sipos, E.F. and Szuhaj, B.F., Lecithins, in *Bailey's Industrial Oil and Fat Products*, 5th ed., Vol. 1, Hue, Y, Ed., John Wiley, New York, 1996, pp. 310–395.
3. List, G.R., Commercial manufacture of lecithin, in *Lecithins: Sources, Manufacture and Uses*, Szuhaj, B.F., Ed., American Oil Chemists' Society, Champaign, IL, 1989, pp. 145–162.
4. Yen, C.E. and Zeisel, S., Choline, phospholipids and cell suicide, in *Choline, Phospholipids, Health and Disease*, Zeisel, S. and Szuhaj, B., Eds., AOCS Press, Champaign, IL, 1998, pp. 11–22.
5. Scholfield, C.R., *Occurrence, Structure, Composition and Nomenclature of Lecithin*, AOCS Monograph, Szuhaj, B.F. and List, G.R., Eds., American Oil Chemists' Society, Champaign, IL, 1985, chap. 1.
6. Suzhaj, B.F. and List, G.R., in *Lecithins*, AOCS Monograph, American Oil Chemists' Society, Champaign, IL, 1985, chaps. 2–4.
7. Dashiell, G.L., *Lecithin in Food Processing Applications*, AOCS Monograph, Szuhaj, B.F., Ed., American Oil Chemists' Society, Champaign, IL, 1989, pp. 213–224.
8. Cherry, J.P. and Kramer, W.H., Plant sources of lecithin, in *Lecithins: Sources, Manufacture and Uses*, AOCS Monograph, Szuhaj, B.F., Ed., American Oil Chemists' Society, Champaign, IL, 1989, pp. 16–31.
9. Kuksis, A., Animal lecithins, in *Lecithins*, AOCS Monograph, Szuhaj, B.F. and List, G.R., Eds., American Oil Chemists' Society, Champaign, IL, 1985, pp. 105–162.

10. Schneider, M., in *Lecithins: Sources, Manufacture and Uses*, AOCS Monograph, Szuhaj, B.F., Ed., American Oil Chemists' Society, Champaign, IL, 1989, pp. 109–131.
11. Nyberg, L., Sphingomyelin from bovine milk, in *Phospholipids: Characterization, Metabolism and Novel Biological Applications*, Cevc, G. and altauf, F., Eds., AOCS Press, Champaign, IL, 1993, pp. 120–125.
12. Siakotos, A.N., Rouser, G., and Fleischer, S., *Lipids*, 4, 239, 1969.
13. *Yearbook and Trading Rules*, National Oilseed Processors' Association, Washington, DC, 1995–1996, p. 96.
14. Flider, F., The manufacture of soybean lecithins, in *Lecithins*, AOCS Monograph, Szuhaj, B.F. and List, G.R., Eds., American Oil Chemists' Society, Champaign, IL, 1985.
15. *Code of Federal Regulations*, 21CFR paragraphs 182, 184, Office of the Federal Register, U.S. Government Printing Office, Washington, DC, 1995.
16. *Food Chemicals Codes*, 3rd ed., National Academy Press, Washington, DC, 1981, pp. 147–148 and 166–167.
17. U.S. Patent 4,235,793, 1980.
18. Schmidt, J.C. and Orthoefer, F., Modified lecithins, in *Lecithins*, Szuhaj, B.F. and List, G.R., Eds., American Oil Chemists' Society, Champaign, IL, 1985, pp. 203–213.
19. Stanley, J., in *Soybeans and Soybean Products*, Vol. 2, Markley, K.S., Ed., Interscience, New York, 1951, pp. 593–647.
20. Zeisel, S.H., Growdon, J.H., Wurtman, R.J., Magil, S.G., and Logue, M., Normal plasma choline responses to ingested lecithin, *Neurology*, 30, 1226, 1980.
21. Zeisel, S.H., Lecithin in health and disease, in *Lecithins*, Szuhaj, B.F. and List, G.R., Eds., American Oil Chemists' Society, Champaign, IL, 1985, pp. 323–347.
22. Bartus, R.T., Dean, R.L., Goods, A.J., and Lippas, A.S., Age related changes in passive avoidance retention: modulation with dietary choline, *Science*, 209, 301–303, 1980.
23. Burt, M.E., Hanin, I., and Brennan, M.F., *Lancet*, 2, 638, 1980.
24. Zeisel, S.H. and Szuhaj, B.F., *Choline, Phospholipids, Health, and Disease*, AOCS Press, Champaign, IL, 1998.
25. Toffano, G., The therapeutic value of phosphatidylserine effect in the aging brain, in *Lecithin: Technological, Biological and Therapeutic Aspects*, Hanin, L. and Ansell, G.B., Eds., Plexum Press, New York, 1987, pp. 137–146.
26. Yates, J., Lecithin works wonders, *Prevention*, 32, 55–59, 1980.
27. Hirsch, M.J., Growdon, J.H., and Wurtman, R.J., Relations between dietary choline and lecithin's intake, serum choline levels and various metabolic indices, *Metabolism*, 27, 953–960, 1978.
28. Wurtman, R.J., Hirsch, M.J., and Growden, J.H., Lecithin consumption raises serum-free choline levels, *Lancet*, July, 68–69, 1977.
29. Izaki, Y., Hashimoto, M., Arita, J., Iriki, M., and Hibno, H., Intraperitoneal injection of 1-oleoyl-2-decosahexnoil phosphatidylcholine enhances discriminatory shock avoidance learning in rats, *Neuroscience Lett.*, 167, 171–174, 1994.
30. Bartus, R.T., Dean, R.L., Pontecorvo, M.J., and Flicker, C., The cholinergic hypothesis: a historical overview, current perspectives and future directions, *Ann. N.Y. Acad. Sci.*, 444, 332–358, 1985.
31. Canty, D.J., Zeisel, S., and Jolitz, A.J., Lecithin and Choline: Research Update on Health and Nutrition, Central Soya Co., Fort Wayne, IN, 1996.
32. Crook, T.H. et al., Effects of phosphatidylserine in age associated memory impairment, *Neurology*, 41, 644–649, 1991.
33. Crook, T.H. et al., Effects of phosphatidylserine in Alzheimer's disease, *Psychopharmacol. Bull.*, 28, 61–66, 1992.
34. Heiss, W.D. et al., Long term effects of phosphatidylserine, pyritinol and cognitive training in Alzheimer's disease, *Cognit. Deteriorat.*, 5, 88–98, 1994.
35. Boushey, D.J., Beresford, S.A.A., Omenn, G.S., and Motulsky, A.G., A qualitative assessment of plasma homocysteine as a risk factor for vascular disease, *JAMA*, 274, 1049–1057, 1995.
36. O'Brien, B.C. and Andrews, V.G., Influence of dietary egg and soybean phospholipids and triacylglycerols on human serum lipoproteins, *Lipids*, 28, 7–12, 1993.
37. Wilson, T.A., Meservey, C.M., and Nicolosi, R.J., The hypocholesterolemic and antiatherogenic effects of soy lecithin in hypercholesterolemic monkeys and hamsters: beyond linoleate, *Atherosclerosis*, 140, 147–153, 1998.

38. Zeisel, S.H., Choline: an important nutrient in brain development, liver function and carcinogenesis, *J. Am. Coll. Nutr.*, 11, 473–481, 1992.

39. Lieber, C.S., Robins, S.J., and Li, J., Phosphatidylcholine protects against fibrosis and cirrhosis in the baboon, *Gastroenterology*, 106, 152–159, 1994.

40. Conlay, L.A., Wurtman, R.J., Blusztajn, L., Coviella, I.L.G., Maher, T.J., and Evoniuke, G.E., Decreased plasma choline concentration in marathon runners, *N. Engl. J. Med.*, 315, 892, 1986.

41. von Allworden, H., Horn, S., and Feldheim, W., The influence of lecithin on the performance and the recovery process of endurance athletes, in *Phospholipids: Characterization, Metabolism and Novel Biological Applications*, Cevc, G. and Paltauf, F., Eds., AOCS Press, Champaign, IL, 1993, pp. 319–325.

42. Davis, S.S. and Illum, L. The Use of Phospholipids in Drug Delivery. *In Phospholipids: Characterization, Metabolism, and Novel Biological Applications*, Cevc, G. and Paltauf, F., Eds., AOCS Press, Champaign, IL, 1993, pp, 67–79.

43. Allison, M.C., Howatson, A.G., Torrance, C.J., Lee, F.O., and Russell, R.I., Gastrointestinal damage associated with the use of nonsteroidal anti-inflammatory drugs, *N. Engl. J. Med.*, 32, 749–754, 1992.

44. Lichtenberger, L.M., Ulloa, C., and Vanous, A., Zuetterionic phospholipids enhance aspirin's therapeutic activity as demonstrated in rodent model systems, *J. Pharmacol. Exp. Therap.*, 277, 1221–1227, 1996.

45. Barbeau, A., Growden, J.H., and Wurtman, R., *Nutrition and the Brian: Choline and Lecithin in Brain Disorders*, Vol. 5, Raven Press, New York, 1979, p. 76.

46. Parnham, M.J. and Mendel, A., Phospholipids and Liposomes: Safety for Cosmetics and Pharmaceutical Use, Rhone-Poulenc Rorer, Cologne, Germany, 1992.

47. Cenacchi, B., Cognitive decline in the elderly: a double-blind, placebo-controlled multicenter study on efficacy of phosphatidylserine administration, *Aging Clin. Exp. Res.*, 5, 123–133, 1993.

48. Heywood, R., Cozens, D.D., and Richold, M., Toxicology of a phosphatidylserine preparation from bovine brain (BC–PS), *Clin. Trials J.*, 24, 25–32, 1987.

49. Orthoefer, F.T., Compressed Lecithin Formulation, U.S. Patent 6,312,703, Nov. 6, 2001.

50. Orthoefer, F.T., Lecithin: Beyond Emulsification, presented at the World Conference and Exhibition on Oilseed and Edible, Industrial and Specialty Oils, American Oil Chemists' Society conference, Istanbul, Turkey, August 12–15, 2002.

29 Centrifugal Partition Chromatography (CPC) as a New Tool for Preparative-Scale Purification of Lipid and Related Compounds

Udaya Wanasundara and Paul Fedec
POS Pilot Plant Corp., Saskatoon, Saskatchewan, Canada

CONTENTS

29.1   INTRODUCTION

The increasing need for cost-effective ways to purify bioactive compounds such as lipids or lipid-derived compounds is generating new approaches to downstream purification. Use of various chromatographic techniques is necessary for the isolation and purification of milligram to gram quantities for structural studies, bioassays, pharmacological tests, reference substances, and standards for qualitative determinations. Therefore, developing techniques for compound isolation and purification has become very significant.

About half a century ago, the concept of partitioning solutes between liquids gave birth to two methods: one was countercurrent distribution chromatography and the other was liquid–liquid partition chromatography. Development in the 1980s of modern countercurrent chromatography based on the fundamental principles of liquid–liquid partition initiated a resurgence of interest in the separation sciences. The advantage of applying continuous liquid–liquid extraction, a process for separation of a multicomponent mixture according to the differential solubility of each component in two immiscible solvents, has long been recognized. In spite of limitations in the traditional countercurrent distribution methods, liquid–liquid partitioning was used successfully to fractionate natural compounds[1]. In recent years, significant improvements have been made to enhance performance and efficiency of liquid–liquid partitioning with different configuration instruments[2].

Centrifugal partition chromatography (CPC) is an emerging liquid chromatographic technique that utilizes liquid–liquid partition, countercurrent distribution of solute mixtures between two liquid phases, in the absence of a solid support, to perform separations of complex mixtures of biological molecules. Although CPC has been used for separation and purification of a broad range of naturally occurring chemical species, this technology offers distinct advantages for the isolation and purification of lipid-based substances. So far, CPC has demonstrated its potential for the separation of natural products at levels ranging from milligrams to several grams. New instrumentation is being developed, in particular to cover the preparative area, and there is no doubt that soon this technique will allow production at kilogram levels.

29.2 CENTRIFUGAL PARTITION CHROMATOGRAPHY

Although the first published experiments on partition chromatography were described in 1906 by Goppelsroeder, the real development of this technique occurred in the 1940s and it is attributed to the work of Martin and Synge[3,4]. Today, a century after its initial trials, chromatography is a powerful tool used in every chemical, biochemical, and biotechnology laboratory and has found multiple applications in the chemical and pharmaceutical industries.

Liquid–liquid partition chromatography was further developed by Martin and Synge[3,4], and eventually they received the Nobel Prize for their work in 1952. According to this concept, one liquid was used as a "sorbent" and a second liquid was allowed to come into contact with the former, until equilibrium was achieved. The solvents may be single chemical substances or complex mixtures, including buffers, among others.

Most chemists are familiar with chromatographic techniques that use a solid support for the so-called stationary phase. The stationary phase is well defined in paper chromatography, thin-layer chromatography, column chromatography, high-performance liquid chromatography (HPLC), or gas–liquid chromatography. In all of these cases the liquid phase adheres to a solid support that can be either columnar or planar.

In liquid–liquid systems, both the stationary and the mobile phases are liquids. Probably the simplest illustration of this kind of separation is the separation of substances between two immiscible phases in a separatory funnel.

The operating principles of extraction are based on the concept of heterogeneous equilibrium in multiphase liquid–liquid partition systems. Extraction processes are primarily concerned with the fraction of a given solute in a given phase at equilibrium. The fraction in a single phase is the portion of the solute molecules that will be removed or transferred at a given step, when that particular phase is separated. This underlying principle in the extraction process is based on the distribution law that was first recognized by Nernst in 1891. The distribution law states that a solute dissolved in one phase in equilibrium with another immiscible phase will distribute itself between the two phases, so that the ratio of the concentration in the two phases is constant at a fixed temperature. This can be represented by the equation $K = C_1/C_2$, where C_1 and C_2 are the concentrations of the solute molecule in the first and second phases, respectively, and K is usually called the "distribution constant" or the "partition coefficient." The partition coefficient is constant for a given substance, in a specific solvent system, and at a given temperature. Therefore, a mixture of several substances with different partition coefficients could be separated.

The fundamentals of the countercurrent distribution (chromatography) process and its associated calculations are best illustrated by an example involving the separation of two solutes using a separatory funnel operation (Figure 29.1). In this instance, successive portions of upper and lower solvent phases are moved over to be equilibrated with equal successive volumes of new lower and

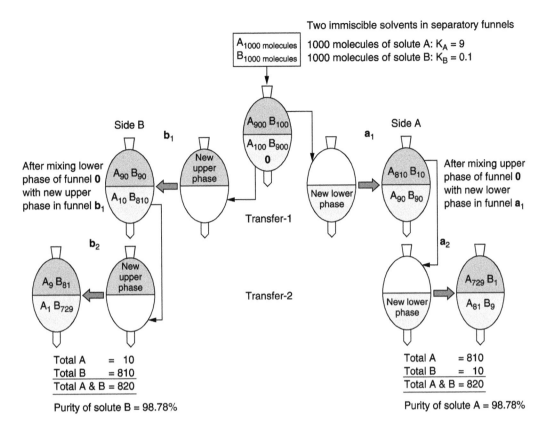

FIGURE 29.1 Countercurrent distribution of solutes A and B.

upper solvent phases. To clarify the relationship of the theory to the apparatus actually used for countercurrent distribution, the lower phase will be held stationary (Figure 29.1). Assume three separatory funnels, numbered **0**, a_1, and a_2, each containing a given volume of lower phase solvent (Figure 29.1, side A). To funnel **0** is added an equal volume of upper phase solvent and 1000 molecules each of solutes A and B. Assume further, for illustrative purposes, that solute A is 9 times more soluble in the upper phase than in the lower phase while solute B is 9 times more soluble in the lower phase than in the upper phase. Therefore, the partition coefficient for solute A is 9 and that for B is 0.1.

Considering for the moment only compound A, the distribution of 1000 molecules of A in funnel **0** after equilibration and settling is calculated. As shown in Figure 29.1 (funnel **0**), 900 molecules will be found in the upper phase and 100 molecules in the lower phase after the first equilibration. After taking the lower phase out, the upper phase of funnel **0** is next transferred to funnel a_1 (with new lower phase) and a volume of new upper phase solvent is replaced in funnel **0** (now this funnel is named b_1). After shaking and settling, the distribution of solute A in funnel b_1 distributes to give 90 molecules in the upper phase and 10 molecules in the lower phase. The 900 molecules transferred to funnel a_1 are now distributed as 810 molecules in the upper phase and 90 molecules in the lower phase. In the next operation the upper phase of funnel a_1 is transferred to funnel a_2 (with new lower phase). The 810 molecules (solute A) transferred to funnel a_2 are now distributed as 729 molecules in the upper phase and 81 molecules in the lower phase.

Solute B, by analogous reasoning, will be found to have the reverse distribution. Therefore, solute B content in funnel b_2 will be distributed as 729 molecules in the lower phase and 81 molecules in the upper phase. When both solutes are now considered, funnel **a₂** is found to contain 81% of solute A of 98.78% purity and funnel **b₂** contains 81% of solute B of 98.78% purity. This separatory funnel operation demonstrates the countercurrent distribution process and also shows the increased separation of solutes A and B over that for the single equilibration where the corresponding purities of solutes A and B are over 98%.

CPC, pioneered by Ito in 1964, is a liquid–liquid chromatography without a sorbent, requiring two immiscible solvent phases. This is basically an outgrowth of countercurrent distribution, as developed by Craig[5]. The most distinct variant of CPC is that one liquid phase remains stationary while the second solvent phase passes through the stationary-phase solvent. The principle of separation involves the partition of a solute between two immiscible solvents (mobile and stationary phases). The relative proportion of solute passing into each of the two solvent phases is determined by the respective partition coefficients. CPC is a liquid–liquid separation technique, which offers distinct advantages for the separation, isolation, and purification of lipids such as fatty acids or their derivatives, phospholipids, and tocopherols compared to traditional liquid–solid separation methods, i.e., normal column chromatography and/or HPLC.

CPC is exclusively a liquid–liquid partition method and has several important advantages:

- It does not require solid supports, which in many cases are very costly, but instead it relies on readily available solvents.
- Elimination of solid supports avoids problems associated with irreversible retention of highly retentive sample components; the chromatographer is virtually assured of almost 100% recovery of the compounds from a chromatogram due to this reason.
- Any two-phase solvent system may be used; many partition systems can be prepared with nontoxic and commonly available solvents.
- The volume ratio of the stationary phase to the total column (rotor) volume is greater in CPC than in conventional liquid chromatography, and therefore large quantities of materials can be retained in the stationary phase.
- Decomposition and denaturation of valuable sample components such as polyunsaturated fatty acids, often encountered with conventional packed chromatographic columns, are virtually nonexistent under the mild operating conditions used in CPC.
- As the stationary phase is retained by centrifugal force, the mobile phase can be pumped at high speeds through the apparatus resulting in reduced separation times.

One additional advantage of CPC is the low solvent consumption and both normal- and reversed-phase elution may be conducted with the same solvent pair. CPC can be readily adapted for large-scale continuous separations[6]. The entire process is performed in the liquid phases, in a closed system. Environmental problems are minimal and solvents may be completely recovered and recycled.

29.3   INSTRUMENTS

Although several minor variants are available, instruments are basically of two types: the CPC instrument manufactured by Sanki Engineering Ltd (Kyoto, Japan) and a multicoil countercurrent system designed by Ito and manufactured by PharmaTech Research (Baltimore, USA). The basic components of the unit manufactured by Sanki Engineering are shown in Figure 29.2. Similar to most of the chromatographic systems, the CPC is comprised of basic components such as pumps for solvent delivery, valves to control solvent delivery, and a sample injector, detector, and recorder. The main component of the instrument is its rotor. The rotor is made of stackable engraved polyphenylene sulfide (PPS) or polychlorotrifluoroethylene (DAIFLON) disks separated by Teflon seal sheets and stainless steel plates (Figure 29.3). The disks consist of channels and ducts on each

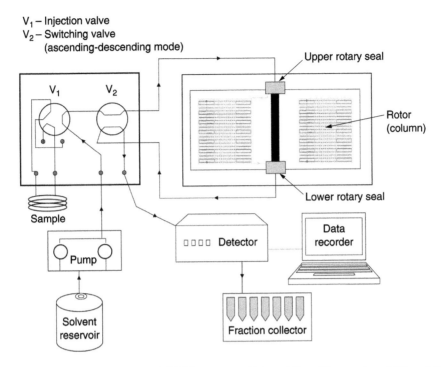

FIGURE 29.2 Components of the Sanki CPC instrument. (Adapted from Wanasundara, U.N. and Fedec, P., *Inform*, 13, 726–730, 2002.)

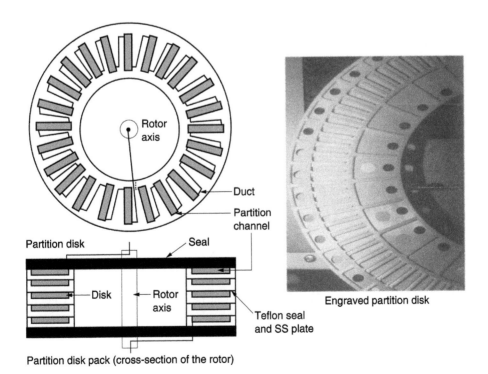

FIGURE 29.3 Inside view of the partition disk of the Sanki CPC.

side. The liquid stationary phase is held in these channels and the centrifugal field generated by the spinning rotor holds the stationary phase in place, enabling a mobile phase to be pumped through.

Recently, another model of the CPC instrument was manufactured by Kromaton Technologies (Angers, France). This model is now available for commercial use. Another commercially available rotating coil instrument is the multicoil countercurrent chromatograph, introduced by Pharma-Tech Research. The multicoil chromatograph consists of two or three identical multilayer coils arranged symmetrically around the rotor frame of the centrifuge, thus eliminating the need for balancing with a counterweight. Each coil column undergoes synchronous planetary motion in such a way that it revolves around the central axis of the centrifuge and simultaneously rotates around its own axis at the same angular velocity[7,8]. The columns are equipped with flow tubes arranged in such a fashion that they do not twist, allowing seal-free operation of the instrument.

29.3.1   SELECTION OF SOLVENT SYSTEMS

The correct choice of a solvent system is of paramount importance for a successful CPC separation. Selection of a two-phase (biphasic) solvent system for CPC is similar to choosing the solvents for other chromatographic methods such as for a normal column or HPLC[2]. Important criteria are the polarity of the sample components and their solubility, their charge state, and their ability to form complexes, among others. The most critical factors in selecting solvent systems for CPC are twofold: one is sample solubility and the other is that the partition coefficients of the molecular species that are to be separated must differ from each other.

The chromatographic literature contains numerous examples of solvent systems used in different countercurrent chromatographic separations[9] and these references may provide leads as to possible systems useful for the separation in question. Alternatively, a classic chloroform/methanol/water (polar) or n-hexane/ethyl acetate/methanol (less polar) system can be selected as the starting point and the proportions of the individual solvents may be changed until the required distribution of the sample between the two phases is obtained[10]. Chloroform-based solvent systems provide large density differences and relatively high interfacial tension between the two solvent phases; consequently they are frequently employed for the separation of natural products by CPC. Because of their short setting times, chloroform/methanol/water systems normally produce satisfactory phase retention. However, one drawback of the chloroform system is that it readily leads to overpressure problems with the Sanki apparatus.

The other approach for selecting proper solvent systems for CPC is by using the solvent phase diagrams. Ternary diagrams for many solvent systems have been compiled by Sorensen and Arlt[11] and two of the diagrams are shown in Figure 29.4. They often consist of two immiscible solvents plus a third solvent that is soluble in the two primary solvents. Most systems conform to type 1, comprising of one solvent miscible with two other immiscible solvents (Figure 29.4A). A typical example of the type 1 ternary system is chloroform/methanol/water. Very few systems are like type 0, made with three solvents fully miscible by pairs, but for which a zone exists in the ternary diagram where a biphasic system occurs when mixing them in a suitable ratio (Figure 29.4B). A good example of a type 0 system is water/dimethyl sulfoxide/tetrahydrofuran.

Foucault[2] has suggested three criteria to follow when ternary diagrams are used for the selection of solvents:

1. Select the solvent(s) in which the sample can be completely dissolved. CPC has a preparative goal, and so the final mixture of solvents must be able to dissolve large amounts of sample and thus it should contain at least one of the "best" solvents that make the sample freely soluble.
2. Aided by the polarities of the solvents (numerous polarity scales may be found in the literature), choose a solvent on each side of the selected best solvent; that is, one less

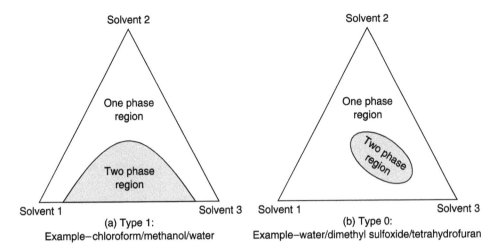

FIGURE 29.4 Ternary diagrams encountered in CPC. (Adapted from Wanasundara, U.N. and Fedec, P., *Inform*, 13, 726–730, 2002.)

polar and one more polar, in order to obtain a biphasic system where the best solvent will partition into the two other solvents.

3. Adjust the ratio of the best solvent to disperse the sample into two phases; that is, the less polar fraction of the sample will preferentially go into the less polar phase and the more polar fraction will preferentially go into the more polar phase, so that the average partition coefficient will stay around 1.

29.3.2 BIPHASIC LIQUID SYSTEM PREPARATION

The required volumes of solvents that comprise the biphasic liquid system are measured and poured together in a container. The mixture is then shaken vigorously for a few minutes. The settling time of the phase, that is, the time needed to obtain a meniscus between two liquid phases, should be established. According to Ito[9], if this time is less than one minute, the system has good interfacial tension and density differences between phases, and it should be easily retained inside the CPC apparatus. If the settling time is much longer, the solvent system will likely not be retained in the CPC rotor. It should still be possible to use this liquid system in a CPC unit having a higher rotational speed.

29.3.3 PREPARATION OF CPC APPARATUS FOR CHROMATOGRAPHY

Operation of a Sanki CPC instrument is enabled by the following sequence. The clean rotor is first filled with the liquid stationary phase while the rotor is spinning at low speed (200 to 300 rpm). The operating mode is selected: *descending* (or head-to-tail), if the mobile phase is denser than the stationary phase, or *ascending* (tail-to-head) in the opposite case. Once the rotor is filled with stationary phase, the speed of centrifugation can be increased to the rate chosen on the basis of the phase density difference. Then the mobile phase pump is turned on at the desired flow rate, which should be lower than the "flooding flow rate"[12]. As long as the liquid stationary phase exits the CPC rotor, equilibrium is not reached, and *the pressure increases*. When the first drop of mobile phase occurs in the rotor, equilibrium is reached and the pressure stabilizes. The volume of the stationary phase displaced is measured; it corresponds to the volume of the mobile phase inside the CPC rotor. The pressure should be noted and if it decreases, it will be a sign of stationary phase loss. If everything is stable, including the baseline of the chromatogram, the rotor is ready and the sample can be injected. The rotational speed of the instrument can be varied, although higher speeds generally lead to a better resolution, and the backpressure from the instrument may be subjected to increase. The most frequently used rotational speed varies between 800 and 1000 rpm, depending on the specific gravity of the solvents.

The separation of compounds present in the injected sample is conceptually simple. When the solute moves with the mobile phase through the stationary phase, the solute is exchanged back and forth between the drops of mobile phase and the stationary phase. If the solute has more affinity for the stationary phase than the mobile phase, it tends to stay in the stationary phase longer and has a longer retention time. Conversely, if the solute has less affinity for the stationary phase, it tends to remain longer in the mobile phase. Spatial separation of different analytes occurs through this partitioning between the stationary and mobile phases.

29.3.4 Application of CPC for Isolation and Purification of Lipid Components

Separation and purification of various lipid classes including free fatty acids (FFA) and fatty acid ethyl esters (FAEE), monoacylglycerols (MAG), diacylglycerols (DAG), triacylglycerols (TAG), phospholipids, and glycosphingolipids (GSL) has been successfully carried out using the CPC method[13]. Fatty acids, MAG, DAG, and TAG were separated under the following conditions: 600 rpm at 10 mL/min flow rate of the solvent system n-hexane/methanol/water (1:1:0.05, v/v) using both reversed descending and normal ascending modes[13]. Otsuka et al.[14] have separated glycolipids, both acidic and neutral, using countercurrent chromatography with the solvent system chloroform/benzene/methanol/water (50:25:65:30, v/v). The details of phospholipid, fatty acid, and tocopherol isomer purification are given in the following sections.

29.3.4.1 Fractionation and Purification of Phospholipids

Centrifugal partition chromatography has been used for isolation and purification of molecular species of phospholipids from various natural sources. Highly purified phospholipids have been isolated and purified from egg yolk using CPC. An ethanol extract of egg yolk yields phosphotidylcholine (PC, lecithin) having 98.3% purity in a single run through CPC. Byproducts of this purification include phosphotidylethanolamine (PE), lysophosphotidylcholine (LPC), sphingomyelin (SPM), and TAG, all of reasonably good purity. Renaud et al.[15] were able to purify different molecular species of PC (PC with different fatty acids), extracted from fish oil using the Kromaton CPC apparatus in combination with the solvent system heptane/ethyl acetate/acetonitrile (1:0.55:1.0, v/v/v) at 500 rpm and 2 mL/min flow rate. They were able to demonstrate that major molecular species of PC contained palmitic acid (C16:0) and docosahexaenoic acid (C22:6n3, DHA). Baudimant et al.[16] applied the CPC technique for purification of PC to homogeneity from a crude phospholipids extract from squid. The starting extracts contained PE (24.1%), phosphotidylinositol (PI, 17.5%), phosphotidylserine (PS, 4.9%), PC (48.3%), and SPM (5.1%). The crude extract was submitted to a first run using n-heptane/ethyl acetate/acetonitrile (1:0.65:1.0, v/v/v) as the solvent system. The first elution performed in a normal ascending mode allowed the removal of all components except PC. The most polar PC was then eluted from the rotor in a reversed descending mode. Baudimant et al.[17] were able to patent the purification process for PC containing long-chain polyunsaturated fatty acids using the CPC method.

29.3.4.2 Fractionation and Purification of Fatty Acids

Bousquet et al.[18,19] have tested CPC separation of eicosapentaenoic acid (EPA) and docosahexaenoic acid (DHA) from microalgal oil and were able to isolate pure EPA and DHA from this oil with excellent yields. The first separation used heptane as the stationary phase and acetonitrile/water (3%) as the mobile phase. The minor fatty acids were removed by this elution, leaving a mixture of four major polyunsaturated fatty acids. As an example, 2.4 g of fatty acid preparation yielded a mixture (1 g) of polyunsaturated fatty acids (i.e., α-linolenic acid, C18:3 n-3 (43%); stearoidonic acid, C18:4 n-3 (7.5%); EPA (45%); and DHA (4.5%)). This mixture was subjected to a second separation, using heptane as a stationary phase and methanol/water as the mobile phase. Under these conditions, a good separation was achieved and pure EPA and DHA were isolated with excellent yields. Du et al.[20] also applied high-speed countercurrent chromatography to separate EPA and DHA ethyl esters with a solvent system comprised of hexane/dichloromethane/acetonitrile (5:1:4, v/v/v). They were able to obtain a mixture of EPA/DHA esters with a proportion of 56.4 and 39.3%, respectively. Unfortunately, these two esters could not be resolved with this solvent system because they have a low partition coefficient under the experimental conditions used. Murayama et al.[6] have successfully separated a mixture of ethyl esters of stearic (C18:0), oleic (C18:1), linoleic (C18:2), and α-linolenic (C18:3) acids because their partition coefficients are distributed over a wide range in the two-phase solvent n-hexane/acetonitrile (1:1, v/v). The ethyl esters of C18:2 and C18:3 were separated during the first normal ascending elution whereas C18:1 was recovered by switching the elution mode.

A CPC method to purify DHA from algal oil was recently developed[21]. The starting algal oil contained 39.7% DHA and 15.2% docosapentaenoic acid (DPA, n-6) along with several other fatty acids. The FFA of algal oil were eluted using a hexane/methanol/water two-phase solvent system in a normal phase ascending mode. Under these conditions it was possible to purify DHA up to 84.6% and DPA to 84.9%. However, under these conditions myristic acid (C14:0) coeluted with DHA, and therefore a higher purity could not be achieved for DHA (Figure 29.5a). In order to isolate ultrapure DHA (fine chemical grade), prepurification of algal oil FFA was carried out by complexing with urea to remove the coeluting C14:0 fatty acid. When the prepurified FFA mixture was subjected to CPC, it resulted in ultrapure DHA (99%) with a high recovery (Figure 29.5b). The same solvent system was used to purify arachidonic acid (ARA); however, linoleic acid (C18:2) was coeluted with ARA due to the very close partition coefficient of these two fatty acids in the selected solvent system (Figure 29.6). The above solvent system and conditions were applied to borage oil FFA to obtain ultra pure γ-linolenic acid (GLA). The starting borage oil contained 21.8% GLA and upon CPC purification, 98.3% GLA was obtained (Figure 29.7).

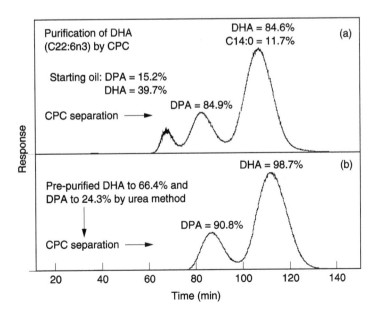

FIGURE 29.5 CPC purification of docosahexaenoic acid (DHA).

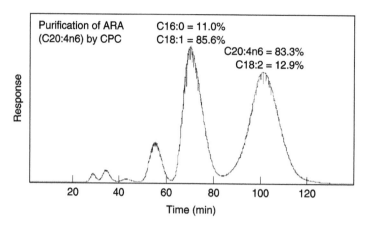

FIGURE 29.6 CPC purification of arachidonic acid (ARA).

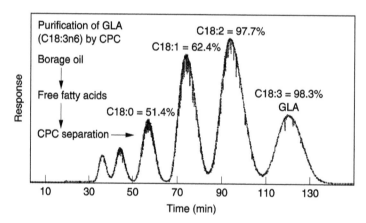

FIGURE 29.7 CPC purification of γ-linolenic acid (GLA).

29.3.4.3   Fractionation and Purification of Natural Tocopherols

Purification of natural tocopherol isomers is possible by applying the CPC method. A starting mixture containing 70% total tocopherols (tocopherol isomers) applied to a Sanki instrument with the solvent system hexane/acetonitrile/ethanol and ascending mode elution resulted in tocopherol isomers being separated with very high purity (Figure 29.8). Using this solvent system, γ- and δ-tocopherols of 97% purity and α-tocopherol of 99% purity were obtained with high recovery. Therefore, it is possible to isolate and purify some isomeric compounds by CPC technology.

29.4   CONCLUSIONS

CPC is a powerful process-scale separation technology which offers distinct advantages for separation, isolation, and purification of lipids and related compounds, and it is complementary to HPLC[22]. Although capital investment for CPC equipment often runs higher than that for HPLC, operating costs are generally an order of magnitude lower when compared to the same separation

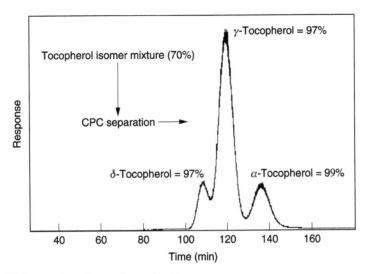

FIGURE 29.8 CPC separation of natural tocopheol isomers.

performed with conventional liquid chromatography or preparative HPLC. Scaling-up problems associated with HPLC due to variability of column packing, chemistry, surface morphology, and geometry are not a consideration in CPC, since there is no solid stationary phase in this system. Currently, the number of CPC manufacturers is limited to only a few companies (i.e., Pharma-Tech, USA; Kromaton Technologies, France; Sanki Engineering, Japan). All companies listed manufacture instruments for isolation of compounds at the gram scale. Recent attempts have been made to design equipment capable of handling kilogram quantities of materials. When these instruments are commercially available, the potential of CPC for separation science will be increased tremendously.

REFERENCES

1. Hostettmann, K. and Marston, A., Liquid–liquid partition chromatography in natural products isolation, *Anal. Chim. Acta*, 236, 63–76, 1990.
2. Foucault, A.P., Solvent systems in centrifugal partition chromatography, in *Centrifugal Partition Chromatography*, Chromatographic Science Series, Vol. 68, Foucault, A.P., Ed., Marcel Dekker, New York, 1994, pp. 71–97.
3. Martin, A.J.P. and Synge, R.L.M., Separation of the higher monoamino acids by countercurrent liquid–liquid extraction: the amino acid composition of wool, *Biochem. J.*, 35, 91–121, 1941.
4. Martin, A.J.P. and Synge, R.L.M., A new form of chromatogram employing two liquid phases: I. A theory of chromatography. II. Application of the micro-determination of the higher monoamino acid in proteins, *Biochem. J.*, 35, 1358–1368, 1941.
5. Craig, L.C. and Post, O., Apparatus for countercurrent distribution, *Anal. Chem.*, 21, 500–504, 1949.
6. Murayama, W., Kosuge, Y., Nakaya, N., Nunogaki, Y., Nunogaki, K., Cazes, J., and Nunogaki, H., Preparative separation of unsaturated fatty acid esters by centrifugal partition chromatography, *J. Liq. Chromatogr.*, 19, 283–300, 1988.
7. Ito, Y. and Chou, F.E., New high-speed countercurrent chromatograph equipped with a pair of separation columns connected in series, *J. Chromatogr.*, 391, 382–386, 1988.
8. Ito, Y., Oka, H., and Slemp, J.L., Improved high-speed countercurrent chromatography with three multilayer coils connected in series: I. Design of the apparatus and performance of semi preparative columns in 2,4-dinitrophenyl amino acid separation, *J. Chromatogr.*, 475, 219–227, 1989.
9. Ito, Y., High-speed countercurrent chromatography, *Crit. Rev. Anal. Chem.*, 17, 65–143, 1986.
10. Marston, A., Borel, C., and Hostettmann, K., Separation of natural products by centrifugal partition chromatography, *J. Chromatogr.*, 450, 91–99, 1988.

11. Sorensen, J.M. and Arlt, W., Ternary systems, in *Liquid–liquid Equilibrium Data Collection*, Chemistry Data Series, Behrens, D. and Eckermann, R., Eds., Dechema, Deutsche Gesellschaft fur Chemisches Apparatewesen, Frankfurt, Germany, 1980.

12. Berthod, A. and Amstrong, D.W., Centrifugal partition chromatography: II. Selectivity and efficiency, *J. Liq. Chromatogr.*, 11, 567–583, 1988.

13. Alvarez, J.G., Cazes, J., Touchstone, J.C., and Grob, R.L., Isolation and purification of lipids by centrifugal partition chromatography, *J. Liq. Chromatogr.*, 13, 3603–3614, 1990.

14. Otsuka, H., Suzaki, A., and Yamakawa, T., Application of droplet countercurrent chromatography (DCC) for the separation of acidic glycolipids, *J. Biochem.*, 94, 2935–2014, 1983.

15. Renaud, C., Roy, P., and Baudimant, G., Separation of molecular species of phosphatidylcholine by centrifugal partition chromatography, Marine Lipids: proceedings of the symposium held in Brest, France, 19–20 November 1998, pp. 15–20.

16. Baudimant, G., Maurice, M., Landrrein, A. Durand, G., and Durand, P., Purification of phosphotidylcholine with high content DHA from squid by centrifugal partition chromatography, *J. Liq. Chromatogr. Rel. Technol.*, 19, 1793–1804, 1996.

17. Baudimant, G., de la Poype, F., and Durand, P., Purification Process of the Phosphotidylcholine With Long Chain Polyunsaturated Fatty Acids by Centrifugal Partition Chromatography, French Patent 9607960, 1996.

18. Bousquet, O., Sellier, N., and Goffic, F.L., Characterization and purification of fatty acids from micro algae by GC-MS and countercurrent chromatography, *Chromatographia*, 39, 40–44, 1994.

19. Bousquet, O. and Goffic, F.L., Countercurrent chromatographic separation of polyunsaturated fatty acids, *J. Chromatogr. A*, 704, 211–216, 1995.

20. Du, Q., Shu, A. and Ito, Y., Purification of fish oil ethyl esters by high-speed countercurrent chromatography using non-aqueous solvent system, *J. Liq. Chromatogr. Relat. Technol.*, 19, 1451–1457, 1996.

21. Wanasundara, U.N. and Fedec, P., Centrifugal partition chromatography (CPC): emerging separation and purification technique for lipids and related compounds, *Inform*, 13, 726–730, 2002.

22. Marston, A. and Hostettmann, K., Countercurrent chromatography as a preparative tool: application and perspectives, *J. Chromatogr. A*, 658, 315–341, 1994.

30 Oilseed Medicinals: Applications in Drugs and Functional Foods*

Robert D. Reichert
Industrial Research Assistance Program, National Research Council of
Canada, Ottawa, Ontario, Canada

CONTENTS

30.1 INTRODUCTION

Ten lipophilic, naturally occurring compounds in edible oils have been developed into prescription and over-the-counter (OTC) drugs for more than 15 medical conditions in some jurisdictions. Eight of these compounds are or have been prescribed and reimbursed through drug insurance plans. In large measure, the use of these compounds is a reflection of the practice of medicine in those jurisdictions (Europe and Japan, particularly) where regulated natural medicines are an integral part of the healthcare system. This chapter describes these compounds and their genesis, therapeutic uses, doses, dietary required intakes where applicable, and their societal relevance. Several of these oilseed medicinals are isolated from refining byproducts. Oilseed medicinals that are prescribed in some jurisdictions are sometimes considered as OTC drugs or dietary supplements in another. Since these compounds already have approved, marketable health benefits, at least in some jurisdictions, some could provide the basis for new functional foods, especially if the cost of these compounds is reduced through molecular farming.

Functional foods are created by incorporating biologically active compounds into food or by using agronomic, genetic engineering, or other techniques to overexpress or introduce these compounds into plants; they have potential in the fight against malnutrition and in disease prevention[1]. Consumption of a functional food is intended to impart a health benefit beyond that conferred by

---

* Portions of this chapter have been reprinted with permission from the following two publications: Reichert, R.D., Oilseed medicinals: in natural drugs, dietary supplements and in new functional foods, *Trends Food Sci. Technol.*, 13, 353–360, 2002; and Mag, T.K. and Reichert, R.D., A new recommended calculation of vitamin E activity: implications for the vegetable oil industry, *INFORM*, 13, 836–839, 2002.

the nutrients that the food normally contains. Approximately six years ago, the U.S. Food and Drug Administration (FDA) approved the use of label health claims for cholesterol-lowering margarine, a nonengineered functional food[2]. The bioactive component in this product was a cholesterol-lowering, reimbursed, prescription drug in the U.S. for more than 25 years[3]; it is isolated from soybean oil or wood processing byproducts, and sometimes esterified or hydrogenated for use in margarines[4]. The history of long-term, effective, and safe use of this drug provided credibility to the idea of using sterols and stanols in margarine.

The purpose of this chapter is to identify and describe those compounds that are both present in edible oils and that have been developed into prescription and OTC drugs. These compounds may have potential in new functional foods.

30.2 GENESIS OF OILSEED MEDICINALS

Oilseeds were among the world's first therapeutics. Flax and sesame, for example, were both used by the ancients as a cough remedy[5] and for liver dysfunction[6], respectively. However, oilseed medicinal is a new term that we use to describe a compound (activity) that is both found in an edible oil, and also cited as the active ingredient in a prescription or OTC drug in a pharmaceutical reference text commonly used by health professionals. Such citations provide concise, practical information on the selection and clinical use of medicines.

Several oilseed medicinals are commercially isolated from byproducts of the refining process (Figure 30.1). Crude edible oil is de*gum*med, de*acid*ified, de*color*ized, and de*odor*ized in order to produce clear, stable, pale, odorless, and bland-tasting oil. In the process, greater than 95% of the phosphatides, sterol glycosides, free fatty acids, and carotenoids and approximately 32 to 61% of the sterols, tocopherols, and tocotrienols are removed. The deodorization temperature largely controls the degree of removal of the latter three classes of compounds[7]. Deodorizer distillate is the principal raw material providing vitamin E and sterols used in pharmaceutical production[8]. Vitamin K activity is also reduced under some refining conditions[9]. Oilseed refining byproducts are fractionated to provide the active ingredients for a variety of prescription (R) and OTC drugs, dietary supplements (DS), and, increasingly, functional foods (FF).

30.3 THERAPEUTIC USES IN HUMANS

Eight oilseed medicinals are or have been prescribed as drugs in some jurisdictions; vitamin E and ricinoleic acid (in castor bean oil) are only marketed as an OTC or a DS (Table 30.1, Figure 30.2). The pharmaceutical firms that market prescribed drugs containing phosphatidylcholine (PC), phosphatidylserine (PS), γ-linolenic acid (GLA), β-carotene (β-C), β-sitosterol (β-S), γ-oryzanol (γ-O), coenzyme Q_{10} (CoQ_{10}), vitamin K_1, and vitamin K_2 as the active ingredient include: Aventis Pharma, Fidia (to 1996), Pharmacia (to 2002), Roche (to 1996), Eli Lilly (to 1982) and Hoyer-Madaus, Nippon Organon, Eisai, Merck, and Eisai, respectively.

Each prescribed drug in Table 30.1 has been approved by a national regulatory agency on the basis of the submission of a body of evidence including results from human clinical efficacy trials. Although oilseed medicinals are only approved for certain indications, they may be effective for other conditions as well. However, since the degree of intellectual property protection afforded by these compounds is low, drug firms are reluctant to spend significant resources proving out other indications.

Except for ricinoleic acid and vitamin E, this chapter deals only with compounds in drug form that are or have been reimbursed through drug insurance plans in some jurisdictions. Drugs achieve reimbursable status only after a panel of healthcare professionals (drug formulary committee) and scientific experts are convinced, on the basis of human clinical trial data, that the drug has sufficient merit. The following active ingredients in drug form are reimbursed: PC and β-S in Germany; γ-O, CoQ_{10}, and vitamin K_2 in Japan; GLA in the U.K. (to 2002), and vitamin K_1 in the

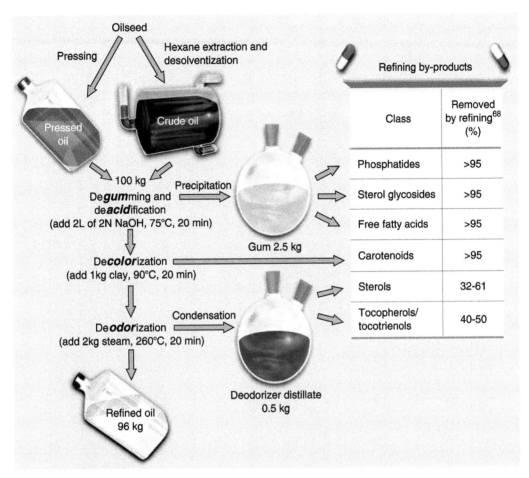

Class	Removed by refining[68] (%)
Phosphatides	>95
Sterol glycosides	>95
Free fatty acids	>95
Carotenoids	>95
Sterols	32-61
Tocopherols/ tocotrienols	40-50

FIGURE 30.1 Refining of vegetable oil creates byproducts that are purified into a variety of medicinal compounds by an industry that interfaces agriculture and medicine.

U.S. These medicinals are likely reimbursable in other jurisdictions as well. PS and β-C were both reimbursed as drugs in Italy and the U.S., respectively, until 1996; however, now as OTC products they are generally not reimbursable.

Oilseed medicinals that are prescribed in some jurisdictions are marketed as an OTC, DS or in a FF in others, continuing a trend to blur the definition of a drug[29]. DSs or FFs containing the same active ingredient as a prescription or OTC drug may or may not confer the same medicinal benefit since differences in composition, dose, and delivery system, affect the bioavailability and pharmacokinetics of the active ingredient. For example, β-carotene is more bioavailable if it is delivered in oil[19]. Several oilseed medicinals that had prescription drug status, including PS, β-C, and β-S, are now becoming less regulated (i.e., ℞ → ℞/OTC or OTC or DS), a trend that has recently also been observed with other prescription drugs[30].

Oilseed medicinals are used therapeutically, often at a relatively high dose, for several chronic conditions. PC, β-S, and γ-O, whose chemical structures are quite dissimilar, are all indicated for alleviating hypercholesterolemia in some jurisdictions; research is needed to determine whether the effects could be additive. PC is indicated for mild forms of hypercholesterolemia only if other nonmedicinal measures and diet (such as physical training and weight reduction) will not give adequate results. Aventis Pharma is not aggressively marketing PC for the latter indication; their

TABLE 30.1
Medicinal Uses of Naturally Occurring Compounds in Edible Oils (Each Has Achieved Prescription or OTC Drug Status)

Active medicinal compound	Indications and usage in humans* (in specific jurisdictions)	Product classification and description** (product, jurisdiction, availability, commercial source) ℞: Prescribed drug; OTC: over-the-counter drug; DS: dietary supplement; FF: functional food
Phosphatides		
Phosphatidylcholine (PC)	Mild forms of hypercholesterolemia[a10]; toxic-nutritive hepatic lesions[a11,12]; chronic hepatitis[a13]; provides choline, a vitamin of the B complex	℞/OTC[a]: Lipostabil® and Essentiale® forte N, Germany, 1988–, soy oil; OTC: PhosChol®, U.S., 1984–, soy oil; DS: Lecithin, U.S., 1950–, soy oil
Phosphatidylserine (PS)	Primary and secondary symptoms of presenile and senile cognitive deterioration[c14]	℞[c]: Bros®, Italy, 1985–96, brain cortex; OTC: NeoBros®, Italy, 1996–, soy oil; DS: Leci-PS™, USA, 1996–, soy oil; FF: Brain Gum™, U.S., 1998–, soy oil
Fatty acids*		
γ-Linolenic acid (GLA)	Symptom relief of cyclical and noncyclical mastalgia (breast pain)[d15,16]; symptomatic relief of atopic eczema[d17]	℞/OTC[d]: Efamast® and Epogam® U.K., 1989–2002, evening primrose oil; DS: Evening primrose, borage and black current oils, U.S.
Ricinoleic acid (RA)	Stimulant laxative	OTC[e]: Purge, U.S., castor bean oil
Carotenoids		
β-Carotene (β-C)	Erythropoietic protoporphyria[b18]; treatment of vitamin A deficiency[b19]	℞[b]: Solatene®, U.S., 1975–96, synthetic; OTC: Lumitene, U.S., 1995–, synthetic; DS: Caromin™, Malaysia, 1996–, palm oil
Sterols		
β-Sitosterol (β-S) (may be hydrogenated and/or esterified in functional foods)	Hypercholesterolemia[b20], hyperbetalipoproteinemia[b]; Benign prostatic hyperplasia[a21]	℞[b]: Cytellin®, U.S., 1955–82, soy oil/wood; ℞/OTC[a]: Harzol®, Germany, 1974–, P. africanum/S. repens; DS: Kholesterol Blocker™, U.S., 1999–, soy oil; FF: Benecol™, Take Control™ margarine, Finland/U.S., 1995/1999–; soy/wood
γ-Oryzanol (γ-O)	Hypercholesterolemia[e22]; menopausal disorders[f23]; irritable bowel syndrome[f24]	℞[e]: γ-OZ®, Japan, 1964–, rice bran oil; DS: Oryzanol, U.S., rice bran oil

TABLE 30.1 (Continued)
Medicinal Uses of Naturally Occurring Compounds in Edible Oils (Each Has Achieved Prescription or OTC Drug Status)

Active medicinal compound	Indications and usage in humans* (in specific jurisdictions)	Product classification and description** (product, jurisdiction, availability, commercial source) Rx: Prescribed drug; OTC: over-the-counter drug; DS: dietary supplement; FF: functional food
Tocopherols		
Vitamin E	Treatment of vitamin E deficiency[b25]	OTC[b]: Nutr-E-Sol® (water-soluble derivative), U.S., 1986–, soy oil; DS: Vitamin E, synthetic or soy oil
Quinones		
Coenzyme Q_{10} (CoQ_{10})	Mild or moderate symptom of congestive heart failure during basic therapy[f26]	Rx[f]: Neuquinon® (Ubidecarenone), Japan 1974–, synth. or fermen. OTC[b], DS: Coenzyme Q_{10}, U.S., fermentation
Vitamin K	Coagulation disorders due to vitamin K deficiency[b27] Improvement of decrease in bone mass and relief of pain in patients with osteoporosis[f28]	Rx[b]: Mephyton® (vitamin K_1, phytomenadione), U.S., 1953–, synth. Rx[f]: Glakay® (vitamin K_2, menatetrenone), Japan, 1995–, synthetic DS: Vitamin K_3 (menadione) is used mainly in animal feeds

* Indications are based on the wording in the physicians' pharmaceutical desk reference texts below. Each prescribed drug has been approved for marketing in a specific jurisdiction by a regulatory agency based on the review of evidence submitted by a drug company. Only key human, clinical references are cited based on references in Martindale[e], the drug or product monograph, or a recommendation from the drug company. Drug monographs are available from the author.

** Drugs designated Rx/OTC are both prescribed by physicians and marketed over-the-counter. Only oral products are considered. The listing of commercial products is exemplary not exhaustive. Drug monographs for Rx and Rx/OTC drugs should be consulted for safety and efficacy data.

*** In medicinal preparations, these fatty acids are usually in the form of triacylglycerols purified from pressed oil in a manner similar to that described in Figure 30.1.

Physicians' Pharmaceutical Desk Reference Texts:

a Rote Liste, Bundesverband der Pharmazeutischen Ind. e.V., Editio Cantor, Frankfurt, Germany (in German, 2001). Lipostabil® and Essentiale® forte N are only being marketed for the indications cited in the table although other uses are also cited in the Rote Liste (Dr. C. Nauert, Aventis Pharma Deutschland GmbH, Cologne, Germany, personal communication).

b Physicians' Desk Reference, Med. Econ. Co., Montvale, NJ, (1980, Cytellin; 1996, Solatene; 2002, PhosChol, Purge, Nutr-E-Sol, Coenzyme Q_{10}, and Mephyton).

c Repertorio Farmaceutico Italiano, OVP Italia S.p.A., Milano, Italy (in Italian, 1993).

d British National Formulary, 43rd ed., British Medical Association and Royal Pharmaceutical Society of Great Britain, Mehta, D., Ed., Pharmaceutical Press, London (2002).

e Martindale, the Complete Drug Reference, 32nd ed., Pharmaceutical Press, London (1999).

f Drugs in Japan, Ethical Drugs, 23rd ed., Japan Pharmaceutical Information Center, Yakugyo Jiho Co. Ltd. Tokyo, Japan (in Japanese, 2000).

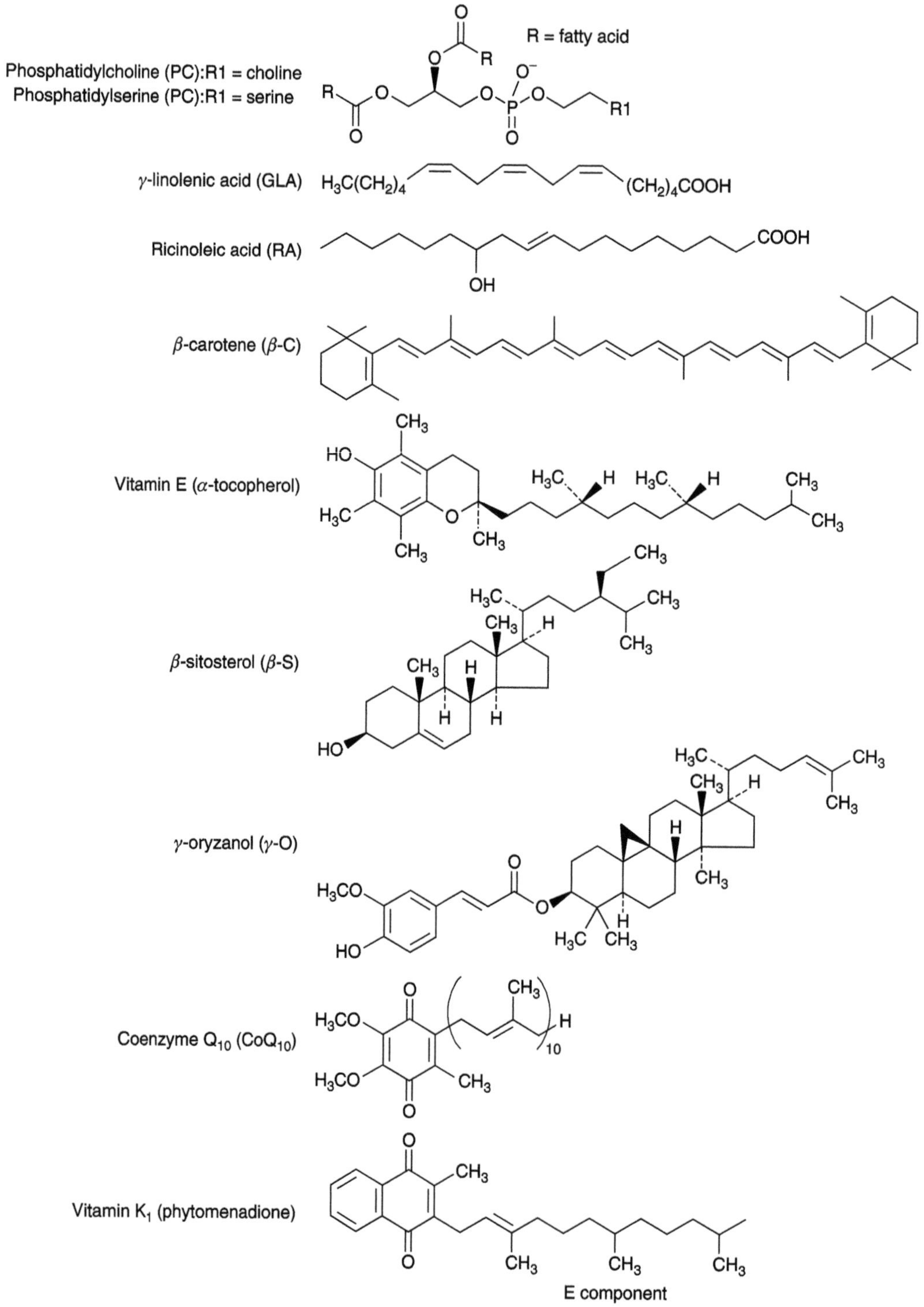

FIGURE 30.2 Structure of medicinal compounds in edible oils. Each lipophilic compound (except for vitamin E and RA) has achieved prescription drug status in some jurisdiction.

main indications for PC are toxic-nutritive hepatic lesions resulting, for example, from chronic alcoholism, and chronic hepatitis (both at 1800 mg/day). In Italy, bovine PS was used until 1996 in the prescribed drug Bros at 100 to 200 mg/day for cognitive deterioration associated with aging; currently, plant-derived PS is extensively marketed as an OTC, DS, and in a FF. Until 1996 in the U.S., β-C was prescribed as the drug Solatene at 30 to 300 mg/day to protect patients with erythropoietic protoporphyria against severe photosensitivity reactions (e.g., burning sensation, skin lesions). β-S in combination with its glucoside (100:1) is commonly prescribed (60 mg/day) in Germany and effective for benign prostatic hyperplasia, a chronic condition resulting in obstructive urinary symptoms experienced to some extent by most men over the age of 50. The latter combination has been shown to stimulate the immune system in humans at very low concentrations, prompting the authors to consider it as an immunomodulatory "vitamin" combination[31]. γ-O and CoQ_{10} are commonly prescribed drugs in Asia, although in the U.S. they are sold as DSs or OTC. In Japan, γ-O is prescribed at 50 mg/day for irritable bowel syndrome, a chronic condition that is highly prevalent in the general population: 14 to 24% of women and 5 to 19% of men[32]. γ-O is also prescribed for hypercholesterolemia and menopausal disorders in Japan. A water-soluble form of vitamin E is used (269 to 806 mg/day of α-tocopherol) in the treatment of vitamin E deficiencies resulting from fat malabsorption syndromes such as childhood cholestasis and short-bowel syndrome[33,34]. Vitamin K activity is provided by vitamin K_1 (present in plants and also commercially synthesized), vitamin(s) K_2 (produced by bacterial fermentation or synthesis), and vitamin K_3 (synthesis). In Japan, vitamin K_2 is prescribed for osteoporosis on the basis of studies including one showing that administration of 45 mg/day to osteoporotic patients for 45 days was associated with a significant increase in mineral bone density and a significant rate of relief from lumbago[28]. In the U.S., vitamin K_1 is prescribed (2.5 to 25 mg/day) for coagulation disorders due to vitamin K deficiency, but the value of vitamin K for bone disorders is still being debated. Osteoporosis afflicts 75 million persons in the U.S., Europe, and Japan, including one third of postmenopausal women[35].

Some fatty acids, usually in the form of triacylglycerols, are also used therapeutically. As a prescribed drug in the U.K. from 1989 to 2002, GLA was provided by evening primrose oil and indicated for mastalgia (breast pain) and atopic eczema, although careful patient selection and timely review were essential to ensure the appropriate use of evening primrose oil in both of these conditions[36]. Mastalgia is a very common condition while atopic eczema constitutes the most frequent inflammatory skin condition in early childhood with an incidence up to 20% of the general population[37]. The recent withdrawal of the marketing authorization for Epogam and Efamast by the Medicines Control Agency was based on the conclusion that the evidence presented did not support the current standard of efficacy required for the authorization of these products as medicines for the treatment of eczema and mastalgia; the safety of evening primrose oil was not an issue. Ricinoleic acid, a hydroxylated fatty acid, is the active ingredient in castor oil, a powerful stimulant laxative marketed for constipation, a relatively infrequent condition[38]. Although eicosapentaenoic acid (EPA) and docosahexaenoic acid (DHA) are not naturally found in oilseeds (they have been genetically engineered into canola[39]), a mixture of their ethyl esters (Omacor®) has recently been approved for marketing and drug reimbursement in the U.K. for two indications: hypertriglyceridemia (1996) and postmyocardial infarction (in 2001) (M. Bryhn, Pronova Biocare, Norway, personal communication).

The therapeutic use of some oilseed medicinals is controversial, especially in the U.S. A particular drug therapy can become controversial when there is conflicting evidence either at the time of drug approval, or when newer therapies become available. The use of Cytellin, for example, which was launched in 1955 for hypercholesterolemia, became questionable when newer therapies became available, finally resulting in withdrawal of the product in 1982.

PC, PS, β-C, β-S, vitamin E, CoQ_{10}, and vitamin K_1 are present in most common edible oils, although the concentration of these medicinals varies widely[9]. γ-O is found in rice oil, while soybean and canola oils are both particularly rich in vitamin K_1[9] and CoQ_{10}[40].

TABLE 30.2
Dietary Reference Intake Values for Oilseed Medicinals (or Pro Forms)[19, 25]; National Academy of Sciences (NAS), 1998, 2000, and 2001

	Men	Women
Recommended dietary allowance (RDA)		
Vitamin A[a] (μg/day)	900 (1000)[b]	700 (800)
Vitamin E (mg/day)	15[c] (10)	15 (8)
Adequate intake (AI)[d]		
Choline[a] (mg/day)	550	425
Vitamin K (μg/day)	120 (80)	90 (65)

[a] β-C and PC provide vitamin A and choline, respectively, via metabolism.
[b] Previous value in parentheses. Choline has no previous AI value.
[c] Only the intake of α-tocopherol contributes to the new RDA.
[d] An AI is set when there is insufficient scientific evidence to establish an RDA.

30.4   INCREASING DEMAND

Global sales of oilseed medicinals exceed $2 billion (U.S.) per year. The vitamin E market is of the order of $1.4 billion with the synthetic version accounting for 70% of these sales[41]. Sales of β-C are approximately $212 million, while sales of all carotenoids are approaching $800 million[42]. CoQ_{10} is one of the top 500 drugs in the world with sales estimated to be $288 million, excluding cosmetic and other uses; U.S. sales climbed rapidly from $20 million in 1995 to $100 million in 1999 and further increases are expected[43]. Sales of Glakay, a Japanese prescribed drug based on vitamin K_2, were $95 million in 1999[44]; it is one of Eisai's leading drugs. PC (lecithin) sales are estimated to be $187 million in the U.S. food industry alone; the recent establishment[10] of an *Adequate Intake* level for choline by the National Academy of Sciences (NAS) may increase the demand for PC (Table 30.2). Essentiale forte N (based on PC) was the fourth largest selling drug in Russia in 2000[45]. Market estimates for PS, GLA, RA, β-S, and γ-O were not readily available.

Several recent developments have prompted an increase in demand for *natural* vitamin E. In 2000 the NAS increased the *Recommended Dietary Allowance* (RDA) for vitamin E (specifically α-tocopherol) by 50% for males and by 88% for females[25]. According to the USDA, 69% of Americans already did not meet their daily required intake of vitamin E using the previous RDA values[46].

For the first time in some 20 years, the NAS also redefined the formula for the calculation of vitamin E activity considering *only* the α-tocopherol content (including the natural form and three synthetic forms), and not the content of β-, γ-, and δ-tocopherols or tocotrienols. This recommendation has resulted in a reordering of the value of vegetable oil with regard to vitamin E content (Table 30.3). Since soybean oil contains high levels of γ-tocopherol, the vitamin E content of soybean oil decreases by approximately half when only α-tocopherol is used to calculate its vitamin E content. Whereas canola oil can now be labeled as a "good source" of vitamin E in Canada, soybean can only be labeled as a "source." Recent research has shown that synthetic vitamin E has only half of the bioactivity of vitamin E extracted from plant sources[25]. In 1997 the American Psychiatric Association recommended the use of large doses of vitamin E for the treatment of moderately impaired Alzheimer's patients[47].

The future demand for pharmaceutical-grade vitamin K may increase considering that the NAS has recently increased the *Adequate Intake* level for vitamin K by approximately 50%[19]

TABLE 30.3
A New Ranking of Commodity Vegetable Oils Based on Vitamin E Content Calculated Using Only the Content of α-Tocopherol

	Method of vitamin E calculation			
Type of oil	Old method (including α-, β-, and γ-tocopherols and tocotrienols) (mg/100 g oil)	New method (including only α-tocopherol) (mg/100 g oil)	% of recommended daily intake in 10 g[a]	Vitamin E claim allowed in Canada
Sunflower	63	62	62	"Excellent Source"
Cottonseed	40	36	36	"Excellent Source"
Corn	33	24	24	"Good Source"
Canola	24	20	20	"Good Source"
Peanut	15	14	14	"Source"
Palm	10	10	10	"Source
Soybean	19	9	9	"Source"

[a] In Canada, vegetable oils can be labeled as a "Source," "Good Source," or "Excellent Source" of vitamin E if they provide (in 10 g) at least 5, 15, and 25%, respectively, of the daily vitamin E requirement (10 mg).
(From Mag, T.K. and Reichert, R.D., *INFORM*, 13, 836–839, 2002.)

(Table 30.2). Demand could further increase based on increased awareness of the high-dose use of vitamin K_2 for osteoporosis in Japan, particularly considering that no adverse effects of high intakes of this vitamin have been reported[19].

The worldwide demand for *sterols and stanols* has increased since cholesterol-lowering margarines were approved for marketing in the U.S., the E.U. and elsewhere. The demand may increase further as a result of practice guidelines released by the National Cholesterol Education Program suggesting an increase in the consumption of plant sterols and stanols[48]. The development of cholesterol-reducing mayonnaise using γ-O may increase the demand for this compound in the future[49].

30.5 POTENTIAL IN FUNCTIONAL FOODS

Since the oilseed medicinals in Table 30.1 already have approved, marketable health benefits, at least in some jurisdictions, some could provide the basis for new functional foods, especially vegetable oils. Oils containing added oilseed medicinals could have a large impact on human health considering the high consumption of vegetable oil in most industrialized nations: 56.9 lb per person in the U.S. in 1999 for example[50]. The need for such "functionalized" oils is already widely recognized by consumers in Japan and, in response, a new Japanese industrial niche has recently emerged including firms such as Kao, Nisshin Oil Mills, and Honen[51]. Some of these oils contain enhanced levels of both vitamin E and β-S, and some have achieved regulatory approval as functional foods in Japan[52]. The trend toward fortified refined oils is analogous to fortification of white wheat flour that began in the U.S. during the 1940s. Although lipophilic functional foods have been the first target for oilseed medicinals, there is also the possibility of their application in hydrophilic foods such as fruit juices. An emerging niche involves the delivery of oilseed medicinals into water-based foods using specialized strategies such as the use of charged adjunct molecules to solubilize the lipophilic compound[53].

Functional foods containing oilseed medicinals at higher levels than naturally present in common foods will require regulatory approval in most jurisdictions. Since food processing, storage, and handling affect the biological activity of medicinals, thorough testing of the safety and efficacy of these functional foods will be required. Labeling and perhaps segregation of such foods in the retail market will help ensure that consumers make educated choices.

Unfortunately, however, the first major functional food in the U.S. (cholesterol-lowering margarine) costs up to three times as much as ordinary margarine[3]. Consequently, Law[20] contends that the 25% risk reduction in heart disease associated with consumption of these margarines largely benefits the affluent minority of society. The active ingredient is a major contributor to the price because relatively large quantities (i.e., 1.3 and 3.4 g per day of sterols or stanol esters, respectively) must be present in each serving in order to qualify for the cholesterol-lowering health claim[2]. These compounds are relatively expensive because of processing costs and since they are present in only low concentration in plant materials; soybean seeds, for example, contain of the order of only 0.07% total sterols[9]. Molecular farming of oilseed medicinals may be required to produce more economical functional foods incorporating these compounds. Some progress has recently been made in this area.

The introduction of the gene *phytoene synthase* into *Brassica napus* (canola) resulted in a 50-fold increase in carotenoids, thus producing "red" canola oil with levels of carotenoids higher than crude palm oil[54,55]. However, the concentration of carotenoids in these seeds is still relatively low (~0.1%). β-C was elevated in white rice to produce seeds with a "golden" endosperm[56,57] but a β-C content of the order of only 1 to 3 μg/g (I. Potrykus, Swiss Federal Institute of Technology, Zurich, Switzerland, personal communication). Shintani and DellaPenna[58,59] elevated the vitamin E activity of arabidopsis lines by a factor of approximately nine by overexpression of the enzyme α-tocopherol methyltransferase that converted most of the γ-tocopherols to α-tocopherol; however, the total tocopherol content did not change significantly. Sterols[60,61], GLA[62,63], and RA[64,65] have also recently been overexpressed in the seeds of prototype plants. In the case of GLA, the transgenic oil has already been analyzed by food scientists[66].

30.6 CONCLUSIONS AND RECOMMENDATIONS

Edible oils are a valuable source of compounds that have been developed into a variety of natural drugs with marketable health benefits in some jurisdictions for some conditions. Functional or medical foods could be developed for those conditions that are chronic and that have a higher incidence if the cost of oilseed medicinals is reduced through molecular farming in plants, i.e., special-purpose crops containing elevated levels of oilseed medicinals. Such an approach should result in lower-priced functional foods accessible to all levels of society and a concomitant reduction in societal health care costs. However, to achieve a practical outcome, this area of research will require very close cooperation among food scientists, molecular biologists, plant breeders, clinicians, edible oil refiners, and food companies.

Special focus is needed in the development of functional oils, since current industrial practice significantly reduces the concentration of several oilseed medicinals in refined products. Processes for less rigorous refining or fortification of oils should be developed and applied. Functional oils containing oilseed medicinals in various amounts and ratios need to be created and clinically tested.

ACKNOWLEDGMENTS

The expert intelligence-gathering skills of Marion Boyd (Canada Institute for Scientific and Technical Information) and Isao Kaneko, Steffen Preusser, and Caroline Martin (Department of Foreign Affairs and International Trade in Japan, Germany, and the U.K., respectively) are gratefully acknowledged.

REFERENCES

1. Blackburn, G.L., Pasteur's quadrant and malnutrition, *Nature*, 409, 397–401, 2001.
2. U.S. Food and Drug Administration, Interim Final Ruling on Food Labeling: Health Claims; Plant Sterol/Stanol Esters and Coronary Heart Disease, *FDA Fed. Reg.*, 65, 54685–54739, 2000.
3. Hollingsworth, P., Hicks, K.B., and Moreau, R.A., Development of cholesterol-fighting foods, *Food Technol.*, 55, 59–67, 2001.
4. Miettenen, T., Wester, I., and Vanhanen, H., Substance for Lowering High Cholesterol Level in Serum and Methods for Preparing and Using the Same, Raisio Benecol Ltd., Raisio, Finland, U.S. Patent, 6, 174, 560, 2001.
5. Nunn, J.F., *Ancient Egyptian Medicine*, University of Oklahoma/British Museum Press, London, 1996.
6. Grant, M., Dieting for an emperor, in *Studies in Ancient Medicine*, Vol. 15, Scarborough, J., Ed., E.J. Brill, New York, 1997.
7. Walsh, L., Winters, R.L., and Gonzalez, R.G., Optimizing deodorizer distillate tocopherol yields, *INFORM*, 9, 78–83, 1998.
8. Ramamurthi, S., McCurdy, A.R., and Tyler, R.T., Deodorizer distillate: a valuable by-product, in *Emerging Technologies, Current Practices, Quality Control, Technology Transfer, and Environmental Issues*, Koseoglu, S.S., Rhee, K.C., and Wilson, R.F., Eds., AOCS Press, Champaign, IL, 1998, pp. 130–134.
9. Scherz, H. and Senser, F., *Food Composition and Nutrition Tables*, 6th ed., Medpharm, Stuttgart, Germany, 2000.
10. Food and Nutrition Board, *Dietary Reference Intakes for Thiamin, Riboflavin, Niacin, Vitamin B_6, Folate, Vitamin B_{12}, Pantothenic Acid, Biotin, and Choline*, National Academy of Sciences, National Academy Press, Washington, DC, 1998.
11. Put, A., Samochowiec, L., Ceglecka, M., Tustanowski, S., Birkenfeld, B., and Zaborek, B., Clinical efficacy of "essential" phospholipids in patients chronically exposed to organic solvents, *J. Int. Med. Res.*, 21, 185–191, 1993.
12. Panos, M.Z., Polson, R., Johnson, R., Portmann, B., and Williams, R., Polyunsaturated phosphatidyl-choline for acute alcoholic hepatitis: a double blind, randomized, placebo-controlled trial, *Eur. J. Gastroenterol. Hepatol.*, 2, 351–355, 1990.
13. Niederau, C., Strohmeyer, G., Heinteges, T., Peter, K., and Gopfert, E., Polyunsaturated phosphatidyl-choline and interferon alpha for treatment of chronic hepatitis B and C: a multi-center, randomized, double-blind, placebo-controlled trial, *Hepato-Gastroenterology* 45, 797–804, 1998.
14. Pepeu, C., Pepeu, I.M., and Amaducci, L., A review of phosphatidylserine pharmacological and clinical effects. Is phosphatidylserine a drug for the aging brain?, *Pharmocol. Res.*, 33, 73–80, 1996.
15. Gateley, C.A., Miers, M., Mansel, R.E., and Hughes, L.E., Drug treatments for mastalgia: 17 years experience in the Cardiff mastalgia clinic, *J. R. Soc. Med.*, 85, 12–15, 1992.
16. Pye, J.K., Mansel, R.E., and Hughes, L.E., Clinical experience of drug treatments for mastalgia, *Lancet*, Aug., 373–377, 1985.
17. Morse, P.F., Horrobin, D.F., Manku, M.S., Stewart, J.C.M., Allen, R., Littlewood, S., Wright, S., Burton, J., Gould, D.J., Holt, P.J., Jansen, C.T., Mattila, L., Meigel, W., Dettke, Th., Wexler, D., Guenther, L., Bordoni, A., and Patirizi, A., Meta-analysis of placebo-controlled studies of the efficacy of Epogam in the treatment of atopic eczema. Relationship between plasma essential fatty acid changes and clinical response, *Br. J. Dermatol.*, 121, 75–90, 1989.
18. Todd, D.J., Erythropoietic protoporphyria, *Br. J. Dermatol.*, 131, 751–766, 1994.
19. Food and Nutrition Board, *Dietary Reference Intakes for Vitamin A, Vitamin K, Arsenic, Boron, Chromium, Copper, Iodine, Iron, Manganese, Molybdenum, Nickel, Silicon, Vanadium, and Zinc*, National Academy of Sciences, National Academy Press, Washington, DC, 2001.
20. Law, M., Plant sterol and stanol margarines and health, *Br. Med. J.*, 320, 861–864, 2001.
21. Berges, R.R., Kassen, A., and Senge, T., Treatment of symptomatic benign prostatic hyperplasia with β-sitosterol: an 18-month follow-up, *Br. J. Urol.* 85, 842–846, 2000.
22. Yoshino, G., Kazumi, T., Amono, M., Tateiwa, M., Yamasaki, T., Takashima, S., Iwai, M., Hatankana, H., and Baba, S., Effects of gamma-oryzanol on hyperlipidemic subjects, *Curr. Therap. Res.*, 45, 543–552, 1989.
23. Kushima, K., Hasegawa, N., and Nakano, S., Drug efficacy evaluation of gamma-oryzanol for menopausal disorders by double-blind test, *Gynecol. Practice* 25, 1099–1111, 1976 (Japanese).

24. Namiki, M. et al., Clinical evaluation of gamma-oryzanol against gastrointestinal psychosomatic disorders; double blind comparative test, *Rinsyo to Kenkyu (Clinic and Research)*, 59, 955–963, 1982 (Japanese).

25. Food and Nutrition Board, *Dietary Reference Intakes for Vitamin C, Vitamin E, Selenium and Carotenoids*, National Academy of Sciences, National Academy Press, Washington, DC, 2000.

26. Overvad, K., Diamant, B., Holm, L., Holmer, G., Mortensen, S.A., and Stenser, S., Co-enzyme Q_{10} in health and disease, *Eur. J. Clin. Nutr.*, 53, 764–770, 1999.

27. Sherer, M.J., Vitamin K, *Lancet*, 345, 229–234, 1995.

28. Orimo, H., Fujita, T., Onomura, T., Inoue, T., Kushida, K., and Shiraki, M., Cinical evaluation of Ea-0167 (Menatrenone) in the treatment of osteoporosis, *Clin. Eval. (Tokyo)* 20, 45–100, 1992 (Japanese).

29. Brower, V., Nutraceuticals: poised for a healthy slice of the healthcare market?, *Nature Biotechnol.*, 16, 728–731, 1998.

30. Grubert, N., *The Impact of Rx-to-OTC Switching on the Pharmaceutical Market*, Decision Resources, Waltham, MA, 2001.

31. Bouic, P.J.D., Etsebeth, S., Liebenberg, R.W., Albrecht, C.F., Pegel, K., and van Jaarsveld, P.P., Beta-Sitosterol and beta-sitosterol glucoside stimulate human peripheral blood lymphocyte proliferation: implications for their use as an immunomodulatory vitamin combination, *Int. J. Immunopharmac.*, 18, 693–700, 1996.

32. Jailwala, J., Imperiale, T.F., and Kroenke, K., Pharmacologic treatment of the irritable bowel syndrome: a systematic review of randomized, controlled trials, *Ann. Intern. Med.*, 133, 136–147, 2000.

33. Sokol, R.J., Heubi, J.E, Butler-Simon, N., McClung, H.J., Lilly, J.R., and Silverman, A., Treatment of vitamin E deficiency during chronic childhood cholestasis with oral d-α-tocopheryl polyethylene glycol-1000 succinate, *Gastroenterology*, 93, 975–985, 1987.

34. Traber, M.G., Schiano, T.D., Steephen, A.C., Kayden, H.J., and Shike M., Efficacy of water-soluble vitamin E in the treatment of vitamin E malabsorption in short-bowel syndrome, *Am. J. Clin. Nutr.*, 59, 1270–1274, 1994.

35. South-Paul, J.E., Osteoporosis: I. Evaluation and assessment, *Am. Fam. Physician*, 63, 867–904, 908, 2001.

36. National Prescribing Centre, Atopic eczema and mastalgia: the place of evening primrose oil, *MeReC Bull.*, 9, 5–8, 1998.

37. Leung, D.Y.M. and Soter, N.A., Cellular and immunologic mechanisms in atopic dermatitis, *J. Am. Acad. Dermatol.*, 44, S1–S12, 2001.

38. Shiller, L.R., The therapy of constipation, *Alim. Pharm. Therapeut.*, 15, 749–763, 2001.

39. Facciotti, D., Metz, J.G., and Lassner, M., Production of Polyunsaturated Fatty Acids by Expression of Polyketide-Like Synthesis Genes in Plants, Calgene Inc., CA, U.S. Patent 6,140,486, 2000.

40. Kamei, M., Fujita, T., Kanbe, T., Sasaki, K., Oshiba, K., Otami, S., Matsui-Yuasa, I., and Morisawa, S., The distribution and content of ubiquinone in foods, *Int. J. Vit. Nutr. Res.*, 56, 57–63, 1986.

41. Patterson, E., *Vitamin E: US Market Poised for a Boom*, Decision Resources, Waltham, MA, 1997.

42. Marz, U., *The Global Market for Carotenoids*, Business Communications, Norwalk, CT, 2000.

43. Challener, C., Nutraceuticals/natural products 2000: specialty dietary supplement ingredients are the hot spot, *Chem. Marketing Rep.*, 258, 8, 2000.

44. Engel Publishing, The top 500 prescription drugs by worldwide sales, *Med Ad News*, 19, 59, 2000.

45. Association of International Pharmaceutical Manufacturers in Russia, Market Bulletin, Moscow, Dec. 2000.

46. Agricultural Research Service, *USDA's 1994–96 Continuing Survey of Food Intakes by Individuals*, U.S. Dept. of Agriculture, Beltsville, MD, 1999.

47. Work Group on Alzheimer's Disease and Related Dementias, Practice guidelines for the treatment of patients with Alzheimer's disease and other dementias of late life, American Psychiatric Association Practice Guidelines, *Am. J. Psychiatry*, 154, 5, 1997.

48. Expert panel on detection, evaluation, and treatment of high blood cholesterol in adults, Executive summary of the third report of the National Cholesterol Education Program (NCEP) expert panel on detection, evaluation, and treatment of high blood cholesterol in adults (Adult Treatment Panel III), *J. Am. Med. Assoc.* 285, 2486–2497, 2001.

49. Imai, Y., Tomonari, O., Chiho, T., Masatoshi, K., and Hidero, T. Food Additive and Use Thereof, Amano Pharmaceutical Co. Ltd, Nagoya, Japan, U.S. Patent 5, 514, 398, 1996.

50. Economic Research Service, *Food Consumption, Prices and Expenditures*, U.S. Dept. of Agriculture, Washington, DC, 2001.
51. Burke, C., Oily success, in *Funct. Foods Nutraceut.*, 3, 32, 2000.
52. Japanese Ministry of Health and Welfare, Foods for Specified Health Uses (FOSHU), Tokyo, 2000.
53. Borowy-Borowski, H., Sikorska-Walker, M., and Walker, P.R., Water-Soluble Compositions of Bioactive Lipophilic Compounds, National Research Council of Canada, Ottawa, Canada, U.S. Patent 6, 045, 826, 2000.
54. Shewmaker, C.K., Methods for Producing Carotenoid Compounds and Specialty Oils in Plant Seeds, Calgene Inc., CA, WO Patent, 98/06862, 1998.
55. Kishore, G.M. and Shewmaker, C.K., Biotechnology: enhancing human nutrition in developing and developed worlds, *Proc. Natl. Acad. Sci. USA*, 96, 5968–5972, 1999.
56. Ye, X., Al-Babili, S., Klöti, A., Zhang, J., Lucca, P., Beyer, P., and Potrykus, I., Engineering the provitamin A (β-carotene) biosynthetic pathway into (carotenoid-free) rice endosperm, *Science*, 287, 303–305, 2000.
57. Beyer, P. and Potrykus, I., Method for Improving the Agronomic and Nutritional Value of Plants, Greenovation Pflanzenbiotechnologie GmbH, Freiburg, Germany, WO Patent 00/53768, 2000.
58. Shintani, D. and DellaPenna, D., Manipulation of Tocopherol Levels in Transgenic Plants, University of Nevada, PCT Patent US99/19483, 2000.
59. Shintani, D. and DellaPenna, D., Elevating the vitamin E content of plants through metabolic engineering, *Science*, 282, 2098–2100, 1998.
60. Venkatramesh, M, Corbin, D.R., Bhat, G.B, Boddupalli, S.S., Grebenok, R.J., Kishore, G.M., Lardizabal, K.D., Rangwala, S.H., and Karunanandaa, B., Transgenic Plants Containing Altered Levels of Sterol Compounds and Tocopherols, Monsanto Co., WO Patent 00061771A2, 2000.
61. Karunanandaa, B., Jaehyuk, Y., and Kishore, G.M., Nucleic acid Molecules and other Molecules Associated with Sterol Synthesis and Metabolism, Pharmacia Corp., WO Patent 00104314A2, 2001.
62. Reddy, A.S. and Thomas, T.L., Expression of a cyanobacterial Δ6-desaturase gene results in γ-linolenic acid production in transgenic plants, *Nature Biotechnol.*, 14, 639–642, 1996.
63. Thomas, T.L., Reddy, A.S., Nuccio, M., and Freyssinet, G.L., Production of Gamma Linolenic Acid by a Δ6-Desaturase, Texas A & M University, U.S. Patent 5, 689, 050, 1997.
64. Somerville, C. and van de Loo, F., Plant Fatty Acid Hydroxylase, Carnegie Institution, Washington, DC, U.S. Patent 6, 028, 248, 2000.
65. Smith, M., Moon, H., and Kunst, L., Production of hydroxy fatty acids in the seeds of *Arabidopsis thaliana*, *Biochem. Soc. Trans.*, 28, 947–950, 2000.
66. Liu, J.W., DeMichele, S., Bergana, M., Bobik, E., Hastilow, C., Chuang, L.T., Mukerji, P., and Huang, Y.S., Characterization of oil exhibiting high α-linolenic acid from a genetically transformed canola strain, *J. Am. Oil Chem. Soc.*, 78, 489–493, 2001.
67. Mag, T.K. and Reichert, R.D., A new recommended calculation of vitamin E activity: implications for the vegetable oil industry, *INFORM*, 13, 836–839, 2002.
68. *Practical Handbook of Soybean Processing and Utilization*, Erickson, D.R., Ed., American Oil Chemists Society/United Soybean Board, Champaign, IL, 1995.

Index